PRENTICE-HALL INTERNATIONAL SERIES IN

THEORETICAL AND APPLIED MECHANICS

Under the General Editorship of N. M. Newmark

Professor and Head of the Department of Civil Engineering

University of Illinois

PRENTICE-HALL, INC.

PRENTICE-HALL INTERNATIONAL, INC., UNITED KINGDOM AND EIRE

PRENTICE-HALL OF CANADA, LTD., CANADA

PRENTICE-HALL SERIES IN

FLUID MECHANICS

Richard Skalak, Editor

PRENTICE-HALL INTERNATIONAL, INC., *London*
PRENTICE-HALL OF AUSTRALIA, PTY., LTD., *Sydney*
PRENTICE-HALL OF CANADA, LTD., *Toronto*
PRENTICE-HALL OF INDIA (PRIVATE) LTD., *New Delhi*
PRENTICE-HALL OF JAPAN, INC., *Tokyo*
PRENTICE-HALL DE MEXICO, S. A., *Mexico City*

Oceanographical

Engineering

PRENTICE-HALL, INC. / ENGLEWOOD CLIFFS, N.J.

ROBERT L. WIEGEL

PROFESSOR OF CIVIL ENGINEERING, UNIVERSITY OF CALIFORNIA, BERKELEY

Oceanographical

Engineering

Preface

OCEANOGRAPHICAL ENGINEERING is expanding and developing rapidly. Although it is a very ancient field, having begun at least as early as 3000 B.C. with the construction of Port A-ur near the mouth of the Nile, significant contributions from oceanography toward the understanding of problems involved in the rational design of a structure, a device, or a system have been made only since World War II. As is the case with many modern fields, oceanographical engineering cuts across the boundaries of several disciplines. In addition, it is often difficult to distinguish clearly between engineering and scientific endeavors in this field. This is quite natural, since both groups have the same ultimate aim: to understand the nature of the ocean and to make use of this understanding for the benefit of mankind through better ports, safer and more economical operations at sea, and a greater use of the oceans' natural resources—food, raw materials and recreation.

Much of oceanography is still descriptive, so that certain portions of the book, pertaining to the distribution of temperatures and salinities in the ocean, are largely descriptive. Other portions are mathematical, such as the sections on waves. A large number of drawings and photographs have been included for two reasons. First, much oceanographical data can only be presented in this form; and second, the author has found in his experience in teaching this subject that these illustrations are necessary to assist the students in a subject that is not well covered in most courses in fluid mechanics.

This book is the result of the gradual development of a set of notes for parts of three graduate courses in the Civil Engineering Department of the University of California, Berkeley, California: River and Harbor Hydraulics, Advanced Design of Hydraulic Structures, and Coastal Engineering. A knowledge of fluid mechanics is necessary for all portions of the book, a background in the motions of sediments in fluids is desirable for some portions, and a concurrent course in structural dynamics is desirable for a more mature understanding of the chapters on wave forces on structures and ship moorings. *Oceanographical Engineering* has been written for use both as a text for coursework and as a reference for practicing engineers and scientists.

Since portions of this book were originally started nearly ten years ago, the author has drawn upon the comments of many of his colleagues and students to modify various statements and sections. The author would like to express his indebtedness to each person who has helped in this way. He would like to thank especially Morrough P. O'Brien (Dean Emeritus of the College of Engineering, University of California), who started the author in this field; both Dr. O'Brien and Thorndike Saville (Dean Emeritus of the College of Engineering, New York University) were probably the founders of coastal engineering in the United States, largely through their long, active association with the Beach Erosion Board, Corps of Engineers, U. S. Army. The author is extremely grateful to his colleagues of many years, Professors J. W. Johnson, H. A. Einstein, John V. Wehausen, and J. D. Isaacs, with whom he discussed many of the ideas expressed in this book and from whom he received useful advice.

ROBERT L. WIEGEL

Berkeley, California

Contents

CHAPTER ONE

Introduction

1. INTRODUCTION

Physical oceanography can be defined as that science treating of the oceans, their forms, physical features and phenomena. This book deals largely with the applications of physical oceanography to civil engineering. It is not intended to be a design manual. To the contrary, considerable emphasis is given to the lack of precise quantitative design information and procedures. Before World War II, the designs of most civil engineering structures built in the ocean were based upon past experience. Because of the stimulus of amphibious operations during World War II, and more recently, because of offshore oil operations there has been a great increase in the knowledge of oceanographical engineering.

Some aspects of oceanographical engineering are new, whereas other aspects go back to antiquity. The design and construction of harbors is one of the oldest branches of engineering. The port of A-ur was located on the Canopic branch of the Nile prior to 3000 B.C., and it is well established that a harbor was built nearby on the open coast of Egypt (a forerunner of Alexandria) about 2000 B.C., called the Port of Pharos (Savile, 1940). This port, apparently built by the Minoan Cretans, had an area of about 300 acres. Among other structures it had a breakwater 8500 ft long which consisted in the main part of two rubble mound structures between 130 and 200 ft apart, each with an upper width of 26 to 40 ft and 20 to 30 ft high. These were made of very large blocks and the space between them was filled with smaller stone. The top width was between 180 and 250 ft.

Another famous harbor of antiquity was Tyre, which was built about 900 to 1000 B.C. Hewn rectangular blocks weighing as much as 15 tons each were used, with many of them being tied together with iron dowels run in with lead (Savile, 1940).

Great harbors were constructed by the Greeks and Romans, the last of the large construction jobs being done by the Romans. The Romans invented a hydraulic cement (pozzuolana) in the third century B.C., and developed pile driving for foundations and cofferdams (Savile, 1940). Using a combination of these, they constructed seawalls of concrete. They had considerable trouble with sanding up of harbors in some locations, such as in the vicinity of the mouth of the Tiber. For example, at Ostia the shoreline advanced about 3300 ft between 634 B.C. and 82 B.C., and about 27,500 ft since 634 B.C. The port of Ostia was built about 43 A.D., and it apparently became inoperable within 75 years because of its sanding up.

2. WAVES AND THEIR EFFECTS

A large portion of this book is devoted to waves, the theory of waves, their generation, propagation, and effects upon sediments and structures. This is due to three reasons: To begin with, waves are important in almost every phase of oceanographical engineering; secondly, the author's main field of research has been on waves and their effects; finally, not as much is known of most of the remaining fields.

There are many examples in the literature on the combined effects of waves, storm tides, and astronomical tides. The great winter coastal storm of 5–9 March, 1962, off the east coast of the United States was responsible for 34 dead and more than 300 injured, together with damage in excess of $200 million (U.S. Congress, 1962). About 1800 houses were destroyed and another 2200 were severely damaged. The storm surge in the North Sea of January 31–February 1, 1953, drowned more than 300 people in England and 1800 people in the Netherlands; 1800 houses were destroyed in the Netherlands along with a total damage to dikes, buildings, agriculture, and livestock of the order of $250 million (Wemelsfelder, 1954). Tsunamis have also caused tremendous damage. The June 15, 1896, tsunami killed more than 27,000 people and destroyed over 10,000 houses in Japan (Leet, 1948); the April 1, 1946, tsunami was responsible for the deaths of 163 people and $25 million damage in the Hawaiian Islands (Shepard, MacDonald, and Cox, 1950).

Waves are also of interest from the standpoint of understanding the physics of fluids. It is well established that, for one fluid motion to be similar to another, the two situations must not only be geometrically and kinematically similar, but also dynamically similar. Dynamical similarity can exist only if the ratio of one prototype force to model force, say, has the same numerical value as the ratio of each other pair of forces; i.e., the ratio of the prototype to model inertia force must have the same numerical value as the ratio of the prototype to model gravitation force, the ratio of the prototype to model viscous force, etc. These can also be transposed so that the ratio of the inertia force to gravity force of the prototype must be equal to the ratio of the inertia to gravity force of the model, etc. These

ratios form the well-known dimensionless numbers of fluid mechanics: Froude number (ratio of inertia to gravity forces), Mach number (ratio of inertia to elastic forces), Weber number (ratio of inertia to surface tension forces), and Reynolds number (ratio of inertia to viscous forces). These ratios can be expressed in a different form. Froude number becomes the ratio of the speed of a disturbance to the speed of a surface gravity wave; Mach number becomes the ratio of the speed of a disturbance to the speed of an acoustic wave; Weber's number becomes the ratio of the speed of a disturbance to the speed of a surface tension wave; Reynolds number ($VD/\nu = VD/c\lambda$, where ν is the kinematic viscosity) becomes the product of the ratio of the speed of the disturbances (V) to the mean molecular speed (c) and the ratio of representative dimension of the disturbance (D) to the mean free path of the molecules (λ). This last relationship was pointed out by Von Karman in 1923 (see Von Karman, 1956, p. 164) for compressible flow.

The theory of periodic waves is covered rather thoroughly, beginning with the linear theory of irrotational waves, both standing and progressive. It is surprising how well linear theory predicts many of the characteristics of uniform periodic waves. Where experimental data were available, they are given, as it is mandatory for an engineer to know the range of validity of the linear theory. Curves and tables are also presented to facilitate the use of theory by practicing engineers. Linear theory can be used, by the principle of linear superposition, to develop equations that express irregular waves which, in some ways, are a better approximation of the sea surface than is a set of regular waves. One example in which linear superposition is used to a considerable extent is the treatment of the two-dimensional power spectra of wind waves and swell. It is the author's opinion that this method must be used with considerable caution. The waves in the generating area are highly nonlinear, and the techniques used for calculating power spectra assume that they are a linear phenomenon. This results in ascribing energy to high frequencies as linear component waves, whereas these frequencies are higher harmonics of the lower-frequency nonlinear waves. Because of simplifications of this sort, wrong impressions are created, and the essential nonlinear processes which govern many aspects of wave generation are neglected.

A considerable literature has been developed in which linear superposition has been used to describe ocean waves. This approach appears to be useful in some engineering problems in deep water, such as the response of ships to waves. Most of the work along these lines has been done by statistical means, relating the power spectrum of the waves to the "power spectrum" of heave, pitch, or roll of the ship. Some work has been done on the co- and quadrature spectra, but the physical significance of the results is not always clear. In deep water, the results are probably more likely to be valid for swell than for seas. In shallow water, some of the nonlinear characteristics of waves start to predominate even for swell. The second, third, and higher harmonics become important in calculating water particle velocities and accelerations which are necessary to calculate wave-induced forces on structures. Under these conditions, use of linear superposition is probably a poorer representation of the waves than is the approximation of replacing a single wave of a group by a nonlinear theoretical wave train of the height and period of the single wave. This is especially true, considering the characteristic of ocean waves for the largest waves to be in groups of three, four five, etc., waves which are nearly periodic, as will be shown in Chapter 9.

Because of the importance of the nonlinear waves to engineering problems, the section on linear theory is followed by the theory for Stokes' waves of finite amplitude, including a section on the highest possible waves of this type. The theory of Stokes' waves for a ratio of wave length to water depth of greater than about 10 is not satisfactory in many respects. In the more shallow water region, the theory of *cnoidal* waves should be more useful, and a section on this type of wave is presented.

In nature, wave trains are not of infinite extent, rather they consist of groups of waves. There are many facets of wave group theory that are still not clear; but the theory as it exists, making use of the principle of linear superposition, is a useful tool in predicting, for example, the time it will take waves to travel from the storm area to the section of coast, or other ocean areas, in question. The phenomenon of wave groups is connected with the dispersive characteristic of waves.

"Water seeks its own level" is a well-known saying. How it does so is a gravity water wave phenomenon. If by some means water is heaped up over an area of the water surface and then released, the distrubance disperses, the water seeks its own level. The heaped-up water may be expressed in Fourier integral form consisting of a combination of all possible wave lengths. If each wave component traveled out at the same velocity, the disturbance wouldn't disperse, it would travel as a concentrated disturbance. If each wave length travels at a different velocity, say, a function of the wave length, then the disturbance will disperse. The longer the wave, the faster it will travel, for a given depth of water. A group of irregular waves will sort itself as it travels, with the longer components gradually moving ahead, and the shorter waves dropping behind.

This is the case, theoretically, except for the peculiar nonlinear phenomenon known as a *solitary wave*. The solitary wave may be considered as a limiting case of wave motion. It is a purely positive wave, i.e., there is no trough. The theory, and observations of this wave are treated in a separate chapter.

Waves are generated impulsively, whether by a gust of wind blowing over the water surface, a ship moving through the water, a ground effect machine moving over the water surface, the low-pressure area of a hurricane moving over the continental shelf, or by a submarine seismic disturbance. The theory of impulsively generated waves is developed and compared with observations. At one limit are waves which behave as predicted by linear theory. As will be shown in a section describing experiments of waves generated impulsively by suddenly adding a volume of water at one end of a wave tank, the other limit is not simply the solitary wave of theory. A series of crests and troughs, all above the initial water level, has been observed, with the group being amplitude-dispersive. Under extreme conditions, bores have been generated.

As has been pointed out, the effects of tsunamis and storm surges on man are great, even devastating at times. Many of the things that are known of these catastrophic phenomena are described, together with some information on predicting possible maximum conditions in certain areas. Because of the close relationship between the storm surges and tsunamis as forcing functions and the response characteristics of shelfs, bays, and harbors, the problem of this type of oscillation is covered in the same chapter.

Beaches, currents, breakwaters, drilling platforms—how do they affect waves? How high do waves run up on a beach or breakwater? How much energy is dissipated, and how much is reflected? Here we must leave theory, except for the simplest of cases, because we can tell the mathematician so little on how wave energy is dissipated at a structure. Results of much laboratory work must be used to answer problems of the type just posed.

Waves moving in shoaling water transform. The speed of a wave depends upon the water depth as well as upon the wave length, with the more shallow the water, the slower the wave speed. Waves moving in shoaling water at an angle to the bottom contours must bend because of this dependence of speed upon water depth. This process is known as *wave refraction*. Graphical methods are presented for the determination of the pattern, and the resulting spatial distribution of wave energy, in complicated coastal regions. Finally, these waves either break along shores, or transform into multiple crests. Methods of predicting wave heights on beaches and in harbors are given. In calculating breaker heights, it is necessary to leave theory and make use of empirical results.

Waves reach areas in the geometric shadow of a breakwater. This phenomenon is known as *wave diffraction*. The theory of this phenomenon and the comparison of model studies with theory are given. Graphical means of predicting diffracted wave heights in the lee of breakwaters, and similar structures or natural obstructions, are given. Connected with diffraction is the Mach-stem type of reflection (or nonreflection). Because of this, wave energy is concentrated in some circumstances, and waves can even be swung around by a curving breakwater. This can lead to conditions that could not be predicted by linear theory. The theory that is available is for blast waves for the simplest of conditions. It is in areas of this sort that the engineer must resort to model studies. Laboratory studies are presented for several problems of this sort in order to develop in the engineer a feeling for some of the more complex phenomena that he may encounter.

Great detail is given on the initial generation of waves by winds and their growth under the action of continuing winds. Statistical representations are presented on the characteristics of wind waves and their relationship to the characteristics of the winds creating them. These are forced waves. When they leave the storm area, they become free gravity waves. They disperse. They transform into what is known as *swell*. The changes in the wave characteristics are given in quantitative terms. Much is known about wave characteristics and the relationships of these characteristics to the winds that cause them. Much is not known. Even in the laboratory using controlled wind speeds and fetches and using the "proper" statistical techniques, there is considerable scatter of results in these relationships. Evidently we do not know enough.

The information on the relationship between wave characteristics and the winds forms the basis of wave forecasting which is of great importance in both the planning and construction of marine structures and in dredging operations. The uncertainty of the relationship between wind and waves, together with the uncertainties of the data on weather maps and the interpretation of the weather maps, results in doubts as to the usefulness of wave forecasting techniques at the present time without relying upon the knowledge of local conditions by a forecaster.

Let us consider an operation in the construction of an offshore structure, such as placing riprap around the base to prevent scouring. One type of information needed is whether or not a barge can be used on a particular day to place the riprap. This depends upon the distance the tug must tow the barge and upon the maximum seas in which this particular operation can be performed. Little information is available on operational limits with respect to sea conditions (see Chapter 17); these data indicate that for a dump barge and tug, a wave 3 ft high would be about the limit with no indication as to the associated wave period. The problem is forecasting in advance whether the waves (significant wave height?) would be less or greater than 3 ft high during the time necessary to tow the barge to the site, dump the rock, and return to port. If it is forecast that the sea conditions will be operationally safe, and the waves are higher than forecast, then there is the possibility of damage to, or loss of, the equipment. If the waves are forecast to be higher than the operational limit, but are actually lower, then the cost of the equipment and crew is lost for that day. A study made by Isaacs and Saville (1949) was based upon the original data of Sverdrup and Munk and hence is no longer completely applicable in detail. Isaacs and Saville made 271 forecasts for one region (a nine-month interval) and 201 forecasts for another region (an eight-month interval). One of their findings is still valid. It was found that 97 per cent of the recorded significant increases in wave heights were forecast, but 23 per cent of the forecast wave trains failed to arrive. According to the authors, the rather large proportion of nonarrivals apparently resulted from the erroneous selection of fetches, frequently because of difficulty in determining the limits of effective angles of the winds with respect to the point at which the waves were recorded. This would result in a number of days in which operations would be canceled needlessly.

When a structure is placed in waves, forces are induced. That these forces can be very large is substantiated by the evidence of failures of many massive structures. The nature of these forces and the available empirical data useful in the prediction of these forces are presented.

The ocean waves to which marine structures are subjected are phenomena imperfectly understood. They are nonuniform in three dimensions and in time; i.e., the vertical distance from trough to crest varies from wave to wave, the horizontal distance from crest to crest varies, the distance along a crest for which a wave may be distinguished varies, and the fluctuation of the surface at a point varies with time. Only in recent years has the engineer had the tools necessary to describe these phenomena approximately; even at the present time it is not possible to describe in detail the wave characteristics in the direction at right angles to their direction of propagation. Statistical terms must be used to describe the characteristics of length, height, and period.

Nearly all measurements of wave characteristics of actual ocean waves have been made of what might be termed their surface features, i.e., a statistical description of their height and period, and the velocity of the wave crests. With the exception of the case of impact, one is not as interested in the phase velocity (velocity of the wave crest) as with the motion of the water particles. As far as the author knows, there are almost no published data on particle velocities or accelerations in the ocean (Inman and Nasu, 1956). Thus one is forced to use theoretical values of these quantities for conditions which have not been checked. In fact there have been only a limited number of measurements in the laboratory, and in these measurements the quantity being measured was so small that experimental error appears to have masked many of the significant characteristics the investigators were seeking, although some were readily apparent.

Waves encountered in the ocean present a more difficult problem than the uniform waves dealt with both theoretically and in most laboratory tests. Because they are nonuniform both in the direction of motion and along the crests, they often have been treated by means of a linear superposition of component waves rather than by the concept of a representative train of waves with one height and length. There appear to be no published data from which to determine the relative errors introduced by using the linear superposition technique or the representative wave approach in computing the wave forces. Further, the so-called design wave will almost always be under the action of wind, so there will be a considerable wind-induced current near the surface. A formula which can be used to predict the force on a pile where a current is present has been developed. At present, however, the lack of quantitative knowledge of these surface currents precludes the use of this formula.

3. TIDES, CURRENTS, SHORES, SEA WATER, AND MIXING

The level of the oceans is continually varying. The effects of astronomical tides are fairly well known and are well documented for many areas. The effects of what have been termed *meteorological tides* are not as well known. Considerable detail is given in Chapter 5 on the heights of the water levels due to winds and barometric pressures. The theory and the results of laboratory measurements of the piling up of water along coasts due to wind stress on the water surface is given on the section on tides. In addition, long-term variations of sea level changes will be described.

Currents may be oceanic in scope, such as the Gulf Stream, or they may be local and temporary, due to the wind blowing over a limited area of water for a short time. The effects of the oceanic currents may be far-reaching, being largely responsible for certain climatic conditions of many portions of the world. The small currents are of considerable importance in regard to the movement of sewage discharged from a sewer outfall, or the movement of oil slicks. Tidal currents are largely responsible for the depths of channels into estuaries. Another type of current is the littoral current which is created by waves breaking at an angle to the coast line. Littoral currents transport vast amounts of sand along coasts, often in the range of five hundred to a thousand cubic yards per day.

Waves, tides, currents, and winds acting on coasts create many severe problems, such as shoaling of harbors and erosion of beaches. The shore and its environment is covered in detail, and the effects of waves, tides, currents, and winds on beach configuration, longshore drift of sediments, and shoaling of estuarial entraces are discussed both qualitatively and quantitatively. As an example of the type of problem confronting the engineer, let us suppose it is desired to make a beach by artificial fill of sand. The proposed beach will have an exposure

of some sort to waves that can be determined. A sieve analysis of the available sand can be made. These two pieces of information can then be used to predict the steepness of the beach, and from this to predict the relative safety of the beach from the standpoint of people wading and swimming in the surf.

Salinity, temperature, and density of the oceans' waters affect the design of most marine structures and hardware. Sometimes, only the mean values of these properties are important; for other designs, one needs the range to be expected over a few days, a few months, or perhaps the long-term changes. Much of the available information has been assembled here to acquaint the engineer with this aspect of the environment. For the more detailed information that is often necessary for design, original sources are cited.

Mixing processes are of major importance in the disposal of waste at sea, as well as for an understanding of some of the physics of the growth and decay of thermoclines and currents. The mixing of turbulent jets, mixing by wind waves, and mixing by currents are described, together with values for the eddy coefficient for both vertical and horizontal mixing. The design of ocean outfall sewers is complicated by the fact that, during certain seasons of the year, the density gradient of the sea water in many locations is such that the mixing takes place while the effluent is completely submerged. During other seasons, the mixture rises to the surface and spreads out. Criteria are established for the prediction of the type of mixing that will occur under a given set of conditions. Which result is preferable is not resolved.

4. FUNCTIONAL DESIGN

With the present state of the art, economics, and science of oceanographical engineering, it is often more difficult to design a structure from the functional than from the structural standpoint. For example, a groin field can be built that will stand up for the required length of time; but how does an engineer decide upon the spacing, length, and height of the groins to trap a maximum amount of sand? Or, even more important in many instances, will a groin field prevent the erosion of a beach in a particular region?

Sometimes the installation of a structure will create problems as severe as the one it was conceived to solve. A case history of such a situation is given, the construction of a harbor, subsequent filling of the harbor together with downdrift beach erosion problems that had to be overcome.

5. MOORINGS

Ships, buoys, and other equipment must be anchored at sea and in harbors. The differential equation describing the motions of a moored structure is nonlinear in its simplest form. Hence in solving many engineering jobs, the engineer must rely upon model studies. The model laws to be followed are described, as are the results of model tests. The reliance that can be placed on the results of model tests to predict prototype action is based upon past experience. Comparison of predictions with prototype observations have been sparse, but they seem to indicate that such model tests are useful in this field.

6. INSTRUMENTS

At some stage in almost every engineering investigation, measurements must be made. The literature on this subject is so large that no attempt has been made to describe instruments in this book.

7. CORROSION AND EROSION

Due to space limitations, some extremely important subjects, such as the corrosion and erosion of materials and marine growths, are not covered. Entire books are available on these subjects (see, for example, Uhlig, 1948; Woods Hole Oceanographic Institute., 1952).

Studies made on the durability of steel sheet piling used in shore structures have shown that the rate of deterioration is strongly dependent upon their environment (Ross, 1948; Rayner and Ross, 1952). Seventy-seven harbor bulkheads, 26 beach bulkheads, and 50 groins were measured at intervals between 1936 and 1946. They were located from Miami Beach, Florida, to Portland, Maine. It was found that in general the rates of loss of web thickness were much lower for harbor bulkheads than for shore structures, with the harbor bulkheads losing an average 0.0033 in. per year (ranging from 0.0023 in. per year in the northern regions to 0.0062 in. per year in the southern regions) and the shore structures losing an average of 0.016 in. per year (ranging from about 0.010 in. per year in the northern regions to 0.018 in. per year in the southern regions). The harbor structures probably were affected only by corrosion, and the foregoing figures are in agreement with corrosion rates from all over the world (La Que, 1948). The rates measured for the shore structures suggest that erosion by sand plays an important part either by direct abrasion, or by removing rust or any other surface layer from the steel so that the bare steel is continually exposed to the salt water or spray.

The location of the portion of the pile tested with respect to the water level was found to be important, as shown in Table 1.1, and the effect of sand, earth, or other cover was also found to be important, as shown in Table 1.2.

It was found that when the piles were painted at least once the rate of loss of web thickness decreased substantially (Rayner and Ross, 1952). The paint was found to

Table 1.1. Effect on Rate of Deterioration of Location of Sample with Respect to Water Level (from Rayner and Ross, 1952)

Zone	Harbor bulkheads (in./yr)	Beach bulkheads (in./yr)	Groins and jetties (in./yr)
8 ft above MHW	0.0049	0.020	0.010
5 to 8 ft above MHW	0.0049	0.022	0.010
2 to 5 ft above MHW	0.0049	0.0081	0.010
Mean high water (MHW)	0.0027	0.0074	0.0055
Mean tide level	0.0024	0.001	0.024
Mean low water	0.0035	0.002	0.028

Table 1.2. Effect on Rate of Deterioration of Sand, Earth, or Other Cover (from Rayner and Ross, 1952)

Cover	Harbor bulkheads (in./yr)	Beach bulkheads (in./yr)	Groins and jetties (in./yr)
No cover on either surface	0.0075	0.0027	0.019
One surface never covered, other covered part time	0.0076	0.020	0.014
One surface only covered	0.0026	0.0094	0.020
One surface always covered, other covered part time	—	0.0065	0.0057
Both surfaces covered part time	—	—	0.017
Both surfaces always covered	—	0.0017	0.0026

last only about six months on groins, however (Ross, 1948).

More detailed studies were made on five experimental groins at Palm Beach, Florida (Ross, 1948). It was found that different results were obtained for steel piles installed in four different zones: (1) the air corrosion zone in which the piles were usually subjected to air and spray; (2) the sand protected zone, i.e., the area at least 0.2 ft beneath the beach surface; (3) the abrasion zone, extending from 0.2 ft beneath to 0.6 ft above the beach surface; and (4) the wetting and drying zone, in which the piles were periodically immersed in sea water and then in air—roughly the zone between mean low water and mean high water. It was found that the rate of decrease in steel thickness was 0.011 in. per year for the air corrosion zone, 0.001 in. per year for the sand protected zone, 0.117 in. per year for the abrasion zone, and 0.005 in. per year for the wetting and drying zone. In the abrasion zone, the sand not only abraded the steel, but also removed the rust covering so as to accelerate the chemical corrosion rate. Some perforations appeared in the abrasion zone within one year, which would indicate some local abrasion at about 0.373 in. per year. The rate at which a large section of a pile might be corroded and abraded would depend upon the cut and fill of a particular beach. If the beach face moved back and forth as is usually the case, no one portion of a pile would be in the abrasion zone continuously. Because of this, it would be difficult to predict the life of a structure, except as an average.

Experience with steel "H" section piles used as

structural members of piers in Southern California have shown that they should be protected in the surf zone to prevent rapid deterioration (Schaufele, 1951). "H" section 8 in. × 8 in., 32-lb piles were given two coats of asphalt emulsion, with a minimum of $\frac{1}{16}$-in. coating. After these had been driven into the bottom (sand on top of shale), the piles in the breaker zone were sheathed with a 14-in. diameter, $\frac{3}{16}$-in. thick steel cylinder which was driven through the sand into the top of the shale. The space between the cylinder and the pile was jetted clean, then filled with a 1:2 cement grout, using a high silica cement. After twenty years, the piles were still in good shape, whereas the "H" piles of a nearby pier which were not protected in this manner had to be either replaced or protected at the end of five years.

Piles seaward of the surf zone merely coated with the asphalt emulsion were also found to be satisfactory, whereas an experiment with a heavier uncoated "H" section (54 lb, 10 in. × 10 in.) was found to be unsatisfactory.

The decking of the piers was made of timber. All bearings and laps were coated with hot creosote; the timber was spaced to provide considerable circulation of air, and all nails, bolts, etc., were galvanized. In twenty years of service only minor repairs were necessary.

It was found that concrete structural members "grew" in the sea water, whether mass or reinforced concrete. There was a case of a 13-ft diameter concrete cylinder which increased 6 in. in diameter and 4 in. in height. In another case, 6-ft diameter steel cylinders made of 1/4-in. plate and filled with concrete, the "growing" had either ruptured welds or caused the plate to fail. Schaufele (1951) states that they had success using high silica cement, then stripping the forms off and coating the concrete as soon as possible to prevent the sea water from getting into the concrete. Additional information on concrete in sea water is available (see Cook, 1951).

Asphalt has been used to protect concrete piles with considerable success (Engineering News Record, 1935). Concrete piles impregnated with asphalt in both Los Angeles harbor, California, and Melville, Rhode Island, were in excellent condition at the end of ten years. Asphalt and asphalt mixes have also been used to protect underwater pipelines (Aldridge, 1956; Timothy, 1956).

For further information on this subject and on marine borers see Voigt's bibliography (1948) on the effect of salt spray on materials, and the proceedings of a conference on the deterioration of materials in the marine environment (National Academy of Sciences., 1951).

Many structures made of iron are protected against corrosion in the region below mean tide level by cathodic devices, such as a magnesium, aluminum or zinc anode, or an impressed direct current using scrap steel (or iron) or graphite anodes.

Most corrosion in sea water occurs because the water is a good electrolyte, and the system acts as a battery.

This is called *galvanic corrosion*. Consider a series of metals listed according to their relative potential in sea water (Table 1.3). If two of these metals, connected by a conductor, are placed in sea water the one closer to the anodic end of the list will go into solution as an ion, releasing electrons which flow through the conductor to the metal closer to the cathodic end where they react with hydrogen ions to form hydrogen gas, or in the presence of oxygen gas to form hydroxide ions. This type of corrosion of the metal closer to the anodic end is known as *bimetallic corrosion*. It is of considerable importance where there are brass screws in iron plates, bronze propellers on steel shafts, bronze fixtures on iron pipes, etc.

The process is complicated by other reactions that take place. If there is little oxygen available, the hydrogen gas that is formed might create a film at the cathode which causes a decrease in corrosion rate; this is called *polarization*. If sufficient oxygen is present, one has $2H^+ + O_2 + 4e \rightarrow 2OH^-$ and little polarization occurs. The rate of corrosion, then, depends upon the amount of mixing of the sea water with air (or other sources of oxygen) and the rate at which water currents move by a structure.

One type of galvanic corrosion of considerable importance to the engineer is that of pipelines and piers. A steel pile is not a bimetallic. Why does it corrode? One reason is the presence of mill scale, which has a lower potential than steel, so that the mill scale acts as a cathode and the steel as an anode (Evans, 1948). Even

Table 1.3. RELATIVE POTENTIAL POSITIONS OF SOME COMMON METALS IN SEA WATER

Anodic (corroded) end

Magnesium
Magnesium alloys
Zinc
Galvanized iron
Aluminum 4 S
Cadmium
Aluminium 17 S–T
Mild steel
Wrought iron
Cast iron
18–8 stainless steel (active)
Solders (lead-tin)
Lead
Tin
Naval brass
Nickel (active)
Brasses
Copper
Bronzes
Nickel (passive)
18–18 stainless steel (passive)
Silver
Graphite
Gold
Platinum

Cathodic (protected) end

if the mill scale is removed, steel piles will corrode because of local chemical or physical differences in the metal. Thus, the galvanic corrosion is in reality "bimetallic."

It is generally concluded that coatings are the most effective means of protecting steel piles either above or in the splash zone or in the zone of tidal action (Humble, 1949; Ryan, 1955; Spencer, 1958). In the zone beneath mean low water, cathodic protection has been used extensively and beneficially, and detailed studies have been made of this problem (La Que, 1948; Humble, 1949; Ryan, 1955). Two methods are most often used. One consists of placing sacrificial anodes of magnesium (or some other metal high up in the anodic region of Table 1.3) which are connected electrically to the steel piles. The magnesium goes into solution as an ion, and electrons flow through the electrical connection to the steel which acts as a cathode. The second system supplies direct current by means of an ac-dc generator (or some other device) to anodes made of scrap iron or graphite which are placed at proper locations in the sea water, with the negative side of the power impressed system being connected electrically to the steel piles to make them cathodic.

The decision about which of the two systems to use depends upon the location of the structure, the type of personnel available, and the cost (Ryan, 1955). The electrical current that must be generated, whichever of the two systems is used, has been found to be about 10 milliamperes per square foot of steel surface to be protected for the first year and about 7 ma after that. Rusty steel requires about 8 ma per square foot the first year and about 6 ma afterwards (Ryan, 1955). Cox (1949) has found that, if cathodic current densities between 100 and 400 ma per square foot of steel are impressed until 4–10 ampere hours per square foot are used, the steel piles will become electroplated directly from the sea water with a magnesium-cement type of coating so that only 3 ma per square foot is needed afterward (Ryan, 1955).

Model studies have proved useful in determining the proper locations to put anodes in order to insure that all parts of a large structure, such as a steel pile pier, have sufficient electrical current supplied (Ryan, 1955).

Stray currents can present considerable problems (Morgan, 1960). One major cause of these currents is grounding of electrical equipment. If the equipment is grounded to the steel pier directly the negative side should be grounded. If a positive ground is necessary, it should be connected to a large metal plate immersed in the sea water (Calahan,—).

8. MARINE BORERS AND FOULING

Marine borers are of major concern to engineers in the design and maintenance of timber structures in sea water. The Teredo and Limnoria are not only the most destructive borers, but are widely distributed; they cause many millions of dollars of damage each year (Horonjeff and Patrick, 1952). Another destructive mollusk is the Martesia; other crustaceans are the Chelura and the Spaeroma. The mollusks enter the timber through small holes, usually as larvae, and they grow to full size while inside the wood. Hence the outside appearance of a piece of timber can be misleading; it may be infested although only small holes are visible. On the other hand, crustaceans destroy the surface of the wood. They burrow very close together, creating a system of interlacing holes which are broken off by the water (U.S. Navy, 1951). Both types attack timber in the region between the mudline and the high water line, with the Teredo being most numerous near the mudline, and the Limnoria in greatest numbers between tide levels, in some regions (Hill and Kofoid, 1927).

Marine borers are a problem only in the marine or estuarial environment. Limnoria find salinities less than 6.5 to 10 parts per 1000 lethal (ocean salinity is about 35 parts per 1000) and do hardly any damage in salinities lower than about 12 to 16 parts per 1000. The *Teredo navalis* finds salinities less than 5 parts per 1000 lethal and requires more than about 9 parts per 1000 to be active (Horonjeff and Patrick, 1952). Domestic sewage does not seem to have any effect on them, but some industrial wastes may cut down marine borer activity (U.S. Navy, 1951). They are not particularly affected by temperatures; the Teredo will attack wood until the water temperature is nearly freezing, and a few Limnoria are present as far north as Kodiak, Alaska. Details of marine borer activity at many important harbors in the world have been accumulated and analyzed by the Clapp Laboratories for the U.S. Navy and have been put into report form (U.S. Navy, 1951).

Borers can make timber structurally unsound in a relatively short time. There have been cases of 16-in. diameter untreated piles severed in six months and treated piles in need of replacement in less than two years. Total destruction by marine borers of sections of piles 12 in. to 15 in. in diameter have been observed in widely separated areas, such as Puerto Rico, Florida, California, New York, Newfoundland, Nova Scotia, and Alaska (U.S. Navy, 1951).

Some tropical woods, such as greenheart from British Guiana, are naturally resistant to attack by marine borers (Wangaard, 1953), and the service life of this wood in many areas is available (Jachowski, 1955). Most woods must be protected. Two common methods of protecting piles against attack by marine borers are creosote pressure treatment and either gunnite or precast concrete jackets. Electric currents impressed by electrodes mounted on piles have been tested, and it was found that alternating currents (6.3 volts) of 0.01 amp and 1.3 amp per pile caused a noticeable decrease in marine borer damage, whereas at intermediate currents damage increased (Hochman and Roe, 1956).

In using a gunnite jacket, an average thickness of 2 in. has been recommended. No. 12 gauge 2 × 2 in. galvanized mesh is installed approximately 1 in. from the surface of the pile first. About one week is allowed for curing (Horonjeff and Patrick, 1952). As marine borers (Pholads) are known to attack low-grade concrete and soft stone, care must be exercised in using this type of casing (Kofoid and Miller, 1923; U.S. Navy, 1951).

For Southern pine, it has been recommended that at least 16 to 24 lb/cu ft of creosote be used and that a 4-in. penetration occur. For Douglas fir, a minimum of 12 lb/cu ft has been recommended, which would be only a $\frac{3}{4}$-in. penetration. As creosote is not effective against some of the Limnoria, work has been done with some copper-based preservatives which are far more effective (University of Miami, 1960).

It has been found that coal-tar creosote impregnated piles which are protected in temperate waters have been subject to considerable marine borer attack in tropical and subtropical waters. It was believed that one type of borer might be responsible (Menzies, 1952). A thorough study was made by the U.S. Navy at Port Hueneme, California (Hochman, et al., 1956). *Limnoria tripunctata*, *Limnoria quadripunctata*, Teredo, Chelura

and Bankia were present in large numbers in the harbor. Creosoted piles that had been in the harbor for from ten to fifteen years were found to be heavily infested with *Limnoria tripunctata*, and with a few Chelura. Test blocks of untreated wood were placed in the harbor, and the untreated wood was soon infested with all types of marine borers. Toxicity tests were made by placing a number of *L. quadripunctata* in a dish of salt water with a chip of creosoted pile in it, and a number of *L. tripunctata* in a similar dish. After twenty-four hours, most of the *L. quadripunctata* were dead whereas all the *L. tripunctata* were living. Other tests of various sorts were run, all of which led the investigators to conclude that *L. tripunctata* are probably responsible for the many early failures of creosoted piles in tropical and subtropical waters.

In using treated wood, all cuts and holes between the mudline and high water should be made before treatment (American Wood-Preservers Association., 1955).

Fouling organisms, such as barnacles, mussels, hydroids, corals, and algae (which includes kelp), will not be covered here (Woods Hole Oceanographic Institution., 1952). Neither, because of the author's unfamiliarity with certain subjects—submarine geology, underwater acoustics, and sea ice, for example—are those topics dealt with in this book.

REFERENCES

Aldridge, Clyde, What's involved in planning and constructing an offshore pipeline, *Oil Gas J.*, **54**, 59 (June 18, 1956), 174–79.

American Wood-Preservers' Association, *Manual of recommended practice* (August, 1955).

Calahan, H. A., *When metals go to sea*, The International Nickel Co., Inc.

Cook, Herbert K., Concrete piles and their protection in harbor construction, *Proc. Wrightsville Beach Marine Conf.*, Nat. Acad. Sci.–Nat. Res. Council. (1951), 8–14.

Cox, George C., Discussion of the cathodic protection of steel piling in sea water, *Corrosion*, **5**, 9 (September, 1949), 300–301.

Engineering News-Record, Bitumized concrete casing for piles in sea water (January, 24 1935), 121–24.

Evans, U. R., An outline of corrosion mechanisms, including the electrochemical theory, *The corrosion handbook*. New York: John Wiley & Sons, Inc., 1948, 3–11.

Hill, C. L., and C. A. Kofoid (edits.), *Marine borers and their relation to marine construction on the Pacific Coast, Final report of the San Francisco Bay Marine Piling Committee*. San Francisco, Calif.: The Committee, 1927.

Hochman, H., H. Vind, T. Roe, Jr., J. Muraoka, and J. Casey, *The role of Limnoria tripunctata in promoting early failure of creosoted piling*, U.S. Naval Civil Eng. Res.

Evaluation Lab., Port Hueneme, Calif., Tech. Memo. M–109 (April, 1956).

Hochman, H. and T. Roe, Jr., *An electrical protection system for wooden piling*, U.S. Naval Civil Eng. Res. Evaluation Lab., Port Hueneme, Calif., Tech. Note N–270 (June, 1956).

Horonjeff, R. and D. A. Patrick, Action of marine borers and protective measures against attack, *Proc. Second Conf. Coastal Eng.*, Berkeley, Calif.: The Engineering Foundation Council on Wave Research, 1952, 86–100.

Humble, H. A., The cathodic protection of steel piling in sea water, *Corrosion*, **5**, 9 (September, 1949), 292–302.

Inman, Douglas L. and Noriyuki Nasu, *Orbital velocity associated with wave action near the breaker zone*, U.S. Army, Corps of Engineers, Beach Erosion Board, Tech. Memo. No. 79 (March, 1956).

Isaacs, J. D. and Thorndike Saville, Jr., A comparison between recorded and forecast waves on the Pacific Coast, *Annals New York Acad. Sci.*, **51**, 3 (May, 1949), 502–510.

Jachowski, Robert A., *Factors affecting the economic life of timber in coastal structures*, U.S. Army, Corps of Engineers, Beach Erosion Board, Tech. Memo. No. 66 (December, 1955).

Kofoid, C. A. and Robert C. Miller, An unusual occurrence of rock boring mollusks in concrete on the Pacific Coast, *Science*, **57**, 1474 (March, 30, 1923), 383–84.

La Que, F. L., Corrosion by sea water, *The corrosion handbook*. New York: John Wiley & Sons, Inc., 1948, 383–429.

Leet, L. Don, *Causes of catastrophe; earthquakes, volcanos, tidal waves, hurricanes*. New York: McGraw-Hill Book Company, Inc., 1948.

Menzies, Robert J., The phylogeny, systematics, distribution, and natural history of Limnoria. Ph. D. thesis, University, Southern California, June, 1952.

Morgan, J. H., *Cathodic protection, its theory and practice in the prevention of corrosion*. New York: Macmillan Company, 1960.

National Academy of Sciences–National Research Council, *Proc. Wrightsville Beach Marine Conf.*, Publ. No. 203, 1951.

Rayner, Albert C. and Culbertson W. Ross, *Durability of steel sheet piling in shore structures*, U.S. Army, Corps of Engineers, Beach Erosion Board, Tech. Memo. No. 12, (February, 1952).

Ross, Culbertson W., *Experimental steel pile groins, Palm Beach, Florida, Ibid.*, Tech. Memo. No. 10, 1948.

Ryan, L. T., Cathodic protection of steel pile wharves, *Dock and Harbour Authority*, **35**, 412 (February, 1955), 303–307.

Savile, Sir Leopold Halliday, Presidential Address, *J. Inst. Civil Engrs.*, 1 (November, 1940), 1–26.

Schaufele, H. J., Erosion and corrosion on marine structures, Elwood, Cal.: *Proc. First Conf. Coastal Eng.*, Berkeley, Calif: The Engineering Foundation Council on Wave Research, 1951, 326–34.

Shepard, F. B., G. A. MacDonald, and D. C. Cox, The tsunami of April 1, 1946, *Bull. Scripps Inst. Oceanog.*, **5**, 6, (1950), 391–528.

Spencer, K. A., Cathodic protection of ships and marine structures, *Dock Harbour Authority*, **38**, 447 (January, 1958), 307–312.

Timothy, P. H., New method for coating offshore pipelines, *Oil Gas J.*, **54**, 35 Jan. 2, 1956, 70–75.

University of Miami, *Final report, May, 1960*, Marine investigations, Contract NOy–81879. (Unpublished.)

U.S. Congress, *Improvement of storm forecasting procedures*, Hearing before the Subcommittee on Oceanography of the Committee on Merchant Marine and Fisheries, House of Representatives, 87th Cong., 2nd Sess., 1962.

U.S. Navy, Bureau of Yards and Docks, *Report on marine borers and fouling organisms in 56 important harbors and tabular summaries of marine borer data from 160 widespread locations*, NAVDOCKS TP-Re-1, April, 1951.

Uhlig, Herbert H. (ed.), *The corrosion handbook*. New York: John Wiley & Sons, Inc., 1948.

Voigt, Lorraine R., Selected bibliography on salt spray testing, 1935–1946, *Corrosion*, **4**, 10 (October, 1948), 492–502.

von Kármán, Theodore, *Collected works of Theodore von Kármán*, vol. 2. London: Butterworth Scientific Publications, 1956.

Wangaard, F. F., Tropical American woods for durable waterfront structures, *Report of the Marine Borer Conference* (April, 1953).

Wemelsfelder, P. J., The disaster in the Netherlands caused by the storm flood of February 1, 1953, *Proc. Fourth Conf. Coastal Eng.*, Berkeley, Calif.: The Engineering Foundation Council on Wave Research, 1954, 258–71.

Woods Hole Oceanographic Institution, *Marine fouling and its prevention*. Annapolis, Md.: U.S. Naval Institute, 1952.

CHAPTER TWO

Theory of Periodic Waves

1. INTRODUCTION

Ocean waves are complex phenomena, difficult, if not impossible, to describe correctly in mathematical terms. Many of their characteristics can, however, be described within acceptable limits. This section develops only the wave theory necessary to several succeeding sections. For a detailed mathematical summary of wave theory see the paper by Wehausen and Laitone (1960).

Stokes (1880) studied two-dimensional oscillatory waves for the case of a fluid which is frictionless, homogeneous, and incompressible, and uniform in depth. In addition, the inertia of the air and the pressure due to a column of air whose height is comparable with that of the waves were neglected, so that the pressure at the upper surface of the fluid could be assumed to be zero, provided that atmospheric pressure be added afterward. In the following section on wave theory the original development is due to Stokes, but the notation follows Lamb (1945). The motion was assumed to have been generated from rest by normal forces and thus irrotational. There will then be a velocity potential ϕ which will satisfy Laplace's equation of continuity

$$\frac{\partial^2 \phi}{\partial x^2} + \frac{\partial^2 \phi}{\partial y^2} = 0 \qquad (2.1)$$

where x and y are the rectangular axes in the plane xy vertical to the fluid surface; x is in the direction of propagation of the waves and y is normal to the surface, being measured positively upwards. The plane xz is horizontal and coincides with the surface of the fluid when in equilibrium. The horizontal and vertical components of water particle velocities at a point x, y are $u = \partial\phi/\partial x$ and $v = \partial\phi/\partial y$. The dynamical equation in its integrated form (Bernoulli's equation) will be

$$\frac{p}{\rho} = -gy - \frac{\partial \phi}{\partial t} - \frac{1}{2}\left[\left(\frac{\partial \phi}{\partial x}\right)^2 + \left(\frac{\partial \phi}{\partial y}\right)^2\right] \qquad (2.2)$$

where p is the pressure, ρ is the density of the incompressible fluid, and g is the acceleration of gravity.

11

The two foregoing equations describe the body of the fluid. To these must be added the boundary conditions. The equation which expresses the fact that the fluid particles on the rigid bottom plane remain in contact with it is

$$\frac{\partial \phi}{\partial y} = 0 \quad \text{when } y = -d \quad (2.3)$$

where d is the water depth measured from the still water level (SWL).

The conditions at the free surface must be that the pressure be zero,

$$\frac{p}{\rho} = 0, \quad (2.4)$$

and the same surface of particles continues to be the free surface throughout the motion

$$\frac{\partial p}{\partial t} + \frac{\partial \phi}{\partial x}\frac{\partial p}{\partial x} + \frac{\partial \phi}{\partial y}\frac{\partial p}{\partial y} = 0 \quad \text{when } p = 0. \quad (2.5)$$

Neglecting small quantities of the second order and substituting the several partials of p found from Eq. 2.2 into Eq. 2.5, the following first approximation is obtained:

$$g\frac{\partial \phi}{\partial y} + \frac{\partial^2 \phi}{\partial t^2} = 0 \quad \text{when } y = 0. \quad (2.6)$$

This equation, together with Eqs. 2.1, 2.3, and 2.4 will give the first approximation.

2. STANDING WAVES, LINEAR THEORY

The most simple supposition that can be made is that ϕ is a simple harmonic function of x. The general case consistent with the assumption made can be obtained by use of Fourier's theorem of superposition (Lamb, 1945). Assume

$$\phi = (P \sin kx)\, e^{i(\sigma t + \epsilon)} \quad (2.7)$$

where P is a function of y only. Equation 2.1 then gives

$$\frac{d^2 P}{dy^2} - k^2 P = 0. \quad (2.8)$$

So

$$P = A\, e^{ky} + B\, e^{-ky}. \quad (2.9)$$

For Eq. 2.7 and 2.9 to satisfy Eq. 2.2 we find

$$A\, e^{-kd} = B\, e^{kd} = \frac{1}{2}\, D. \quad (2.10)$$

So

$$\phi = \frac{1}{2}\, D[e^{k(y+d)} + e^{-k(y+d)}](\sin kx)\, e^{i(\sigma t+\epsilon)}$$
$$= D[\cosh k(y+d)](\sin kx)\, e^{i(\sigma t+\epsilon)}. \quad (2.11)$$

For the case of simple harmonic motion, Eq. 2.6 can be written

$$\sigma^2 \phi = g\frac{\partial \phi}{\partial y}. \quad (2.12)$$

From Eqs. 2.11 and 2.12,

$$\sigma^2 = gk \tanh kd. \quad (2.13)$$

Surface equation. The equation for the surface, y_s, is from Eq. 2.2 for $p = 0$

$$y_s = -\frac{1}{g}\left(\frac{\partial \phi}{\partial t}\right)_{y=y_s} \quad (2.14)$$

or, to the first approximation,

$$y_s = -\frac{1}{g}\left(\frac{\partial \phi}{\partial t}\right)_{y=0}. \quad (2.15)$$

Substituting from Eq. 2.11 into Eq. 2.15, we find

$$y_s = -\frac{i\sigma D}{g}(\cosh kd \sin kx)\, e^{i(\sigma t+\epsilon)} \quad (2.16)$$

or, putting it in the terms of the wave amplitude a, where

$$a = \frac{\sigma D}{g}\cosh kd \quad (2.17)$$

and retaining only the real part

$$y_s = a \sin kx \sin(\sigma t + \epsilon); \quad (2.18)$$

Eq. 2.18 represents a system of "standing waves," of wave length $L = 2\pi/k$. The relation between the wave length and the wave period $T = 2\pi/\sigma$ is given by Eq. 2.13. The wave amplitude, a, is equal to one-half the *wave height*, which is the vertical distance between the wave crest and trough.

Potential and stream functions. The potential function is the real part of Eq. 2.11,

$$\phi = \frac{ga}{\sigma}\frac{\cosh k(y+d)}{\cosh kd}\sin kx \cos(\sigma t + \epsilon) \quad (2.19)$$

and the corresponding stream function ψ is

$$\psi = -\frac{ga}{\sigma}\frac{\sinh k(y+d)}{\cosh kd}\cos kx \cos(\sigma t + \epsilon). \quad (2.20)$$

Particle motions. In order to find the motions of the individual particles, letting ξ, η be the coordinates of a particle relative to the mean position x_0, y_0, we use

$$\frac{d\xi}{dt} = u = \frac{\partial \phi}{\partial x}, \qquad \frac{d\eta}{dt} = v = \frac{\partial \phi}{\partial y}, \quad (2.21)$$

neglecting small quantities of the second order. ξ and η can be found by obtaining $\partial \phi/\partial x$ and $\partial \phi/\partial y$ from Eq. 2.19 and integrating with respect to t. It is found that

$$\xi = a\frac{\cosh k(y_0+d)}{\sinh kd}\cos kx_0 \sin(\sigma t + \epsilon), \quad (2.22)$$

$$\eta = a\frac{\sinh k(y_0+d)}{\sinh kd}\sin kx_0 \sin(\sigma t + \epsilon). \quad (2.23)$$

The particle motions are rectilinear, with the direction of motion being vertical beneath the crests and troughs $(kx = (m+\frac{1}{2})\pi)$ and horizontal beneath the nodes $(kx = m\pi)$. The amplitudes of both vertical and horizontal components of motion decrease with distance below the water surface.

For the case of a wave length, small compared with the water depth (deep water), Eqs. 2.22 and 2.23 simplify as

$$\frac{\cosh k(y+d)}{\sinh kd} \cong \frac{\sinh k(y+d)}{\sinh kd} \cong e^{ky}.$$

Then

$$\xi = a\, e^{ky} \cos kx_0 \sin(\sigma t + \epsilon), \qquad (2.24)$$

$$\eta = a\, e^{ky} \sin kx_0 \sin(\sigma t + \epsilon), \qquad (2.25)$$

The orbital motions decrease rapidly with increasing distance below the water surface. At a distance below the surface equal to one-half a wave length, ξ and η are 1/23.1 their surface values and at a distance below the water surface equal to a wave length ξ and η are only 1/535 their surface values.

The water particle component velocities are

$$u = \frac{gak}{\sigma} \frac{\cosh k(y+d)}{\cosh kd} \cos kx \cos(\sigma t + \epsilon), \qquad (2.26)$$

$$v = \frac{gak}{\sigma} \frac{\sinh k(y+d)}{\cosh kd} \sin kx \cos(\sigma t + \epsilon). \qquad (2.27)$$

Energy. The energy of the standing wave system per wave length per unit width of wave is given by

$$E = \frac{\gamma H^2 L}{16}, \qquad (2.28)$$

and the energy per unit area of the water surface is

$$E_s = \frac{\gamma H^2}{16}, \qquad (2.29)$$

where half the energy is kinetic and half is potential. γ is the specific weight of the water. It will be seen later that this is one-half the energy per unit area of water of a progressive wave of the same height.

3. PROGRESSIVE WAVES, LINEAR THEORY

If it is assumed that the solution is of the form (Lamb, 1945)

$$\phi = P\, e^{i(\sigma t - kx)}. \qquad (2.30)$$

Then the equation for the surface ordinate y_s is

$$y_s = a \sin(kx - \sigma t),$$

or by shifting the origin and substituting other symbols for a, k, and σ,

$$y_s = \frac{H}{2} \cos 2\pi \left(\frac{x}{L} - \frac{t}{T} \right), \qquad (2.31)$$

where H is the wave height, measured vertically from trough to crest, L is the wave length, and T is the wave period. Graphical descriptions of the terms used in these equations for progressive waves, together with the coordinate system used in the equations, are shown in Fig. 2.1. Comparisons of measured surface time histories (laboratory) with theory are shown in Fig. 2.2 for relatively deep water. In shallow water, the agreement between theory and measured values is not nearly so

good. More detailed comparisons for a greater range of conditions will be covered in the section on irrotational waves of finite amplitude.

The potential is given by

$$\phi = \frac{ga}{\sigma} \frac{\cosh k(y+d)}{\cosh kd} \cos(kx - \sigma t),$$

or by shifting the origin to correspond to the origin of Eq. 2.31,

$$\phi = \frac{ga}{\sigma} \frac{\cosh k(y+d)}{\cosh kd} \sin(kx - \sigma t). \qquad (2.32)$$

The phase velocity, C, is given by the equation

$$C = \frac{\sigma}{k} = \frac{gT}{2\pi} \tanh \frac{2\pi d}{L} = \left(\frac{gL}{2\pi} \tanh \frac{2\pi d}{L} \right)^{1/2}. \qquad (2.33)$$

Then

$$\phi = \frac{1}{2} HC \frac{\cosh 2\pi(y+d)/L}{\sinh 2\pi d/L} \sin 2\pi \left(\frac{x}{L} - \frac{t}{T} \right). \qquad (2.34)$$

In deep water ($d/L > \frac{1}{2}$ for many practical purposes), the wave velocity becomes

$$C_0^2 = \frac{gL_0}{2\pi} \qquad (2.35)$$

or, as for periodic waves,

$$L = CT \qquad (2.36)$$

$$C_o = \frac{gT}{2\pi} \qquad (2.37)$$

where the subscript o refers to deep water conditions. In shallow water ($d/L < \frac{1}{25}$ for many practical purposes) $\tanh 2\pi d/L \cong 2\pi d/L$ and Eq. 2.33 becomes

$$C^2 = gd. \qquad (2.38)$$

The equation for expressing wave length in terms of wave period and water depth is

$$L = \frac{gT^2}{2\pi} \tanh \frac{2\pi d}{L}. \qquad (2.39)$$

In deep water, this becomes

$$L_o = \frac{gT^2}{2\pi}. \qquad (2.40)$$

C and L as functions of T and d are plotted in Fig. 2.3.

Useful functions. The ratio of wave velocity in water of any depth to the wave velocity in deep water is

$$\frac{C}{C_o} = \frac{\begin{cases} C = \sigma/k = gT/2\pi \tanh 2\pi d/L \\ \quad = [gL/2\pi \tanh (2\pi d/L)]^{1/2} \end{cases}}{C_o = gT/2\pi} \qquad (2.41)$$

$$= \tanh \frac{2\pi d}{L}$$

and the ratio of the wave length in water of any depth to the length in deep water is

$$\frac{L}{L_o} = \frac{[L = (gT^2/2\pi) \tanh 2\pi d/L]}{[L_o = gT^2/2\pi]} = \tanh \frac{2\pi d}{L}. \qquad (2.42)$$

Equations 2.41 and 2.42 are very useful if the wave

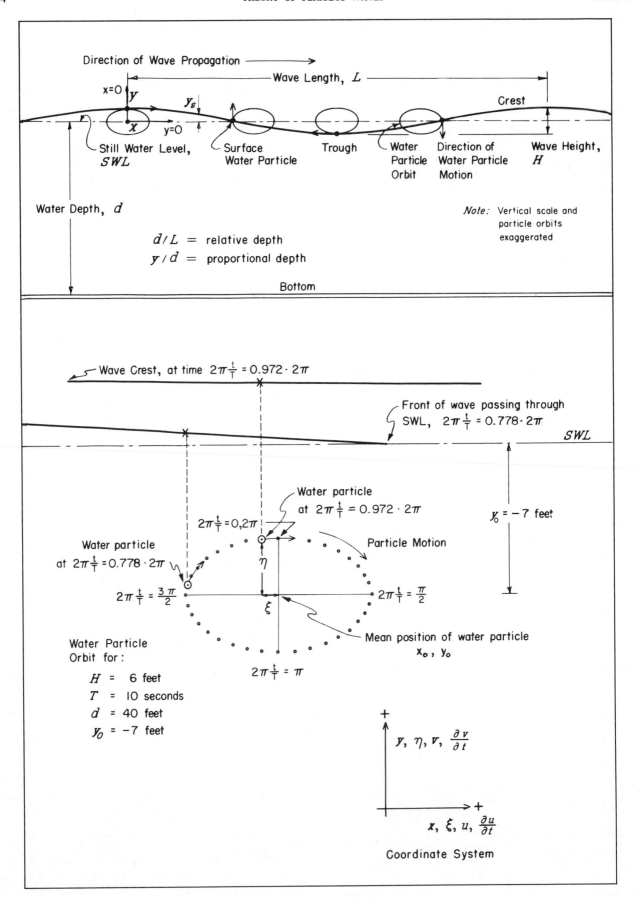

Fig. 2.1. Graphical description of terms

d/L = 0.32   H/L = 0.025

T   = 1.16 sec.   H   = 2.17 ft.

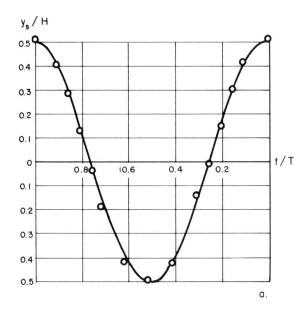

O - Experimental   ——— Theory

d/L = 0.22   H/L = 0.016

T   = 1.43   d   = 2.17 ft.

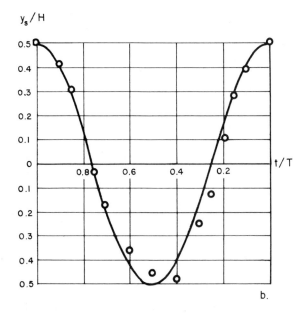

Fig. 2.2. Comparison of laboratory water surface time history with linear theory

period is an independent variable. Then for any particular water depth, within the limits of the linear theory, the deep water wave velocities and wave lengths can be obtained from Eqs. 2.37 and 2.40. Values of d/L can be obtained as a function of even increments of d/L_o by successive approximations using the following equation:

$$\frac{d}{L} \tanh \frac{2\pi d}{L} = \frac{d}{L_o}. \qquad (2.43)$$

The term d/L has been tabulated as a function of d/L_o, together with many other useful functions of d/L, such as $2\pi d/L$, $\tanh 2\pi d/L$, etc. (Wiegel, 1954) in Appendix 1. Some of these functions can also be seen in Fig. 2.4.

Particle motions. The horizontal component of water particle velocity is

$$u = \frac{\partial \phi}{\partial x} = \frac{gak}{\sigma} \frac{\cosh k(y+d)}{\cosh kd} \cos (kx - \sigma t)$$

$$= \frac{gHT}{2L} \frac{\cosh 2\pi(y+d)/L}{\cosh 2\pi d/L} \cos 2\pi \left(\frac{x}{L} - \frac{t}{T}\right) \qquad (2.44)$$

and as

$$C = \frac{gT}{2\pi} \tanh \frac{2\pi d}{L}$$

$$u = \frac{\pi H}{T} \frac{\cosh 2\pi(y+d)/L}{\sinh 2\pi d/L} \cos 2\pi \left(\frac{x}{L} - \frac{t}{T}\right). \qquad (2.45)$$

The vertical component of water particle velocity is

$$v = \frac{\partial \phi}{\partial y} = \frac{gak}{\sigma} \frac{\sinh k(y+d)}{\cosh kd} \sin (kx - \sigma t)$$

$$= \frac{gHT}{2L} \frac{\sinh 2\pi(y+d)/L}{\cosh 2\pi d/L} \sin 2\pi \left(\frac{x}{L} - \frac{t}{T}\right), \qquad (2.46)$$

$$= \frac{\pi H}{T} \frac{\sinh 2\pi(y+d)/L}{\sinh 2\pi d/L} \sin 2\pi \left(\frac{x}{L} - \frac{t}{T}\right). \qquad (2.47)$$

These are component water particle velocities at any point x, y within the fluid.

In deep water (effectively $d/L > \frac{1}{2}$)

$$\frac{\cosh 2\pi(y+d)/L}{\sinh 2\pi d/L} \simeq \frac{\sinh 2\pi(y+d)/L}{\sinh 2\pi d/L} \simeq e^{2\pi y/L},$$

$$u = \frac{\pi H}{T} e^{2\pi y/L} \cos 2\pi \left(\frac{x}{L} - \frac{t}{T}\right) \qquad (2.48)$$

$$v = \frac{\pi H}{T} e^{2\pi y/L} \sin 2\pi \left(\frac{x}{L} - \frac{t}{T}\right). \qquad (2.49)$$

The horizontal and vertical components of water particle velocities can be seen to be of equal magnitude in deep water.

Some comparisons of laboratory measurements of the horizontal components of water particle velocities with theory for relatively deep water are shown in Fig. 2.5. The measurements are for the velocities in a vertical plane through the wave crest and a vertical plane through the wave trough. The measurements were obtained from motion pictures taken through a glass wall section of a wave tank of the motions of small particles of a fluid.

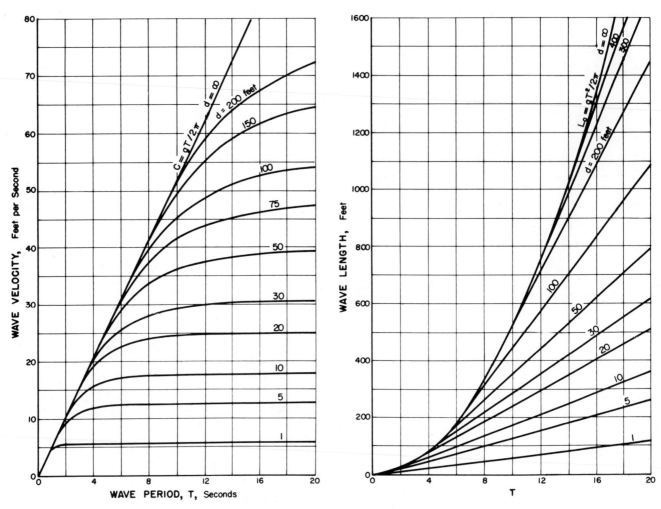

Fig. 2.3. Relationship between wave period, wave velocity (left), and wave length (right) and water depth as given by

$$C^2 = \left(\frac{gL}{2\pi}\right) \tanh\left(\frac{2\pi d}{L}\right) \quad \text{and} \quad L = CT$$

The fluid particles were a mixture of carbon tetrachloride and xylene with zinc oxide added so that they would be easily visible in the film. The proportions of the components were adjusted so that the mixture had as near the same specific gravity as water; some drops would rise slowly, however, and some drops would sink slowly. It can be seen that the linear theory correctly describes the horizontal component of water particle velocity even for waves of appreciable steepness (H/L) for water depth to length ratios (called the relative depth) greater than about 0.2. The effects of decreasing relative depth on actual water particle motions are discussed in more detail in the section on irrotational waves of finite amplitude.

A few comparisons of the theoretical maximum vertical component of particle velocity (Eq. 2.47) and measured values are in Fig. 2.5. The theory predicts the velocities within the range of experimental error for values of d/L as low as about 0.10 for waves of appreciable steepness. It can be seen that the more shallow the water depth, for a given wave length, the greater the deviation between

measured and theoretical values with the difference being of the order of plus 100 per cent, referred to the theoretical value, for a d/L of 0.035. The data do not show any noticeable trend for the upward velocities being of greater or less magnitude than the downward velocities.

To the first approximation, the component velocities of the individual water particles as they move in their orbits are given by the same equations, provided that the particle still water coordinates x_0, y_0 are substituted for x and y in the equations. It is evident that the results are different for the horizontal components. In the Euler form, the equation for the water particle velocity shows it is greater when under the crest (at maximum elevation y with respect to its still water elevation y_0) than when under the trough (at minimum elevation y with respect to its still water elevation y_0). In the Lagrange form, the horizontal component of velocity is the same under the crest as it is under the trough.

The horizontal component of water particle local acceleration at any point x, y within the fluid is

$$\frac{\partial u}{\partial t} = gak\frac{\cosh k(y+d)}{\cosh kd}\sin(kx - \sigma t), \qquad (2.50)$$

$$\frac{\partial u}{\partial t} = \frac{2\pi^2 H}{T^2}\frac{\cosh 2\pi(y+d)/L}{\sinh 2\pi d/L}\sin 2\pi\left(\frac{x}{L} - \frac{t}{T}\right). \qquad (2.51)$$

The vertical component of water particle local acceleration is

$$\frac{\partial v}{\partial t} = -gak\frac{\sinh k(y+d)}{\cosh kd}\cos(kx - \sigma t), \qquad (2.52)$$

$$\frac{\partial v}{\partial t} = -\frac{2\pi^2 H}{T^2}\frac{\sinh 2\pi(y+d)/L}{\sinh 2\pi d/L}\cos 2\pi\left(\frac{x}{L} - \frac{t}{T}\right). \qquad (2.53)$$

There will also be field accelerations, but in the linear theory these are neglected. Measurements of particle accelerations have been almost nil. The few data that are available will be discussed in the section on irrotational waves of finite amplitude.

The motion of the individual particles from their mean position x_0, y_0 can be obtained from $d\xi/dt = u$ and $d\eta/dt = v$ which neglects as second-order effects the differences in velocities at x_0, y_0, and $x_0 + \xi$ and $y_0 + \eta$.

$$\xi = -a\frac{\cosh k(y_0+d)}{\sinh kd}\sin(kx_0 - \sigma t), \qquad (2.54)$$

$$= -\frac{H}{2}\frac{\cosh 2\pi(y_0+d)/L}{\sinh 2\pi d/L}\sin 2\pi\left(\frac{x_0}{L} + \frac{t}{T}\right), \quad (2.55)$$

$$\eta = a\frac{\sinh k(y_0+d)}{\sinh kd}\cos(kx_0 - \sigma t), \qquad (2.56)$$

$$= \frac{H}{2}\frac{\sinh 2\pi(y_0+d)/L}{\sinh 2\pi d/L}\cos 2\pi\left(\frac{x_0}{L} - \frac{t}{T}\right). \qquad (2.57)$$

The water particles move in closed orbits; i.e., each particle returns to its original position each wave cycle. In infinitely deep water, the orbits are circular, but in water of finite depth, they are elliptical—the more shallow the water, the flatter will be the ellipse (Fig. 2.6). Also, in water of finite depth, the greater the distance below the surface, the flatter will be the orbit. Comparison of theory with laboratory measurements for a particular value of d/L is shown in Fig. 2.7. It will be noted in the photographs that the orbits are not closed. This is owing to mass transport and is explained in a later

Fig. 2.4. Progressive wave characteristics in transitional and shallow water relative to deep water-linear theory

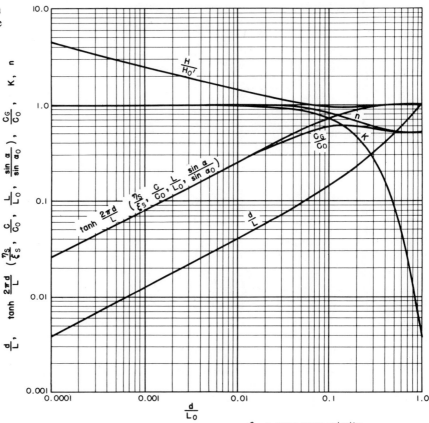

H = wave height
L = wave length
C = wave phase velocity
α = angle of wave crest with bottom
d = water depth
K = pressure response factor at bottom

C_G = wave group velocity
n = ratio of wave group velocity to phase velocity
$^{\prime}$ = superscript refers to waves not affected by refraction
$_0$ = subscript refers to deep water conditions
η_S, ξ_S = vertical and horizontal components surface water particle displacements

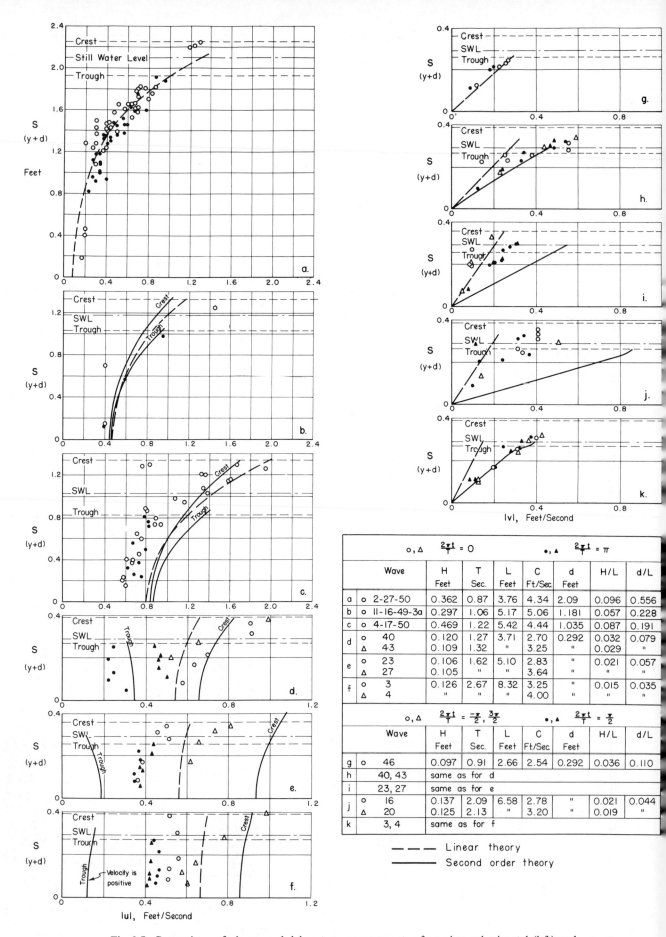

Fig. 2.5. Comparison of theory and laboratory measurements of maximum horizontal (left) and vertical (right) components of particle velocities (after Morison and Crooke, 1953)

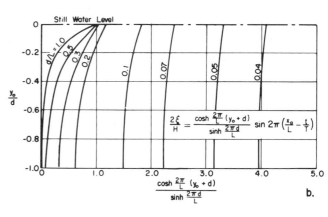

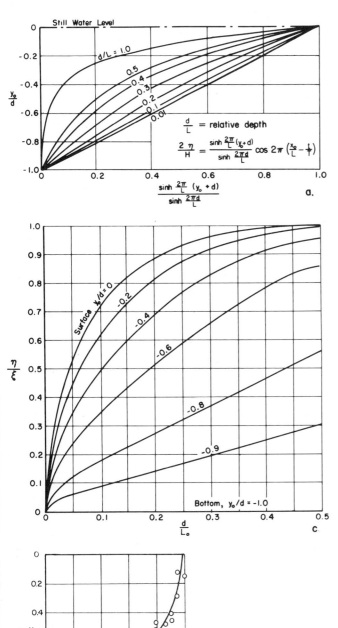

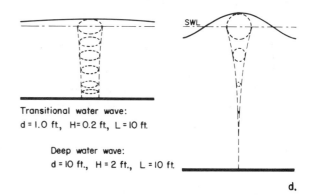

Fig. 2.6. *above:* Ratios of vertical (a) and horizontal (b) double amplitudes of water particle orbits to wave height related to relative depth and proportional depth for water of finite depth (after Wiegel and Johnson, 1951)
left: Ratio of vertical to horizontal particle orbit diameters
below: Water particle orbits for two wave conditions

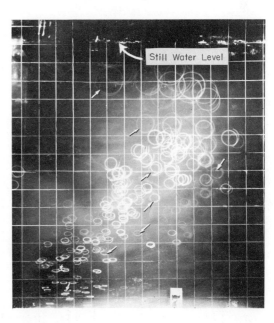

Transitional water wave:
d = 1.0 ft, H = 0.2 ft, L = 10 ft.

Deep water wave:
d = 10 ft., H = 2 ft., L = 10 ft.

d.

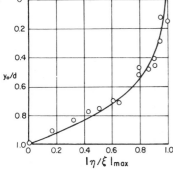

(a) Comparison of the ratios of measured orbit lengths with theory for $d/L = 0.39$

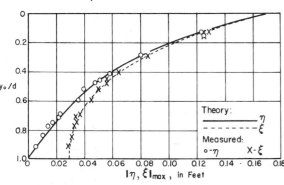

(b) Comparison of measured orbit semi-major and semi-minor lengths with theory for $d/L = 0.39$

(c) Photograph of water particle orbits for a wave of the dimensions: $d = 2.50$ ft, $H = 0.339$ ft, $L = 6.42$ ft, $T = 1.12$ sec, $d/L = 0.39$

Fig. 2.7.

section where second-order effects are considered. Some additional data are shown in Fig. 2.7.

Subsurface pressure. The subsurface pressure can be obtained from Eqs. 2.2 and 2.25. Neglecting terms higher than first-order gives

$$p + \rho g y = \rho g a \frac{\cosh k(y + d)}{\cosh kd} \cos (kx - \sigma t), \quad (2.58)$$

$$\frac{p}{\rho g} + y = \frac{H}{2} \frac{\cosh 2\pi(y + d)/L}{\cosh 2\pi d/L} \cos 2\pi \left(\frac{x}{L} - \frac{t}{T}\right). \quad (2.59)$$

The term on the right-hand side of the equation is known as the *dynamic pressure*. It is positive under the wave crest and negative under the wave trough.

The ratio

$$\frac{\cosh 2\pi(y + d)/L}{\cosh 2\pi d/L}$$

is known as the *pressure response factor K*. This ratio has been plotted as a function of y/d and d/L in Fig. 2.8.

Wave energy and power. The total energy of a wave per unit width of crest is the sum of its kinetic and potential energies with the energies being half kinetic and half potential. The development shown here is after Rayleigh (1877) and Lamb (1945).

The kinetic energy in an element of the fluid is

$$\frac{\gamma}{2g} q^2 \, dx \, dy \, dz$$

where q^2 is $u^2 + v^2 + w^2$; $w = 0$, however, so the kinetic energy in an element of the fluid is $(\gamma/2g)(u^2 + v^2)dx \, dy \, dz$ where u and v are given by Eqs. 2.30 and 2.30b. The energy in a wave per unit width of crest is

$$E_k = \int_0^L \int_0^{-d} \frac{\gamma}{2g} (u^2 + v^2)dx \, dy, \quad (2.60)$$

$$= \int_0^L \int_0^{-d} \frac{\gamma}{2g} \left[\left(\frac{gak}{\sigma} \frac{\cosh k(y + d)}{\cosh kd} \cos (kx - \sigma t)\right)^2 \right.$$

$$\left. + \left(\frac{gak}{\sigma} \frac{\sinh k(y + d)}{\cosh kd} \sin (kx - \sigma t)\right)^2 \right] dx \, dy$$

$$= \frac{\gamma a^2 L}{4} = \frac{\gamma H^2 L}{16} \quad (2.61)$$

where γ is the unit weight of water (ρg).

The potential energy in an element of fluid displaced a distance y from the still water level is

$$y\gamma \, dx \, dy \, dz.$$

The total potential energy in a wave per unit width of crest is

$$E_p = \int_0^L \int_0^{y_s} \gamma y \, dx \, dy \quad (2.62)$$

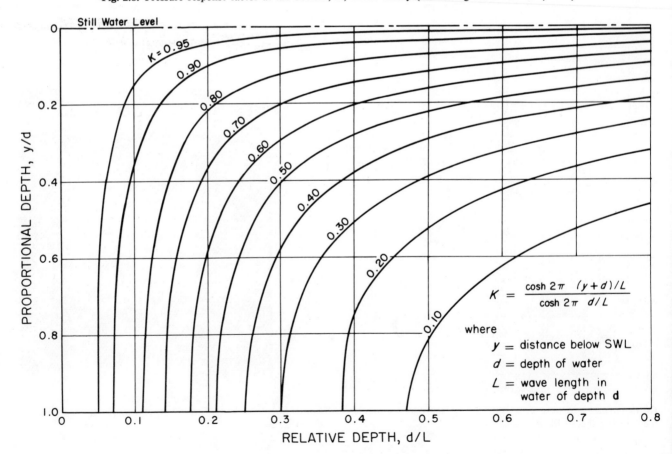

Fig. 2.8. Pressure response factor at the bottom, K, linear theory (after Wiegel and Johnson, 1951)

where y_s is given by Eq. 2.31, where $a = H/2$.

$$E_p = \int_0^L \gamma y_s^2 \, dx$$

$$= \frac{\gamma a^2}{2} \int_0^L \cos^2 (kx - \sigma t) \, dx$$

$$= \frac{\gamma a^2 L}{4} = \frac{\gamma H^2 L}{16} . \qquad (2.63)$$

The total energy in a wave per unit width of crest is

$$E = E_k + E_p = \frac{\gamma H^2 L}{8} . \qquad (2.64)$$

Power is the product of the force acting on a plane in the fluid and the velocity of the flow across this plane; the force acting on a fluid plane is the product of the pressure and the area. In considering the rate at which energy is transmitted in the direction of wave propagation across a vertical plane, the first approximation of the mean power per unit width of wave crest can be expressed as (Reynolds, 1877, and Rayleigh, 1877)

$$P = \frac{1}{T} \int_0^T \int_0^{-d} (p + \rho g y) u \, dt \, dy \qquad (2.65)$$

where $p + \rho g y$ is the variable pressure (the mean power associated with the static pressure being zero) and is given by Eq. 2.58.

$$P = \frac{1}{T} \int_0^T \int_0^{-d} \left[\rho g a \frac{\cosh k(y + d)}{\cosh kd} \cos (kx - \sigma t) \right]$$

$$\left[\frac{gak}{\sigma} \frac{\cosh k(y + d)}{\cosh kd} \cos (kx - \sigma t) \right] dt \, dy$$

$$= \frac{1}{T} \frac{\rho g^2 a^2 k}{\sigma \cosh^2 kd} \int_0^T \int_0^{-d} \cosh^2 k(y + d)$$

$$\cos^2 (kx - \sigma t) dt \, dy.$$

Integrating the preceding equation and substituting $C/\tanh kd$ for g/σ, we get

$$P = \frac{1}{2} \rho g a^2 C \cdot \frac{1}{2} \left[1 + \frac{2kd}{\sinh 2kd} \right], \qquad (2.66)$$

$$= \frac{\gamma H^2 L}{8T} \frac{1}{2} \left[1 + \frac{2kd}{\sinh 2kd} \right] . \qquad (2.67)$$

The term $\gamma H^2 L/8$ is the expression for the total energy in a wave per unit width of crest and

$$\frac{1}{2} \left[1 + \frac{2kd}{\sinh 2kd} \right] = \frac{1}{2} \left[1 + \frac{4\pi d/L}{\sinh 4\pi d/L} \right] = \frac{C_G}{C} = n \qquad (2.68)$$

is the ratio of the group velocity to the phase velocity (C_G/C). Equation 2.67 can be written

$$P = \frac{nE}{T} . \qquad (2.69)$$

In deep water, n is equal to $\frac{1}{2}$ and in shallow water it is equal to 1.

Equation 2.66 has led to confusion in the literature, with some authors stating that all of the wave energy travels with the group velocity and others stating that half the wave energy travels with the phase velocity. Nothing in the development of the equation can be used to substantiate either claim over the other, or that either claim is correct. Equation 2.66 is the equation of power transmission in an infinite train of waves: it does not mark and follow a little "bit" of energy. That the equation is essentially correct has been shown for the case of waves in a channel where the waves are generated at one end by a moving piston and the transmitted power absorbed at the other end (Snyder, Wiegel, and Bermel, 1958). The mean power necessary to operate the piston to sustain a train of uniform periodic waves was almost exactly that computed by Eq. 2.67 for the measured values of H, L, T, and d.

If E is given in foot-pounds and T in seconds, the mean wave power may be expressed as horsepower by

$$P = \frac{nE}{550T} \text{ hp.} \qquad (2.70)$$

Table 2.1 gives the energy horsepower per foot of crest width of waves of various heights and periods in deep water. It should be emphasized that the table shows figures for waves which are of much greater amplitude than that for which the linear theory is applicable.

Limiting condition, linear theory. Stokes pointed out that, for the linear theory to be valid, the wave steepness must be small, and in addition $L^2 H/2d^3$ must be small. This has also been discussed by Ursell (1953). The limiting value of $L^2 H/2d^3$ has not been determined experimentally, but Longuet-Higgins (1956) states that

$$\frac{L^2 H}{2d^3} \ll \frac{16\pi^2}{3} . \qquad (2.71)$$

4. STANDING WAVES, FINITE AMPLITUDE

Miche (1944) has developed the equations for irrotational cylindrical standing waves in water of finite depth to the second order for zero mass transport. Carry (1953), as a part of a mathematical study on partial standing waves also developed equations to the second order. Sekerz-Zen'Kovic (1951) developed equations to the first, second, and third orders. There are a few differences between the three sets of equations. When these differences occurred, the equations of Miche were followed. In general, only second-order theory is presented, although some of the details from one higher-order theory will be given. In the literature, the term *clapotis* is often used for standing wave, especially when referring to the standing wave system resulting from a progressive wave train reflecting from a vertical wall.

The potential is given by

$$\phi = \frac{HL}{2T} \frac{\cosh 2\pi(y + d)/L}{\sinh 2\pi d/L} \sin \frac{2\pi x}{L} \cos \frac{2\pi t}{T}$$

$$- \frac{3\pi H^2}{32T} \frac{\cosh 4\pi(y + d)/L}{\sinh^4 2\pi d/L} \cos \frac{4\pi x}{L} \sin \frac{4\pi t}{T} \qquad (2.72)$$

Table 2.1. THE ENERGY AND POWER OF WAVES, PER FOOT OF CREST, IN DEEP WATER

A. $E_o = \dfrac{\gamma H_o^2 L_o}{8} = \dfrac{64}{8} H_o^2 L_o = 8 H_o^2 L_o$ for salt water

T (sec)	L_o (ft)	E_o, wave energy per foot width of wave crest (ft-lb/ft) H_o (ft)									
		2	4	6	8	10	15	20	30	40	50
2	20.5	656									
4	81.9	2,621	10,500	23,600	41,900	65,500					
6	184	5,890	23,600	53,000	94,200	147,000	331,000	589,000			
8	328	10,500	42,000	94,500	168,000	262,000	590,000	1,050,000	2,360,000	4,200,000	
10	512	16,400	65,500	148,000	262,000	410,000	922,000	1,640,000	3,690,000	6,550,000	10,200,000
12	737	23,600	94,300	212,000	377,000	590,000	1,330,000	2,360,000	5,310,000	9,430,000	14,700,000
14	1004	32,100	129,000	289,000	514,000	803,000	1,810,000	3,210,000	7,230,000	12,900,000	20,100,000
16	1310	41,900	168,000	377,000	671,000	1,050,000	2,360,000	4,190,000	9,430,000	16,800,000	26,200,000
18	1658	53,100	212,000	478,000	849,000	1,330,000	2,880,000	5,310,000	11,900,000	21,200,000	33,200,000
20	2047	65,500	262,000	590,000	1,050,000	1,640,000	3,680,000	6,550,000	14,700,000	26,200,000	40,900,000

B. $P_o = \dfrac{nE_o}{550T}$ horsepower per foot width of wave crest

T (sec)	L_o (ft)	P_o, wave power per foot width of wave crest (hp/ft) H_o (ft)									
		2	4	6	8	10	15	20	30	40	50
2	20.5	0.3									
4	81.9	0.6	2.4	5.4	9.5	15.9					
6	184	0.9	3.6	8.0	14.3	22.3	50.2	89.2			
8	328	1.2	4.8	10.7	19.1	29.8	67.1	120	268	477	
10	512	1.5	6.0	13.4	24.8	37.2	83.8	149	335	597	931
12	737	1.8	7.1	16.1	28.6	44.7	101	179	402	715	1,120
14	1004	2.1	8.3	18.8	33.4	52.2	117	207	469	835	1,300
16	1310	2.4	9.5	22.4	38.1	59.5	134	238	536	953	1,490
18	1658	2.7	10.7	24.1	43.9	67.0	151	268	603	1,070	1,670
20	2047	3.0	11.9	26.8	47.6	74.4	168	298	700	1,190	1,860

where H, L, and T are the height, length, and period of the standing wave. This point is emphasized, as in oceanographic problems standing waves are sometimes the result of the combination of the waves incident to, and reflected from, walls, in which case the incident wave height is often used in calculations—this difference must be kept in mind. If the standing wave system results from waves reflected from a vertical wall, the coordinate system will be such that the wall is located at $x = L/4$.

To the second order (Sekerz-Zen'Kovic, 1951), the wave length is given by the same equation as to the first order; that is,

$$L = \frac{gT^2}{2\pi} \tanh \frac{2\pi d}{L}$$

which is also the relationship among L, T, and d for progressive waves to the second order.

The free surface is expressed by

$$y_s = \frac{H}{2} \sin \frac{2\pi x}{L} \sin \frac{2\pi t}{T} - \frac{\pi H^2}{4L} \coth \frac{2\pi d}{L}$$

$$\left\{ \sin^2 \frac{2\pi t}{T} - \frac{3 \cos 4\pi t/T + \tanh^2 2\pi d/L}{4 \sinh^2 2\pi d/L} \right\} \cos \frac{4\pi x}{L}. \quad (2.73)$$

In deep water, this becomes

$$y_s = \frac{H}{2} \sin \frac{2\pi x}{L} \sin \frac{2\pi t}{T} - \frac{\pi H^2}{4L} \cos \frac{4\pi x}{L} \sin^2 \frac{4\pi t}{T}. \quad (2.74)$$

The parametric equations for the free surface (in Lagrangian coordinates) are

$$x = x_0 + \frac{H}{2} \coth \frac{2\pi d}{L} \cos \frac{2\pi x_0}{L} \sin \frac{2\pi t}{T}$$
$$- \frac{\pi H^2}{2L} \frac{\sin 4\pi x_0/L}{\sinh^2 2\pi d/L} \left[\sin^2 \frac{2\pi t}{T} + \frac{\cosh 4\pi d/L}{4 \sinh^2 2\pi d/L} \right.$$
$$\left. \left(3 \cos \frac{4\pi t}{T} + \tanh^2 \frac{2\pi d}{L} \right) \right], \quad (2.75)$$
$$y = \frac{H}{2} \sin \frac{2\pi x_0}{L} \sin \frac{2\pi t}{T} + \frac{\pi H^2}{2L} \frac{\sinh 4\pi d/L}{\sinh^2 2\pi d/L}$$
$$\left[\sin^2 \frac{2\pi t}{T} + \frac{\cos 4\pi x_0/L}{4 \sinh^2 2\pi d/L} \left(3 \cos \frac{4\pi t}{T} + \tanh^2 \frac{2\pi d}{L} \right) \right].$$

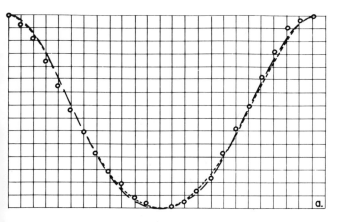

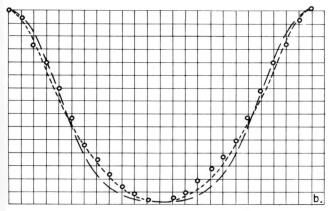

T, sec.	L, cm.	d, cm	H̄/2,cm	H̄/L	d/L
a.	1	137	30	7.8	0.057	0.22
b.	1.2	179	30	15.4	0.086	0.17

   o Experimental points
\- \- \- \- Eulerian equations (to the second order)
\- \- \- \- - Lagrangian " " " "

Fig. 2.9. Comparison of measured and theoretical standing wave profiles (after Suquet and Wallet, 1953)

A few laboratory measurements have been made by Suquet and Wallet (1953) of standing wave profiles. These data are shown in Fig. 2.9. It can be seen that Eq. 2.75 predicts the profiles with a high degree of precision within the range of conditions tested. In regard to the maximum steepness of a standing wave, Miche considered that the wave would break when $\partial x/\partial x_0 = 0$, and that this condition occurred when

$$1 - 2\left(\frac{\pi H}{2L}\coth\frac{2\pi d}{L}\right) - 2\left(\frac{\pi H}{2L}\coth\frac{2\pi d}{L}\right)^2$$
$$\left[\frac{\frac{3}{4} + \sinh^2 2\pi d/L}{\sinh^2 2\pi d/L \cosh^4 2\pi d/L}\right] = 0. \tag{2.76}$$

In deep water, this reduces to

$$\frac{H}{L} = \frac{\tanh 2\pi d/L}{\pi} \simeq \frac{1}{\pi} = 0.318. \tag{2.77}$$

Danel (1952) studied this phenomenon in a wave channel, creating standing waves by generating progressive waves which were reflected from a vertical wall. At the same time, he also studied the limiting steepness of progressive waves. The data are shown in Fig. 2.10. The data for the standing waves are in terms of $H/2L$, as he plotted the steepness in terms of the progressive wave heights. It can be seen that the maximum steepness in relatively deep water is considerably less than $\frac{1}{2} \times 0.318 = 0.159$.

Penney and Price (1952) have made a theoretical study of standing waves. They state that they were not able to

Fig. 2.10. Laboratory determination of limiting steepness of progressive and standing waves in a channel (after Danel, 1952)

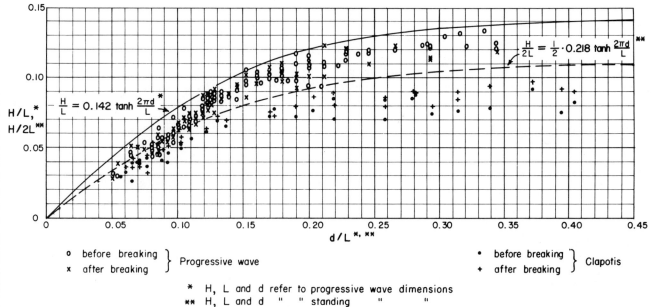

prove mathematically the existence of truly periodic standing waves beyond doubt, but that they would appear to exist. They found that in infinitely deep water the maximum wave steepness to the fifth order would be

$$\left(\frac{H}{L}\right)_{\max} = 0.218 \qquad (2.78)$$

and that the maximum crest height above the still water level and the minimum trough depth below the still water level would be

$$\left(\frac{H_c}{L}\right)_{\max} = 0.141,$$

$$\left(\frac{H_t}{L}\right)_{\max} = 0.077. \qquad (2.79)$$

Laboratory experiments by Taylor (1953) showed these to be very nearly correct (Fig. 2.11). Taylor found however that for $H_0/L > 0.125$ the waves became three-dimensional in character and a mass of water moved vertically upward from the wave.

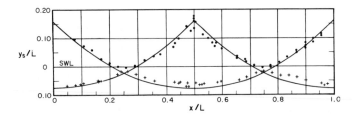

Measurements with probes: x, measured from one end of tank; y, measured from water surface before starting wave maker; $\cdot$, downward pointing probe; $+$, upward pointing probe

Fig. 2.11. Comparison of theory and laboratory measurements of standing waves of limiting steepness in deep water (after Taylor, 1953)

Penney and Price did not solve the problem of maximum steepness of standing waves in water of finite depth. Some theoretical studies they made showed that it would depend upon the water depth and would tend to zero as the depth decreased. In Fig. 2.10 an equation is plotted similar to the one for the limiting steepness of progressive waves in water of any depth

$$\frac{H}{L} = 0.218 \tanh \frac{2\pi d}{L}. \qquad (2.80)$$

In Fig. 2.10, one-half this value is plotted in conformance with the way Danel plotted his data. The curve appears to provide an acceptable upper limit to Danel's data.

Penney and Price found that to the fifth order the water surface was never perfectly flat, but at its minimum deviation from a flat surface, with respect to time, was

$$y_s = \frac{L}{2\pi}\left(\frac{1}{7} A^4 \cos \frac{4\pi x}{L}\right) \qquad (2.81)$$

and the surface at a time $2\pi t/T = \pi$ later was

$$y_s = \frac{L}{2\pi}\left[\left(A + \frac{1}{32} A^3 - \frac{47}{1344} A^5\right) \cos \frac{2\pi x}{L}\right.$$
$$+ \left(\frac{1}{2} A^2 - \frac{79}{672} A^4\right) \cos \frac{4\pi x}{L}$$
$$+ \left(\frac{3}{8} A^3 - \frac{12563}{59136} A^5\right) \cos \frac{6\pi x}{L}$$
$$\left. + \frac{1}{3} A^4 \cos \frac{8\pi x}{L} + \frac{295}{768} A^5 \cos \frac{10\pi x}{L}\right]. \qquad (2.82)$$

The mean position of the water surface, averaged with respect to time was found to be

$$y_m = \frac{L}{2\pi}\left[\left(\frac{1}{4} A^2 + \frac{1}{16} A^4\right) \cos \frac{4\pi x}{L} + \frac{1}{8} A^4 \cos \frac{8\pi x}{L}\right] \qquad (2.83)$$

rather than a horizontal plane. Further, it was found that there were no points on the water surface which had no vertical movement at all. At the points which would be true nodes in the linear theory, it was found that the water surface would vary between $-\frac{1}{7}A^4$ and $-\frac{1}{2}A^2 + \frac{101}{224}A^4$.

It is reasonable to expect that this phenomenon will be even more prominent for shallow water standing waves. An example of a shallow water standing wave in a long narrow channel is shown in Fig. 2.12. The ends of the channel were not vertical, but their slopes were steep compared with the wave length and wave steepness. It can be seen that the surface time histories at the two antinodes have a single periodicity whereas the surface time history at the node has a frequency of twice that of the antinodes. This is also apparent at the quarter points.

The height of the wave crest above the still water level is

$$H_c = \frac{L}{2\pi}\left(A + \frac{1}{2} A^2 + \frac{13}{32} A^3 + \frac{145}{672} A^4 + \frac{2021}{17484} A^5\right), \qquad (2.85a)$$

and the depth of the wave trough below the still water level is

$$H_t = \frac{L}{2\pi}\left(-A + \frac{1}{2} A^2 - \frac{13}{32} A^3 + \frac{145}{672} A^4 - \frac{2021}{17484} A^5\right). \qquad (2.85b)$$

If the relationship between the linear wave period and length is given by

$$T_l = \sqrt{\frac{2\pi}{g} L_l} \qquad (2.86)$$

then the relationship between the fifth-order wave period to the linear wave period for the same wave length is

$$\left(\frac{T_l}{T}\right)^2 = 1 - \frac{1}{4} A^2 - \frac{13}{128} A^4. \qquad (2.87)$$

This shows that for a given wave length (say, a standing wave formed between two vertical walls), the greater the wave height, the longer the period.

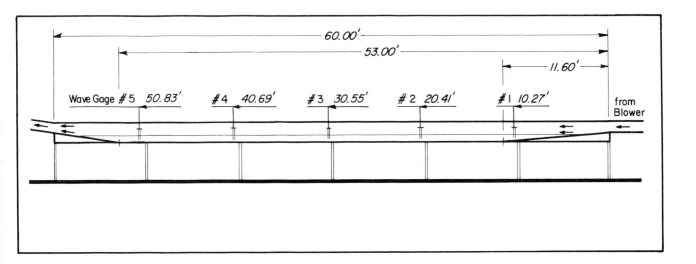

(Run 16:   V = 15,   d = 0.2098')

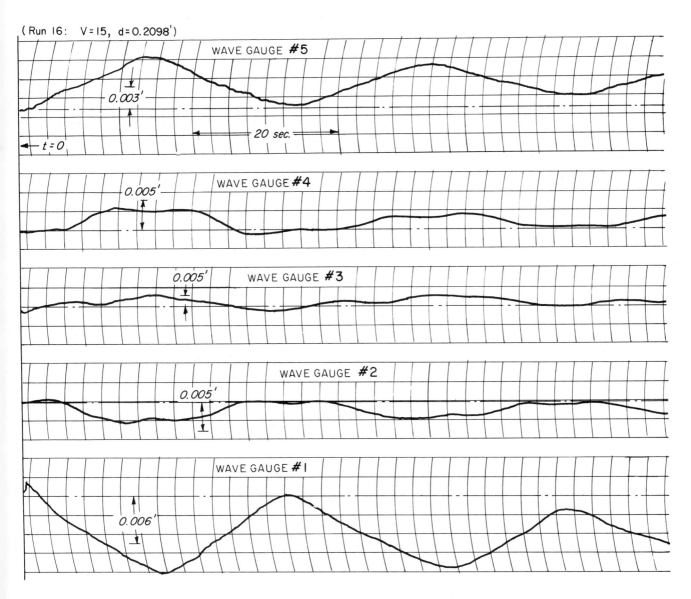

Fig. 2.12. Measured characteristics of impulsively generated standing wave set up in channel by wind starting to blow over water surface (after Tickner, 1958)

Table 2.2. RELATIONSHIP BETWEEN RATIO OF LINEAR STANDING WAVE PERIOD
TO FIFTH-ORDER PERIOD AND WAVE STEEPNESS
(from Taylor, 1953)

A	0	0.1	0.2	0.3	0.4	0.5	0.592
T_l/T	1.000	0.9987	0.9949	0.9883	0.9785	0.9650	0.9487
H_c/L	0	0.0168	0.0356	0.0570	0.0816	0.1103	0.1411

Taylor (1953) has tabulated T_l/T and H_c/L as a function of A as shown in Table 2.2.

When considering the case of a progressive wave train reflecting from a vertical wall, the periods of progressive waves and the standing waves are identical, so the wave length must be different. To the third order, in deep water, for relatively low waves, Penney and Price found the ratio of the progressive wave length, L_p, to the resulting standing wave length, L_s, is

$$\frac{L_p}{L_s} = 1 + 8 \left(\frac{\pi H_p}{L_p} \right)^2. \quad (2.88)$$

For the case of $H_p/L_p = 0.05$, it is found that $L_p/L_s = 1.20$; that is, the resulting standing wave is 20 per cent shorter than the progressive wave. It is difficult to see how such a standing wave can exist in terms of what we normally think of as a standing wave.

A few laboratory measurements are shown in Fig. 2.13. It can be seen that the lengths of the standing waves formed by progressive waves reflecting from a vertical wall are a little shorter than predicted from linear theory.

The mean level of the standing wave in its plane of

Fig. 2.13. Comparison of measured and theoretical standing wave lengths, standing wave formed by reflected progressive wave from vertical wall in wave channel

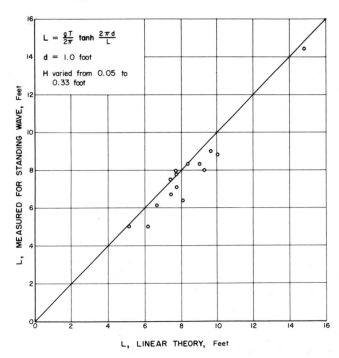

maximum vertical motion (the antinode) is above the still water level by the amount Δh (Miche, 1944).

$$\Delta h = \frac{\pi H^2}{4L} \left(1 + \frac{3}{4 \sinh^2 2\pi d/L} - \frac{1}{4 \cosh^2 2\pi d/L} \right)$$
$$\cdot \coth \frac{2\pi d}{L}. \quad (2.89a)$$

As $\cosh 2\pi d/L > \sinh 2\pi d/L$, $\Delta h > (\pi H^2/4L) \coth 2\pi d/L$ where

$$\frac{\pi H^2}{4L} \coth \frac{2\pi d}{L} \quad (2.89b)$$

is a commonly used expression for Δh. The relationships among $\Delta h/H$, H/L, and d/L as expressed by Eq. 2.89a can be seen in Table 2.3. Blank squares indicate that the standing wave would have broken before these particular combinations of H/L and d/L could have been reached, according to the data of Danel (1952).

Miche presented some data which tended to substantiate the validity of Eq. 2.89. These data, shown in Table 2.4, are for standing waves formed at vertical breakwaters. Laboratory studies made at the University of California show that Eq. 2.89a predicts Δh more closely than does Eq. 2.89b (Fig. 2.14), at least within the range of experimental data.

The pressure p any point x, y in the fluid is

$$\frac{p}{\rho g} + y = \frac{H}{2} \frac{\cosh 2\pi (y + d)/L}{\cosh 2\pi d/L} \sin \frac{2\pi x}{L} \sin \frac{2\pi t}{T}$$
$$- \frac{\pi H^2}{8L} \frac{\cos^2 2\pi t/T}{\sinh 2\pi d/L \cosh 2\pi d/L}$$
$$\left[\cosh \frac{4\pi}{L}(y + d) + \cos \frac{4\pi x}{L} - 1 \right] \quad (2.90)$$
$$+ \frac{3\pi H^2}{16L} \frac{\cosh 4\pi(y + d)/L}{\sinh^3 2\pi d/L \cosh 2\pi d/L} \cos \frac{4\pi x}{L}$$
$$\cos \frac{4\pi t}{T} + \frac{\pi H^2}{4L} \tanh \frac{2\pi d}{L} \cos \frac{4\pi t}{T}.$$

In Fig. 2.15 are presented the data used by Miche to show the validity of Eq. 2.90. These data were obtained by Stucky (1934) in some laboratory tests. Miche points out that this expression contains one item which is significantly different from the second-order equation for progressive waves. There is one term which does not attenuate with depth.

$$\frac{\pi H^2}{4L} \tanh \frac{2\pi d}{L} \cos \frac{4\pi t}{T} \longrightarrow \frac{\pi H^2}{4L} \cos \frac{4\pi t}{T}$$
$$\text{as} \quad \frac{d}{L} \to \infty.$$

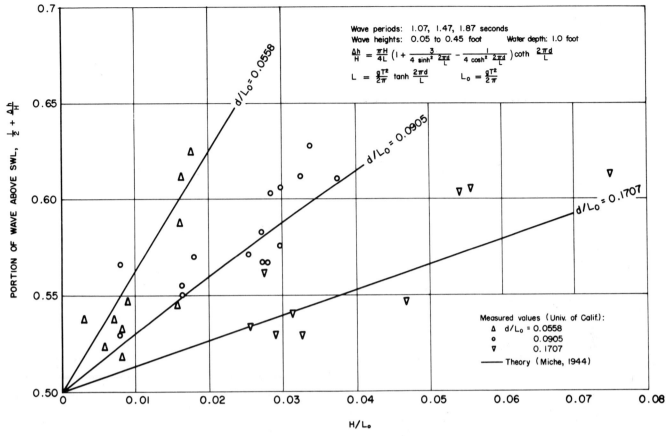

Fig. 2.14. Portion of standing wave above still water level

Table 2.3. RATIO OF MEAN WAVE LEVEL TO STANDING WAVE HEIGHT AT ANTINODE, AS A FUNCTION OF d/L AND H/L

d/L \ H/L	0.005	0.010	0.020	0.050	0.075	0.100	0.150	0.200
0.5	0.004	0.008	0.016	0.040	0.060	0.079	0.119	0.158
0.4	0.004	0.008	0.016	0.040	0.060	0.081	0.121	0.161
0.3	0.004	0.009	0.017	0.043	0.065	0.086	0.130	0.173
0.2	0.006	0.011	0.023	0.056	0.084	0.113	0.169	0.225
0.1	0.018	0.035	0.070	0.176	0.264	0.352		
0.08	0.030	0.060	0.120	0.300				
0.06	0.063	0.127	0.254					
0.04	0.198	0.395						
0.02								
0.01								

Table 2.4. ELEVATION OF MEAN WAVE LEVEL ABOVE STILL WATER LEVEL FOR STANDING WAVE
(from Miche, 1944, p. 67)

Structure location	Depth d (meters)	Standing wave ht. H (meters)	Wave length L (meters)	Wave steepness H/L	L/d	Elevation of mean standing wave level above SWL (meters)		
						obs.	Miche (Eq. 2.89a)	Sainflou (Eq. 2.89b)
Gênes (Umberto's principle)	1.5	10	110	1/11	7.3	1.73	1.71	1.02
Catane	18	16.50	155	1/9.4	8.6	4.75	4.40	2.18
Algiers (Mustapha)	21	21.50	185	1/8.6	8.8	4.25	6.47	3.15

At very great depths in deep water

$$\frac{p}{\rho g} + y \cong \frac{\pi H^2}{4L} \cos \frac{4\pi t}{T}. \qquad (2.91)$$

This is a dynamic pressure which has a period of one-half that of the standing wave. The significance of this with respect to the formation of microseisms in the ocean is of considerable scientific interest. Cooper and Longuet-Higgins (1951) have made laboratory measurements of the pressure in water deep enough that the normal dynamic pressure was very small compared with the dynamic pressure expressed by the term of Eq. 2.91. For the case of a water depth of 1.352 ft, a standing wave of 0.065 ft in height and 0.461 sec in period ($d/L = 1.244$ and $H/L = 0.060$) pressure measurements were made at distances below the still water surface of $y/d = 0.524$, 0.644, 0.765, and 0.887. The ratio of measured pressure to theoretical pressure as given by Eq. 2.91 was found to be 0.97, 0.96, 0.99, and 1.03, respectively, for the given values of y/d. Observations were also made for different positions of x under the wave, and it was found that these pressures were independent of x.

The components of water particle velocities at any point x, y in the fluid are

$$u = \frac{\pi H}{T} \frac{\cosh 2\pi(y + d)/L}{\sinh 2\pi d/L} \cos \frac{2\pi x}{L} \cos \frac{2\pi t}{T}$$

$$+ \frac{3}{8} \left(\frac{\pi H}{T}\right)\left(\frac{\pi H}{L}\right) \frac{\cosh 4\pi(y + d)/L}{\sinh^4 2\pi d/L} \qquad (2.92)$$

$$\cdot \sin \frac{4\pi x}{L} \sin \frac{4\pi t}{T},$$

$$v = -\frac{\pi H}{T} \frac{\sinh 2\pi(y + d)/L}{\sinh 2\pi d/L} \sin \frac{2\pi x}{L} \cos \frac{2\pi t}{T}$$

$$+ \frac{3}{8} \left(\frac{\pi H}{T}\right)\left(\frac{\pi H}{L}\right) \frac{\sinh 4\pi(y + d)/L}{\sinh^4 2\pi d/L} \qquad (2.93)$$

$$\cdot \cos \frac{4\pi x}{L} \sin \frac{4\pi t}{T}.$$

The components of water particle velocities of particles with still water positions x_0, y_0 are

$$u = \frac{\pi H}{T} \frac{\cosh 2\pi(y_0 + d)/L}{\sinh 2\pi d/L} \cos \frac{2\pi x_0}{L} \cos \frac{2\pi t}{T}$$

$$- \frac{1}{4} \left(\frac{\pi H}{T}\right)\left(\frac{\pi H}{L}\right) \frac{1}{\sinh^2 2\pi d/L} \qquad (2.94)$$

$$\left[1 - \frac{3}{2} \frac{\cosh 4\pi(y_0 + d)/L}{\sinh^2 2\pi d/L}\right] \sin \frac{4\pi x_0}{L} \sin \frac{4\pi t}{T},$$

$$v = -\frac{\pi H}{T} \frac{\sinh 2\pi(y_0 + d)/L}{\sinh 2\pi d/L} \sin \frac{2\pi x_0}{L} \cos \frac{2\pi t}{T}$$

$$- \frac{1}{4} \left(\frac{\pi H}{T}\right)\left(\frac{\pi H}{L}\right) \frac{\sinh 4\pi(y_0 + d)/L}{\sinh^2 2\pi d/L}$$

$$\left[1 - \frac{3}{2} \frac{\cos 4\pi x_0/L}{\sinh^2 2\pi d/L}\right] \sin \frac{4\pi t}{T}. \qquad (2.95)$$

It is important to note that these velocities are approximately twice the velocities of the progressive waves that

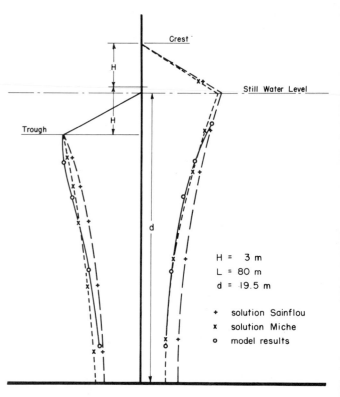

Fig. 2.15. Pressure distribution of standing wave on vertical wall (after Miche, 1944)

would form a standing wave by reflecting from a vertical wall which is normal to the direction of advance of the progressive waves.

The components of water particle local accelerations at any point x, y in the fluid are

$$\frac{\partial u}{\partial t} = -2 \left(\frac{\pi^2 H}{T^2}\right) \frac{\cosh 2\pi(y + d)/L}{\sinh 2\pi d/L} \cos \frac{2\pi x}{L} \sin \frac{2\pi t}{T}$$

$$+ \frac{3}{2} \left(\frac{\pi^2 H}{T^2}\right)\left(\frac{\pi H}{L}\right) \frac{\cosh 4\pi(y + d)/L}{\sinh^4 2\pi d/L}$$

$$\cdot \sin \frac{4\pi x}{L} \cos \frac{4\pi t}{T}, \qquad (2.96)$$

$$\frac{\partial v}{\partial t} = 2 \left(\frac{\pi^2 H}{T^2}\right) \frac{\sinh 2\pi(y + d)/L}{\sinh 2\pi d/L} \sin \frac{2\pi x}{L} \sin \frac{2\pi t}{T}$$

$$+ \frac{3}{2} \left(\frac{\pi^2 H}{T^2}\right)\left(\frac{\pi H}{L}\right) \frac{\sinh 4\pi(y + d)/L}{\sinh^4 2\pi d/L}$$

$$\cdot \cos \frac{4\pi x}{L} \cos \frac{4\pi t}{T}. \qquad (2.97)$$

The components of water particle local accelerations with respect to their still water positions x_0, y_0 are

$$\frac{\partial u}{\partial t} = -2 \left(\frac{\pi^2 H}{T^2}\right) \frac{\cosh 2\pi(y_0 + d)/L}{\sinh 2\pi d/L} \cos \frac{2\pi x_0}{L} \sin \frac{2\pi t}{T}$$

$$- \left(\frac{\pi^2 H}{T^2}\right)\left(\frac{\pi H}{L}\right) \frac{1}{\sinh^2 2\pi d/L} \qquad (2.98)$$

$$\left[1 - \frac{3}{2} \frac{\cosh 4\pi(y_0 + d)/L}{\sinh^2 2\pi d/L}\right] \sin \frac{4\pi x_0}{L} \cos \frac{4\pi t}{T},$$

$$\frac{\partial v}{\partial t} = 2\left(\frac{\pi^2 H}{T^2}\right)\frac{\sinh 2\pi(y_0 + d)/L}{\sinh 2\pi d/L}\sin\frac{2\pi x_0}{L}\sin\frac{2\pi t}{T}$$

$$-\left(\frac{\pi^2 H}{T^2}\right)\left(\frac{\pi H}{L}\right)\frac{\sinh 4\pi(y_0 + d)/L}{\sinh^2 2\pi d/L}$$

$$\left[1 - \frac{3}{2}\frac{\cos 4\pi x_0/L}{\sinh^2 2\pi d/L}\right]\cos\frac{4\pi t}{T}. \tag{2.99}$$

The displacements of the water particles from their still water positions x_0, y_0 are

$$\xi = \frac{H}{2}\frac{\cosh 2\pi(y_0 + d)/L}{\sinh 2\pi d/L}\cos\frac{2\pi x_0}{L}\sin\frac{2\pi t}{T} - \frac{\pi H^2}{8L}$$

$$\cdot\frac{\sin 4\pi x_0/L}{\sinh^2 2\pi d/L}\left[\sin^2\frac{2\pi t}{T} + \frac{\cosh 2\pi(y_0 + d)/L}{4\sinh^2 2\pi d/L}\right.$$

$$\left.\left(3\cos\frac{4\pi t}{T} + \tanh^2\frac{2\pi d}{L}\right)\right], \tag{2.100}$$

$$\eta = \frac{H}{2}\frac{\sinh 2\pi(y_0 + d)/L}{\sinh 2\pi d/L}\sin\frac{2\pi x_0}{T}\sin\frac{2\pi t}{T} + \frac{\pi H^2}{8L}$$

$$\cdot\frac{\sinh 4\pi(y_0 + d)/L}{\sinh^2 2\pi d/L}\left[\sin^2\frac{2\pi t}{T} + \frac{\cos 4\pi x_0/L}{4\sinh^2 2\pi d/L}\right.$$

$$\left.\left(3\cos\frac{4\pi t}{T} + \tanh^2\frac{2\pi d}{L}\right)\right]. \tag{2.101}$$

5. STOKES' WAVES, FINITE AMPLITUDE

Stokes (1880) studied waves of small, but finite, wave steepness. The method used by Stokes and many later investigators was to expand the velocity potential about the still water level, obtaining a nonlinear surface condition for the potential on the plane of the still water level which consists of an infinite series containing partial derivatives of the potential. The solution is obtained by successive approximations. The solutions require a great amount of detailed calculations of coefficients, etc., and small errors often occur. Because of this, the final results of the various investigators are not always the same. The proof of the convergence of the series was made by Levi-Civita (1925) for infinitely deep water and by Struik (1926) for water of finite depth. Certain corrections to Struik's work have been made by Wolf (1944) and Hunt (1953).

Hunt points out that the inequalities used in the work of Levi-Civita restricts his existence proof to waves with steepnesses less than 1:200, although with certain refinements the limit can be stretched to a wave steepness of 1:98. Wilton (1914) did work on Stokes' series to the twelfth order in deep water and concluded that although the series for the wave profile became divergent in the neighborhood of the crest of the highest wave, it held right up to this point. For water of finite depth, the restriction on wave steepness is more severe. Stokes was the first to point this out (Stokes, 1880, p. 325). He stated that the approximation was slower for the case of finite depth. In deep water, the profile of the wave lends itself to the type of series developed by Stokes

whereas in shallow water the nearly isolated and widely separated crests would require a comparatively large number of terms to express the wave form relatively accurately. In the extreme case, the fact that the wave is one of a series would have little to do with water particle motion in the vicinity of the crests, the only portion of the wave where the motion is appreciable.

A comparison has been made between cnoidal and Stokes' waves by De (1955). It appears that the cnoidal wave theory is better when considering waves in shallow water than is the theory of Stokes' waves. De extended Stokes' theory to the fifth order. He showed, by comparing certain features of the theories, that Stokes' theory (to the fifth order) should not be used for d/L less than about 0.125, the minimum value depending upon the value of H/L, with the greater the values of H/L, the greater the value of d/L at which Stokes' theory becomes unreliable.

Second order. The equations as presented herein are, in general, compatible with the equations developed by Miche (1945) and Biesel (1952).

To the second approximation, the potential function is

$$\phi = aC\frac{\cosh k(y + d)}{\sinh kd}\sin(kx - \sigma t) + \frac{3}{4}\frac{\pi a^2 C}{L}$$

$$\cdot\frac{\cosh 2k(y + d)}{\sinh^4 kd}\sin 2(kx - \sigma t), \tag{2.102}$$

$$= \frac{HL}{2T}\frac{\cosh 2\pi(y + d)/L}{\sinh 2\pi d/L}\sin 2\pi\left(\frac{x}{L} - \frac{t}{T}\right)$$

$$+ \frac{3\pi H^2}{16T}\frac{\cosh 4\pi(y + d)/L}{\sinh^4 2\pi d/L}\sin 4\pi\left(\frac{x}{L} - \frac{t}{T}\right). \tag{2.103}$$

To the second order, the equations of wave velocity and wave length are the same as for the linear theory; that is,

$$C = \frac{gT}{2\pi}\tanh\frac{2\pi d}{L} \quad \text{and} \quad L = \frac{gT^2}{2\pi}\tanh\frac{2\pi d}{L}.$$

The free surface is expressed by

$$y_s = \frac{H}{2}\cos 2\pi\left(\frac{x}{L} - \frac{t}{T}\right) + \frac{\pi H^2}{8L}$$

$$\cdot\frac{\cosh 2\pi d/L[2 + \cosh 4\pi d/L]}{\sinh^3 2\pi d/L}\cos 4\pi\left(\frac{x}{L} - \frac{t}{T}\right)$$

$$= \frac{H}{2}\cos 2\pi\left(\frac{x}{L} - \frac{t}{T}\right) + \frac{\pi H^2}{4L} \tag{2.104}$$

$$\left(1 + \frac{3}{2\sinh^2 2\pi d/L}\right)\coth\frac{2\pi d}{L}\cdot\cos 4\pi\left(\frac{x}{L} - \frac{t}{T}\right).$$

In Fig. 2.16 are plotted the theoretical profiles together with profiles measured in a wave tank. The laboratory data are from Morison and Crooke (1953).

This shows that the mean level of the wave surface lies above the still water level by the amount

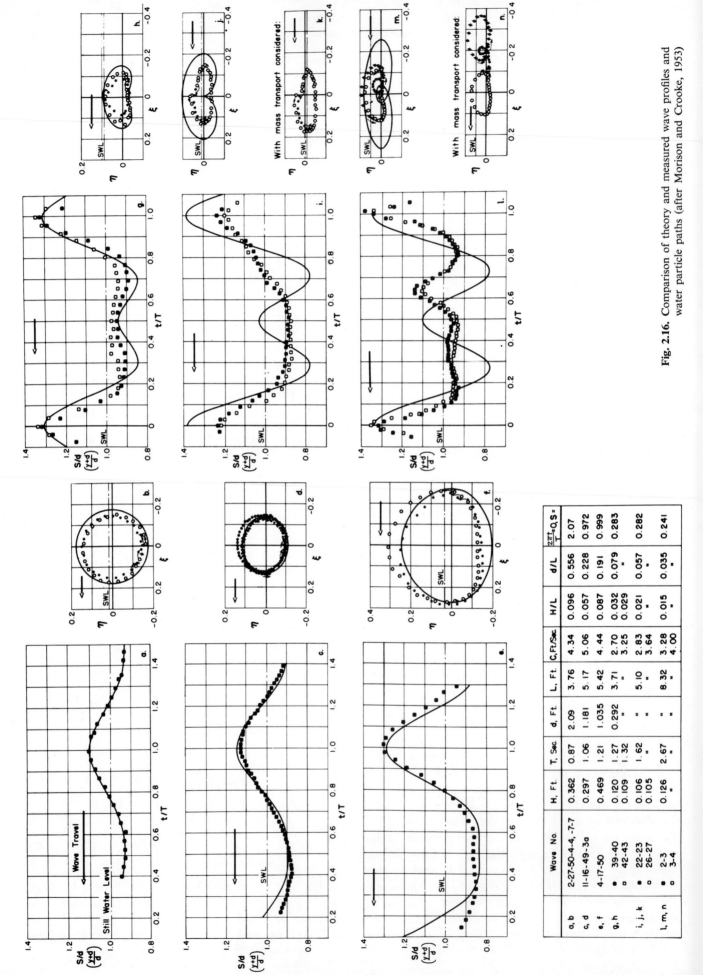

Fig. 2.16. Comparison of theory and measured wave profiles and water particle paths (after Morison and Crooke, 1953)

Wave No.		H, Ft.	T, Sec.	d, Ft.	L, Ft.	C, Ft/Sec.	H/L	d/L	$\frac{2\pi t}{T}=\theta, S=$
a, b	2-27-50-4-4, -7-7	0.362	0.87	2.09	3.76	4.34	0.096	0.556	2.07
c, d	11-16-49-3a	0.297	1.06	1.181	5.17	5.06	0.057	0.228	0.972
e, f	4-17-50	0.469	1.21	1.035	5.42	4.44	0.087	0.191	0.999
g, h	39-40	0.120	1.27	0.292	3.71	2.70	0.032	0.079	0.283
	42-43	0.109	1.32	"	"	3.25	0.029	"	"
i, j, k	22-23	0.106	1.62	"	5.10	2.83	0.021	0.057	0.282
	26-27	0.105	"	"	"	3.64	"	"	"
l, m, n	2-3	0.126	2.67	"	8.32	3.28	0.015	0.035	0.241
	3-4	"	"	"	"	4.00	"	"	"

30

Table 2.5. Displacement of Mean Wave Level Above Still Water Level

d/L_0	0.5	0.4	0.3	0.2	0.1	0.08	0.06	0.04	0.02	0.01	0.005
d/L	0.502	0.405	0.312	0.225	0.141	0.123	0.104	0.0833	0.0576	0.0403	0.0284
$1 + \dfrac{3}{2 \sinh^2 2\pi d/L}$	1.01	1.04	1.13	1.40	2.482	3.06	4.02	5.98	12.0	23.9	47.9
$\dfrac{\Delta h_2}{H} \cdot \dfrac{L}{H} = \dfrac{\pi}{4} \coth \dfrac{2\pi d}{L}$	0.79	0.80	0.82	0.87	1.11	1.21	1.37	1.64	2.10	3.17	4.46
$\dfrac{\Delta h_1}{H} \cdot \dfrac{L}{H} = \dfrac{\pi}{4} \coth \dfrac{2\pi d}{L} \left(1 + \dfrac{3}{2 \sinh^2 2\pi d/L}\right)$	0.80	0.82	0.84	1.23	2.86	3.71	5.48	9.80	25.1	75.6	213

$$\Delta h_1 = \frac{\pi H^2}{4L}\left(1 + \frac{3}{2 \sinh^2 2\pi d/L}\right) \coth \frac{2\pi d}{L} \quad (2.105)$$

which, as Miche points out, is in excess of the commonly used value of

$$\Delta h_2 = \frac{\pi H^2}{4L} \coth \frac{2\pi d}{L}. \quad (2.106)$$

Equation 2.105 predicts a greater portion of the wave above the still water level than does Eq. 2.106. Modifications of Eq. 2.105 and 2.106 have been tabulated in Table 2.5. Comparisons of Eq. 2.106 with laboratory measurements have been inconclusive due to the scatter of the data.

The waves are symmetrical about vertical planes drawn through their crests and troughs. The crests are steeper and the troughs are flatter than is the case for the linear theory.

For some combinations of H/L and d/L, the second harmonic will be such that a secondary crest will appear at the trough. The limiting condition for a null trough has been given by Miche (1944) as

$$\frac{H}{L} = \frac{\sinh^2 2\pi d/L}{3\pi} \tanh \frac{2\pi d}{L}. \quad (2.107)$$

The value of wave steepness for various values of d/L is shown in Table 2.6.

For deep water ($d/L > \frac{1}{2}$ for many practical purposes), the equation for the free surface becomes

$$y_{s_0} = \frac{H_0}{2} \cos 2\pi \left(\frac{x}{L_0} - \frac{t}{T}\right) + \frac{\pi H_0^2}{4L_0} \cos 4\pi \left(\frac{x}{L_0} - \frac{t}{T}\right). \quad (2.108)$$

The subsurface pressure at any point x, y in the fluid is given by

$$\frac{p}{\rho g} + y = \frac{H}{2} \frac{\cosh 2\pi (y + d)/L}{\cosh 2\pi d/L} \cos 2\pi \left(\frac{x}{L} - \frac{t}{T}\right)$$
$$+ \frac{3}{8} \frac{\pi H^2}{L} \frac{\tanh 2\pi d/L}{\sinh^2 2\pi d/L} \left(\frac{\cosh 4\pi (y + d)/L}{\sinh^2 2\pi d/L} - \frac{1}{3}\right)$$
$$\cdot \cos 4\pi \left(\frac{x}{L} - \frac{t}{T}\right) - \frac{1}{8} \frac{\pi H^2}{L} \frac{\tanh 2\pi d/L}{\sinh^2 2\pi d/L}$$
$$\cdot \cosh \frac{4\pi}{L} (y + d). \quad (2.109)$$

It should be noted that there is a nonperiodic pressure which attenuates rather slowly with the depth below the still water surface.

The components of water particle velocities at any place x, y in the fluid are given by

$$u = \frac{\partial \phi}{\partial x}$$
$$= \frac{\pi H}{T} \frac{\cosh 2\pi (y + d)/L}{\sinh 2\pi d/L} \cos 2\pi \left(\frac{x}{L} - \frac{t}{T}\right) + \frac{3}{4}$$
$$\left(\frac{\pi H}{T}\right)\left(\frac{\pi H}{L}\right) \frac{\cosh 4\pi (y + d)/L}{\sinh^4 2\pi d/L} \cos 4\pi \left(\frac{x}{L} - \frac{t}{T}\right), \quad (2.110)$$

$$v = \frac{\partial \phi}{\partial y}$$
$$= \frac{\pi H}{T} \frac{\sinh 2\pi (y + d)/L}{\sinh 2\pi d/L} \sin 2\pi \left(\frac{x}{L} - \frac{t}{T}\right) + \frac{3}{4}$$
$$\left(\frac{\pi H}{T}\right)\left(\frac{\pi H}{L}\right) \frac{\sinh 4\pi (y + d)/L}{\sinh^4 2\pi d/L} \sin 4\pi \left(\frac{x}{L} - \frac{t}{T}\right). \quad (2.111)$$

The displacements of the water particles from their still water positions are

$$\xi = -\frac{1}{2} H \frac{\cosh 2\pi (y_0 + d)/L}{\sinh 2\pi d/L} \sin 2\pi \left(\frac{x}{L} - \frac{t}{T}\right)$$
$$+ \frac{\pi H^2}{8L \sinh^2 2\pi d/L} \left[1 - \frac{3}{2} \frac{\cosh 4\pi (y_0 + d)/L}{\sinh^2 2\pi d/L}\right]$$
$$\cdot \sin 4\pi \left(\frac{x_0}{L} - \frac{t}{T}\right) + \frac{\pi H^2}{4L} \frac{\cosh 4\pi (y_0 + d)/L}{\sinh^2 2\pi d/L}$$
$$\cdot \frac{2\pi t}{T}, \quad (2.112)$$

$$\eta = \frac{1}{2} H \frac{\sinh 2\pi (y_0 + d)/L}{\sinh 2\pi d/L} \cos 2\pi \left(\frac{x_0}{L} - \frac{t}{T}\right)$$
$$+ \frac{3\pi H^2}{16L} \frac{\sinh 4\pi (y_0 + d)/L}{\sinh^4 2\pi d/L} \cos 4\pi \left(\frac{x_0}{L} - \frac{t}{T}\right)$$
$$+ \frac{\pi H^2}{8L} \frac{\sinh 4\pi (y_0 + d)/L}{\sinh^2 2\pi d/L}. \quad (2.113)$$

Comparisons of the water particle orbits as measured in the laboratory (Morison and Crooke, 1953) and the

Table 2.6. H/L Versus d/L For Null Wave Trough

d/L	0.01	.02	.03	.04	.05	.06	.07	.08	.09	.10	.11	.12	.13	.14	.15	.16
H/L	0.00003	.00021	.00071	.0017	.0033	.0057	.0090	.014	.019	.027	.036	.040	.059	.074	.093	.113

preceding equations are made in Fig. 2.16. The theoretical curves are not quite the same as given by Morison and Crooke. The measured data of Morison and Crooke had the mass transport component motion subtracted, so the mass transport portion of Eq. 2.112 has been left out of the theoretical curves. The effect of the mass transport in an extreme case can be seen in Fig. 2.16(n) where it is actually in the direction opposite to the direction of wave travel because the wave tank was a closed system. This phenomenon is discussed in detail later.

The components of velocities of the individual particles with respect to their still water positions x_0, y_0 are

$$u = \frac{\pi H}{T} \frac{\cosh 2\pi(y_0 + d)/L}{\sinh 2\pi d/L} \cos 2\pi \left(\frac{x_0}{L} - \frac{t}{T}\right)$$

$$+ \left(\frac{\pi H}{T}\right)\left(\frac{\pi H}{L}\right) \frac{3}{4 \sinh^2 2\pi d/L} \left[-\frac{1}{2} + \frac{3}{4}\right.$$

$$\left. \cdot \frac{\cosh 4\pi(y_0 + d)/L}{\sinh^2 2\pi d/L}\right] \cos 4\pi \left(\frac{x_0}{L} - \frac{t}{T}\right)$$

$$+ \frac{1}{2}\left(\frac{\pi H}{T}\right)\left(\frac{\pi H}{L}\right) \frac{\cosh 4\pi(y_0 + d)/L}{\sinh^2 2\pi d/L}, \qquad (2.114)$$

$$v = \frac{\pi H}{T} \frac{\sinh 2\pi(y_0 + d)/L}{\sinh 2\pi d/L} \sin 2\pi \left(\frac{x_0}{L} - \frac{t}{T}\right)$$

$$+ \frac{3}{4}\left(\frac{\pi H}{T}\right)\left(\frac{\pi H}{L}\right) \frac{\sinh 4\pi(y_0 + d)/L}{\sinh^4 2\pi d/L}$$

$$\cdot \sin 4\pi \left(\frac{x_0}{L} - \frac{t}{T}\right). \qquad (2.115)$$

Some comparisons of theory with laboratory data are shown in Fig. 2.5. The mass transport of the horizontal component has been removed from both theory and measured data.

It can be seen in Eqs. 2.112 and 2.114 that there is a nonperiodic drift in the direction of wave advance. This drift is called *mass transport*. The mass transport velocity is (Stokes, 1880)

$$\bar{U} = \frac{1}{2}\left(\frac{\pi H}{T}\right)\left(\frac{\pi H}{L}\right) \frac{\cosh 4\pi(y_0 + d)/L}{\sinh^2 2\pi d/L}. \qquad (2.116)$$

For deep water this reduces to

$$\bar{U}_o = \left(\frac{\pi H}{T}\right)\left(\frac{\pi H}{L}\right) e^{4\pi y/L}. \qquad (2.117)$$

Ursell (1953) has presented a mathematical proof for the mass transport velocity for a null net transport of water. His equation is the same as obtained by Miche (1944) and is

$$\bar{U}_n = \frac{1}{2}\left(\frac{\pi H}{T}\right)\left(\frac{\pi H}{L}\right)$$

$$\cdot \frac{\cosh 4\pi(y + d)/L - (L/4\pi d)\sinh 2\pi d/L}{\sinh^2 2\pi d/L}. \qquad (2.118)$$

In deep water this reduces to

$$\bar{U}_{n_o} = \left(\frac{\pi H}{T}\right)\left(\frac{\pi H}{L}\right)\left(e^{4\pi y/L} - \frac{L}{4\pi d}\right) \qquad (2.119)$$

which is different from the equation given by either Ursell or Longuet-Higgins (1953). It is the same, however, as the equation given by Mitchim (1940) who obtained Eq. 2.119 from Eq. 2.117 by assuming that, in a closed wave channel, the water moved to one end of the channel by mass transport will return in a flow that is steady and with a uniform velocity distribution. This gives a very simple physical explanation of Eq. 2.119, and also Eq. 2.118. The return flow having a uniform velocity distribution results in the waves still being irrotational in character. A few measurements of mass transport associated with deep water waves in a laboratory tank by Mitchim (1940) confirmed Eq. 2.119. Other laboratory experiments (Mason, 1941) have also confirmed Eq. 2.119, which is for deep water.

The equation for mass transport in relatively shallow water has been found to be invalid (Mason, 1941; Bagnold, 1947). A proper solution must include viscous terms. This has been done by Longuet-Higgins (1953), and it explains some of the findings of Bagnold (1947). These findings, together with those of Russell and Osorio, (1958) will be discussed in a later section. The most important effect of the viscosity is that the direction of the movement of the mass transport near the bottom is always in the direction of wave advance.

The components of water particle local accelerations at any point x, y in the fluid are

$$\frac{\partial u}{\partial t} = 2\left(\frac{\pi^2 H}{T^2}\right) \frac{\cosh 2\pi(y + d)/L}{\sinh 2\pi d/L} \sin 2\pi \left(\frac{x}{L} - \frac{t}{T}\right)$$

$$+ 3\left(\frac{\pi^2 H}{T^2}\right)\left(\frac{\pi H}{L}\right) \frac{\cosh 4\pi(y + d)/L}{\sinh^4 2\pi d/L}$$

$$\cdot \sin 4\pi \left(\frac{x}{L} - \frac{t}{T}\right), \qquad (2.120)$$

$$\frac{\partial v}{\partial t} = -2\left(\frac{\pi^2 H}{T^2}\right) \frac{\sinh 2\pi(y + d)/L}{\sinh 2\pi d/L} \cos 2\pi \left(\frac{x}{L} - \frac{t}{T}\right)$$

$$- 3\left(\frac{\pi^2 H}{T^2}\right)\left(\frac{\pi H}{L}\right) \frac{\sinh 4\pi(y + d)/L}{\sinh^4 2\pi d/L}$$

$$\cdot \cos 4\pi \left(\frac{x}{L} - \frac{t}{T}\right). \qquad (2.121)$$

The components of local accelerations of the individual water particles with respect to their still water positions x_0, y_0 are

$$\frac{\partial u}{\partial t} = 2\left(\frac{\pi^2 H}{T^2}\right) \frac{\cosh 2\pi(y_0 + d)/L}{\sinh 2\pi d/L} \sin 2\pi \left(\frac{x_0}{L} - \frac{t}{T}\right)$$

$$+ \left(\frac{\pi^2 H}{T^2}\right)\left(\frac{\pi H}{L}\right) \frac{2}{\sinh^2 2\pi d/L} \left[-1 + \frac{3}{2}\right.$$

$$\left. \cdot \frac{\cosh 4\pi(y_0 + d)/L}{\sinh^2 2\pi d/L}\right] \sin 4\pi \left(\frac{x_0}{L} - \frac{t}{T}\right), \qquad (2.122)$$

$$\frac{\partial v}{\partial t} = -2\left(\frac{\pi^2 H}{T^2}\right) \frac{\sinh 2\pi(y_0 + d)/L}{\sinh 2\pi d/L} \cos 2\pi \left(\frac{x_0}{L} - \frac{t}{T}\right)$$

$$- 3\left(\frac{\pi^2 H}{T^2}\right)\left(\frac{\pi H}{L}\right)\frac{\sinh 4\pi (y_0 + d)/L}{\sinh^4 2\pi d/L}$$

$$\cdot \cos 4\pi \left(\frac{x_0}{L} - \frac{t}{T}\right). \qquad (2.123)$$

Elliott (1953) made few measurements of the horizontal components of the velocities and accelerations of immiscible fluid particles with the same specific gravity as the water which were placed at various distances below the water surface (Figs. 2.17 and 2.18). The results were compared with Eqs. 2.114 and 2.122. As can be seen in Fig. 2.17, the magnitudes of the maximum values were good in all three cases; the position of the maximum forward acceleration, however, led the wave crest by a considerably smaller time than Eq. 2.122 predicts. It should be stated that Eq. 2.122 is the equation for the horizontal component of particle local acceleration, whereas the measured values were of the horizontal component of total acceleration.

Third order. Stokes extended his work to the third order, as have several other investigators. Many of the equations of the third order shown here, however, are due to Skjelbreia (1959). The equation for the free water surface is

$$y_s = \frac{1}{2} a \cos 2\pi \left(\frac{x}{L} - \frac{t}{T}\right) + \frac{\pi a^2}{L} f_2 \left(\frac{d}{L}\right)$$

$$\cdot \cos 4\pi \left(\frac{x}{L} - \frac{t}{T}\right) + \frac{\pi^2 a^3}{L^2} f_3 \left(\frac{d}{L}\right)$$

$$\cdot \cos 6\pi \left(\frac{x}{L} - \frac{t}{T}\right) \qquad (2.124)$$

where $f_2 \left(\dfrac{d}{L}\right) = \dfrac{(2 + \cosh 4\pi d/L)\cosh 2\pi d/L}{2 \sinh^3 2\pi d/L}$,

$$f_3 \left(\frac{d}{L}\right) = \frac{3}{16} \frac{1 + 8 \cosh^6 2\pi d/L}{\sinh^6 2\pi d/L},$$

and the relationship between the wave height H and a is given by

$$H = 2a + 2\frac{\pi^2}{L^2} a^3 \cdot f_3 \left(\frac{d}{L}\right). \qquad (2.125)$$

For d/L large (say $> \frac{1}{2}$), $f_2(d/L) \to 1$ and $f_3(d/L) \to \frac{3}{2}$, and

$$y_{s_0} = a \cos 2\pi \left(\frac{x}{L} - \frac{t}{T}\right) + \frac{\pi a^2}{L} \cos 4\pi \left(\frac{x}{L} - \frac{t}{T}\right)$$

$$+ \frac{3\pi^2 a^3}{2L^2} \cos 6\pi \left(\frac{x}{L} - \frac{t}{T}\right).$$

The value of L to be used in these third-order equations is given by

$$L = \frac{gT^2}{2\pi} \tanh \frac{2\pi d}{L} \left[1 + \left(\frac{2\pi a}{L}\right)^2 \frac{14 + 4 \cosh^2 4\pi d/L}{16 \sinh^4 2\pi d/L}\right]. \qquad (2.126)$$

It can be seen that the wave length depends upon the wave height as well as the water depth and wave period. This makes the use of the third-order equation difficult

as the various functions must be tabulated in relation to both H/d and $2\pi d/gT^2$.

The wave velocity is given by

$$C^2 = \frac{gL}{2\pi} \tanh \frac{2\pi d}{L} \left[1 + \left(\frac{\pi a}{L}\right)^2 \frac{8 + \cosh 8\pi d/L}{8 \sinh^4 2\pi d/L}\right]$$

$$= \frac{gL}{2\pi} \tanh \frac{2\pi d}{L} \left[1 + \left(\frac{\pi a}{L}\right)^2 \frac{14 + 4 \cosh^2 4\pi d/L}{16 \sinh^4 2\pi d/L}\right]. \qquad (2.127)$$

This corresponds to the equation developed by Hunt (1953), but differs from several other expressions given in the literature. The equation for wave velocity to the third order as developed by Stokes is

$$C^2 = \frac{gL}{2\pi} \tanh \frac{2\pi q}{CL} \left[1 + \left(\frac{2\pi a}{L}\right)^2 \frac{12 + S_4 + 2S_2}{D_1^4}\right] \qquad (2.128a)$$

or $C^2 = \dfrac{gL}{2\pi} \tanh \dfrac{2\pi q}{CL} \left[1 + \left(\dfrac{2\pi a}{L}\right)^2 \dfrac{10 + S_2^2 + 2S_2}{D_1^4}\right]$ (2.128b)

where $D_n = 2 \sinh \dfrac{2n\pi q}{CL}$ and $S_n = 2 \cosh \dfrac{2n\pi q}{CL}$,

and $\dfrac{q}{CL} = \dfrac{d}{L} - \dfrac{\pi H^2}{4L^2} \coth \dfrac{2\pi q}{CL}$.

Using this relationship, the wave velocity was found independently by Hunt (1953) and Skjelbreia (1959) to be as shown in Eq. 2.127. In several places in the literature, Eq. 2.128 has been used with d/L replacing q/CL, which is incorrect to the third order.

The ratio

$$\frac{4 \cosh^2 (4\pi d/L) + 14}{16 \sinh^4 2\pi d/L}$$

is shown in its relation to d/L in Fig. 2.22. This ratio grows large without bounds as $d/L \longrightarrow 0$, for $\cosh^2 4\pi d/L \longrightarrow 1$ and $\sinh^4 2\pi d/L \longrightarrow 0$. This growing function is counterbalanced by the function $(2\pi a/L)^2$ which grows very small as $d/L \longrightarrow 0$.

In deep water, Eq. 2.127 reduces to

$$C_0^2 = \frac{gL_0}{2\pi} \left[1 + \left(\frac{\pi a}{L_0}\right)^2\right]. \qquad (2.129)$$

Some measurements of the velocity of laboratory waves have been made by Wiegel (1950), Morison (1951), Suquet and Wallet (1953), and Savage (1954), but the scatter of data were of the same magnitude as the difference between third-order and linear theory. Some of these data are shown in Fig. 2.19. The measurements show that the waves of large amplitude traveled faster than the waves of small amplitude. The work of Suquet and Wallet includes no data for d/L less than 0.2, but the work of Morison does. In the region of small d/L, where the third-order equation for wave velocity blows up, the measured values of wave velocity were not large; the combinations of wave heights and lengths and water depths used in the various laboratory

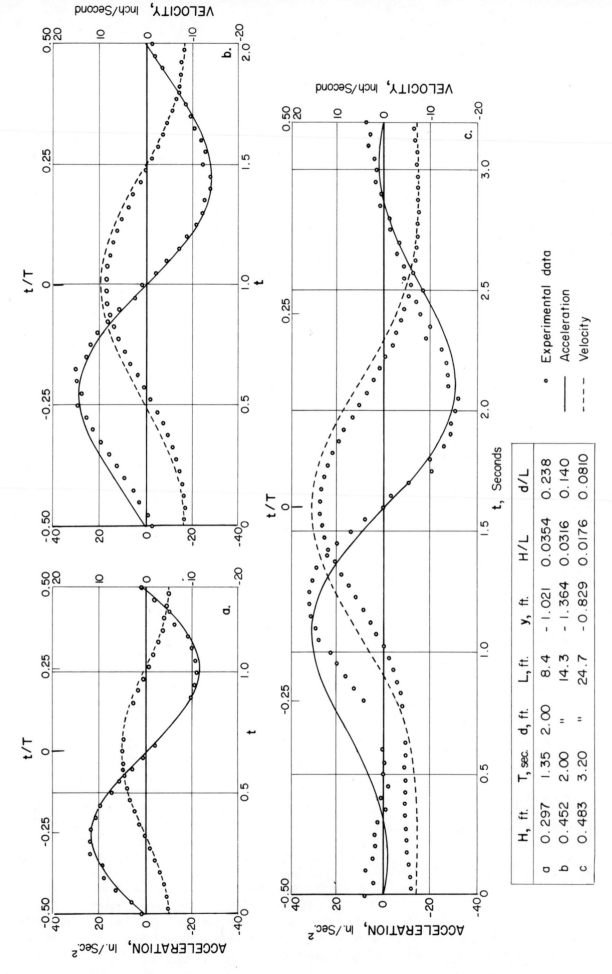

	H, ft.	T, sec.	d, ft.	L, ft.	y, ft.	H/L	d/L
a	0.297	1.35	2.00	8.4	-1.021	0.0354	0.238
b	0.452	2.00	"	14.3	-1.364	0.0316	0.140
c	0.483	3.20	"	24.7	-0.829	0.0176	0.0810

Fig. 2.17. Comparison of horizontal velocity and acceleration with Stokes' wave theory (after Elliott, 1953)

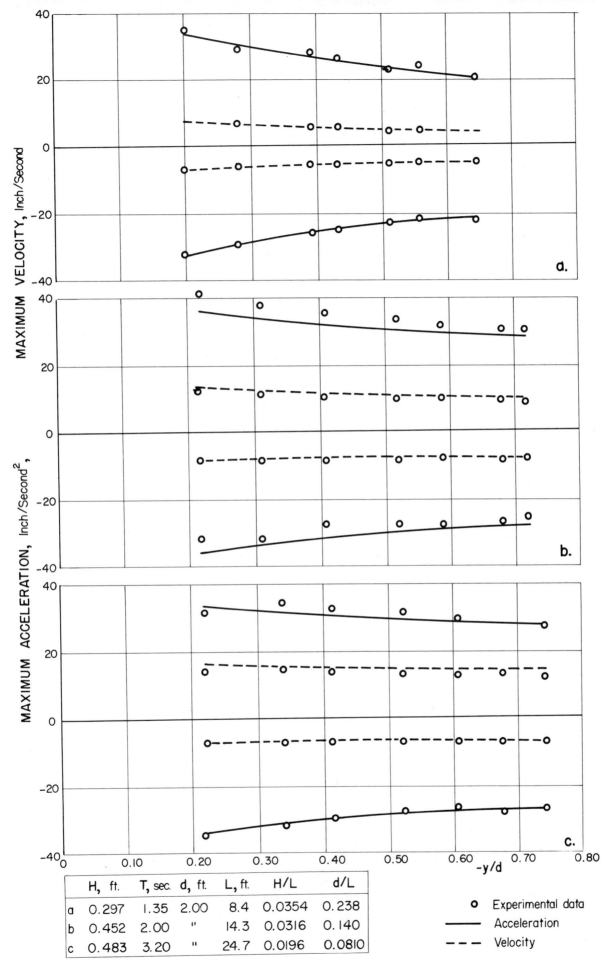

	H, ft.	T, sec.	d, ft.	L, ft.	H/L	d/L
a	0.297	1.35	2.00	8.4	0.0354	0.238
b	0.452	2.00	"	14.3	0.0316	0.140
c	0.483	3.20	"	24.7	0.0196	0.0810

o   Experimental data
――― Acceleration
‒ ‒ ‒ Velocity

Fig. 2.18. Comparison of maximum horizontal velocity and acceleration with Stokes' wave theory (after Elliott, 1953)

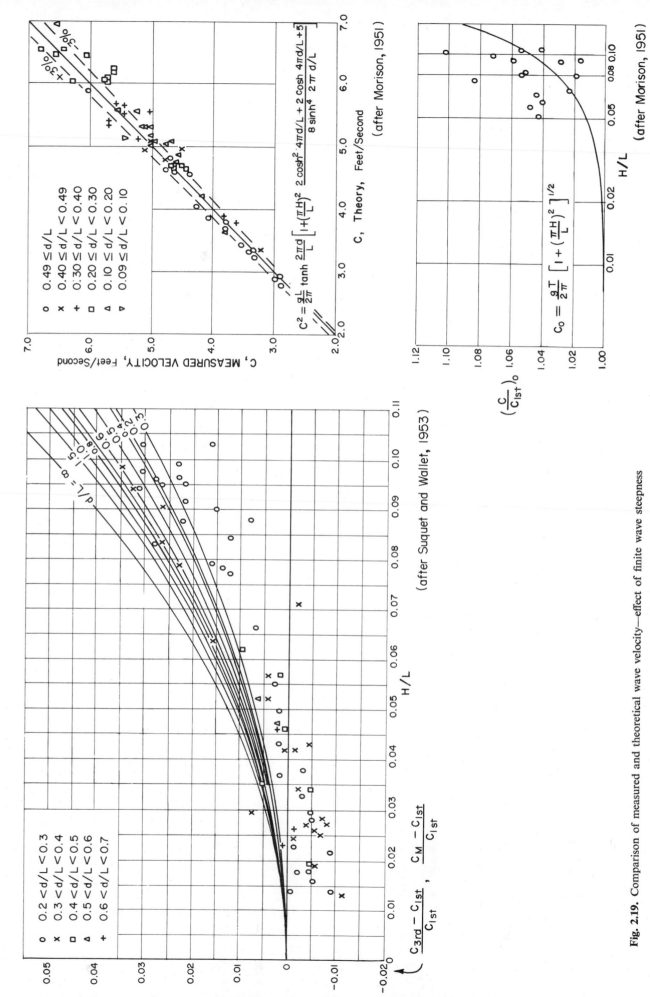

Fig. 2.19. Comparison of measured and theoretical wave velocity—effect of finite wave steepness

measurements were not sufficient, however, to test it thoroughly. The theoretical wave velocity C_T used by Morison is Eq. 2.128b, with d/L used in place of qC/L and where first-order L is used in place of third-order L. The error introduced by this is small, however, and would not invalidate the general conclusions.

Equation 2.124 can be written (Skjelbreia, 1959)

$$\frac{y_s}{L} = A_1 \cos 2\pi \left(\frac{x}{L} - \frac{t}{T}\right) + A_2 \cos 4\pi \left(\frac{x}{L} - \frac{t}{T}\right)$$
$$+ A_3 \cos 6\pi \left(\frac{x}{L} - \frac{t}{T}\right) \quad (2.130)$$

and Eq. 2.125 can be written as

$$\frac{H}{d} = \frac{L}{d}\left[2A_1 + 2\pi^2 A_1^3 \cdot f_3\left(\frac{d}{L}\right)\right] \quad (2.131)$$

where $\quad A_1 = \frac{a}{L}, \quad A_2 = \pi A_1^2 \cdot f_2\left(\frac{d}{L}\right),$

and $\quad A_3 = \pi^2 A_1^3 \cdot f_3\left(\frac{d}{L}\right).$ $\quad (2.132)$

Equation 2.126 can be written

$$\frac{2\pi d}{gT^2} = \frac{d}{L} \tanh \frac{2\pi d}{L}\left[1 + (2\pi A_1)^2 \frac{8 + \cosh 8\pi d/L}{8 \sinh^4 2\pi d/L}\right]. \quad (2.133)$$

In order to use this theory, it is necessary to tabulate the various functions of H, T and, d. This has been done by Skjelbreia (1959) using Eqs. 2.131 and 2.133. The functions A_1, A_2, and A_3 (as defined by Eq. 2.132) and F_1, F_2, and F_3 have been tabulated, where

$$F_1 = \frac{2\pi a}{L} \cdot \frac{1}{\sinh 2\pi d/L},$$

$$F_2 = \frac{3}{4}\left(\frac{2\pi a}{L}\right)^2 \cdot \frac{1}{\sinh^4 2\pi d/L}, \quad (2.134)$$

$$F_3 = \frac{3}{64}\left(\frac{2\pi a}{L}\right)^3 \cdot \left(\frac{11 - 2\cosh 4\pi d/L}{\sinh^7 2\pi d/L}\right).$$

These functions can be computed after obtaining d/L as a function of H/d and $2\pi d/gT^2$ from Figs. 2.20 and 2.21 which have been plotted from the tables of Skjelbreia.

The velocity potential is

$$\phi = \frac{CL}{2\pi}\left[F_1 \cosh \frac{2\pi}{L}(y + d) \sin 2\pi \left(\frac{x}{L} - \frac{t}{T}\right)\right.$$

$$+ \frac{1}{2}F_2 \cosh \frac{4\pi}{L}(y + d) \sin 4\pi \left(\frac{x}{L} - \frac{t}{T}\right)$$

$$\left.+ \frac{1}{3}F_3 \cosh \frac{6\pi}{L}(y + d) \sin 6\pi \left(\frac{x}{L} - \frac{t}{T}\right)\right] \quad (2.135)$$

The components of water particle velocities at any point x, y in the fluid can be easily found from the velocity potential, and are

$$\frac{u}{C} = F_1 \cosh \frac{2\pi}{L}(y + d) \cos 2\pi \left(\frac{x}{L} - \frac{t}{T}\right)$$

$$+ F_2 \cosh \frac{4\pi}{L}(y + d) \cos 4\pi \left(\frac{x}{L} - \frac{t}{T}\right) \quad (2.136)$$

$$+ F_3 \cosh \frac{6\pi}{L}(y + d) \cos 6\pi \left(\frac{x}{L} - \frac{t}{T}\right),$$

$$\frac{v}{C} = F_1 \sinh \frac{2\pi}{L}(y + d) \sin 2\pi \left(\frac{x}{L} - \frac{t}{T}\right)$$

$$+ F_2 \sinh \frac{4\pi}{L}(y + d) \sin 4\pi \left(\frac{x}{L} - \frac{t}{T}\right) \quad (2.137)$$

$$+ F_3 \sinh \frac{6\pi}{L}(y + d) \sin 6\pi \left(\frac{x}{L} - \frac{t}{T}\right).$$

The components of water particle local acceleration at any point x, y in the fluid are

$$\frac{\partial u}{\partial t} = \frac{2\pi C}{T} F_1 \cosh \frac{2\pi}{L}(y + d) \sin 2\pi \left(\frac{x}{L} - \frac{t}{T}\right)$$

$$+ \frac{4\pi C}{T} F_2 \cosh \frac{4\pi}{L}(y + d) \sin 4\pi \left(\frac{x}{L} - \frac{t}{T}\right)$$

$$+ \frac{6\pi C}{T} F_3 \cosh \frac{6\pi}{L}(y + d) \sin 6\pi \left(\frac{x}{L} - \frac{t}{T}\right),$$

$$(2.138)$$

$$\frac{\partial v}{\partial t} = -\frac{2\pi C}{T} F_1 \sinh \frac{2\pi}{L}(y + d) \sin 2\pi \left(\frac{x}{L} - \frac{t}{T}\right)$$

$$- \frac{4\pi C}{T} F_2 \sinh \frac{4\pi}{L}(y + d) \sin 4\pi \left(\frac{x}{L} - \frac{t}{T}\right)$$

$$- \frac{6\pi C}{T} F_3 \sinh \frac{6\pi}{L}(y + d) \sin 6\pi \left(\frac{x}{L} - \frac{t}{T}\right).$$

$$(2.139)$$

Using

$$u_1 = F_1 \cosh \frac{2\pi(y + d)}{L}, \quad u_2 = F_2 \cosh \frac{4\pi(y + d)}{L},$$

$$u_3 = F_3 \cosh \frac{6\pi(y + d)}{L}$$

$$v_1 = F_1 \sinh \frac{2\pi(y + d)}{L}, \quad v_2 = F_2 \sinh \frac{4\pi(y + d)}{L},$$

$$v_3 = F_3 \sinh \frac{6\pi(y + d)}{L}$$

the components of water particle total acceleration are

$$\frac{du}{dt} = \frac{2\pi C}{T}\left[\left(u_1 - \frac{u_1 u_2}{2} - \frac{v_1 v_2}{2}\right) \sin 2\pi \left(\frac{x}{L} - \frac{t}{T}\right)\right.$$

$$+ 2\left(u_2 - \frac{u_1^2}{4} + \frac{v_1^2}{4}\right) \sin 4\pi \left(\frac{x}{L} - \frac{t}{T}\right) \quad (2.140)$$

$$\left.+ 3\left(u_3 - \frac{u_1 u_2}{2} + \frac{v_1 v_2}{2}\right) \sin 6\pi \left(\frac{x}{L} - \frac{t}{T}\right)\right]$$

$$\frac{dv}{dt} = \frac{2\pi C}{T}\left[u_1^2 - \left(v_1 - \frac{u_1 v_2}{2} - \frac{u_1 u_2}{2} - u_2 v_1\right)\right.$$

$$\cdot \cos 2\pi \left(\frac{x}{L} - \frac{t}{T}\right) - 2v_2 \cos 4\pi \left(\frac{x}{L} - \frac{t}{T}\right)$$

$$\left.- 3\left(v_3 + \frac{u_2 v_1}{6}\right)\cos 6\pi \left(\frac{x}{L} - \frac{t}{T}\right)\right] \quad (2.141)$$

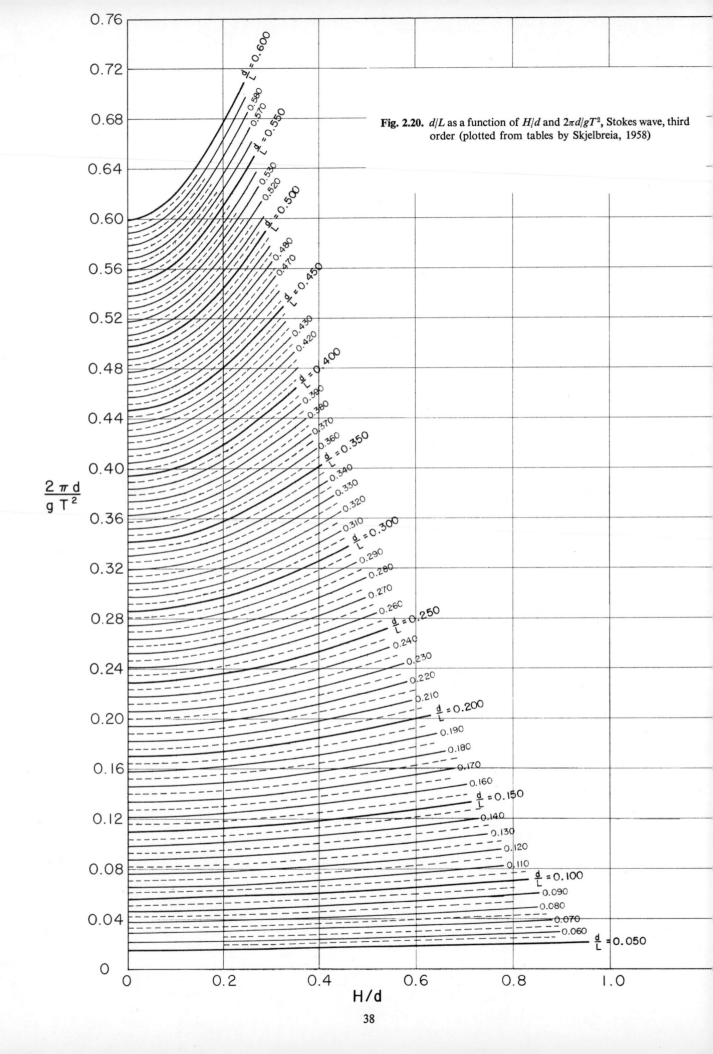

Fig. 2.20. d/L as a function of H/d and $2\pi d/gT^2$, Stokes wave, third order (plotted from tables by Skjelbreia, 1958)

$$\frac{2\pi d}{gT^2}$$

H/d

38

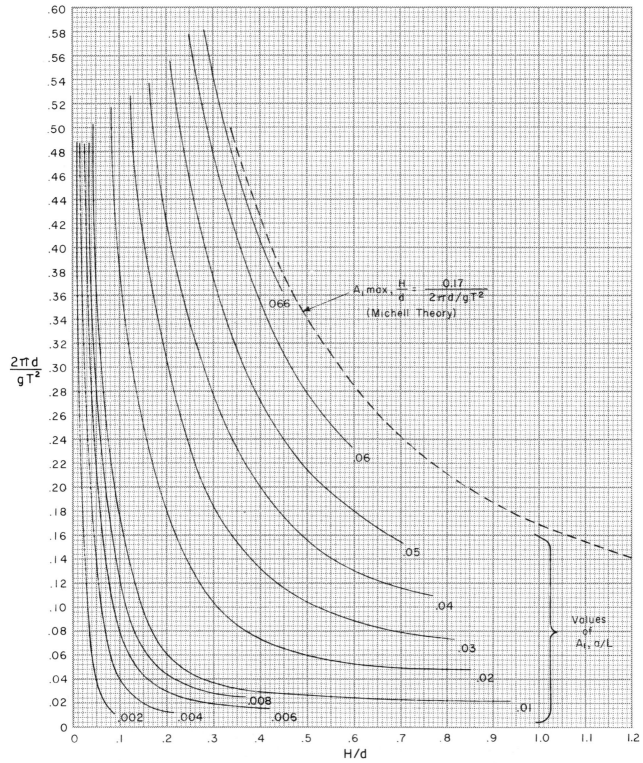

Fig. 2.21. A_1 as a function of $2\pi d/gT^2$ and H/d, Stokes wave, third order (plotted from tables by Skjelbreia, 1959)

The displacements of the water particles from their still water positions x_0, y_0 are

$$
\begin{aligned}
\xi = &-\frac{L}{2\pi}\left[F_1\left(1 - \frac{F_1^2}{8}\right)\cosh\frac{2\pi}{L}(y_0 + d)\right.\\
&\left.+ \frac{F_1}{8}(3F_1^3 + 10F_2)\cosh\frac{6\pi}{L}(y_0 + d)\right]\\
&\cdot\sin 2\pi\left(\frac{x_0}{L} - \frac{t}{T}\right) - \frac{L}{4\pi}\left[-\frac{1}{2}F_1^2 + F_2\right.\\
&\left.\cdot\cosh\frac{4\pi}{L}(y_0 + d)\right]\sin 4\pi\left(\frac{x_0}{L} - \frac{t}{T}\right)\\
&- \frac{L}{6\pi}\left[\frac{1}{4}F_1(F_1^2 - 5F_2)\cosh\frac{2\pi}{L}(y_0 + d)\right.\\
&\left.+ F_3\cosh\frac{6\pi}{L}(y_0 + d)\right]\sin 6\pi\left(\frac{x_0}{L} - \frac{t}{T}\right)\\
&+ \frac{Ct}{2}F_1^2\left[\cosh\frac{4\pi}{L}(y_0 + d) - F_1\cosh\frac{2\pi}{L}(y_0 + d)\right.\\
&\left.\cdot\cosh\frac{4\pi}{L}(y_0 + d)\cdot\cos 2\pi\left(\frac{x_0}{L} - \frac{t}{T}\right)\right],
\end{aligned}
$$
(2.142)

$$
\begin{aligned}
\eta = &\frac{L}{2\pi}\left[F_1\left(1 - \frac{3}{8}F_1^2\right)\sinh\frac{2\pi}{L}(y_0 + d)\right.\\
&\left.+ \frac{F_1}{8}(F_1^2 + 6F_2)\sinh\frac{6\pi}{L}(y_0 + d)\right]\\
&\cdot\cos 2\pi\left(\frac{x_0}{L} - \frac{t}{T}\right)\\
&+ \frac{L}{4\pi}F_2\sinh\frac{4\pi}{L}(y_0 + d)\cos 4\pi\left(\frac{x_0}{L} - \frac{t}{T}\right)\\
&+ \frac{L}{6\pi}\left[-\frac{3}{4}F_1F_2\sinh\frac{2\pi}{L}(y_0 + d)\right.\\
&\left.+ F_3\sinh\frac{6\pi}{L}(y_0 + d)\right]\cos 6\pi\left(\frac{x_0}{L} - \frac{t}{T}\right)\\
&- \frac{Ct}{2}F_1^3\left[\sinh\frac{2\pi}{L}(y_0 + d)\cdot\cosh\frac{4\pi}{L}(y_0 + d)\right.\\
&\left.\cdot\sin 2\pi\left(\frac{x_0}{L} - \frac{t}{T}\right)\right].
\end{aligned}
$$
(2.143)

6. LIMITING STEEPNESS OF PROGRESSIVE WAVES

The form of the progressive wave of maximum steepness in deep water has been studied by several people. Stokes (1880) concluded that, for any wave whose crest angle was greater than 120 degrees, his series would cease to be convergent and the wave form would be discontinuous. But the possibility of a wave existing with a crest angle equal to 120 degrees was not shown until later. Wilton (1914) worked along lines similar to Stokes and found that the wave maximum steepness would have a crest angle of 120 degrees. Michell (1893) found that the theoretical limit was $H_0/L_0 = 0.142$ and Havelock (1918) found it to be 0.1418. Michell found the velocity of the highest wave in deep water to be 10 per cent greater than the velocity of a wave of infinitesimal height

Table 2.7. SURFACE PROFILE OF PROGRESSIVE WAVE OF MAXIMUM STEEPNESS IN DEEP WATER

x/L	0.000	0.049	0.092	0.149	0.246	0.334	0.418	0.500
y_s'/L	0.143	0.116	0.094	0.068	0.035	0.015	0.004	0.000

(linear theory). The form of the wave of maximum steepness as developed by Michell and extended by Keulegan (1950) is shown in Table 2.7. (The table shown here is due to Keulegan and it differs slightly from the values given by Michell.)

In this table, $x = 0$ is at the wave crest and y_s' is measured positively upward from the wave trough. The surface profile of a wave of maximum steepness as photographed in a wave tank by Suquet and Wallet (1953) was found to compare favorably with the theoretical values of Table 2.7. The wave was not as high as theoretically possible, but was similar in shape.

For progressive waves in water of finite depth, Miche (1944) gives the limiting steepness as

$$
\left(\frac{H}{L}\right)_{\max} = \left(\frac{H_0}{L_0}\right)_{\max}\tanh\frac{2\pi d}{L} = 0.142\tanh\frac{2\pi d}{L}.
$$
(2.144)

This curve has been plotted in Fig. 2.10. The data of Danel (1952) show that it is satisfactory from an engineering standpoint.

7. CNOIDAL WAVES

Mathematical arguments show that Stokes' waves are most nearly valid in water deeper than about $d/L > \frac{1}{8}$ to 1/10 (Keulegan, 1950; De, 1955). In shallower water, the theory for a wave type known as *cnoidal* appears to be more satisfactory. Keulegan points out that the solution has been developed for the square of the inclination of the water surface small in comparison to unity. The development of the theory for use by engineers as given herein is essentially from Wiegel (1960).

Korteweg and DeVries (1895) developed a theory of cnoidal waves. One limiting case of this theory is the solitary wave; another limiting case is the linear wave theory (sinusoidal waves). Keulegan and Patterson (1940) also studied the cnoidal wave. Keller (1948) treated the problem using nonlinear shallow water theory and obtained formulas which were similar to those of Korteweg and DeVries. Littman (1957) has shown the existence of permanent periodic waves of this type. The approximate region of validity of the cnoidal wave theory is shown in Fig. 2.22 (Wehausen, 1963). Benjamin and Lighthill (1954) have advanced the theory considerably in regard to the formation of hydraulic jumps, and Iwasa (1955) has also considered it. The formula for the wave profile is expressed in terms of Jacobian elliptic function $cn\ u$; hence the term *cnoidal*, analogous to sinusoidal.

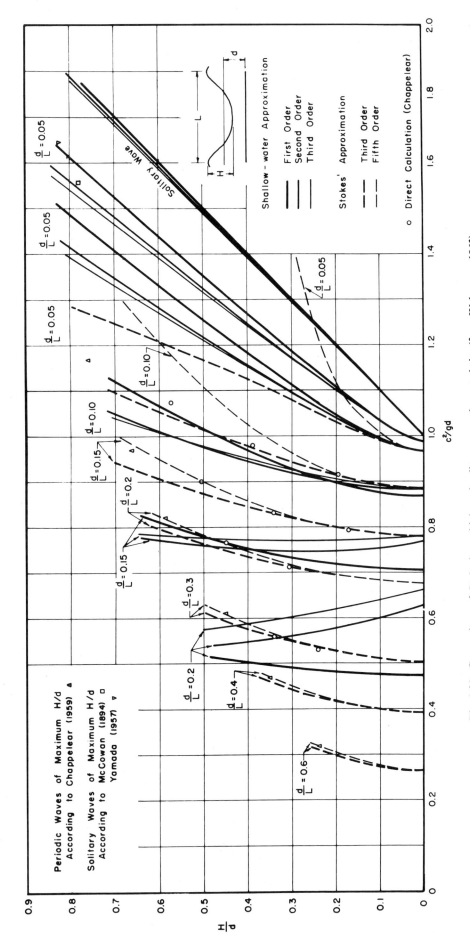

Fig. 2.22. Comparison of Stokes', cnoidal, and solitary wave characteristics (from Wehausen, 1963)

41

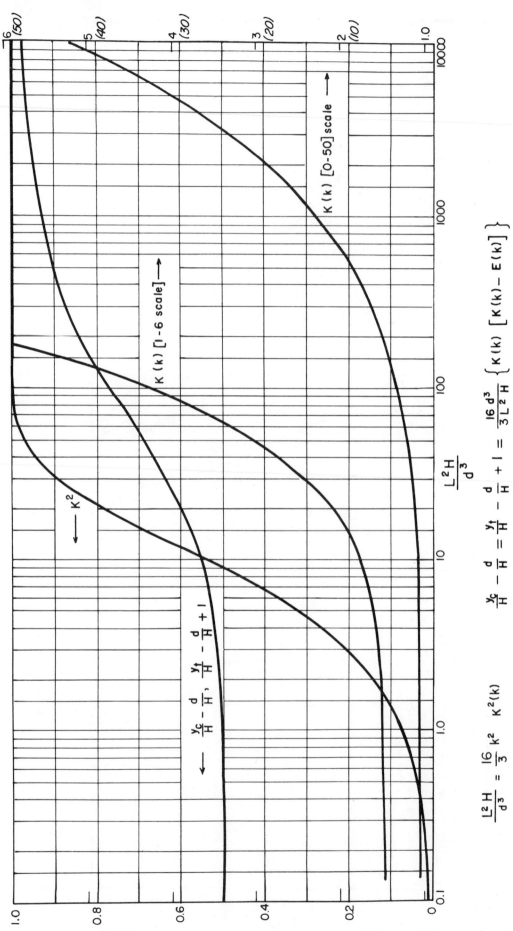

Fig. 2.23. Relationship among L^2H/d^3 and the square of the elliptic modulus (k^2), y_c/H, y_t/H, and $K(k)$ (after Wiegel, 1959)

$$\frac{y_c}{H} - \frac{d}{H} = \frac{y_t}{H} - \frac{d}{H} + 1 = \frac{16}{3}\frac{d^3}{L^2 H}\left\{K(k)\left[K(k) - E(k)\right]\right\}$$

$$\frac{L^2 H}{d^3} = \frac{16}{3}k^2\ K^2(k)$$

Korteweg and DeVries, Keulegan and Patterson, and Keller use different symbols; but the critical formulas obtained by them are essentially the same. The wave length is given by

$$\frac{L}{d} = \frac{4}{\sqrt{3}} K(k) \left(2\bar{L} + 1 - \frac{y_t}{d}\right)^{-1/2} \quad (2.145)$$

where d is the still water depth, $K(k)$ is the complete elliptic integral of the first kind of modulus k (it should be noted that $K(k)$ is sometimes designated as $F_1(k)$); y_t is the distance from the ocean bottom to the wave trough; and $\bar{L}$ and k are defined by the following two equations:

$$k^2 = \frac{(y_c/d) - (y_t/d)}{2\bar{L} + 1 - (y_t/d)}, \quad (2.146)$$

$$\left(2\bar{L} + 1 - \frac{y_t}{d}\right) E(k) = \left(2\bar{L} + 2 - \frac{y_c}{d} - \frac{y_t}{d}\right) K(k), \quad (2.147)$$

where y_c is the distance from the ocean bottom to the wave crest and $E(k)$ is the complete elliptic integral of the second kind of modulus k. The following inequalities must also hold:

$$2\bar{L} + 1 > \frac{y_c}{d} > \frac{y_t}{d} \quad \text{and} \quad 0 < k^2 \le 1, \quad (2.148)$$

In using tables of elliptical integrals and functions, it is cautioned that they are often tabulated as functions of the parameter m, where $m = k^2$. Equation 2.146 can be written

$$\left(2\bar{L} + 1 - \frac{y_t}{d}\right) = \frac{(y_c/d) - (y_t/d)}{k^2} = \frac{H/d}{k^2}. \quad (2.149)$$

Substituting this into Eq. 2.145 and squaring gives

$$\frac{L^2 H}{d^3} = \frac{16}{3} [k \, K(k)]^2. \quad (2.150)$$

Both Keller (1948) and Littman (1957) also obtain this relationship as their approximate solution. L^2H/d^3 is plotted as a function of k^2 in Figs. 2.23 and 2.24. If the wave length, wave height, and water depth are known, then the many formulas of the cnoidal wave theory can be used as they are expressed in terms of various functions of the square of the modulus k. The terminology of the elliptic functions and integrals as used herein are as used by Milne-Thompson (1950).

The wave length is

$$L = \sqrt{\frac{16d^3}{3H}} \cdot k \, K(k). \quad (2.151)$$

Equation 2.147 can be written as

$$\left(2\bar{L} + 1 - \frac{y_t}{d}\right) E(k) = \left(2\bar{L} + 1 - \frac{y_t}{d}\right) K(k)$$

$$+ \left(1 - \frac{y_c}{d}\right) K(k), \quad (2.152)$$

which, upon rearranging, becomes

$$E(k) - K(k) = \frac{[1 - (y_c/d)]}{[2\bar{L} + 1 - (y_t/d)]} K(k). \quad (2.153)$$

Substituting Eq. 2.145 into Eq. 2.153 gives

$$\frac{y_c}{d} = \frac{16d^2}{3L^2} \{K(k)[K(k) - E(k)]\} + 1; \quad (2.154)$$

or, multiplying by d/H, we get

$$\frac{y_c}{H} - \frac{d}{H} = \frac{16d^3}{3L^2 H} \{K(k)[K(k) - E(k)]\}. \quad (2.155)$$

y_t can be obtained from the relationship

$$\frac{y_t}{d} = \frac{y_c}{d} - \frac{H}{d}$$

$$= \frac{16d^2}{3L^2} \{K(k)[K(k) - E(k)]\} + 1 - \frac{H}{d} \quad (2.156a)$$

or $\quad \dfrac{y_t}{H} - \dfrac{d}{H} + 1 = \dfrac{16d^3}{3L^2 H} \{K(k)[K(k) - E(k)]\}.$
$$(2.156b)$$

These equations have been plotted in Figs. 2.23 and 2.24. The relationships among k^2, $K(k)$, and $E(k)$ have been tabulated up to $k^2 = 1 - 10^{-6}$ by Kaplan (1946, 1948) and partially by Hayashi (1930, 1933) and Airey (1935). In order to extend these functions to the range needed for the study of waves (up to $k^2 = 1 - 10^{-40}$) the following equations were used (Jahnke and Emde, 1945):

$$K(k) = \Lambda + \frac{\Lambda - 1}{4} k'^2 + \frac{3}{16}\left(\Lambda - \frac{7}{16}\right) k'^4$$

$$+ \frac{25}{256}\left(\Lambda - \frac{37}{30}\right) k'^6 + \cdots, \quad (2.157a)$$

$$E(k) = 1 + \frac{1}{2}\left(\Lambda - \frac{1}{2}\right) k'^2 + \frac{3}{16}\left(\Lambda - \frac{13}{12}\right)k'^4$$

$$+ \frac{15}{128}\left(\Lambda - \frac{6}{5}\right) k'^6 + \cdots, \quad (2.157b)$$

where $k' = \sqrt{1 - k^2}$ and $\Lambda = ln\,(4/k')$.

The wave profile is

$$y_s = y_t + Hcn^2\left[2K(k)\left(\frac{x}{L} - \frac{t}{T}\right), \, k\right], \quad (2.158)$$

where cn is one of the Jacobian elliptic functions. The function cn is singly periodic as long as k is a real number and $0 \le k < 1$. The period becomes infinite when $k = 1$ (in which case we have the solitary wave). Whereas the period of the cn function is $4K(k)$, the period of the cn^2 function is $2K(k)$. The cn^2 $(2K(k)\,(x/L - t/T),\, k)$ is plotted in Figs. 2.25 and 2.26 as a function of x/L, t/T, with k^2 as the parameter. Values of the cn function are available over a limited range of k^2 (Milne-Thompson, 1950; Spenceley and Spenceley, 1947; Schuler and Gabelin, 1955). The extension of this function above $k^2 = 0.95$ to three decimal places was made using the following series (Milne-Thompson, 1950):

$$cn(\bar{u}|k^2) = \text{sech}\,\bar{u} - \tfrac{1}{4}k'^2 \tanh \bar{u}$$

$$\cdot \text{sech}\,\bar{u}(\sinh \bar{u} \cosh \bar{u} - \bar{u}), \quad (2.159)$$

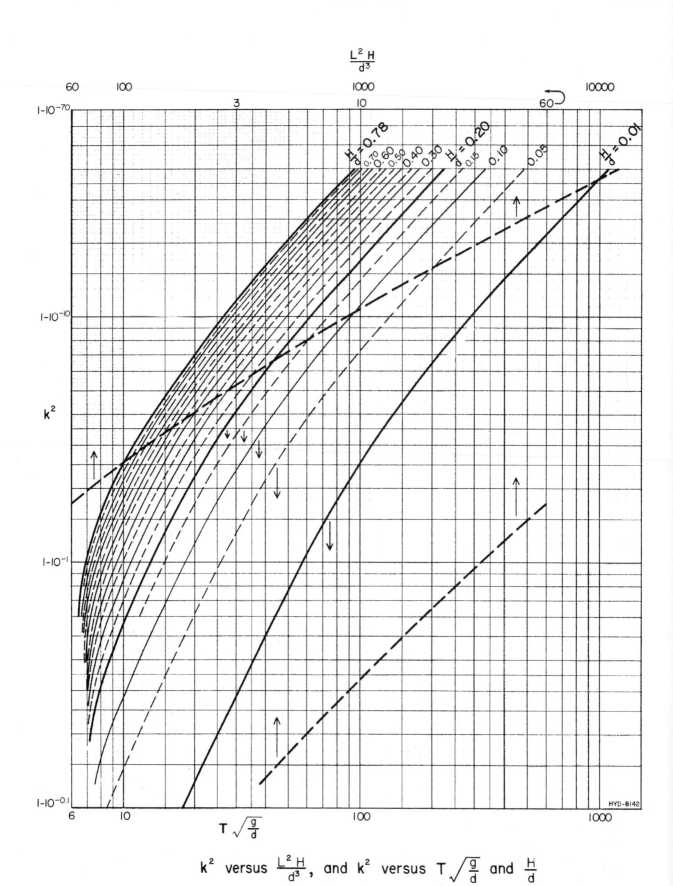

$$k^2 \text{ versus } \frac{L^2 H}{d^3}, \text{ and } k^2 \text{ versus } T\sqrt{\frac{g}{d}} \text{ and } \frac{H}{d}$$

Fig. 2.24. k^2 versus L^2H/d^3, and k^2 versus $T\sqrt{g/d}$ and H/d (from Wiegel, 1960)

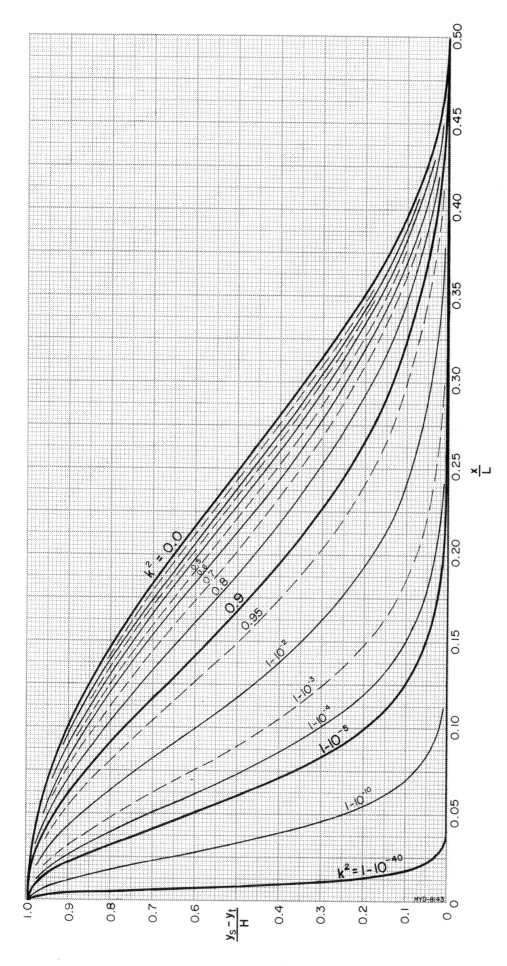

Fig. 2.25. Surface profile of the cnoidal wave (from Wiegel, 1960)

45

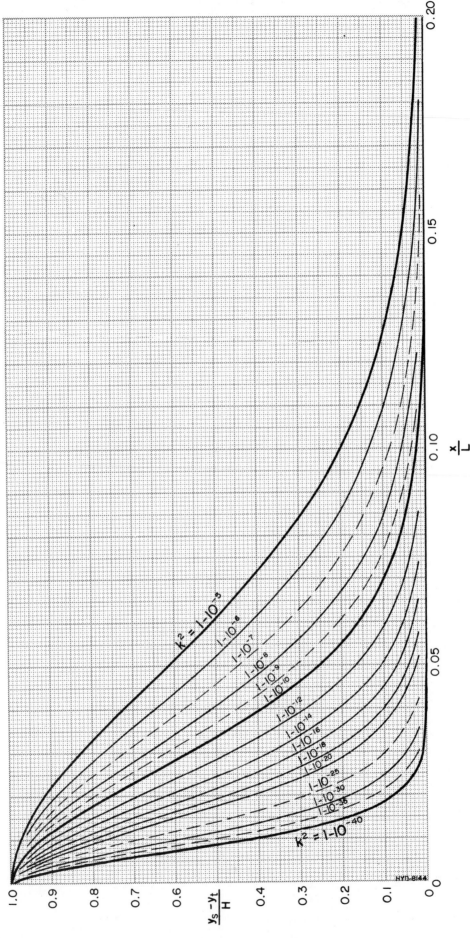

Fig. 2.26. Surface profile of the cnoidal wave (from Wiegel, 1960)

46

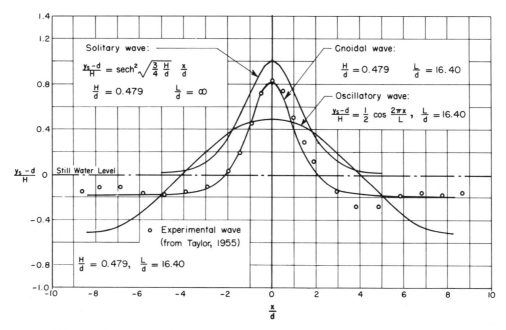

Fig. 2.27. Comparison of measured and theoretical wave profiles (from Wiegel, 1960)

where $\bar{u}$ is the incomplete elliptic integral of the first kind and sinh, cosh, tanh, and sech are hyperbolic functions. x/L is the same as $\bar{u}/2K(k)$ for the cn^2 function. $\bar{u}$ has been used rather than the more commonly accepted symbol u, because u has been used in this book as the horizontal component of water particle velocity. In Fig. 2.27 a comparison of the theoretical surface profile is compared with some measurements made by Taylor (1955). It can be seen that the cnoidal theory predicts the wave profile very well.

It is interesting to note that when the modulus k is zero, $cn\,(\bar{u}|k) = cn\,(\bar{u}|0) = \cos \bar{u}$ and $K(k) = \pi/2$; hence $4K(k) = 2\pi$, and we have the trignometric functions. When $k = 1$, $cn\,(\bar{u}|1) = \operatorname{sech} \bar{u}$ and we have the hyperbolic function with $K(k) = \infty$; hence, the period becomes infinite and we then have the solitary wave. It can be seen that, when k is reduced from 1 to 0.9999, the period $4K$ is reduced from infinity to about 7π, whereas the further reduction of k to 0 reduces $4K$ to only 2π. Because of this, one limiting case of the cnoidal wave, the solitary wave, can be thought of as having a finite period for many practical purposes.

The wave velocity (using Stokes' second definition of wave velocity, which is the velocity of the propagation of the wave form when the horizontal momentum of the liquid has been reduced to zero by the addition of a uniform motion) is

$$C = \sqrt{gd}\left[1 + \frac{H}{d}\cdot\frac{1}{k^2}\left(\frac{1}{2} - \frac{E(k)}{K(k)}\right)\right], \quad (2.160a)$$

or, as $(16d^3/3L^2H)\,K^2(k) = 1/k^2$,

$$C = \sqrt{gd}\left[1 + \frac{16d^2}{3L^2}\,K^2(k)\left(\frac{1}{2} - \frac{E(k)}{K(k)}\right)\right]. \quad (2.160b)$$

Equation 2.160a has been plotted in Fig. 2.32

For one limiting case (the solitary wave), k^2 of unity, $E(k)$ is unity and $K(k)$ is infinity; hence

$$C = \sqrt{gd}\left(1 + \frac{H}{2d}\right). \quad (2.160c)$$

This approximation to

$$C = \sqrt{gd\left(1 + \frac{H}{d}\right)} \quad (2.160d)$$

is higher by a maximum of only 2 per cent even for the case of the solitary wave of maximum steepness ($H/d = 0.78$).

For the other limiting case (the linear theory, where $k^2 \to 0$), $E(k)/K(k) \to$ unity and

$$C = \sqrt{gd}\left(1 - \frac{2\pi^2 d^2}{3L^2}\right), \quad (2.160e)$$

which is an approximation of

$$C = \sqrt{gd\left(1 - \frac{4\pi^2 d^2}{3L^2}\right)}. \quad (2.160f)$$

Now in the linear theory

$$C = \sqrt{\frac{gL}{2\pi}\tanh\frac{2\pi d}{L}}. \quad (2.160g)$$

But the first two terms in the expansion of $\tanh 2\pi d/L$ are

$$\frac{2\pi d}{L} - \frac{8\pi^2 d^3}{3L^3}.$$

So

$$C \cong \sqrt{gd\left(1 - \frac{4\pi^2 d^2}{3L^2}\right)}, \quad (2.160h)$$

which is in agreement with Eq. 2.160f.

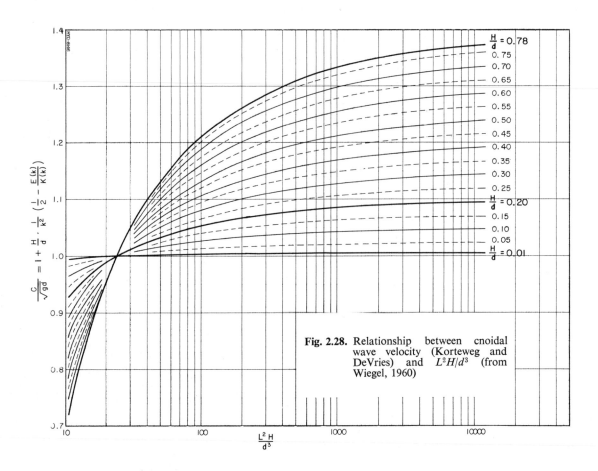

Fig. 2.28. Relationship between cnoidal wave velocity (Korteweg and DeVries) and L^2H/d^3 (from Wiegel, 1960)

The equation for wave velocity as given by Keulegan and Patterson (1940), and Littman (1957), which is apparently the velocity of the wave crest with respect to fixed coordinates, can be written

$$C^2 = gd\left\{1 + \frac{H}{d}\left[-1 + \frac{1}{k^2}\left(2 - 3\frac{E(k)}{K(k)}\right)\right]\right\} \cdot \quad (2.161)$$

In Fig. 2.29 are shown comparisons of the wave velocity as obtained from Eq. 2.161 and some measurements of waves in the laboratory.

As cnoidal waves are periodic and of permanent form, $L = CT$, $T = L/C$; the wave period is given by

$$T\sqrt{\frac{g}{d}} = \sqrt{\frac{16d}{3H}}\left\{\frac{kK(k)}{\left[1 + \frac{1}{2}\frac{H}{dk^2} - \frac{E(k)}{K(k)}\right]}\right\}, \quad (2.162)$$

using the velocity as given in Eq. 2.160a.

From this equation, k^2 can be determined as a function of $T\sqrt{g/d}$ and H/d, and from this L^2H/d^3 can be determined. This has been done and plotted in Fig. 2.30.

Using the velocity as given in Eq. 2.161 results in

$$T\sqrt{\frac{g}{d}} = \sqrt{\frac{16d}{3H}}\left\{\frac{kK(k)}{\sqrt{1 + \frac{H}{d}\left[-1 + \frac{1}{k}\left(2 - 3\frac{E(k)}{K(k)}\right)\right]}}\right\},$$

$$(2.163)$$

which has been plotted in Fig. 2.24 as a function of k^2 and H/d. When this equation is plotted as a function of L^2H/d^3 and H/d, it shows that a wave of a given period and height, and for a given water depth, can have two possible lengths. The physical significance of this is not apparent.

The pressure at any distance y above the bottom has been shown by Keller (1948) to the second approximation to be

$$p = \rho g(y_s - y), \quad (2,164)$$

where y_s is given by Eq. 2.156. Laitone (1960), however, found that it held only to the first approximation, and not to the second approximation.

The horizontal and vertical components of water particle velocities at any point x, y within the fluid can be obtained from the following equations of Keulegan and Patterson (1940):

$$\frac{u}{\sqrt{gd}} = \left[\frac{h}{d} - \frac{h^2}{4d^2} + \left(\frac{d}{3} - \frac{y^2}{2d}\right)\frac{\partial^2 h}{\partial x^2}\right],$$

$$\frac{v}{\sqrt{gd}} = -y\left[\left(\frac{1}{d} - \frac{h}{2d^2}\right)\frac{\partial h}{\partial x} + \frac{1}{3}\left(d - \frac{y^2}{2d}\right)\frac{\partial^3 h}{\partial x^3}\right],$$

$$h = y_s - d$$

$$= -d + y_t + H\,cn^2\left[2K(k)\left(\frac{x}{L} - \frac{t}{T}\right), k\right].$$

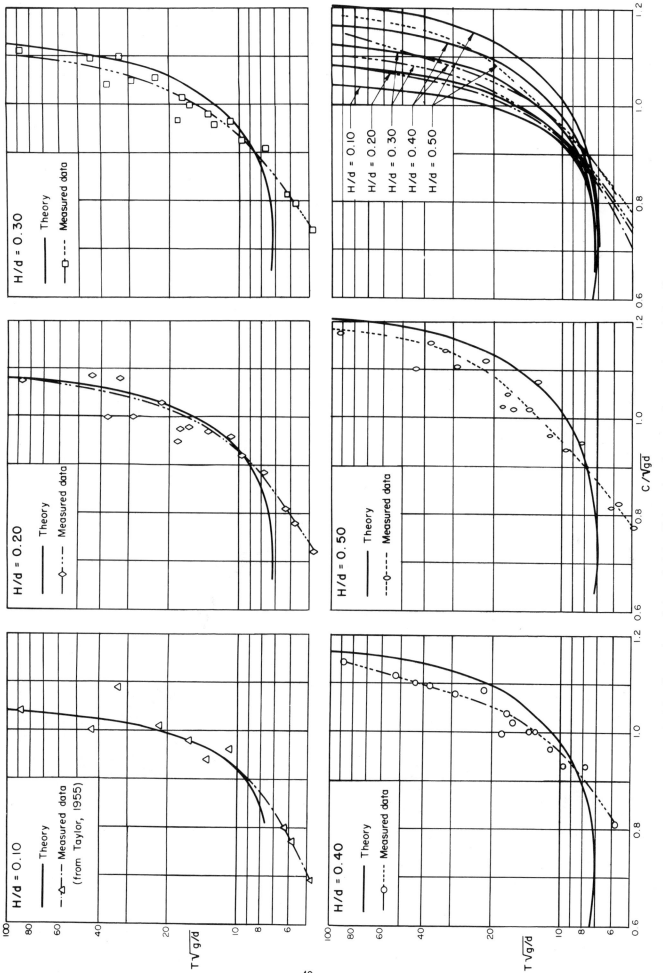

Fig. 2.29. Comparison of theoretical (cnoidal wave) and measured wave phase velocity (from Wiegel, 1960)

49

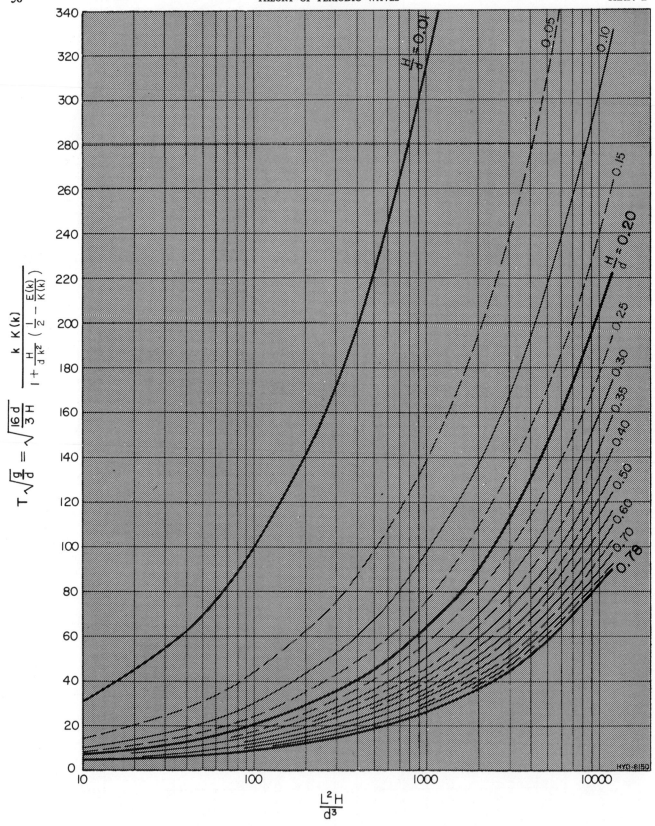

Fig. 2.30. Relationships among $T\sqrt{g/d}$, L^2H/d^3, and H/d (Korteweg and DeVries) (from Wiegel, 1960)

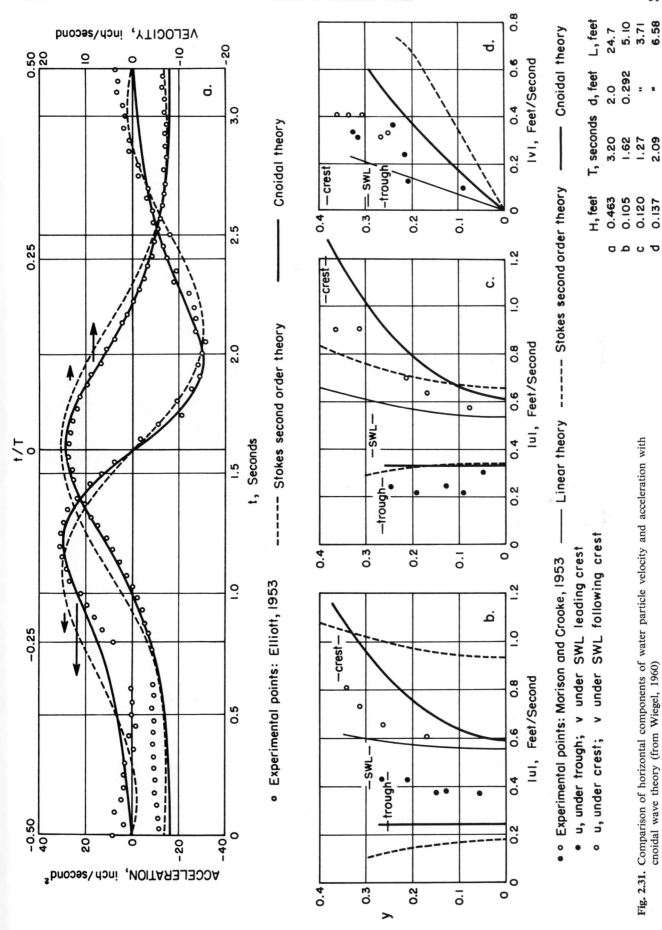

Fig. 2.31. Comparison of horizontal components of water particle velocity and acceleration with cnoidal wave theory (from Wiegel, 1960)

These equations become.

$$\frac{u}{\sqrt{gd}} = \left[-\frac{5}{4} + \frac{3y_t}{2d} - \frac{yt^2}{4d^2} + \left(\frac{3H}{2d} - \frac{y_t H}{2d^2} \right) cn^2(\ ) \right.$$
$$- \frac{H^2}{4d^2} cn^4(\ ) - \frac{8HK^2(k)}{L^2} \left(\frac{d}{3} - \frac{y^2}{2d} \right) \qquad (2.165)$$
$$(-k^2 sn^2(\ ) cn^2(\ ) + cn^2(\ ) dn^2(\ )$$
$$\left. - sn^2(\ ) dn^2(\ )) \right],$$

$$\frac{v}{\sqrt{gd}} = y \cdot \frac{2HK(k)}{Ld} \left[1 + \frac{y_t}{d} + \frac{H}{d} cn^2(\ ) \right.$$
$$+ \frac{32K^2(k)}{3L^2} \left(d^2 - \frac{y^2}{2} \right)(k^2 sn^2(\ ) \qquad (2.166)$$
$$\left. - k^2 cn^2(\ ) - dn^2(\ )) \right] sn(\ ) cn(\ ) dn(\ ),$$

where $sn(\ )$ refers to $sn[2K(k)(x/L - t/T), k]$, etc. The local accelerations are

$$\frac{\partial u}{\partial t} = \sqrt{gd} \cdot \frac{4HK(k)}{Td} \left[\left(\frac{3}{2} - \frac{y_t}{2d} \right) - \frac{H}{2d} cn^2(\ ) \right.$$
$$+ \frac{16K^2(k)}{L^2} \left(\frac{d^2}{3} - y^2 \right)(k^2 sn^2(\ ) \qquad (2.167)$$
$$\left. - k^2 cn^2(\ ) - dn^2(\ )) \right] sn(\ ) cn(\ ) dn(\ ),$$

$$\frac{\partial v}{\partial t} = y\sqrt{gd} \cdot \frac{4HK^2(k)}{LTd} \left\{ \left[1 + \frac{y_t}{d} \right][sn^2(\ ) dn^2(\ ) \right.$$
$$- cn^2(\ ) dn^2(\ ) + k^2 sn^2(\ ) cn^2(\ )]$$
$$+ \frac{H}{d} [3sn^2(\ ) dn^2(\ ) - cn^2(\ ) dn^2(\ )$$
$$+ k^2 sn^2(\ )] cn^2(\ ) - \frac{32K^2(k)}{3L^2} \left[d^2 - \frac{y^2}{2} \right]$$
$$\cdot [9k^2 sn^2(\ ) cn^2(\ ) dn^2(\ ) - k^2 sn^4(\ )$$
$$\cdot (k^2 cn^2(\ ) + dn^2(\ )) + k^2 cn^4(\ )(k^2 sn^2(\ )$$
$$\left. + dn^2(\ )) + dn^4(sn^2(\ ) - cn^2(\ ))] \right\} \qquad (2.168)$$

In order to use the equations for water particle velocities, the necessary numbers can be obtained from Figs. 2.23, 2.24, 2.25, and 2.26, as

$$\bar{u}|k \equiv 2K(k)\frac{x}{L}, \qquad (2.169)$$

$$sn^2(\bar{u}|k) \equiv 1 - cn^2(\bar{u}|k), \qquad (2.170)$$

$$dn^2(\bar{u}|k) \equiv 1 - k^2[1 - cn^2(\bar{u}|k)]. \qquad (2.171)$$

A few comparisons of theory with laboratory measurements are shown in Fig. 2.31. In considering the vertical velocity, it should be noted that the curve plotted for the cnoidal theory is for the phase of maximum vertical velocity and this occurs before the wave profile goes through the still water level.

Many of the necessary functions are available in tabulated form (Masch and Wiegel, 1961). Second-order

$H_R/H_I = 0.00$, pure swell

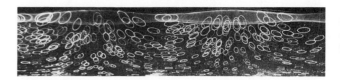

$H_R/H_I = 0.24$

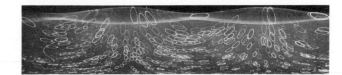

$H_R/H_I = 0.38$

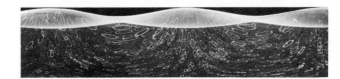

$H_R/H_I = 0.53$

$H_R/H_I = 0.71$

$H_R/H_I = 0.85$

$H_R/H_I = 1.00$, pure standing wave

Fig. 2.32. Water particle motions of pure swell, partially reflected waves, and pure standing wave (after Wallet and Ruellan, 1950)

solutions have also been obtained which should be used for many calculations (Laitone, 1960; 1962).

8. PARTIALLY REFLECTED WAVES

The incident wave may be given by

$$y_{SI} = \frac{H_I}{2} \cos 2\pi \left(\frac{x}{L} - \frac{t}{T} \right),$$

and the reflected wave given by

$$y_{SR} = \frac{H_R}{2} \cos 2\pi \left(\frac{x}{L} - \frac{t}{T} \right)$$

where H_I is the height of the incident wave and H_R is the height of the reflected wave. The resultant water surface is

$$y_s = \frac{1}{2}(H_I + H_R) \cos \frac{2\pi x}{L} \cos \frac{2\pi t}{T}$$
$$+ \frac{1}{2}(H_I - H_R) \sin \frac{2\pi x}{L} \sin \frac{2\pi t}{T} \quad (2.172)$$

The envelope of y_s has a maximum $H_{\max}$ of $H_I + H_R$ at $x = 0$, $L/2$, L, etc., and a minimum $H_{\min}$ of $H_I - H_R$ at $x = L/4$, $3L/4$, etc. The reflection coefficient is

$$\frac{H_R}{H_I} = \frac{H_{\max} - H_{\min}}{H_{\max} + H_{\min}}. \quad (2.173)$$

The equations for the water particle motions will not be given here. A classic example of the motions of the water particles is given however, in Fig. 2.32 (Wallet and Ruellan, 1950) which shows the complete range of conditions from $H_R = 0$ to $H_R = H_I$.

The problem of waves partially reflecting from a boundary is difficult, especially when the amount of energy either dissipated or transformed at many boundaries is unknown. Some laboratory tests of the reflection from an impervious slope have been made by Schoemaker and Thijsse (1949) and are shown in Fig. 2.33, as replotted by Iribarren and Nogales y Olano (1950). As the slope decreases from the vertical, the wave reflection decreases quite rapidly, with the amount of energy being dissipated by the wave breaking increasing rapidly. Schoemaker and Thijsse comment that, as the slope decreased from the vertical, a phase shift between incident and reflected waves developed.

Iribarren and Nogales y Olano presented an equation to relate the slope to the wave steepness. A more general equation has been developed by Miche (1953) and is

$$\left(\frac{H_o}{L_o} \right)_{\max} = \sqrt{\frac{2\alpha}{\pi}} \cdot \frac{\sin^2 \alpha}{\pi}, \quad (2.174)$$

where H_o is the deep water wave height, L_o is the deep water wave length and α is the slope measured from the horizontal (in radians). This equation is plotted in Fig. 2.34.

Miche obtained a solution for the theoretical reflection coefficient, R',

$$R' = \frac{(H_o/L_o)_{\text{max-theory}}}{(H_o/L_o)_{\text{actual}}}, \quad \text{where } R' < 1. \quad (2.175)$$

This is shown in Fig. 2.35. The actual reflection coefficient R is

$$R = \frac{H_R}{H_I} = \rho R'. \quad (2.176)$$

Comparing Eq. 2.175 with the measurements of Schoemaker and Thijsse, Miche concluded that for a smooth impervious beach $\rho \cong 0.8$ and for the stepped slope $\rho \cong 0.33$. For regular slopes of rocks, Miche recommends $\rho \cong 0.3$ to 0.6.

9. SHORT-CRESTED WAVES

Waves generated by winds blowing over the water surface are short-crested, whether these waves are created by strong winds blowing over the surface of the ocean, or gentle winds blowing over the water in a small laboratory tank (Fig. 2.36). They are waves of finite lateral extent and, as such, differ from the theory that follows, as diffraction is an important phenomenon associated with them. The theory of short-crested waves is applicable for the wave field resulting from a system of waves of infinite crest length reflecting from a vertical wall placed at an angle with the direction of wave advance.

The theory of short-crested waves was developed by Jeffreys (1924) and extended considerably by Fuchs (1952). The following equations are after Fuchs.

The first-order velocity potential (linear theory) is

$$\phi = \frac{ga \cosh r(d+y)}{mC \cosh rd} \cos m(x - Ct) \cos ny, \quad (2.177)$$

where m is $2\pi/L$, n is $2\pi/L'$, L is the wave length, L' is the crest length (see Fig. 2.37 for a pictorial representation of the terms), a is the wave amplitude ($H/2$ in the linear theory), and

$$r^2 = m^2 + n^2. \quad (2.178)$$

The coordinate system can be seen in Fig. 2.37; y is positive upward in the following equations, rather than downward as given by Fuchs.

The surface profile (linear theory) is

$$y_s = a \sin m(x - Ct) \cos nz. \quad (2.179)$$

The wave velocity is

$$C^2 = \frac{gr}{m^2} \tanh rd \geq - \tanh md$$
$$= \frac{gL}{2\pi} \sqrt{1 + \left(\frac{L}{L'}\right)^2} \tanh \left[\frac{2\pi d}{L} \sqrt{1 + \left(\frac{L}{L'}\right)^2} \right]$$
$$\geq \frac{gL}{2\pi} \tanh \frac{2\pi d}{L}, \quad (2.180)$$

so that the short-crested wave travels faster than a long-crested wave of the same length L in the same water depth.

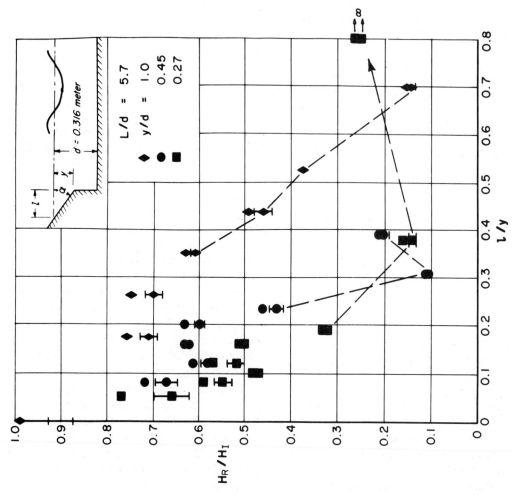

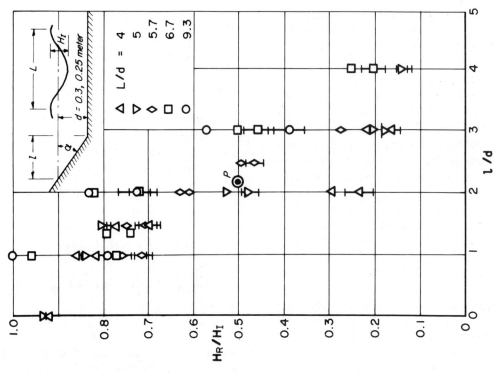

Fig. 2.33. Wave reflection from impervious slope (after Iribarren and Nogales y Olano, 1950; data from Schoemaker and Thijsse, 1949)

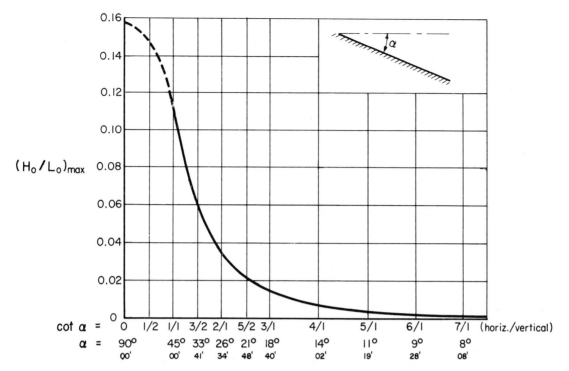

Fig. 2.34. Slope of the impervious beach (after Miche, 1951)

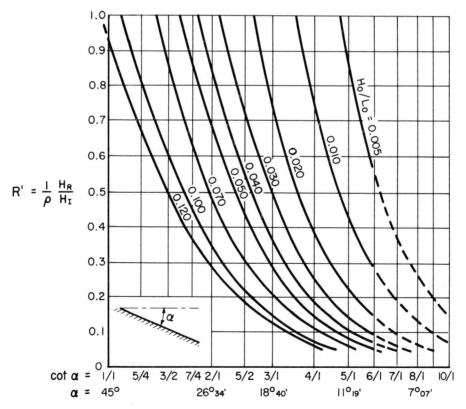

Fig. 2.35. Reflecting capacity of the impervious beach (after Miche, 1951)

Fig. 2.36. Wind-generated waves, Pt. Reyes, California, June 1, 1945

The appearance of short-crested waves refracting in shoaling water will be different than long-crested waves, as the refraction angle will be different. In addition, they will become more long-crested as they move into shallow water.

The orbits of the water particles are ellipses, with one axis in the direction of wave advance, and the plane of the ellipse makes an angle β with the vertical, where

$$\tan \beta = \frac{n \tan nz}{r \tanh r(d + y)}. \qquad (2.181)$$

The components of water particle velocities are

$$u = \frac{ga}{C} \frac{\cosh r(d + y)}{\cosh rd} \sin m(x - Ct) \cos nz, \qquad (2.182a)$$

$$v = -\frac{gar}{mC} \frac{\sinh r(d + y)}{\cosh rd} \cos m(x - Ct) \cos nz, \qquad (2.182b)$$

$$w = \frac{gan}{mC} \frac{\cosh r(d + y)}{\cosh rd} \cos m(x - Ct) \sin nz. \qquad (2.182c)$$

The pressure is given by

$$\frac{p}{\rho g} = -y + a \frac{\cosh r(d + y)}{\cosh rd} \sin mx \cos nz \qquad (2.183)$$

The pressure response factor K at the bottom is

$$K = \frac{\Delta p}{\rho g} = \frac{H}{\cosh rd}. \qquad (2.184)$$

This function has been plotted in Fig. 2.37(c) for three values of L/L'. It is evident that, if the pressure response factor for long-crested waves is applied to the measurements made by a wave pressure recorder under short-crested waves, the wave height will be underestimated.

The potential for the second order is

$$\begin{aligned} \phi = {} & A \cosh r(d + y) \cos m(x - Ct) \cos nz \\ & + B \cosh 2r(d + y) \cdot \sin 2m(x - Ct) \cos 2nz \\ & + D \cosh 2m(d + y) \sin 2m(x - Ct), \qquad (2.185) \end{aligned}$$

where

$$A = \frac{am}{r} \frac{\cosh r(d + y)}{\sinh rd} \cos nz,$$

$$B = -\frac{3}{16} \frac{A^2 r^2}{Cm \sinh^2 rd},$$

$$D = \frac{A}{8mC} \frac{4n^2 \cosh^2 rd - 3r^2}{2 \cosh 2md - (m \sinh 2md/r \tanh rd)}.$$

The wave profile is

$$\begin{aligned} y_s = {} & a \sin m(x - Ct) \cdot \cos nz + \frac{a^2 r}{4 \sinh 2rd} \\ & \left(2 \sinh^2 rd - 1 + \frac{3 \cosh 2rd}{\sinh^2 rd} \right) \cdot \cos 2m(x - Ct) \\ & \cdot \cos 2nz + \frac{a^2(m^2 - n^2 \cosh 2rd)}{4r \sinh 2rd} \cdot \cos 2nz \\ & + \frac{a^2}{4r \sinh 2rd} \left[\frac{(3r^2 - 4n^2 \cosh^2 rd) \cosh 2md}{\cosh 2md - (m \sinh 2md/2r \tanh rd)} \right. \\ & \left. - m^2 \cosh^2 rd + 3r^2 \sinh^2 rd + n^2 \cosh^2 rd \right] \\ & \cdot \cos 2m(x - Ct). \qquad (2.186) \end{aligned}$$

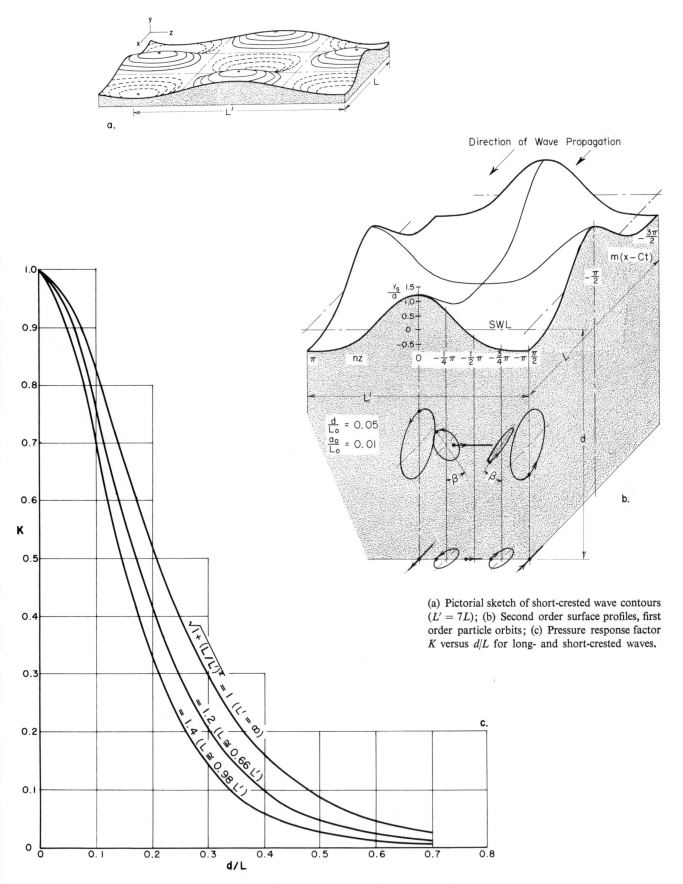

(a) Pictorial sketch of short-crested wave contours ($L' = 7L$); (b) Second order surface profiles, first order particle orbits; (c) Pressure response factor K versus d/L for long- and short-crested waves.

Fig. 2.37. Short crested wave (after Fuchs, 1952)

This equation has been plotted in Fig. 2.37(b).

The mass transport velocity is

$$\bar{U} = \frac{a^2 Cm^2 \coth^2 rd}{4r^2} [r^2(\tanh^2 rd + 1)$$

$$+ \cos 2nz(r^2 \tanh^2 rd + m^2 - n^2)]. \qquad (2.187)$$

The mass transport velocity is a maximum along lines through the wave crests and troughs in the direction of wave advance and a minimum along lines midway between these. This would probably result in circulation cells in the vicinity of the breaker zone along a beach.

10. ROTATIONAL WAVES

The trochoidal wave theory for deep water was developed by Gerstner (1802), Froude (1862), and Rankine (1863). The theory has been used widely by naval architects and civil engineers in their studies. One advantage of the theory is its exactness. It appears to represent the actual wave profiles as well as satisfying the pressure conditions at the surface, and the continuity condition. It requires rotation of the particles in a sense opposite to that of the orbital paths, however, and the orbits are closed. It is difficult to see how this type of wave could be generated impulsively by normal forces (which would result in an irrotational wave) or by the tangential stress of winds (which would create a rotation in a sense opposite to that predicted by trochoidal theory). The conditions might be nearly fulfilled by swells running counter to a current which varies with distance below the surface, such as a drift current. The equations of the surface profile (see Fig. 2.38 for definitions of terms) are

$$y_s = R - r \cos \theta,$$
$$x = R\theta - r_s \sin \theta. \qquad (2.188)$$

It can be seen that the wave length L_o is equal to $2\pi R$, whereas the wave height H is equal to $2r_s$, where r_s is the value for the surface orbit. In order to plot the equation of wave shape in dimensionless form with the origin of the coordinates at the crest and the vertical dimension measured negatively downward, these equations may be transformed to

$$\frac{y'_s}{H} = \frac{1}{2}(1 - \cos \theta),$$

$$\frac{x'}{L_o} = 1 - \left(\frac{\operatorname{rad}\theta}{2\pi} - \frac{H}{2L_o}\sin \theta\right), \qquad (2.189)$$

where y'_s and x' are measured from the wave crest. These have been plotted in Fig. 2.39 with H/L_o as the parameter. As H/L_o approaches zero, the curve approaches a sine wave and the surface is nearly that developed in the irrotational theory of Stokes for waves of small amplitude.

The positions of the crest and trough relative to the still water level are

$$\text{height of crest} = H - \left(r_s - \frac{r_s^2}{2R}\right) = \frac{H}{2} + \frac{\pi H^2}{4L_o}, \qquad (2.189a)$$

$$\text{depth of trough} = \frac{2\pi R}{L_o}\left(r_s - \frac{r_s^2}{2R}\right) = \frac{H}{2} - \frac{\pi H^2}{4L_o}. \qquad (2.189b)$$

Thus, the crest is more than half the wave height above the still water level, while the trough is less than half the wave height below this level. Experiments performed by the Beach Erosion Board (1941) have verified this relationship. Note that the data verify the equations of Stokes (1880) as well.

The paths described by the water particles during one cycle are circles with the radii decreasing exponentially with depth. This is expressed as

$$a' = b' = r_s e^{2\pi y/L_o} = \frac{H}{2} e^{2\pi y/L_o}. \qquad (2.190)$$

The energy of a trochoidal wave per unit width of crest is half potential and half kinetic and is expressed by

Fig. 2.38. Trochoidal wave (after Gaillard, 1935)

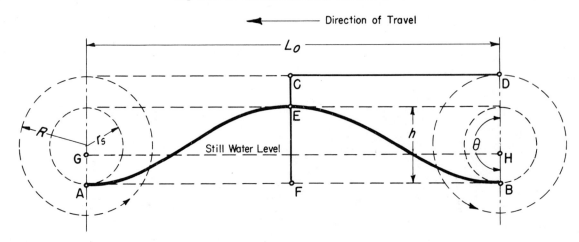

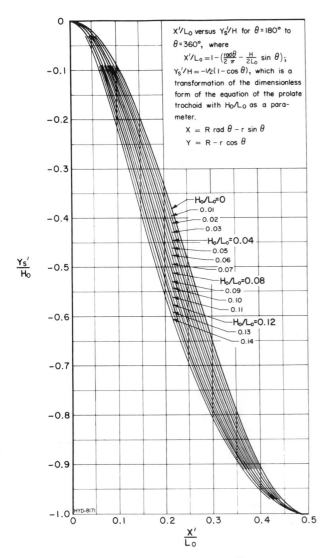

X'/L_0 versus Y_s'/H for θ = 180° to θ = 360°, where

$X'/L_0 = 1 - \left(\frac{\text{rad}\,\theta}{2\pi} - \frac{H}{2L_0}\sin\theta\right)$;

$Y_s'/H = -\frac{1}{2}(1-\cos\theta)$, which is a transformation of the dimensionless form of the equation of the prolate trochoid with H_0/L_0 as a parameter.

$X = R\,\text{rad}\,\theta - r\sin\theta$

$Y = R - r\cos\theta$

$H_0/L_0 = 0$
0.01
0.02
0.03
$H_0/L_0 = 0.04$
0.05
0.06
0.07
$H_0/L_0 = 0.08$
0.09
0.10
0.11
$H_0/L_0 = 0.12$
0.13
0.14

Fig. 2.39. Trochoidal wave profile

$$E = \frac{\gamma H^2 L_0}{8}\left[1 - \frac{1}{2}\left(\frac{\pi H}{L_0}\right)^2\right]. \qquad (2.191)$$

Stokes (1880) has shown that the rotation (vorticity) of the particles of a trochoidal wave is expressed by

$$\omega = \frac{-C_0(2\pi^3 H^2/L_0^3)\,e^{4\pi y/L_0}}{1 - (\pi H/L_0)^2\,e^{4\pi y/L_0}}. \qquad (2.192)$$

This would require a horizontal current of the following distribution with depth:

$$U' = -C_0\left(\frac{\pi H}{L_0}\right)^2 e^{4\pi y/L_0}; \qquad (2.193)$$

this is, in the direction opposite to that of the wave advance. It is interesting to note that, in deep water, the equation for the mass transport for irrotational waves is

$$\overline{U}_0 = C_0\left(\frac{\pi H}{L_0}\right)^2 e^{4\pi y/L_0}, \qquad (2.194)$$

the only difference between this being a difference in sign.

When the exact equations of the surface profile of the

trochoidal theory are expanded in series form, as has been done by Stokes (1880), it is found that up to the third order it is the same as for a Stokes wave.

Gerstner found that the wave length was dependent upon the wave period, but not upon the wave height. The wave length is given by

$$L_0 = \frac{gT^2}{2\pi}, \qquad (2.195)$$

and the wave velocity is given by

$$C_0 = \frac{L_0}{T} = \frac{gT}{2\pi}. \qquad (2.196)$$

It is interesting to note that these are the same equations as for the linear theory of waves in deep water, but are different from the equations for irrotational waves of finite amplitude.

Additional work has been done on rotational waves by Dubrail-Jacotin (1934), Miche (1944), and Kravtchenko and Daubert (1957).

11. SURFACE TENSION WAVES

The restoring force in the waves discussed so far has been gravity. There is another type of wave for which surface tension is the restoring force. Lamb (1945) has shown that the surface profile for a progressive wave system, of infinite extent, is given by

$$y_s = a\cos(kx - \sigma t). \qquad (2.197)$$

The wave velocity is given by

$$C = \sqrt{\frac{\tau k}{\rho}} = \sqrt{\frac{2\pi\tau}{\rho L}}, \qquad (2.198)$$

where the density of the air has been neglected owing to its small magnitude compared with the density of water, ρ. τ is the surface tension, in pounds per foot. It can be seen that the shorter the wave length the greater is the velocity of a surface tension wave.

In contrast to the gravity waves for which the group velocity in deep water is one-half the phase velocity for surface tension waves, it is three-halves the phase velocity. Thus, in a wave group, new waves form in the group front and die out in the rear of the group.

If both gravity and surface tension are considered, the celerity of a surface wave in deep water is (Milne-Thompson, 1955)

$$C = \sqrt{\left(\frac{gL}{2\pi} + \frac{2\pi\tau}{\rho L}\right)\tanh\frac{2\pi d}{L}}, \qquad (2.199a)$$

which in deep water becomes

$$C_0 = \sqrt{\frac{gL_0}{2\pi} + \frac{2\pi\tau}{\rho L_0}}. \qquad (2.199b)$$

This equation shows that there is a minimum celerity, which occurs for a wave length of about 0.06 ft. Equation 2.199b has been plotted in Fig. 2.40, along with Eq.

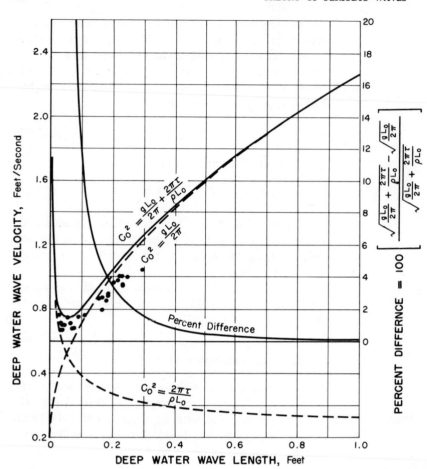

Fig. 2.40. Effect of surface tension on deep water wave velocity in fresh water at 70°F (experimental data after Chinn, 1949)

2.26b. It can be seen that for waves longer than about 0.1 ft the effect of surface tension can generally be neglected. In this figure, some of the data obtained by Chinn (1949) in a ripple tank have been plotted.

One aspect of a surface tension wave is very interesting. It is not possible for a negative solitary gravity wave (that is, a depression of the water surface) to exist. Korteweg and Vries (1895) have shown that a negative solitary surface tension wave can exist; however, the water depth must be very small, of the order of a quarter-inch.

12. SOME EFFECTS OF VISCOSITY

Viscous damping. The effect of viscous damping of water waves of small amplitude and small viscosity has been studied mathematically by Lamb (1932) for waves in deep water, and by Hough (1896) for waves in water of finite depth (assuming the bottom to be perfectly smooth). The modulus of decay (the time necessary for the wave height to be reduced in the ratio of $e:1$) is given by

$$t_\nu = \frac{L^2}{8\pi^2 \nu}, \qquad (2.200)$$

where ν is the kinematic viscosity of water. Very short waves die out rapidly, but the damping is extremely small for waves of appreciable length. For example, a wave 1 ft long would be reduced to $e:1$ in 0.35 hr, but it would take a wave 50 ft long almost 900 hr to be reduced to the same extent.

Mass transport in fluid of small viscosity: In deep water the fact that water is viscous has little effect upon the mass transport as has been shown by laboratory measurements. The few measurements made in shallow water (Caligny, 1878; Mason, 1941; Bagnold, 1947) indicated that Stokes' theory was not satisfactory in shallow water. Longuet-Higgins (1953) developed a theory for null net transport of water for a fluid of small viscosity. An equation was derived which contained what he called *conduction* and *convection terms*. The conduction terms were those that described the diffusion of the vorticity from the bottom, and surface boundary layers into the body of the fluid by the viscous action of the fluid; the convection terms were those that described the movement of the vorticity by the mass transport current into the body of the fluid from the end boundaries of the wave system (for example, the wave generator and the beach in a laboratory tank, or the wave generating area and the beach, reef, or cliff in nature). By assuming the convection terms to be negligible compared with the conduction terms, Longuet-Higgins derived the following equation for progressive waves:

$$\overline{U}_{nv} = \frac{1}{4}\left(\frac{\pi H}{T}\right)\left(\frac{\pi H}{L}\right) \cdot \frac{1}{\sinh^2 2\pi d/L}$$

$$\left\{2\cosh\left[\frac{4\pi d}{L}\left(-\frac{y}{d}-1\right)\right] + 3 \right.$$

$$+ \frac{2\pi d}{L}\left(3\frac{y^2}{d^2} + 4\frac{y}{d} + 1\right)\sinh\frac{4\pi d}{L}$$

$$\left. + 3\left(\frac{\sinh 4\pi d/L}{4\pi d/L} + \frac{3}{2}\right)\left(\frac{y^2}{d^2}-1\right)\right\}. \qquad (2.201)$$

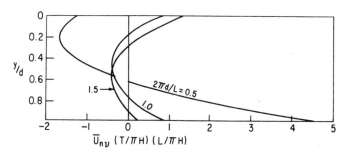

Fig. 2.41. Graphs of $\overline{U}_{nv}(T/\pi H)(L/\pi H)$ when $2\pi d/L = 0.5$, 1.0, and 1.5, representing the profile of the mass-transport velocity in the interior of the fluid in a progressive wave (conduction solution) (after Longuet-Higgins, 1953)

By neglecting the conduction terms, Longuet-Higgins found that the solution was indeterminate for progressive waves. No general solution, including both sets of terms, was given.

The relationship among the dimensionless quantities $\overline{U}_{nv}(T/\pi H)(L/\pi H)$, d/L, and y/d are shown in Fig. 2.41. There are two outstanding features of this theory and its partial experimental verification: first, the transport near the bottom is always in the direction of wave advance; second, for relatively shallow water, the surface transport is in the opposite direction to the direction of wave advance.

Although the argument necessary to the solution of the conduction equations leading to Eq. 2.201 requires that the wave amplitude be very much smaller than the boundary layer thickness, which is small; laboratory measurements made by Russell and Orsorio (1958) show it to be useful for waves of appreciable amplitude (Fig. 2.42).

Russell and Orsorio found that for relatively deep water ($2\pi d/L > 2.1$ in their experiments) the equation of Longuet-Higgins was not as satisfactory as the equa-

tion of Stokes; it gave surface velocities which were many times greater than the measured velocities. An extensive test showed that steady state conditions existed. For the range $0.7 < 2\pi d/L < 1.5$ they found that Longuet-Higgins' conduction equation predicted their results rather well. For values of $2\pi d/L < 0.3$, the mass transport velocity profiles varied in an unsystematic manner. It was found that as long as the measurements were made more than 1 in from the channel walls and not too close to the beach they were consistent.

In an appendix to the paper of Russell and Orsorio, Longuet-Higgins (1958) has shown that the equation should apply in general terms for turbulent motion as well as for laminar motion.

Scattering by turbulence: The water particle velocities associated with turbulence must be considered in addition to the velocity field of pure wave motion. The sizes

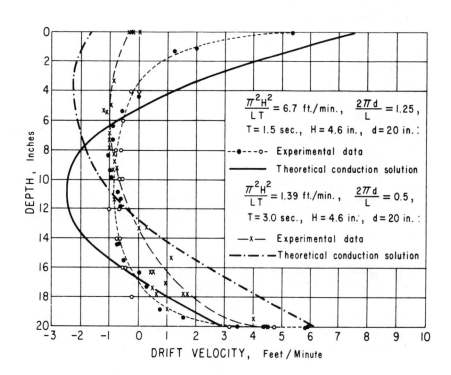

Fig. 2.42. Comparison between measured and theoretical null net mass transport velocity profiles in water of uniform depth (from Russell and Orsorio, 1958)

of the eddies in the ocean may range from a fraction of an inch to at least several hundred yards and thus are of the same order of magnitude as the waves. One of the effects of this turbulence is the scattering of wave energy, with the resulting attenuation of the original wave train.

Phillips (1959) has studied this problem mathematically. He concluded that, for wave lengths in excess of about six feet, the effect of scattering by turbulence would be more important in attenuating the waves than would be the effect of viscous damping.

REFERENCES

Airey, J. R., Toroidal functions and the complete elliptic integrals, *Phil. Mag.*, 7th ser., **19**, 124 (January, 1935), 177–88.

Bagnold, R. A., Sand movement by waves: some small-scale experiments with sand at very low density, *J., Inst. Civil Engrs.*, **27**, 4 (February, 1947), 447–69.

Benjamin, T. B. and M. J. Lighthill, On cnoidal waves and bores, *Proc. Roy. Soc.* (London), ser. A, **224**, 1159 (July, 1954), 448–60.

Biésel, F., General second-order equations of irregular waves, *La Houille Blanche*, Supplement to No. 3, 1952.

Caligny, A. de, Expériences sur les mouvements des molécules liquides des ondes courantes, considérées dans leur mode d'action sur la marche des navires, *Compt. Rend. Acad. Sci.* (Paris), **87**, (1878), 1019–23.

Carry, C., Clapotis partiel, *La Houille Blanche*, No. 4 (1953), 482–94, 494.

Chappelear, J. E., *On the theory of the highest waves*, U.S. Army, Corps of Engineers, Beach Erosion Board, Tech. Memo. No. 116 (July, 1959).

Chinn, Allen J., The effect of surface tension on wave velocities in shallow water, M.S. thesis in Civil Engineering, Univ. of Calif., Berkeley, June, 1949.

Cooper, R. I. B., and M. S. Longuet-Higgins, An experimental study of the pressure variations in standing water waves, *Proc. Roy. Soc.* (London), ser. A, **206** (1951), 424–35.

Danel, Pierre, *On the limiting clapotis, Gravity Waves*, National Bureau of Standards Circular No. 521 (November, 1952), 35–38.

De, S. C., Contributions to the theory of Stokes' waves, *Proc. Camb. Phil. Soc.*, **51** (1955), 713–36.

Dubreil-Jacotin, L., Sur le détermination rigoureuse des ondes permanentes périodiques d'ampleur finie, *J. Math.*, **13** (1934), 267–91.

Elliott, John G., *Interim Report*, Hydrodynamics Laboratory, Calif. Inst. of Tech., Contract NOy–12561 U.S. Navy, Bureau Yards and Docks, July, 1953.

Froude, W., On the rolling of ships, *Trans. Inst. Naval Architects*, **3** (1862), 45–62.

Fuchs, Robert A., On the theory of short-crested oscillatory waves, *Gravity Waves*, National Bureau of Standards Circular No. 521 (November, 1952), 187–200.

Gaillard, D. D., *Wave action in relation to engineering structures*, reproduced from 1935 ed. Fort Belvoir, Va: The Engineer School, 1945.

Gerstner, F., Theorie der wellen, *Abhandlungen der Koniglichen böhmischen Gesellschaft der Wissenschaften*, Prague, 1802. Also, Gilbert *Annalen der Physik*, **32** (1809), 412–25.

Havelock, E. T., Periodic irrotational waves of finite height, *Proc. Roy. Soc.* (London), ser. A, **95** (1918), 38–51.

Hayashi, Keiichi, *Tafeln der Besselschen, theta—   , kugel, und Anderer Funktionen.* Berlin: Springer 1930.

————, *Tafeln für die Differenzenrechnung sowie für die hyperbel, Besselschen, elliptischen, und anderer Funktionen*, Berlin: Springer 1933.

Hough, S. S., On the influence of viscosity of waves and currents, *Proc. London Math. Soc.*, **28** (1896), 264–88.

Hunt, J. N., A note on gravity waves of finite amplitude, *Quart. J. Mech. Applied Math.*, **6** (1953), 336–43.

Iribarren, Ramon Cavanillel, and Casto Nogales y Olano, Talud limite entre la rotura y la reflexion de las olas, *Rev. Obras Publicas* (Madrid), February, 1950.

Iwasa, Yodhiaki, Analytical considerations on cnoidal and solitary waves, *Memoire of Faculty of Engineering*, Kyoto Univ. (Japan), **17**, 4, 1955.

Jahnke, Eugene and Fritz Emde, *Tables of functions with formulae and curves*, 4th ed., New York: Dover Publications, Inc., 1945.

Jeffreys, H., On water waves near the coast, *Phil. Mag.*, ser. 6, **48** (1924), 44–48.

Kaplan, E. L., Auxiliary table of complete elliptic integrals, *J. Math. Physics*, **25**, 1 (February, 1946), 26–36.

————., Auxiliary table for the incomplete elliptic integrals, *J. Math. Physics*, **27**, 1 (April, 1948), 11–36.

Keller, Joseph B., The solitary wave and periodic waves in shallow water, *Comm. Applied Math.*, **1**, 4 (December, 1948), 323–39.

Keulegan, Garbis H. and George W. Patterson, Mathematical theory of irrotational translation waves, *J. Res., National Bur. Standards*, **24**, 1 (January, 1940), 47–101.

————., Wave motion, in *Engineering hydraulics, proc. of the fourth hydraulics conf.*, ed. Hunter Rouse. New York: John Wiley & Sons, Inc., 1950, pp. 711–768.

Korteweg, D. J. and G. de Vries, On the change of form of long waves advancing in a rectangular canal, and on a new type of long stationary waves, *Phil. Mag.*, 5th *ser.* **39** (1895), 422–43.

Kravtchenko, J. and A. Daubert, La houle á trajectoires fermées en profondeur finie, *La Houille Blanche*, **12**, 3 (July–August, 1957), 408–29.

Laitone, E. V., The second approximation to cnoidal and solitary waves, *J. Fluid Mech.* **9** (1960), 430–44.

———., Limiting conditions for cnoidal and Stokes' waves, *J. Geophys. Res.*, **67**, 4 (April, 1962), 1555–64.

Lamb, Sir Horace, *Hydrodynamics*, 6th ed. New York: Dover Publications, Inc., 1945.

Levi-Civita, T. Determination rigoureuse des ondes d'ampleur finie, *Math. Annalen*, **93** (1925), 264–314.

Littman, Walter, On the existence of periodic waves near critical speed, *Comm. Pure Applied Math.*, **10** (1957), 241–69.

Longuet-Higgins, M. S., Mass transport in water waves, *Phil. Trans., Roy. Soc.* (London), ser. A, **245**, 903 (March 31, 1953), 535–81.

———., The refraction of sea waves in shallow water, *J. Fluid Mech.*, **1**, Part 2 (July, 1956) 163–76.

———., The mechanics of the boundary-layer near the bottom in a progressive wave, an appendix to the paper by Russell and Orsorio, 1958, *Proc. Sixth Conf. Coastal Eng.* Berkeley, Calif.: The Engineering Foundation Council on Wave Research (1958), 184–93.

Masch, Frank D. and R. L. Wiegel, *Cnoidal waves: tables of functions.* Berkeley, Calif.: The Engineering Foundation Council on Wave Research, 1961.

Mason, Martin A., *A study of progressive oscillatory waves in water*, U.S. Army, Corps of Engineers, Beach Erosion Board, Tech. Rept. No. 1, 1941.

McCowan, J., On the highest wave of permanent type, *Phil. Mag.*, **38** (1894), 1351–358.

Miche, Robert, Mouvements ondulatoires des mers en profondeur constante ou décroissante, *Annales des Ponts et Chaussées* (1944), 25–78; 131–64; 270–92; 369–406.

———, The reflecting power of maritime works exposed to action of the waves, Bulletin of the Beach Erosion Board, Corps of Engineers, U.S. Army, **7**, 2 (April, 1953), 1–7. (NOTE: This is a translation of the original article that appeared in 1951.)

Michell, J. H., On the highest waves in water, *Phil. Mag.*, 5th ser. **36** (1893), 430–37.

Milne-Thompson, L. M., *Jacobian elliptic function tables.* New York: Dover Publications, Inc., 1950.

———., *Theoretical hydrodynamics*, 3rd ed. New York: The Macmillan Company, 1955.

Mitchim, C. F., Oscillatory waves in deep water, *Military Eng.* (March–April 1940), 107–109.

Morison, J. R., The effect of wave steepness on wave velocity, *Trans. Amer. Geophys. Union*, **32**, 2 (April, 1951), 201–206.

——— and R. C. Crooke, *The mechanics of deep water, shallow water, and breaking waves*, U.S. Army, Corps of Engineers, Beach Erosion Board, Tech. Memo. No. 40, March, 1953.

O'Brien, M. P. and Martin A. Mason, *A summary of the theory of oscillatory waves*, U.S. Army, Corps of Engineers, Beach Erosion Board, Tech. Report No. 2, 1942.

Penny, W. G. and A. T. Price, Finite periodic stationary gravity waves in a perfect liquid, Part II "Some gravity wave problems in the motion of perfect liquids", *Phil. Trans. Roy. Soc.* (London), ser. A, **244**, 882 (March, 1952), 254–84.

Phillips, O. M., The scattering of gravity waves by turbulence, *J. Fluid Mech.*, **5**, Part 2 (February, 1959), 177–92.

Rankine, W. J. M., On the exact form of waves near the surface of deep water, *Phil. Trans., Roy. Soc.* (London), (1863), 127–38.

Rayleigh, Lord, On progressive waves, *Proc. London Math. Soc.*, **9** (November, 1877), 21–26.

Reynolds, Osborne, On the rate of progression of groups of waves and the rate at which energy is transmitted by waves, *Nature*, **36** (August, 1877), 343–44.

Russell, R. C. H. and J. D. C. Orsorio, An experimental investigation of drift profiles in a closed channel, *Proc. Sixth Conf. Coastal Eng.*, Berkeley, Calif.: The Engineering Foundation Research on Wave Research (1958), 171–83.

Savage, Rudolph P., *A statistical study of the effect of wave no steepness on wave velocity*, U.S. Army Corps of Engineers, *Bulletin of the Beach Erosion Board* **8**, 4 (October, 1954), 1–10.

Schoemaker, H. J. and J. Th. Thijsse, Investigations of the reflection of waves, *Third Meeting, Intern. Assoc. Hyd. Structures Res.*, I–2 (September, 1949).

Schuler, M. and H. Gebelein, *Eight-and nine-place tables of elliptical functions.* Berlin: Springer Verlag, 1955.

Sekersh-Zen'Kovich, Y. I., On the theory of standing waves of finite amplitude on the surface of a heavy liquid of finite depth, *Doklady Akad. Nauk SSSR*, new ser., **58** (1947), 551–53.

Skjelbreia, Lars, *Gravity waves, Stokes' third order approximation; tables of functions.* Berkeley, Calif.: The Engineering Foundation Council on Wave Research 1959.

Snyder, C. M., R. L. Wiegel and C. J. Bermel, Laboratory facilities for studying water gravity wave phenomena, *Proc. Sixth Conf. Coastal Eng.*, Berkeley, Calif.: The Engineering Foundation Council on Wave Research 1958, 231–51.

Spenceley, G. W. and R. M. Spenceley, *Smithsonian elliptic functions tables.* Washington, D.C.: Smithsonian Institution, 1947.

Stokes, G. G., On the theory of oscillatory waves, *Mathematical and Physical Papers*, **I**, Cambridge: Cambridge University Press, 1880.

Struick, D. J., Determination rigoureuse des ondes irrotationelles periodiques dans un canal a profundeur finie, *Math. Annalen*, **95** (1926), 595–634.

Stucky, M. A., Contribution á l'étude de l'action des vagues sur une paroi verticale, *Bull. Tech. Suisse Romande*, **60**, 20 (September, 1934), 229–33.

Suquet, F. and A. Wallet, Basic experimental wave research, *Proc. Minn. Intern. Hydraulics Convention, International Assoc. Hyd. Res. Hyd. Div., ASCE.* Minneapolis, Minn.:

St. Anthony Falls Hydraulic Laboratory (August, 1953), pp. 173–91.

Taylor, Donald Charles, An experimental study of the transition between oscillatory and solitary waves, M. S. thesis, Massachusetts. Institute of Technology, November, 1955.

Taylor, Sir Geoffrey, An experimental study of standing waves, *Proc. Roy. Soc.* (London), ser. A, **218**, 1132, (June, 1953), 44–59.

Ursell, F., The long-wave paradox in the theory of gravity waves, *Proc. Camb. Phil. Soc.*, **49** (1953), 685–94.

——., Mass transport in gravity waves, *Proc. Cambridge Phil. Soc.*, **49**, Part 1 (January, 1953), 145–50.

Wallet, A. and F. Ruellan, Trajectories of particles within a partial clapotis, *La Houille Blanche*, No. 4 (July–August, 1950), 483–89.

Wehausen, J. V. and E. V. Laitone, Surface waves, *Handbuch der Physik*, **9** (1960), 446–778.

Wehausen, J.V., Recent developments in free surface flows, Univ. of Calif., Berkeley, Calif., *Inst. Eng. Res. Tech. Rept. NA–63–5* (June, 1963).

Wiegel, R. L., Experimental study of surface waves in shoaling water, *Trans. Amer. Geophys. Union*, **31**, 3 (June, 1950), 377–85.

——., *Gravity waves, tables of functions*. Berkeley, Calif.: The Engineering Foundation Council on Wave Research, 1954.

——., A presentation of cnoidal wave theory for practical application, *Jour. Fluid Mech.*, **7**, Part 2 (1960), 273–86.

Wilton, J. R., On deep water waves, *Phil. Mag.*, 6th ser. **27** (1914), 385–940.

Wolf, F., *Memorandum on the theory of oscillatory waves in a rectangular channel of finite depth*, Univ. Calif., Dpt. of Eng., Tech. Rept. HE–116–21, November, 1944.

Yamada, Hikoji, On the highest solitary wave, *Reports of Research Institute for Applied Mechanics*, **5**, 18 (1957), 53–67.

CHAPTER THREE

The Solitary Wave

1. INTRODUCTION

The waves considered in Chapter 2 were oscillatory, or approximately oscillatory; i.e., the water particles moved forward and backward as the waves passed by. The linear theory described purely oscillatory waves, whereas the higher-order theories showed that, for waves of finite amplitude there was a mass transport in the direction of wave advance. Thus, for waves of finite amplitude, the water particles move back and forth, but each particle moves farther forward than it does backward, and the particle is translated a small net amount forward with the passage of each wave. In closed systems where there must be null net mass transport, the water particles in some layers are translated forward and in other layers they are translated backward. When the water particles move only in the direction of wave advance, the wave is called a *wave of translation*. The solitary wave is of this type. The pure solitary wave is difficult to form although it is rather easy to form an approximate solitary wave, one with a tail of small dispersive waves.

The true solitary wave is entirely above the original water level. It can be generated by adding water to a body of water, dropping a body into water, or pushing forward the end of a wave tank, etc. As will be shown in Chapter 4, however, other types of waves can be formed by the same means.

2. THEORY AND EXPERIMENT

The solitary wave is a limiting case of the cnoidal wave. When $k^2 = 1$, $K(k) = \infty$ and $cn^2\,(\bar{u}\,|\,1) \approx \text{sech}^2\,\bar{u}$, $y_t = d$, and Eq. 2.156 becomes

$$y_s = d + H\,\text{sech}^2\left[\sqrt{\frac{3}{4}\frac{H}{d^3}}(x - Ct)\right], \qquad (3.1)$$

considering Eq. 2.149 for the wave length. The coordinate system is shown in Fig. 3.1. This is the first approximation of Boussinesq (1872), Laitone (1959, 1960), and many others. The speed is given by Eq. 2.158c, and is

$$C = \sqrt{gd}\left(1 + \frac{1}{2}\frac{H}{d}\right). \qquad (3.2a)$$

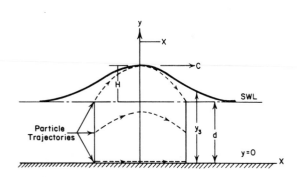

Fig. 3.1. Coordinate system for the solitary wave (after Perroud, 1957)

This is the same as the first approximation of Laitone (1959, 1960) and is nearly equal to

$$C = \sqrt{gd\left(1 + \frac{H}{d}\right)}, \qquad (3.2b)$$

which is the equation determined empirically by Russell (1844), and as a first approximation in the theoretical studies of Boussinesq (1872), Rayleigh (1876), and McCowan (1891).

Laboratory measurements of the speed have been made by Daily and Stephan (1952, 1953), as well as many other investigators. Some data are shown in Fig. 3.2, and it can be seen that Eq. 3.2b is satisfactory for many practical purposes. Equation 3.1 predicts the wave profile for appreciable values of H/d, but theory and measurements deviate for large values of H/d (Fig. 3.3).

To the first approximation, the horizontal and vertical components of water particle speeds are given by Eqs. 2.163 and 2.164 (Laitone, 1959)

$$\frac{u}{\sqrt{gd}} = \frac{H}{d} \operatorname{sech}^2\left[\sqrt{\frac{3}{4}\frac{H}{d^3}}(x - Ct)\right], \qquad (3.3)$$

$$\frac{v}{\sqrt{gd}} = \sqrt{3}\left(\frac{H}{d}\right)^{3/2}\frac{y}{d}\operatorname{sech}^2\left[\sqrt{\frac{3}{4}\frac{H}{d}}(x - Ct)\right]$$
$$\cdot \tanh\left[\sqrt{\frac{3}{4}\frac{H}{d}}(x - Ct)\right]; \qquad (3.4)$$

Fig. 3.2. Comparison of measured solitary wave speed and theory (data, Daily and Stephan, 1953)

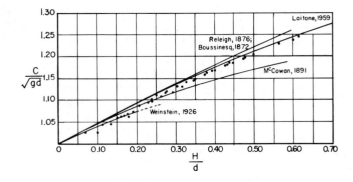

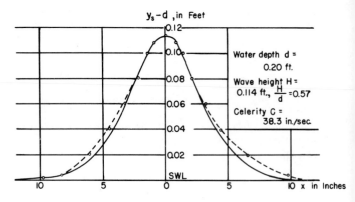

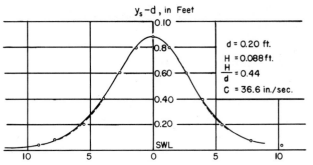

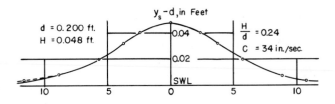

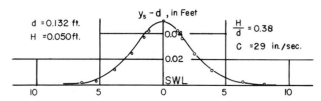

Fig. 3.3. Solitary wave profiles (after Perroud, 1957)

u is always positive; v is positive in front of the wave crest and negative behind the wave crest.

The pressure is given by

$$p = \rho g(y_s - y) \qquad (3.5)$$

Laitone extended the first-order shallow water approximation to obtain the second and third approximation of the shallow water theory in order to get the exact second approximation to the cnoidal wave and the solitary wave. For the limiting case, the solitary wave, the depth of water beneath the trough and the undisturbed water depth become identical so that the equations are easy to apply. These equations are presented instead of the equations of other investigators because the results agree more favorably with laboratory measurements of several wave characteristics than is the case with the other theoretical results. For example, Laitone predicts that the ratio of the maximum height of a solitary wave to the undisturbed water depth, H/d, is $5/7 = 0.714$, whereas other theories predict values of 0.78, and greater. Ippen and Kulin (1955) generated several hundred solitary waves and were never able to obtain a value of H/d greater than 0.72 in water of constant depth. In addition, in McCowan's work (1891), the pressure on the surface is not constant, and is atmospheric only at the crest and mean level.

For the wave speed, Laitone gives

$$\frac{C}{\sqrt{gd}} = 1 + \frac{1}{2}\frac{H}{d} - \frac{3}{20}\left(\frac{H}{d}\right)^2 + \cdots, \quad (3.6a)$$

which is a little smaller than the value due to Boussinesq (1872) and Rayleigh (1876) of

$$\frac{C}{\sqrt{gd}} \approx \sqrt{1 + \frac{H}{d}} \approx 1 + \frac{1}{2}\frac{H}{d} - \frac{1}{8}\left(\frac{H}{d}\right)^2 + \cdots. \quad (3.6b)$$

The measurements of Daily and Stephan (1953) are given in Fig. 3.2. Considering the entire range of values of H/d, Laitone's equation compares most favorably with the measurements.

For the wave profile, Laitone gives

$$y_s = d + H \operatorname{sech}^2(A\omega x - Ct) - \frac{3}{4}H\left(\frac{H}{d}\right)$$
$$\cdot \operatorname{sech}^2(A\omega x - Ct)[1 - \operatorname{sech}^2(A\omega x - Ct)], \quad (3.7)$$

where

$$A\omega x = \frac{x}{d}\sqrt{\frac{3}{4}\frac{H}{d}}\left[1 - \frac{5}{8}\frac{H}{d}\right] \quad (3.8)$$

and C is given by Eq. 3.6a. Experimental data have

shown that Eq. 3.7 predicts the profile accurately for a large value of H/d (Laitone, 1959; 1960).

The water particle velocities are

$$\frac{u}{\sqrt{gd}} = \frac{H}{d}\left[1 + \frac{1}{4}\frac{H}{d} - \frac{3}{2}\frac{H}{d}\frac{y^2}{d^2}\right]\operatorname{sech}^2(A\omega x - Ct)$$
$$+ \frac{H^2}{d}\left[-1 + \frac{9}{4}\frac{y^2}{d^2}\right]\operatorname{sech}^4(A\omega x - Ct) \quad (3.9)$$

$$\frac{v}{\sqrt{gd}} = \sqrt{3}\left(\frac{H}{d}\right)^{3/2}\frac{y}{d}\operatorname{sech}^2(A\omega x - Ct)$$
$$\cdot \tanh(A\omega x - Ct)\left[1 - \frac{3}{8}\frac{H}{d} - \frac{1}{2}\frac{H}{d}\frac{y^2}{d^2}\right.$$
$$\left. + \frac{H}{d}\left(-2 + \frac{3}{2}\frac{y^2}{d^2}\right)\operatorname{sech}^2(A\omega x - Ct)\right]$$
$$(3.10)$$

These expressions for u and v differ from those given by Eqs. 2.165 and 2.166 when $k^2 = 1$, although Eqs. 3.9 and 2.165 give the same numerical values when $\operatorname{sech}(\ ) \approx 1$.

These solutions are not adequate near the water surface in the vicinity of the crest for large values of H/d, as some additional terms including $(H/d)^3$ and $(H/d)^4$ would have to be included. For example, if $H/d = 0.714$ and $y/d = 1.714$, $u/C = 1.14$, where C is calculated using Eq. 3.6a. This results in $u > C$, which cannot occur without the wave breaking. There are very few data available with which to check the existing theories for water particle velocities, as reliable measurements are difficult to obtain. These data (Dailey and Stephan,

Fig. 3.4. Comparison of components of local fluid element velocity according to McCowan with components determined by experiment. Heavy isovels show equal vertical velocity. Light isovels show equal horizontal velocity. $H/d = 0.509$, $d = 0.259^1$, $C = 3.502$ ft/sec. Values of isovels are ratios of element velocities to wave celerity (after Daily and Stephan, 1952)

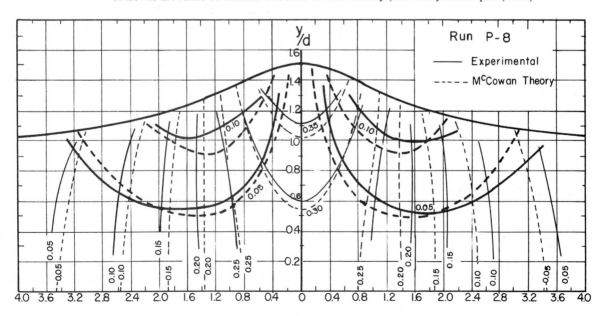

1952, 1953) indicate that the equations due to McCowan (1891) are the most reliable (Fig. 3.4, for example). It is surprising that this is the case, as McCowan's theory does not predict the other wave characteristics as well as some other theories. The equations are

$$\frac{u}{C} = N\left\{\frac{1 + \cos(My/d)\cosh Mx/d}{[\cos(My/d) + \cosh(Mx/d)]^2}\right\} \quad (3.11)$$

$$\frac{v}{C} = N\left\{\frac{\sin My/d \cdot \sinh Mx/d}{[\cos(My/d) + \cosh(Mx/d)]^2}\right\} \quad (3.12)$$

where M and N are as used by Munk (1949) and are given by

$$\frac{H}{d} = \frac{N}{M}\tan\frac{1}{2}\left[M\left(1 + \frac{H}{d}\right)\right] \quad (3.13)$$

$$N = \frac{2}{3}\sin^2\left[M\left(1 + \frac{2}{3}\frac{H}{d}\right)\right] \quad (3.14)$$

M and N have been calculated by Munk (1949) and are shown in Fig. 3.5. Equation 3.9 predicts lower values of u/C under the wave crest ($x = 0$) than the theory of Boussinesq (see Munk, 1949), but higher than the theory of McCowan. A comparison of Eqs. 3.9 and 3.11 for $x = 0$ is given in Fig. 3.6. The two theories compare favorably for H/d less than about 0.3.

The pressure is given by

$$p = \rho g\left\{y_s - y - \frac{3}{4}\frac{H^2}{d}\left[2\left(\frac{y}{d} - 1\right) + \left(\frac{y}{d} - 1\right)^2\right]\right.$$

$$\left. \cdot[2\,\text{sech}^2\,(A\omega x - Ct) - 3\,\text{sech}^4\,(A\omega t - Ct)]\right\} \quad (3.15)$$

Waves traveling must lose energy due to internal friction. The problem of solitary waves moving in a wave tank has been studied theoretically by Keulegan (1948). It was found that

Fig. 3.5. Values of parameters M and N defined by Eqs. 3.13 and 3.14 (after Munk, 1949)

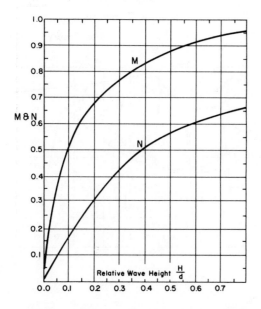

$$\left(\frac{H}{d}\right)^{-1/4} - \left(\frac{H_x}{d}\right)^{-1/4} = \left[\frac{1}{12}\left(1 + \frac{2d}{B}\right)\sqrt{\frac{v}{g^{1/2}d^{3/2}}}\right]\frac{x}{d} = K\frac{x}{d} \quad (3.16)$$

where H_x is the wave height after it has traveled a distance x ft, v is the kinematic viscosity of the liquid, ft²/sec, and B is the width of the tank, in feet. Experimental studies by Dailey and Stephan (1952, 1953) verified the form of Eq. 3.16 for relatively smooth bottoms and walls, but found that the value of K was less than predicted by theory. For water depths of 0.2, 0.3, and 0.4 ft K was found to be 0.00372, 0.00327, and 0.00318 compared with the theoretical values of 0.00480, 0.00379, and 0.00335, respectively. Ippen, Kulin, and Raza (1955) and Ippen and Mitchell (1957) have made extensive studies of the damping of solitary waves moving over a rough bottom.

3. SHOALING WATER

It has been said that a wave in shoaling water just prior to breaking might be treated as a solitary wave as an approximation (see McCowan, 1891). This approximation has proved to be useful under some circumstances (Bagnold, 1947; Munk, 1949; Wiegel and Beebe, 1957; Housley and Taylor, 1957). It was found, for example, that the solitary wave theory would predict approximately the height of a breaker. Laboratory studies by Ippen and Kulin (1955) of the characteristics of solitary waves in shoaling water, however, led them to conclude that the observed breaking height–water depth ratio of oscillatory waves and the theoretical maximum solitary wave value was fortuitous. As the cnoidal wave theory is available, it is suggested that the solitary wave theory no longer be employed to approximate shallow water waves.

The experimentally determined breaking characteristics of the solitary wave could not be reconciled with available results on long oscillatory waves. For example, the ratio of breaker height to wave height in the constant depth portion of the channel decreases with steepening beach slope for the same incident wave, whereas for oscillatory waves this ratio increases with increasing beach slope. The breaker characteristics of solitary waves can be seen in Figs. 3.7 and 3.8. It is evident that the ratio $H/d = 0.71$ for limiting solitary wave steepness does not hold for waves in shoaling water, as the ratio H/d was always well in excess of this value.

Ippen and Kulin (1955) performed their experiments on slopes of intermediate steepness, ranging from 0.023 to 0.065 (1°19′ to 3°43′). They found that, on the 0.023 slope, the waves tended to spill, but on the steeper slopes the waves tended to plunge. On the 0.023 slope, it was found that, when the bottom was roughened, the spilling became more pronounced. Caldwell (1949) studied the action of solitary waves formed in water 5 in. deep running up smooth impermeable slopes ranging

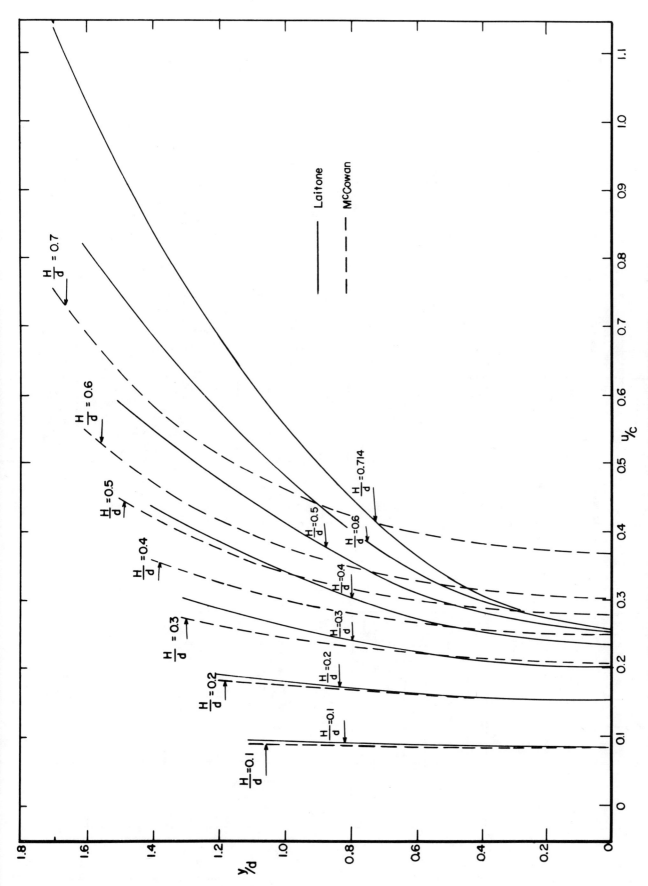

Fig. 3.6. Comparison of Laitone and McCowan theories for u/c under wave crest ($x = 0$)

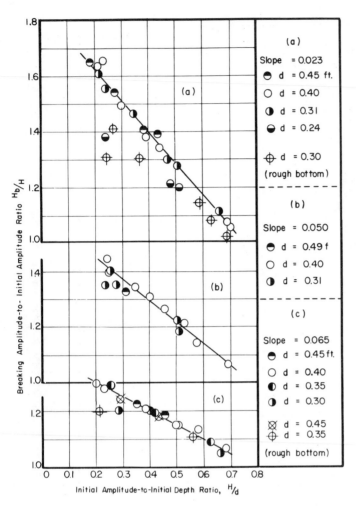

Fig. 3.7. Breaking amplitudes (after Ippen and Kulin, 1955)

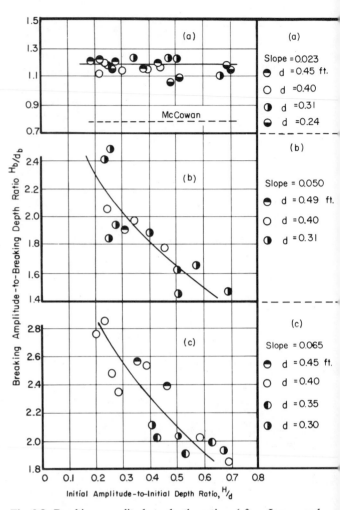

Fig. 3.8. Breaking amplitude-to-depth ratios (after Ippen and Kulin, 1955)

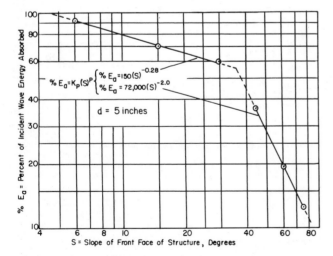

Fig. 3.9. Relationship between wave energy absorbed and slope of impermeable structure (after Caldwell, 1949)

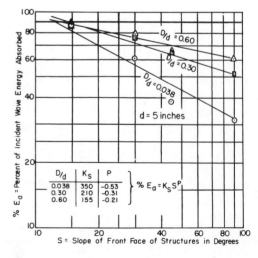

Fig. 3.10. Relationships among wave energy absorbed, slope, and relative permeability of permeable structure (after Caldwell, 1949)

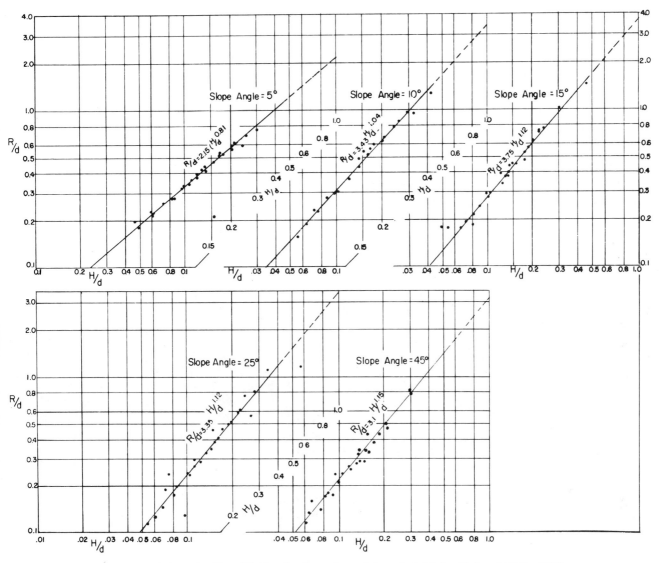

Fig. 3.11. Relation of R/d to H/d for the five values of slope tested (after Hall and Watts, 1953)

from 0.105 to 3.732 (6° to 75°). It was found that the waves never broke, and that the action was a gentle ride-up of the wave on the slope.

Caldwell (1949) measured the solitary wave dimensions incident to the slope and after the waves had been reflected in order to determine the amount of energy absorbed in the process. The portion of the energy absorbed was calculated from

$$E_a = \frac{(\frac{3}{4} H_I d)^{3/2} - (\frac{4}{3} H_R d)^{3/2}}{(\frac{4}{3} H_I d)^{3/2}} \tag{3.17}$$

where H_I is the incident solitary wave height and H_R is the reflected solitary wave height. The results are shown in Fig. 3.9. It appears that negligible energy is reflected from a slope less than about four degrees (slope of 0.070, that is, 1 in 14.3). A permeable slope was found to be a more effective absorber of wave energy, as can be seen in Fig. 3.10. Here D is the median diameter of the rock composing the permeable slope and d is the

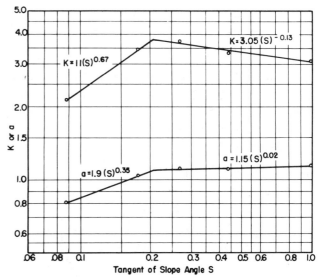

Fig. 3.12. Relations of K and a to tangent of slope angle (after Hall and Watts, 1953)

water depth at the toe of the structure (the water depth of the horizontal portion of the wave tank).

Hall and Watts (1953) made laboratory studies of the vertical rise, R, of solitary waves on impermeable slopes for water depths in the horizontal bottom portion of the wave tank ranging from 0.50 ft. to 2.25 ft. The results are summarized in Fig. 3.11, with the coefficients K and a given in Fig. 3.12 for the equation

$$\frac{R}{d} = K \left(\frac{H}{d}\right)^a \qquad (3.18)$$

R is the vertical distance above the still water level to which the solitary wave rises on the impermeable slope. It can be seen that K is a function of the slope S.

4. REFLECTION—THE MACH-STEM EFFECT

Little work has been done on the reflection of solitary waves approaching a slope at other than normal incidence. Perroud (1957) made laboratory studies of the reflection of solitary waves from a vertical wall. It was found that three types of patterns were obtained; one pattern, however, was a particular case of another of the patterns. The critical angle of incidence, i, separating these two types of patterns appeared to be 45 degrees. (Chen, 1961, found the critical angle to be somewhat under 40 degrees.)

For *incident angles* (the angle between the direction of wave advance and the vertical wall) greater than 45 degrees the reflection pattern is "normal" (Fig. 3.13). The incident and reflected waves are slightly disturbed near the wall, but the angle of reflection is equal to the angle of incidence, and the reflected wave height is only slightly less than the incident wave height. The reflected wave is followed by a trough, however, except for the case of the incident wave making an angle of approximately 90 degrees. It is interesting to note that in the latter case the wave height at the wall was 20 per cent higher than double the incident wave height. For angles of incidence less than 45 degrees, the reflection appears to be of the type called a *Mach reflection* in acoustics (Lighthill, 1949).

For angles of incidence less than 20 degrees, the wave crest bends so that it becomes perpendicular to the wall (Fig. 3.13) and no reflected wave appears. When the angle of incidence is greater than 20 degrees but less than 45

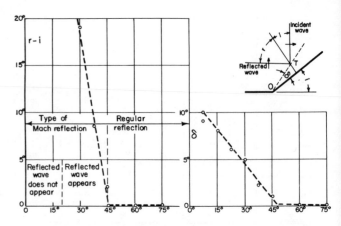

Fig. 3.14. Oblique reflection of a solitary wave, experimental results; water depth $d = 0.132$ ft (after Perroud, 1957)

degrees, three waves are present: the incident wave I; a reflected wave R; and a wave M approximately perpendicular to the wall, the width of which seems to grow as the wave travels (Fig. 3.13). The reflected wave height is smaller than the incident wave height and the angle of reflection (r) is greater thsn the angle of incidence. The height of the portion of the wave perpendicular to the wall (called the *Mach stem*), is greater than the incident wave height, and is at its maximum height at the wall. The reflected and Mach-stem waves are followed by a trough.

The relationships between reflected and incident wave angle is shown in Fig. 3.14; the relationships among incident, reflected, and Mach-stem heights and incident wave angle are shown in Fig. 3.15. In these figures, δ is

Fig. 3.15. Oblique reflection of a solitary wave, experimental results (after Perroud, 1957)

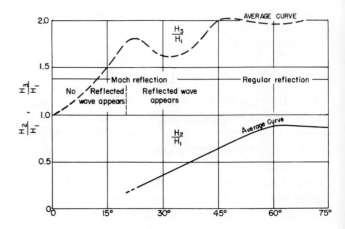

Fig. 3.13. Reflection patterns of solitary wave (after Perroud, 1957)

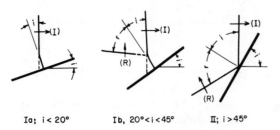

Ia; $i < 20°$ Ib, $20° < i < 45°$ II; $i > 45°$

the angle between the wall and the curve along which point T moves. T is the intersection of the three wave crests for incident wave angles between 20 and 45 degrees, and it is the intersection of the perpendicular to the wall (the Mach stem) and the prolongation of the undisturbed incident wave crest for incident wave angles less than 20 degrees. For the case of incident waves with incident angles greater than 45 degrees, H_3 refers to the wave height at the vertical wall.

Most coastal areas and structures have slopes. Experiments made with smooth impermeable slopes (Chen, 1961) showed that for nearly vertical walls the phenomenon looked the same as for the case of a vertical wall ($\beta = 90$ degrees). As the wall slope was decreased, a large horizontal eddy was seen to form over the slope, and as the slope was decreased further, the wave broke along the slope; the slope angle at which this occurred, β, was found to depend upon the angle of incidence, i.

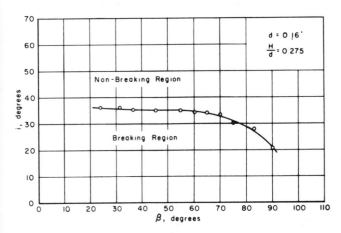

Fig. 3.16. Angle of incidence separating breaking and non-breaking regions (after Chen, 1961)

In Fig. 3.16 are shown the values of i and β which determine whether or not the wave breaks over the slope for a specific value of H/d. For small values of H/d, no break would occur in the vicinity of $\beta = 90$ degrees. It was not possible to determine whether a wave would break for β in the vicinity of 90 degrees for intermediate values of H/d because of the small size of the tank. Friedlander (1946) has shown theoretically for a sound pulse incident to a wedge (less than 90-degree wedge) that the pulse must travel a considerable distance along the wedge before the pressure at the wedge builds up to its maximum value of twice the pressure of the incident pulse, and that the smaller the wedge angle (hence the smaller i) the greater this distance must be. For example, for a 90-degree wedge the pressure builds up to about 90 per cent of its final value within six pulse lengths, whereas for a 30-degree wedge the pressure pulse must travel nearly sixty pulse lengths to build up to 90 per cent of its final value.

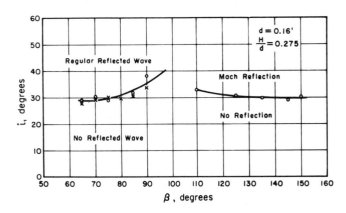

Fig. 3.17. Incident angle i beyond which regular reflected wave occurs for different slopes β, and i at which Mach reflection starts for each negative slope ($\beta > 90°$) (after Chen, 1961)

Waves incident to an overhanging wall ($\beta > 90$ degrees) behaved in a manner similar to a vertical wall except that the stem did not grow with distance from the start as was found by Perroud (1957) for the vertical wall. At least, the stem did not appear to grow within the limits of the experimental facilities. The regions of i and β for which various types of reflections occur are shown in Fig. 3.17.

The author has observed periodic waves in shallow water along some jetties and seawalls that had the characteristics of the Mach-stem phenomenon. In some of these cases, waves encountered the structure at small angles of incidence at the seaward end of the structures (Wiegel, 1961). The structures gradually curved up to angles as great as 90 degrees with the direction of incident wave advance. This condition was simulated in a wave tank by bending a piece of sheet metal as shown in Fig. 3.18 (Sigurdsson and Wiegel, 1962). A 2-ft long straight section extended from the tank wall making an angle of 12 degrees with it. This straight section was connected tangentially to a 66-degree segment of a circle of 40-in. radius. The plan view of the solitary wave crest used in the experiment is shown for several positions of the wave as it advances along the barrier in Fig. 3.18, together with the elevations along the crest. Tests made with a barrier with a smooth wavy surface (corrugated aluminum, with corrugations 0.021 ft deep by 0.135 ft long, about 1/2 and 3 times the incident solitary wave height, respectively) and with a rough barrier showed similar results. That the amplitudes were greater in valleys than at the ridges of the corrugations was unexpected.

It was found that close to the end of the barrier the wave stem began to separate from the main wave. In this region there were essentially two separate waves: the "incident wave" and the wave stem advancing at an appreciable angle to the "incident wave." These two waves were connected by a transition zone, and in this zone the incident and stem waves were superimposed. In

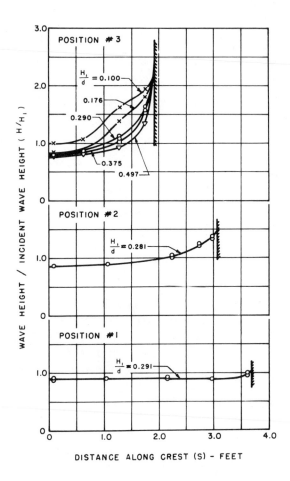

Fig. 3.18. Crest position and crest elevation series *A* (from Sigurdsson and Wiegel, 1962)

this region there was considerable variation in the profiles of the crest. Figure 3.19 shows that the maximum wave height at the barrier occurred before its end. From this study, it was not possible to determine whether this occurred because of end effects or because the "incident wave" had already struck the end of the barrier and had started to retreat when the stem portion arrived.

A model pervious breakwater built of $\frac{3}{8}$-in. gravel to the same plan as the impervious barriers, but with a 36-degree side slope showed no indication of either a buildup of wave height at the structure or a reflected wave.

5. DIFFRACTION BY BREAKWATER GAP

Perroud (1957) also made some observations in the laboratory of a solitary wave moving through a breakwater gap, i.e., the phenomenon of diffraction. The incident wave was normal to the vertical wall breakwater. It was found that the diffracted wave was followed by a trough so that it was no longer a true solitary wave. Furthermore, just inside the breakwater a rather strong vortex formed that persisted for a considerable length of time.

REFERENCES

Bagnold, R. A., Sand movement by waves: some small-scale experiments with sand of very low density, *Inst. Civil Engrs.*, **27**, 4, Paper No. 5554 (February, 1947), 447–69.

Boussinesq, J., Theorie des onde et de remous qui se propagent le long d'un canal rectangulaire horizontal, en communiquant au liquide contenu dans ce canal des vitesses sensiblement pareilles de la surface au fond, *J. Math. Pures Appliquées*, ser. 2, **17** (1872), 55–108.

Caldwell, Joseph M., *Reflection of solitary waves*, U.S. Army

Corps of Engineers, Beach Erosion Board, Tech. Memo. No. 11 (November, 1949).

Chen, T. C., *Experimental study on the solitary wave reflection along a straight sloped wall at oblique angle of incidence*, U.S. Army, Corps of Engineers, Beach Erosion Board, Tech. Memo. No. 124, March, 1961.

Dailey, James W. and Samuel C. Stephan, Jr., Characteristics of the solitary wave, *Trans. ASCE*, **118** (1953), 575–87.

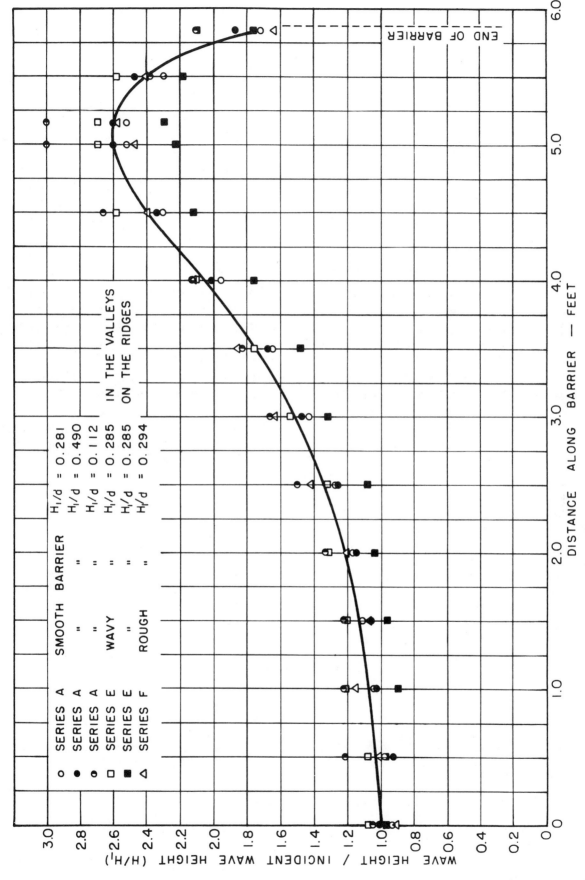

Fig. 3.19. Maximum wave elevation at vertical barriers (from Sigurdsson and Wiegel, 1962)

————. *The solitary wave, its celerity, profile, internal velocities and amplitude attenuation*, Mass. Institute of Technology, Hydro. Lab., Tech. Rept. No. 8, June 1952. (Unpublished.)

Friedlander, F. G., The diffraction of sound pulses. I. Diffraction by a semiinfinite plane. II. Diffraction by an infinite wedge. III. Note on an integral occurring in the theory of diffraction by a semi-infinite screen. IV. On a paradox in the theory of reflexion, *Proc. Roy. Soc.* (London), ser. A, **186**, 1006 (Sept. 24, 1946), 322–67.

Hall, Jay V., Jr., and George M. Watts, *Laboratory investigation of the vertical rise of solitary waves on impermeable slopes*, U.S. Army, Corps of Engineers, Beach Erosion Board, Tech. Memo. No. 33, March, 1953.

Housley, John G., and Donald C. Taylor, Application of the solitary wave theory to shoaling oscillatory waves, *Trans. Amer. Geophys. Union*, **38**, 1 (February, 1957), 56–61.

Ippen, Arthur T. and Gershon Kulin, The shoaling and breaking of the solitary wave, *Proc. Fifth Conf. Coastal Eng.*, Berkeley, Calif.: The Engineering Foundation Council on Wave Research (1955), 27–49.

————., ————, and Mir A. Raza, *Damping characteristics of the solitary wave*, Mass. Inst. of Tech., Hydro. Lab., Tech. Rept. No. 16, April, 1955. (Unpublished.)

————., and Melvin M. Mitchell, *The damping of the solitary wave from boundary shear measurements*, Mass. Inst. of Tech., Hydro. Lab., Tech. Rept. No. 23, June, 1957. (Unpublished.)

Keulegan, G. H., Gradual damping of solitary waves, *J. Res. National Bur. Standards*, **40**, Research Paper No. RP 1895 (June, 1948), 487–98.

Laitone, E. V., *Water waves. IV, shallow water waves*, Univ. Calif., Inst. Eng. Res., Tech. Rept. 82–11, November, 1959.

————., The second approximation to cnoidal and solitary waves, *J. Fluid Mech.*, **9**, Part 3 (1960), 430–44.

Lighthill, M. J., The diffraction of blast. I, *Proc. Roy. Soc.* (London), ser. A, **198**, 1055 (Sept. 7, 1949), 454–70.

McCowan, J., On the solitary wave, *London, Edinburgh, Dublin Phil. Mag. J. Sci.*, **32**, 5 (1891), 45–58.

Munk, Walter H., The solitary wave theory and its application to surf problems, ocean surface waves, *Annals New York Acad. Sci.*, **51**, 3 (May, 1949), 376–423.

Perroud, Paul Henri, "The solitary wave reflection along a straight vertical wall at oblique incident," Ph. D. thesis, Univ. Calif.,; also, Inst. Eng. Res., Tech. Rept. No. 99–3, September, 1957. (Unpublished.)

Rayleigh, Lord, On waves, *London, Edinburgh, Dublin Phil. Mag. J. Sci.*, **1**, 4 (April, 1876), 257–79.

Russell, J. Scott, "Report on waves," *Fourteenth Meeting* Brit. Assoc. Advance. Sci., (1844), pp. 311–390.

Sigurdsson, Gunnar and R. L. Wiegel, Solitary wave behavior at concave barriers, Calcutta, *The Port Engineer*, **9**, 4, Oct., 1962, 4–8.

Wiegel, R. L. and K. E. Beebe, The design wave in shallow water, *J. Waterways Division, Proc. ASCE*, **82**, WW1, Paper 910, March, 1956.

————., Research related to tsunamis performed at the Hydraulic Laboratory, University of California, Berkeley, *Proc. Tsunami meetings*, Pacific Science Congress, 1961. Inter. Union Geodesy and Geophys., Paris (1963).

CHAPTER FOUR

Impulsively Generated and Other Waves

1. PROPAGATION OF WAVE GROUPS

The concept of the movement of a group of waves is basic to the understanding of impulsively generated waves. The simplest type of group results from the superposition of two trains of waves of the same amplitude, of approximately the same wave length, and extending to infinity in both directions. For the two-dimensional case, the surface profile is given by (Rayleigh, 1877) as

$$
\begin{aligned}
y_s &= a \sin (kx - \sigma t) + a \sin (k'x - \sigma' t) \\
&= 2a \cos \left[\tfrac{1}{2}(k - k')x - \tfrac{1}{2}(\sigma - \sigma')t \right] \\
&\quad \cdot \sin \left[\tfrac{1}{2}(k + k')x - \tfrac{1}{2}(\sigma + \sigma')t \right],
\end{aligned}
\tag{4.1}
$$

where the symbols are the same as those used in Chapter 2. As $k - k'$ and $\sigma - \sigma'$ are very small, the cosine term varies slowly with x and t so that the superimposed wave trains consist of waves in the form of sines in which the amplitude varies gradually from 0 to $2a$, back to 0, etc. The system has the appearance of a series of groups of waves.

The distance between centers of groups of waves is $2\pi/(k - k')$ and the time necessary for the wave system to shift this distance is $2\pi/(\sigma - \sigma')$, so the speed at which the group moves, C_G, is

$$
C_G = \frac{\sigma - \sigma'}{k - k'} = \frac{d\sigma}{dk}
\tag{4.2}
$$

in the limit. In terms of wave speed (which is also called *phase speed*),

$$
C_G = \frac{d(kC)}{dk} = C - L \frac{dC}{dL}.
\tag{4.3}
$$

Substituting the value of C given by Eq. 2.33,

$$
C_G = \frac{1}{2} C \left(1 + \frac{2kd}{\sinh 2kd} \right).
\tag{4.4}
$$

This is the same as Eq. 2.68, which is the speed at which the wave energy of an infinitely long train of uniform periodic waves is propagated.

If the effect of surface tension is included, the equation for group velocity in deep water is

$$C_G = \frac{1}{2} C \left[1 + \frac{1}{C^2} \frac{d(kC^2)}{dk} \right] = \frac{1}{2} C \frac{g + 3k^2 T_s}{g + k^2 T_s} \quad (4.5)$$

where T_s is the surface tension. When surface tension can be neglected, $C_G = \frac{1}{2}C$. When surface tension predominates, $C_G = \frac{3}{2}C$, and the wave group moves faster than the individual waves. When $k^2 T_s = g$, $C_G = C$, which is the case of the minimum speed of propagation of gravity–surface tension waves.

Because these single groups of waves and the energy of waves in infinite trains of uniform periodic waves are propagated at the same speed, it has been thought that Eq. 4.4 should predict the speed at which a train of waves of finite extent propagates. The problem becomes difficult because the dispersive nature of waves causes the spreading of a finite wave train and the segregation of the wave components with respect to their lengths. These difficulties will be described in more detail in the following sections.

2. INITIAL ELEVATIONS

The classic solution to the waves generated by one type of initial elevation, without particle velocities at $t = 0$, was due to Cauchy and Poisson (Lamb, 1945, p. 238). For the two-dimensional case, the elevation and potential of this particular initial disturbance are given by

$$y_s = \frac{1}{\pi} \int_0^\infty \cos \sigma t \, dk \int_{-\infty}^\infty f(\alpha) \cos k(x - \alpha) \, d\alpha \quad (4.6)$$

$$\phi = \frac{g}{\pi} \int_0^\infty \frac{\sin \sigma t}{\sigma} e^{ky} \, dk \int_{-\infty}^\infty f(\alpha) \cos k(x - \alpha) \, d\alpha \quad (4.7)$$

The disturbance is considered to consist of the superposition of an infinite number of trains of regular waves extending to infinity so that all of their phases agree at the origin, but at all other points the phases are such that the wave trains interfere and result in zero displacement. The initial elevation is confined to the immediate neighborhood of the origin and $f(\alpha)$ vanishes except for infinitesimal values of α. It is assumed that

$$\int_{-\infty}^\infty f(\alpha) \, d\alpha = 1 \quad (4.8)$$

$$\phi = \frac{g}{\pi} \int_0^\infty \frac{\sin \sigma t}{\sigma} e^{ky} \cos k \times dk. \quad (4.9)$$

One solution to this as given by Lamb is

$$y_s = \frac{t}{x} \sqrt{\frac{g}{2\pi x}} \left[C(u) \cdot \cos \left(\frac{gt^2}{4x} \right) + S(u) \cdot \sin \left(\frac{gt^2}{4x} \right) \right], \quad (4.10)$$

where $C(u)$ and $S(u)$ are the Fresnel integrals, taking $u = t\sqrt{g/2\pi x}$. $C(u)$ and $S(u)$ have been tabulated by Jahnke and Emde and by Van Wyngaarden and Schien (see Chapter 8). Equation 4.10 appears to be dimensionally nonhomogeneous. This is because in Eq. 4.8 the area of the disturbance was taken as unity. The periods

and lengths of the individual waves are given by $T = 4\pi x/gt$ and $L = 8\pi x^2/gt^2$. Equation 4.10 shows that, at a fixed place, the wave period and length continually decrease with increasing time while the amplitude continually increases, whereas at a fixed time there is a continual decrease in wave period and length in the direction of the origin and a continued increase in wave amplitude.

At any time $t > 0$ there will be no complete agreement in phase of all the components, but there will be a large number of component trains that will nearly agree in phase. If a very large number of component wave trains of nearly equal phase velocity are in agreement in phase at any point, the sum of them must determine approximately the water surface displacement at that point. Thus, at some given distance from the origin and at some given time, waves will be passing the point that can be described as having a length L. If we define the group velocity as $C_G = x/t$ for waves of this length, Green (1909) has shown that this value of C_G is expressed by Eq. 4.2. The physical significance of the group velocity (or as Green prefers, *stationary phase velocity*) discussed in Section 4.1 is thus more broad than originally expected.

The initial disturbance previously described is for the case of an initial elevation which has an infinite amount of energy, and as such is of only limited interest. Two cases that are more enlightening from a physical standpoint are due to Kelvin (1906). Both of these cases are for a finite amount of energy. The equations are not repeated here, but the general results of one case are shown in Fig. 4.1. Rather than using the natural value for g, Kelvin chose a value that would simplify his calculations. As a result, $\sqrt{\pi}$ is the period of an infinite train of waves of wave length 2.

Green (1909), using the foregoing concept of group velocity, computed the positions of the maximums shown in Fig. 4.1 as a function of time. The location of the crest was taken as the product of the group velocity of the crest and the time, with the group velocity being that of a wave whose length was taken as twice the horizontal distance between the zero crossings preceding and following the crest in question. By the time $t = 8\sqrt{\pi}$, the values computed by Green were very close to those shown in the figure.

This illustration of an approximate physical situation shows that, owing to dispersion, the leading waves are relatively long and the trailing waves are short, and eventually become almost imperceptible.

Another case of interest is an initial elevation of finite constant height that extends over a finite length of water surface. Unoki and Nakano (1953a) have developed the theory for both the two-dimensional and three-dimensional cases for water of infinite depth. The initial elevation is $y_s = h$ and extends over the region $-\lambda \leq x \leq \lambda$, and $y_s = 0$ for $x > \lambda$. The solution for $x \gg \lambda$ is

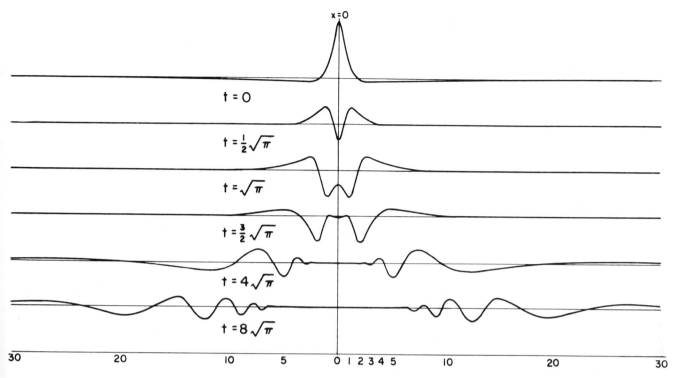

Fig. 4.1. Dispersion of initial water surface elevation (after Kelvin, 1906)

$$y_s = \frac{4h}{t} \sqrt{\frac{x}{\pi g}} \sin\left(\frac{gt^2}{4x} \cdot \frac{\lambda}{x}\right) \cos\left(\frac{gt^2}{4x} - \frac{\pi}{4}\right) \quad (4.11)$$

This shows individual waves with characteristics

$$T = \frac{4\pi x}{gt}; \qquad L = \frac{8\pi x^2}{gt^2}; \qquad C = \sqrt{\frac{gL}{2\pi}} = \frac{2x}{t} \quad (4.12)$$

with the group, or amplitude envelope, with characteristics

$$T_g = \frac{4\pi x}{gt} \cdot \frac{x}{\lambda}; \qquad L_g = \frac{8\pi x^2}{gt^2} \cdot \frac{x}{2\lambda};$$

$$C_g = \frac{x}{t} = \text{constant.} \quad (4.13)$$

The group velocity is equal to one-half the phase velocity, as in the other analyses for deep water waves.

Prins (1958a, 1958b) studied experimentally waves generated by an initial surface elevation. The initial surface elevation (or depression) was obtained using an airtight plexiglass box placed in the tank with the front of the box consisting of a sliding wall. The box was partially evacuated to cause an initial surface elevation, and compressed air was added to obtain an initial surface depression. When the sliding wall was raised suddenly, the vertical wall of water collapsed, generating waves in the process. When the length (λ) and height (h) of the initial elevation, or depression, were not large compared with the water depth, waves were generated which in general had the characteristics described by Eq. 4.11. As an example of the accuracy of Eq. 4.11 in predicting the group (amplitude envelope zero-points) speeds, see

Fig. 4.2, in which the surface time histories at 0, 5, 15, 25, and 35 ft from the origin are shown for an initial depression and an initial elevation of the water surface. The prediction of phase speed is about as reliable. It can be seen that the waves resulting from an initial depression are the negative of the waves resulting from an initial elevation. When the initial elevation was of a size to generate "deep water" waves, the phase period was found to be nearly that predicted by Eq. 4.11, but the wave height was about 25 per cent less that predicted by Eq. 4.11.

When the length and height of the initial elevation (or depression) become large relative to the water depth, Eq. 4.11 is no longer valid, although the theory of Ursell (1958) will be valid for the range of transitional water waves. Upon further increase in λ or h, or both, the resulting waves become nonlinear. In all the cases where the initial elevation generated nonlinear waves, the initial depression generated waves that were dispersive, and similar to the linear case, except that the mean of the generated waves was depressed below the initial water level in the tank, as shown in Fig. 4.3(a). For the largest values of λ and h relative to the undisturbed water depth in the tank, a bore formed first, and gradually dissipated energy as it traveled down the tank, becoming a series of complex "solitary waves," as shown in Fig. 4.3(b). For values of λ and h a little smaller than this, "solitary" or complex "solitary waves" formed.

The types of waves generated by an initial elevation as a function of h/d and λ/d are shown in Fig. 4.4.

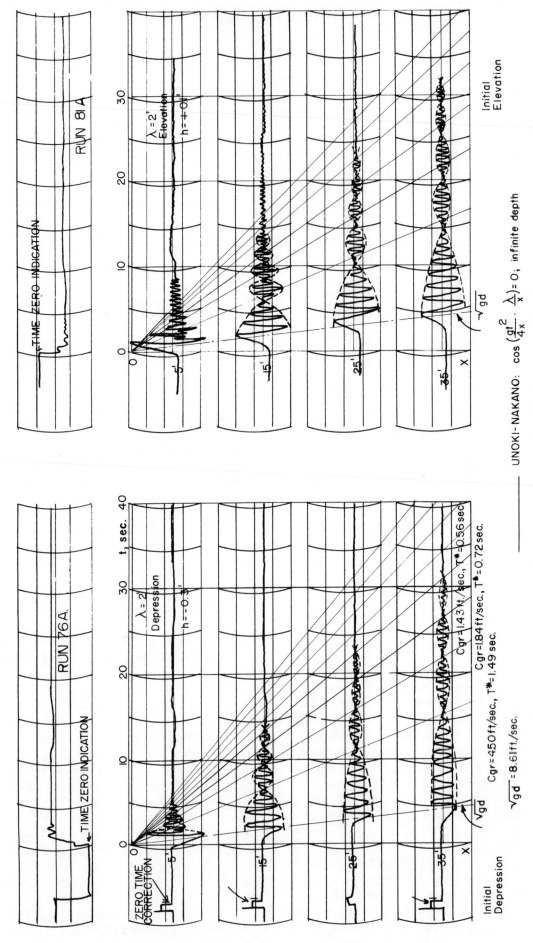

Fig. 4.2. Amplitude envelope zero-points, $d = 2.3$ ft (from Prins, 1958)

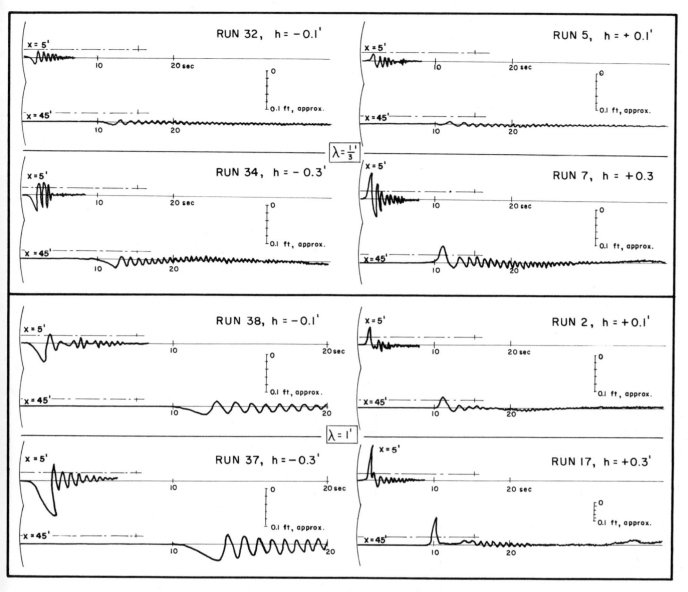

(a) Leading parts of the wave patterns at depth $d = 0.5$ ft (after Prins, 1958)

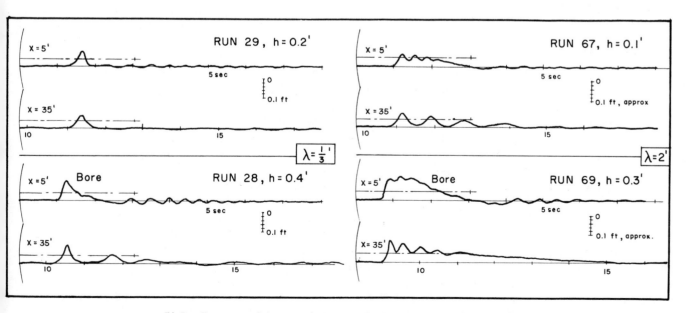

(b) Leading parts of the wave patterns at depth $d = 0.2$ ft (after Prins, 1958)

Fig. 4.3.

81

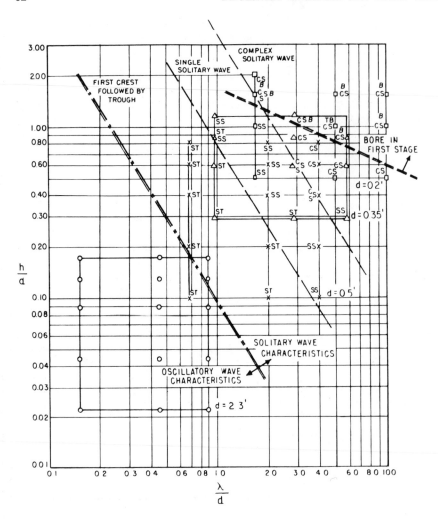

Fig. 4.4. Relation between λ/d, h/d and the characteristics of the leading wave in the case of elevation (after Prins, 1958)

Theories have been developed for the three-dimensional case (Penney, 1950; Unoki and Nakano, 1953b; Kranzer and Keller, 1955).

3. INITIAL IMPULSE

The case of an initial concentrated impulse at the water surface was developed by Cauchy and Poisson (see Lamb, 1945, p. 239). Impulses spread over an area have been studied theoretically for both the two- and three-dimensional cases (Penney, 1950; Unoki and Nakano, 1953b, 1953c; Kranzer and Keller, 1955).

For the two-dimensional case in deep water, Unoki and Nakano give

$$y_s = \frac{2P}{\rho\sqrt{\pi g x}} \sin\left(\frac{gt^2}{4x} \cdot \frac{\lambda}{x}\right) \sin\left(\frac{gt^2}{4\pi} - \frac{\pi}{4}\right), \quad (4.14)$$

where P is the initial impulse acting uniformly on the water surface in the region $-\lambda < x < \lambda$. As in the case of the initial elevation, the resulting waves are a beat phenomenon. The period and length of the beat depends upon the extent of the impulse. For the three-dimensional case, Unoki and Nakano obtained

$$y_s = -\frac{P\lambda t}{\rho r^2\sqrt{2}} e^{-\mu t} \sin\left(\frac{gt}{4r}\right) \cdot J_1\left(\frac{gt^2}{4r} \cdot \frac{\lambda}{r}\right), \quad (4.15)$$

where λ is the radius of the initial impulse, r is the distance from the center of the source, μ is a coefficient of internal friction (given by Unoki and Nakano as 2.75×10^{-4} sec^{-1}, and J_1 is a Bessel function.

Some three-dimensional laboratory tests have been made of an approximation to the type of initial surface impulse considered theoretically (Johnson and Bermel, 1949). Steel disks were dropped from various elevations into water about one foot deep; these results, however, have not been compared with theory.

Waves generated by underwater explosions have been studied theoretically by considering the impulse to consist of a submerged bubble, which may pulsate, if desired (Kirkwood and Seegar, 1950; Penney, 1950; Fuchs, 1952). Measurements of the waves generated by underwater explosions have been given by Bryant (1950), and compared with the theory of Penney with some success.

Another type of impulse is the sudden movement of a submerged body for a short interval of time. This may be considered to be representative of a submarine landslide. The effect of the length of such a disturbance, as determined in a two-dimensional model study, is shown in Fig. 4.5, together with the type of waves generated in relatively deep water (Wiegel, 1955). The amount of energy in the wave disturbance was found to be of the order of 1 per cent of the initial potential energy (net, submerged) of the body, as is shown in Fig. 4.6.

A more realistic model of a landslide is a section of material moving down a slope, rather than vertically. The results of an experiment with this type of mechanism are shown in Fig. 4.7. The period of the main wave increases with increasing slope, with the rate of increasing period becoming large for slopes less steep than about one in three ($18\frac{1}{2}$ degrees).

It is interesting to note that the relationship between the length of the vertically falling body and the initial period of the waves, when extrapolated considerably, indicates that an underwater disturbance of the order of a few thousand feet could generate waves with an initial period of the order of 10 to 15 minutes. If the body were sliding down a one in three slope the dimensions would be much less. Because of the great length of such a wave, it could travel a considerable distance without being affected greatly by its dispersive qualities, because it would be traveling a short distance as measured in wave lengths. In fact, it would have the general appearance of the tsunamis observed in nature. Slides of the dimensions necessary to generate small tsunamis have occurred at the

Fig. 4.5. Effect of body length (from Wiegel, 1955)

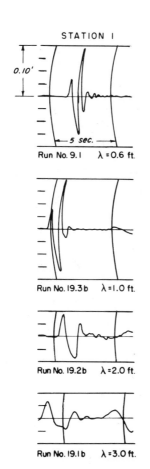

(a) Effect of length of body on wave characteristics, constant water depth; net weight varied from $30\frac{1}{4}$ to $32\frac{1}{2}$ pounds; Series II

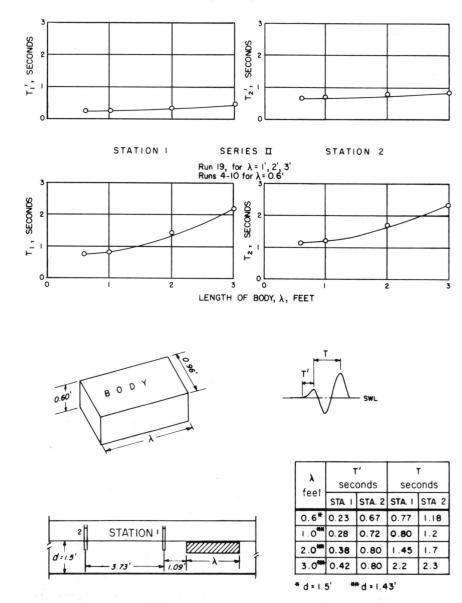

λ feet	T' seconds		T seconds	
	STA. 1	STA. 2	STA. 1	STA. 2
0.6*	0.23	0.67	0.77	1.18
1.0**	0.28	0.72	0.80	1.2
2.0**	0.38	0.80	1.45	1.7
3.0**	0.42	0.80	2.2	2.3

* d = 1.5'   ** d = 1.43'

(b) Relationship between length of vertically falling body and wave period

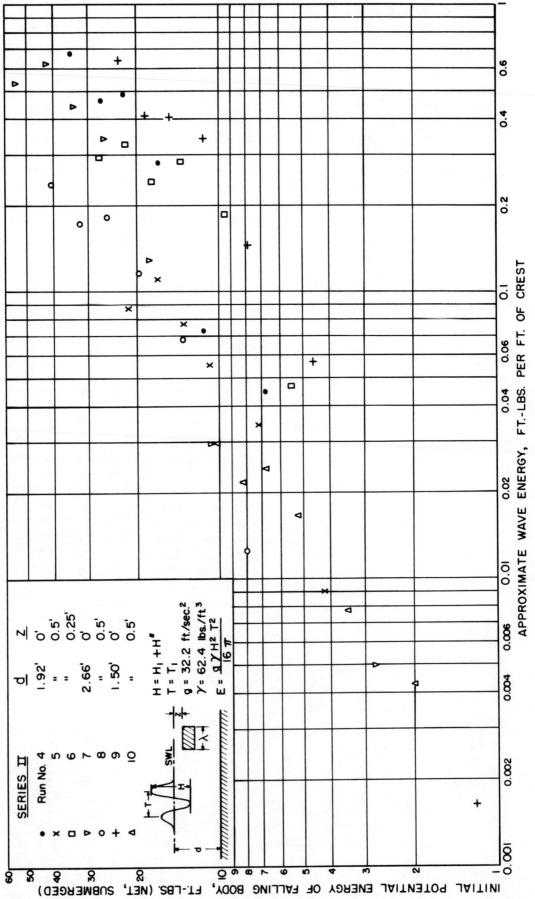

Fig. 4.6. Relationship between energy of wave disturbance and initial potential energy (net, submerged) of a vertically falling body (after Wiegel, 1955)

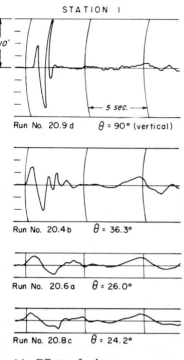

(a) Effect of slope on wave characteristics, constant water depth, and constant body dimensions; Series II

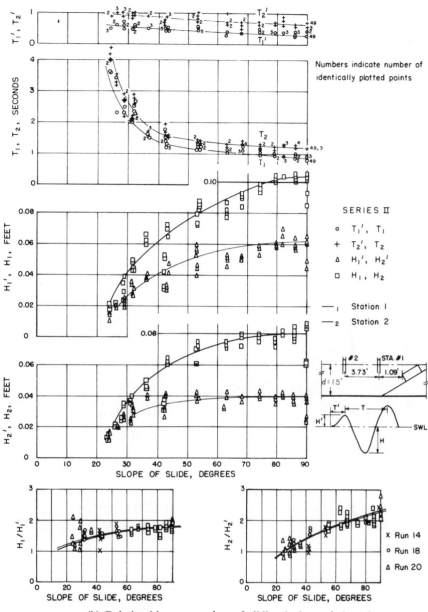

Numbers indicate number of identically plotted points

Fig. 4.7. Effect of incline angle (after Wiegel, 1955)

(b) Relationships among slope of sliding body, period, and amplitude of waves

boundary to the ocean (Miller, 1960), and there is little reason to expect that they would not occur in such submarine regions as the Aleutian Trench.

A classic example of the magnitude of a wave that can be caused by a large body of rock sliding into a bay occurred in Lituya Bay, Alaska. The results of such a slide have been studied by Miller (1960), who also lists many similar phenomena in various parts of the world. The potential energy of the landslide was about 3.5×10^{14} ft-lb, whereas the energy of the wave was about 6×10^{12} ft-lb, which is approximately 2 per cent of the slide energy. The slide occurred along one side of the bar of a T-shaped bay, causing water to surge up the opposite side to a maximum elevation of 1200 ft above the water level and a wave, nearly solitary in

form, about 200 ft high moved down the main part of the bay, and out to sea. The evidence of its height consisted mainly of the scoured border of the bay (Fig. 4.8), where trees about 50 ft high were removed as if they were match sticks. Model studies made by the author reproduced the major features of the wave, and computations of the hydrodynamic forces exerted on the trees by such a wave showed the forces to have been nearly ten times as great as necessary to snap the tree or uproot it.

There have been many such large waves caused by landslides in the fjords of Norway, with loss of life usually resulting: large waves generated by landslides have occurred in Tafjord in 1718, 1755, 1805, 1868, and 1934 (Jørstad, 1956). In the 1934 slide, 41 persons were

Fig. 4.8. Lituya Bay, Alaska: (*top*) September, 1954; (*bottom*) August, 1958 (after Miller, 1960; courtesy of U. S. Geological Survey)

killed by the waves. Tafjord is about six-tenths of a mile wide and about 600 feet deep in the vicinity of the landslide. About one million cubic meters of an overhang fell, with the top of the slide being 730 meters above the water level in the fjord (Holmson, 1936). The wave apparently was from 90 to 180 feet high in the vicinity of its origin, about 30 feet high several miles away. Jørstad (1956) has estimated that a landslide must be of at least 130,000 cubic yards to cause a devastating wave. There is good evidence of the occurrence in historic times of a landslide into Tafjord of about sixteen million cubic yards (Jørstad, 1956). Landslides have apparently reached speeds in excess of 325 ft/sec (Holmson, 1936).

4. TRAVELING IMPULSE

A hurricane (typhoon) is a small, roughly circular tropical disturbance containing poor weather, heavy rains, a very low-pressure area, and strong counter-clockwise winds (in the Northern Hemisphere). This is a traveling impulse, with winds. The low-pressure area in the center is often 2 to 3 in. (of mercury) lower than the surrounding air. Maximum winds often exceed 100 knots. The speed of advance of a hurricane and its path vary considerably; the speed may, however, exceed 30 knots in the more northern latitudes, as can be seen in Table 4.1. There is evidence (Dunn, 1951; Unoki and

Table 4.1. AVERAGE FORWARD SPEED (KNOTS) 2 HR BEFORE AND AFTER ENTERING THE COAST, BY REGIONS (UNITED STATES HURRICANES WITH CENTRAL PRESSURES BELOW 29.00 IN.) (From U.S. Dept. of Commerce, Weather Bureau, 1957).

Texas 1900–49	Mid-Gulf 1877–1955	Florida peninsula	Florida keys	Atlantic (north of Fla.–south of Hatteras) 1893–1955	Atlantic (Cape Hatteras and northward) 1893–1955
4	6	6	8	7	16
8	7	6	8	9	18
8	7	7	9	9	22
8	9	8	10	12	23
10	10	8	10	14	24
10	10	8	11	15	29
11	11	10	13	16	30
11	11	10	13	16	33
11	14	10	16	17	40
11	16	11		18	47
12	16	11		21	
13	16	11		22	
14	21	11		28	
15	23	11			
15	25	12			
	28	13			
		13			
		14			
		14			
		17			
Mean					
11	14	11	11	16	28

Nakano, 1955; Harris, 1956) that long water gravity waves are associated with some hurricanes.

Lord Kelvin (1906) developed a theory for a diffused line pressure source moving over a body of infinitely deep water. Havelock (1909, 1922, 1925–26), working along similar lines developed the theory for water of any depth, with the same assumptions being made as in the development of the linear theory.

Kelvin and Havelock found that the moving pressure source would generate transverse waves with phase velocities equal to the velocity of the moving pressure source. The wave periods would be the period associated with the particular wave velocity and water depth. The group of waves would form to the rear of the disturbance as the group velocity of gravity waves is less than the phase velocity.

Havelock (1909) studied a diffused line pressure of the following sort:

$$p = f(x) = \frac{P}{\pi} \frac{\alpha}{\alpha^2 + x^2}, \qquad (4.16)$$

where
$$P = \int_{-\infty}^{\infty} f(x)\, dx = \text{constant}, \qquad (4.17)$$

with x measured in the direction of motion, and p the pressure. α expresses the degree of diffusion (it has the dimension of length, the "width" of the pressure line) and should be small compared with the normal wave dimensions. The amplitude of the waves formed by such a system is given by

$$a = \frac{K(2\pi/L)p\, e^{-2\pi\alpha/L}}{1 - [(4\pi d/L)/(\sinh 4\pi d/L)]}, \qquad (4.18)$$

where L = wave length, d = water depth, and K is a constant (probably $2/\gamma$ where γ is the specific weight of water). The term $1 - [(4\pi d/L)/(\text{sihn } 4\pi d/L)]$ expresses the effect of the difference between group velocity and phase velocity upon the accumulation of energy in the waves.

Some model studies have been made of a two-dimensional low-pressure area moving over the water surface at a series of constant forward speeds for different constant water depths (Wiegel, Snyder, and Williams, 1958). The relationship between the wave height (twice the wave amplitude, a) and the wave velocity has been plotted in Fig. 4.9 for a low pressure area 2 ft long traveling over water 1.06 ft in depth, taking α equal to one-half the length of the box which represents the "hurricane" in the series of tests. On the same graph, the experimental points of wave height versus velocity of the pressure area have been plotted. It is apparent that the solution, which depends upon linear wave theory, becomes invalid as the wave speed approaches $\sqrt{gd}$ because the wave amplitude becomes very large, which violates the assumption of the linear theory.

Examples of waves generated in the laboratory for several values of $V/\sqrt{gd}$, where V is the forward speed of the low-pressure area are shown in Fig. 4.10.

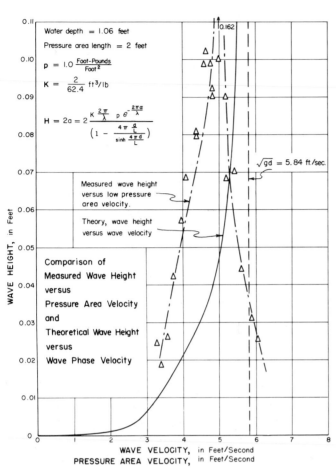

Fig. 4.9. Comparison of measured wave height versus pressure area velocity and theoretical wave height versus wave phase velocity (After Wiegel, Snyder, and Williams, 1958)

It is evident from these tests that waves can be formed in shoal water with relatively long periods. It is necessary to determine whether hurricanes do move with the necessary velocity over shoal water to generate such waves. The path of the center of Hurricane Carol (August 30, 31, 1954) is plotted in Fig. 4.11. An examination of U.S. Coast and Geodetic Survey Chart No. 1000 (Cape Sable to Cape Hatteras) showed that the hurricane traveled over shallow water for an appreciable distance with a forward speed of approximately 30 to 35 knots (between Cape Hatteras, North Carolina, and Long Island, New York). The shallow water wave velocities ($\sqrt{gd}$) associated with water depths of 5, 10, 15, 20, and 25 fathoms are 18.4, 26.1, 31.9, 36.8, and 41.2 knots. It is evident that Hurricane Carol should have been able to generate the type of long wave that would be known as a *surge*, especially in the vicinity of Long Island. This is in agreement with the data presented by Harris (1956) which have also been shown in Fig. 4.11.

The forward speed of Hurricane Carol was not unusually great for the coastal area north of Cape Hatteras as can be seen from Table 4.1.

A more realistic mathematical model, one circular in shape, was developed by Havelock (1922). In this theory

$$p = f(r) = \frac{Aa}{(a^2 + r^2)^{3/2}}, \qquad (4.19)$$

where r is the distance from the center of the disturbance,

Fig. 4.10. Surface time histories of shallow water waves generated by the 6-foot low pressure area for a water depth of 0.50 foot (after Wiegel, Snyder, and Williams 1958)

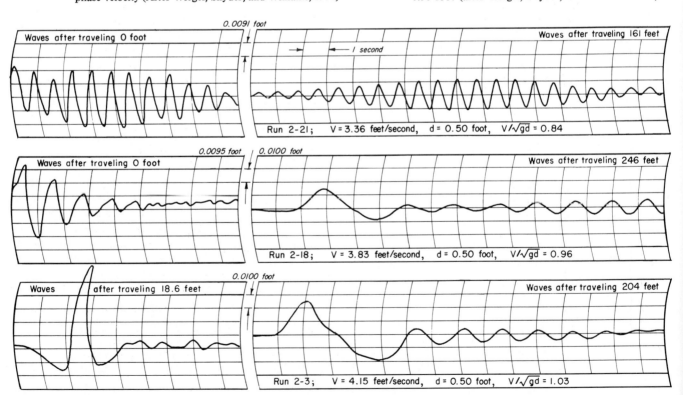

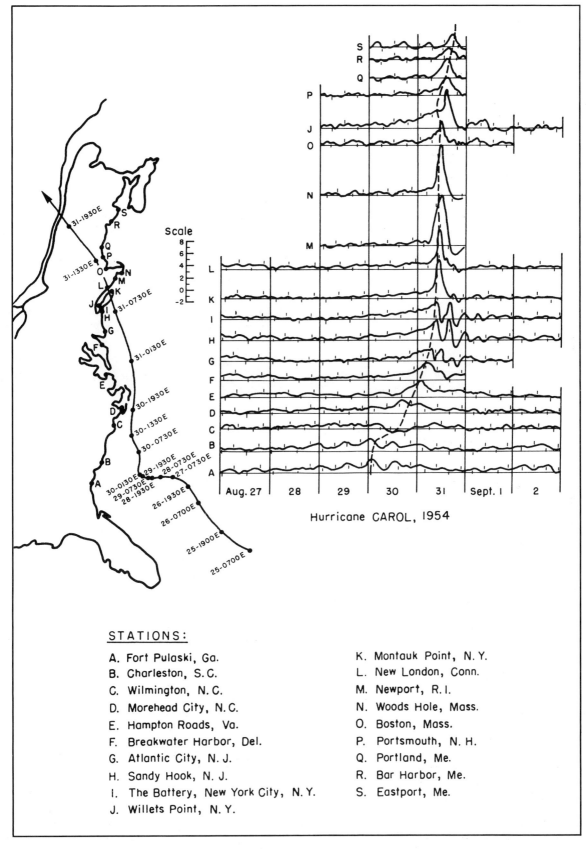

Hurricane CAROL, 1954

STATIONS:

A. Fort Pulaski, Ga.
B. Charleston, S.C.
C. Wilmington, N.C.
D. Morehead City, N.C.
E. Hampton Roads, Va.
F. Breakwater Harbor, Del.
G. Atlantic City, N.J.
H. Sandy Hook, N.J.
I. The Battery, New York City, N.Y.
J. Willets Point, N.Y.

K. Montauk Point, N.Y.
L. New London, Conn.
M. Newport, R.I.
N. Woods Hole, Mass.
O. Boston, Mass.
P. Portsmouth, N.H.
Q. Portland, Me.
R. Bar Harbor, Me.
S. Eastport, Me.

Fig. 4.11. Hourly storm surge height (observed minus predicted sea level), Atlantic coast tide stations, August 27–September 2, 1954 (after Harris, 1956)

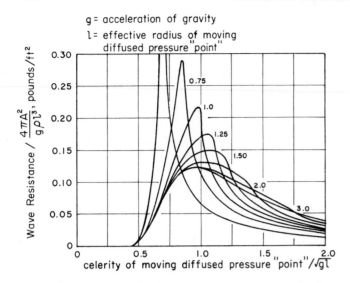

g = acceleration of gravity

l = effective radius of moving diffused pressure "point"

Fig. 4.12. Relationship between wave resistance and ratio of speed of moving diffused pressure "point" to $\sqrt{gL}$ (theory of Havelock, 1922; from Brandmair and Moretti, 1962)

maximum pressure, p_{max}, is at the center of the disturbance, and thus from Eq. 4.19, $A = a^2 p_{max}$. The wave resistance, R, is

$$R = \frac{4\pi A^2 V^2}{\rho} \int_{\phi_o}^{\pi/2} \frac{k^3 e^{2ka} \sec \phi \, d\phi}{g(V^2 - gd \sec^2 \phi) \sec^2 \phi + k^2 V^4 d},$$

(4.20)

where $V^2 = (g/k) \tanh kd \cdot \sec^2\phi$, and the lower limit ϕ_o is given by

$$\phi_o = 0, \quad \text{for } V^2 < gd; \quad \text{and}$$
$$\phi_o = \arccos (gd/V^2)^{1/2} \quad \text{for } V^2 > gd.$$

(4.21)

Havelock made numerical calculations for a range of conditions, and results from his theory are shown in Fig. 4.12, when $D = 2a$, the diameter of the pressure area, and $\gamma = \rho g$, the specific weight of the water.

The curves in Fig. 4.13(a) have been constructed by multiplying the values of $V/\sqrt{gd/2}$ for which the maximums occur by $\sqrt{D/2}/\sqrt{d}$, and plotting the product as a function of D/d, with the asymptotic value of the shallow water curve being obtained from a theoretical study by Inui (1936), and the slope of the deep water maximum curve at $D/d = 0$ being obtained from Havelock (1926). Values obtained in laboratory studies

A is a constant with the dimension of force, and a is a constant with the dimension of length which may be considered to be the radius of the disturbance. The

Fig. 4.13. Maximum wave height in axis of wave pattern as a function of velocity and diameter of pressure disturbance (from Abraham, 1961)

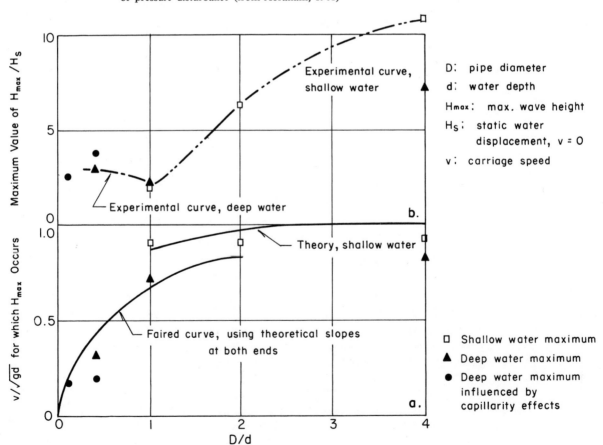

D: pipe diameter

d: water depth

Hmax: max. wave height

Hs: static water displacement, v = 0

v: carriage speed

☐ Shallow water maximum

▲ Deep water maximum

● Deep water maximum influenced by capillarity effects

are shown in the figure (Abraham, 1961). The ratios of the maximum values of the wave heights to the static water displacement (which is the same as the difference between the pressure in the center of the disturbance and the normal surrounding atmospheric pressure) as a function of D/d are shown in Fig. 4.13b. Abraham also found that a *phase shift* (the distance between the center of the pressure disturbance and the first wave crest) occurred as a function of $V/\sqrt{gd}$, with the shift being gradual for $D/d \approx 1$, but rapid for $D/d \approx 4$, with the rapid shift occurring in the vicinity of $V/\sqrt{gd}$ between 0.9 and 1.0.

The data in Fig. 4.13(b) show that large long waves can be generated in shallow water. These data are the steady state value, however, and a major problem is whether hurricanes in nature are over shallow water for a long enough time for the large coupled wave to build up. Abraham (1961) studied the results of several actual natural hurricanes and other large-scale atmospheric conditions and concluded that their durations were of adequate length to permit the formation of these coupled waves. Indeed, the records show waves of similar characteristics.

5. SHIP WAVES

In their simplest form, ship waves are those described in the previous section. In fact, the theories cited in that section were developed to increase the understanding of the wave-making resistance of ships. As there is an extensive literature on ship resistance, much of it concerned with the caluclation of ship resistance for actual ship forms, it will be considered only briefly here. In practice, the wave resistance of ships is usually measured while towing models in a towing tank.

In the case of a moving pressure point, the controlling parameter is a Froude number based upon the water depth, $F = V/\sqrt{gd}$. For values of F less than 1, two sets of waves are generated, transverse and diverging (see, for example, Havelock, 1909; Inui, 1936). Where these two sets of waves intersect, a large superimposed wave is noticed. The angle formed by a line through the intersections (called the *cusp locus*) and the centerline (α) depends upon F, being $19°\,18'$ for $F \to 0$, and rising sharply to $90°$ for $F = 1$. For values of F greater than 1, no transverse waves form (Fig. 4.14).

Most naval architect studies of wave resistance are concerned with ships in deep water. In harbors and canals, however, shallow water effects are of considerable importance, not only because of wave resistance, but because of the effects on shores, boats, and structures of the waves generated in this manner. Johnson (1958) has made a study of the waves generated by several ship models, in both deep and shallow water, the study including the change in the heights of the waves as these moved away from the source. The relationship between

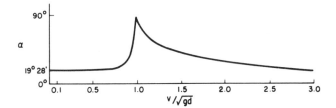

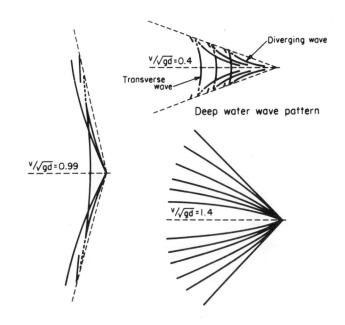

v/√gd	α	v/√gd	α
0.38	19° 28'	0.96	59° 27'
0.42	19° 28'	0.99	78°
0.5	19° 29'	1.0	90°
0.55	19° 30'	1.005	84°
0.6	19° 37'	1.41	45°
0.7	20° 18'	1.73	35°
0.82	23° 42'	2.0	30°
0.92	39° 19'	3.0	19° 28'

v: velocity of disturbance
d: water depth
a: half angle of aperture of wedge within which the wave pattern is contained

Fig. 4.14. Wave patterns caused by point disturbance of pressure advancing over water of finite depth (influence of surface tension neglected) (after Havelock, 1908)

the actual measurements of α and the theoretical values of Kelvin (1906a) was quite close (Fig. 4.15). An example of the decrease in wave height with distance from the ship model is shown in Fig. 4.16, where λ is the length of the ship model, D is the draft of the ship model, V is the ship speed, H is the wave height, and d is the water depth.

6. BULKHEAD WAVE GENERATOR

Many problems encountered by engineers in the design and operation of structures subject to waves require the

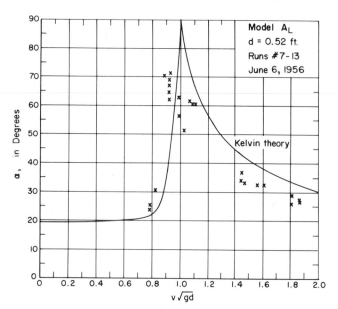

Fig. 4.15. Relationships between α, (angle between the cusp locus and the sailing line), and the Froude Number, $V\sqrt{gd}$ (from Johnson, 1958)

the occurrence of groups of nearly periodic waves, these waves usually being relatively high. It is therefore desirable to be able to design a mechanism to generate periodic waves. In addition, if the proper relationships between a moving mechanism and the waves generated by such a mechanism are available, it is possible to design a device to generate irregular waves corresponding to those measured in the ocean.

There are three types of wave generators in wide use: (1) plunger; (2) piston, or rigid flap; and (3) pneumatic. A variety of other types are employed on occasions but are not in general use (Ross and Bowers, 1953). The plunger type has been studied by Sibul and Pister (1954), and the pneumatic type has been studied in detail (Caldwell, 1955; Brownell, Asling and Marks, 1957).

The piston, or rigid flap, type is in most common use. Furthermore, the necessary theory and empirical data are available with which to design a reliable wave generator. The theory has been developed by Biesel (1951) and comparisons between theory and practice are available (Snyder, Wiegel, and Bermel, 1958; Ursell, Dean, and Yu, 1959). The work by Ursell *et al.*, includes the theory of the effect of multiple reflections as well as experimental verification of the reliability of the theory. Data on the peak power requirements and other information necessary in choosing the proper motor and for designing the gear box, etc., are also available (Caldwell, 1955).

use of models. It is necessary to generate waves of the proper characteristics to do this. The wind-generated waves of the ocean are irregular, as is discussed in detail in Chapter 9. One of their characteristics, however, is

Fig. 4.16. Relationships among the ship wave heights, ship draft, and ratio of distance from the sailing line to ship length (x/λ) (from Johnson, 1958)

Fig. 4.17. Computed stroke compared with measured stroke (from Snyder, Wiegel, and Bermel, 1958)

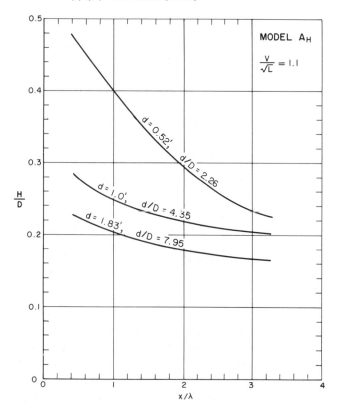

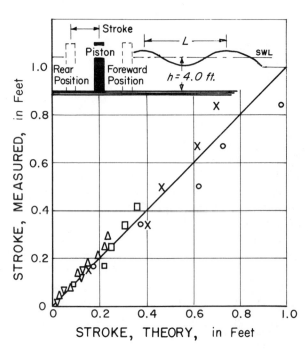

Biesel's theory shows, as would be expected, that, for waves which are relatively long compared with the water depth, the piston type of wave generator is superior to the rigid flap higed at the bottom, whereas for waves which are relatively short compared with the water depth, the opposite is true. Both types of wave generators require the wave to travel a certain distance before it acts as an entirely free wave rather than as a forced wave. Biesel calculated that, after the waves have traveled a distance of three times the water depth, any effect of either of these types of wave generators on the free waves is negligible.

In determining the stroke of a piston type of wave generator, Biesel developed the equation:

$$\text{stroke} = \frac{\text{wave height}}{2K} \qquad (4.22)$$

where

$$K = \frac{2 \sinh^2 2\pi d/L}{(2\pi d/L) + [\sinh (2\pi d/L)] \cdot \cosh 2\pi d/L} \qquad (4.23)$$

and the stroke is measured from the mean position of the piston (i.e., one-half its total motion). The results of comparison of the measured values of the stroke and the computed values are shown in Fig. 4.17; comparisons of the theoretical values of K and the values computed from the measured wave characteristics were good. The average power needed can be obtained from Eq. 2.70; the peak horsepower needed is about three times the average horsepower (Snyder, Wiegel, and Bermel, 1958).

REFERENCES

Abraham, G., Hurricane storm surge considered as a resonance phenomenon, *Proc. Seventh Conf. Coastal Eng.*, Berkeley, Calif.: The Engineering Foundation Council on Wave Research, 1961, pp. 585–604.

Biesel, F., Etude théorique d'un certain type d'appareil à houle, *La Houille Blanche*, **6**, No. 2 (March–April, 1951), 156–65.

Brandmaier, H. E., and F. J. Moretti, A note on Havelock's shallow water wave-resistance curves, *J. Aerospace Sci.*, **29**, 3 (March, 1962), 357–58.

Brownell, W. F., W. L. Asling, and W. Marks, A 51-ft pneumatic wavemaker and a wave absorber, *Trans. Eleventh General Meeting of Amer. Towing Tank Conf.*, U.S. Navy, David Taylor Model Basin, Report No. 1099, **III** (September, 1957), 709–42.

Bryant, A. R., Surface waves produced by underwater explosions. Comparisons of the theory of W. G. Penney with experimental results for a 32-lb. charge, *Underwater Explosion Research*, U.S. Navy, ONR, **II** (1950), 701–706.

Caldwell, J. M., The design of wave channels, *Proc. First Conf. Ships and Waves*, The Engineering Foundation Council on Wave Research and the Society of Naval Architects and Marine Engineers, 1955, pp. 271–87.

Dunn, Gordon E., Tropical cyclones, in *Compendium of meteorology*, ed. Thomas F. Malone. American Meteorological Society, 1951, pp. 887–901.

Fuchs, R. A., *Theory of surface waves produced by underwater explosions*, Univ. Calif., IER, Tech. Rept. No. 3-335, May, 1952. (Unpublished.)

Green, George, On group-velocity and on the propagation of waves in a dispersive medium, *Proc. Roy. Soc.* (Edinburgh), **29** (1909), 445–70.

Harris, D. Lee, *Some problems involved in the study of storm surges*, U.S. Dept. of Commerce, Weather Bureau, National Hurricane Research Project Report No. 4, December, 1956.

Havelock, T. H., The propagation of groups of waves in a dispersive media, with application to waves on water produced by a travelling disturbance, *Proc. Roy. Soc.* (London), ser. A, **81** (1908), 398–430.

———, The wave-making resistance of ships: a theoretical and practical analysis, Proc. Roy. Soc. (London), ser. A, **82** (1909), 276–300.

———, The effect of shallow water on wave resistance, *Proc. Roy. Soc.* (London), ser. A, **100**, A 706 (Feb. 1, 1922), 499–505.

———, Some aspects of the theory of ship waves and wave resistance, *Trans., Northeast Coast Inst. Engrs. Shipbuilders.* **XLII** (1925–26), 71–86.

Holmson, Gunnar, De siste bergskred i Tafjord of Loen, Norge, *Svensk Geog. Arsbok*, **12** (1936), 171–90.

Inui, T., On deformation, wave patterns and resonance phenomenon of water surface due to a moving disturbance, *Proc. Physico-Mathematical Soc. Japan*, ser. 3, **18**, 2 (February, 1936), 60–113.

Johnson, J. W., Ship waves in navigation channels, *Proc. Sixth Conf. Coastal Eng.*, Berkeley, Calif.: The Engineering Foundation Council on Wave Research, 1958, pp. 660–90.

——— and K. J. Bermel, Impulsive waves in shallow water as generated by falling weights, *Trans. Amer. Geophys. Union*, **30**, 2 (April, 1949), 223–30.

Jorstad, Finn A., Fjellskredet ued Tjelle; et 200-års minne, *Naturen*, **80**, 6 (1956), 323–33.

Kelvin, Lord, Deep sea ship-waves, *Phil. Mag.*, 6th ser., **11**, 61 (January, 1906a), 1–25.

———, Initiation of deep-sea waves of three classes: (1) from a single displacement: (2) from a group of equal and similar displacements; (3) by a periodically varying surface-pressure, *Proc. Roy. Soc.* (Edinburgh), **26** (November, 1906b), 399–436.

Kirkwood, J. G. and R. J. Seeger, Surface waves from an underwater explosion, *Underwater Explosion Research*, **II** U.S. Navy, ONR (1950), 707–60.

Kranzer, H. C. and J. B. Keller, *Water waves produced by explosions*, New York University, Institute of Math. Sci, IMM-NYU-222, September, 1955.

Lamb, Sir Horace, *Hydrodynamics*, 6th ed. New York: Dover Publications, Inc., 1945.

Miller, Don J., *Giant waves in Lituya Bay, Alaska*, U.S. Geological Survey Professional Paper 354–C, 1960.

Penney, W. G., Gravity waves produced by surface and underwater explosions, *Underwater Explosion Research*, U.S. Navy, ONR, **II** (1950), 679–700.

Prins, J. E., Characteristics of waves generated by a local disturbance, *Trans. Amer. Geophys. Union*, **39**, 5 (October, 1958a), 865–74.

———, Water waves due to a local disturbance, *Proc. Sixth Conf. Coastal Eng.*, Berkeley, Calif.: The Engineering Foundation Council on Wave Research, 1958b, pp. 147–62.

Rayleigh, Lord, On progressive waves, *Proc. London Math. Soc.*, **9** (November, 1877), 21–26.

Ross, James and C. E. Bowers, *Laboratory surface wave equipment*, Univ. Minn., St. Anthony Falls Hyd. Lab., Contract Nonr–710(05), November, 1953. (Unpublished.)

Sibul, O. and K. Pister, *Propagation of radial waves generated by an oscillating body*, Univ. Calif., Inst. Eng. Res., Tech. Rept. 3–367, June, 1954. (Unpublished.)

Snyder, C. M., R. L. Wiegel, and K. J. Bermel, Laboratory facilities for studying water gravity wave phenomena, *Proc. Sixth Conf. Coastal Eng.*, Berkeley, Calif.: The Engineering Foundation Council on Wave Research 1958, pp. 231–51.

U.S. Dept. of Commerce, Weather Bureau, *Survey of meteorological factors pertinent to reduction of loss of life and property in hurricane situations*, National Hurricane Research Project Report No. 5, March, 1957.

Unoki, S. and M. Nakano, On the Cauchy-Poisson waves caused by the eruption of a submarine volcano (1st paper), *Oceanographical Mag.*, **4**, 4 (1953a), 119–141.

——— and ———, On the Cauchy-Poisson waves caused by the eruption of a submarine volcano (2nd paper), *Oceanographical Mag.*, **5**, 1 (1953b), 1–13.

——— and ———, On the Cauchy-Poisson waves caused by the eruption of a submarine volcano, *Oceanographical Mag.*, **4**, 3–4 (1953c), 139–50.

——— and ———, A note on forecasting ocean waves caused by typhoons, *Records oceanographic works in Japan*, **2**, 1 (March, 1955), 151–61.

Ursell, F., On the waves generated by a local surface disturbance (personal communication, July, 1958).

———, R. G. Dean, and Y. S. Yu, Forced small-amplitude water waves: a comparison of theory and experiment, *J. Fluid Mech.*, **7**, Part 1 (1959), 33–52.

Wiegel, R. L., Laboratory studies of gravity waves generated by the movement of a submerged body, *Trans. Amer. Geophys. Union*, **36**, 5 (October, 1955), 759–74.

———, C. M. Snyder, and J. B. Williams, Water gravity waves generated by a moving low pressure area, *Trans. Amer. Geophys. Union*, **39**, 2 (April, 1958), 224–36.

CHAPTER FIVE

Tsunamis, Storm Surges, and Harbor Oscillations

1. TSUNAMIS

The term *tidal wave*, which is often used for the water gravity waves associated with submarine seismic disturbances, is now seldom used in the technical literature as the waves are not related to the tides. The Japanese word *tsunami* is usually used instead.

Tsunamis can be generated by several mechanisms. It is known that the explosion of an underwater volcano can cause one (Ogawa, 1924), or an island exploding, such as Krakatoa (Wharton and Evans, 1888). Most tsunamis are associated with submarine seismic disturbances. But many submarine seismic disturbances, although large, have not caused tsunamis. Leet (1948) states that a catalog made in 1861 reported only 124 tsunamis during the interval that 15,000 earthquakes were observed along coastlines; a later catalog listed 1098 earthquakes along the west coast of South America with only 19 accompanied by tsunamis. There are probably two

reasons for this: not all earthquakes result in tsunamis; secondly small tsunamis probably go unnoticed. It has been found that earthquakes which have been accompanied by tsunamis are always followed by aftershocks, that the earthquakes have a magnitude in excess of about 6, and that they have a relatively shallow focal depth (Iida, 1958, 1963b). Tsunamis associated with large earthquakes with focal depths greater than 80 km can hardly be observed and are small for shocks with focal depths between 50 and 80 km. Most tsunamis that have been observed were associated with earthquakes with focal depths between 0 and 40 km. It was found that for noticeable tsunamis to occur, the magnitude of the earthquake, M, must be greater than $6.3 + 0.01H$, where H is the focal depth in kilometers. It was also found that, for disastrous tsunamis to occur, M must be greater than $7.75 + 0.008H$.

Heck (1947) prepared a list of the geographical distribution of tsunamis which is given in Table 5.1 (see also Cueller, 1953). This list is only partially complete;

Table 5.1. GEOGRAPHICAL DISTRIBUTION OF TSUNAMIS (after Heck, 1947)

Number	Location	Number	Location
2	Azores	2	Hawaiian Islands
2	West coast, Europe	5	Philippine Islands
12	Italy and west Mediterranean	47	Japan to Kamchatka
13	East Mediterranean	5	Alaska
2	Bay of Bengal	4	California
47	Dutch East Indies	11	Mexico
3	New Guinea	2	Central America
4	Solomons and New Hebrides	34	Caribbean
3	Rest of Oceania	1	Newfoundland
2	New Zealand		

Table 5.2. TSUNAMIS REACHING THE HAWAIIAN ISLANDS; RUN-UP AND DAMAGE AT HILO, HAWAII
(after Cox, 1962)

Year	Date	Source	Period of initial wave (minutes)	Maximum run-up (feet)	Earthquake magnitude	Damage (dollars)
1837	Nov. 7	Chile	?	20?		Many houses destroyed
1841	May 17	Kamchatka	?	3		
1854	Dec. 23	Japan	No record at Hilo			
1868	Apr. 2	Hawaii	?	10		Minor
1868	Aug. 13	Peru, Chile	?	15		Small
1869	July 25	Peru, Chile	No record at Hilo			
1872	Aug. 23	Hawaii ?	6	4		
1877	May 10	Chile	?	16		Moderate (14,000)
1878	Jan. 20	Chile ?	?			
1883	Aug. 27	Krakatoa eruption				
1896	June 15	Japan	6 ?	8	7.6	
1901	Aug. 9	Japan	?	4	7.7	
1906	Jan. 31	Colombia	30	3	8.6	
1906	Aug. 16	Chile	?	5	8.4	
1913	Oct. 11	New Guinea	No record at Hilo			
1914	May 26	New Guinea	No record at Hilo		7.9	
1917	May 1	Kermadec Is.	No record at Hilo		8	
1917	June 25	Samoa Is.	No record at Hilo		8.3	
1918	Aug. 15	Philippine Is.	No record at Hilo		8.3	
1918	Sept. 7	Kurile Is.	?	5	8.3	
1919	Apr. 30	Tonga Is.	?	2	8.3	
1922	Nov. 11	Chile	?	7	8.3	
1923	Feb. 3	Kamchatka	?	20	8.3	Considerable
1923	Apr. 13	Kamchatka	?	Small wave	7.2	
1927	Nov. 4	California	20	1.5	7.3	
1927	Dec. 28	Kamchatka	25	0.2	7.3	
1928	June 17	Mexico	22	0.8	7.8	
1929	Mar. 6	Aleutian Is.	15	0.6	8.1	
1931	Oct. 3	Solomon Is.	13	0.3	7.9	
1932	June 3	Mexico	18	0.6	8.1	
1932	June 18	Mexico	?	0.1	7.8	
1932	June 22	Mexico	?		6.9	
1933	Mar. 2	Japan	?	2	8.3	
1938	Nov. 10	Alaska	No record at Hilo		8.3	
1943	Apr. 6	Chile	No record at Hilo		7.9	
1944	Dec. 7	Japan	No record at Hilo		8.0	
1946	Apr. 1	Aleutian Is.	?	27	7.4	Great (26,000,000)
1946	Dec. 20	Japan	No record at Hilo		8.2	
1948	Sept. 8	Tonga Is.	No record at Hilo		7.8	
1951	Aug. 21	Hawaii	12	0.2	6.9–7.0	
1952	Mar. 3	Japan	16	0.2	8.2	
1952	Nov. 4	Kamchatka	18	12	8.3	Moderate (300,000)
1953	Sept. 14	Fiji		0.2	6.8	
1956	Mar. 30	Kamchatka	No record at Hilo			
1957	Mar. 9	Aleutian Is.	18	14	8.0	Moderate (400,000)
1957	Oct. 31	?				
1958	Nov. 6	Kurile Is.	20	0.6	8.3	
1959	May 4	Kamchatka	22	0.5	8.2	
1960	May 23	Chile	34	35	8.5	Great (22,000,000)

Table 5.3. TSUNAMI ENERGY
(after Iida, 1958)

	Date	Tsunami	Earthquake energy		Tsunami energy		Tsunami energy / Earthquake energy
			(ergs)	(ft-lb)	(ergs)	(ft-lb)	
1.	Mar. 3, 1933	Sanriku	20.0×10^{23}	14.7×10^{16}	17×10^{22}	12.5×10^{15}	0.085
2.	Nov. 3, 1936	Fukushima	2.2	1.6	0.2	0.15	0.0091
3.	May 23, 1938	Fukushima	0.28	0.21	0.04	0.03	0.011
4.	Nov. 5, 1938	Fukushima	2.2	1.6	0.2	0.15	0.0091
5.	Dec. 7, 1944	Tonankai	6.4	4.7	7.9	5.8	0.12
6.	Feb. 10, 1945	Fukushima	0.56	0.41	0.04	0.03	0.007
7.	Dec. 21, 1946	Nankaido	9.0	6.6	8.0	5.9	0.098
8.	Mar. 2, 1952	Tokachi	9.0	6.6	8.0	5.9	0.098
9.	Nov. 4, 1952	Kamchatka	13.0	9.6	15	11	0.11
10.	Nov. 25, 1953	Boso	1.0	0.7	0.7	0.5	0.07

for example, 49 tsunamis have been noticed in the Hawaiian Islands up to 1962 (Cox, 1962), whereas Heck lists only two. The origin, characteristics, and magnitude of the associated earthquakes for the tsunamis that have reached the Hawaiian Islands are tabulated in Table 5.2.

The literature on tsunamis shows considerable disagreement as to the mechanism by which these water gravity waves are generated. It appears that they can be caused by submarine fault movements or by submarine landslides (Gutenberg, 1939). Model studies have shown that both types can cause waves with the characteristics of tsunamis (Takahasi, 1934; 1963a; Wiegel, 1955) provided that the disturbances are of large enough magnitude. The wave created by a large mass of rock sliding into the bay that was discussed in Chapter 4, had an energy of about 6×10^{12} ft-lb, which was about 2 per cent of the potential energy of the landslide. The tsunami of March 9, 1957, had an energy of about 9.5×10^{14} ft-lb (Van Dorn, 1959) and this appears to be of the order of magnitude of many other tsunamis (Iida, 1958, 1963a, 1963b). Thus, there has been a wave that is known to

have been generated by a landslide that had 1 per cent of the energy of a fairly large tsunami.

Iida (1958; 1963a; 1963b) studied the relationship between the energy of tsunamis and the energy of the associated earthquakes. He found that the tsunami energy was from one-tenth to one-hundredth of the total energy of the seismic waves. It was further found that the larger the tsunami, the larger the ratio of tsunami energy to the energy of the associated seismic waves, as is shown in Table 5.3 (Iida, 1963a).

Iida (1963b) studied the relationship between the depth of water over the epicenter of an earthquake and the magnitude (m) of the accompanying tsunami for about 100 tsunamis occurring in Japan between 1700 and 1960, most of these being generated in the region between Taiwan and Kamchatka. The results are shown in Fig. 5.1. It appears that large tsunamis must be generated in deeper water than small tsunamis. The relationships between the tsunami magnitude classification and the energy and run-up of the tsunamis in Japan are given in Table 5.4.

Table 5.4. MAGNITUDE, ENERGY, AHD RUN-UP ELEVATION OF TSUNAMIS IN JAPAN
(after Iida, 1963b)

Tsunami magnitude classification	Tsunami energy		Maximum run-up elevation	
(m)	ergs	(ft-lb)	(meters)	(ft)
5	25.6×10^{23}	18.9×10^{16}	>32	>105
4.5	12.8	9.4	24–32	79–105
4	6.4	4.7	16–24	52.5–79
3.5	3.2	2.4	12–16	39.2–52.5
3	1.6	1.2	8–12	26.2–39.2
2.5	0.8	0.59	6–8	19.7–26.2
2	0.4	0.29	4–6	13.1–19.7
1.5	0.2	0.15	3–4	9.9–13.1
1	0.1	0.074	2–3	6.6–9.9
0.5	0.05	0.037	1.5–2	4.9–6.6
0	0.025	0.018	1–1.5	3.2–4.9
−0.5	0.0125	0.0092	0.75–1	2.5–3.2
−1	0.006	0.0044	0.50–0.75	1.6–2.5
−1.5	0.003	0.0022	0.30–0.50	1.0–1.6
−2	0.0015	0.0011	<0.30	<1.0

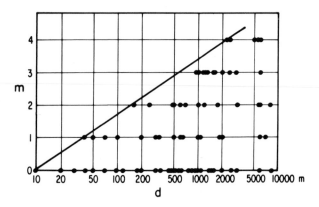

Fig. 5.1. The relationship between tsunami magnitude m and water depth d at the epicenter of earthquake (after Iida, 1961 b)

The transformation of waves generated by an initial surface displacement or impulse has been described in Chapter 4, and certain of these observations apply to tsunamis. For example, the travel time of the initial wave disturbance can be obtained using $C = \sqrt{gd}$. For the April 1, 1946, tsunami it was found that the average wave speeds ranged from 375 to 490 miles per hour, depending upon the ocean depths between the origin and the station. The arrival times were computed to within a few minutes for the stations relatively near the epicenter (1100 statute miles with a travel time of just under 3 hr) and within 30 min for a distant station (8000 miles with a travel time just over 18 hr) (Green, 1946). Travel times to the Hawaiian Islands have been computed for tsunamis originating in several areas, Fig. 5.2. A chart of

Fig. 5.2. Travel times of tsunamis to Honolulu, Hawaii (U. S. Coast and Geodetic Survey) (from Zetler, 1947)

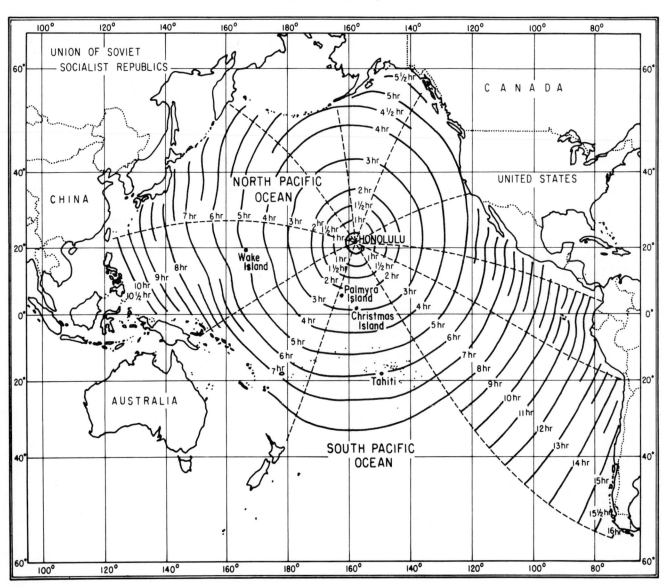

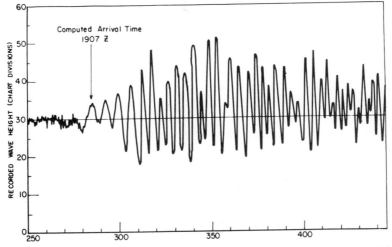

(a) Tsunami of March 9, 1957, as recorded at Wake Island (after Van Dorn, 1959)

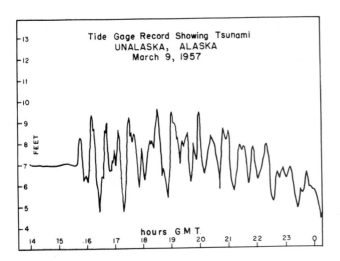

(b) Period 1st to 2nd crest; 27 minutes; 3rd wave highest (after Salsman, 1959)

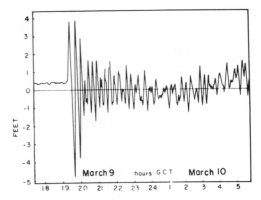

(d) Record of the tsunami of March 9, 1957, from the Hilo tide gage. Time is 10 hours ahead of Hawaiian time. Period 1st to 2nd crest; 19 minutes. 1st wave highest (Courtesy, U.S.C. & G.S.)

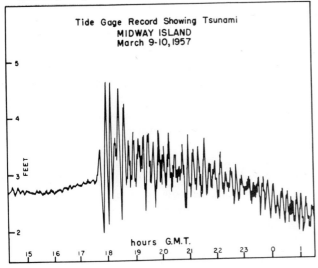

(c) Period 1st to 2nd crest; 12 minutes; 1st wave highest (after Salsman, 1959)

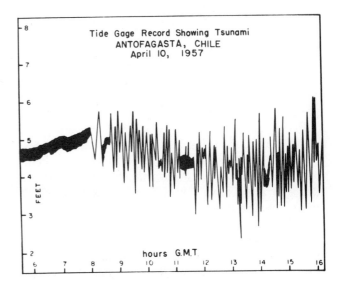

(e) Period 1st to 2nd crest; 14 minutes; 7th wave highest (after Salsman, 1959)

Fig. 5.3.

this type is very useful in setting up a tsunami warning system, as the tsunamis take a much longer time to reach an area than do the seismic waves.

The heights and periods of the waves in the open ocean have not been measured. The nearest thing to such a measurement has been made with a specially designed long period wave recorder on Wake Island in the Pacific Ocean (Van Dorn, 1959). The size and hydrography of this island are such that it is similar to a large pile rising from the floor of the ocean. The diameter of the island is about 5 miles whereas the lengths of the early waves in the tsunami train were from 20 to 30 miles. It is believed that such an island may give reliable records for long period waves, from the standpoint of measuring their open ocean characteristics (Munk, Snodgrass and Tucker, 1959). The record of the March 9, 1957, tsunami is shown in Fig. 5.3(a). The highest wave was only 1.3 ft and the longest period was only 7 min (first crest) followed by a dispersive train with periods degrading in several hours from 5 to 2½ minutes. This is at considerable variance with the periods recorded on the tide gauges in many bays, as shown in Fig. 5.3(b–e) where the initial periods were anywhere from 8 to 27 min (GMT—Greenwich Mean Time—is used on these records, whereas GCT—Greenwich Civil Time—is used on other records. GMT replaced GCT in official records in the United States on January 1, 1953). Some of the initial periods, for tide records not shown in the figure, were nearly one hour long (Apra Harbor, Guam, for one). If the periods of the waves measured at Wake Island are the periods of the waves in the open ocean, and if the periods of the waves are close to the periods associated with the peak of the energy density spectrum (as is probably the case) then one is faced with the difficulty of trying to explain the much longer waves observed in the various bays. Perhaps the long period forcing function is associated with a second-order effect due to varying mass transport of the fluctuating waves in the open ocean. It is not clear how the energy spectrum of the leading portion of a tsunami train should be obtained or used as the waves are essentially a group phenomenon and cannot be considered to be a stationary process.

Examples are shown in Table 5.5 of the initial period and the maximum height obtained from tide gauge records of the tsunami of November 4, 1952, originating off the east coast of the Kamchatka Peninsula, U.S.S.R., (lat. 52 1/2° N., long. 159° E. (see Savarensky et al. 1958 for a complete description of the effect of this tsunami in the U.S.S.R.). An examination of the data for 71 tide stations given by Zerbe (1953) shows almost no correlation between the height and period of the first waves with distance from Kamchatka. This suggests that the phenomenon being measured by the tide gauges is a complex oscillation of the bay in which the tide gauge is located at one or more of the bay's natural periods, this oscil-

Table 5.5. Some Initial Periods and Maximum Heights of Tsunami Originating Near Kamchatka on November 4, 1952, as Obtained from Tide Gauge Records (after Zerbe, 1953)

Tide station	Initial wave period first to second crest (min)	Initial rise (ft)	Maximum rise (or fall) (ft)
Adak, Alaska	48	1.4	6.9
Tolfino, B. C.; Canada	28	0.6	2.0
Crescent City, Calif.	25	1.7	6.8
Avila, Calif.	20	1.4	9.5
Los Angeles Harbor, Calif.	55	0.4	3.6
La Paz, Mexico	38	0.2	1.6
Salina Cruz, Mexico	35	0.3	4.0
Apra Harbor, Guam	53	0.5	0.9 (fall)
Wake I.	12	1.2	1.7
Midway I.	8	1.9	6.6
Hilo, Hawaii	20	4.0	7.9 (fall)
Johnson I.	29	1.2	1.4 (fall)
Yap I.	55	0.2	0.4 (fall)
Canton I.	13	0.3	0.7
Pago Pago, Samoa	18	0.9	6.0 (fall)
La Libertad, Ecuador	33	1.5	6.2 (fall)
Antofugasta, Chile	17	1.3	4.7 (fall)
Caldera, Chile	20	2.0	9.3

lation being excited by the tsunami. There is some suggestion that regions, such as the area between the coast of Southern California and the offshore islands, may also be set to oscillating at a natural period of the area by a tsunami (Munk, 1953) and that islands have a natural period, or periods, associated with them. In regard to the spectra of tsunamis, one study found that the spectra are similar at one tide gauge for different tsunamis, and dissimilar at different tide gauges for one tsunami, which seems to show the controlling role of local resonance (Munk, Snodgrass, and Tucker, 1959). This study shows that records obtained in closed harbors are not of much use in studying the character of tsunamis on even nearby coasts, and that even long period wave recorders on the open coasts may not be satisfactory. Takahasi (1963b), on the other hand, found that, although the response of a particular location was strongly dependent upon the local natural periods of oscillation (see Section 4 of this chapter) the peak energy would occur at a different natural period for one tsunami than for another, with the greater the magnitude of the earthquake associated with the tsunami, the longer the period at which the peak energy occurred. He found a linear relationship between tsunami peak energy period and the earthquake magnitude given by $T_e = 0.57 M - 2.85$, where T_e is the effective tsunami period in minutes and M is the magnitude of the associated earthquake.

Tsunamis are of low height in the open ocean, but can run up to substantial elevations on shores as can be seen in Fig. 5.4. Use of refraction diagrams alone does not permit the correct prediction of distribution of wave

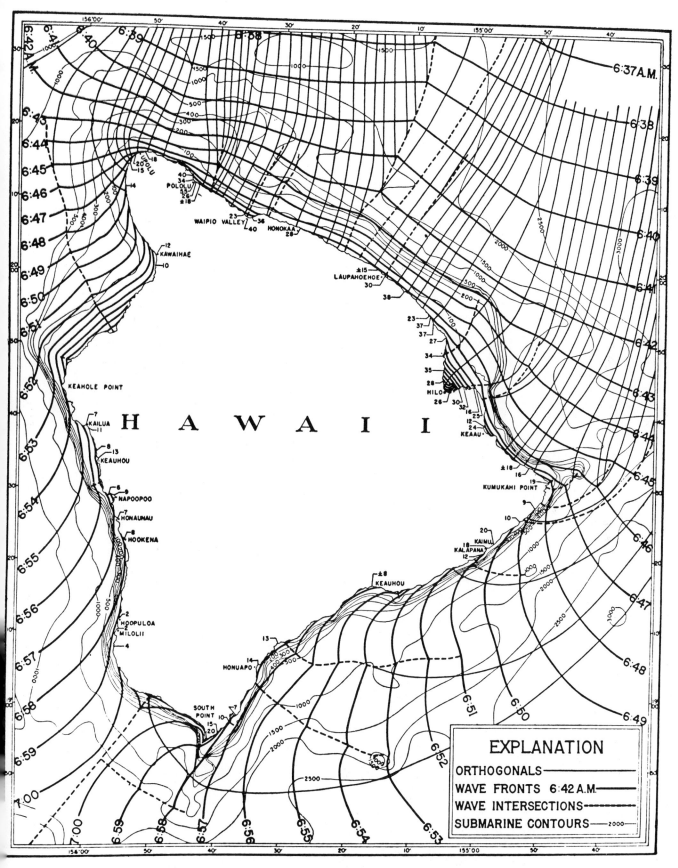

Fig. 5.4. Map of the island of Hawaii, showing heights (in feet above lower low water) reached by the water during the tsunami, wave fronts, orthogonals, and submarine contours (in fathoms). Times refer to computed time of arrival of first wave (from Shepard, MacDonald and Cox, 1950)

101

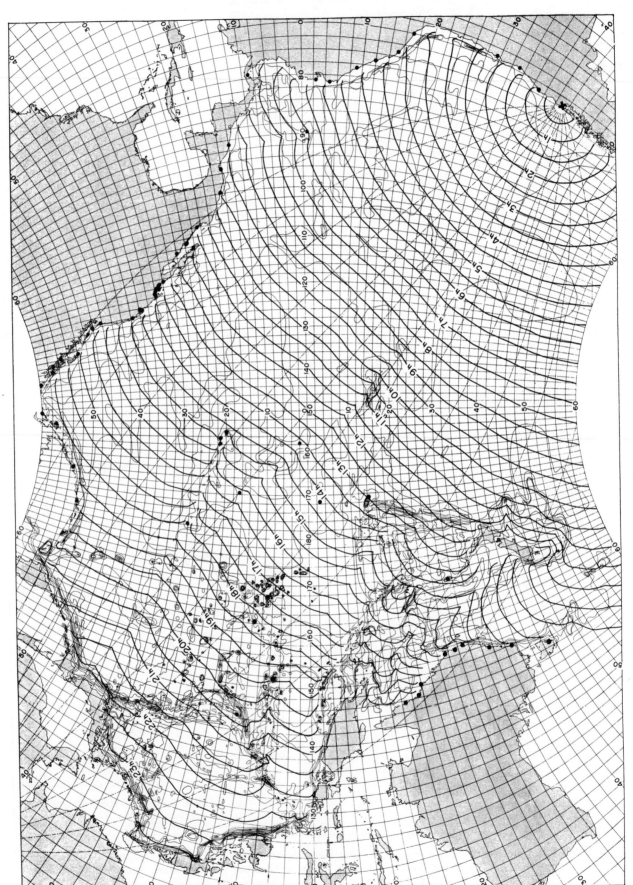

Fig. 5.5. Wave refraction chart for the Chilean tsunami of May 23, 1960 (after Wadati, Hirono, and Hisamoto, 1963)

heights around an island, but the diffraction theory of sound waves incident to a circular cylinder (Wiener, 1947; see also Chapter 11) permits an approximate prediction when used in conjunction with refraction drawings, such as the detailed one for a coastal area as shown in Fig. 5.4 and one for the whole ocean as shown in Fig. 5.5. Omer and Hall (1949) found that the pattern of heights for the 1946 tsunami at Kauai Island, Hawaii, was very similar to the pattern predicted by diffraction theory (see also Van Dorn, 1959).

It was suggested to the author by Professor J.D. Isaacs that tsunamis sometimes have the characteristics of a Mach-stem phenomenon (see Chapter 3). The April 1, 1946, tsunami originating in the Aleutian Island area appeared to behave in this manner as it moved along the cliffs forming the western boundary of the bay at Hilo, Hawaii. The experimental results presented in Chapter 3 showed that, once such a phenomenon is formed, the portion of the wave immediately adjacent to the boundary will be swung around by a boundary that curves, and that small irregularities of the coastline will not affect substantially the amplitude build-up of the portion of the wave adjacent to the boundary. This concept of the wave's motion appears to be substantiated by an examination of the run-up and damage pattern (photograph and observations, Child, 1962).

As will be discussed in Chapters 6 and 17, the characteristics of a wave on a slope depend upon the slope and the wave steepness in deep water, which in this case can refer only to the wave steepness in oceanic depths rather than in deep water because of the lengths of tsunamis. Most shore slopes are relatively steep when compared with waves as long as tsunamis; hence, they act in some ways as the tides in that they form breakers, or bores, only under certain limited conditions. The run-up on a shore can be many times the height of the wave in the oceanic depths. There is not much information on this phenomenon (see Van Dorn, 1959, for a discussion of it). A set of model studies was made for two shore slopes for a relatively large range of wave steepnesses (Kaplan, 1955). If the laboratory measurements are extrapolated as shown in Fig. 5.6, the heights reached by actual tsunamis can be predicted approximately, provided that the shore slopes are nearly the same as those used in the model study. In this study, the run-up, R, refers to the maximum elevation reached by the wave above the initial water level. The wave length, L, was the length in the uniform water depth portion of the tank. As this depth will not, in general, be the same as the oceanic depth relative to the wave length, considerable care must be used in applying these data to a specific set of conditions.

The effect of shore slope and terminal shore conditions, such as a seawall, are apparent. This confirms the findings on the effect of slopes in nature where a slope of one in fifty was associated with a run-up seven times as high as for a shore slope of one in five hundred (Matuo, 1934). A method for extending the limited data that are available to other slopes and wave steepnesses has been proposed by Van Dorn (1959).

The elevation reached by tsunamis on shores depends upon the offshore characteristics of the waves, diffraction, the slope and configuration of the shore, and resonance. Resonance problems of bays are considered in detail in another part of this chapter; a general feature of resonance, however, is that the greater the build-up of an oscillation in a harbor, the greater the number of waves necessary to cause the peak response of a bay. Thus, if the highest waves occur shortly after arrival of the first wave, then damping is of considerable importance. At Hilo, for example, the highest wave was the third (or perhaps it was the fifth or sixth, depending on how one interprets the record) for the May, 1960, tsunami (Eaton, Richter, and Ault, 1961), and this was apparently the case for that of April, 1946, also (Shepard et al., 1950), whereas in both the 1952 (MacDonald and Wentworth, 1954) and the 1957 tsunamis (Fraser, Eaton and Wentworth, 1959) the first wave was the highest. It would appear that, for this bay, damping is of considerable importance. In at least two cases a bore was observed, which ran ashore and did extensive damage, an indication of great damping of the energy of the waves.

It was found in Japan (Takahasi, 1963c) that the heights of the tsunamis at the heads of triangular bays are considerably higher than at the mouths of the bays. This relationship, however, appears to be a function of the ratio of period of the tsunamis to the period of the bay, and for tsunamis with periods several times as long as the periods of the bays there was little increase in the amplitude at the head of the bay.

Some peculiar observations of tsunami heights in open and hook bays have been made. Along the California coast, the bays that had the most protection from the direction of approach of the April 1, 1946, tsunami had the largest rise in water level (Bascom, 1946). For example, at Monterey Bay there was practically no wave at the south side of the bay, but the water level rose 10 ft at Santa Cruz on the north side of the bay.

Tsunamis are important because of the loss of life and vast property damage that have resulted from the larger ones. One of the worst occurred in Japan during the evening of June 15, 1896. The wave rushing onto land was between 75 and 100 ft in height, engulfing entire villages. More than 27,000 persons were killed and over 10,000 houses destroyed (Leet, 1948). Many other tsunamis have caused appreciable, but less, loss of life. For example, the April 1, 1946, wave killed more than 150 persons, injured badly another 163 persons, and caused about $25 million in property damage in

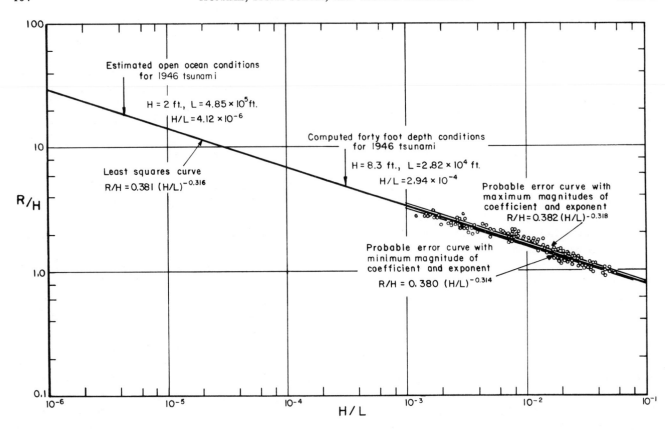

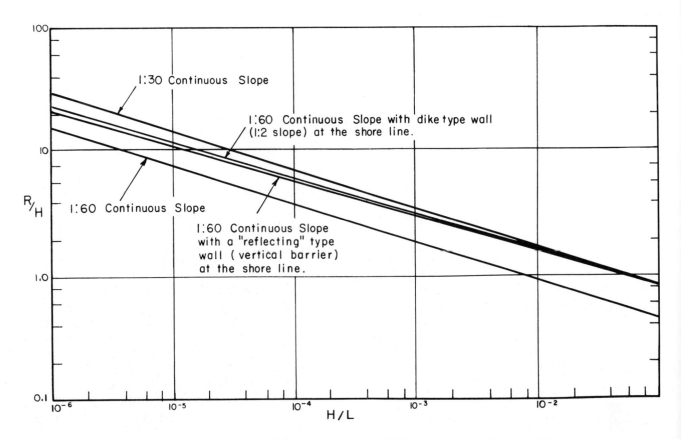

Fig. 5.6. Relative run-up (R/H) vs wave steepness (H/L) for waves (after Kaplan, 1955)

the Hawaiian Islands, largely in Hilo (Shepard, Mac-Donald, and Cox, 1950). An example of the damage caused by this same tsunami, but in a different location, is shown in Fig. 5.7. The base of the radio mast that was knocked down by the wave (or waves) was at an elevation of +106 ft, and the elevation of the foundation of the lighthouse was +45 feet.

On May 23, 1960, another disastrous tsunami struck Hilo. Some of the damage that was done can be seen in Fig. 5.8. Frame buildings were largely destroyed, but reinforced concrete buildings were only damaged. The force of the wave can be estimated from the bent parking meters shown in the photograph. The turf in the area between the parking lot and the concrete building in the background remained in excellent condition. Fig. 5.9 is included to show the inundated area and a measure of the extent of severe damage (60 per cent or more structural damage) for both the April 1, 1946, and the May 23, 1960, tsunamis.

Extensive reports have been written on several tsunamis which include information on the damage and a partial evaluation of protective systems and measures. The reader is referred to them for details (Wharton and Evans, 1888; Tokyo Imperial University, 1934; Shepard, MacDonald, and Cox, 1950; Savarensky et al., 1958; U.S. Army, Corps of Engineers, 1960; Matlock, Reese, and Matlock, 1961; Solov'ev and Ferchev, 1961; Committee for Field Investigations of the Chilean Tsunami of 1960, 1961). The type of damage can be classified in several ways. One system is as follows: A, the damage was approximately that which would be expected from an equal tidal inundation without surf; B, intermediate between A and C probably due to currents; C, damage

relatively great compared with what would be expected from a tidal inundation of similar height, probably due to a bore. In Type A the damage was usually due to the effect of water on goods or due to the effect of hydrostatic pressure only; in Type B scouring by currents may cause foundation failures; in both Types B and C damage can be done by currents carrying drift material (such as timber, boats, etc.) which can cause extensive damage to a structure (Horikawa, 1961). In Type C, the damage is caused by dynamic pressures as well as by currents and hydrostatic pressures, with considerable inertial forces being exerted on large structures as the front of the bore passes.

In regard to scouring action, it has been found that buildings with continuous footings generally did not sustain much structural damage (Architectural Institute of Japan, 1961). It has been found that groves of trees are beneficial in that they offer protection against drift materials (Horikawa, 1961; 1963). In the 1960 tsunami, it was found at Hilo, Hawaii, which was subjected to a bore (Type C damage), that although all light frame buildings and most heavy timber structures were demolished, good standard reinforced concrete structures withstood the force of the tsunami (Matlock et al., 1961). It was further found that the reinforced concrete structures that stood tended to shield weaker structures in their lee. The range of dynamic pressures was from 400 psf to 1800 psf with a norm of about 750 psf. These figures were based in part on an examination of the damage.

Statistical studies of tsunami frequencies are of value to the engineer in assessing the economic and social values of structures or other protective schemes being

Fig. 5.7. Effect of tsunami of April 1, 1946, on Scotch Cap Lighthouse, Unimak Island, Alaska. Foundation of lighthouse +45 ft; foundation of upper radio mast +106 ft; upper plateau +115 ft (courtesy U.S. Coast Guard)

(a) Before

(b) After

(a) Before

(b) After

Fig. 5.8. Damage at Hilo, Hawaii due to tsunami of May 23, 1960 (courtesy Larry Kadooka, *Hilo Tribune-Herald*)

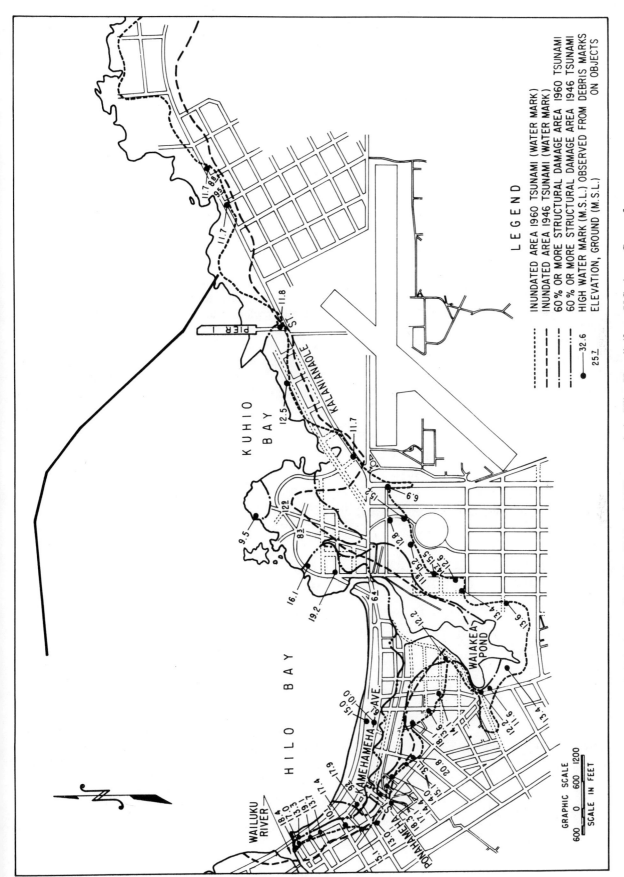

Fig. 5.9. Areas affected by 1946 and 1960 tsunamis in Hilo, Hawaii (from U.S. Army Corps of Engineering, 1960)

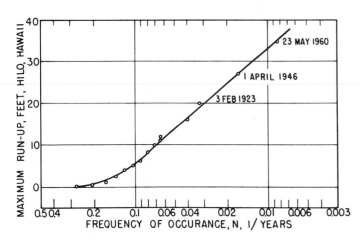

Fig. 5.10. Probability of tsunami maximum run-up at Hilo, Hawaii, exceeding a given value in a given duration

considered for a given location for protection against damage by tsunamis. The frequency distributions for Hilo, Hawaii, are given in Fig. 5.10 in which N is the number of times per year that a particular value of tsunami maximum run-up is equaled or exceeded. Following the method used by Wemelsfelder (1961) for storm tides along the Netherlands coast, the probabilities of tsunami run-up of a given elevation occurring in a given length of time have been computed from Fig. 5.10 and are given in Fig. 5.11. The curves of Fig. 5.11 were constructed from the curve in Fig. 5.10 according to Poisson's law

$$q = 1 - e^{-n}, \tag{5.1}$$

where q is the chance of a given height being exceeded in D years, and where

$$n = ND. \tag{5.2}$$

Thus there is a 10 per cent chance that a tsunami run-up equal to, or greater than, 33 ft will occur at Hilo in a

Fig. 5.11. Probability of tsunami maximum run-up at Hilo, Hawaii, exceeding a given value in a given duration

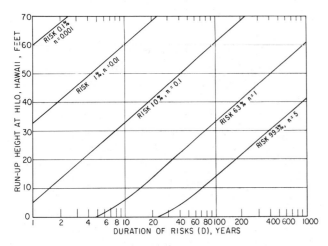

10-year interval, and a 63 per cent chance that such a tsunami will occur in a 100-year interval. It should be emphasized here that little is known of the statistics of geophysical maximums, and that the extension of run-up heights above those that have been measured must be done with caution; for example, stating that there is a 10 per cent chance of a tsunami run-up of 60 ft in a 100 year interval.

2. STORM SURGES

When violent storms, such as hurricanes, move from the ocean along, or across, the coast, the sea level rises above the normal tide level. Part of this rise is due to the rise in the surface resulting from a low barometric pressure, but most of the rise is due to coupled long waves and to the wind setup (wind tide). The coupled wave phenomenon has been discussed in Chapter 4 (traveling impulse generated waves). The wind setup is caused by winds dragging the surface water to the shoreline. The mechanism and measurements of wind setup will be discussed in the chapter on tides. This section describes some of the characteristics of the total phenomenon, together with some details of the type of damage that has been caused by it. A good deal of the damage has been caused, not by storm surges directly, but rather by the large wind-generated waves which accompany such conditions, since they act upon land and structures not normally exposed to them.

The term *surge* will be used to describe a more or less transient motion of a portion of a body of water, although the surge may consist of more than one variation in the surface (and associated water particle motions) of the body of water. A surge, which may have a semiperiodic "tail," may reflect and rereflect from the boundaries of the body of water. Thus, a tsunami may be classified as a surge.

Several mechanisms are known to cause surges, and undoubtedly there are other mechanisms that have not yet been recognized. A large mass of rock sliding into the water causes a surge, underwater seismic disturbances sometimes cause surges, hurricanes moving over relatively shallow water cause surges. These same hurricanes moving over deep water also generate coupled waves, but these are of much smaller amplitude, and are periodic, being of the same general class as ship waves. Strong winds drag surface water onto shore, and this phenomenon sometimes is called a surge. Pressure areas moving along a coast generate edge waves, which also act as a surge (Ursell, 1952; Munk, Snodgrass, and Carrier, 1956).

Surges are not restricted to the oceans. On June 26, 1954, a severe squall line moved across Lake Michigan. There was a rapid pressure jump associated with this, being of the order of about 0.1 ft of water in a few minutes. This pressure jump line moved across the lake at a speed of approximately 66 miles per hour, or a little less. A coupled wave of the resonant type could be formed if this pressure area moved over water about 290 ft deep, this being the depth for which the forward speed of the pressure area equals the long wave speed, $\sqrt{gd}$. The cross-hatched portion of Fig. 5.12 shows the region of the lake for which this condition holds approximately (Ewing, Press, and Donn, 1954). Surges 3–6 ft high were observed at Michigan City and surges 3–8 ft high were observed at Chicago an hour or so later. Seven persons were drowned by the surge (Harris, 1957). The surge at Chicago was probably the reflected wave from Michigan City. The characteristics of the observation and recordings of water level changes at various stations along the shore of Lake Michigan showed that the high amplitudes were not due to seiching of the lake (Harris, 1957). (Seiche refers to the free oscillations of an enclosed body of water.) Laboratory studies of waves generated by a moving pressure area (Wiegel *et al.*, 1958; Abraham, 1961) and the run-up characteristics of long waves (Kaplan, 1955) are sufficient to explain the heights of the surges generated by a pressure disturbance of the order of only 0.1 ft of water pressure difference.

With the possible exception of tsunamis, the greatest loss of life and property damage is associated with storm surges accompanying hurricanes (called typhoons in the Western Pacific Ocean). Between 1900 and 1955 there were over 11,750 deaths from hurricanes and property damage was in excess of $500 million for each of just two hurricanes. The worst storm surge in the United States, from the standpoint of loss of life, occurred in September, 1900, when more than 6000 persons were drowned, most of them in Galveston, Texas (Price, 1956). Between 1940 and 1953, the loss of life in all countries was 3744; and of this total, 590 deaths occurred in the United States. It has been estimated that three-fourths of the loss of life by drowning in the United States resulted from inundations by the sea and inland lakes (U.S. Dept. of Commerce, Weather Bureau, 1957).

Fig. 5.12. Lake Michigan depth contours; area of generation and paths of the June 26, 1954 wave (after Ewing, Press, and Donn, 1954)

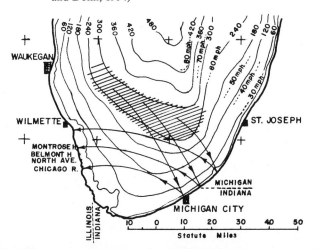

The number of tropical storms in the western Atlantic Ocean and the Gulf of Mexico has averaged about eight per year for the past 75 years, nine per year for the past 40 years, and ten per year for the past 20 years, and about 58 per cent of these reach hurricane intensity (Dunn, 1958). Of these hurricanes, about two per year arrive along the United States coastline. An average of seven typhoons per year enter the South China Sea from the western Pacific Ocean (Republic of the Philippines). Japan had 60 major storm surges between 1900 and 1953 (Wadati and Hirono, 1955). In addition, this type of storm occurs in other localities, such as Mexico, various Pacific islands, etc.

Some of the characteristics of the storm surge resulting from a hurricane can be seen in Fig. 4.17 (see Harris, 1956 and 1958 for many other examples). The maximum height of the storm surge occurred some miles to the right of the path of the center of the hurricane as it crossed the coastline. This is not because of local conditions. An examination of tide gauge records, supplemented by field observations, for five hurricanes along the eastern coast of the United States also showed that the maximun height of the surge occurred to the right of storm center, as can be seen in Fig. 5.13 (Redfield and Miller, 1957). The same general trend has been found for hurricanes in the Gulf of Mexico, although the maximum is not displaced so much to the right (Harris, 1958).

Although the height of the storm surge depends upon the steepness of the surge, the slope of the shore, and the resonance characteristics of the area in question (discussed in Chapter 4), there appears to be an average relationship between the height of the surge and the central pressure in the hurricane, Fig. 5.14 (Harris, 1958). The central pressure given in this figure was determined by a method described by Myers (1954).

The height of the storm surges along the Altantic Coast was greatest within certain embayments, but was also relatively high along the open coast; in Japan the highest water levels occurred in bays, or inland seas. It has been suggested that the difference might occur because the coast of southern New England has a relatively wide and shallow shelf, whereas the outer coast of Japan is bold, rising from relatively deep water (Redfield and Miller, 1957). An example of the distribution of maximum water levels, and the time of their occurrence is given in Fig. 5.15.

In Japan (Wadati and Hirono, 1955), it has been found that there are two types of storm surges, one type occurring along the coasts adjacent to shallow seas (the Japan Sea, for example) and the other type typical of a coastline along a deep sea (the Pacific Ocean). Along the coasts of the shallow seas, the storm surges consist of a gradual rise in sea level and then a slow return to normal tide levels, occasionally with a large oscillation superimposed upon the gradual rise. The maximum rise

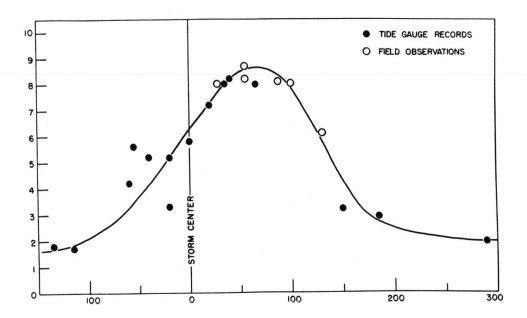

above normal tide level has been found to be from 6 to 10 ft. Along coasts adjacent to deep seas the gradual rise is insignificant, but the coast has been subject to several huge waves as high as 30 ft.

Munk, Snodgrass, and Tucker (1959), in their study of the spectra of low-frequency ocean waves, have shown that resonance characteristics of bays and other bodies of water, including shelves, are of some importance with respect to the heights and periods of water level variations due to storm surges, and other waves that can force oscillations.

There are numerous accounts of the water level falling, rising again, falling, rising, and so on, many hours after a hurricane has passed. If a bay has been set to oscillating in one or more modes, the reason is clear. These later high waters, however, also occur in areas that are not still oscillating. The term *resurgence* has been used to describe the general class of these phenomena (Redfield and Miller, 1957). Some of the observed waves occurring after the hurricanes on the East Coast of the United States have some characteristics of edge waves (Munk, Snodgrass, and Carrier, 1956). Other resurgences appear to have some of the characteristics of solitary or cnoidal waves. It appears that some of the resurgence characteristics, of the type which has occurred in the North Sea, can be explained by considering at least one part of a storm surge to be a Kelvin wave (Charnock and Crease, 1957; see also Chapter 12). Rossiter (1958) has found that negative surges in the North Sea caused by southerly winds are followed by a positive surge which moves progressively down along the east coast of England, easterly along the Netherlands, Germany, and Denmark, and then up along the coast of Norway. In this same study, it appeared that a positive surge was not accompanied by a seiching action, but was heavily

Fig. 5.13. Distribution of maximum water levels, recorded by tide guages or reliable field observations along outer coast, relative to distance from path of storm center. *Ordinate:* Elevation above predicted tide (ft). *Abscissa:* Shortest distance between station and storm center track (*n* mi). Figure based upon five storms (after Redfield and Miller, 1957)

Fig. 5.14. Storm surge height as a function of the central pressure in the storm (after Harris, 1958)

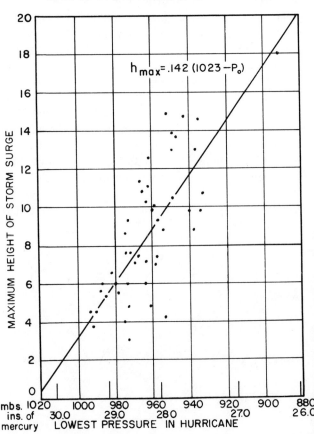

$$h_{max} = .142 (1023 - P_o)$$

MAXIMUM HEIGHT OF STORM SURGE

mbs.
ins. of
mercury   LOWEST PRESSURE IN HURRICANE

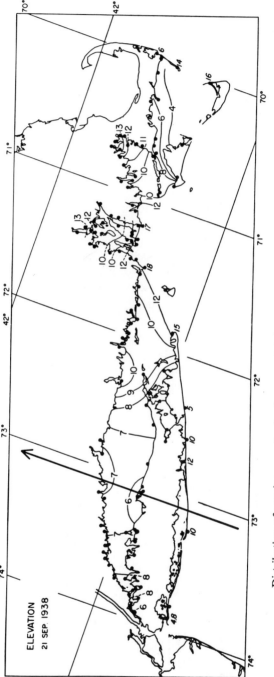

Distribution of maximum water levels reported from shores of southern New England and Long Island during the hurricane of September 21, 1938. *Contours:* Elevation above predicted tide (ft). *Solid circles:* Points of observation. *Arrow:* Storm center trajectory

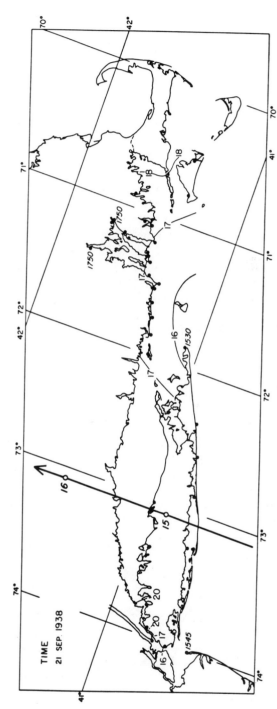

Time of maximum water level reported from shores of southern New England and Long Island during the hurricane of September 21, 1938. *Contours:* Isochrones, EST. *Solid circles:* Points of observation. *Arrow:* Storm center trajectory

Fig. 5.15. (after Redfield and Miller, 1957)

damped with a considerable phase lag, whereas a negative surge appeared to be accompanied by a seiching action of the North Sea.

The author has observed some phenomena in a laboratory study that should be of help in understanding resurgence. These studies were concerned with the generation of wind tides in shallow water (Sibul, 1955; Tickner, 1958). A fan was used to blow air over the water surface in a wind-wave tunnel. The wind shear on the water surface dragged water along the surface until it reached a sloping boundary (beach) at the downwind end, where it piled up until a sufficient head was developed to force a hydraulic flow of equal quantity in the opposite direction, the head being called the *wind tide*. When the fan was stopped, the water in the tank eventually reached a new equilibrium. The process of reaching a new equilibrium occurred in four different manners, depending primarily upon the water depth and setup: (1) A seiche occurred, the entire body of water in the tank oscillating with a gradual dissipation of energy; (2) a solitary wave ran back and forth between the two ends of the tank (both ends were slopes rather than vertical walls, or sometimes a bore was initiated which gradually lost energy and became a solitary wave; (3) a wave train, with gradually increasing period, formed and ran back and forth, with the number of waves increasing with time due to dispersion; (4) a solitary wave ran from the setup side to the upwind side, and upon reflection turned into a series of waves, this wave train gradually increasing in length and number of wave due to dispersion.

The detailed theory of many aspects of forced and free waves with respect to particular boundary conditions is beyond the scope of this book, and the reader is referred to the original work on the subject (for example, see Proudman, 1953; Reid, 1957; Kajiura, 1959). Details on a possible method of the prediction of hurricane storm surges have been given by Wilson (1959).

The low-lying countries bordering the North Sea are subject to severe storm surges (including the wind tide). The worst, from a standpoint of the loss of life, occurred long ago; it has been estimated that 100,000 lives were lost in England and the Netherlands in each one of three such cases, these storm surges occurring in 1099, 1421, and 1446 (Peters, 1954). More recently, during the storm tide of January 31–February 1, 1953, more than 300 lives were lost in England, and over 160,000 acres of agricultural land flooded by salt water. The same storm surge drowned nearly 1800 people in the Netherlands, flooded about 800,000 acres of land with salt water, and damaged 47,300 houses, 9215 of them either badly or irreparably (Wemelsfelder, 1954).

The gale that caused this storm surge was extremely severe and of long duration. It moved across the North Sea in a southeasterly direction. The lowest pressure recorded was 970 millibars, which was measured in the Orkney Islands. Some 5 hours after the low had passed, gusts up to 125 miles per hour, with a mean wind speed of 90 miles per hour were recorded in the Orkney Islands. Gusts of 101 mph were recorded near Aberdeen, Scotland, and gusts over 90 mph were recorded as far south as Suffolk. The gale was probably the most severe to occur over the western and central North Sea in hundreds of years (Peters, 1954). The general path of the storm is shown in Fig. 5.16.

The maximum water levels, above astronomical high water, are shown in Fig. 5.17 as a function of geographical location. The highest waters, and the greatest

Fig. 5.16. Track of the depression and position of the storm center at the moment of the beginning of the disaster (after Wemelsfelder, 1954)

Fig. 5.17. Raise in water level above astronomical tide high water (after Wemelsfelder, 1954)

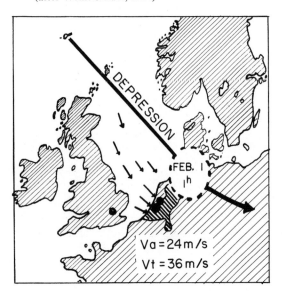

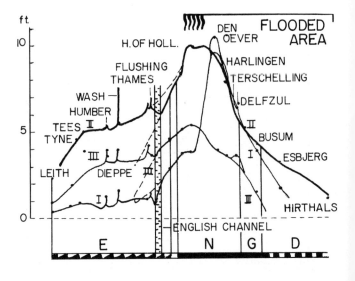

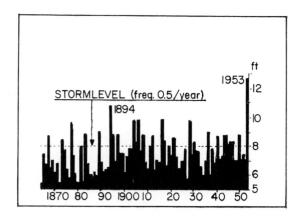

Fig. 5.18. Hook of Holland 90 years higher high water in every year since 1864 (after Wemelsfelder, 1954)

damage, occurred in the Netherlands. The maximum storm surge did not coincide with maximum astronomical tide, and this apparently was true for all other storm surges. It is not known whether this is fortuitous, or whether there is some physical reason for it (Farquharson, 1954). It may be due to some sort of nonlinear

Fig. 5.19. Frequency curve of high water at Hook of Holland (after Wemelsfelder, 1961)

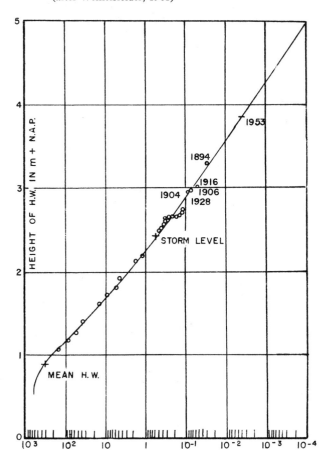

interaction of surge and tide of the sort studied by Rossiter (1961) in the Thames.

An idea of the relative magnitude of the water level can be seen in Fig. 5.18, and the statistical chance of it occurring (once in about four hundred years) can be seen in Fig. 5.19. Wemelsfelder makes use of statistical techniques to transform the data of Fig. 5.19 into a form that is more meaningful to the design engineer. This was done by treating the data as a two-dimensional problem, considering both the duration and the probability of an event occurring through use of Eqs. 5.1 and 5.2. The result is shown in Fig. 5.20. It is believed that the terms in the figure are self-explanatory. The levels

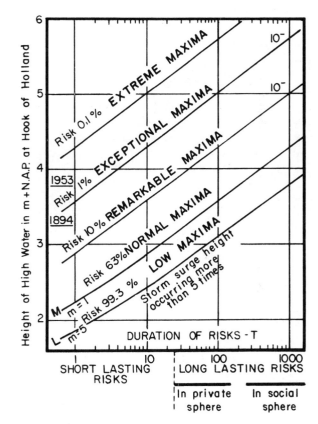

Fig. 5.20. Probability as a function of period and risk (after Wemelsfelder, 1961)

of the two highest storm tides in historical times (1894 and 1953) are shown in the figure.

Some photographs of the damage in the Netherlands can be seen in Figs. 5.21–5.23. Details of the damage done in the British Isles are given in a publication by the Institution of Civil Engineers (1954; also Spalding, 1954).

With the advent of high-speed digital computers and electric analogy computers, considerable headway is now being made in the theoretical study of storm surges with complicated bottom and lateral boundaries (Hansen, 1956; Ishiguro, 1959).

Fig. 5.22. Road destroyed by currents in the Netherlands, Feb. 1, 1953 (from Wemelsfelder, 1954)

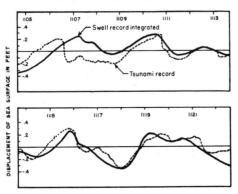

Fig. 5.21. A dike that failed in the Netherlands, Feb. 1, 1953 (from Wemelsfelder, 1954)

Fig. 5.23. Homes destroyed by currents in the Netherlands, Feb. 1, 1953 (from Wemelsfelder, 1954)

3. SURF BEATS

In Chapter 9 data are given on the characteristics of ocean waves. One of these characteristics is that there are groups of high waves and groups of low waves, as can be seen in Fig. 9.4. Associated with the motion of waves is a relatively small transport of water in the direction of wave advance (see Chapter 2). This transport increases as the water depth decreases and becomes relatively large. It is common experience that the water level in the surf zone gradually rises and falls, at intervals in the vicinity of 2 to 5 minutes. This is especially noticeable on a flat beach. The phenomenon has been termed *surf beat*.

Munk (1949) measured long period waves with a tsunami recorder about 1000 feet from shore, and measured simultaneously the swell (periods of the order of 10 seconds) about 2000 feet from shore. It was found that generally the groups of large swell were accompanied by an increase in the amplitude of the 2-minute waves on the tsunami recorder. The time lag between the two recorders was very close to the time computed for the swell to move to shore break, and a long period wave

to travel back out to sea past the tsunami recorder. The amount of mass transport associated with the swell was computed, and then an amplitude assigned to a long period wave to account for this mass transport. A comparison of the integrated swell record with the surf beat record obtained from the tsunami recorder is shown in Fig. 5.24. It was found that approximately 1 per cent of the incoming wave energy (swell) was returned in the form of long period waves.

The general characteristics of the surf beat, and its

Fig. 5.24. Surf beat (after Munk, 1949)

relationship with the grouping of wind waves and swell have been confirmed (Tucker, 1950), although there is some difference in opinion as to the method by which the mechanism of mass transport generates the surf beat. An example of the relationship between the amplitude of the swell height and long period wave heights are shown in Fig. 5.25. Tucker's measurements indicated the best correlation between high groups of waves above the wave recorder and the trough of long period waves. This seems to be explained theoretically by the work of Biesel (1952) and Longuet-Higgins (National Oceanographic Council, 1962) which showed that groups of high wind waves resulted in decreased or even negative mass transport.

Once surf beats have been generated, and they reflect seaward, total refraction may occur in the manner described by Williams and Isaacs (1952), and the waves turn back to land. Some coasts may be so shaped that, when combined with the offshore hydrography, they act as a trap for such waves (Biesel, 1955). In such a region, one could expect strong oscillations in harbors, and in fact these do occur.

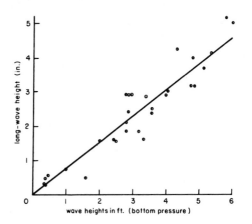

Fig. 5.25. Relation between long-wave height and ordinary-wave height (after Tucker, 1950)

4. HARBOR OSCILLATIONS

One often hears the term *harbor surging* or *seiching* used to describe the oscillations of a harbor. The term *harbor oscillation* will be used here instead, as surge denotes a "once only" motion, and *seiche* more properly refers to the free oscillations of a closed body of water, such as a lake. The phenomenon being described may be the oscillation of the harbor at one or more of its harmonic frequencies, or it may be oscillating at the frequency of the long period wave train exciting it. Usually the vertical motions are low, but when the oscillations are excited by a tsunami or storm surge they may be quite large. The exciting (forcing) mechanism may be tsunamis, storm surges, variable winds, air oscillations, or surf beat (Wilson, 1957), and it

has been shown that currents of sufficient strength moving past the entrance of a narrow bay may set up oscillations within the bay (Nakano and Abe, 1958). In the latter case, the immediate mechanism is the set of unsymmetric von Karman vortices that form.

These oscillations may be relatively common in some ports, but rare in others. At a given port, they may be more common at certain times of the year, and they may be numerous in one year and rare in another. The number of occurrences reported depends upon the natural periods of the harbors with respect to the energy available in the exciting mechanisms, and upon whether the occurrences are easily observed. The noticeable periods of harbor oscillation depend upon the size of the basin, ranging from about 2–3 minutes at Terminal Island, California, through 7–11 minutes for several harbors in the Netherlands, to 15–40 minutes in some Japanese ports. The relative importance of the various exciting mechanisms are given for a series of Japanese harbors by Amano (1957).

Proof that the problem of harbor oscillation is connected with one or more of the natural periods of the harbor has been provided by Wemelsfelder (1957). Measurements were made of the period of oscillations for four harbors as a function of tide stage, and the periods were found to be a function of tide stage (hence, water depth) provided that the harbor was oscillating at one of the natural periods.

When the vertical motions are large, it is easy to notice harbor oscillations, but oscillations of small amplitudes are more difficul to observe. One reason for the importance of an understanding of oscillations of harbors and the causative mechanism is that a small vertical motion is accompanied by a relatively large horizontal motion of the water. When the period of horizontal water motion coincides with the natural period of surge, sway, or yaw of a moored ship, a further resonance phenomenon occurs which results in a considerable motion of a moored ship (see Chapter 19). For example, on quite a few occasions damage occurred to ships, mooring lines, fenders, and piles in certain areas of Los Angeles Harbor. Observations showed that the damage often occurred when the moored ships were moving longitudinally (surging) or transversely (swaying) up to 5 ft or more (Knapp and Vanoni, 1945; Carr, 1953). At the time of these motions the wind waves were of negligible size, and the swell was of about 15-second period and less than 1 ft high in the harbor. The ship heaved with a period of about 14 sec, but surged and swayed with periods of from $1\frac{1}{2}$ to $3\frac{1}{2}$ min. Measurements made with special recorders showed that at these times the harbor was oscillating with about a 3-min period, and with vertical motions of the order of 0.1 to 0.2 ft, but with horizontal displacements of approximately 5 ft.

Consider a rectangular pan, with vertical walls and a flat bottom, partially filled with water. Tilt the pan

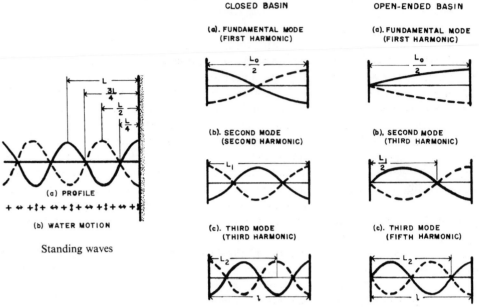

Fig. 5.26. (after Carr, 1953)

Surface profiles for oscillating waves

slightly along one of its axes, and then set it flat again. The water will oscillate back and forth in the pan. Now, if the water at one end increases in elevation while the water at the other end decreases in elevation, it is oscillating at its fundamental frequency, and the distance between the two reflecting walls is one-half the length of the standing wave. In the case of the first harmonic, the water goes up and down at the two ends and also at the middle (Fig. 5.26). The natural period is obtained from Eq. 2.36, $L = CT$, as $2\lambda/n = L$, where λ is the length of the basin and n is the order (first, second, etc., where $n = 1$ is the first harmonic), and C is the speed of an equivalent progressive wave (at period T) in the depth of water corresponding to the water depth in the basin (Rayleigh, 1876).

$$T_n = \frac{2\lambda}{nC} \qquad (5.3)$$

The theory predicts correctly the measurements made by Guthrie (1875). If one end of the basin were open and connected to a large body of water in such a manner that flow could occur across the opening, a node could occur at this end rather than an antinode, and the natural periods would be given by

$$T_n = \frac{4\lambda}{nC} \qquad (5.4)$$

where n has values of 1, 3, 5, etc.; this does not mean that, if a basin has an open end, the type of oscillation will be of this sort. In fact, it seems that it is easier to excite the first type with a standing wave antinode forming at the harbor entrance. C is given by Eq. 2.33, or better by Eq. 2.158b in general; we are, however, dealing

with waves that are long compared with the water depth, so that for most cases $C^2 = gd$ will suffice. Knowing the dimensions of a port, many of the various natural periods can be computed.

A basin can also oscillate between the other two walls, and can oscillate between both sets of walls simultaneously (Rayleigh, 1876; 1945, sec. 197). This oscillation results in complex patterns, for real harbors are not of uniform depth, the walls do not usually form a rectangle, they are not necessarily even approximately vertical, and there is an opening to the sea through which incident long waves move.

If a rectangular basin oscillates in both its longitudinal and transverse directions, the period equation is (Lamb, 1945, sec. 190)

$$T_{mn} = \frac{2\lambda}{C\sqrt{m^2 + (\lambda n/B)^2}} \qquad (5.5)$$

where B is the breadth of the basin, and m and n are integers of possible values of 0, 1, 2, ... defining the nodality of the oscillation in its longitudinal and transverse axes, respectively. The equation for the free surface is

$$y_s = a_0 \sin\frac{2\pi t}{T} \cos\frac{m\pi x}{\lambda} \cos\frac{n\pi z}{B} \qquad (5.6)$$

where x and z are the horizontal coordinates with the origin at one corner, with z the direction of incident wave advance and a_0 a coefficient defining the amplitude of the oscillations. Examples of the types of oscillations for various values of m and n for a square tank are shown in Fig. 5.27.

A more general solution can be obtained using a series of terms. For example, taking $m = 1$ and $n = 2$,

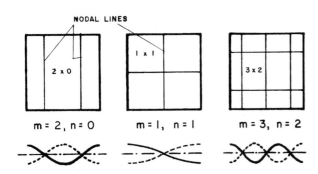

Fig. 5.27. (after McNown, 1951)

Fig. 5.28. (after McNown, 1951)

one such combination for a square tank (B = λ) might be

$$y_s = \left(A_0 \cos\frac{\pi x}{\lambda} \cos\frac{2\pi z}{\lambda} + A_1 \cos\frac{2\pi x}{\lambda} \cos\frac{\pi z}{\lambda} \right) \sin\frac{2\pi t}{T}$$

$$(5.7)$$

The resulting oscillations for various values of A_0 and A_1 are shown in Fig. 5.28. For the case of $m = 2$ and $n = 4$, a similar equation would show the oscillations given in Fig. 5.29.

A basin with an entrance can be forced to oscillate at one of its natural periods by waves entering the harbor through its entrance. Horizontal motions can occur across the entrance or, in the resonance case, an antinode can exist at the entrance, with standing waves outside the harbor as well as in it, the waves outside the harbor being the exciting waves (McNown, 1951, 1952; McNown and Danel, 1952).

Oscillations induced in models of harbors by periodic long waves incident to the harbor entrance have been studied by several investigators (U.S. Army, Corps of Engineers, Waterways Experiment Station, 1938, 1949; Knapp and Vanoni, 1945; Carr, 1953; Biesel, 1955). It has been possible to induce both simple and complex oscillations. An example of the second harmonic from side to side is shown in Fig. 5.30, and an example of the third harmonic from side to side, together with the second harmonic of the front to back mode, is shown in Fig. 5.31. The amplitudes shown in the figures are of relative magnitude, and the dimensions are in the terms

of the prototype, with the 6-minute and 3-minute wave periods being the values calculated to be harmonics of the harbor. The model was subjected to periodic waves of a range of frequencies, and the maximum amplitudes in the harbor measured. The magnification factor (ratio of maximum amplitude inside harbor to amplitude outside harbor) as a function of incident wave period is shown in Fig. 5.32, where it can be seen that the resonant peaks occur at approximately the values of the first few calculated harmonic periods of the harbor. Higher harmonics were found to be relatively unimportant because the energy in these frequencies was dissipated rather rapidly in the process of reflecting from the boundaries of the harbor.

The problems associated with models for long period waves and harbors are difficult. Often harbors are within bays which have natural periods of oscillations, so that primary amplification or reductions of the long waves or surges arriving from the ocean may occur within the bay first, and then further amplifications may occur within the harbor (Biesel, 1955; Wilson, 1957). Thus, even though the greatest magnification found in many of the model studies (see Fig. 5.32) was under 2, the over-all magnification factor may be much more. Problems connected with distorted models have been considered in detail by Biesel (1955), and the reader who wishes to make such a model study is referred to his work.

Considerable insight into the mechanism of harbor

Fig. 5.29. (after McNown, 1951)

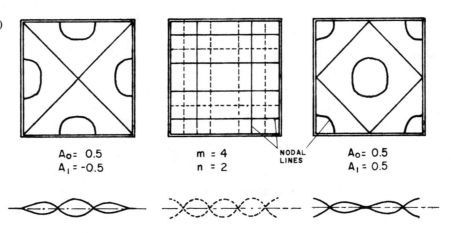

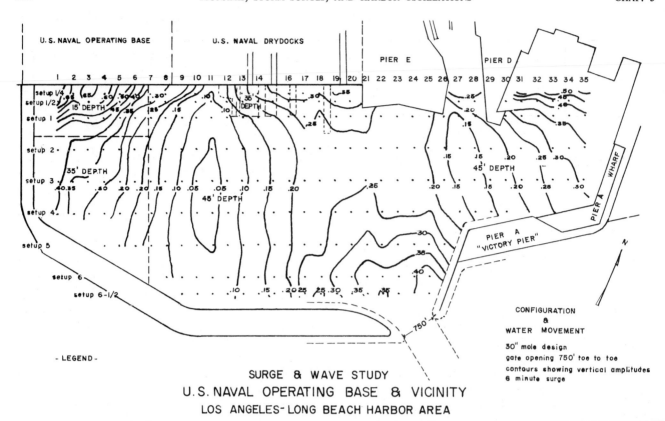

Fig. 5.30. Vertical water movements caused by 6-minute surge (after Knapp and Vanoni, 1945)

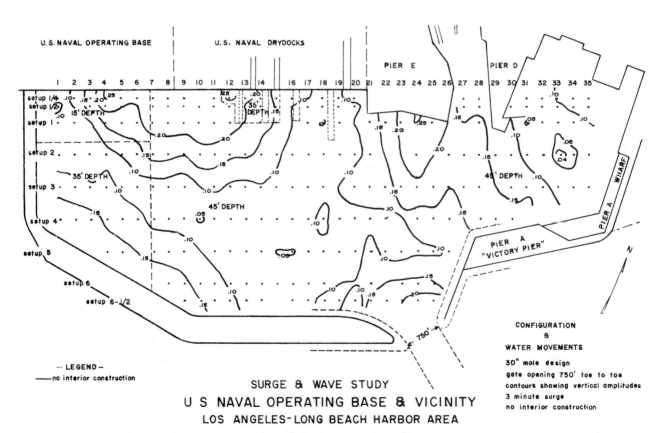

Fig. 5.31. Vertical water movements caused by 3-minute surge (after Knapp and Vanoni, 1945)

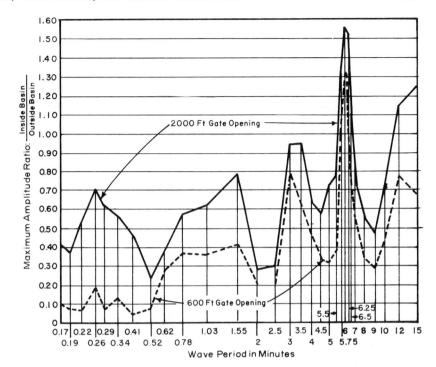

Fig. 5.32. Frequency response for model of Terminal Island mole basin, Los Angeles Harbor. Values given in terms of the prototype (after Carr, 1953)

oscillations is afforded by the theoretical and laboratory investigations of McNown (1951, 1952; McNown and Danel, 1952). A circular and a square basin, both with constant water depths and vertical walls, were studied.

Consider first a closed circular cylindrical basin. Rayleigh (1876) developed the theory for this case and found that the undamped free oscillations could be described by a Bessel function, and his theory predicted correctly the experimental values obtained by Guthrie (1875). As for the case of the rectangular basin, the boundary condition was taken that there be no flow of water in and out of the boundary so that an antinode (loop) of the standing wave must be at the boundary. The water surface elevation is given in cylindrical coordinates by

$$y_s = a \frac{kT}{2\pi} \sin \frac{2\pi t}{T} \cos n\theta \cdot J_n(kr), \qquad (5.8)$$

where the origin is at the center of the basin, r is the radius, a is the amplitude, θ is the angle, n is any integer, and J_n is the Bessel function of order n. The wave number k is obtained from the condition that there can be no water particle motion normal to the walls at the walls, and this condition is fulfilled when $J'_n(kr) = 0$ at $r = R$, where R is the radius of the circular cylinder.

For the case of $n = 0$, the motion is symmetrical about the center and the waves have annular crests and troughs. The lowest roots of $J'_0(kR) = 0 \ [-J_1(kR) = 0]$ are $kR/\pi = 1.2197, 2.2330, 3.2383, \ldots$, and the corresponding wave lengths are $1.640\,R, 0.896\,R, 0.618\,R, \ldots$. An example of this, and two other modes of oscillations is shown in Fig. 5.33. The physical significance of this is that $kR/\pi = t/T$, where t is the time it would take a

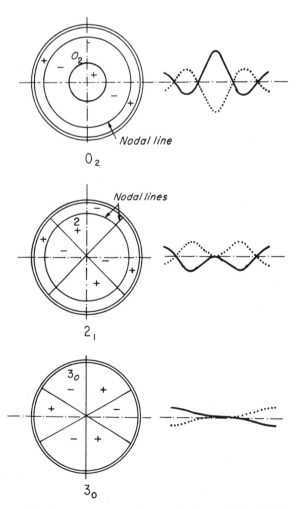

Fig. 5.33. Some resonant motion in a circular port (after McNown, 1951)

progressive wave of the same length to travel from the center of the circular cylinder to the wall and then back to the center; that is, it is a statement of $L = CT$ (Lamb, 1945, sec. 191).

McNown considered two conditions for both the circular and square cylindrical harbors. One was termed the *resonant* case; the other, the *nonresonant* case. Both types of harbors had openings to the sea, so that the incident waves could transmit power into the harbors. In the resonant case, the incident wave had a period that was one of the natural periods of the harbor so that a standing wave developed in a short time, and the standing wave pattern was such that an antinode formed at the entrance to the harbor. Physically then, the harbor with an opening to the sea was similar to a closed cylinder. When the incident waves first arrive, no standing wave forms at the entrance, so the energy density inside the harbor increases. Eventually a standing wave condition is obtained. As power transmitted by a wave is the integrated product of pressure and the horizontal component of water particle velocity in the direction of power transmission, no power can be transmitted across through the entrance, and even in the absence of dissipative forces and scattering mechanisms, the wave height within the harbor will build up to only a finite value. In reality, some energy is dissipated or transferred to other frequencies by nonlinear processes at the boundaries so

that an almost-standing wave exists at the entrance to the harbor, and this permits the feeding into the harbor of the relatively small amount of power needed to sustain the oscillations. This is not inconsistent with the boundary condition that there be no motion normal to the wall inside the harbor as wave energy is diffracted along the wall.

For the resonant case, McNown found theoretically that the maximum motion within a circular harbor was

$$y_s = a_0 \frac{J_n(kr)}{J_n(kR)} \cos n\theta \qquad (5.9)$$

where a_0 is the amplitude of the standing wave at the entrance (twice the amplitude of the incident wave). Equation 5.9 is identical to the case of a closed basin. Comparisons between theory and actual measurements are shown in Fig. 5.34. The opening to the harbor was appreciable, being one-sixteenth of the circumference, but it was not large compared with the wave length.

McNown also studied the effect of a smaller entrance and found that reducing the size of the entrance did not reduce the amplitude of oscillations within the harbor. A study by LeMehauté (1955) of a rectangular harbor indicates that, in general, the smaller the entrance, the larger the motions within the harbor. When the entrance becomes too small, however, frictional effects dominate, and this is no longer true. This will be discussed later.

Fig. 5.34. Comparison of theory and measured amplitudes of oscillations in a circular cylindrical harbor for the first, second and third harmonics (after McNown, 1951)

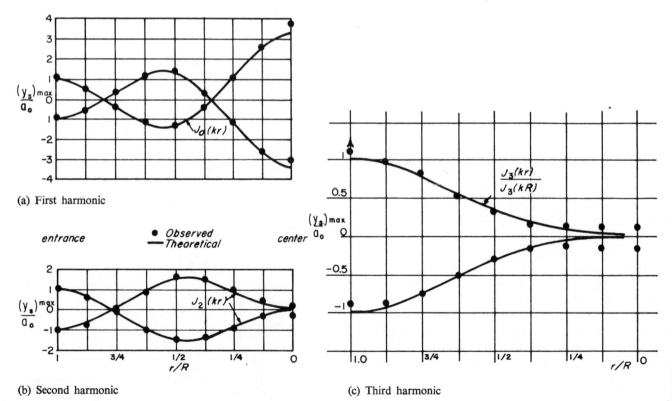

(a) First harmonic

(b) Second harmonic

(c) Third harmonic

The nonresonant case is more difficult to work with as it involves a series of Bessel functions. McNown has computed the amplitudes for several cases, and a comparison of one case with measurements is shown in Fig. 5.35.

McNown also studied two cases of the rate at which the amplitudes of oscillations decreased after the exciting force (the incident waves) had ceased. In one case, the harbor entrance was closed, and the rate of decrease of the oscillations as a function of the dissipative and scattering mechanisms within the harbor was measured. The other case was for the incident waves to be stopped, but with the harbor entrance left open. It can be seen in Fig. 5.36 that the motions in the harbor decreased more rapidly for the latter case as power was radiated through the entrance and out to sea.

For square, or rectangular, harbors it was found that, in regions where the amplitude of oscillations should be negligible, vertical motions would occur with frequencies of two or three times that of the principal motion (McNown and Danel, 1952). In order to generate oscillations comparable with theory, the entrance had to be placed in a position where antinodes should occur. Furthermore, relatively large entrances would extend past positions of antinodes, and nonresonant motions would occur. This emphasizes the difficulties in studying a real harbor. When the experiment was set up properly, fairly good agreement between theory (Kravtchenko and McNown, 1955) and measurements were obtained, with the exception of the higher-frequency oscillations mentioned above. Examples of two cases are shown in Fig. 5.37.

Apté (1955; Apté and Marcou, 1955) continued the studies on resonant oscillations of ports, and also worked on the problem of nonresonant oscillations (see also Gaillard, 1960 for nonresonant oscillations). For a rectangular basin of given horizontal dimensions and water depth, resonant periods can be calculated for various values of m and n. About sixty of the combinations of m and n were tested from the set of possible values of m and n, ranging from 0 to 8, each. The periods for the experimental setup ranged from 0.679 to 1.718 sec. The incident waves in the channel leading to the harbor entrance, which was the same width as the entrance, were generated with each of these periods. The entrance was not placed symmetrically with respect to the harbor, but was displaced so that one edge was 1.65 meters in from one side of the harbor and the other edge of the channel was 1.25 meters in from the other side (Fig. 5.38). It was found that standing wave conditions at the entrance could be obtained for almost all the combinations of m and n for which antinodes of the n modes should occur at the entrance (for example, $n = 0$). It is remarkable that resonant conditions were obtained with well-developed standing waves even when the opening was of

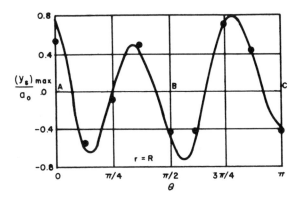

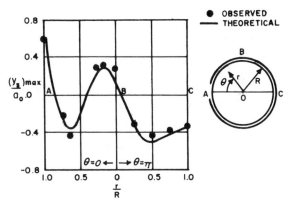

Fig. 5.35. Comparison of results of experiment and theory for nonresonant movement in a circular port (after McNown, 1951)

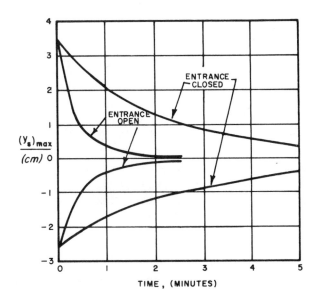

Fig. 5.36. Diminution of amplitude with time in the absence of stimulation by the external wave (after McNown, 1951)

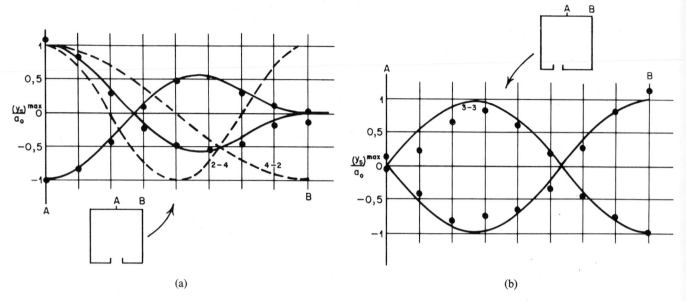

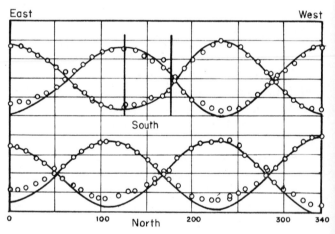

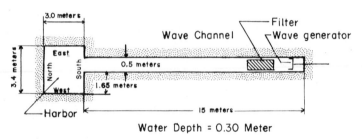

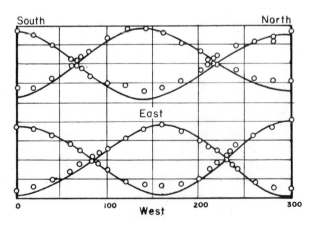

(a)   (b)

Fig. 5.37. (a) Square basin, with $m = 2$ and $n = 4$, entrance at center of one wall; (b) square basin, with $m = 3$ and $n = 3$, entrance offset from center of one wall but located at one antinode (after McNown, 1951)

Fig. 5.38. Laboratory set-up (from Apté, 1955)

the same order as the standing wave length in the transverse direction (for example, $m = 5$, $n = 0$). It was found that some of the amplitudes inside the harbor were often 2 to 3 times the amplitude of the standing wave in the channel (4 to 6 times the amplitude of the incident wave) for $m > 0$. The theory developed by Apté (1955) was found to predict the wave profiles within the harbor. Figure 5.39 shows a comparison of theory and laboratory results for the case of $m = 3$ and $n = 2$, so that a node occurs at the entrance for the n mode. Under conditions such as these, resonance was still obtained inside the harbor, and the disturbance at the entrance extended to about 10 to 15 ft down the channel before normal incident waves were evident. It was also found that completely stationary waves did not exist within the harbor, but that the movements of the nodal lines were small compared with the amplitude of the waves. This was probably due to the nonlinearity of the standing waves (see Chapter 2).

In some experimental work, LeMehauté (1955) found that the amplitudes in a rectangular harbor could be increased to more than 5 times the incident wave height

Fig. 5.39. Examples of data (from Apté, 1955)

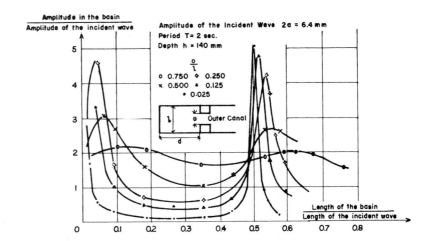

Fig. 5.40. Agitation in a basin of restricted opening, limited by two jetty heads, as a function of the opening and of the relative length of the basin (after LeMéhauté, 1955)

by reducing the size of the entrance sufficiently, and that the length of time to build up to maximum conditions increased rapidly as the entrance was made very small. Finally, the entrance was made so small that friction effects predominated, and the amplitude of oscillations then decreased with decreasing size of entrance. An example of this is shown in Fig. 5.40. That a resonant peak occurs at a ratio of basin length to wave length of 0.5 is as expected. Another peak at a value of the ratio of about one-thirtieth was found as predicted by the theoretical work of LeMehauté. Additional peaks are predicted at half wave length multiples. Other configurations were studied, including changes in depths as well as in planforms. Two examples, which are of interest in the tsunamis at Hilo, Hawaii, are shown in Figs. 5.41 and 5.42.

Another theoretical study (Miles and Munk, 1961)

also shows that the smaller the entrance width with respect to the size of the bay, the greater will be the ratio of the amplitude of standing wave motion within the bay to the incident wave. This study was for the case of an infinitely long ocean shoreline so that the bay and the opening were small compared with the ocean. A discussion of this work (LeMehauté, 1962) points out that as the entrance width decreases, the frequency bandwidth decreases as the magnification factor increases; hence, the probability of its occurring in nature decreases. In another discussion of the paper (Wilson, 1962), an example of a prototype was cited, namely, Duncan Basin in Table Bay Harbor, Cape Town, South Africa. By decreasing the entrance width from 750 to 400 ft (which decreased the ratio of the entrance to basin width from 0.13 to 0.07) the amplitudes of motion within the basin were reduced by one-third.

Fig. 5.41. Agitation in a basin of expanded opening limited by a deepening, as a function of the relative length of the basin (after Le Méhauté, 1955)

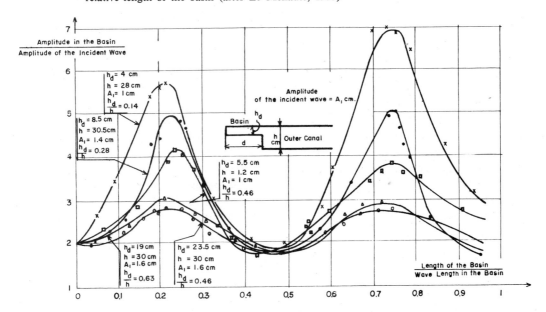

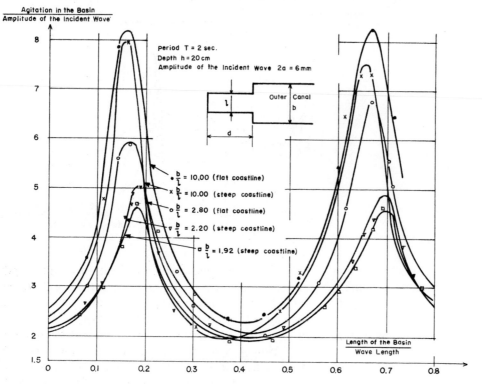

Fig. 5.42. Agitation in a basin of expanded opening limited by a widening, as a function of the relative length of the basin (after Le Méhauté, 1955)

Linearized friction has been considered for a basin of width small compared with its length for both horizontal and uniformly sloping bottoms, and for the case of uniformly slowly converging walls (Dorrestein, 1961). Basin lengths of less than one-quarter wave length only were considered. For the case of no friction, or small friction, it was found that the basin of uniform width resulted in greater amplification than for the case of slowly converging walls. When the friction became appreciable, the reverse was true. The theory assumed that the influence of the bay opening on the wave level along the coast was negligible so that the opening would have to be relatively small.

The effect of a bay opening to an ocean bounded by cliffs so that diffraction of energy into the bay by the waves reflected from the cliffs is important has been studied theoretically by Nakano (1963). The bay was rectangular in plan and of uniform depth. With friction neglected, it was found that the smaller the width of the bay with respect to its length, the greater will be the energy density of the standing waves in the bay; and the shallower the bay, the greater will be the increase in energy density in the bay.

What is the effect on harbor oscillations of an incident single surge? Some of them certainly must be able to set up oscillations, just as hitting a drum with a stick causes the drum membrane to vibrate. It would appear from the studies of LeMehauté that such an exciting force might be important in a harbor with a wide entrance, but of little importance for a harbor with a narrow entrance as there would not be sufficient time to increase the energy density in the harbor to a high level.

If one considers the response of a harbor to be a simple linear phenomenon, then the mean square harbor oscillation

$$\int_0^\infty S_2(f)\,df$$

can be obtained from the power spectral density of the incident waves, $S_1(f)$, by multiplying the amplification relationship $A^2(f)$ by $S_1(f)$ to get the power spectral density of the waves inside the harbor, $S_2(f)$ (Miles and Munk, 1961). Here, f is the wave frequency. In the light of the work of Apté where he was able to generate many modes of oscillations (various values of m and n), each one occurring for a precise value of incident wave period, it is not clear how one obtains the amplification relationship. It seems that a considerable amount of work should be done both in the laboratory and in prototype on the response of a harbor to various inputs $S_1(f)$ rather than to uniform periodic incident waves.

REFERENCES

Abraham, G., Hurricane storm surge considered as a resonance phenomenon, *Proc. Seventh Conf. Coastal Eng.*, Berkeley, Calif.: The Engineering Foundation Council on Wave Research, 1961, pp. 585–602.

Amano, Ryokichi, Long period waves in Japanese ports, *Permanent Intern. Assoc. Navigation Congresses, 19th Congress* (London), Sect. 2, Comm. 1, 1957, pp. 145–66.

Apté, Achyut S., "Recherches théoriques et expérimentales sur les mouvements des liquides pesants avec surface libre," D.Sc. thesis, Univ. Grenoble, 1955.

————, and C. Marcou, Seiches in ports, *Proc. Fifth Conf. Coastal Eng.*, Berkeley, Calif.: The Engineering Foundation Council on Wave Research, 1955, pp. 85–94.

Architectural Institute of Japan, *General state of damage of building, The Chilean Tsunami of 1960.* Tokyo: Maruzen Co., Ltd., 1961, pp. 151–64.

Bascom, W., *Effect of seismic sea wave on California coast*, Univ. Calif., IER, Tech. Rept. 3–204, April 16, 1946. (Unpublished.)

Biesel, F., Equations générales au second ordre de la houle irrégulière, *La Houille Blanche*, **7**, 3 (May–June, 1952), 372–76.

————, The similitude of scale models for the study of seiches in harbors, *Proc. Fifth Conf. Coastal Eng.*, Berkeley, Calif.: The Engineering Foundation Council on Wave Research, 1955, pp. 99–118.

Carr, John H., Long-period waves or surges in harbors, *Trans. ASCE*, **118**, Paper no. 2556 (1953), 588–603.

Charnock, H., and J. Crease, North Sea surges, *Physical Oceanography, Science Progress*, **45** (1957), 494–511.

Child, M. C., Personal communication, Hilo, Hawaii (April, 1962).

Committee for Field Investigations of the Chilean Tsunami of 1960, *Report on the Chilean tsunami of May 24, 1960, as observed along the coast of Japan.* Tokyo: Maruzen Co., Ltd., 1961.

Cox, Doak, Tsunami records at Hilo, Hawaii, Hawaii Institution of Geophysics, 1962. (Unpublished.)

Cuellar, Marcial P., *Annotated bibliography on tsunamis*, U.S. Army, Corps of Engineers, Beach Erosion Board, Tech. Memo, No. 30, February, 1953.

Dorrestein, Richard, Amplification of long waves in bays, *Eng. Prog. Univ. Florida*, **15**, 12, December, 1961.

Dunn, Gordon E., Hurricanes and hurricane tides, *Proc. Sixth Conf. Coastal Eng.*, Berkeley, Calif.: The Engineering Foundation Council on Wave Research, 1958, pp. 19–29.

Eaton, J. P., D. H. Richter, and W. U. Ault, The tsunami of May 23, 1960, on the Island of Hawaii, *Bull. Seis. Soc. Amer.*, **51**, 2 (April, 1961), 135–57.

Ewing, Maurice, Frank Press, and William L. Donn, An explanation of the Lake Michigan wave of 26 June 1954, *Science*, **120**, 3122 (Oct. 29, 1954), 1–2.

Farquharson, W. I., Storm surges on East Coast of England, *Conf. North Sea Floods 31 January/1 February, 1953*, Inst. Civil Engrs. (London), 1954, pp. 14–27.

Fraser, George D., Jerry P. Eaton, and Chester K. Wentworth, The tsunami of March 9, 1957, on the Island of Hawaii, *Bull. Seis. Soc. Amer.*, **49**, 1 (January, 1959), 79–90.

Gaillard, P., Des oscillations non linéaires des eaux portuaires, *La Houille Blanche*, **15**, 2 (March–April, 1960), 164–72.

Green, C. K., Seismic sea wave of April 1, 1946, as recorded on tide gages, *Trans. Amer. Geophys. Union*, **27**, 4 (1946), 490–500.

Gutenberg, B., Tsunamis and earthquakes, *Bull. Seis. Soc. Amer.*, **29**, 4 (October, 1939), 517–26.

Guthrie, Frederick, On stationary liquid waves, *Phil. Mag.*, ser. 4, **50** (1875), 290–302; 377–89.

Hansen, W., Theorie zur Errechnung des Wasserstandes und der Strömungen in Randmeeren nebst Anwendungen, *Tellus*, **8**, 3 (August, 1956), 287–300.

Harris, D. Lee, *Some problems involved in the study of storm surges*, U.S. Dept. of Commerce, Weather Bureau, National Hurricane Research Project Report No. 4, December, 1956.

————, The effect of a moving pressure disturbance on the water level on a lake, *Interaction of Sea and Atmosphere* 1, Meteorological Monograph, **2**, 10 (June, 1957), 46–57.

————, The hurricane surge, *Proc. Sixth Conf. Coastal Eng.*, Berkeley, Calif.: The Engineering Foundation Council on Wave Research, 1958, pp. 96–114.

Heck, N. H., List of seismic sea-waves, *Doc. Scient. Comm. Raz de Marée, IUGG*, 1947.

Horikawa, Kiyoshi, Tsunami phenomena in the light of engineering viewpoint, *The Chilean Tsunami of 1960.* Tokyo: Maruzen Co., Ltd., 1961, pp. 136–50.

————, Evaluation of tsunami protection measures, *Proc. Tsunami Meetings Associated with the Tenth Pacific Science Congress*, Monograph No. 24, IUGG (July, 1963), 250–62.

Iida, K., Magnitude and energy of earthquakes accompanied by tsunami and tsunami energy, *J. Earth Sciences*, (Nagoya Univ.), **6**, 2 (December, 1958), 101–112.

————, On the estimation of tsunami energy, *Proc. Tsunami Meetings Associated with the Tenth Pacific Science Congress*, Monograph No. 24, IUGG (July, 1963a), 167–73.

————, Magnitude, energy, and generation of tsunamis, and catalogue of earthquakes associated with tsunamis, *Proc. Tsunami Meetings Associated with the Tenth Pacific Science Congress*, Monograph No. 24, IUGG (July, 1963b), 7–18.

Institution of Civil Engineers (London), *Conference on the North Sea floods of 31 January/1 February 1953*, 1954.

Ishiguro, S., A method of analysis for long-wave phenomena in the ocean, using electronic network models. I, The earth's rotation ignored, *Phil. Trans., Roy. Soc.* (London), ser. A, **251**, 996 (March, 1959), 303–340.

Kajiura, Kinjiro, *A theoretical and empirical study of storm induced water level anomalies*, Agricultural and Mechanical College of Texas, Texas Agr. and Mech. Research Foundation, Reference 59–23F, December, 1959. (Unpublished.)

Kaplan, Kenneth, *Generalized laboratory study of tsunami run-up*, U.S. Army, Corps of Engineers, Beach Erosion Board, Tech. Memo. No. 60, 1955.

Knapp, Robert T., and Vito A. Vanoni, *Wave and surge*

study for the Naval Operating Base, Terminal Island, California, Calif. Inst. Tech., Hyd. Structures Lab., January, 1945.

Kravtchenko, Julien, and John S. McNown, Seiche in rectangular ports, *Quar. Applied Math.*, **13**, 1 (April, 1955), 19–26.

Lamb, Sir Horace, *Hydrodynamics*, 6th ed. New York: Dover Publications, Inc., 1945.

Leet, L. Don, *Causes of catastrophe: earthquakes, volcanos, tidal waves, hurricanes.* New York: McGraw-Hill, Inc., 1948.

LeMehauté, B., Two-dimensional seiche in a basin subjected to incident waves, *Proc. Fifth Conf. Coastal Eng.*, Berkeley, Calif.: The Engineering Foundation Council on Wave Research, 1955, pp. 119–50.

———, Discussion of "Harbor Paradox," *J. Waterways Harbors Div.*, *Proc. ASCE*, **88**, WW2 (May, 1962), 173–85.

MacDonald, Gordon A., and Chester K. Wentworth, The tsunami of November 4, 1952, on the Island of Hawaii, *Bull. Seis. Soc. Amer.*, **44**, 3 (July, 1954), 463–69.

McNown, John S., Sur l'entretien des oscillations des eaux portuaires sous l'action de la haute mer, Thèse, Faculté des Sciences, Univ. Grenoble, August, 1951.

———, Waves and seiche in idealized ports, *Gravity Waves*, U.S. Dept. of Commerce, National Bureau of Standards Circular 521, 1952, pp. 153–64.

———, and P. Danel, Seiche in harbours, *Dock and Harbour Authority*, **33**, 384 (October, 1952), 177–80.

Matlock, Hudson, Lymon C. Reese, and Robert B. Matlock, *Analysis of structural damage from the 1960 tsunami at Hilo, Hawaii*, University of Texas, Structural Mechanics Laboratory, October, 1961. (Unpublished.)

Matuo, Haruo, Estimation of energy of tsunami and protection of coasts, Papers and Reports on the Tsunami of 1933 on the Sanriku Coast, *Japan, Bull. Earthquake Inst. of Japan*, Supplementary vol. I, Tokyo Imperial University (March, 1934), 55–64.

Miles, John, and Walter Munk, Harbor paradox, *J. Waterways Harbors Div.*, *Proc. ASCE*, **87**, WW3 (August, 1961), 113–130.

Munk, W. H., Surf beats, *Trans. Amer. Geophys. Union*, **30**, 6 (December, 1949), 849–54.

———, Small tsunami waves reaching California from the Japanese earthquake of 4 March 1952, *Bull. Seis. Soc. Amer.*, **43**, 3 (July, 1953), 219–22.

———, F. Snodgrass, and G. Carrier, Edge waves on the continental shelf, *Science* (January, 1956), 127–32.

———, ———, and M. J. Tucker, Spectra of low frequency ocean waves, *Bull. Scripps Inst. Oceanogr.*, **7**, 4 (1959), 283–362.

Myers, Vance A., *Characteristics of United States hurricanes pertinent to levee design for Lake Okeechobee, Florida,*

U.S. Dept. of Commerce, Weather Bureau, Hydrometeorological Report No. 32, 1954.

Nakano, Masito, A theory of growth of tsunami in a bay, *Proc. Tsunami Meetings Associated with the Tenth Pacific Science Congress*, Monograph No. 24, IUGG (July, 1963), 125–28.

———, and T. Abe, Standing oscillation of bay water induced by currents, *Geophy. Mag.*, **28**, 3 (March, 1958), 375–97.

National Oceanographic Council, *Annual report of the National Oceanographic Council 1960–61.* Cambridge: Cambridge University Press, 1962.

Ogawa, T., Notes on the volcanic and seismic phenomena in the volcanic district of Shimabara, with a report on the earthquake of December 8, 1922, Kyoto Imperial Univ., *Mem. Coll. Sci.*, ser. B, **1** (1924), 219–24.

Omer, Guy C., Jr., and Harold H. Hall, The scattering of a tsunami by a cylindrical island, *Bull. Seis. Soc. Amer.*, **39**, 4 (October, 1949), 257–60.

Peters, Sidney Percival, Some meteorological aspects of North Sea floods, with special reference to February, 1953, *Conf. North Sea Floods 31 January/1 February 1953*, *Inst. Civil Engrs.* (London), 1954, pp. 28–36.

Price, W. Armstrong, *Hurricanes affecting the coast of Texas from Galveston to Rio Grande*, U.S. Army, Corps of Engineers, Beach Erosion Board, Tech. Memo. No. 78, March, 1956.

Proudman, J., *Dynamical Oceanography.* London: Methuen & Co., Ltd., 1953.

Rayleigh, Lord, On waves, *Phil. Mag.*, 5th ser., **1**, 4 (April, 1876), 257–79.

———, *The Theory of Sound.* New York: Dover Publications, Inc., 1945.

Redfield, Alfred C., and A. R. Miller, Water levels accompanying Atlantic Coast hurricanes, *Interaction of Sea and Atmosphere, Meteorological Monographs*, **2**, 10 (June, 1957), 1–23.

Reid, Robert O., *Forced and Free Surges in a Narrow Basin of Variable Depth and Width: A Numerical Approach*, Agricultural and Mechanical College of Texas, Texas Agr. and Mech. Research Foundation, Reference 57–25 T, August, 1957. (Unpublished.)

Republic of the Philippines, Weather Bureau Climatological Division, *Tropical Cyclones of* (year). (Annual).

Rossiter, J. R., Storm surges in the North Sea, 11 to 30 December 1954, *Phil. Trans. Roy. Soc.* (London), ser. A, **251**, 991 (4 Dec., 1958), 139–60.

———, Interaction between tide and surge in the Thames, *Geophys. J. Roy. Astron. Soc.*, **6**, 1 (1961), 29–53.

Salsman, Garrett G., *The tsunami of March 9, 1957, as recorded at tide stations*, U.S. Dept. of Commerce, Coast and Geodetic Survey, Tech. Bull. No. 6, July, 1959.

Savarensky, E. F., V. G. Tischchenko, and A. E. Sviatlovsky, The tsunami of 4–5 November 1952, *Bull. Council Seismology, Acad. Sci. USSR*, no. 4 (1958) 1–60. Wilvan G.

Van Campen, trans., Univ. of Hawaii, East-West Center, Hawaii Inst. Geophys., *Trans. series* no. 10, 1961.

Shepard, F. B., G. A. MacDonald, and D. C. Cox, The tsunami of April 1, 1946, *Bull. Scripps Inst. Oceanog*, **5**, 6 (1950), 391–528.

Sibul, O., *Laboratory study of wind tides in shallow water*, U.S. Army, Corps of Engineers, Beach Erosion Board, Tech. Memo. No. 61, August, 1955.

Solov'ev, S. L., and M. D. Ferchev, Summary of data on tsunamis in the U.S.S.R., *Bull. Council Seismology, Acad. Sci. USSR*, no. 9, pp. 23–55, 1961. Wilvan G. Van Campen, trans., Univ. of Hawaii, East-West Center, Hawaii Inst. Geophys., *Trans. Series* no. 9, 1961.

Spalding, John Victor, A general survey of the damage done and action taken, *Conf. North Sea Floods 31 January/1 February 1953, Inst. Civil Engrs.* (London), 1954, 5–13.

Takahasi, Ryutaro, A model experiment on the mechanism of seismic sea wave generation. Part I, Tokyo Imperial University Earthquake Research Institute, Bulletin Supplement no. 1 (1934), 152–78.

——, On some model experiments on tsunami generation, *Proc. Tsunami Meetings Associated with the Tenth Pacific Science Congress*, Monograph No. 24, IUGG (July, 1963a), 235–48.

——, On the spectra and the mechanism of generation of tsunamis, *Proc. Tsunami Meetings Associated with the Tenth Pacific Science Congress*, Monograph No. 24, IUGG (July, 1963b), 19–25.

——, A summary report on the Chilean tsunami of May 24, 1960, as observed along the coast of Japan, *Proc. Tsunami Meetings Associated with the Tenth Pacific Science Congress*, Monograph No. 24, IUGG (July, 1963c), 77–86.

Tickner, E. G., Transient wind tides in shallow water, M. S. thesis in Mechanical Engineering, Univ. Calif., August, 1958.

Tokyo Imperial University, Papers and reports on the tsunami of 1933 on the Sanriku Coast, Japan, *Bull. Earthquake Inst. of Japan*, supplementary vol. I, March, 1934.

Tucker, M. J., Surf beats: sea waves of 1 to 5 min. period, *Proc. Roy. Soc.* (London), ser. A, **202**, 1071 (August, 1950), 565–73.

U.S. Army, Corps of Engineers, Honolulu District, *Hilo Harbor, Hawaii: Report on survey for tidal wave protection and navigation*, 15 Nov., 1960.

——, Waterways Experiment Station, *Hydraulic model study for reducing surge conditions in Hilo Harbor, Hawaii*, 1938.

——, *Wave and surge action, Monterey harbor, Monterey, California* (*Model investigation*), Tech. Memo No. 2–301, Sept., 1949.

U.S. Department of Commerce, Weather Bureau, *Survey of meteorological factors pertinent to reduction of loss of life and property in hurricane situations*, National Hurricane Research Project Report No. 5, March, 1957.

Ursell, F., Edge waves on a sloping beach, *Proc. Roy. Soc.* (London), ser. A, **214**, 1116 (7 Aug., 1952), 79–97.

Van Dorn, William G., *Impulsively generated waves, Report No. II*, Univ. of Calif., Scripps Inst. Oceanogr., Contract Nonr 233(35), August, 1959. (Unpublished.)

Wadati, K., and T. Hirono, Storm tides caused by typhoons, *Proc. UNESCO Symposium Typhoons*, Japanese National Commission for UNESCO, 1955, pp. 31–48.

——, ——, and S. Hisamoto, On the tsunami warning service in Japan, *Proc. Tsunami Meetings Associated with the Tenth Pacific Science Congress*, Monograph No. 24, IUGG (July, 1963), 138–45.

Wemelsfelder, P. J., The disaster in the Netherlands caused by the storm flood of February 1, 1953, *Proc. Fourth Conf. Coastal Eng.*, Berkeley, Calif.: The Engineering Foundation Council on Wave Research, 1954, pp. 258–71.

——, Origin and effects of long period waves in ports, *Permanent Intern. Assoc. Navigation Congresses, 19th Congress* (London), Sect. 2, Comm. 1 (1957), pp. 167–76.

——, On the use of frequency curves of stormfloods, *Proc. Seventh Conf. Coastal Eng.*, Berkeley, Calif.: The Engineering Foundation Council on Wave Research, 1961, pp. 617–32.

Wharton, W. J. L., and Sir F. J. Evans, The eruption of Krakatoa and subsequent phenomena. Part III, On the seismic sea waves caused by the eruptions of Krakatoa, August 26th and 27th, 1883, Royal Society of London (1888), pp. 89–151.

Wiegel, R. L., Laboratory studies of gravity waves generated by the movement of a submerged body, *Trans. Amer. Geophys. Union*, **36**, no. 5 (October, 1955), 759–74.

——, C. M. Snyder, and J. E. Williams, Water gravity waves generated by a moving low pressure area, *Trans. Amer. Geophys. Union*, **39**, 2 (April, 1958), 224–36.

Wiener, F. M., Sound diffraction by rigid spheres and cylinders, *J. Acoust. Soc. Amer.*, **19**, 3 (May, 1947), 444–51.

Williams, E. Allan, and John D. Isaacs, The refraction of groups and of the waves which they generate in shallow water, *Trans. Amer. Geophys. Union*, **33**, 4 (August, 1952), 523–30.

Wilson, Basil W., Origin and effects of long period waves in ports, *Permanent Intern. Assoc. Navigation Congresses, 19th Congress* (London), Sect. 2, Comm. 1 (1957), pp. 13–61.

——, *The prediction of hurricane storm-tides in New York Bay*, Agricultural and Mechanical College of Texas, Texas Agr. and Mech. Research Foundation, Reference 59–20F, October 1959. (Unpublished.)

——, Discussion of "Harbor Paradox," *J. Waterways Harbors Div.*, *Proc. ASCE*, **88**, no. WW2 (May, 1962), 185–94.

Zerbe, W. B., *The tsunami of November 4, 1952, as recorded at tide stations*, U.S. Dept. of Commerce, Coast and Geodetic Survey, Special Publication No. 300, 1953.

Zetler, Bernard D., Travel times of seismic sea waves to Honolulu, *Pacific Science*, **1**, 3 (July, 1947), 185–88 plus plate.

CHAPTER SIX

Effect of Structures on Waves

1. BEACHES

Beaches can be classed loosely as structures that can adjust themselves to waves. If a beach is very steep and the wave is very flat, a wave approaching nearly normal to the beach will be almost totally reflected, with some energy loss due to bottom friction and percolation. At the other limit are relatively flat beaches and steep waves; in which case, the wave dissipates its energy gradually in a spilling type of breaking. Details of this are given in Chapter 7. For details of the effect of bottom friction and percolation on waves see Putnam (1949) and Putnam and Johnson (1949). A nonrigid bottom also contributes to the change in wave characteristics and loss of energy (Gade, 1958); for the case of a clay or mud bottom, the decrease in wave height is of considerable importance. Gade refers to a location on the Louisiana coast where the bottom mud acts as a viscous fluid and the damping action on waves is so pronounced that, in rough weather, the local fishing boats use it as an emergency harbor.

Gade has computed the wave attenuation caused by a layer of mud 1.5 ft thick and 1000 ft long: an 8 sec wave 2.0 ft high traveling in 4 ft of water will be only 37 per cent as high at the end of the stretch of mud as at the start.

In the laboratory, it is sometimes not convenient to use sand or gravel to absorb wave energy. One type of material that has been used with considerable success is stainless steel or aluminum borings. These borings are not packed tightly or too much wave energy would be reflected. The "beach" has a large internal surface area, so that the wave's energy is scattered and ultimately damped by skin friction and eddy viscosity. It should be noted that beaches are not efficient absorbers of the energy of very long waves, and the energy of these waves is largely reflected. Some tests performed using aluminum borings showed that the reflected wave height was only 6 per cent of the incident wave height for waves with periods less than 1.7 sec, but then the reflected wave height increased rapidly with increasing wave period, being 25 per cent for a wave period of 2.2 sec (Paulling, 1957).

Many other types of highly porous wave absorbers have been tried. For a detailed study of many of these, the reader is referred to Straub, Bowers, and Herbich (1958).

At the other extreme are slopes consisting of impervious material, such as a board in a wave tank. Some laboratory measurements on the relationships among wave steepness, slope, and reflection characteristics are given in Fig. 2.33.

2. GRAVITY STRUCTURES

The most common type of gravity structure is the rubble mound, masonry, or similar type of breakwater. Waves may be reflected, or they may break on this type of structure, depending upon the slope of the walls, the wave steepness, and the relative water depth. Some design criteria on this phenomenon are given in Fig. 2.33.

Sunken ships have been used as gravity breakwaters, for example, during the World War II invasion of Normandy. Owing to the shape of a ship's hull, it is less stable against tipping than a flat-bottom caisson which will do the same job. The caisson may be constructed of reinforced concrete, floated to the site, and filled with water until it rests on the bottom. The most famous unit of this type was the Phoenix, used in the Mulberry harbors during the Normandy invasion of World War II (Jellett, 1948; Wood, 1948). These units

were approximately 60 ft wide by 60 ft deep by 200 ft long. The primary effect of this type of structure on waves is to reflect them completely, although some loss of wave energy may occur if the waves break over the top of the caissons at high tide.

3. VERTICAL RIGID THIN BARRIER

One possible type of breakwater consists of a thin rigid vertical barrier extending from above the water surface to some distance below the water surface, such as siding mounted on a pile structure. One of the problems in the design of such a structure is the determination of the distance below the free surface to which such a structure must extend in order to function effectively.

Ursell (1947) developed a theory for the partial transmission and partial reflection of uniform periodic water waves in deep water for a fixed vertical, infinitely thin barrier extending from the water surface to some depth below this surface. The solution was

$$\frac{H_T}{H_I} = \frac{K_1(2\pi h/L)}{\sqrt{\pi^2 I_1^2(2\pi h/L) + K_1^2(2\pi h/L)}}, \qquad (6.1)$$

where H_T is the transmitted wave height, H_I is the incident wave height, I_1 $(2\pi h/L)$ and K_1 $(2\pi h/L)$ are modified Bessel functions; h is the distance from the water surface to the bottom of the structure, and L is the wave length. The ratio H_T/H_I is plotted in Fig. 6.1 together with the experimental data.

Wiegel (1960) developed a theory for water of any depth, based upon the assumption that the power being transmitted by a wave between the bottom of the vertical

Fig. 6.1. Comparison of measured and theoretical wave height transmission coefficient for a vertical rigid thin barrier immersed to various distances below the water surface (from Wiegel, 1960)

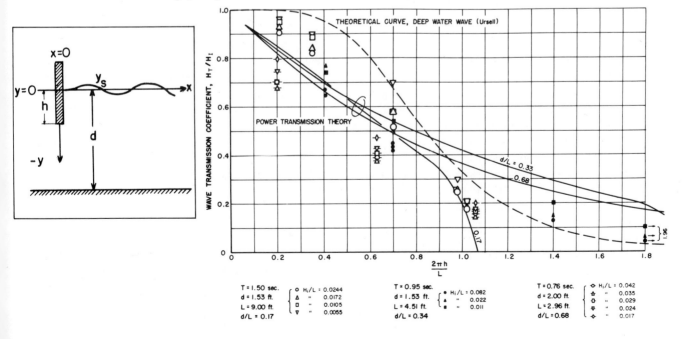

barrier and the ocean bottom, were the structure not there, will be the power transmitted past the structure. The solution is in terms of modified Bessel functions, but the higher-order terms account for only a small percentage of the total power transmitted. A simplified solution is

$$\frac{H_T}{H_I} = \sqrt{\frac{P_T}{P_I}} = \sqrt{\frac{\dfrac{4\pi(y+d)/L}{\sinh 4\pi d/L} + \dfrac{\sinh 4\pi(y+d)/L}{\sinh 4\pi d/L}}{1 + \dfrac{4\pi d/L}{\sinh 4\pi d/L}}}.$$

$$(6.2)$$

This equation is plotted in Fig. 6.2. In order to compare this with Eq. 6.1, take $y = -h$.

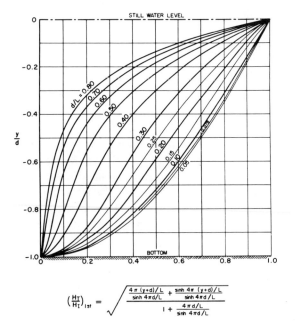

$$\left(\frac{H_T}{H_I}\right)_{1st} = \sqrt{\frac{\dfrac{4\pi(y+d)/L}{\sinh 4\pi d/L} + \dfrac{\sinh 4\pi(y+d)/L}{\sinh 4\pi d/L}}{1 + \dfrac{4\pi d/L}{\sinh 4\pi d/L}}}$$

Fig. 6.2. Ratio of transmitted wave height to incident wave height as a function of rigid barrier depth below still water level; first approximation (from Wiegel, 1960)

Comparison of theory and measurements made in the laboratory is shown in Figs. 6.1 and 6.3. It appears that the power transmission theory is useful to the engineer, but that improvements in the theory are needed. The trend of decreasing values of the transmission coefficient (H_T/H_I) with increasing wave steepness probably occurs because, all other conditions being equal, the water particle velocity increases as the wave steepness increases so that the energy loss due to separation at the bottom of the barrier would increase with increasing wave steepness resulting in a decrease of the transmission coefficient.

Except under certain special circumstances, a rigid barrier extending only a short distance below the surface does not appear to be useful as a breakwater. This may be attributed to the fact that the ratio of the transmitted wave height to incident wave height is proportional to

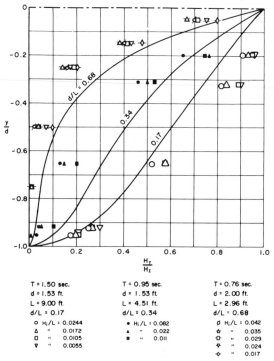

Fig. 6.3. Comparison of measured transmitted wave height and the first order power transmission theory (from Wiegel, 1960)

the square root of the ratio of the transmitted wave power to the incident wave power. Because of this, if only 4 per cent of the power is transmitted, 20 per cent of the wave height is transmitted.

Stelzriede and Carr (1951) made a model study (1:40 scale) with a system consisting of three vertical rigid thin barriers rigidly connected one to the other with the horizontal distance between the first and second barriers being 73 ft, prototype, and 127 ft between the second and third barriers. The two compartment lengths, 73 ft and 127 ft, are in the ratio $1:\sqrt{3}$, an irrational number, This was chosen because previous tests on floating structures of similar design had shown that resonant wave oscillations could occur in the compartments and it was desired to eliminate any overtones between the two compartments. The distances from the bottom of the wave tank and the three vertical barriers were 15, 5, and 15 ft, respectively, prototype. The relationship between H_T/H_I and λ/L (λ being 200 ft) was exponential, with H_T/H_I being approximately 0.55, 0.38, 0.15, 0.075, 0.025 for λ/L of 0.25, 0.5, 0.75, 1.0, and 2.0, respectively.

A modification of the continous vertical barrier just considered has been studied in the laboratory (De Long Corporation, 1959). The device consists of a series of louvers mounted horizontally. One model had fixed louvers and another model had swinging louvers. The effectiveness of both types depended upon the ratio of their areas to the spacing between the louvers. In order to be effective, they had to resemble a vertical wall.

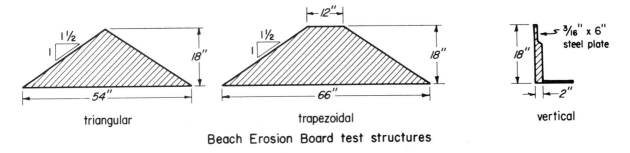

Beach Erosion Board test structures

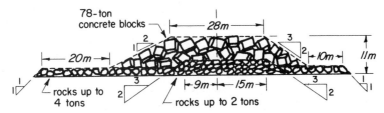

typical profile of Stucky and Bonnard structure (prototype dimensions)

(a)

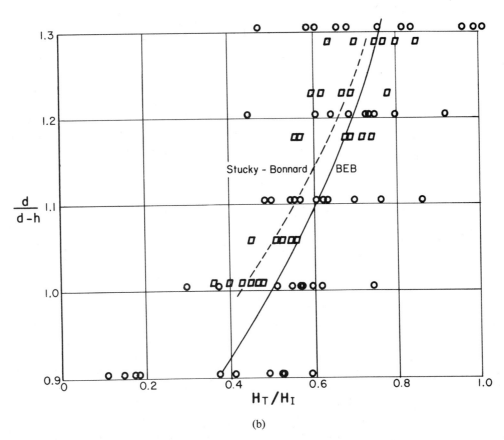

(b)

Fig. 6.4. (a) Types of submerged breakwaters tested; (b) Comparison of Beach Erosion Board (○) and Stucky and Bonnard (□) data with trapezoidal structures (after Hall and Hall, 1940)

4. SUBMERGED BREAKWATER

A *submerged breakwater* is one with its top below the still water level. Part of the wave energy is reflected seaward. The remaining energy is largely dissipated in a breaker, transmitted shoreward as a multiple crest system (Johnson, Fuchs, and Morison, 1951), or transmitted shoreward as a simple wave system.

The simplest case of a submerged barrier is a thin rigid vertical barrier extending from the bottom to a distance h beneath the still water level. Dean (1945) studied theoretically the case for water infinitely deep and found the transmission coefficient H_T/H_I to be a function of $4\pi^2 h/gT^2$ (which is also $2\pi h/Lo$). The results are given in Table 6.1.

Table 6.1. TRANSMISSION COEFFICIENT FOR THIN RIGID VERTICAL BARRIER EXTENDING FROM BOTTOM TO A DISTANCE h BENEATH THE SURFACE IN WATER INFINITELY DEEP

$\dfrac{4\pi^2 h}{gT^2}, \dfrac{2\pi h}{Lo}$	0	0.0125	0.025	0.05	0.1	0.25	0.5	0.75	1.00
$\dfrac{H_T}{H_I}$	0	0.573	0.637	0.710	0.792	0.900	0.964	0.992	0.999

For water that is not infinitely deep, Eq. 6.2 can be modified to give the ratio of transmitted to incident wave height for the case of a submerged breakwater. It is

$$\frac{H_T}{H_I} = \sqrt{1 - \frac{\dfrac{4\pi(y+d)/L}{\sinh 4\pi d/L} + \dfrac{\sinh 4\pi(y+d)/L}{\sinh 4\pi d/L}}{1 + \dfrac{4\pi d/L}{\sinh 4\pi d/L}}}. \tag{6.3}$$

In order to compare results from this equation with Table 6.1 take $y = -h$.

It can be verified easily that this is the same as the equation due to Johnson, Fuchs, and Morison (1951). A few laboratory tests have been made by Hall and Hall (1940), but they were too few to check the theory, except roughly. These results were very interesting in that tests were made of vertical, triangular, and trapezoidal submerged breakwaters (Fig. 6.4), and within the limits of the tests, the thin vertical structure appeared to be a little more effective than the other two forms. The scatter of data was considerable, as can be seen in Fig. 6.4. A good deal of the scatter may be attributed to the use of several values of d/L and H/L in the tests; there were not enough runs, however, for any one constant value of each of the parameters to allow an analysis from this viewpoint. In this figure, the height of the top of the structure above the bottom is $d - h$.

Johnson, Fuchs, and Morison (1951) presented the results of model tests of a submerged rectangular breakwater. One of the experimental setups was as shown in Fig. 6.5(a). The data are compared with Eq. 6.3 and the theory of Jeffreys (1944). Equation 6.3 does not take into

consideration the width of the submerged structure, whereas Jeffreys' theory does; Jeffreys' theory, however, is for shallow water only. It is interesting to note in Fig. 6.5(b) that the data for the flatter wave lie close to the theory of Jeffreys. There was no experimental evidence of the sharp peak of Jeffreys' theory; it is a sharply tuned peak, however, and it might exist.

Hamilton (1950) made laboratory studies of the effect of fixed inclined barriers on waves. The results of the relationship between wave transmission and geometry of the barrier are shown in Fig. 6.6.

The effectiveness of the foregoing types of submerged breakwaters is due to their ability to reflect wave energy, to dissipate wave energy, and to transfer wave energy to other frequencies. Another mechanism of a submerged structure (which might well be just a pile of coarse sand) is utilization of refraction in a manner to concentrate wave energy in an area where it is not harmful and to cause the associated area of low energy density to be at the entrance to a harbor. Costa and Perestrelo (1959) present some thoughts on this subject. Laboratory studies by Mobarek (1962) have shown that their conclusions are essentially valid, but that the submerged structure would have to be so close to the surface to be effective that it would probably be a navigation hazard.

5. PILE ARRAY

Under some conditions, it is desirable to build a combination pile-supported pier and breakwater. If this is done, what would be the effect of the piles alone? Consider a single row of piles D ft in diameter with a distance b between piles. If the assumption is made that the portion of the power transmitted through the pile structure is proportional to the portion of gaps between the piles, then

$$\frac{H_T}{H_I} = \sqrt{\frac{P_T}{P_I}} = \sqrt{\frac{b}{D+b}}. \tag{6.4}$$

As an example, if 14-in. diameter piles were placed with 10 in. between piles, $H_T/H_I = 0.63$. For a spacing of 3 in. between the same sized piles, $H_T/H_I = 0.42$, and for a spacing between the piles of only $1\frac{1}{2}$ in. $H_T/H_I = 0.31$. A 1:12 scale model study for these spacings showed the values of H_T/H_I to be 0.83, 0.53, and 0.43, respectively (Wiegel, 1961a). The measured transmitted wave height was almost 25 per cent greater than the transmitted wave height predicted by this simple theory (analogous to the geometric optics approach) because of diffraction effects (see Fig. 8.18). A few tests made with the line of piles at a 45-degree angle with the waves showed this configuration to be a less effective breakwater than for the case of the pile line parallel to the wave crests.

When an array of piles is used which has more than

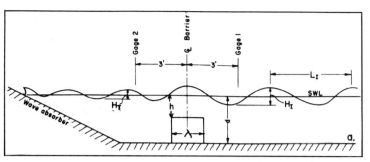

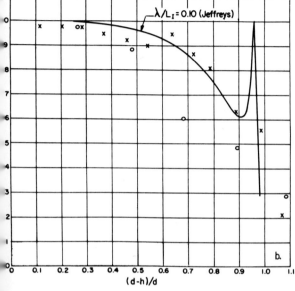

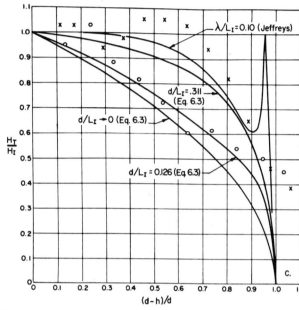

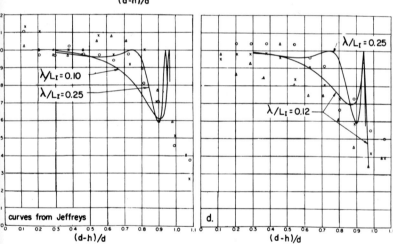

Fig. 6.5. (a) Experimental set-up
(b) Effect of wave steepness on wave action over a rectangular underwater barrier
$\times$: $H_I/L_I = 0.0295$, $d/L_I = 0.137$, $\lambda/L_I = 0.091$;
$\bigcirc$: $H_I/L_I = 0.0583$, $d/L_I = 0.144$, $\lambda/L_I = 0.11$.
(c) Effect of relative depth on wave action over a rectangular underwater barrier
$\times$: $H_I/L_I = 0.052$, $d/L_I = 0.311$, $\lambda/L_I = 0.101$;
$\bigcirc$: $H_I/L_I = 0.055$, $d/L_I = 0.126$, $\lambda/L_I = 0.906$.
(d) Effect of relative width of a structure on wave action over a rectangular underwater barrier (*left*) $H_I/L_I = 0.0195$, $d/L_I = 0.106$, $\lambda/L_I = 0.039$ ($\bigcirc$), 0.07 ($\times$), 0.105 ($\triangle$); (*right*) $H_I/L_I = 0.0927$, $d/L_I = 0.371$, $\lambda/L_I = 0.12$ ($\bigcirc$), 0.24 ($\times$), 0.36 ($\triangle$).
(after Johnson, Fuchs and Morison, 1951)

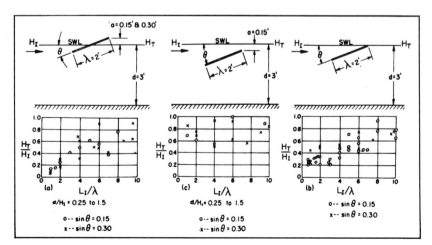

Fig. 6.6. Transmission coefficients for submerged, fixed, inclined barriers (after Hamilton, 1950)

133

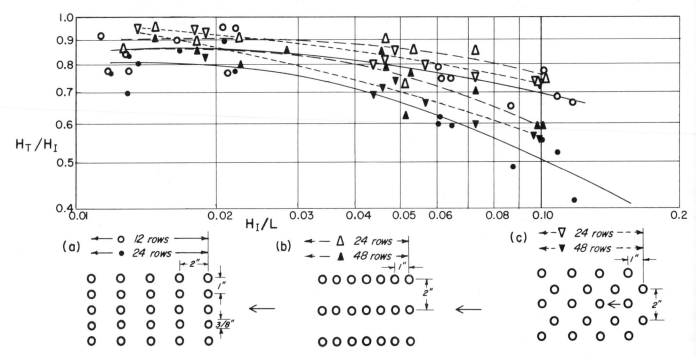

Fig. 6.7. Effect of incident wave steepness on wave transmission (from Costello, 1952)

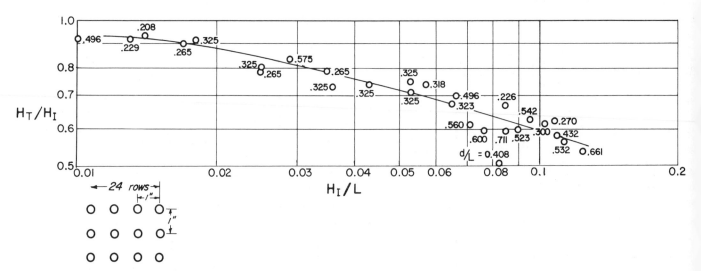

Fig. 6.8. Relationship among transmission coefficient, wave steepness and relative depth (from Costello, 1952)

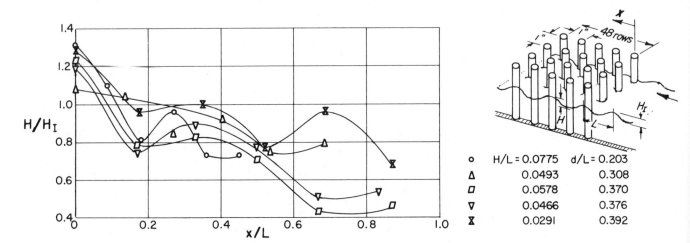

	H/L = 0.0775	d/L = 0.203
○	0.0775	0.203
△	0.0493	0.308
▢	0.0578	0.370
▽	0.0466	0.376
✕	0.0291	0.392

Fig. 6.9. Wave attenuation within a pile array (after Costello, 1952)

one row, the phenomenon becomes complicated. The second row of piles transmits a portion of the energy that gets through the first row of piles and reflects a portion of the energy, the amounts depending upon the geometry of the pile layout. The reflected portion is partially transmitted back through the first row of piles and partially reflected by it, but with a phase lag between this and the parent wave, this "wave" now being transmitted through the second row of piles and partially reflected, etc. In addition, a portion of the energy is scattered, the scattered wave having a frequency that depends upon the pile diameter, and a portion of the energy is dissipated by skin drag and form drag. Costello (1952) measured the effect of various arrays of piles on the transmission of wave energy (Fig. 6.7). The total number of piles in Fig. 6.7(b) is the same as in Figs. 6.7(a) and 6.7(c), even though the number of rows of piles in the direction of wave travel is only one-half the number of rows in the other two figures, because of the pile spacing. For a given number of piles, there does not appear to be any appreciable difference in the effect of the various array configurations upon the effectiveness of the structure as a breakwater.

The effect of wave steepness on the transmission coefficient can be seen in Figs. 6.7 and 6.8. In the setup shown in Fig 6.8, $37\frac{1}{2}$ per cent of the space parallel to the wave front is occupied by the front row of piles. This one row of piles should transmit approximately 80 per cent of the wave height, so it appears that, for waves of low steepnesses, the additional rows of piles have little effect for this configuration. The effect of wave steepness is considerable, as can be seen in the figure. When the number of rows of piles was doubled, the effect of the additional rows became appreciable (Fig. 6.7), but the effectiveness of the system as a breakwater is not good. The effect of the individual rows of piles upon the transmission coefficient can be partially assessed from a study of Fig. 6.9. It is important to note that the wave height immediately seaward of the first row of piles is higher than the incident wave by from 10 to 30 per cent, this being due to reflections. This results in a partial standing wave at the pile, with the accompanying increase in water motions at the bottom.

The type of structure just considered might be generalized as a porous structure. Many such types were experimented with before the invasion of Normandy during World War II. The results of large laboratory experiments on a large number of types of such breakwaters are given in Table 6.2, the dimensions being projected to the prototype (Todd, 1948).

6. FIXED SURFACE STRUCTURE

Stoker (1957, p. 432) (also see Stoker, Fleishman, and Weliczker, 1953) has treated the case of a rigid board of length λ fixed at the still water surface in shallow water, using linear wave theory. It was found that

$$\frac{H_T}{H_I} = \frac{1}{\sqrt{1 + (\pi\lambda/L)^2}}, \tag{6.5}$$

$$\frac{H_R}{H_I} = \frac{\pi\lambda/L}{\sqrt{1 + (\pi\lambda/L)^2}}, \tag{6.6}$$

where H_R is the reflected wave height. For $\lambda/L = 0.5$, 1.0, and 2.0, $H_T/H_I = 0.54$, 0.30, and 0.16, respectively.

The pressure under the board is

$$p(x, t) = p_1(x) \cos \frac{2\pi t}{T} - p_2(x) \sin \frac{2\pi t}{T}, \tag{6.7}$$

where

$$p_1(x) = \frac{1}{2} \rho g H \left[b_1(x) \sin \frac{\pi\lambda}{L} + b_2(x) \cos \frac{\pi\lambda}{L} \right]$$

$$p_2(x) = \frac{1}{2} \rho g H \left[b_2(x) \sin \frac{\pi\lambda}{L} - b_1(x) \cos \frac{\pi\lambda}{L} \right]$$

$$b_1(x) = \frac{(\pi\lambda/L)^2}{1 + (\pi\lambda/L)^2} \cdot \frac{2x}{\lambda} + 1$$

$$b_2(x) = \frac{\pi\lambda/L}{1 + (\pi\lambda/L)^2} \cdot \frac{2x}{\lambda}$$

where $x = 0$ is located at the midpoint of the rigid board with $x = +\lambda/2$ being the edge of the board facing the incident waves. The pressure varies linearly in x at any time t, the pressure variation at the leading edge is greater than the pressure variation at the trailing edge, and the pressure amplitude is linearly related to the wave height.

The foregoing equations are for shallow water waves and an infinitely thin board. It is interesting to compare Eq. 6.7 with the pressure variation under a board of considerable thickness in water which is not shallow (Fig. 6.10). The board was placed in a wave tank so that the bottom of the board was below the wave troughs. The pressures, P, shown in Fig. 6.10(a–c) are the maximum total pressures in feet of water, occurring at each location during one cycle, divided by the incident wave height. Several conclusions can be drawn: (1) the pressure at the leading edge can be higher than the incident wave height since part of the wave energy is reflected; (2) the pressure for waves of small steepness is lower than predicted from Eq. 6.7 because the bottom of the board is beneath the still water surface; (3) the higher waves produce a higher ratio of P/H, all other conditions being equal. The data in Fig. 6.10(d–f) show that the pressure is linearly related to wave height.

Although the measurements were made for relatively deep water, most structures would be placed in shallow water. The total lift on such a structure could be obtained from Eq. 6.7 by integrating the equation with respect to x.

Carr, Healy, and Stelzriede (1950; see also Carr, 1952) developed an equation for shallow water based upon

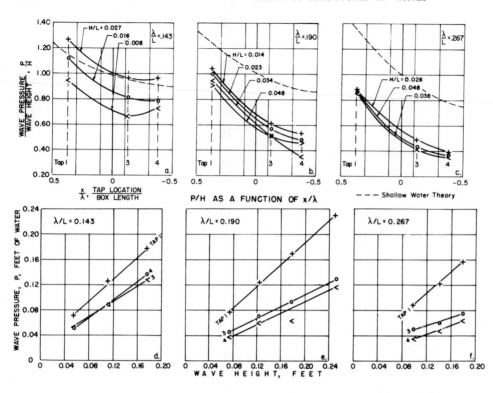

Fig. 6.10. Wave pressure as a function of wave height for constant λ/L (data from Beebe, 1954)

Table 6.2. PARTICULARS OF WAVE BREAKER MODELS
(after Todd, 1948)

Principal Features	Wave length (feet)	Wave height trough to crest (feet)		Remarks
		Ahead of breaker	Behind breaker	
1 Floating grids of brushwood, logs, etc.	160	6–10	6–10	Grids moved with passage of wave and had no noticeable damping effect.
2 Triangular section, both faces of brushwood	160	10.0	0.75	Practically no wave motion behind breaker; both sides of breaker were moving in and out, or "breathing," and there was some disturbance in consequence.
3A Seaward face of brushwood, rear face open at top, solid at bottom	160	10.0	2.5	This breaker had considerable damping effect, but was not so efficient as No. 2.
3B As 3A, but with rear face solid to top	160	10.0	0.3	There was little movement behind the breaker—it was due to the structure "breathing" a little.
4A Lower part solid, upper part open, with grids, on both faces	160	10.0	4.0	The wave motion immediately behind breaker was much reduced, but farther back the long swell appeared again, and reached a height of 6–7 ft.
4B As 4A, with grids on rear face covered, and rear face open at bottom	160	10.0	3.5	Confused sea immediately behind breaker, re-forming into 6–7 ft waves again.
5A Triangular supporting structure at rear, with vertical series of horizontal slots built up in front	160	10.0	5.0	Wave reduced to about half height, and confused sea caused by spilling effect.
5B As 5A, but with forward face of triangle boarded in	160	10.0	2.0	No waves could now pass through, since face was now solid. Waves came over top and caused small waves behind.
6A Simple triangular cross section, completely submerged	160	10.0	9.0	Wave crests tended to break just as they passed over breaker, but very little diminution of height.

Table 6.2. (Cont.)

Principal Features	Wavelength (feet)	Wave height trough to crest (feet)		Remarks
		Ahead of breaker	Behind breaker	
6B As 6A with vertical flap hinged to top edge of triangular section, the movement of flap being controlled by buoyant cylinders attached to it on rearward side	160	10.0	1.67	Flap leaning backward, and water at rest 20 in. above bottom edge. The movement of flaps was restrained to a large extent by the presence of the cylinders behind, in the relatively calm water, and this small movement generated only small waves behind the breaker. Standing waves of some height were formed in front of the breaker due to reflection.
7A Solid face and back, with forward-curved parapet at top	160	10.0	nil	Waves reflected from face, no disturbance behind. Breaking water at crests was thrown back and did not go over breaker.
7B As 7A, but with circular holes in sea and rearward faces	160	10.0	0.25	Water spilled through first row of holes above water, causing ripples behind breakwater. On face, large quantities of water were thrown out of holes when crest was next to breaker.
8A Beach type with slots	160	10.0	5.0	Immediately behind breaker there was a confused sea. The waves re-formed and at 500 ft back there was a swell similar to that in front, and 5–7 ft high.
8B As 8A, but with lower rows of slots filled in to give complete beach	160	10.0	3.0 increasing to 5 ft aft of breaker.	Similar to 8A.
9 Similar to 8 with equal slope each side	160	10.0	3.0	Waves reduced by breaker, but re-formed into swell again on moving away.
10A Two vertical slotted walls	160	10.0	7.0	Waves reduced, but re-formed farther on.
10B As 10A, with interior of wall filled with brushwood	160	10.0	3.3 increasing aft of breaker.	Considerable reduction, but brushwood was evidently not packed tightly enough, and moved about between the walls.
10C As 10B, with brushwood more firmly packed	160	10.0	Small to 6.7 ft.	Waves re-formed to about two-thirds of original height.
11 Triangular base with vertical wall on top	160	10.0	9.0	Practically no reduction in wave height.

John's (1949) work, which is similar to Eq. 6.5, but which includes a term to express the effect of finite draft.

$$\frac{H_T}{H_I} = \frac{1}{\sqrt{1 + \left[\frac{\pi\lambda}{L}(1 + S)\right]^2}}, \qquad (6.8)$$

where S is the ratio of the depth of the bottom of the immersed body below the undisturbed water level to the vertical distance from the bottom of the immersed body and the bottom of the water. H_R/H_I can be obtained from this, using the relationship that

$$\left(\frac{H_T}{H_I}\right)^2 + \left(\frac{H_R}{H_I}\right)^2 = 1, \qquad (6.9)$$

which holds, provided that no energy is lost or transferred to other frequencies in the process. A comparison with laboratory measurements is shown in Fig. 6.11, which gives measured data on both transmitted and reflected wave heights. The measured results do not fulfill Eq. 6.9, difference being due to energy dissipated in the process.

7. FLOATING STRUCTURES

The most common floating structure is a ship; because of its shape, however, its effect on waves is one of the most difficult to develop theoretically. Hence, mathe-

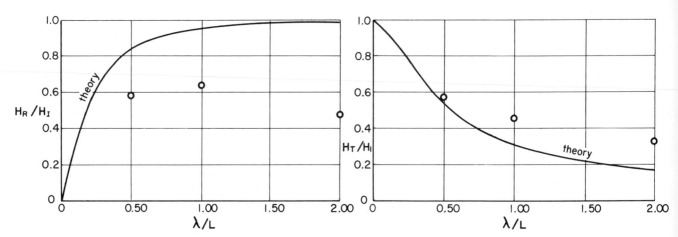

Fig. 6.11. Reflection and transmission coefficients related to the ratio of length of fixed body to wave length (after Carr, Healy, and Stelzriede, 1950)

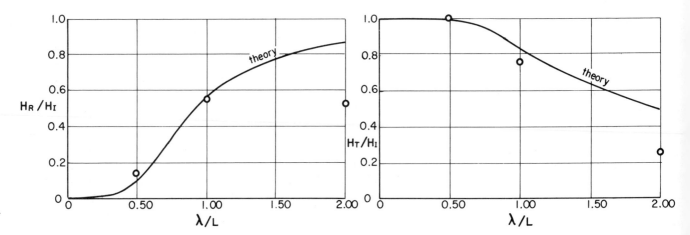

(a) Theoretical and experimental reflection coefficients   (b) Transmission coefficients

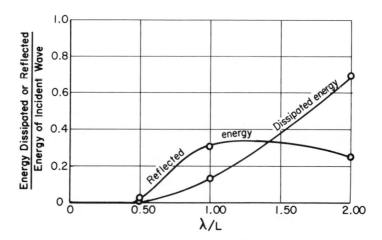

(c) Comparison of reflected and dissipated wave energy

Fig. 6.12. Effect of freely floating body on waves (after Carr, Healy, and Stelzriede, 1950)

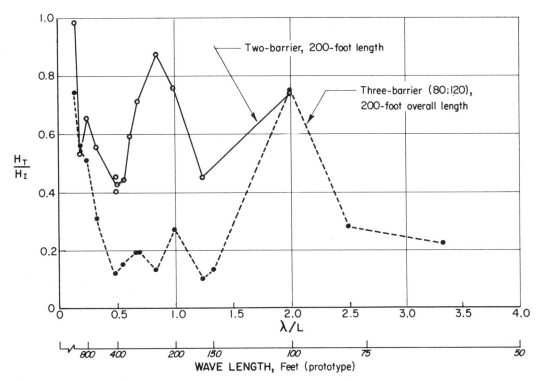

Fig. 6.13. Effect of freely floating spaced barriers on wave transmission; bottom clearance: 5 feet (after Stelzriede, 1951)

matical studies have usually been made of simpler shapes. John (1949) treated the case of cylinders floating at the surface under the action of waves represented by the linear theory. The case of a flat structure was taken to be a section of cylinder of infinite diameter. In shallow water, the coefficient for an infinitely thin structure is

$$\frac{H_R}{H_I} = \frac{(\pi\lambda/L)^5}{45\left\{\left[\left(1 - 0.4\left(\frac{\pi\lambda}{L}\right)^2\right)^2 + \left(\frac{\pi\lambda}{L}\right)^2\left(1 - \frac{(\pi\lambda/L)^2}{15}\right)^2\right]\left[\left(1 - \frac{(\pi\lambda/L)^2}{3}\right)^2 + \left(\frac{\pi\lambda}{L}\right)^2\right]\right\}^{\frac{1}{2}}}$$

(6.10)

where H_T/H_I can be obtained from Eq. 6.9.

The results of laboratory experiments by Carr, Healy, and Stelzriede (1950) are compared with theory in Fig. 6.12a,b. The energy associated with the measured reflected wave is shown in Fig. 6.12(c), together with the curve representing the difference between the incident wave energy and the sum of the transmitted and reflected wave energies, this difference being dissipated by some mechanism, or mechanisms, in the process. For the conditions tested, a considerable portion of the incident wave energy was dissipated for large values of λ/L.

Stelzriede (1951) and Stelzriede and Carr (1951) made a model study (1 : 40 scale model) with a system consisting of rigid vertical barriers connected one to another, as described in Section 6. In one set of tests, two vertical barriers were used 200 ft apart (prototype), the barriers

having sufficient freeboard to prevent waves from going over the top of them. This resulted in a secondary peak transmission at a ratio of λ/L of about unity (Fig. 6.13). A third barrier was placed 80 ft from the leading edge, with the result that the original peak no longer existed, but with the addition of another peak at a smaller wave length.

Many other tests were conducted, including one with a reduced freeboard so that waves would either break over the top or flow over the top. No significant trends were noted.

The effects of various moorings of the system are shown in Fig. 6.14, together with a rigid and a freely floating system. It is difficult to make comparisons as not enough information on elasticity of the moorings, natural periods of the systems, etc., was given.

Model tests have been made of a similar structure, including the measurement of forces in the mooring system by Ross (1957).

A floating breakwater to be effective must extend deep enough into the water that little wave power can be transmitted beneath it, and it must have the mass and damping characteristics necessary to prevent it from moving extensively so it will not transmit wave power as if it were a portion of the water. In order to fulfill the latter requirement, the floating breakwater must have natural periods that are large compared with the wave periods to which it will be subjected. This means that its mass should be large, its viscous and form damping large,

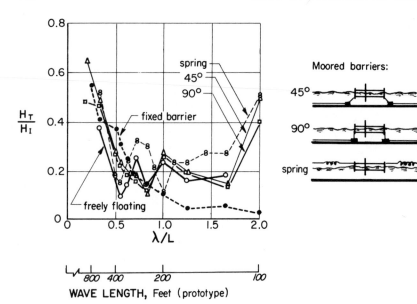

Fig. 6.14. Effect of moored floating spaced barriers on wave transmission (after Stelzriede and Carr, 1951)

and its "elasticity" small. Its "elasticity" is represented by its change in buoyancy as it heaves, rolls, and pitches. In order to make the "elasticity" small and the mass large at the same time the bulk of the breakwater must be below the water surface at all times. A moored structure has an additional elastic restraining force due to the mooring lines, and these are primarily effective with respect to the surging, swaying, and yawing motions. The mass of the structure to be considered is its virtual mass, which includes the added mass term (see Chapter 11). A solution of this problem is the Bombardon floating breakwater designed for, and used in, the Normandy invasion of World War II (Lochner, Faber, and Penney, 1948). This breakwater was constructed of steel plate in sections and was cruciform in cross section. The buoyancy of the structure was such that the cross arms always remained below the water surface. This was done by partially filling the structure with water. The controlling factor in the use of such a structure appears to be the moorings. The storm that commenced on June 19, 1944, was nearly the worst in history for that time of year and the waves were so high that the moorings were stressed to from six to eight times the values considered in their design, and they failed (Lochner, 1948). The other problem was the large torsional stresses induced in the

structure by waves approaching at a considerable angle rather than the wave crests being parallel to the structure (Lee, 1948).

The case of a beam of finite specific weight and finite elasticity in shallow water has been studied mathematically by Stoker, Fleishman, and Weliczke (1953; see also Stoker, 1957). Linear wave theory was used. The resulting equations were solved numerically for a number of conditions so that some information could be obtained on the effect of the various parameters on the wave field. The case of a semi-infinite beam was considered first, extending from $x = 0$ to $x = -\infty$. The numerical results in Table 6.3 were calculated from tabulations in Stoker's book for the case of water 40 ft deep. Almost all of the wave height is transmitted.

A relatively lightweight floating structure does not appear to be very useful as a reflector of waves, nor does one of small moment of inertia. It is interesting to note the relative ineffectiveness of a semi-infinite beam as a wave reflector compared with a beam of finite length. Table 6.4 presents the transmission coefficients for two wave lengths. The irregularity of the coefficient stands out. If these tables predict correctly the behavior of an actual floating structure, the effectiveness in a wave system of various wave lengths as a breakwater is small.

Table 6.3. REFLECTION COEFFICIENT AS A FUNCTION OF WEIGHT OF SECTION, MOMENT OF INERTIA, AND MODULUS OF ELASTICITY IN WATER 40 FT DEEP

Wave period, T, (sec)	3.14	4.19	6.28	8.00	15.0	8.00	15.0	8.00	15.0	8.00	15.0
H_T/H_I	0.90	0.92	0.97	0.98	0.99	0.66	0.86	0.92	0.97	≈ 1	≈ 1
W (lb/ft³ per foot width)	85	85	85	85	85	384	384	85	85	85	85
I (ft⁴)	0.20	0.20	0.20	0.20	0.20	0.20	0.20	2.0	2.0	0	0
E (lb/ft²)	4.37×10^9	4.37×10^9	4.37×10^9	4.37×10^9	4.37×10^9	4.37×10^9	4.37×10^9	4.37×10^9	4.37×10^9	0	0

Table 6.4. Reflection Coefficient As a Function of Beam Length

$W = 85$ lb/ft³ per ft width, $I = 0.20$ ft⁴
$E = 4.37 \times 10^9$ lb/ft², and $d = 40$ ft

$2\pi/T = 0.785$

L/λ	14.3	5.08	3.44	2.54	1.77	1.27	0.86	0.56	0.48	0.38	0.29	0
$\dfrac{H_T}{H_I}$	1.00	0.36	1.00	0.66	0.99	1.00	0.94	0.95	0.99	0.99	0.95	0.98

$2\pi/T = 0.418$

L/λ	3.45	2.55	1.72	1.13	0.86	0.77	0.581	0.579	0.54	0
$\dfrac{H_T}{H_I}$	0.98	0.85	0.99	0.44	0.67	0.78	≈ 1	≈ 1	0.78	0.99

A study of Table 6.3 would lead one to expect that a thin lightweight sheet of flexible plastic floating on the water surface would be ineffective as a breakwater. This, however does not appear to be the case. Studies were made at the suggestion of E. R. Vorenkamp (Wiegel and Friend, 1958) using sheets of plastic floating on the water surface in a wind-wave tank, with the waves being generated by wind blowing over the water surface. The tank was about 60 ft long; sheets of plastic 1, 2.5, 5, or 10 ft in length were mounted in about the center of the tank with the plastic sheets floating on the surface and extending the width of the tank. The water depth was about $\frac{1}{2}$ ft. The advantage of using wind-generated waves is that they have the same type of spectrum that exists in the ocean. In presenting the data, the significant wave length and height are used. The effect of absolute length of the sheet of plastic is shown in Fig. 6.15, but the scatter is such that no trend is evident. The effect of different materials is shown in Fig. 6.16. It appears that to reduce the wave height to $\frac{1}{4}$ of the incident wave height, a sheet of plastic about twice as long as a sheet of plywood would be needed. The mechanism by which the sheets of plastic affect the wave field is not understood at present.

Another device which acts as a breakwater is a plastic bag filled with water, the ensemble being of zero buoy-

Fig. 6.15. Effect of length of floating plastic sheet on wave transmission (from Wiegel and Friend, 1958)

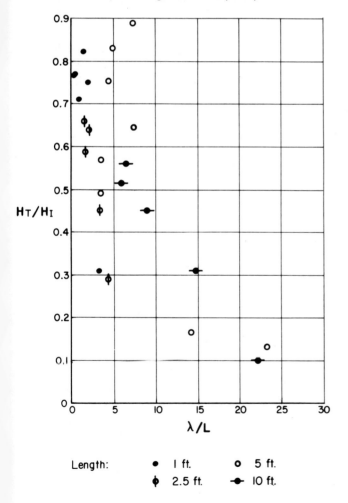

Length: ● 1 ft.   ○ 5 ft.
        ϕ 2.5 ft.   ●─ 10 ft.

Fig. 6.16. Effect of different types of floating sheets on wave transmission (from Wiegel and Friend, 1958)

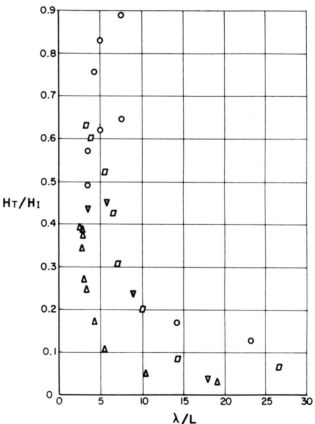

○ plastic   ▽ foam rubber
△ plywood   ▱ laminated Ensolite & foam rubber

5 feet in length, with holes

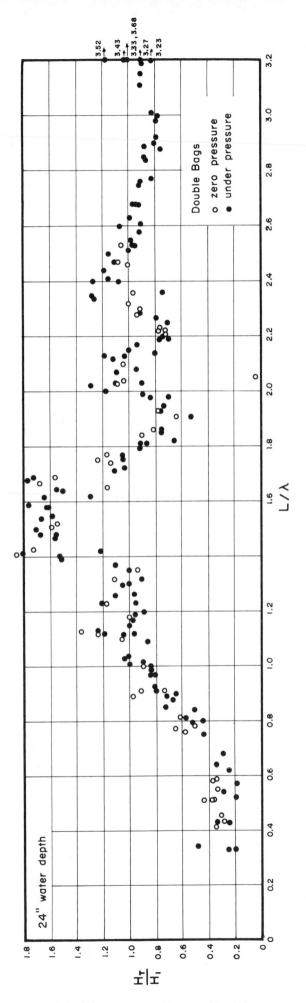

Fig. 6.17. Effect of water depth and ratio of wave length to water mattress length on wave transmission coefficient

ancy. A bag 10 ft by 10 ft by 4 in. thick was moored in a model basin and subjected to uniform, periodic waves (Wiegel, 1959). The wave heights were measured about 5 ft to either side of the bag at the front, and about 5 ft to the rear of the bag along the centerline of the bag. The effect of two bags, one on top of the other, is shown in Fig. 6.17. One set of points is for the water in the bags under zero pressure gauge, whereas the other set of points is for the water in the bags under pressure. It appears that a bag about twice the wave length should be fairly effective as a breakwater.

Laboratory tests of the forces in the mooring lines show that a single bag extending throughout the entire depth of water would be better than a bag that does not touch bottom, from a standpoint of mooring the unit.

Six bags 4 ft by 4 ft in cross section by 20 ft long were lashed side by side to form a 24 ft by 20 ft by 4 ft deep breakwater (Wiegel, Shen, and Cumming, 1962). The bags (called a *hovering breakwater*) were moored in San Francisco Bay and filled with bay water. The water depth was 7 ft below MLLW, and the tide range during the tests was approximately 6 ft. The bags were subjected to wind waves with periods from 1 to 2 sec and heights from $\frac{1}{2}$ to $1\frac{1}{2}$ ft. Reduction of the data was done by an IBM 704 computer using share subroutine #574 "CS TUKS." An example of the output is given in Table 6.5, and the spectral energy density of incident and transmitted waves are shown in Fig. 6.18.

The results were computed from $N = 1201$ observations at time intervals $\Delta t = 1/8$ sec. $M = 45$ spectral estimates were made. The symbols used in Table 6.5 are defined as follows:

A = primary record auto covariance
X = primary record energy spectrum
B = secondary record auto covariance
y = secondary record energy spectrum
E = in phase cross correlation
Z = in phase energy spectrum, or co-spectrum
F = out of phase cross correlation
W = out of phase energy spectrum, or qua-spectrum
P = normalized co-spectrum
Q = normalized qua-spectrum
R — coherence
PHI 1 = phase lead of secondary to primary wave

The energy spectrum gives the energy for a frequency band width of $1/2\ M\ \Delta t$ centered about the frequency $f_k = K/2M\ \Delta t$. The first number in each value shows the number of places the decimal is located from the left side of the number (for example, the first value under

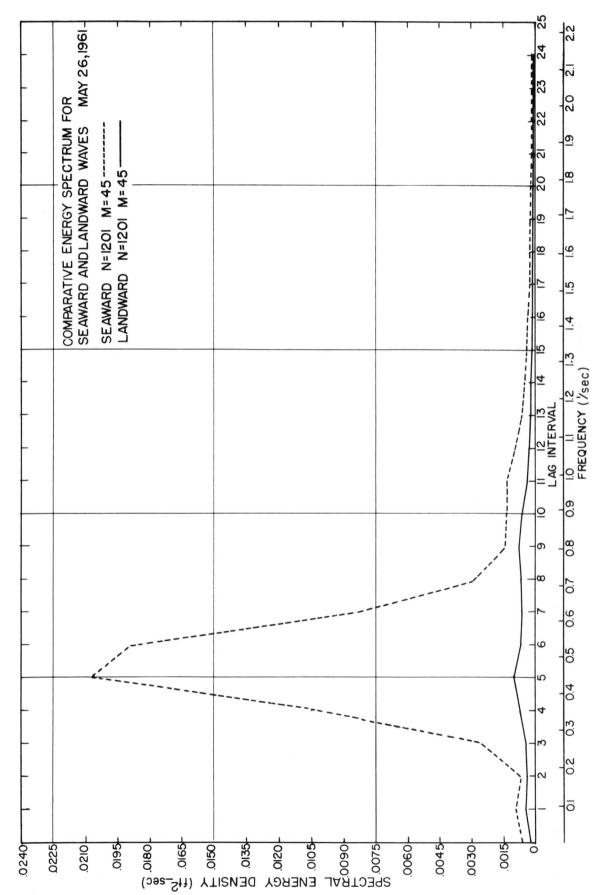

Fig. 6.18. Comparative energy spectrum for seaward and landward waves, May 26, 1961 (from Wiegel, Shen, and Cumming, 1962)

Table 6.5. SAMPLE SPECTRAL ANALYSIS OF FLOATING BREAKWATER TESTS
(from Wiegel, Shen, and Cumming, 1962)

K	A	X	log X	B	Y	log Y	E	Z	F	W	P	Q	R	R*R	PHI1
	3 7446	1 6206	20.79	2 8238	1 2033	20.31	2 1922	1 1356	0000	0000	.38	.00	.38	.15	.0
1	3 6504	1 9282	20.97	2 6360	1 4723	20.67	2 1795	1 2518	1 8379	4135	.38	.06	.39	.15	9.3
2	3 4689	1 7183	20.86	2 3538	1 3948	20.60	2 1481	1 1258	2 1585	−6615	.24	−.12	.27	.07	−27.7
3	3 2378	2 2487	21.40	1 9195	1 4072	20.61	1 9735	−5559	2 2226	−1 8039	−.06	−.01	.06	.00	−171.8
4	1497	3 1035	22.01	1−8417	1 7447	20.87	1 3643	8170	2 2446	1 5338	.03	.19	.19	.04	81.3
5	3−2104	3 2059	22.31	2−1671	1 8426	20.93	1−1951	1 5719	2 2293	1 9491	.14	.23	.27	.07	58.9
6	3−3703	3 1880	22.27	2−1867	1 7091	20.85	1−5868	1 6036	2 1893	1 6353	.17	.17	.24	.06	46.5
7	3−4672	2 8413	21.92	2−1456	1 6073	20.78	1−7413	1 1158	2 1289	1 3192	.05	.14	.15	.02	70.1
8	3−4984	2 2861	21.46	1−7419	1 6468	20.81	1−7655	−5195	1 4426	1 2141	−.04	.16	.16	.03	103.6
9	3−4652	2 1467	21.17	1−1350	1 7602	20.88	1−6370	5050	1−6163	−6434	.05	−.06	.08	.01	−51.9
10	3−3764	2 1279	21.11	1 2406	1 6095	20.78	1−3843	9572	2−1483	1−1774	.11	−.20	.23	.05	−61.7
11	3−2425	2 1230	21.09	1 2061	1 3533	20.55	1−2450	6358	2−2049	−1153	.10	−.02	.10	.01	−10.3
12	2−8257	1 8855	20.95	−9771	1 2287	20.36	9992	−1823	2−2220	5374	−.04	.12	.13	.02	108.7
13	2 8248	1 6601	20.82	1−4188	1 1578	20.20	1 5882	−2354	2−2002	−1−4719	−.07	−.01	.07	.01	−168.7
14	3 2235	1 4198	20.62	1−4633	1 1161	20.06	2 1096	−1 4612	2−1550	−1 4995	.02	.02	.03	.00	47.3
15	3 3234	1 2747	20.44	1−1740	1 1100	20.04	2 1516	−1−7929	1−9466	1362	−.05	.08	.09	.01	120.2
16	3 3692	1 2297	20.36	1 3303	1 1184	20.07	2 1700	−2319	1−3413	−1−6026	−.14	−.04	.15	.02	−165.4
17	3 3617	1 1814	20.26	1 7493	1 1076	20.03	2 1726	−2167	1 2039	−1 5597	−.16	.04	.16	.03	165.5
18	3 3048	1 1580	20.20	2 1068	7648	19.88	2 1630	−1 3899	1 4157	−2 3020	.04	.00	.04	.00	4.4
19	3 2159	1 1673	20.22	2 1187	6088	19.78	2 1188	2404	1 3833	−2525	.24	−.25	.35	.12	−46.4
20	3 1103	1 1414	20.15	2 1124	5354	19.73	1 3466	−1 2814	1 4904	−2231	.03	−.26	.26	.07	−82.8
21	1 2719	1 1144	20.06	1 7845	4212	19.62	1−6134	−1575	1 7646	−1−6793	−.23	−.10	.25	.06	−156.7
22	2−8993	1 1134	20.05	1 3834	3998	19.60	2−1238	−1−9412	2 1026	−1−4039	−.14	−.06	.15	.02	−156.8
23	3−1577	9842	19.99	−7808	3732	19.57	2−1438	−3 2601	2 1468	−1−2501	.00	−.04	.04	.00	−89.4
24	3−1975	7857	19.90	1−5497	3230	19.51	2−1494	−1 4147	2 1757	−1−3780	.08	−.08	.11	.01	−42.3
25	3−2079	7857	19.90	1−7322	2274	19.36	2−1397	−2 7393	2 1888	−1−4982	.02	−.12	.12	.01	−81.6
26	3−1887	7975	19.90	1−5976	1601	19.20	2−1122	−2−4343	2 1691	−1 3192	−.01	.09	.09	.01	97.7
27	3−1476	8355	19.92	1−4422	1960	19.29	1−7982	−1 5097	2 1034	−1 8709	.13	.22	.25	.06	59.7
28	2−9196	7517	19.88	1−3192	1892	19.28	1−3455	−1 6526	1331	−1 6518	.17	.17	.24	.06	45.0
29	2−3206	5828	19.77	1−3126	1275	19.11	4475	−1 1479	1−8919	−2 1079	.05	.00	.05	.00	4.2
30	2 2684	5849	19.77	1−5005	1165	19.07	1 3876	−1−1961	2−1409	−1−2564	−.08	−.10	.12	.02	−127.4
31	2 8012	6319	19.80	1−7549	1643	19.22	1 6991	−1−2459	2−1258	−1 1654	−.08	.05	.09	.01	146.1
32	3 1259	5798	19.76	1−9744	1647	19.22	1 9883	−1−4017	1−9471	−1 4374	−.13	.14	.19	.04	132.6
33	3 1489	5444	19.74	2−1263	1494	19.17	2 1035	−1−4364	1−6210	−3−2439	−.15	−.00	.15	.02	−179.7
34	3 1473	5305	19.72	2−1396	1768	19.25	1 7150	−2 3103	1−3674	−1−2719	.01	−.09	.09	.01	−83.5
35	3 1217	5236	19.72	2−1317	1815	19.26	1 2005	−1 4941	−8647	−1 2703	.16	.09	.18	.03	28.7
36	2 7762	5403	19.73	1−9300	1676	19.22	1−2683	−1 4881	1 1695	−1 5524	.16	.18	.24	.06	48.5
37	2 2185	5166	19.71	1−4570	1464	19.17	1−5076	−1 2314	1 4208	−1 4020	.08	.15	.17	.03	60.1
38	2−3377	4672	19.67	1−1655	1290	19.11	1−6438	−1 1701	1 6507	−1 3067	.07	.12	.14	.02	61.0
39	2−7825	4791	19.68	7989	1338	19.13	1−6515	−2 5496	1 9122	−1 1956	.02	.08	.08	.01	74.3
40	3−1074	5002	19.70	1 3635	1291	19.11	1−5500	−1−1989	1 8604	−1 1072	−.08	.04	.09	.01	151.7
41	3−1179	5619	19.75	1 3931	1081	19.03	1−3343	−1−1469	1 7674	−3−7294	−.06	−.00	.06	.00	−177.2
42	3−1113	5985	19.78	1 2244	−1 9528	18.98	−2989	−2−3085	1 6263	−1−2182	−.01	−.09	.09	.01	−98.0
43	2−8792	4853	19.69	−5040	−1 9601	18.98	1 1479	−2−1075	1 5238	−1−1110	−.00	−.05	.05	.00	−95.5
44	2−5529	4008	19.60	1−2791	1233	19.09	6685	−1 1436	1 3257	−1 3310	.06	.15	.16	.03	66.5
45	2−1621	2028	19.31	1−3834	−1 7413	18.87	−1 4868	−1 1293	6210	0000	.11	.00	.11	.01	.0

A is 744.6). The degree of freedom for each spectral estimate, $2N/M$ is approximately 53.

The hovering breakwater was effective, as can be seen from the spectral energy densities of the incident and transmitted waves (Fig. 6.18). The linear coherence was negligible for nearly all frequencies, except for the very lowest, and the phase difference jumped around. An examination of the normalized co- and quadrature spectra and the coherence show that the incident and transmitted wave records are unrelated, if they are considered to be linear. They are related, of course, but through a nonlinear process.

The mechanism, or mechanisms, by which the hovering breakwater attenuates the waves was not determined, but reflection of wave energy was not significant. A portion of the energy was transferred to waves of other frequencies.

8. RESONANT STRUCTURE REFLECTORS

Waves can be reflected by a resonator tuned to the wave period, the structure being installed on one side or both sides of a channel (Fig. 6.18). The length of the resonator λ must be one-quarter the length of the wave

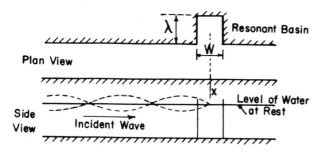

Fig. 6.19. Standing wave created by resonator tuned to the wave period (after Valembois, 1953)

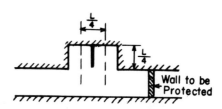

Fig. 6.20. System of two resonant basins one-quarter wave length apart (after Valembois, 1953)

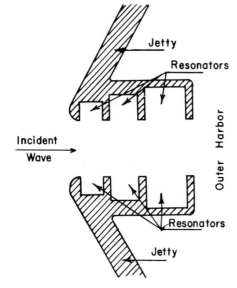

Fig. 6.21. Protection of an outer harbor (after Valembois, 1953)

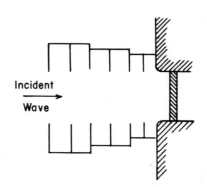

Fig. 6.22. Protection of a reflecting structure (after Valembois, 1953)

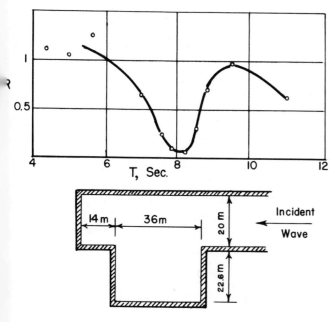

Fig. 6.23. Action of a resonant basin on the clapotis at a wall (after Valembois, 1953)

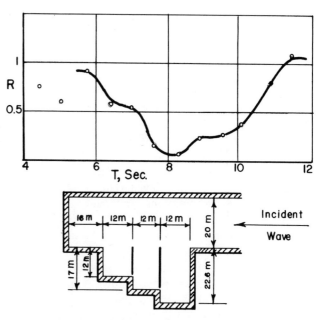

Fig. 6.24. Action of a group of three resonant basins on the clapotis at a wall (after Valembois, 1953)

145

to be reflected, and the width W should be about one-half the wave length (Valembois, 1953). The effectiveness of a resonator is dependent upon the width of the channel in front of the resonator. If there is a wall at the end of the channel, the resonator might be ineffective, depending upon the ratio of the wave length and the distance between the resonator and the wall. One way of getting around this difficulty is by installing two adjacent resonators, the centerlines of which are one-quarter of a wave length apart, Figs. 6.19 and 6.20.

In order to protect a harbor from waves with periods ranging from T to $2T$, a series of resonators (as shown in Figs. 6.21 and 6.22) should be used.

Some results of laboratory tests on two models are shown in Figs. 6.23 and 6.24 for the case of a wall at the end of the channel. In these figures, R is the ratio of the wave height at the wall without the resonators to the wave height at the wall with the resonators.

9. PNEUMATIC AND HYDRAULIC BREAKWATERS

The mechanism by which pneumatic and hydraulic breakwaters affect waves is not related to the actual structure, which is merely a device to generate currents in the water. A brief discussion of these devices is included in this chapter because they are used in place of structures. The literature on harbor protection contains a considerable number of articles on the pneumatic breakwater (see, for example, Green, 1961, which contains an extensive list of references on the subject, and a discussion of this paper by Schijf, 1961), many of which offer conflicting viewpoints of the feasibility of the breakwaters.

The pneumatic breakwater preceded the hydraulic breakwater, and its first use was probably by Brasher (*Engineer, 1916*). This breakwater (U.S. Patent, 1907) consisted of a submerged pipeline containing spaced holes and supplied with compressed air. More recent ones have been installed in England (Laurie, 1952; 1955) and in Japan (Kurihara, 1955; 1956). Many of the claims for the effectiveness of this type of breakwater stem from early stories of experiments at a pier at El Segundo, California. It was concluded from these studies, however, that the apparatus was of no utility or benefit and therefore abandoned (U.S. District Court, 1923), but these conclusions apparently were never published.

It has been claimed that the screen of rising air bubbles is responsible for the reduction in the height of waves moving past the breakwater. That this is not the case in shallow water, at least, has been shown theoretically by Schiff (1949a; 1949b). He found that a screen of air bubbles one-half wave length thick would reduce the height of a wave transmitted through this screen by only 1/2 per cent. Laboratory experiments by Carr (1950) showed

that an aerated region one-half a wave length thick provided about the same wave attenuation as a narrow aerated region. In addition, Kurihara (1956) observed both in model and in prototype experiments that the attenuation of wave heights occurred before the waves entered the region of the air bubbles. It was found that the waves broke in the region of strongest horizontal flow. This was also observed by Hensen (1955). It is also doubtful that large-scale turbulence plays an appreciable role in the process as in some experiments on transitional and shallow water waves. Carr (1950) used flow rates so high that they caused violent surface disturbances without any appreciable effect on the waves. Furthermore, Carr has found that high volume, low-velocity discharge was more effective (i.e., relatively less large-scale turbulence, but greater "pumping action"). There is, however, some evidence that large eddies in the region of the horizontal surface current are important (Kurihara, 1956). Radionov (1958) has stated that one of the effects of the interaction of the rising air-water column and the waves is to distort the motion of the rising column in such a manner that it acts as a generator of waves of the same frequency as the incident waves, provided that the manifold is the proper distance below the water surface for a given wave length. If the phase relationship between the generated and the incident wave is proper, a reinforcement occurs on the seaward side of the manifold, and a cancellation occurs on the lee side.

The principal mechanism by which a pneumatic breakwater attenuates waves is apparently the horizontal surface current generated by the rising air bubbles (Schijf, 1940). Air bubbles are discharged from a manifold some distance below the water surface. The bubbles move with the water, creating a two-phase fluid mixture which is less dense than the surrounding single-phase fluid. This results in a density pump, with the water flowing in toward the manifold at the bottom, mixing with the air bubbles, the mixture rising to the surface, the air escaping from the water to the atmosphere, and a horizontal current of water moving along the surface. Taylor (1955) developed theoretically an equation for the prediction of the maximum horizontal current, $U_{\max}$

$$U_{\max} = 1.9\,(qg)^{1/3} \qquad (6.11)$$

or

$$q = 0.00454\,U_{\max}^3, \qquad (6.12)$$

in which q is the air discharge rate per unit length of manifold. The thickness of the horizontal surface current at the start, b, was

$$b = 0.28\,D \qquad (6.13)$$

where D is the distance below the water surface of the manifold.

The method by which a current in deep water completely stops the transmission of wave energy was developed by Unna (1942) who found that, when the current velocity was greater than one-quarter of the deep water wave velocity, no energy could be transmitted upstream. Yu (1952) showed this to be so experimentally. For other than uniform current distribution with depth, it is probable that some of the wave energy can be transmitted upstream unless the surface current is quite high.

Manifolds discharging water horizontally at the water surface have also been used to attenuate waves (Wetzel, 1955; Snyder, 1959). The mechanism by which the hydraulic breakwater attenuates waves has been shown to be the current (Williams, 1958). This seems to be further proof that the principal mechanism of the pneumatic breakwater for attenuating waves is the horizontal surface current it generates.

Why should there be a discrepancy between the observations in various model studies and claims made for the protoype pneumatic breakwater? It is believed that one of the reasons is partly real (Kurihara, 1962) and partly psychological. Waves that are still in the generating area are steep, many of them breaking due to their steepness, and many of them nearly breaking. Beacause of this, an opposing current, which will cause the waves to steepen, will force many of them to break and dissipate wave energy. In addition, waves in the ocean are irregular and for many purposes can be described by an energy spectrum. On the other hand, most laboratory tests are performed with periodic waves of uniform height. For many purposes, the laboratory waves can be, and have been, associated with the portion of the energy spectrum in the vicinity of the peak energy density which in turn is closely related to the significant wave (Wiegel, 1961b). Now, suppose a surface current is generated by a pneumatic breakwater, or some other means, that is neither thick enough nor fast enough to stop the longer component waves in the spectrum associated with the maximum energy density, but which can stop the relatively short wave components. Most of the wave energy will be transmitted into the lee of the breakwater, but it will look much smoother than the original wave system because the short steep wave components will have been either reflected by the current or greatly attenuated. This is the psychological part— the wave system no longer looks as high as it did before.

If irregular wave systems can be treated to a certain extent as a superposition of linear wave trains, then a current might be able to affect the wave components in the selective manner just described. Some field data obtained by Kurihara (1956) suggested that this was the case, and experiments by Williams (1961) on a two-frequency wave system showed that it definitely occurred for a simple system. Laboratory measurements of the effect of a hydraulic breakwater on wind-generated waves by Williams and Wiegel (1963) have shown that it does occur.

REFERENCES

Beebe, K. E., *Experimental determination of pressures exerted by waves on a rigidly supported box of small draft*, Univ. Calif., Inst. Eng. Res., Tech. Rept. 61–8, November, 1954. (Unpublished.)

Carr, John H., *Mobile breakwater studies*, Calif. Inst. Tech., Hydro. Lab., Rept. N–64.2, December, 1950. (Unpublished.)

———, Mobile breakwaters, *Proc. Second Conf. Coastal Eng.*, Berkeley, Calif., The Engineering Foundation Council on Wave Research, 1952, pp. 281–95.

———, J. J. Healy, and Marshall E. Stelzriede, *Reflection and transmission of water waves by floating and fixed rigid surface barriers*, Progress Report for July, 1950, Calif. Inst. Tech., Hydro. Lab., Contract NOy-12561 July, 1950. (Unpublished.)

Costa, F. Vasco, and J. Fuiza Perestrelo, Modification of the sea bed with a view to concentration and dispersal of sea water, *Dock and Harbour Authority*, **39**, 460 (February, 1959), 305–306.

Costello, R. D., Damping of water waves by vertical circular cylinders, *Trans. Amer. Geophys. Union*, **33**, 4 (August, 1952), 513–19.

Dean, W. R., On the reflection of surface waves by a submerged plane barrier, *Proc. Cambridge Phil. Soc.*, **41**, Part 3 (October, 1945), 231–36.

De Long Corporation, *Mobile self-elevating platform with breakwater used to form a multi-purpose harbour unit*, Contract NBy-3153 with U.S. Naval Civil Engineering Laboratory, Port Hueneme, Calif., November, 1959.

Engineer, The Brasher air breakwater, **121** (May 19, 1916), 414–15.

Gade, Herman G., Effects of a nonrigid, impermeable bottom on plane surface waves in shallow water, *J. Marine Res.*, **16**, 2 (1958), 61–82.

Green, James L., Pneumatic breakwater to protect dredges, *J. Waterways Harbors Div.*, Proc. ASCE- **87**, WW2 (May, 1961), 67–87.

Hall, William C., and Jay V. Hall, *A model study of the effect of submerged breakwaters on wave action*, U.S. Army, Corps of Engineers, Beach Erosion Board, Tech. Memo. No. 1, May, 1940.

Hamilton, W. S., Forces exerted by waves on a sloping board, *Trans. Amer. Geophys. Union*, **31**, 6 (December, 1950), 849–55.

Hensen, Walter, Model tests with pneumatic breakwater, *Dock and Harbour Authority*, **36**, 416 (June, 1955), 57–61.

Jeffreys, H., *Note on the offshore bar problems and reflection from a bar*, Gt. Brit. Ministry of Supply, Wave Report No. 3, June, 1944.

Jellett, John Holmes, The lay-out, assembly, and behaviour of the breakwaters at Arromanches Harbour (Mulberry 'B'), *The Civil Engineer in War*. vol, 2, *Docks and Harbours*. Inst. Civil Engrs., 1948, pp. 291–312.

John, F., On the motion of floating bodies, *Comm. Pure Applied Math.*, **2**, 1 (March, 1959), 13–57.

Johnson, J. W., R. A. Fuchs, and J. R. Morison, The damping action of submerged breakwaters, *Trans. Amer. Geophys. Union*, **32**, 5 (October, 1951), 704–718.

Kurihara, Michinori, Discussion of pneumatic breakwaters to protect dredges, *J. Waterways Harbors Div., Proc. ASCE*, **88**, WW3 (August, 1962), 171–74.

———, Pneumatic breakwater (II), *Proc. Second Conf. on Coastal Eng. in Japan, November, 1955*, K. Horikawa, trans. Univ. Calif., IER Tech. Rept. 104–5, August, 1958.

———, Pneumatic breakwater (III), *Proc. Third Conf. on Coastal Eng. in Japan*, November, 1956, K. Horikawa, trans. Univ. Calif., IER., Tech. Rept. 104–6, November, 1958.

Laurie, A. H., Pneumatic breakwater, *Dock and Harbour Authority*, **33**, 379 (May, 1952), 11–13.

———, Pneumatic and other breakwaters, *Dock and Harbour Authority*, **36**, 422 (December, 1955), 265.

Lee, Donovan H., Discussion on Mulberry breakwaters, *The Civil Engineer in War*, vol. 2, *Docks and Harbours*. Inst. Civil Engrs., 1948, pp. 321–25.

Lochner, R., Discussion on Mulberry breakwater, *The Civil Engineer in War*, vol. 2, *Docks and Harbours*, Inst. Civil Eng., 1948, pp. 327–32.

———, Oscar Faber, and William G. Penney, The Bombardon floating breakwater, *The Civil Engineer in War*, vol. 2, *Docks and Harbours*, Inst. Civil Eng., 1948, pp. 256–90

Mobarek, Ismail, *Effect of bottom slope on wave diffraction*, Univ. Calif., IER, Tech. Rept. 89–8, 1962.

Paulling, J. R., Jr., Beach reflection studies, *Trans. Eleventh General Meeting Amer. Towing Tank Conf.* U.S. Navy, Bureau Ships, David Taylor Model Basin Report 1099, **3**, (September, 1957), 751–56.

Putnam, J. A., Loss of wave energy due to percolation in a permeable sea bottom, *Trans. Amer. Geophys. Union*, **30**, 3 (June, 1949), 349–56.

——— and J. W. Johnson, The dissipation of wave energy by bottom friction, *Trans. Amer. Geophys. Union*, **30**, 1 (February, 1949), 67–74.

Radionov, S. I., *Wave dissipation by compressed air (pneumatic breakwater)* published by Ocean Transport, Moscow, 1948. Hildegard Arnesen, trans. Univ. Calif., IER, Tech. Rept. 104–9, 1960.

Ross, Culbertson W., *Model tests on a triple-bulkhead type of floating structure*, U.S. Army, Corps of Engineers, Beach Erosion Board, Tech. Memo. No. 99, September, 1957.

Schiff, Leonard I., *Air bubble breakwaters*, Calif. Inst. Tech., Hydro. Lab., Rept. N–64.1, June, 1949(a). (Unpublished.)

———, *Gravitational waves in a shallow compressible liquid*, Calif. Inst. Tech., Hydro. Lab., Rept. N–64, May, 1949(b).

Schijf, J. B., Het vernietigen van golven door bet inspuiten van lucht (pneumatische golfbrekers), *De Ingenieur*, **55**, (1940), 121–25.

———, Discussion of pneumatic breakwaters to protect dredges, *J. Waterways Harbors Div., Proc. ASCE*, **87**, WW4 (November, 1961), 127–36.

Snyder, C. M., Model study of a hydraulic breakwater over a reef, *J. Waterways Harbors Div., Proc. ASCE*, **85**, WW1, Paper No. 1979 (March, 1959), 41–68.

Stelzriede, Marshall, *Mobile breakwater study, interim report*, October, 1951, Calif. Inst. Tech., Hydro. Lab., Contract NOy–12561, October, 1951. (Unpublished.)

———, and John H. Carr, *Mobile breakwater study, interim report, December, 1951*, Calif. Inst. Tech., Hydro. Lab. Contract NOy–12561, December, 1951. (Unpublished.)

Stoker, J. J., *Water waves*. New York: Interscience Publishers, Inc., 1957.

———, B. Fleishman, and L. Weliczkes, *Floating breakwaters in shallow water*, New York Univ., Inst. Math. Mech., IMM-NYU–192, February, 1953. (Unpublished.)

Straub, Lorenz G., C. E. Bowers, and John B. Herbich, Laboratory tests of permeable wave absorbers, *Proc. Sixth Conf. Coastal Eng.*, Berkeley, Calif.: The Engineering Foundation Council on Wave Research, 1958, pp. 729–42.

Stucky, A., and D. Bonnard, Contribution to the experimental study of marine rock fill dikes, *Bull. Tech. Suisse Romande*, August, 1937.

Taylor, Sir Geoffrey, The action of a surface current used as a breakwater, *Proc. Roy. Soc.* (London), ser. A, **231**, 1187, (September, 1955), 466–78.

Todd, F. H., Model-experiments on different designs of breakwaters, *The Civil Engineer in War*, vol. 2, *Docks and Harbours*, Inst. Civil Engrs., 1948, pp. 243–55.

United States District Court, Northern District of California, Case at Law No. 16879, April 16, 1923.

U.S. Patent, Protecting objects from wave action, U.S. Patent No. 843,926, Feb. 12, 1907.

Unna, P. J. H., Waves and tidal streams, *Nature*, **149**, 3773, (February, 1942), 219–20.

Ursell, F., The effect of a fixed barrier on surface waves in deep water, *Proc. Cambridge Phil. Soc.*, **43**, Part 3 (July, 1947), 374–82.

Valembois, J., Investigation of the effect of resonant structures on wave propagation, *Proc. Minn. Intern. Hydraulics Convention, Intern. Assoc. for Hyd. Res. and Hyd. Div. ASCE*, St. Anthony Falls Hyd. Lab., Minneapolis, Minn., September, 1953, pp. 193–99. Also Translation No. 57–6, U.S. Army, Corps of Engineers, Waterways Experiment Station.

Wetzel, J. M., *Experimental studies of pneumatic and hydraulic breakwaters*, Univ. Minn., St. Anthony Falls Hyd. Lab., Project Rept. No. 46, May, 1955. (Unpublished.)

Wiegel, R. L., *Floating breakwater survey to August 15, 1959*, Univ. Calif., Inst. Eng. Res., Tech. Rept. 140–2, Contract NBy–3139 August, 1959. (Unpublished.)

————, Transmission of waves past a rigid vertical thin barrier, *J. Waterways Harbors Div., Proc. ASCE*, **86**, WW1 Paper 2413, March, 1960.

————, Closely spaced piles as a breakwater, *Dock and Harbour Authority*, **42**, 491 (September, 1961a), 150.

————, Wind waves and swell, *Proc. Seventh Conf. Coastal Eng.*, Berkeley, Calif.: The Engineering Foundation, Council on Wave Research, 1961b, 1–40.

————, and R. A. Friend, *Model study of wind wave abatement*, Univ. of Calif., Inst. Eng. Res., WRL Field Rept. No. 49, Contract with Shell Oil Co., August, 1958. (Unpublished.)

————, H. W. Shen, and J. D. Cumming, Hovering breakwater, *J. Waterways Harbors Div., Proc. ASCE*, **88**, WW2, (May, 1962), 23–50.

Williams, John A., *Scale effects of models of hydraulic breakwaters*, IER Tech. Rept. 104–7, Univ. Calif., Berkeley, October, 1958. (Unpublished.)

————, *Small amplitude water waves and their superposition*, Report for C. E. 299, Univ. Calif., Berkeley, Dept. of Civil Eng., 1961. (Unpublished.)

————, and R. L. Wiegel, Final report on the hydraulic breakwater, attenuation of wind waves by a hydraulic breakwater, *Proc. Eighth Conf. Coastal Eng.*, Berkeley, Calif.: The Engineering Foundation, Council on Wave Research, 1963, 500–20.

Wood, Cyril Raymond James, Phoenix, *The Civil Engineer in War*, vol. 2, *Docks and Harbors*, Inst. Civil Enginrs., 1948, pp. 336–68.

Yu, Yi-Yuan, Breaking of waves by an opposing current, *Trans. Amer. Geophys. Union*, **33**, 1 (February, 1952), 39–41.

CHAPTER SEVEN

Waves in Shoaling Water

1. TRANSFORMATION OF UNIFORM PERIODIC WAVES

The theory presented in Chapter 2 is for waves in water of constant depth. How does one handle the problem of waves in shoaling water? There have been some papers which give mathematical solutions for the case of a rigid impermeable, sloping bottom. A thorough discussion of this problem is presented by Stoker (1957). The solutions are complicated and probably will not be used by engineers for some time.

A simpler approach, and one that is easy to use is due to Rayleigh (1911). The assumption made is that there is no loss of wave energy and no reflection. The power being transmitted by the wave train in water of any depth is set equal to the power being transmitted by the wave system in deep water. It is assumed that the wave period remains constant in water of any depth, whereas the wave length, velocity, and height vary. The wave power is given by Eq. 2.67 for linear theory.

The ratio of wave height in water of any depth, H, to the wave height in deep water unaffected by refraction, H_o' is found to be

$$\frac{H}{H_o'} = \sqrt{\frac{1}{2n} \cdot \frac{C_o}{C}} \qquad (7.1)$$

where n is defined by Eq. 2.68. The ratio of the wave speed in water of any depth, C, to the wave speed in deep water, C_0, and the ratio of wave length in water of any depth, L, to the wave length in deep water, L_0, are given by

$$\frac{C}{C_o} = \frac{L}{L_o} = \tanh\frac{2\pi d}{L} \qquad (7.2)$$

These are shown in graphical form as functions of d/L_0 in Fig. 2.4 and are tabulated in Appendix 1. There has been no similar development for nonlinear waves to the author's knowledge.

As an example, a wave train in deep water has the characteristics of $H_o = 10.0$ ft and $T = 12.0$ sec. What

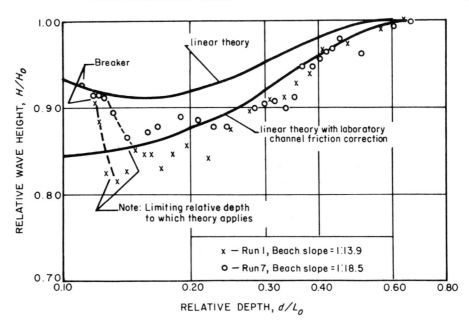

Fig. 7.1. Experimental confirmation of wave height transformation on a shoaling beach (from Iversen, 1951)

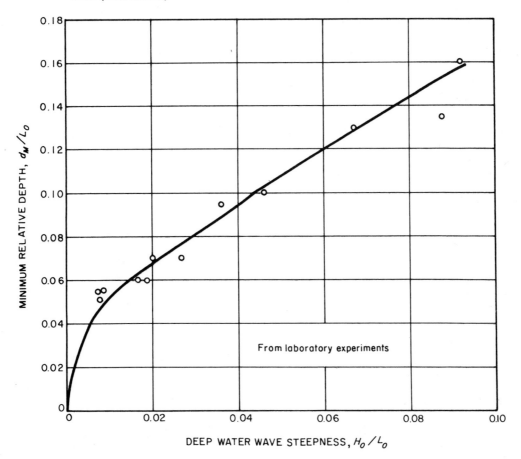

Fig. 7.2. Minimum depth to which wave theories apply for wave height transformations in shoaling water (from Iversen)

d_M — Minimum depth

L_O — Deep water wave length

H_O — Deep water wave height

is the length, speed, and height in water 25.0 ft deep?
$L_0 = (g/2\pi)\ T^2 = 737$ ft; $C_0 = (g/2\pi)\ T = 61.4$ ft/sec; $d/L_0 = 25.0/737 = 0.0340$. From Fig. 2.4 or Appendix 1, $C/C_0 = L/L_0 = 0.446$, and $H/H_0' = 1.098$, $L = 0.446 \times 737 = 329$ ft, $C = 0.446 \times 61.4 = 22.4$ ft/sec, and $H = 10.0 \times 1.098 = 10.98$ ft.

Laboratory studies have shown that Eq. 7.1 is adequate in many respects, except near the breaker zone, provided that frictional losses are considered, (see Fig. 7.1). (See Iverson (1952) for details on friction losses.) Iversen (1951) established experimentally the limiting value of d_M/L for which Eq. 7.1 appeared to be valid, as a function of the wave steepness H_0/L_0. This relationship is given in Fig. 7.2. (Remember that the theory and experiments were for uniform periodic waves.)

2. TRANSFORMATION OF IRREGULAR WAVES

The transformation of irregular waves (which ocean waves are) is a complex process which is not yet fully understood. One method of treating the problem which has been partially successful is to represent the actual system by a series of sinusoidal waves of different heights, periods, phases, and directions. Such a system would have a two-dimensional energy spectrum. Any resultant wave crest deforms as it moves, and it is difficult to define a wave length, period, height, or velocity, as these are merely instantaneous representations. The rate of deformation of one of the elevations depends upon the variation of component wave velocities and the angular divergence of direction of advance.

The travel time can be obtained for waves of permanent form from a knowledge of the refraction angle and the wave velocity; for irregular waves, however, the problem is difficult. Some complications of this and other problems of irregular waves have been discussed by Fuchs (1952, 1955), and it appears that large-amplitude crests generally deform at a slower rate than small-amplitude crests. The less the water depth, the slower is the deformation because the less the depth, the less the effect of wave period. For beaches of slopes less than about 1/10 the effect of slope on wave speed is very small. Because of selective refraction of the different wave component periods of an irregular wave train, the transformation in shoaling water is complex. This is especially true for waves of small period and large refraction angles.

Assuming that the speed of a wave crest depends upon the water depth immediately below the crest and on the wave period, according to linear theory

$$t = \int_{x_1(d_1)}^{x_2(d_2)} \frac{x'}{C\cos\alpha} = \frac{T}{\beta} \int_{d_1/L_0}^{d_2/L_0} \frac{(d/L_0)'}{C\cos\alpha/C_0}$$

$$= \frac{T}{\beta}\left[S_{\alpha_0}\left(\frac{d_1}{L_0}\right) - S_{\alpha_0}\left(\frac{d_2}{L_0}\right)\right] \qquad (7.3)$$

$$S_{\alpha_0}\left(\frac{d}{L_0}\right) = \int_{d/L_0}^{d/L_0=1/2} \frac{(d/L_0)'}{C\cos\alpha/C_0} \qquad (7.4)$$

where x' and $(d/L_0)'$ refer to differentiation, $d/L_0 = 1/2$ refers to deep water conditions, α is the angle between a wave crest and the bottom contour, β is the beach slope (constant), t is the travel time (sec) from d_1 to d_2 with angles α_0, β, and wave period T, and x is the horizontal distance (ft) measured perpendicular to the parallel bottom contours with origin at the deep water point for which $d/(g/2\pi)\ T^2 = \frac{1}{2}$.

Assuming the wave speed is independent of beach slope, and that refraction follows Snell's law (see section on wave refraction), the integrals were evaluated by Fuchs (1954) using finite difference methods (Table 7.1). Using this table of $S_{\alpha_0}\ (d/L_0)$ permits the rapid computation of travel time, t.

As an example of the use of Table 7.1 consider the problem of predicting the travel time necessary for waves to travel between two locations, such as Recorder Site No. 1 and Recorder Site No. 2 in Fig. 7.3. One recorder (Site No. 1) was of the pressure type located a few feet

Table 7.1.–TABLE OF S_{α_0}
(from Wiegel and Fuchs, 1955)

d/L_0	S_{α_0}				
	70°	50°	30°	15°	0°
0.50	0.	0	0	0	0
0.45	0.142	0.078	0.058	0.052	0.050
0.40	0.280	0.155	0.116	0.104	0.101
0.35	0.413	0.232	0.174	0.157	0.152
0.30	0.539	0.309	0.233	0.210	0.203
0.25	0.655	0.386	0.293	0.264	0.256
0.20	0.762	0.463	0.355	0.321	0.311
0.19	0.782	0.478	0.367	0.332	0.322
0.18	0.802	0.494	0.380	0.344	0.333
0.17	0.822	0.509	0.393	0.356	0.345
0.16	0.841	0.525	0.406	0.368	0.357
0.15	0.860	0.540	0.419	0.381	0.369
0.14	0.879	0.556	0.433	0.393	0.381
0.13	0.898	0.572	0.447	0.406	0.394
0.12	0.917	0.588	0.461	0.420	0.407
0.11	0.936	0.604	0.475	0.433	0.420
0.10	0.955	0.621	0.490	0.447	0.434
0.09	0.974	0.638	0.505	0.462	0.449
0.08	0.993	0.655	0.521	0.477	0.464
0.07	1.013	0.674	0.538	0.493	0.480
0.06	1.033	0.692	0.556	0.510	0.496
0.05	1.054	0.712	0.575	0.528	0.514
0.04	1.077	0.734	0.595	0.548	0.534
0.03	1.101	0.757	0.618	0.571	0.556
0.02	1.129	0.785	0.644	0.597	0.582
0.01	1.164	0.819	0.678	0.631	0.616
0.009	1.168	0.823	0.682	0.634	0.620
0.008	1.173	0.828	0.687	0.639	0.624
0.007	1.177	0.832	0.691	0.644	0.629
0.006	1.183	0.837	0.696	0.649	0.634
0.005	1.188	0.843	0.702	0.654	0.640
0.004	1.194	0.849	0.708	0.660	0.646
0.003	1.201	0.856	0.715	0.667	0.652
0.002	1.209	0.864	0.723	0.675	0.660
0.001	—	—	—	—	0.671

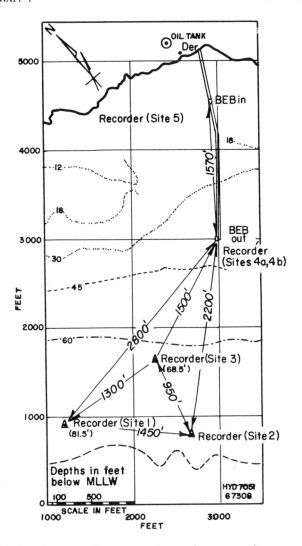

Fig. 7.3. Location of wave recorders and survey stations, Davenport, California; depth contours from USC & GS chart; seasonal changes or scouring had increased the depth at the end of the pier to 46 ft at the time the gages were installed (after Wiegel and Fuchs, 1955)

Fig. 7.4. Comparison of wave records obtained 4200 ft apart in water 85.5 and 16.5 ft deep (after Wiegel and Fuchs, 1955)

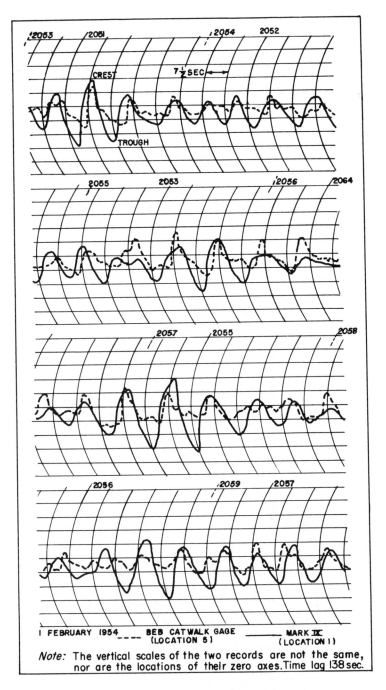

above the bottom in 85.5 ft of water and the other (Site No. 5) was of the surface type located in water 16.5 ft deep, which was in the immediate vicinity of the surf zone. The distance between the two recorders was 4200 ft. The average bottom slope between the two sites was 1/64. An example of the two records, one displaced in time with respect to the other for best visual correlation is given in Fig. 7.4. The displacement was found to be 138 sec. Using the average wave period of 13 sec and a deep water wave angle of almost 90 degrees (obtained from a study of weather maps) the values in Table 7.1 were extrapolated to the approximate 90-degree case and a travel time of 123 sec was computed. This compares favorably with the measured 138 sec.

3. LINEAR LEAST-SQUARES METHOD OF PREDICTION OF WAVE MOTION

The prediction of the fluctuating time history of the surface elevation or subsurface pressure at one place from the record at another place can be done by a linear least-squares technique. The procedure is complex and the reader is referred to the paper by Putz (1955). An example of the results of this procedure is given in Fig. 7.5.

4. WAVE REFRACTION

a. *Introduction.* In the linear theory of progressive water gravity waves, the phase velocity is given by Eq.

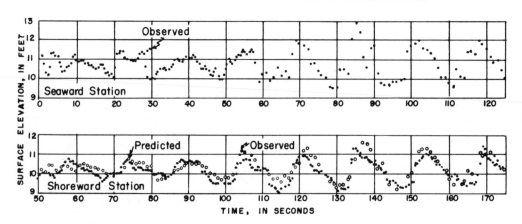

Fig. 7.5. Least-squares surface wave prediction over 1570 ft distance (four-term kernel based upon approximate spectral information from both positions) (after Putz, 1955)

2.33. This phase velocity depends upon both the water depth and the wave length. The relative importance of water depth and wave length depends upon the ratio of these two independent variables, d/L. When d/L is large, say, greater than $\frac{1}{2}$, the phase velocity is a function of only the wave length, or wave period. When d/L is small, say, less than 1/25, Eq. 2.33 reduces to Eq. 2.38 and the phase velocity is a function of only the water depth.

When the wave height is not small with respect to either the wave length or the water depth, the equation for the phase velocity becomes more complicated and the phenomenon is nonlinear. This will not be treated here.

For waves (whether linear or not) in water depths which are small enough compared with the wave length ($d/L < 1/2$ for most practical purposes), each part of a wave travels with a phase velocity that is dependent upon the water depth under it; if the water depth is not constant, the wave must bend. This bending of the wave is known as *wave refraction* (Fig. 7.6). In treating the phenomenon of wave refraction from an engineering standpoint, the assumption has been made in the past that what is known in optics as Snell's law can be applied

$$\frac{\sin \alpha_2}{\sin \alpha_1} = \frac{C_2}{C_1} \qquad (7.5)$$

where α_1 and α_2 are angles between adjacent wave front positions and the respective adjacent bottom contours. The ratio C_2/C_1 can be obtained for given water depths and wave periods by use of tabulated wave functions.

It should be emphasized that waves may also be refracted by currents (Unna, 1942; Johnson, 1947).

The height, period, and direction of waves in deep water may be forecast either from synoptic or prognostic weather maps or by direct measurements with

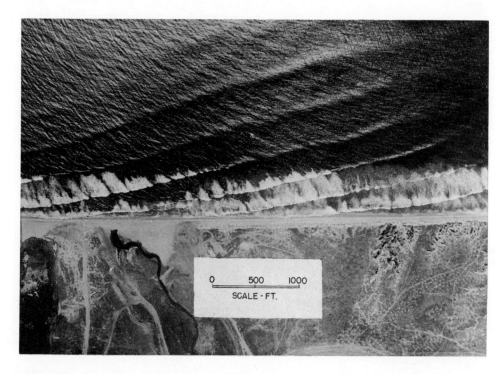

Fig. 7.6. Aerial photograph of swell, breakers, and surf north of Oceanside, Calif., Aug. 17, 1945 (Utitity Squadron 12)

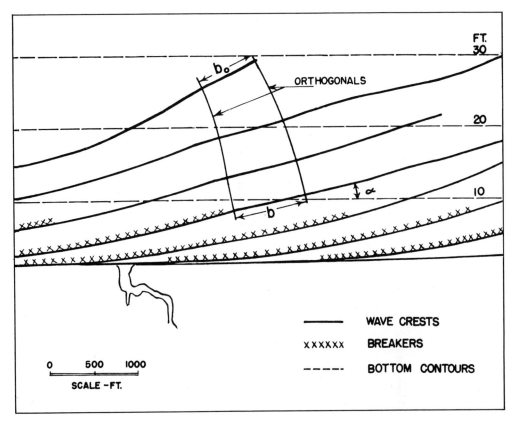

Fig. 7.7. Wave pattern from aerial photograph (from Johnson, O'Brien, and Isaacs, 1948)

suitable instruments. The design and construction of marine structures often depends upon wave heights, periods, and directions in coastal regions where they are affected by the hydrography. In order to forecast these quantities, refraction must be taken into account. Some simplified cases which were solved analytically gave very accurate results, but unfortunately were not applicable to most natural situations. The changes due to refraction are best estimated by the construction of refraction diagrams. Such diagrams can be prepared entirely from aerial photographs, as was done in Fig. 7.7, but are generally constructed graphically. The graphical means of constructing refraction diagrams for complex hydrography has been used for some time. Graduate students working under Professor M. P. O'Brien constructed them as far back as 1937 (Lapsley, 1937), but the theory was not published until a later date (O'Brien, 1942).

Refraction diagrams are constructed in two ways. The first, known as the *wave-front method*, is essentially a map showing the wave crests at a given time or the successive positions of a particular wave crest as it moves shoreward. A second set of lines, everywhere perpendicular to the "wave crests" is then constructed on the "map." These new lines are known as *orthogonals* (Fig. 7.7). It is considered that constant power is transmitted between any two orthogonals so that it is possible to estimate variations in wave height due to refraction. In

the second method, known as the *orthogonal* method, the orthogonals are drawn directly.

Sometimes the hydrography is such that adjacent orthogonals cross. There are two general types of crossing. The simpler of the two occurs when the wave crest is severed and the severed sections later cross each other. An idealized example of this situation, shown in Fig. 7.8, illustrates the case of a wave train passing over a corner-shaped ledge, where its speed is reduced to one-half its deep water speed. As a result, the wave train

Fig. 7.8. Example of crossed wave trains (after Pierson, 1951)

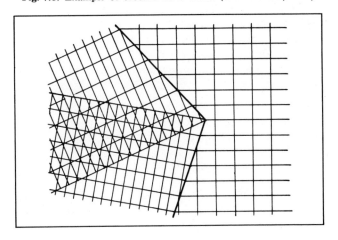

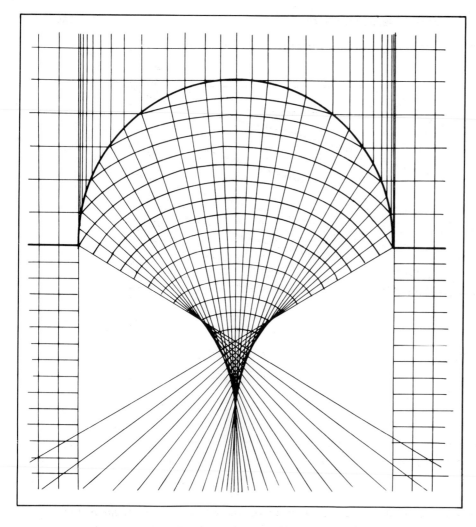

Fig. 7.9. Wave crest-orthogonal pattern for the example of a caustic curve (after Pierson, 1951)

breaks into two trains which cross each other. The more complex situation occurs where the orthogonals form a caustic envelope. This is illustrated in Fig. 7.9 (Pierson, 1951). The convergence of adjacent orthogonals would indicate, on the basis of the simple theory, that the waves become infinitely high which, of course, is not the actual case. In nature, conditions similar to these (bars at river mouths with gently sloping areas to the seaward, for example) do occur. The waves sometimes peak up and break; often a chaotic sea surface results.

Although the refraction techniques discussed in this section are not applicable to all conditions, they are usually valuable in obtaining reliable quantitative information for large portions of the area under study and will give qualitative information about the complex areas.

b. *Theory*. The power transmitted by a train of long-crested sinusoidal waves ("linear theory") is

$$P = C_G b H^2 (\gamma/8) \qquad (7.6)$$

where C_G is the group velocity, γ is the specific weight of water, b is the length of crest (perpendicular to the local direction of travel), and H is the wave height.

Ocean waves are not sinusoidal in form, and their departure from this form increases in shoal water, and especially as they approach the condition of breaking in shoal regions; this formula, however, is sufficient for estimating wave height in many cases and often can be corrected by empirical results. If no energy flows laterally along the wave crest, the same power should be transmitted past all positions between two orthogonals.

$$P = P_o \qquad (7.7)$$

$$\frac{H}{H_o} = \sqrt{\frac{C_{Go}}{C_G}} \sqrt{\frac{b_o}{b}}, \qquad (7.8)$$

where the subscript, o, refers to deep water conditions.

The quantity $\sqrt{b_o/b}$ is the refraction coefficient, and is designated as K_d. The quantity $\sqrt{C_{Go}/C_G}$ represents the effect of a change in depth on the wave height (shoaling coefficient), and is designated as D_d. Equation 7.8 can be written

$$H = H_o D_d K_d. \qquad (7.9)$$

Methods for determining K_d, the refraction coefficient, are presented in this section. Values of the shoaling coefficient, D_d, which depends on the water depth and wave length, have been tabulated in Appendix 1, where it is designated as H/H_o' rather than D_d. For reference,

Table 7.2: THEORETICAL COEFFICIENT OF SHOALING,
D_d, LINEAR THEORY

d/L_o	0.002	0.005	0.007	0.01	0.02	0.04	
D_d	2.12	1.69	1.57	1.45	1.23	1.06	
D/L_o	0.056	0.08	0.1	0.15	0.2	0.3	0.4
D_d	1.0	0.94	0.92	0.91	0.92	0.93	0.96

some values of D_d are presented in Table 7.2. For the case of predicting conditions in the breaker zone along a beach, it should be cautioned that the values of D_d are valid only up to several wave lengths seaward of the breaker zone. Empirical values must be used in the vicinity of the breakers. Some of these values have been presented in the previous section and some will be presented later.

The graphical or analytical determination of wave refraction coefficients assumes (1) that the velocity of the wave crest depends only upon the wave length and the still water depth under the crest at each point; (2) that the elements of the wave crest advance in a direction perpendicular to the crest line; (3) that the wave energy is confined between orthogonals, (4) that the waves are long crested; (5) that the period is constant.

The assumption that Snell's law can be used to describe the refraction of water gravity waves has been shown mathematicaly to be proper for linear wave theory in shallow water (Lowell, 1949), but it has not yet been proved to be legitimate for deeper water. Mathematical solutions have been obtained for certain simplified submarine topography (Pocinki, 1950; Arthur, 1951; and Pierson, 1951) and graphical methods have been developed for application to the complex ocean bottoms of nature (Johnson, O'Brien, and Isaacs, 1948). A theoretical study has also been made of the refraction of short-crested waves (Fuchs, 1952) and of nonuniform short-crested waves (Longuet-Higgins, 1956). These solutions are based upon the condition that Snell's law correctly predicts the refraction of water gravity waves. It has been generally accepted that Snell's law does correctly predict the refraction of water gravity waves except for certain conditions. This acceptance has been based upon qualitative observations, largely of aerial photographs taken of the ocean surface. In addition, a few laboratory tests have been made (Chien, 1954; Ralls, 1956; Wiegel and Arnold, 1957).

Wiegel and Arnold (1957) studied the refraction of uniform periodic waves over a submerged shoal of constant slope with straight parallel contours (slopes ranging from 1:10.6 to vertical) in a model basin 2 1/2 ft deep by 64 ft wide by 150 ft long. Waves of several periods and amplitudes for initial wave front positions with the shoal of 10 through 70 degrees were used. The limits of reliability of Snell's law were determined for these conditions. It was found that Snell's law was valid over a large range of wave periods even for the case of refraction over a vertical reef separating bodies

of water of two different depths (Fig. 7.10). It appeared that the formation of multiple wave crests was associated to some extent with the combination of parameters for which Snell's law became inaccurate in consistently predicting the refracted wave angles.

Model studies by Chien (1954) and Ralls (1956) also showed that Snell's law was applicable over a wide range of conditions. Some of Chien's data are shown in Fig. 7.11.

The cases considered in the following sections are for waves coming shoreward from deep water. Depending upon the conditions, a little, or much, of the wave energy is reflected. Under certain conditions, the refraction of the waves heading seaward swings them shoreward again, and the waves are "trapped" in the shore area; this refraction process has been called *total reflection* (Isaacs, Williams, and Eckart, 1951). It is believed that this mechanism is important from the standpoint of long waves originating as surf beat (Chapter 5): some types of bays could act as traps for these long waves. These waves could be the cause of oscillations in a harbor developed in such a bay.

c. *Straight shore with parallel contours.* For the case of a shoreline and offshore contours which are straight and parallel, refraction may be treated analytically by utilizing Snell's law directly:

$$\frac{\sin \alpha}{\sin \alpha_o} = \frac{C}{C_o} \qquad (7.10)$$

where α is the angle between the wave crest (see Fig. 7.7) and the shoreline, α_o is the angle between the deep water wave and the shoreline, C is the wave velocity, and C_o is the deep water wave velocity. The change in angle determines the increase in crest length, and thus the value of K_d is fixed by the depth (which determines C for a particular C_o) and α_o, the angle in deep water. The value of K_d may be computed from the relationship:

$$\frac{b_o}{\cos \alpha_o} = \frac{b}{\cos \alpha} \qquad (7.11)$$

or

$$K_d = \sqrt{\frac{b_o}{b}} = \sqrt{\frac{\cos \alpha_o}{\cos \alpha}} \qquad (7.12)$$

where

$$\alpha = \sin^{-1}\left(\frac{C \sin \alpha_o}{C_o}\right). \qquad (7.13)$$

For example, if $\alpha_o = 45°$ and the depth and period at the point in question are such that $C/C_o = 0.5$, ($d/L_o = 0.044$), then

$$\alpha = \sin^{-1}(0.5 \times 0.707) = 20.8 \text{ degrees}$$

$$\cos \alpha = 0.935 \quad \text{and} \quad \cos \alpha_o = 0.707$$

$$K_d = 0.87.$$

For convenience, the relationships between α, α_o, depth, period, and K_d are summarized in Fig. 7.12.

A thorough understanding of the nature and magnitude of refraction effects along straight coastlines is helpful in constructing refraction diagrams for complex

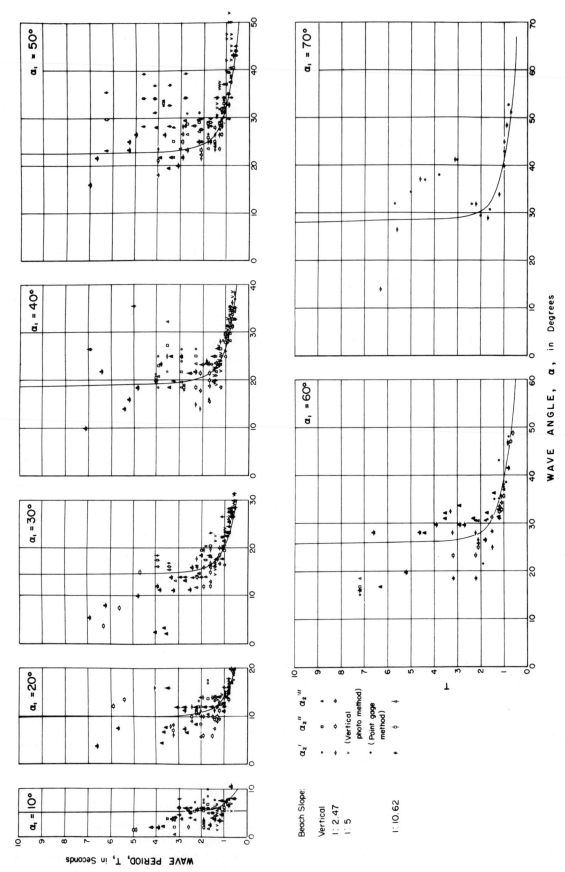

Fig. 7.10. Wave refraction in the model basin (from Wiegel and Arnold, 1957)

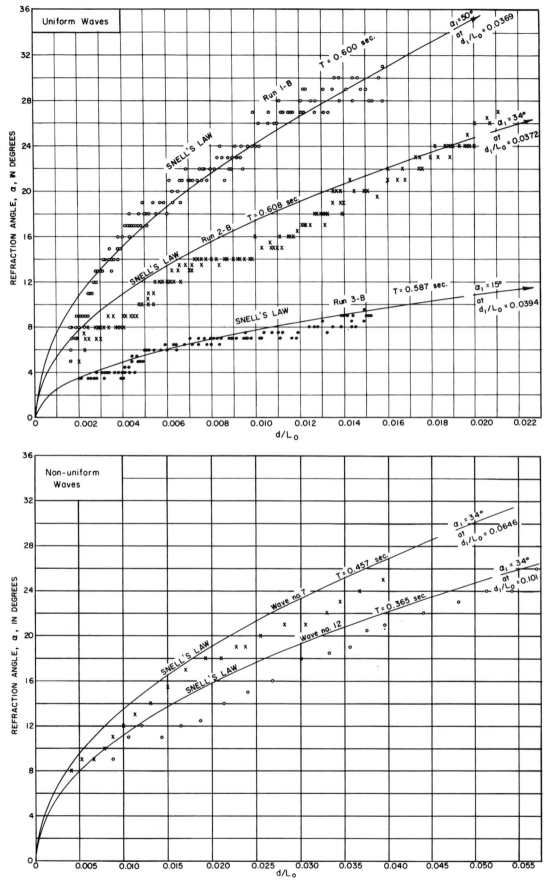

Fig. 7.11. Laboratory study of wave refraction (from Chien, 1954)

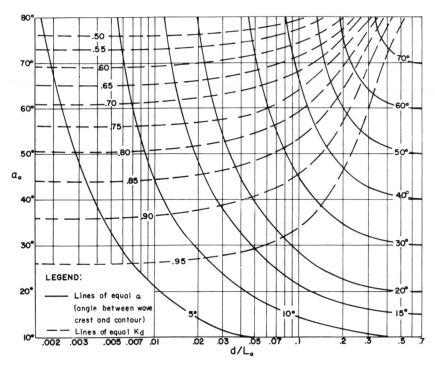

Fig. 7.12. Change in wave direction and height due to refraction on beaches with straight, parallel depth contours (from Johnson, O'Brien, and Isaacs, 1948)

hydrography. It is noteworthy that along a straight shoreline the reduction in wave height by refraction is less than 10 per cent when the initial angle in deep water is less than 36 degrees.

    d. *Islands and shoals with concentric circular contours.* An analytical solution for the refraction of waves

Fig. 7.13. Polar-coordinate representation of an orthogonal associated with wavecrests (after Arthur, 1946)

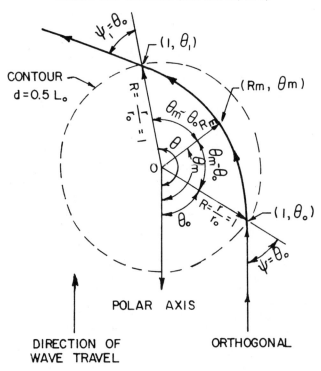

around an island with concentric circular contours has been obtained by the application of Fermat's principle by Arthur (1946); i.e., light waves will travel in a path such that the travel time is a minimum. The problem was one of determining the path between two points for which the following integral has a minimum value:

$$I = \int_{\theta_o}^{\theta} (C_o/C) \sqrt{r^2 + (dr/d\theta)^2} \, d\theta \qquad (7.14)$$

where the initial conditions to be satisfied are $C = C_o$ for $r = r_o$; a given orthogonal passes through point (r_o, θ_o); and a given orthogonal is parallel to the polar axis of the polar coordinate system at point (r_o, θ_o), (See Fig. 7.13 for definition of symbols.)

    Application of the method of the calculus of variation led to the Euler-Lagrange condition which Arthur solved for this case. The solution was

$$\frac{dR}{R\sqrt{(C_o R/C \sin \theta_o)^2 - 1}} = \pm d\theta \qquad (7.15)$$

where $R = r/r_o$ and where the region $0 \leq R \leq 1$ was considered. Solutions of this can be obtained when the function $C = C(R)$ is known. Certain simplified solutions were obtained such as assuming that $R = C/C_o$, etc. The different profiles of the 'islands' studied are shown in Fig. 7.14.

    Case 1A: Point island. Assume $R = C/C_o$, where $0 \leq R \leq 1$. The solution is

$$R = \frac{1}{e^{(\theta - \theta_o) \operatorname{ctn} \theta_o}} \qquad (7.16)$$

This is the equation of a logarithmic spiral; all orthogonals converge towards the center, $R = 0$. Figure 7.15 shows this case with only the orthogonals entering one

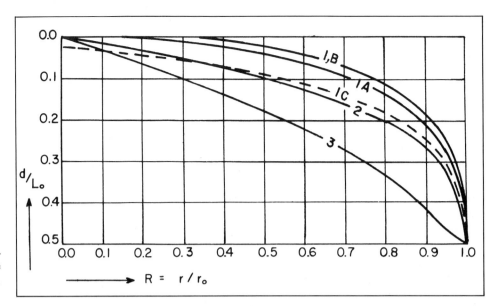

Fig. 7.14. Cross-sections of bottom-slopes for five cases (after Arthur, 1946)

quadrant shown. This is the limiting case between an island and a shoal and is known as a *point island*. There is extreme convergence at the center and there are no refracted waves in the lee beyond the contour $d = 0.5L_o$.

Case 1B: Circular island. Assume $(R - n)/(1 - n) = C/C_o$, where $0 < n < 1$ and $n \leq R \leq 1$.

The results for one quadrant are given in Fig. 7.16. All orthogonals meet the shore of the island ($R = n$,

in this case $n = 0.2$). In addition to the orthogonals, the refraction coefficient, K_d, has been plotted for $d = O$. Actually the waves would break just offshore; the actual value of K_d, however, would be nearly as given in the sketch. The zone in the lee of the island, beyond $d = 0.5\,L_o$, is free of refracted waves. The curve of K_d shows that the minimum value is 0.6; interference between wave trains refracted around each side of the

Fig. 7.15. Orthogonals for a "point island" off which the slope changes from very gentle to very steep; case 1A (after Arthur, 1946)

Fig. 7.16. Orthogonals and *K*-variation at shore for a circular island off which the slope changes from very gentle to very steep; case 1B (after Arthur, 1946)

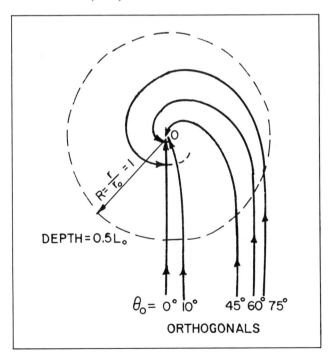

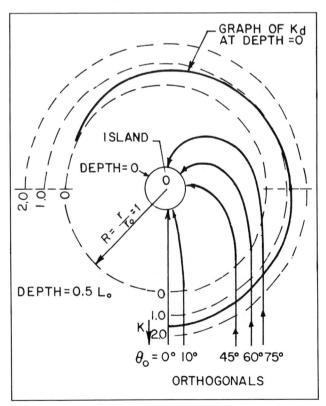

island, however, would produce waves higher than the incident waves all along the shore. Hence, such an island would have no "lee shore" as far as protection is concerned.

Case 1C: Shoal. Assume $(1 - n) R + n = C/C_o$ where $O < n < 1$ and $O \leq R \leq 1$.

For example $n = 0.4$ represents a shoal, with the wave velocity at the most shallow point being $C = 0.4\,C_o$ and the depth at this point being $d = 0.028\,L_o$. This shows that orthogonals associated with an undivided wave train can intersect, and that this is such a case. At such an intersection point there is not what might be called a complete "focusing" effect and K_d is therefore not infinite because each orthogonal is associated with a different wave crest. If waves are refracted about both sides of the island, a very confused sea would exist in the lee owing to the intersection of four sets of crests.

Case 2: Point island. Assume $R = (C/C_o)^2$, where $O < R < 1$. The cross section Fig. 7.14 shows a moderate bottom slope changing to a steep slope near $d = 0.5\,L_o$. It is found that the height of the refracted waves is greatest for that portion of the wave crest turned through the largest angle.

Case 3: Point island. Assume $R\,(2 - R) = (C/C_o)^2$, where $O \leq R \leq 1$. As is shown in Fig. 7.14, the bottom slope is nearly linear. The curve of K shows a decrease in height of refracted waves for increasing angle and hence the decrease in wave height is rapid in the lee of this island. The waves refracted through greater angles are lower.

e. *Caustic.* Analytical and experimental studies of refraction in the region of a caustic have been made by Pierson (1951). A case, similar to the shoal studied by Arthur, was examined in detail. Although the two cases were not identical, it was thought possible to represent the orthogonal pattern over the submerged clock glass (shoal) in a ripple tank schematically from Arthur's theoretical considerations.

Figure 7.17 shows the orthogonal pattern over the clock glass. The wave crests shown beyond the caustic are not as observed, but rather were constructed by Pierson (1951) after the method of Arthur. Figure 7.18 shows the pattern of wave-crest positions as found by the geometric optics approximation which predicts a phase shift through the caustic. Fig. 7.19 shows the actual crest-orthogonal pattern for waves passing over a clock glass. This illustration was drawn by Pierson from a photograph of the light and dark shadows on a screen, caused by the focusing effect of the wave crests on a light shining up through the bottom of a transparent ripple tank.

Fig. 7.17. Theoretical wave crest-orthogonal pattern for waves passing over a clock glass; no phase shift (from Pierson, 1951)

Fig. 7.18. Theoretical wave crest-orthogonal pattern for waves passing over a clock glass; phase shift (from Pierson, 1951)

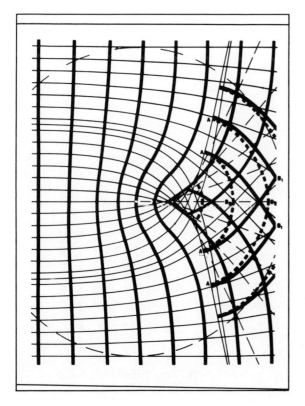

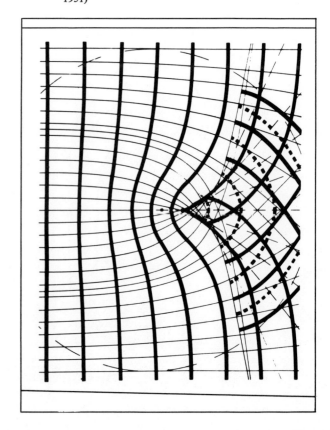

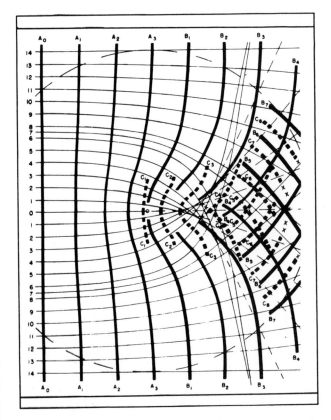

Fig. 7.19. Actual wave crest-orthogonal pattern for waves passing over a clock glass (from Pierson, 1951)

At present, no quantitative information may be obtained for the area beyond the caustic.

f. Graphical construction of refraction diagrams by the wave-front method for long-crested waves. In deep water long-crested uniform waves move forward with their crests parallel, whereas in transitional and shallow water, the reduction in wave speed causes the crest to swing around in the direction which will decrease the angle between the crest and the bottom contour. The meaning

of these terms may be examined in the light of refractive effects by a study of Figures 2.4 and 7.12

The angle through which a wave crest will turn between deep water and some particular value of d/L, over straight parallel bottom contours, may be obtained from Fig. 7.12. It can be seen that the limits of transitional water depend upon the angle α_o and the accuracy to which the diagram is to be constructed. If $\alpha_o = 70°$ and if an accuracy of ± 1 degree is wanted, the diagram should be started in depths even greater than $d = 0.5 L_o$, but if α_o is only 10 degrees, a negligible error is introduced if the diagram is started from $d = 0.3 L_o$. It is usually possible to start the construction of refraction diagrams from straight wave crests in a depth equal to half the deep water wave length.

The initial form of the wave is a straight line in the deep water area. Graphical construction of a refraction diagram is performed by moving each point of a wave crest in a direction perpendicular to the crest by a distance equal to the wave velocity times the time interval selected. Figure 7.20 shows a scale constructed in such a manner as to give the advance of the wave crest at any value of d/L_o on a chart of any scale S.

On the best available hydrographic chart of the area, contours are drawn with an interval which will represent accurately the details of the bottom topography. For many locations, it may be necessary to secure additional soundings. Each contour on the map is converted to mean sea level—or any other desired stage of the tide—adding the proper constant to the chart soundings. For many areas, the work sheets from which the hydrographic charts were prepared are available (U.S. Coast and Geodetic Survey work sheets within the United States) which are generally larger-scale and have more soundings than the charts. The deep water wave length is computed for each wave period to be studied. The contour values, as given in feet below the tide stage selected, are divided by the deep water wave length (in feet) to give contours in terms of d/L_o (Fig. 7.21).

Fig. 7.20. Scale for preparation of wave refraction diagrams (from Johnson, O'Brien, and Isaacs, 1948)

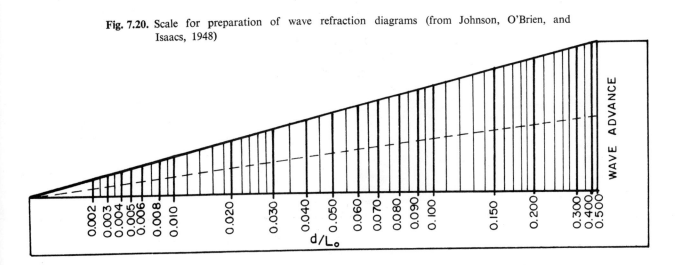

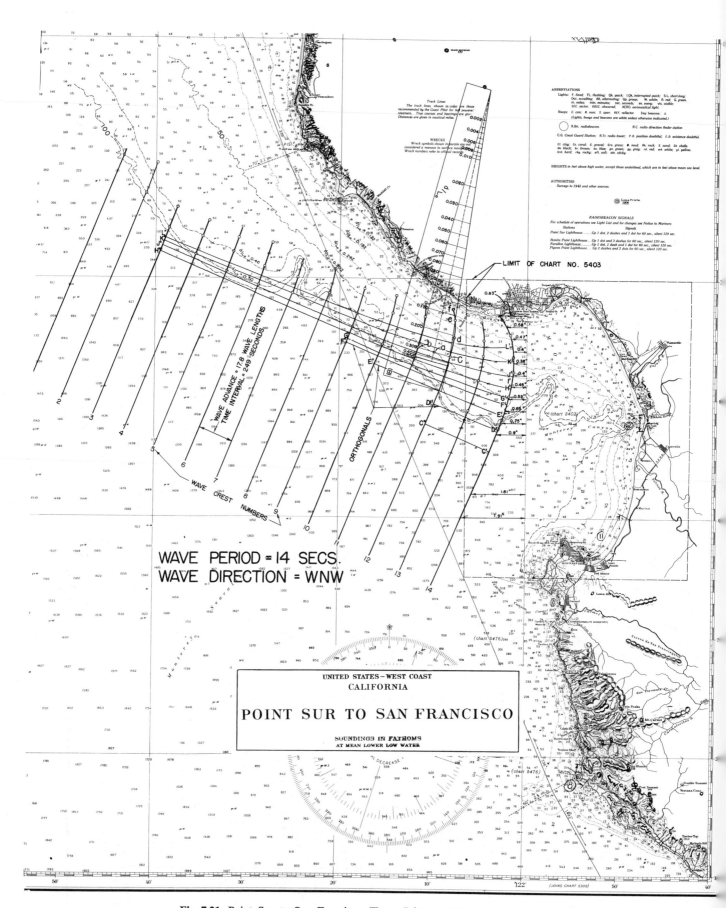

Fig. 7.21. Point Sur to San Francisco (From Johnson, O'Brien, and Isaacs, 1948)

It is necessary to draw only every nth crest, the value of n used depending upon the scale of the chart and the complexity of the bottom topography. The crest interval does not have to be an even value, and usually it will not be the same for the entire chart, since it is necessary to draw more crests where the bottom topography is complex.

It is necessary to draw a refraction diagram in several steps, each to a different scale. The over-all pattern for a long stretch of coastline is developed on a relatively small-scale chart, from deep water to within a few thousand feet from shore; then the end wave is transferred to a larger-scale chart, and the detailed wave pattern is constructed of the areas of particular importance. If the tidal range is large, it may be necessary to construct diagrams for several stages of the tide. It is necessary to draw refraction patterns for a series of wave periods and directions to cover the range to be expected.

Example-refraction diagram, Monterey Bay, California. A refraction diagram for Monterey Bay has been constructed as an example, (from Johnson, O'Brien, and Isaacs, 1948), and the values of K_d determined for points along the coast from Monterey to Santa Cruz for a mean tide condition of 2 ft above MLLW (Mean Lower Low Water), a direction of wave advance in deep water from W.N.W., and a wave period of 14 sec. Thus, $L_o = 5.12T^2 = 5.12\ (14)^2 = 1000$ ft, and the refraction drawing is started at a depth of $L_o/2 = 500$ ft.

The U.S. Coast and Geodetic Survey charts used were No. 5402, scale 1:214,000, and No. 5403, scale 1:50,000. The contours appearing on these charts are in fathoms. In terms of d/L_o these contours are

$$100 \text{ fathoms}; \quad \frac{d}{L_o} = \frac{(6)(100) + 2}{1000} = 0.602,$$

and in a similar manner for 50 fathoms, $d/L_o = 0.302$; for 40 fathoms, $d/L_o = 0.242$; for 30 fathoms, $d/L_o = 0.182$; for 20 fathoms, $d/L_o = 0.122$; for 15 fathoms, $d/L_o = 0.092$; for 10 fathoms, $d/L_o = 0.062$; and for 5 fathoms, $d/L_o = 0.032$.

The wave was carried from deep water to the limits of chart No. 5402. The wave front then had to be transferred to the larger-scale chart, No. 5403. In practice, the refraction diagrams would be drawn on tracing paper overlays placed over the hydrographic chart; for illustration in this report, however, the diagrams were drawn directly on the charts.

Figure 7.21 shows a portion of U.S. Coast and Geodetic Survey Chart 5402 with 14 wave crests marked 1, 2 . . . 14 drawn on it. Crest 1 lies in deep water and was drawn as a straight line. The southerly portion of crests 1–14 remain in deep water whereas the northerly portions of the crests advance into shallow water and bend. The position of each crest is determined from that of the crest behind it by locating a few points on the

new crest. The points were located by means of a scale of the type shown in Fig. 7.20 used as illustrated in Fig. 7.21. As an example, consider the problem of locating point c on crest 12: (1) Lay the scale on the chart so that the dashed centerline and the line of $d/L_o = 0.302$ on the scale intersect with the contour $d/L_o = 0.302$ on the chart (point a); (2) then move the scale so that condition (1) remains satisfied and at the same time the lower side of the scale is tangent to crest 11 on the diagram at the end of the $d/L_o = 0.302$ line on the scale (point b); (3) mark point c where the $d/L_o = 0.302$ line on the scale reaches the upper side of the scale. The other points, such as d, e, and f, are found by making use of the d/L_o contours of 0.242, 0.182, and 0.122, respectively. The process is repeated until enough points are found to determine the position of crest 12.

The other crests are located in a similar manner until the diagram is carried into a locality which is within the limits of a larger-scale chart. For this case, crest 14 is within the limits of the area shown on Chart 5403. This crest is transferred from Chart 5402 by taking offsets from a convenient longitude (122°), correcting for scale ratio, and replotting on Chart 5403 (Fig. 7.22). Care must be taken in transferring a wave from one chart to another, as slight errors here may cause large errors in K_d. Starting with wave crest 14 on Chart 5403, crests 15–23 were plotted by the method previously illustrated.

For a few localities where the bottom configurations were irregular, intermediate crests on Chart 5403 were added. These localities are between crests 16 and 17 near a tributary canyon of the Monterey Canyon, and shoreward from crest 21. Even this spacing is probably inadequate to describe the refraction which probably occurs at certain locations, such as Santa Cruz Harbor, Monterey Harbor, and at the head of Monterey Canyon at Moss Landing. For greater detail in these areas, it is necessary to use the original U.S. Coast and Geodetic Survey work sheets.

After the wave crests have been drawn, orthogonals are constructed on the diagrams. The orthogonals are usually started at the shore and carried seaward as perpendiculars to the wave crests until deep water is reached. One method of constructing orthogonals is to cut about a 4-in. square from acetate or other thin transparent drafting plastic (Fig. 7.23). A straight line is then drawn diagonally across the square and two narrow slots cut perpendicular to the line to accommodate a pencil point. Then the line is held tangent to the first wave front and a light line drawn on the forward slot from a point halfway to the next wave front, back toward the first wave front. The square is then moved forward so that the inked line is tangent to the second wave front with the pencil line from the first wave front terminating

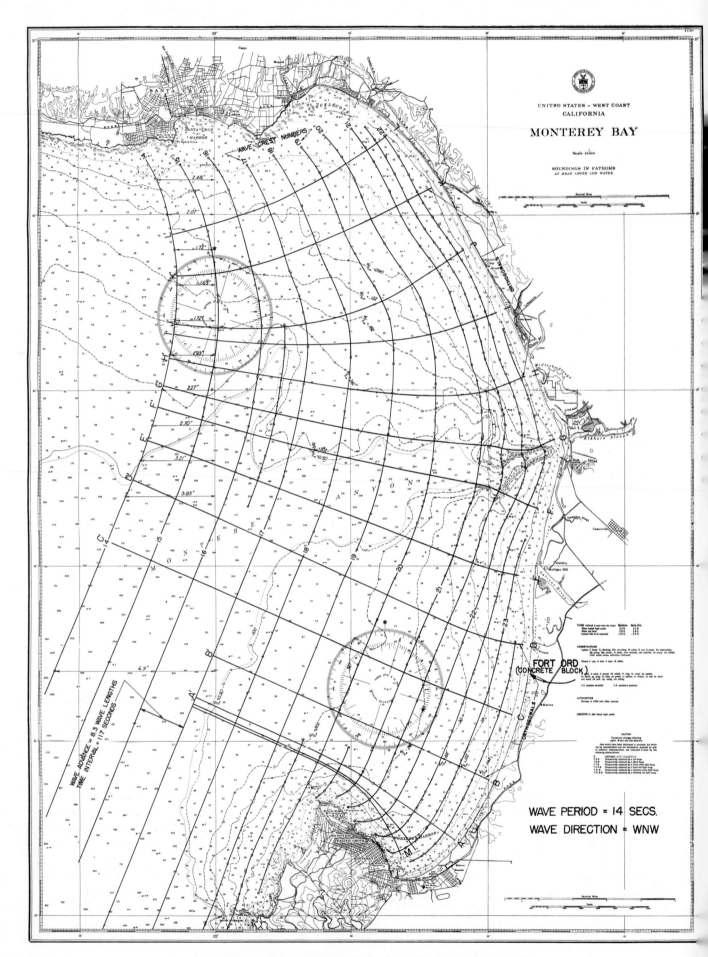

Fig. 7.22. Monterey Bay (From Johnson, O'Brien, and Isaacs, 1948)

166

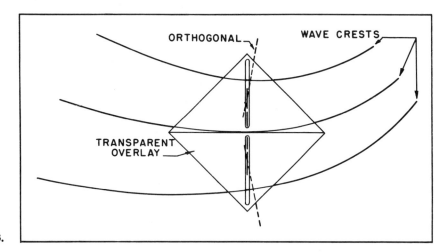

Fig. 7.23.

in the rear slot. A light line is then drawn from a point halfway to the third wave front and back through the rear slot to the first line. The positions of those light lines are usually satisfactory for orthogonals. If the wavefront curvature is great, the orthogonals can be drawn through their intersections with wave fronts. If desired, a smooth curve could be drawn through the points where the perpendiculars cross the crests.

The orthogonals shown on Fig. 7.22 ended at crest 14, and then had to be transferred to Fig. 7.14. In practice, not all the orthogonals can be carried to deep water on the smaller-scale chart nor is it necessary. It was not necessary to carry orthogonal I seaward from crest 14 as orthogonals H and J gave a measure of the refraction for this portion of the wave.

Consider one example of the computation of the refraction coefficient, K_d. It was desired to obtain K_d for a point on the 5-fathom contour midway between orthogonals K and L. On the original drawings of Fig. 7.22, the distance between orthogonals K and L at the 5-fathom line was measured as 2.9 in. and at crest 14 the distance was 1.72 in., whereas on the original drawings of Fig. 7.21 the distance between orthogonals K' and L' was 0.49 in. at crest 14 and 0.12 in. at crest 1. The refraction coefficient for the point midway between orthogonals K and L at the 5-fathom line is

$$K_d = \sqrt{\frac{1.72}{2.9} \times \frac{0.12}{0.49}} = 0.38.$$

On some refraction diagrams, the crests may divide and

Fig. 7.24. Soldiers Beach, Monterey, refraction coefficients

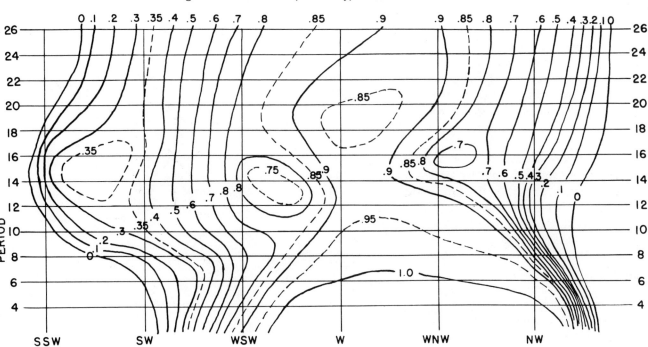

intermingle shoreward of an offshore island or shoal, or underwater ridge, etc. The problem should then be considered as one of two separate wave trains.

g. *Presentation of refraction coefficients.* Refraction diagrams should be prepared for various periods and several deep water wave directions, and the refraction coefficient should be summarized in convenient table or graph form. One of the easiest ways to plot a complex refraction coefficient pattern is similar to drawing contours on topographic maps from spot elevations. A range of periods may be tested as ordinates and a range of different directions as abscissas. The refraction coefficients are then placed on the diagram in appropriate places and the contours drawn by linear interpolation and inspection of the over-all refraction diagrams (See Fig. 7.24).

This information may also be plotted as a polar plot instead of in a rectangular plot with the origin at the point on shore, the various wave periods as concentric circles, and the different directions as radial lines in their true directions. Contours are interpolated as for the rectangular plot. If forecasts are being made for several points along a coast, a convenient means of summarizing the data is a graph which shows K_d factors plotted versus distance along the coast for various wave periods. A separate graph for each wave direction is necessary. Another possible method of summarizing data is to show a map of an area with contours of equal K_d values indicated. A map must be prepared for each wave direction and period.

The assumption of constant wave energy between orthogonals does not apply after a wave breaks as energy is dissipated. If waves pass over a submerged reef, it is necessary to examine this area critically to determine whether the waves break at some, or all, stages of the tide. If the waves break, then the wave heights beyond the reef would be lower than that determined by use of K_d factors from a refraction diagram. In addition, waves passing over a reef transform into several crests and the further refraction of the wave may not be simple. Use of a hydraulic model would probably be necessary for this case.

h. *Graphical construction of refraction diagrams directly by orthogonals.* Another system of determining wave refraction by graphical means is the construction of orthogonals directly without first drawing the wave fronts. This method has the advantage of eliminating an entire graphical step and the drafting errors associated with it. The method is carried out by use of a special protractor which incorporates the proper scales. The protractor is manipulated in steps from contour to contour and at each step indicates the direction of the orthogonal. One orthogonal is drawn from deep water to shore in each series of operations.

The assumptions made in developing this method were (Johnson, O'Brien, and Isaacs, 1948): (1) Contours can be drawn at every abrupt change in slope of the chart. (2) The depth varies linearly between adjacent contours. (3) The wave length and velocity may be considered to vary linearly (the usual assumption) between adjacent contours. (4) The radius of curvature of the orthogonal between contours may be considered constant (a circular arc). (5) The angle of the arc is equal to the change of angle of the orthogonal. (6) The undulations of small magnitude of most contours are more likely to be a measure of observational inaccuracies magnified by rigorous drafting than an indication of the direction of level bottom. The bottom features with dimensions small compared with the wave length do not influence the motion of the wave to any appreciable extent. This justifies the smoothness of the contours. (7) The angle of convergence or divergence of orthogonals at differential intervals is small compared to the angle of refraction. (8) A line drawn through any point midway between two contours and making equal angles with the adjacent contours is approximately the direction of a level line at that point—provided that the requirement implied in assumption (1) is accomplished.

The theory of the orthogonal method of constructing wave refraction drawings is as follows (see Fig. 7.25):

$$\tan \Delta\alpha \approx \sin \Delta\alpha \approx \Delta\alpha = \frac{R_{c2} - R_{c1}}{BB'}$$

$$AA' = OO' = BB' = d \quad \text{(a differential distance)}$$

so $$\Delta\alpha = \frac{R_{c2} - R_{c1}}{d}.$$

For two-place accuracy $\Delta\alpha < 13°$ and for three-place accuracy $\Delta\alpha < 6°$. Now, let C' be the effective velocity

Fig. 7.25. (from Johnson, O'Brien, and Isaacs, 1948)

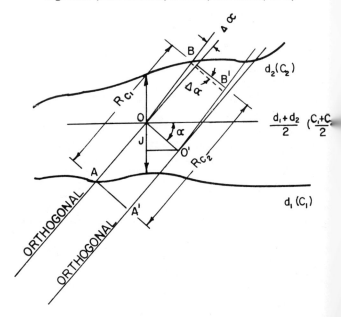

from A to B, C'' be the effective velocity from A' to B', and t be the time required for the wave front to move from AA' to BB'.

Then

$$R_{c2} = C''t; \qquad R_{c1} = C't; \qquad \Delta\alpha = \frac{C't}{d}$$

$$C'' = C' + \Delta C, \quad \text{and} \quad \frac{\Delta C}{C_1 - C_2} = \frac{d \tan \alpha}{J \cos \alpha} = \frac{d \sin \alpha}{J}$$

$$C'' = C' \frac{C_1 - C_2}{J} d \sin \alpha.$$

Then

$$\Delta\alpha = \frac{C_1 - C_2}{J} t \sin \alpha,$$

but

$$t = \frac{R_{C_1}}{C'} \quad \text{and} \quad \frac{C_1 + C_2}{2} = C'.$$

So

$$C' = \frac{R_{C_1}}{J} \frac{C_1 - C_2}{\frac{1}{2}(C_1 + C_2)} \sin \alpha$$

$$\Delta\alpha = \frac{R_c}{J} \frac{\Delta C}{C_{\text{ave}}} \sin \alpha.$$

Hence,

$$\Delta\alpha = \frac{R_c}{J} \frac{\Delta L}{L_{\text{ave}}} \sin \alpha, \qquad (7.17)$$

when $\alpha > 80$ degrees. For the general case, $R_C/J = \sec \alpha$:

so

$$\Delta\alpha = \frac{\Delta L}{L_{\text{ave}}} \tan \alpha. \qquad (7.18)$$

These two equations are independent of the scale of the chart and are independent of the wave period because of the use of the dimensionless ratio $\Delta L/L_{\text{ave}}$.

Equation 7.17 can be used in cases where $\Delta\alpha$ is less than some predetermined limit which depends upon the accuracy desired. Good results are usually obtained as long as $\Delta\alpha$ is less than $13°$, and in practice, it has been found that $\Delta\alpha$ rarely approaches this limit.

It is easier to use Eq. 7.18 than Eq. 7.17 under ordinary conditions. The limitation of Eq. 7.18 stems from the fact that as α approaches 90 degrees, $\tan \alpha$ becomes infinite. This situation may occur but, as it is instantaneously altered by refraction, α becomes less than 90 degrees. The practical difficulty is that the application of Eq. 7.14 normally necessitates crossing an entire contour interval at each step. When α approaches 90 degrees, $\Delta\alpha$ changes rapidly so there is a great change in the rate of refraction, and the interval must be crossed in a series of shorter steps. Equation 7.18 should be used whenever α exceeds about 80 degrees as this equation lends itself to crossing a contour interval in partial steps, provided that R_c (the distance of wave advance) is selected properly.

A special protractor has been constructed to give a graphical solution of Eq. 7.14; it has a table for use when α exceeds 80 degrees which adapts Eq. 7.17 to the graph on the protractor. Since, for almost all ordinary cases, there is no need to refer to this table, it has been made very simple. The graph requires the measurement of only one factor, α, whereas use of the table requires the measurement of three factors, α, R_c, and J. (For details on use of this method the reader is referred to the original publication.)

Other types of protractors can be constructed which are easier to use but more difficult to construct (see Palmer, 1957).

In constructing a refraction drawing directly by orthogonals, first, contours are drawn upon the hydrographic chart at intervals that will adequately represent the details of the bottom topography and which are consistent with assumption (1), just as for the reliability of wave-front method both methods depend upon the amount and accuracy of the hydrographic data. Details on using the orthogonal method for constructing orthogonals from the shore to the sea are given in papers by Saville (1951) and Kaplan (1952).

(i) *Discussion of the methods.* Explanations of two graphical methods, the wave-front and the orthogonal, for constructing refraction diagrams have been presented. Also, two methods of summarizing the refraction coefficients have been described. The presentation for a particular circumstance is largely a choice based upon several factors. If time permits, both methods should be tried to determine which is preferable. In any case, a few diagrams should be checked by both methods to see whether they are in reasonable agreement. As an aid in choosing the basic method to be used for the refraction study, the comparative advantages of each method are presented; circumstances of the specific job will dicate the final choice.

Advantages of the wave-front method

1. Contours need not be drawn over the base hydrography, but they are desirable as an aid to visual interpretation of the results. The wave fronts may be drawn by knowing only the average depth over which the front passes and this depth can be determined from the plotted soundings.

2. Wave crests are shown. Thus the action of the wave in the ocean becomes easier to visualize. In certain cases, the wave crests may be correlated directly with aerial photographs.

3. The method is easy to understand and apply. A person need not understand the complicated wave theory and mathematics involved in the development of the wave-front method to be able to draw the wave fronts as it can be done by following normal drafting and graphical construction methods.

4. When wave fronts are properly drawn, crossed orthogonals are easier to interpret. In fact, they can usually be predicted on the basis of the wave fronts.

5. The method is easy to supervise. A trained supervisor, by close scrutiny of the wave fronts and hydrography, can usually perform a satisfactory detailed visual check of the diagram. If any irregularities are indicated, he can then easily check them with the scales.

Fig. 7.26. Examples of wave refraction: (*top*) Point Pinos, Calif., waves moving over a submarine ridge concentrate to give large wave heights on a point; (*bottom*) Halfmoon Bay, Calif., note the increasing width of the surf zone with increasing degree of exposure to the south

6. Wave-front refraction diagrams are easy to explain to nontechnical people. Progressive positions of a wave front have been indicated and from their shape and curvature most people can visualize whether they are converging or diverging.

7. In certain cases, insignificant wave trains may be eliminated before the wave fronts are carried all the way to shore.

Disadvantages of the wave front method

1. Wave crests are often smoothed in cases when the crest should be severed to indicate crossed orthogonals. Many published reports show that even experienced people have not severed the crests on the refraction diagrams when the hydrography clearly indicated such severance and aerial photos have shown that such severance existed. Nothing in the foregoing presentation of the wave-front method prohibits the drawing of severed crests.

2. Different scales or different chart markings must be used for each wave period considered. Because the scales are not readily changeable, an unreasonable scale might be used. Thus regardless of the scale of the chart, if the hydrography is regular and slopes are gradual, larger wave advances are permissible than if the hydrography is irregular or slopes are steep.

Advantages of the orthogonal method

1. Orthogonals are constructed in a single operation in shorter time. For the majority of cases, only the refraction coefficients are desired and so only orthogonals would be needed.

2. The dimensionless protractor which is used is independent of the scale of the chart and the period of the wave. Thus only one scale need be used for an entire series of refraction diagrams and the rate of advance of the orthogonals may be adjusted to suit the hydrography.

3. Wave fronts may be drawn if needed. Since by definition orthogonals are everywhere perpendicular to wave fronts, the fronts may be easily drawn in wherever needed for illustration.

Disadvantage of the orthogonal method

1. A higher degree of training is required. This method is less straightforward in the graphical stages and requires not only a draftsman to apply it, but also someone with a working knowledge of the fundamental equations.

j. *Examples of refraction.* Various aerial photographs and refraction diagrams are of interest in providing typical examples. Some aerial photographs are shown in Fig. 7.26 and Figs. 14.6–14.8; some examples of refraction drawings are shown in Figs. 14.9–14.11,

k. *Refraction by currents.* The length, steepness, and velocity of waves moving through still water will change when they encounter a current, moving with, against, or at an angle with, the wave direction. When the waves meet a current at an angle, the waves also change their direction. The waves are refracted by currents to an extent which depends upon the initial wave velocity and direction and the strength of the current. The two common conditions treated here are only for deep-water waves (Johnson, 1947).

After waves have run from still, deep water into a deep sound or river where a current runs directly with, or against the advancing waves, the wave period remains constant while the wave length, velocity, and height

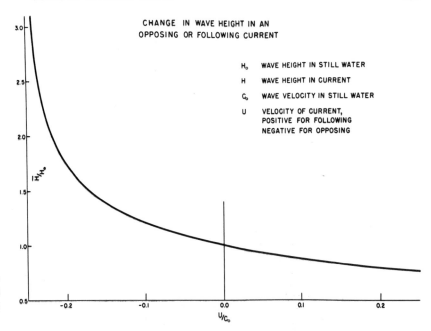

Fig. 7.27. Change in wave height in an opposing or following current (from Scripps Institution of Oceanography, 1944)

change (Scripps Institution of Oceanography, 1944). The wave velocity $\bar{C}$ and wave length $\bar{L}$ relative to the water are given by

$$\frac{\bar{L}}{L_o} = \left(\frac{\bar{C}}{C_o}\right)^2. \qquad (7.19)$$

As the wave velocity over the bottom is equal to $(\bar{C} + U)$, where U is the velocity of the current and

$$T = \frac{L_o}{C_o} = \frac{\bar{L}}{\bar{C} + U} \qquad (7.20)$$

$$\frac{\bar{C}}{C_o} = \frac{1}{2}\left[1 \pm \sqrt{1 + 4\left(\frac{U}{C_o}\right)}\right]. \qquad (7.21)$$

The plus sign must be taken since $\bar{C}$ must equal C_o where $U = O$. Thus if,

$$\sqrt{1 + 4\left(\frac{U}{C_o}\right)} = a$$

$$\frac{\bar{C}}{C_o} = \frac{1}{2}(1 + a),$$

and

$$\frac{\bar{L}}{L_o} = \left(\frac{1 + a}{2}\right)^2. \qquad (7.22)$$

It can be seen that the effect of a following current is to increase the wave length, and the effect of an opposing current is to decrease the wave length.

Consideration of the linear wave power equation gives the ratio of the wave heights as

$$\frac{\bar{H}}{H_o} = \sqrt{\frac{2}{a(1 + a)}} \qquad (7.23)$$

This equation has been plotted in Fig. 7.27 to show the change in wave height in an opposing or following current. The wave steepness is obtained from a combination of Eqs. 7.22 and 7.23,

$$\frac{\bar{H}}{\bar{L}} = \frac{H_o}{L_o}\sqrt{\frac{2}{a(1 + a)}} \cdot \left(\frac{a}{1 + a}\right)^2 \qquad (7.24)$$

where the wave steepness $\bar{H}/\bar{L}$ in the sound is seen to depend upon the initial steepness H_o/L_o and upon the ratio U/C_o.

When a wave travels from the position ABC in deep still water across a current discontinuity to a position $A'B'C'$, the wave length, height, and steepness change (Fig. 7.28). Johnson (1947) has shown that the effect on wave length is

$$\frac{\bar{L}}{L_o} = \frac{1}{[1 - (U/C_o)\sin\alpha]^2} \qquad (7.25)$$

Fig. 7.28. Relative positions of wave crests before and after refraction by a current (from Johnson, 1947)

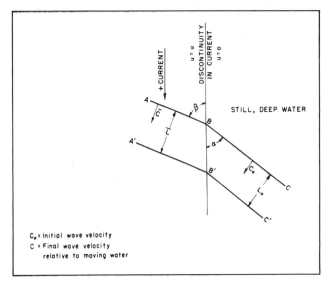

C_o = Initial wave velocity
C = Final wave velocity relative to moving water

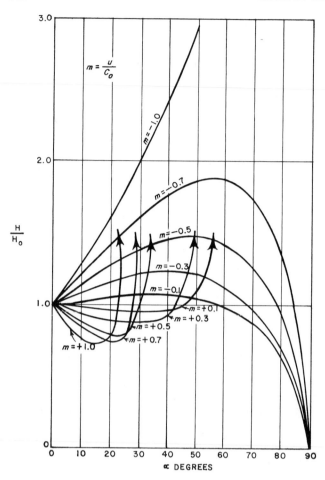

Fig. 7.29. The effect on wave height in the refraction of waves by currents

and that the effect on wave steepness is

$$\frac{\bar{H}}{\bar{L}} = \frac{H_o}{L_o} \sqrt{\frac{\cos \alpha}{\cos \beta} \left\{ \frac{[1 - (U/C_o) \sin \alpha]^6}{1 + (U/C_o) \sin \alpha} \right\}} \quad (7.26)$$

The results of Eq. 7.26 have been plotted in Fig. 7.29, which shows the change in wave height as a function of the strength and direction of the current compared to that of the incident wave. It can be seen that waves which either enter an oncoming or a following current may increase in height (and therefore steepness) and then break. Examination of Fig. 7.29 shows that, when the value of $U/C_o = +0.1$ (a following current), for example, all waves will rapidly increase in height (and therefore break) if the incident angle is larger than about 58 degrees. The waves will break at smaller incident angles for larger values of U/C_o. Similar equations could be developed for waves in transitional water depths.

(1) *Difficulties encountered in the construction of refraction diagrams.* One should constantly refer to the basic assumptions in the methods for constructing refraction diagrams to see whether they agree reasonably with the practical factors. One should also use every means to check the refraction picture which has been

developed graphically. Aerial photos and visual observations are two of the best practical checks of the diagrams. Some of the irregularities which may be encountered are discussed later.

Refraction theory is based upon the assumption that waves are long crested. If the waves are short crested, velocities are higher than predicted, the refraction effect is larger, and if the angle of incidence of the wave to the contours is large, the refraction coefficients are likely to be too small for extreme cases of diverging orthogonals or too large for similarly extreme cases of convergence.

Wave velocities are computed using the formula based on the irrotational theory for waves of infinitely small height. These velocities are somewhat too small for waves of finite height. In general, the greater the deep water steepness of the wave and the shallower the depth, the greater the variation of velocities computed from the two theories. Thus if the wave front is convex and orthogonals are diverging, the measured refraction coefficient will be too small; if the wave front is concave (convergence), the refraction coefficient will be too large.

Refraction diagrams are drawn for wave trains of constant period, whereas in nature the periods of the waves vary in a statistical manner. A thorough discussion of this problem, from the standpoint of linear theory, has been made by Longuet-Higgins (1955).

Refraction diagrams are drawn for waves in deep water from specific directions, while the spectra of waves are two-dimensional. Small changes in direction from the directions investigated should be examined visually to determine whether more diagrams are required from slightly different directions in the case of, say, offshore islands (Arthur, 1951).

Refraction by unknown currents of transverse set may occur in the case of a train of waves progressing into a current at an angle. Very few ocean currents are known which have stable boundaries and known velocity distributions across them. Idealized theoretical studies have been made of the additional refraction effect of currents and many actual observations of the effect of currents on waves have been made (Johnson, 1947; Arthur, 1950).

Diffraction along the wave crest may occur. An extreme convergence of orthogonals indicates an excessive wave height in that area, but if some of the wave energy flows laterally along the crest, the wave height will not be as high as indicated by the refraction study. Similarly, a zone of extreme divergence may indicate wave heights lower than actual.

Crossed orthogonals are a phenomenon which has occupied much attention in published reports (see Pierson, 1951, for example). They probably represent either a severance of the wave crest and subsequent intermingling of the resulting pair of wave trains or a peaking up and breaking of the waves. The only difficulties which should arise over the question of wave-crest severance are the relative sizes of the underwater obstacles, rapidly changing hydrography, or the relative size of islands which will cause an appreciable intermingling of crests. Bottom discontinuities, such as breakwaters, ledges, abrupt changes in slope, etc., if of appreciable size and extent may have a large effect on the refraction of the waves and should be examined carefully. Quantitative information as to the relationships of the size of these discontinuities to the size of the waves is not now available and these factors are at present largely a matter of reasonable judgment. A model study of such a region would be very valuable for an actual case.

The hydrographic charts must be examined carefully to determine their reliability. One should be suspicious of each chart until he has checked its source, date of survey, and general character to see whether it rings true. He should also check it against every other known bit of information available. Sometimes spot checks can be made by limited surveys, reconnaisance, or aerial photos. Adjacent areas to the coast under consideration should be examined to see whether rivers are dumping sediments into the ocean or nearby areas are being eroded rapidly in such a manner as to change the hydrography.

The irregularities in the developed refraction patterns which have just been described may seem formidable. They may seem to provide reason for not developing the refraction picture at all. But remember that they will probably never all be combining at once to distort the refraction picture. Whether or not they are known or suggested by the conditions in a particular case can help to establish the reliability of the refraction picture. They should always be investigated and evaluated as closely as possible. Even if all the foregoing factors combine to form the most unfortunate conditions, and even if hydrographic information is scant, a refraction picture of some sort should always be developed.

m. *Group refraction.* Wave speed decreases continuously with decreasing water depth, but the group speed, C_G, at first increases and then decreases (Fig. 2.4). As C_G is the speed at which energy is propagated, one would expect that wave crests which were not infinitely long would have a shift in the energy density in such a manner as to cause the waves to have an echelon type of distribution (Isaacs, Williams, and Eckart, 1951). Trains of irregular waves would present some interesting characteristics. These have been studied to some extent by Williams and Isaacs (1952), but the physical significance is not clear.

5. BREAKERS AND SURF

Linear theory predicts and measurements confirm that waves moving from deep into shoal water continually decrease in length and speed, whereas the height at first decreases and then increases. It is said that the waves "feel bottom" in shoal water. Because of this, the waves lose energy due to bottom friction, percolation, and nonrigidity of the bottom. This, however, is not the most important thing that occurs to the wave. When waves within a certain range of steepness (H/L) move over a beach within a certain range of slopes, and these waves enter water approximately as deep as the wave is high, the waves become unstable and break. The depth of water in which the waves break depends upon what the wave has done before; not only is the individual wave unstable, but the regime is unsteady. When waves peak up and break on a beach of very small slope, there is no appreciable reflection, and it appears that most of the energy in a wave length (not dissipated by bottom friction) has become concentrated in the kinetic and potential energy of the crest and this mechanical energy is reduced to heat by the turbulence induced by the break. At the other extreme are waves which encounter a vertical wall in water deep enough so that they are reflected with very small loss of energy. Between these two extremes are waves in which a portion of the energy is dissipated by the break and in which another portion of the energy is reflected.

There are three types of breakers: spilling, plunging, and surging (Fig. 7.30). Spilling breakers are characterized by the appearance of "white water" at the crest; they break gradually. Plunging breakers are characterized by a curling over of the top of the crest and a plunging down of this mass of water; the front of the crest first becomes steep and then concave. Surging breakers peak up as if to break in the manner of a plunging breaker; but then the base of the wave surges up the beach face with the resultant disappearance of the collapsing wave crest; it is similar to a standing wave.

During World War II, a relationship was obtained relating the breaker height H_b and the depth of water in which the wave broke to the unrefracted deep water height H_o' and the deep water wave steepness H_o/L_o (U.S. Navy Hydrographic Office, 1955). This relationship was obtained primarily from measurements of waves on ocean beaches of medium steepness. It soon became apparent, when working on steep beaches, that beach slope was also a factor and a laboratory study was performed (Iversen, Crooke, Larocco, and Wiegel, 1950; Hamada, 1951; Iversen, 1952). The results of these tests are shown in Fig. 7.31. In addition, other useful data were obtained such as the elevation of the wave crest above the bottom at the point of breaking.

As shown in Fig. 7.32, for the steeper deep water waves, H_o/L_o greater than about 0.06, all waves tend to spill. For H_o/L_o between about 0.06 and 0.03, the waves tend to spill on the flatter beaches and plunge on the steeper beaches. For H_o/L_o between 0.03 and 0.009, all waves tend to plunge. For H_o/L_o flatter than 0.009, the waves tend to surge on the steeper beaches and to plunge

Spilling breaker

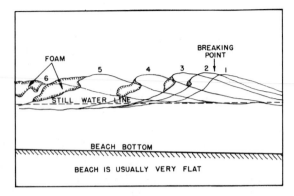

General character of spilling breakers

Plunging breaker

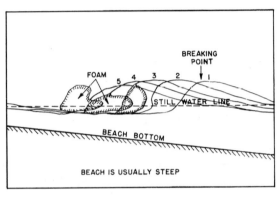

General character of plunging breakers

Surging breaker

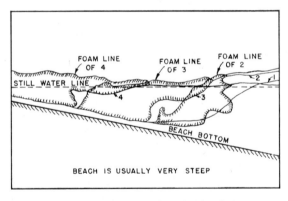

General character of surging breakers

Fig. 7.30. The three types of breakers. The sketches consist of a series of profiles of the wave form as it appears before breaking, during the breaking, and after breaking; the numbers opposite the profile lines indicate the relative times of the occurrences (after Patrick and Wiegel, 1955)

on the flatter beaches. In general, spilling breakers are associated with steep deep water waves and flat beaches, plunging breakers are associated with waves of intermediate steepness and the steeper beaches, and surging breakers are associated with flat waves and steep beaches. Breakers are classified in the foregoing three classes for convenience, but they shade from one type into another with an infinite number of variations.

A re-analysis of the laboratory data was made by Wiegel and Beebe (1956) who found that one factor was

apparently independent of the beach slope or the deep water wave steepness. This factor was the ratio of the portion of the wave height above the undisturbed water level to the wave height $(Y_b - d_b)/H_b$. This factor was about 0.78. This agrees quite well with the rule of thumb used in taking soundings through the surf zone that the trough is a quarter of the breaker height below the undisturbed water level.

Changes in the character of surf in the ocean may occur in a few hours because of the effects of tide. Thus,

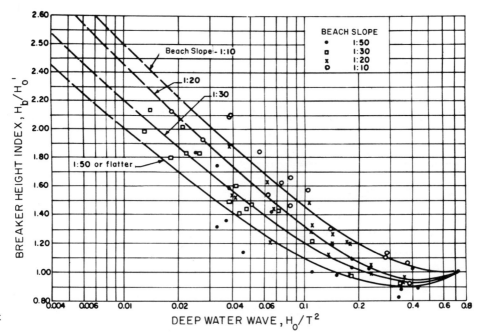

(a) Breaker height

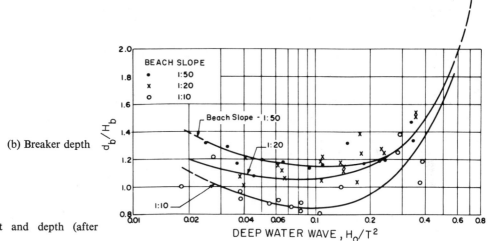

(b) Breaker depth

Fig. 7.31. Breaker height and depth (after Iversen, 1953)

Fig. 7.32. Variation of breaker characteristics with deep water wave steepness and offshore beach steepness (from Patrick and Wiegel, 1955)

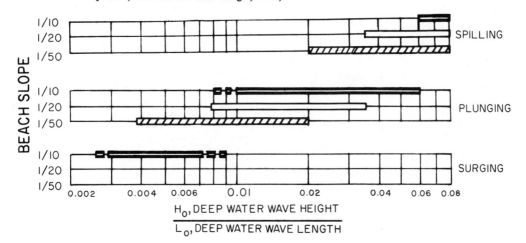

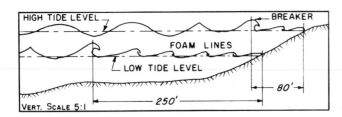

(a) Example of effect of tide on width of surf zone. At high tide the beach has the character of a steep beach with a narrow surf zone; at low tide it appears as a flat beach with a wide surf zone.

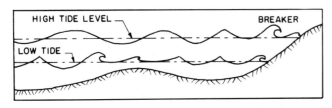

(b) Effect of tide on character of surf zone in the presence of a bar. At high tide there is a single line of breakers on the beach; at low tide there are two lines—one on the bar and one on the beach.

Fig. 7.33. Effect of tide on surf zone (from Patrick and Wiegel, 1955)

for the beach profile illustrated in Fig. 7.33(a), at high tide the beach has the character of a steep beach with a narrow surf zone; whereas at low tide it appears as a flat beach with a wide surf zone. For a beach with an offshore bar, the effect of this tide on the character of the surf zone is illustrated in Fig. 7.33(b). Thus, at high tide there is a single breaker line, whereas at low tide the surf zone has two breaker lines: one on the bar and one at the beach face. On some beaches, particularly on a wide flat beach, several longshore bars are often present, in which case several lines of breakers exist. On such beaches, the waves re-form (with a lower height) and break again on each successive bar with areas of relative calm between them.

Iverson (1952) studied the motions of water particles in breakers. A sample of the data is given in Fig. 7.34. Motion pictures were taken of the breaker, with small white particles of the same specific gravity as water placed in the waves. The motion pictures were analyzed frame by frame and the displacement of the particles divided by the time between frames. The velocities obtained in this manner were plotted as vectors as shown in Fig. 7.34(c).

The data of Iversen were re-analyzed by Wiegel and Beebe (1956), using the suggestions of Munk (1949) in regard to the possible prediction of breaker characteristics. The controlling parameter in the solitary wave theory is the ratio of wave height to undisturbed water depth, γ. The maximum value this parameter can have is 0.78 as the wave becomes unstable for this value. The waves we are discussing are not solitary and the trough lies below the still water level. Munk (1949) suggested that γ be taken as the ratio of wave height to the water depth below the trough. Wiegel and Beebe found that the measured values were always greater than 0.78. It was found that a more reliable criterion was $\gamma' = (Y_b - d_b)/H_b$ as mentioned previously. It appears that, insofar as water particle velocities directly

under the crest are concerned, a modified solitary wave theory permits some useful predictions.

Using a series of velocity vector plots of the type shown in Fig. 7.34(c), Wiegel and Skjei (1958) constructed vectors of water particle accelerations. The measured accelerations were compared with the modified solitary wave theory and the correlation was generally poor.

In the past, attempts have been made to predict breaker characteristics from solitary wave theory, and certain useful criteria have been developed. A solitary wave is not oscillatory in character, however, in that there is no reversal of flow of the water particles because it is a wave of translation; nor is there a trough, all of the wave being above the undisturbed water level. The solitary wave is merely the limiting case of a general class known as *cnoidal waves*. As much of the theory of this class of waves is now in usable form (see Chapter 2), it would be desirable to develop breaker characteristics from the cnoidal wave theory.

6. MULTIPLE CRESTS

Another phenomenon that occurs in shoaling water is the transformation of a wave into a multiple-crested wave. One circumstance under which it has been observed is the case of a wave passing over a reef. This case might be described as one in which the amount of momentum transferred over the reef edge is too great for the shallow water over the reef to transmit, but not enough to cause the wave to break. The relationship between breaking and transforming into multiple crest might be similar to a hydraulic jump and an undular jump in open channel flow.

A similar phenomenon has been observed when a relatively long wave reflects from a beach without breaking. In this case, the reflected wave transforms into a multiple-crest system.

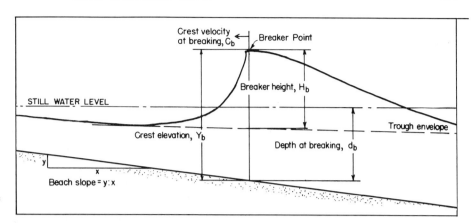

(a) Terminology for a breaking wave
(after Iversen, 1952)

(b) Photograph of breaker in laboratory

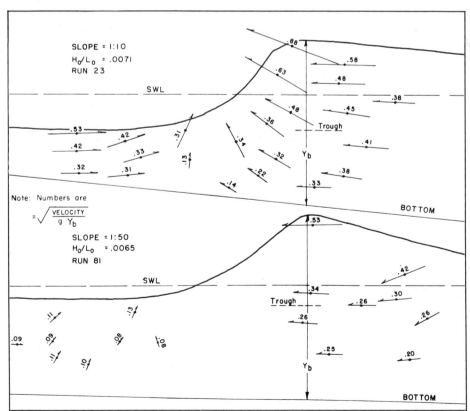

(c) Kinematics of a breaking wave
(after Iversen, 1952)

Fig. 7.34.

REFERENCES

Arthur, R. S., Refraction of waves by islands and shoals with circular bottom contours, *Trans. Amer. Geophys. Union*, **27**, 2 (April, 1946), 168–77.

———, Refraction of shallow water waves: the combined effects of currents and underwater topography, *Trans. Amer. Geophys. Union*, **31**, 4 (August, 1950), 549–52.

———, The effect of islands on surface waves, *Bull. Scripps Inst. Oceanog.*, **6**, 1 (1951), 1–26.

Chien, Ning, Ripple tank studies of wave refraction, *Trans. Amer. Geophys. Union*, **35**, 6 (December, 1954), 897–904.

Fuchs, R. A., On the theory of short-crested oscillatory waves, *Gravity Waves*, U.S. Dept. of Commerce, National Bureau of Standards, Circular No. 512 (November, 1952), 187–200.

———, *Travel time for periodic waves on beaches of small constant slope*, Univ. Calif., Berkeley, IER, Tech. Rept. 29–55, 1954. (Unpublished.)

———, On the theory of irregular waves, *Proc. First Conf. Ships and Waves*, Berkeley, Calif.: The Engineering Foundation Council on Wave Research, 1955, 1–10.

Hamada, Tokuichi, Breakers and beach erosions, *Rept. Transportation Tech. Res. Inst.*, Tokyo, Report No. 1, December, 1951.

Inman, D. L. and Noriyuki Nasu, *Orbital velocity associated with wave action near the breaker zone*, U.S. Army, Corps of Engineers, Beach Erosion Board, Tech. Memo. 79, March, 1956.

Isaacs, J. D., The refraction of surface waves by currents: a discussion, *Trans. Amer. Geophys. Union*, **29**, 5 (October, 1948), 739–42.

———, E. Allan Williams and Carl Eckart, Reflection of surface waves by deep water, *Trans. Amer. Geophys. Union*, **32**, 1 (February, 1951), 37–40.

Iversen, H. W., "Waves in shoaling water," in *Manual of Amphibious Oceanography*, ed. R. L. Wiegel, Univ. Calif., Contract N7onr–29535 Office of Naval Research, 1951.

———, Waves and breakers in shoaling water, *Proc. Third Conf. Coastal Eng.*, Berkeley, Calif.: The Engineering Foundation Council on Wave Research, 1953, 1–12.

———, *Laboratory study of breakers*, U. S. National Bureau of Standards, Circular No. 521 (November, 1952), 9–31.

———, R. C. Crooke, M. J. Larocco and R. L. Wiegel, *Beach slope effect on breakers and surf forecasting*, Univ. Calif., Berkeley, IER, Tech. Rept. 29–38, December, 1950.

Johnson, J. W., The refraction of surface waves by currents, *Trans. Amer. Geophys. Union*, **28**, 6 (December, 1947), 867–74.

———, Engineering aspects of diffraction and refraction, *Proc. ASCE*, **118**, Paper No. 2556 (1953), 617–52.

———, M. P. O'Brien and J. D. Isaacs, *Graphical construction of wave refraction diagrams*, U.S. Navy Hydro. Office Pub. No. 605, January, 1948.

Kaplan, Kenneth, A method for drawing orthogonals seaward

from shore—discussion, *Bull. Beach Erosion Board*, U.S. Army, Corps of Engineers, **6**, 1 (January, 1952), 18–21.

Lapsley, W. W., Sand movement and beach erosion, M.S. thesis in Civil Engineering, Univ. Calif., Berkeley. 1937. (Unpublished.)

Longuet-Higgins, M. S., The refraction of sea waves in shallow water, *J. Fluid Mech.*, **1**, part 2 (July, 1956), 163–76.

Lowell, S. C., The propagation of waves in shallow water, *Comm. Pure Applied Math.*, **2**, 2/3 (1949), 275–91.

Munk, Walter H., The solitary wave theory and its application to surf problems, *Ocean Surface Waves*, *Annals New York Acad. Sci.*, **51**, Art. 3 (May, 1949), 376–423.

O'Brien, M. P., *A summary of the theory of oscillatory waves*, U.S. Army, Corps of Engineers, Beach Erosion Board, Tech. Rept. No. 2, 1942.

Palmer, R. Q., Wave refraction plotter, *Bull. Beach Erosion Board*, U.S. Army, Corps of Engineers, **11**, 1 (July, 1957), pp. 13–16.

Patrick, D. A. and R. L. Wiegel, Amphibian tractors in the surf, *Proc. First Conf. on Ships and Waves*, The Engineering Foundation Council on Wave Research and the American Society of Naval Architects and Marine Engineers, 1955, 397–422.

Pierson, Willard J., Jr., *The interpretation of crossed othogonals in wave refraction phenomena*, U.S. Army, Corps of Engineers, Beach Erosion Board, Tech. Report No. 21, January, 1951.

Pocinki, Leon S., The application of conformal transformation to ocean wave refraction problems, *Trans. Amer. Geophys. Union*, **31**, 6 (December, 1950), 856–66.

Putz, R. R., Prediction of wave motion, *Proc. First Conf. Ships and Waves*, The Engineering Foundation Council on Wave Research and the Society of Naval Architects and Marine Engineers, 1955, 33–54.

Ralls, G. C., A ripple tank study of wave refraction, *J. Waterways Div.*, ASCE, **82**, Paper No. 911, March, 1956.

Rayleigh, Lord, Hydrodynamical notes, *Phil. Mag.*, ser. 6, **21**, 122 (February, 1911), 177–87.

Saville, Thorndike, Jr., A method for drawing orthogonals seaward from shore, *Bull. Beach Erosion Board*, U.S. Army, Corps of Engineers, **5**, 4 (October, 1951), 1–6.

Scripps Institution of Oceanography, *On wave heights in straits and sounds when incoming waves meet a strong tidal current*, Scripps Inst. Oceanog., Apr. 20, 1944. (Unpublished.)

Stoker, J. J., *Water waves*. New York: Interscience Publishers, Inc., 1957.

U.S. Navy Hydrographic Office, *Breakers and surf: principles in forecasting*, H. O. No. 234, November, 1944.

Unna, P. J. H., Waves and tidal streams, *Nature*, **149**, 3773, (1942), 219.

Wiegel, R. L. and A. L. Arnold, *Model study of wave refraction*, U.S. Army, Corps of Engineers, Beach Erosion Board, Tech. Memo. No. 103, December, 1957.

——— and K. E. Beebe, The design wave in shallow water, *J. Waterways Harbors Div.*, ASCE, **82**, WW1, Paper No. 910, March, 1956.

——— and R. A. Fuchs, Wave transformation in shoaling water, *Trans. Amer. Geophys. Union*, **36**, 6 (December, 1955), 975–84.

——— and R. E. Skjei, Breaking wave force prediction, *J. Waterways Harbors Div.*, ASCE, **84**, WW2, Paper No. 1573, March, 1958.

Williams, E. Allan and John D. Isaacs, The refraction of groups and of the waves which they generate in shallow water, *Trans. Amer. Geophys. Union*, **33**, 4 (August, 1952), 523–30.

CHAPTER EIGHT

Wave Diffraction

1. INTRODUCTION

Consider a wave system interrupted by an impervious structure such as a breakwater (Fig. 8.1). The portion of the waves incident to the structure will be reflected, or break, or both, whereas the portion moving past the tip of the structure will be the source of a flow of energy in the direction essentially along the wave crest and into the region in the lee of the structure. The "end" of the wave will act somewhat as a potential source and the wave in the lee of the breakwater will spread out in approximately a circular arc with the amplitude decreasing exponentially along this arc (see Fig. 8.4). The same phenomenon will also occur in the reflected portion of the wave. This complicates the physical picture considerably, as part of the wave energy associated with the "radial" wave being generated from the "end" of the reflected wave will move into the harbor region. The two sets of waves, cylindrical and radial, reinforce and cancel each other in such a manner as to cause an irregular wave height in this region. This physical phenomenon

is known as *diffraction*. There are mathematical solutions of some specific problems which have been taken from the theory of acoustic and light waves and applied to water waves for the case of water of constant depth, and for impermeable rigid structures. Several solutions of considerable practical importance are presented in this chapter.

2. SEMI-INFINITE BREAKWATER

Penny and Price (1944; 1952) showed that the Sommerfeld solution of the diffraction of light (polarized in a plane parallel to the edge of a semi-infinite screen) is also a solution of the water wave diffraction phenomenon. The water surface elevation (linear theory) can be expressed as

$$y_s = \frac{Aik\,C}{g} \, e^{ikCt} \cosh kd \cdot F(x, z) \qquad (8.1)$$

where the real part of the expression on the right is used.

Fig. 8.1. Diffraction of ocean waves at Morro Bay, California

For the case of progressive waves traveling in the direction of the x axis with no structures present, $F(x, z) = e^{-ikx}$, so

$$y_s = \frac{Aik\,C}{g}\,e^{ik(Ct-x)}\cosh kd$$
$$= a\sin k(Ct - x). \qquad (8.2)$$

The diffraction coefficient, K', is defined as the ratio of the wave height in the area affected by diffraction to the wave height in the area unaffected by diffraction; it

Fig. 8.2. Nomenclature for wave diffraction analysis at breakwater tip

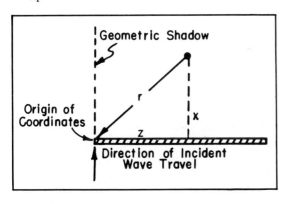

is the ratio of the amplitude of Eq. 8.1 to 8.2. It is given by the modulus of $F(x, z)$ for the diffracted wave

$$K' = |F(x, z)| \qquad (8.3)$$

as the modulus of $F(x, z)$ for the incident wave is $|e^{-ikx}| = 1$. The phase difference is given by the argument of $F(x, z)$ for the diffracted wave plus kx.

Now consider an infinitely thin, vertical, rigid, impermeable, semi-infinite breakwater located as shown in Fig. 8.2. Sommerfeld (1896) showed (see Penny and Price 1952) that

$$F(x, z) = \frac{1 + i}{2}\left\{ e^{-ikx}\int_{-\infty}^{\sigma} e^{-\pi iu^2/2}\,du \right.$$
$$\left. + e^{ikx}\int_{-\infty}^{-\sigma'} e^{-\pi iu^2/2}\,du \right\}. \qquad (8.4)$$

Here

$$\sigma^2 = \frac{4}{L}(r - x) \quad \text{and} \quad \sigma'^2 = \frac{4}{L}(r + x) \qquad (8.5)$$

where $r^2 = x^2 + z^2$. The signs of σ and σ' are to be taken as shown in Fig. 8.3.

Equation 8.4 can be transformed into a form that allows the use of tabulated functions

$$\frac{1 + i}{2}\int_{-\infty}^{\sigma} e^{-\pi iu^2/2}\,du$$

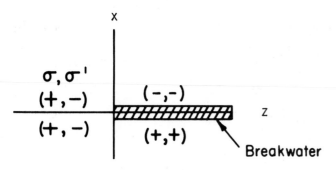

Fig. 8.3. Signs of σ and σ^1

$$= \frac{1 + i}{2}\left[\int_{-\infty}^{0} e^{-\pi i u^2/2}\, du + \int_{0}^{\sigma} e^{-\pi i u^2/2}\, du\right]$$

$$= \frac{1 + i}{2}\left[\frac{1}{2}\int_{-\infty}^{\infty} e^{-\pi i u^2/2}\, du + \int_{0}^{\sigma} e^{-\pi i u^2/2}\, du\right]$$

where σ is the value of u used in the upper limit of the definite integral. Bierens de Haan (1959, p.55) gives

$$\int_{-\infty}^{\infty} e^{-\pi i u^2/2}\, du = (1 - i).$$

Also,

$$\int_{0}^{\sigma} e^{-\pi i u^2/2}\, du = \int_{0}^{u=\sigma} \cos \frac{1}{2}\pi u^2\, du - i\int_{0}^{u=\sigma} \sin \frac{1}{2}\pi u^2\, du$$

$$= \int_{0}^{t=\pi\sigma^2/2} \frac{e^{-it}}{\sqrt{2\pi t}}\, dt = C - iS$$

where $t = \pi u^2/2$, and C and S are Fresnel integrals (Jahnke and Emde, 1945, p. 36). C and S are tabulated by Jahnke and Emde, and in great detail (0.01 steps, from 0.00 to 20.00) by Van Wijngaarden and Scheen (1949). Finally,

$$f(\sigma) = \frac{1 + i}{2}\int_{-\infty}^{\sigma} e^{-\pi i u^2/2}\, du$$

$$= \tfrac{1}{2}[(1 + C + S) - i(S - C)], \quad \text{for } \sigma \text{ positive} \tag{8.6a}$$

$$f(-\sigma) = \frac{1 + i}{2}\int_{-\infty}^{-\sigma} e^{-\pi i u^2/2}\, du$$

$$= \tfrac{1}{2}[(1 - S - C) + i(S - C)], \quad \text{for } \sigma \text{ negative}$$

$$= U + iW \tag{8.6b}$$

and $\qquad\qquad f(\sigma) + f(-\sigma) = 1. \tag{8.6c}$

Equation 8.6b has been computed and put into graphical form by Blue (1948; also, Wiegel, 1962).

Using Eq. 8.6c and the appropriate signs for σ and σ', it is found that

$$F(x, z) = e^{-ikx} - e^{-ikx}f(-\sigma) + e^{ikx}f(-\sigma'), \quad \text{for } z \leq 0$$

$$= e^{-ikx}f(-\sigma) + e^{ikx}f(-\sigma'), \quad \text{for } \begin{cases} z \geq 0 \\ x \geq 0 \end{cases}$$

$$= e^{-ikx} + e^{ikx} - e^{-ikx}f(-\sigma) - e^{ikx}f(-\sigma'),$$

$$\text{for } \begin{cases} z \geq 0 \\ x \leq 0. \end{cases}$$

$$\tag{8.7}$$

Obtaining numerical results from the preceding solution is time-consuming, as can be seen by considering only one case:

$$K' = |F(x, z)| = |e^{-ikx}f(-\sigma) + e^{ikx}f(-\sigma')| \tag{8.8a}$$

$$K' = \{[(U_1 + U_2)\cos kx + (W_1 - W_2)\sin kx]^2$$
$$+ [(W_1 + W_2)\cos kx + (U_2 - U_1)\sin kx]^2\}^{1/2}$$
$$\tag{8.8b}$$

where $f(-\sigma) = U_1 + iW_1$ and $f(-\sigma') = U_2 + iW_2$.

For any x/L, z/L, σ, and σ' are computed, U_1, W_1, U_2, and W_2 calculated or obtained from published graphs, $\cos kx$ and $\sin kx$ obtained, and K' computed. This is done for a large number of x/L, z/L, and contours of equal K' drawn. The calculations can be simplified by choosing values of z/L as 0, 1/4, 1/2, 3/4, 1, etc., so that $\sin kx = 0$, $\cos kx = \pm 1$, or $\sin kx = \pm 1$, $\cos kx = 0$. Then

$$K' = \sqrt{(U_1 + U_2)^2 + (W_1 + W_2)^2}, \quad \frac{x}{L} = 0, \frac{1}{2}, 1, \text{ etc.}$$
$$\tag{8.8c}$$

$$K' = \sqrt{(W_1 - W_2)^2 + (U_2 - U_1)^2}, \quad \frac{x}{L} = \frac{1}{4}, \frac{3}{4}, \frac{5}{4}, \text{ etc.}$$
$$\tag{8,8d}$$

For example, choose $z/L = 5$ and $x/L = 0$. $(r/L)^2 = (z/L)^2 + (x/L)^2 = 25$; $\sigma^2 = 4 \times 5 = 20$; $\sigma'^2 = 4 \times 5 = 20$; $U_1 = U_2 = +0.036$; $W_1 = W_2 = -0.034$; $K' = [(0.036 + 0.036)^2 + (-0.034 - 0.034)^2]^{1/2} \approx 0.1$. Choose $z/L = 5$ and $x/L = 5/4$. $(r/L)^2 = 5^2 + (5/4)^2 = 26.6$, $r/L = 5.15$; $\sigma^2 = 4(5.15 - 1.25) = 15.6$; $\sigma'^2 = 4(5.15 + 1.25) = 25.6$; $U_1 = +0.057$; $U_2 = -0.043$; $W_1 - 0.008$; $W_2 = +0.006$; $K' = [(-0.008 - 0.006)^2 + (-0.043 - 0.057)^2]^{1/2} = 0.10$.

In a similar manner,

$$K' = |F(x, z)| = |e^{-ikx} - e^{-ikx}f(-\sigma) + e^{ikx}f(-\sigma')|$$
$$= \{[(1 - U_1 + U_2)\cos kx + (-W_1 + W_2)\sin kx]^2$$
$$+ [(-W_1 + W_2)\cos kx + (1 + U_1 + U_2)\sin kx]^2\}^{1/2}$$
$$\tag{8.9a}$$

which reduces to

$$K' = \{(1 - U_1 + U_2)^2 + (-W_1 + W_2)^2\}^{1/2},$$
$$\text{for } \frac{x}{L} = 0, \frac{1}{2}, 1, \text{ etc.,} \tag{8.9b}$$

$$K' = \{(-W_1 + W_2)^2 + (1 + U_1 + U_2)^2\}^{1/2},$$
$$\text{for } \frac{x}{L} = \frac{1}{4}, \frac{3}{4}, \frac{5}{4}, \text{ etc.} \tag{8.9c}$$

The phase of the waves can be obtained from arg $F(x,z)$.

A plot of the values of K' and the locations of the wave fronts are in Fig. 8.4. As the graph coordinates are in terms of wave lengths, the graph can be used as an overlay on any chart by enlarging it or reducing it to the appropriate scale.

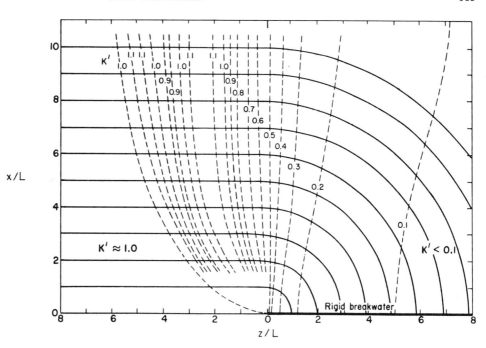

Fig. 8.4. Wave fronts and lines of equal diffraction coefficients for a semi-infinite rigid impervious breakwater, with the incident waves normal to the breakwater (after Penney and Price, 1952)

For the more general case of waves approaching the breakwater at any angle θ_0 (Fig. 8.5) Penny and Price (1952) show

$$F(r, \theta) = f(\sigma) e^{-ikr \cos(\theta - \theta_0)} + f(\sigma') e^{-ikr \cos(\theta + \theta_0)} \quad (8.10)$$

where

$$\sigma = 2\sqrt{\frac{kr}{\pi}} \sin\frac{1}{2}(\theta - \theta_0)$$

and   (8.11)

$$\sigma' = -2\sqrt{\frac{kr}{\pi}} \sin\frac{1}{2}(\theta + \theta_0).$$

In Region S, the numerical values of both σ and σ' are negative for any θ between 0 and π, so using Eqs. 8.6b, 8.6c, 8.10, and 8.11, it can be shown that

Fig. 8.5. Notation for the regions Q, R, and S for oblique incidence of waves on a breakwater (after Penney and Price, 1952)

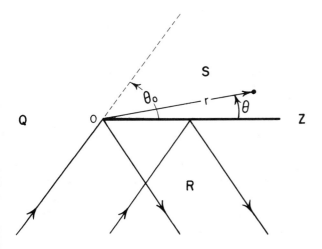

$$F(r, \theta)_s = \{U_1 \cos[kr \cos(\theta - \theta_0)]$$
$$+ U_2 \cos[kr \cos(\theta + \theta_0)]$$
$$+ W_1 \sin[kr \cos(\theta - \theta_0)]$$
$$+ W_2 \sin[kr \cos(\theta + \theta_0)]\}$$
$$+ i\{W_1 \cos[kr \cos(\theta - \theta_0)] \quad (8.12a)$$
$$+ W_2 \cos[kr \cos(\theta + \theta_0)]$$
$$- U_1 \sin[kr \cos(\theta - \theta_0)]$$
$$- U_2 \sin[kr \cos(\theta + \theta_0)]\}$$
$$= A + iB$$
$$|F(r, \theta)_s| = \sqrt{A^2 + B^2}.$$

In Region Q, the numerical value of σ is positive and the numerical value of σ' is negative; hence,

$$F(r, \theta)_Q = \{\cos[kr \cos(\theta - \theta_0)]$$
$$- U_1 \cos[kr \cos(\theta - \theta_0)]$$
$$- W_1 \sin[kr \cos(\theta - \theta_0)]$$
$$+ U_2 \cos[kr \cos(\theta + \theta_0)]$$
$$+ W_2 \sin[kr \cos(\theta + \theta_0)]\}$$
$$+ i\{-\sin[kr \cos(\theta - \theta_0)] \quad (8.12b)$$
$$- W_1 \cos[kr \cos(\theta - \theta_0)]$$
$$+ U_1 \sin[kr \cos(\theta - \theta_0)]$$
$$+ W_2 \cos[kr \cos(\theta + \theta_0)]$$
$$- U_2 \sin[kr \cos(\theta + \theta_0)]\}$$
$$= E + iF$$
$$|F(r, \theta)_Q| = \sqrt{E^2 + F^2}.$$

In Region R, the numerical values of σ and σ' are both

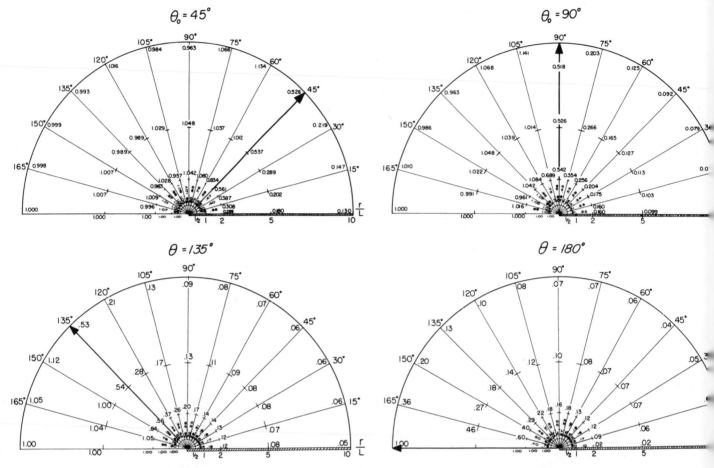

Fig. 8.6. Diffraction coefficients as a function of incident wave angle for a semi-infinite rigid impermeable breakwater (after Wiegel, 1962)

positive; hence,

$$F(r, \theta)_R = \{\cos [kr \cos (\theta - \theta_0)]$$
$$+ \cos kr [\cos (\theta + \theta_0)]$$
$$- U_1 \cos [kr \cos (\theta - \theta_0)]$$
$$- U_2 \cos [kr \cos (\theta + \theta_0)]$$
$$- W_1 \sin [kr \cos (\theta - \theta_0)]$$
$$- W_2 \sin [kr \cos (\theta + \theta_0)]\}$$
$$+ i \{\sin [kr \cos (\theta - \theta_0)] \qquad (8.12c)$$
$$- \sin [kr \cos (\theta + \theta_0)]$$
$$- W_1 \cos [kr \cos (\theta - \theta_0)]$$
$$- W_2 \cos [kr \cos (\theta + \theta_0)]$$
$$+ U_1 \sin [kr \cos (\theta - \theta_0)]$$
$$+ U_2 \sin [kr \cos (\theta + \theta_0)]\}$$
$$= C + iD$$

$$|F(r, \theta)_R| = \sqrt{C^2 + D^2}.$$

Some numerical results are shown in Fig. 8.6. There does not appear to be any method of simplifying the

computations appreciably for the exact solution except for the cases of $\theta_0 = 0$, $\pi/2$, and π. These equations reduce to Eq. 8.7 for $\theta_0 = \pi/2$. For $\theta_0 = 0$, the diffraction coefficient is unity for all values of r and θ. An example of its use is shown in Fig. 17.3. Table 8.1 gives values for other initial wave directions.

Approximations can be made for values of kr greater than 2; this is useful in large harbors (see Putnam and Arthur, 1948). The reliability of the approximations has been verified experimentally (Putnam and Arthur, (1948), as shown in Fig. 8.7. (See Fig. 8.7(d) for coordinate description.) For small harbors, however, the region of small kr is of great interest and the complete equation must be used.

3. SINGLE BREAKWATER GAP

Two theories have been developed for a single gap in a very long breakwater. One theory is for a relatively large gap; the other, for a relatively small gap.

The theory for the large gap was developed by Penny and Price (1944, 1952) for waves approaching at 90 degrees. The wave heights at any point are affected by

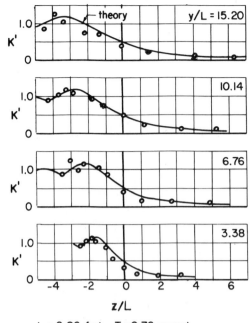

L = 2.96 feet, T = 0.76 second

(a) Incident waves 90° with respect to breakwater

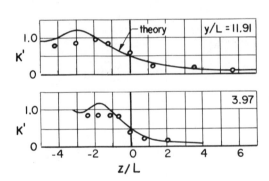

L = 2.52 feet, T = 0.70 second

(b) Incident waves 45° with respect to breakwater

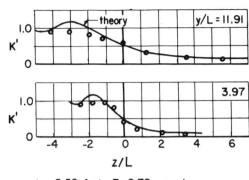

L = 2.52 feet, T = 0.70 second

(c) Incident waves 135° with respect to breakwater

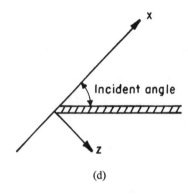

(d)

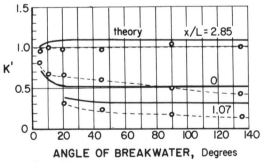

L = 2.52 feet, T = 0.70 second, x/L = 7.94

(e) Effect of angle of breakwater

Fig. 8.7. Comparison of laboratory data and theory of wave diffraction for a semi-infinite breakwater (after Putnam and Arthur, 1948)

Table 8.1. DIFFRACTION COEFFICIENTS, K', AS A FUNCTION OF INCIDENT WAVE ANGLE, θ_0, AND POSITION, r/L AND θ, (from Wiegel, 1962)

Value of r/L	Value of θ (degrees)												
	0	15	30	45	60	75	90	105	120	135	150	165	180
	$\theta_0 = 15°$												
1/2	0.49	0.79	0.83	0.90	0.97	1.01	1.03	1.02	1.01	0.99	0.99	1.00	1.00
1	0.38	0.73	0.83	0.95	1.04	1.04	0.99	0.98	1.01	1.01	1.00	1.00	1.00
2	0.21	0.68	0.86	1.05	1.03	0.97	1.02	0.99	1.00	1.00	1.00	1.00	1.00
5	0.13	0.63	0.99	1.04	1.03	1.02	0.99	0.99	1.00	1.01	1.00	1.00	1.00
10	0.35	0.58	1.10	1.05	0.98	0.99	1.01	1.00	1.00	1.00	1.00	1.00	1.00
	$\theta_0 = 30°$												
1/2	0.61	0.63	0.68	0.76	0.87	0.97	1.03	1.05	1.03	1.01	0.99	0.95	1.00
1	0.50	0.53	0.63	0.78	0.95	1.06	1.05	0.98	0.98	1.01	1.01	0.97	1.00
2	0.40	0.44	0.59	0.84	1.07	1.03	0.96	1.02	0.98	1.01	0.99	0.95	1.00
5	0.27	0.32	0.55	1.00	1.04	1.04	1.02	0.99	0.99	1.00	1.01	0.97	1.00
10	0.20	0.24	0.54	1.12	1.06	0.97	0.99	1.01	1.00	1.00	1.00	0.98	1.00
	$\theta_0 = 45°$												
1/2	0.49	0.50	0.55	0.63	0.73	0.85	0.96	1.04	1.06	1.04	1.00	0.99	1.00
1	0.38	0.40	0.47	0.59	0.76	0.95	1.07	1.06	0.98	0.97	1.01	1.01	1.00
2	0.29	0.31	0.39	0.56	0.83	1.08	1.04	0.96	1.03	0.98	1.01	1.00	1.00
5	0.18	0.20	0.29	0.54	1.01	1.04	1.05	1.03	1.00	0.99	1.01	1.00	1.00
10	0.13	0.15	0.22	0.53	1.13	1.07	0.96	0.98	1.02	0.99	1.00	1.00	1.00
	$\theta_0 = 60°$												
1/2	0.40	0.41	0.45	0.52	0.60	0.72	0.85	1.13	1.04	1.06	1.03	1.01	1.00
1	0.31	0.32	0.36	0.44	0.57	0.75	0.96	1.08	1.06	0.98	0.98	1.01	1.00
2	0.22	0.23	0.28	0.37	0.55	0.83	1.08	1.04	0.96	1.03	0.98	1.01	1.00
5	0.14	0.15	0.18	0.28	0.53	1.01	1.04	1.05	1.03	0.99	0.99	1.00	1.00
10	0.10	0.11	0.13	0.21	0.52	1.14	1.07	0.96	0.98	1.01	1.00	1.00	1.00
	$\theta_0 = 75°$												
1/2	0.34	0.35	0.38	0.42	0.50	0.59	0.71	0.85	0.97	1.04	1.05	1.02	1.00
1	0.25	0.26	0.29	0.34	0.43	0.56	0.75	0.95	1.02	1.06	0.98	0.98	1.00
2	0.18	0.19	0.22	0.26	0.36	0.54	0.83	1.09	1.04	0.96	1.03	0.99	1.00
5	0.12	0.12	0.13	0.17	0.27	0.52	1.01	1.04	1.05	1.03	0.99	0.99	1.00
10	0.08	0.08	0.10	0.13	0.20	0.52	1.14	1.07	0.96	0.98	1.01	1.00	1.00
	$\theta_0 = 90°$												
1/2	0.31	0.31	0.33	0.36	0.41	0.49	0.59	0.71	0.85	0.96	1.03	1.03	1.00
1	0.22	0.23	0.24	0.28	0.33	0.42	0.56	0.75	0.96	1.07	1.05	0.99	1.00
2	0.16	0.16	0.18	0.20	0.26	0.35	0.54	0.69	1.08	1.04	0.96	1.02	1.00
5	0.10	0.10	0.11	0.13	0.16	0.27	0.53	1.01	1.04	1.05	1.02	0.99	1.00
10	0.07	0.07	0.08	0.09	0.13	0.20	0.52	1.14	1.07	0.96	0.99	1.01	1.00
	$\theta_0 = 105°$												
1/2	0.28	0.28	0.29	0.32	0.35	0.41	0.49	0.59	0.72	0.85	0.97	1.01	1.00
1	0.20	0.20	0.24	0.23	0.27	0.33	0.42	0.56	0.75	0.95	1.06	1.04	1.00
2	0.14	0.14	0.13	0.17	0.20	0.25	0.35	0.54	0.83	1.08	1.03	0.97	1.00
5	0.09	0.09	0.10	0.11	0.13	0.17	0.27	0.52	1.02	1.04	1.04	1.02	1.00
10	0.07	0.06	0.08	0.08	0.09	0.12	0.20	0.52	1.14	1.07	0.97	0.99	1.00
	$\theta_0 = 120°$												
1/2	0.25	0.26	0.27	0.28	0.31	0.35	0.41	0.50	0.60	0.73	0.87	0.97	1.00
1	0.18	0.19	0.19	0.21	0.23	0.27	0.33	0.43	0.57	0.76	0.95	1.04	1.00
2	0.13	0.13	0.14	0.14	0.17	0.20	0.26	0.36	0.55	0.83	1.07	1.03	1.00
5	0.08	0.08	0.08	0.09	0.11	0.13	0.16	0.27	0.53	1.01	1.04	1.03	1.00
10	0.06	0.06	0.06	0.07	0.07	0.09	0.13	0.20	0.52	1.13	1.06	0.98	1.00
	$\theta_0 = 135°$												
1/2	0.24	0.24	0.25	0.26	0.28	0.32	0.36	0.42	0.52	0.63	0.76	0.90	1.00
1	0.18	0.17	0.18	0.19	0.21	0.23	0.28	0.34	0.44	0.59	0.78	0.95	1.00
2	0.12	0.12	0.13	0.14	0.14	0.17	0.20	0.26	0.37	0.56	0.84	1.05	1.00
5	0.08	0.07	0.08	0.08	0.09	0.11	0.13	0.17	0.28	0.54	1.00	1.04	1.00
10	0.05	0.06	0.06	0.06	0.07	0.08	0.09	0.13	0.21	0.53	1.12	1.05	1.00
	$\theta_0 = 150°$												
1/2	0.23	0.23	0.24	0.25	0.27	0.29	0.33	0.38	0.45	0.55	0.68	0.83	1.00
1	0.16	0.17	0.17	0.18	0.19	0.22	0.24	0.29	0.36	0.47	0.63	0.83	1.00
2	0.12	0.12	0.12	0.13	0.14	0.15	0.18	0.22	0.28	0.39	0.59	0.86	1.00
5	0.07	0.07	0.08	0.08	0.08	0.10	0.11	0.13	0.18	0.29	0.55	0.99	1.00
10	0.05	0.05	0.05	0.06	0.06	0.07	0.08	0.10	0.13	0.22	0.54	1.10	1.00

Table 8.1. (cont.)

$\theta_o = 165°$

1/2	0.23	0.23	0.23	0.24	0.26	0.28	0.31	0.35	0.41	0.50	0.63	0.79	1.00
1	0.16	0.16	0.17	0.17	0.19	0.20	0.23	0.26	0.32	0.40	0.53	0.73	1.00
2	0.11	0.11	0.12	0.12	0.13	0.14	0.16	0.19	0.23	0.31	0.44	0.68	1.00
5	0.07	0.07	0.07	0.07	0.08	0.09	0.10	0.12	0.15	0.20	0.32	0.63	1.00
10	0.05	0.05	0.05	0.06	0.06	0.06	0.07	0.08	0.11	0.11	0.21	0.58	1.00

$\theta_o = 180°$

1/2	0.20	0.25	0.23	0.24	0.25	0.28	0.31	0.34	0.40	0.49	0.61	0.78	1.00
1	0.10	0.17	0.16	0.18	0.18	0.23	0.22	0.25	0.31	0.38	0.50	0.70	1.00
2	0.02	0.09	0.12	0.12	0.13	0.18	0.16	0.18	0.22	0.29	0.40	0.60	1.00
5	0.02	0.06	0.07	0.07	0.07	0.08	0.10	0.12	0.14	0.18	0.27	0.46	1.00
10	0.01	0.05	0.05	0.04	0.06	0.07	0.07	0.08	0.10	0.13	0.20	0.36	1.00

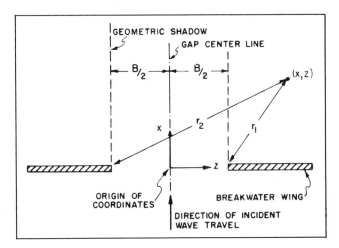

Fig. 8.8. Nomenclature for breakwater gap problem (after Blue and Johnson, 1949)

Fig. 8.9. (Equation 8.13)

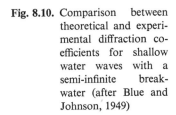

Fig. 8.10. Comparison between theoretical and experimental diffraction coefficients for shallow water waves with a semi-infinite breakwater (after Blue and Johnson, 1949)

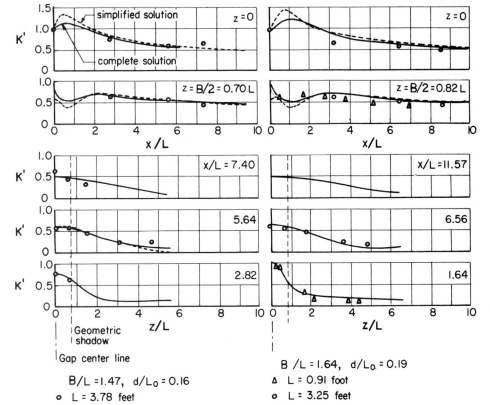

both tips of the breakwater, Fig. 8.8. The solutions for $F(r,\theta)$ are as shown in Fig. 8.9 (Eq. 8.13), where

$$f_1 = e^{-ikx} f(-\sigma_1), \quad f_2 = e^{-ikx} f(-\sigma_2),$$

$$g_1 = e^{-ikx} f(-\sigma_1') \quad g_2 = e^{-ikx} f(-\sigma_2'),$$

$$\sigma_1 = -\sqrt{\frac{4(r_1 - x)}{L}}, \quad \sigma_2 = -\sqrt{\frac{4(r_2 - x)}{L}},$$

$$\sigma_1' = -\sqrt{\frac{4(r_1 + x)}{L}}, \quad \sigma_2' = -\sqrt{\frac{4(r_2 + x)}{L}},$$

and $f(-\sigma)$ and $f(-\sigma')$ are defined by Eq. 8.6b. The diffraction coefficient K' is $|F(r,\theta)|$. Penny and Price concluded from a mathematical analysis that the solution as given by Eq. 8.13 is a good approximation as long as the width of the breakwater gap is greater than 1 wave length.

Laboratory studies by Blue and Johnson (1949; also Blue, 1948) verified the reliability of the solution from a practical standpoint (Fig. 8.10). In addition, experiments were performed with the breakwater wings inclined at 60 degrees and 45 degrees, with the gap centerline. Although these two cases, as well as the normal case (straight breakwater) showed lower than theoretical diffraction coefficients, no significant difference due to orientation of the breakwater wings was apparent.

One phenomenon of particular interest was observed: There were zones of apparent wave-crest discontinuity in the lee of the breakwater. In reality there was no discontinuity, but rather zones of considerably reduced wave heights.

Johnson (1952; 1953) presented graphs and tables for rapidly obtaining the diffraction coefficients for a series of breakwater gap widths, in terms of wave lengths.

An example of the values of K' and the locations of the wave crests is given in Fig. 8.11. Graphs for one-half of the breakwater area are presented in Fig. 8.12, the solution being symmetrical about the centerline of the breakwater gap. For breakwater gaps greater than 5 wave lengths, the semi-infinite solution should suffice and Figs. 8.4 and 8.6 can be used for each portion of the breakwater.

Johnson (1952; 1953) made the assumption that for many practical purposes the solution for waves approaching a breakwater gap at 90 degrees could be used for angles of approach different from this up to a certain limit, provided that a modified "gap width" B' be used, as shown in Fig. 8.13. In Fig. 8.14 it is compared with the exact theory of Morse and Rubenstein as developed by Carr and Stelzriede (1952).

Fig. 8.11. Diffraction of waves at a breakwater gap—contours of equal diffraction coefficient (after Johnson, 1952)

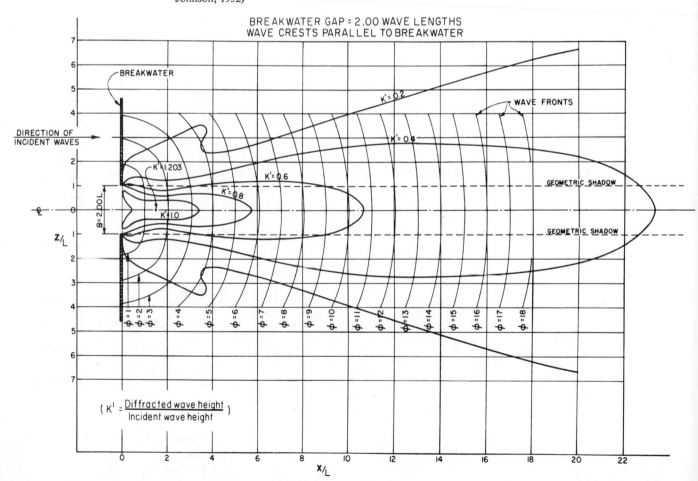

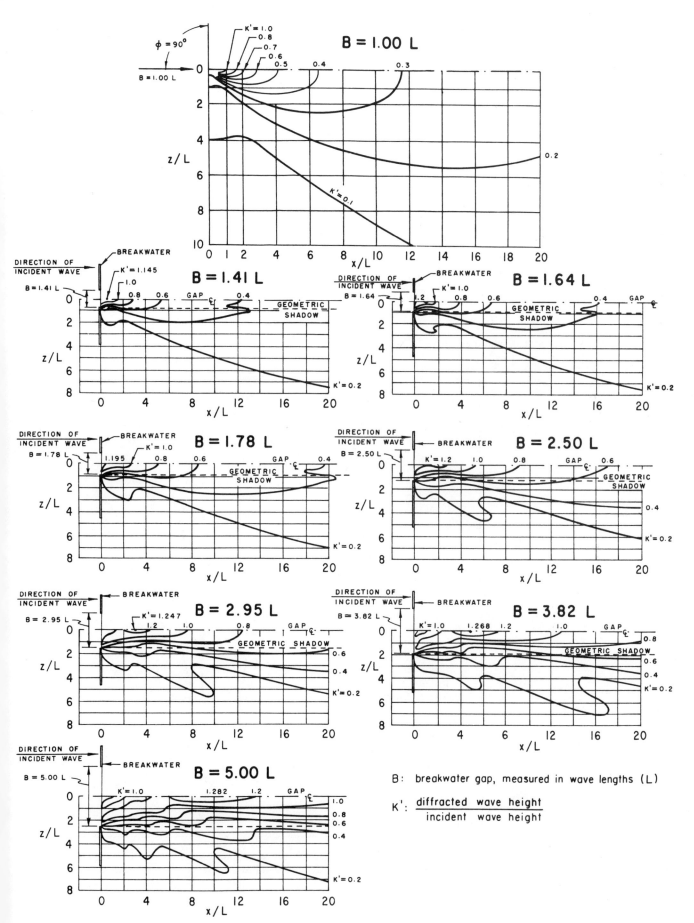

Fig. 8.12. Diffraction of waves at breakwater gap—contours of equal diffraction coefficient (after Johnson, 1952)

189

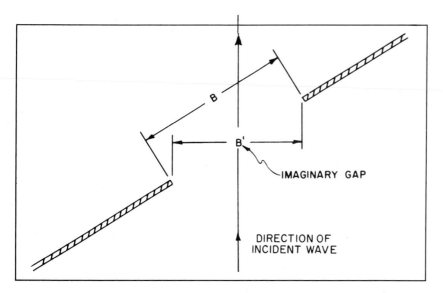

Fig. 8.13. Modified gap (from Johnson, 1952)

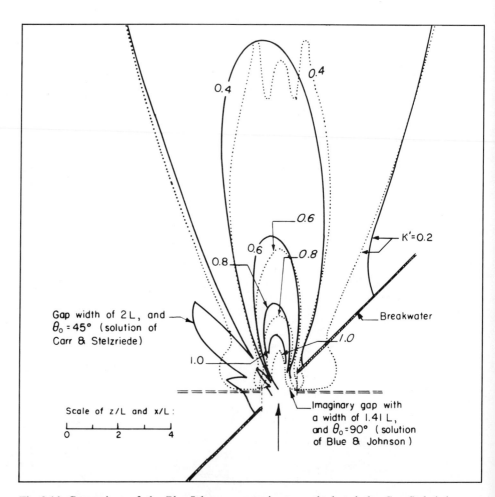

Fig. 8.14. Comparison of the Blue-Johnson approximate method and the Carr-Stelzriede exact method for obtaining diffraction coefficients in the lee of a breakwater gap (after Johnson, 1952)

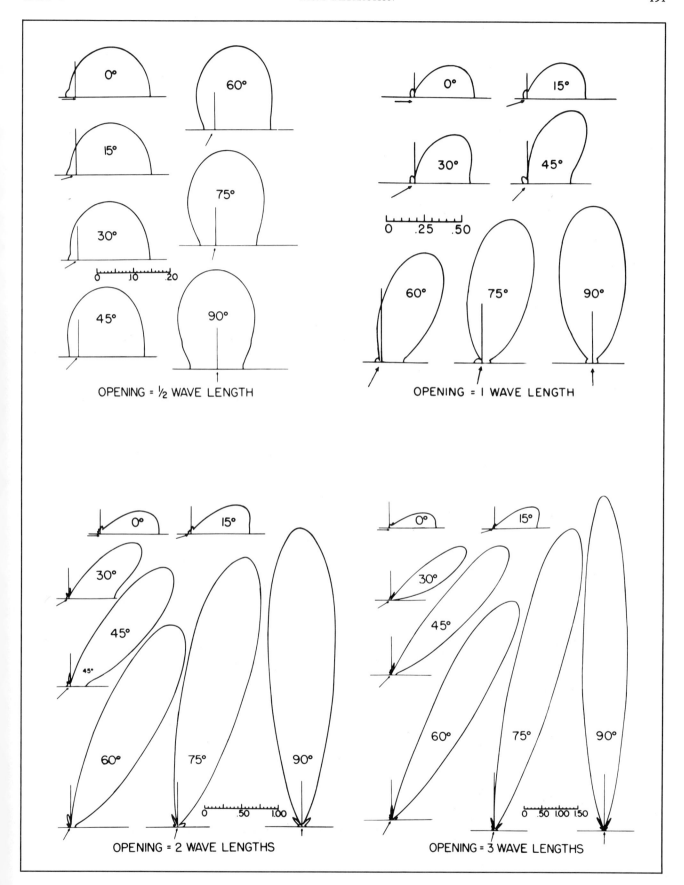

Fig. 8.15. Polar plots of intensity factors (vertical face straight breakwaters) (after Carr and Stelz-riede, 1952)

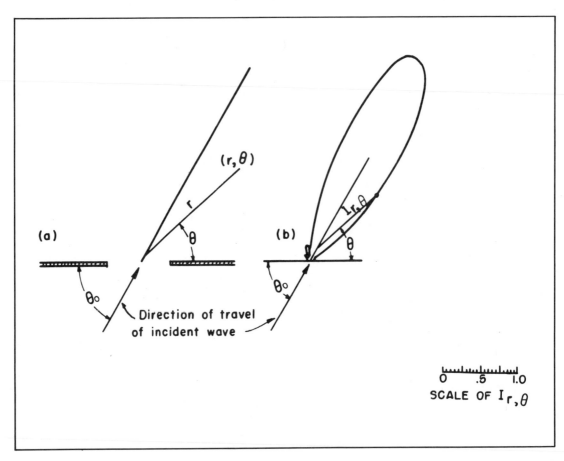

Fig. 8.16. Description of symbols

Exact theories for the diffraction of waves by ribbons (detached breakwater) and slits (breakwater gap) were developed by Morse and Rubenstein (1938). For numerical calculations one must use a solution in terms of a series of Mathieu functions. This series converges rapidly for small slits. It is exact for incident waves of any angle of approach.

The theory was applied by Carr and Stelzriede (1952; also, California Institute of Technology, 1952) to breakwaters and water waves. Many of the necessary computations were made, the results of which are summarized in Fig. 8.15, using the nomenclature shown in Fig. 8.16. The intensity factor $I(r,\theta)$ can be used to obtain the wave diffraction coefficient K' by use of the equation for definitions of symbols:

$$K' = \sqrt{\frac{I(r,\theta)L}{r}}. \qquad (8.14)$$

Johnson (1952), using the results of Carr and Stelzriede, prepared a complete set of curves for the case of $B/L = 1$. These are given in Fig. 8.17. These values, together with values for other B/L's, have been tabulated (California Institute of Technology, 1952). These are considered to be accurate for $r > 2\,B$ and $r > 3\,L$. Errors introduced for r smaller than these values have not been determined (*Ibid.*).

Of interest in certain practical problems is the function that relates the ratio of the energy transmitted through a slit to the energy that would be transmitted if geometric optics applied. A transmission factor has been computed (Morse and Rubenstein, 1938) which is the ratio of the energy transmitted by a wave of angle of incidence θ_0 to the energy that would be predicted from geometric optics for a wave of 90 degree angle of incidence, Fig. 8.18. The limiting case for an opening small compared with the wave length has been solved by Lamb (1945, p. 533) in which the transmission factor is given by

$$\frac{\frac{1}{4}\pi^2}{(\pi B/L)\{[\ln(\pi B/4L) + \gamma]^2 + \frac{1}{4}\pi^2\}}, \qquad (8.15)$$

where γ is Euler's number (0.577) and the incident waves are normal to the breakwater. For L/B of 10, 100, and 1000, the transmission factor is 1.24, 3.80, and 17.2. In practice, the value of the transmission factor would not continue to rise with decreasing ratio of opening width to wave length, as the energy transformed into eddies at the two edges of the opening would predominate for very small openings.

The physical reason for the increase in the transmissions factor with decreasing slit width is that the amount of energy transmitted through the slit by the circular waves generated by the two ends of the reflected waves

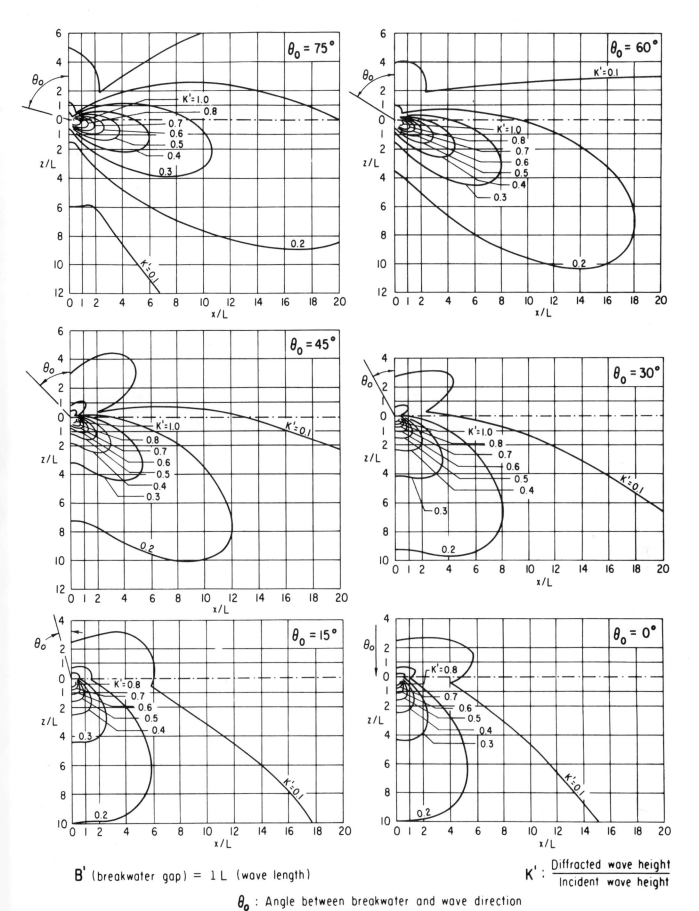

B' (breakwater gap) $= 1\,L$ (wave length)

K' : $\dfrac{\text{Diffracted wave height}}{\text{Incident wave height}}$

θ_0 : Angle between breakwater and wave direction

Fig. 8.17. Diffraction of waves at breakwater gap—contours of equal diffraction coefficient (after Johnson, 1952)

193

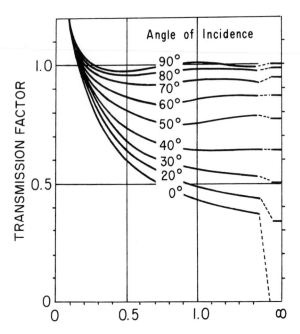

Fig. 8.18. Theoretical transmission factors (after Morse and Rubenstein, 1938)

becomes large relative to the energy associated with the "cylindrical" wave (the geometric optical wave) transmitted through the slit.

4. MISCELLANEOUS

Detached breakwater. One that allows the waves to diffract around both ends. In the Morse-Rubenstein theory this is called a *ribbon*. Detailed graphs for this case have not been prepared. In practice, however, if such a breakwater were to be used it would be many wave lengths long. In this case, each end could be considered to be the end of a semi-infinite breakwater and Figs. 8.3 and 8.4 could be used.

Vertical circular cylinder. The diffraction theory for vertical circular cylinders has been treated in Chapter 11, Section 8. An approximate theory for a series of equally spaced circular cylinders has been given by Lamb (1945, p. 537).

Mach-stem effect. The Mach-stem effect for solitary waves has been covered in Chapter 3. The same phenomenon has been observed for periodic waves both in the laboratory and in the ocean (Nielsen, 1962).

REFERENCES

Bierens de Haan, D., *Nouvelles tables d'intégrales définies*, ed. 1867, corrected. New York: G. E. Stechert & Co., 1939.

Blue, Frank Lee, Jr., Diffraction of water waves passing through a breakwater gap, Ph. D. thesis in Civil Engineering, Univ. Calif., 1948.

————, and J. W. Johnson, Diffraction of water waves passing through a breakwater gap, *Trans. Amer. Geophys. Union*, **30**, 5 (October, 1949), 705–18.

California Institute of Technology, *Wave protection aspects of harbor design*, Hydro. Lab., Hyd. Structures Div., Report No. E–11, August, 1952.

Carr, J. H., and M. E. Stelzriede, *Diffraction of water waves by breakwaters*, U.S. National Bureau of Standards, Circular No. 521 (November, 1952), pp. 109–25.

Jahnke, Eugene and Fritz Emde, *Tables of functions with formulae and curves*, 4th ed. New York: Dover Publications, Inc., 1945.

Johnson, J. W., Generalized wave diffraction diagrams, *Proc. Second Conf. Coastal Eng.*, Berkeley, Calif.: The Engineering Foundation Council on Wave Research, 1952, pp. 6–23.

————, Engineering aspects of diffraction and refraction, *Trans. ASCE*, **118** (1953), 617–48.

Lacombe, H., *The diffraction of a swell. A practical approximate solution and its justification*, U.S. National Bureau of Standards, Circular No. 521 (November, 1952), pp. 129–40.

Lamb, Sir Horace, *Hydrodynamics*. 6th ed. New York: Dover Publications, Inc., 1945.

Morse, P. M. and P. J. Rubinstein, The diffraction of waves by ribbons and slits, *Physical Rev.*, **54** (December, 1938), 895–98.

Nielsen, Arne Hasle, *Diffraction of periodic waves along a vertical breakwater for small angles of incidence*, Univ. Calif., Inst. Eng. Res., Tech. Rept.: HEL–1–2, December, 1962.

Penny, W. G. and A. T. Price, *Diffraction of sea waves by breakwaters*, Directorate of Miscellaneous Weapons Development, Technical History No. 26, Artificial Harbors, Sec. 3D, 1944.

————, The diffraction theory of sea waves by breakwaters, and the shelter afforded by breakwaters, *Phil. Trans., Roy. Soc.* (London), ser. A, **244** (March, 1952), 236–53.

Putnam, J. A. and R. S. Arthur, Diffraction of water waves by breakwaters, *Trans. Amer. Geophys. Union*, **29**, 4 (August, 1948), 481–90.

Sommerfeld, A., Mathematische Theorie der Diffraction, *Math. Ann.* **47**, (1896), 317–74.

Stelzriede, M. E., Discussion of "Engineering aspects of diffraction and refraction," *Trans. ASCE*, **118** (1953), 649–51.

Van Wijngaarden, A. and W. L. Scheen, *Table of Fresnel integrals*, Report R 49, Computation Dept., Mathematical Center, Amsterdam, 1949.

Wiegel, R. L., Diffraction of waves by semi-infinite breakwater, *J. Hyd. Div., Proc. ASCE*, **88**, HY1 (January, 1962), 27–44.

CHAPTER NINE

Wind Waves and Swell

Winds blowing over the water surface generate waves. In general, the higher the wind velocity, the longer the fetch over which it blows; and the longer it blows, the higher and longer will be the average waves. Waves still under the action of the winds that created them are called *wind waves*, or a *sea*. They are forced waves rather than free waves. They are variable in their direction of advance (Arthur, 1949). They are irregular in the direction of propagation (Fig. 2.36). The flow is rotational due to the shear stress of the wind on the water surface and it is quite turbulent, as observation of dye in the water indicates. After the waves leave the generating area, their characteristics become somewhat different. Principally they are smoother, losing their rough appearance due to the disappearance of the multitude of smaller waves on top of the bigger ones and the whitecaps and spray. When running free of the storm, the waves are known as *swell*. In Fig. 9.1 are shown some photographs taken in the laboratory of waves still rising under the action of wind and this same wave system after it has left the

windy section of the wind-wave tunnel. It can be seen that the freely running swell has a smoother appearance than the waves in the windy section. The motion of the swell is nearly irrotational and nonturbulent, unless the swell runs into other regions where the water is in turbulent motion. Turbulence is a property of the fluid rather than of the wave motion. After the waves have traveled a distance from the generating area, they have lost some energy owing to air resistance, internal friction, and large-scale turbulent scattering if they run into other storm areas; the rest of the energy has become spread over a larger area because of the dispersive and angular spreading characteristics of water gravity waves. All these mechanisms lead to a decrease in energy density. Thus, the waves become lower in height. In addition, owing to their dispersive characteristic, the component wave periods tend to segregate in such a way that the longest waves lead the main body of waves and the shortest waves form the tail of the main body of waves. Finally, the swell may travel through areas

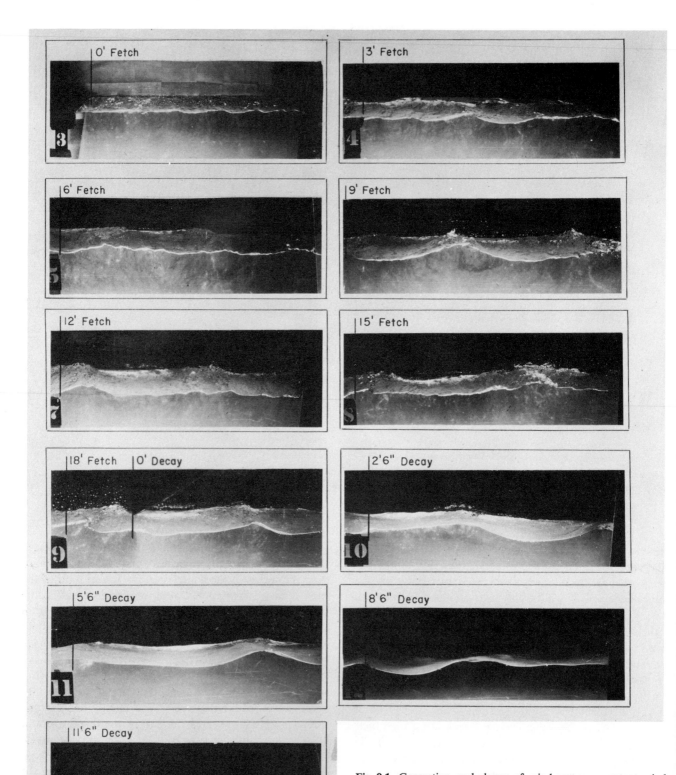

Fig. 9.1. Generation and decay of wind waves; constant wind velocity in the fetch, no wind in the decay area; steady state conditions; photographs taken in the University of California Wave Channel

where winds are present, adding new wind waves to old swell, and perhaps directly increasing or decreasing the size of the old swell.

1. WAVE CHARACTERISTICS

An observer stationed high above the area in which wind waves pass from the fetch into an area of calm (called the *decay area*) would notice that the waves in both regions vary in height, length, and breadth. If he were to follow a particular wave crest, he would notice that it would gradually disappear; he would also notice that new crests form.

An observer watching the crests leaving the fetch and measuring the time intervals between the successive crests as well as the heights would have a true picture of the surface waves at a particular point without having a true picture of the phenomenon. This is so because the surface phenomenon is a result of other complex phenomena. Because there is a spectrum of lengths (or periods) and heights present, there must be some sort of a group phenomenon; i.e., there are no permanent wave forms. Instead, wave crests and troughs gradually appear and disappear. The longer wave components of the group, traveling with greater speeds than the shorter wave components, gradually move ahead, with the shortest wave components dropping behind. Hence, a spreading of the wave system occurs.

In order to understand what happens in an actual case, where the generating areas vary in size, the winds vary in speed and direction and exist for different lengths of time, it is necessary first to consider the simplified case of a stationary storm of constant dimensions with winds that immediately spring up to a constant speed and remain at that speed for a short time, long enough to generate a considerable number of waves, and then die down immediately. In addition, the decay distance must be long enough for complete segregation of wave components to take place. Shortly after the wind starts blowing over the entire generating area, the waves will be short, but probably close to maximum steepness ($H/L = \frac{1}{7}$ in deep water). At some distance downwind from the start of the fetch, the waves will gradually grow in height and length as time increases. The maximum wave dimensions at this point will be obtained when all the waves generated upwind of this point have reached it, this time depending upon the group velocity of the waves. When this has occurred, a quasi–steady-state condition has been reached.

An observer traveling with the wave system in the decay area would notice that the long wave components would gradually move in front of the system and short wave components would drop to the rear to the system. The longer the wave system has traveled, the greater would be this segregation. If the group were to travel many thousands of miles, and there were no other disturbances, this segregation and stretching of the system would become complete. An observer at a fixed distance from the storm would notice a steady decrease in wave period with increasing time.

Actually, the duration of the storm and the relatively short decay distances (even several thousands of miles) are such that complete segregation never takes place. In local storms almost no segregation occurs and the lengths and periods are relatively short, even if the winds are great, the fetch long, and the duration long; stretching of wave system does not occur so that the energy density is high. As the decay distance increases, the segregation increases; hence the long waves, often called *forerunners of storms*, reach the coast before the main body of waves. For a particular storm, the longer the decay distance, the greater is the time between these forerunners and the main body of waves (high energy density).

The normal case is more complicated. For example, suppose the storm lasts for two days and that it takes only one day for the medium length wave component to reach the section of coast under consideration. The longer waves are being continually generated as are all the shorter wave periods. Thus, even as the first of the shortest waves are arriving at the coast, the longer waves which were being formed after the shorter waves have left the generating area, overtake the shorter waves, and arrive at the same time.

It can be seen then that a section of coast a considerable distance from a storm will be subject to long, low waves first with the mean "period" of the waves (with the *period* being defined as the time necessary for two successive crests to pass a fixed point) decreasing with time but with the period spectrum width about the mean value increasing. The wave "heights" (with the *height* being defined as the vertical distance between a trough and the following crest) will be increasing because the greatest energy density is concentrated in the waves with medium periods. As the last of the waves reach the coast, an abrupt decrease in wave period and height will be observed. The average period of even these short waves, at the coast, will always be longer than the period observed at the end of the fetch because of the spreading phenomenon and because the smallest waves will have been either captured by the large waves or dissipated.

The actual phenomenon is more complex than has been described because the winds gradually rise to a maximum speed, then decrease again, always fluctuating, with the wind field varying in size and the storm moving.

As an example, consider the wave spectra at Pendeen, England, during March 14–15, 1945, which have been presented in Fig. 9.2 (Barber and Ursell, 1948). The wave record at Pendeen, the bottom pressure type, was analyzed by a frequency analyzer so that the component period spectra were obtained. It can be seen that the forerunners of a new wave system first appeared at 1300

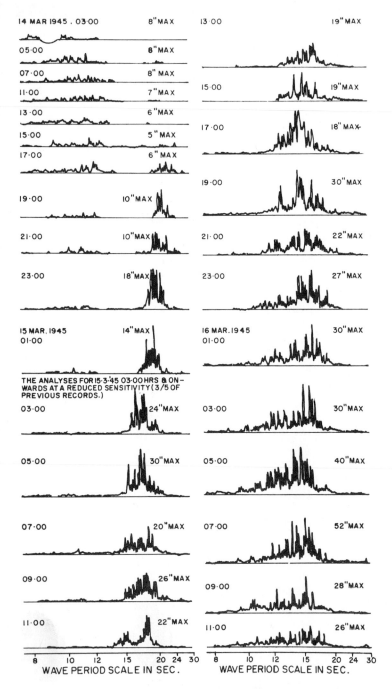

Fig. 9.2. Wave spectra at Pendeen, March 14–16, 1945 (after Barber and Ursell, 1948)

and ocean conditions. The statistical techniques used in obtaining the information that will be presented in the following sections neglect the time sequence of the phenomena as they assume random phase. Thus, one entire class of information is often thrown away in the analysis of waves. From an engineering standpoint, it is the groups of several periodic waves, which are almost always the highest waves in a wave system, that are the most effective in causing structural damage. Donn and McGuinness (1959) report that the waves they measured in a relatively open ocean area occurred in striking sinusoidal groups in contrast to the far more irregular patterns observed in the shallow water off coasts in the area.

One of the main problems connected with describing waves is that of definition: Should every small bump be considered a wave? For some purposes, such as scattering of radar or the reflection and scattering of light, the answer should be yes. For many other purposes the answer should be no. One concept that is useful in this respect is to neglect the very small waves and to measure the highest one-third of the remaining waves, the average of the highest one-third being called the *significant* or *characteristic* wave height. This concept was developed during studies of landing craft operation in the surf in World War II. It was found that the wave height estimated by observers corresponded to the average of the highest 20 to 40 per cent of the waves (Scripps Institution of Oceanography, 1944). Originally, the term *significant wave* was attached to the average of these observations, the highest 30 per cent of the waves, but has evolved to become the average of the highest $33\frac{1}{3}$ per cent (designated as H_S or $H_{1/3}$).

As can be seen in Fig. 9.3, the higher waves often occur in groups which are nearly periodic. The average period of these high groups in a record was termed the *significant period* (designated at T_S or $T_{H_{1/3}}$).

In order for the concept of significant wave height and period to be of more value, studies were made of the distribution of wave heights and periods about their mean values (Ehring, 1940; Seiwell, 1948; Wiegel, 1949; Rudnick 1951; Munk and Arthur, 1951; Darbyshire, 1952; Putz, 1952; Pierson and Marks, 1952; Watters, 1953; Yoshida, Kajiura, and Hidaka, 1953; and Darlington, 1954). These papers advanced the concept of the "statistical nature" of ocean waves. Many of the data were summarized by Longuet-Higgins (1952) in his paper on the statistical distribution of heights of ocean waves.

Almost all the data were obtained from bottom recorders of the pressure type with their inherent practical and theoretical limitations (Folsom, 1949; Pierson and Marks, 1952; Fuchs, 1955; Gerhardt, Jehn, and Katz, 1955; Neuman, 1955). Similar studies were made with both surface recorders and bottom pressure recorders by Wiegel and Kukk (1957), some of the results of which are shown in Fig. 9.4. It was found that the statistical

on March 14, 1945. The mean of the periods gradually decreased while at the same time the width of the spectra increased.

In scientific works on ocean waves the terms *Gaussian* and *random* are often used when referring to the heights and periods of wind waves and swell. These terms must be used with caution as they oversimplify the true complex nature of waves and at the other extreme they are not descriptive to the layman of the many groups of nearly periodic waves that exist. That these groups exist can be seen in Fig. 9.3 in which records of waves in the generating area are presented for laboratory, lake,

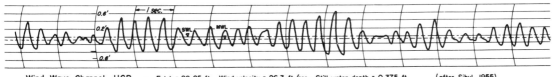

Wind Wave Channel, UCB - Fetch = 22.85 ft., Wind velocity = 26.3 ft./sec., Still water depth = 0.375 ft.   (after Sibul, 1955)

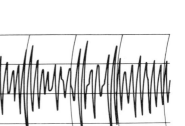

Abbotts Lagoon, California - Fetch = 2970 ft., Av. wind velocity = 44.8 ft./sec., Water depth = 6 ft.   (after Johnson, 1950)

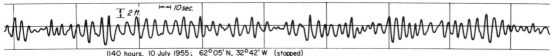

1140 hours, 10 July 1955; 62°05' N, 32°42' W (stopped)

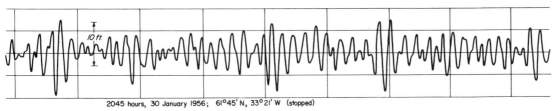

2045 hours, 30 January 1956; 61°45' N, 33°21' W (stopped)

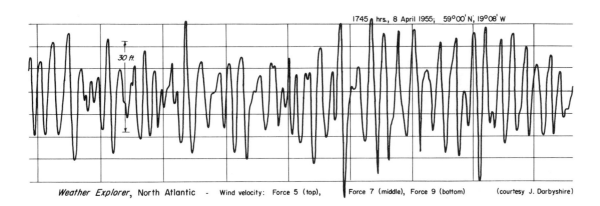

Weather Explorer, North Atlantic - Wind velocity: Force 5 (top),   Force 7 (middle), Force 9 (bottom)   (courtesy J. Darbyshire)

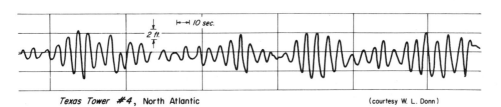

Texas Tower #4, North Atlantic   (courtesy W. L. Donn)

Fig. 9.3. Sample records of waves unaffected by refraction

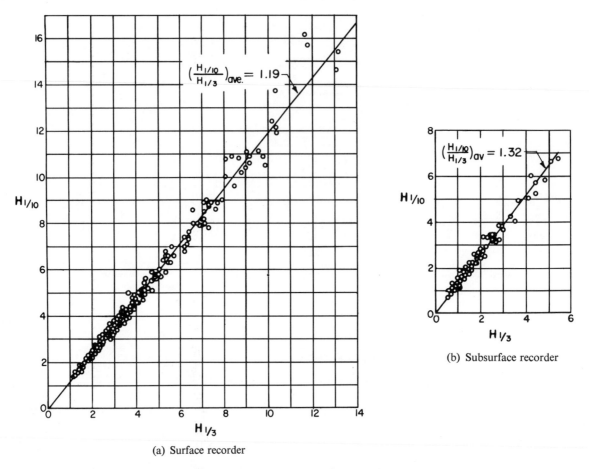

(a) Surface recorder

Fig. 9.4. Wave height relationships (after Wiegel and Kukk, 1957)

ratios obtained by measuring the waves at the surface differed from the ratios obtained from the subsurface pressure recorders (Table 9.1). It is believed that the difference, and the sign of the difference, can be explained by the method of analyzing the records of the subsurface pressure recorders combined with the fact that the subsurface dynamic pressures decrease more rapidly with depth for the shorter period waves than for the longer period waves (Fig. 2.8). Of particular importance is the fact that the surface measurements in the ocean of wind waves and swell of Wiegel and Kukk (1957) resulted in nearly the same ratios as those of the surface measurements by Sibul (1955) of wind waves generated in a wind-wave tunnel. Although many of the measurements of Sibul were for waves in relatively shallow water, the author states that no observable shallow water effect was noticed in the wave-height ratios. It should be noted that the wave-height ratios obtained by Wiegel and Kukk were for swell with wind waves often superimposed, whereas the ratios obtained by Sibul were only for wind waves. It appears then that the wave height distribution about a mean value is almost the same for swell as for wind waves.

Several investigators were able to fit existing mathematical curves to the empirical wave height data, curves

of the "random distribution" type (Putz, 1952; Longuet-Higgins, 1952; and Waters, 1953). The work of Longuet-Higgins is in most general use (it predicts nearly the same values as do the curves of Putz). It allows the prediction of the most probable maximum wave height for a given number of waves, provided that the mean wave height (or some other measure of wave height) is known. These values are presented in Table 9.2. In this table, N is the number of consecutive waves considered and $\mu(a_{max})$ is the most probable maximum wave amplitude (half the wave height) that will occur in N consecutive waves if the sample has a root mean square wave height of $2\ \bar{a}$. In order to find the most probable maximum wave to expect in N waves if the significant height is known, rather than the root mean square wave height, it is only necessary to divide the value $\mu(a_{max})$ by 1.416 (Table 9.3). For example, if $N = 500$ waves, from Table 9.2, $\mu(a_{max})/\bar{a} = 2.509$, from Table 9.3 $a^{(0.333)}/\bar{a} = 1.416$, and the most probable maximum wave height/significant wave height $= 2.509/1.416 = 1.77$. If the significant wave was 10.0 ft, then the most probable maximum wave height would have been 17.7 ft.

The entire spectrum of wave heights can be obtained by use of Table 9.3.

In Table 9.3, $p = 0.1$ refers to the average of the

Table 9.1. MEASURED STATISTICAL RATIOS FOR OCEAN WAVES
(after Wiegel and Kukk, 1957)

Location and type of wave recorder	Statistical ratios							Remarks
	$\dfrac{H_{max}}{H_{1/10}}$	$\dfrac{H_{max}}{H_{1/3}}$	$\dfrac{H_{max}}{H_{mean}}$	$\dfrac{H_{1/10}}{H_{1/3}}$	$\dfrac{H_{1/10}}{H_{mean}}$	$\dfrac{H_{1/3}}{H_{mean}}$	$\dfrac{H_{3/10}}{H_{mean}}$	
1a. Davenport, Calif., surface recorder		1.40	1.90	1.19	1.61	1.37		11 mo, 12–20 min interval every 12 hr, water depth 46′ MLLW (Wiegel and Kukk, '57)
1b. Davenport, Calif., bottom-pressure recorder		1.64	2.64	1.32	2.09	1.48		5mo, 12–20 min interval every 12 hr, water depth 46′ MLLW (Wiegel and Kukk, 1957)
2. N. Atlantic Ocean NIO shipborne recorder			2.40			1.60		138 records, 26–124 waves per record (mean of 69 waves) every 3 hr (Darlington, 1954)
3. Pt. Arguello, Calif., bottom-pressure recorder	1.42	1.85		1.30				3 mo, 20 min every 8 hr; figures refer to daily av. and max., recorder in 75′ of water MLLW (Wiegel, 1949)
4. Pt. Sur, Calif., bottom-pressure recorder	1.46	1.85		1.27				14 mo, 20 min interval every 8 hr; figures refer to daily av. and max.; recorder in 68′ of water MLLW (Wiegel, 1949)
5. Heceta Head, Ore., bottom-pressure recorder	1.47	1.91		1.30				14 mo, 20 min interval every 8 hr; figures refer to daily av. and max.; recorder in 50′ of water MLLW (Wiegel, 1949)
6. Cuttyhunk, Mass., bottom-pressure recorder						1.57		10 mo, 20 min interval every 6 hr; wave recorder in 75′ of water (Seiwell, 1948)
7. Bermuda, bottom-pressure recorder						1.57		4 mo, 20 min interval every 2 hr; recorder in 120′ of water (Seiwell, 1948)
8. Cape Cod, Mass.							1.56	1 record, 14 waves; 1 record, 28 waves; waves only 2–3 sec period (Gibson, 1944)
9. LaJolla, Calif., bottom-pressure recorder		1.63				1.49		46 waves (Munk and Arthur, 1951)
10. North Sea, pressure recorder hanging on cable below float		1.5				1.85 (1.58)		688 waves (every bump counted as a wave); wind waves (Ehring, 1940) 517 waves (neglecting all waves less than 14.8 cm high in above sample (Harney, Saur, and Robinson, 1949)
11. Pacific Coast, U.S.A., and Guam, M.I., bottom-pressure recorder						1.63		25 records, 20 min interval (Putz, 1952)
12. Greymouth, N.Z., bottom-pressure recorder				1.24	1.94	1.58		109 records, 17 min interval, every 2 hr (Watters, 1953)
13. Hachijo I. Japan, (surface?)			2.0			1.5		11 records, 10 min interval, various locations (Yoshida, Kajiura, and Hidaka, 1953)
14. Long Branch, N.J. BEB surface recorder		1.29	1.93			1.50		1 record of 102 waves (Pierson, Neumann, and James, 1955)
15. Bermuda, NIO shipborne recorder				1.24		1.61		6 records, 12–18 min interval (Farmer, 1956)
16. Oceanside, Calif., bottom-pressure recorder				1.25				1 mo, 20 min interval every 8 hr (Wiegel and Kukk, 1957)

Table 9.2. VALUE OF $E(a_{\max})/\bar{a}$ AND $\mu(a_{\max})/\bar{a}$ FOR DIFFERENT VALUES OF N, FOR NARROW SPECTRUM (After Longuet-Higgins, 1952)

N	$(\log N)^{1/2}$	$E(a_{\max})/\bar{a}$ exact expression	asymptotic expression	$\mu(a_{\max})/\bar{a}$
1	0.000	0.886	—	0.707
2	0.833	1.146	—	1.030
5	1.269	1.462	—	1.366
10	1.517	1.676	1.708	1.583
20	1.731	1.870	1.898	1.778
50	1.978	—	2.124	2.010
100	2.146	—	2.280	2.172
200	2.302	—	2.426	2.323
500	2.493	—	2.609	2.509
1000	2.628	—	2.738	2.642
2000	2.757	—	2.862	2.769
5000	2.918	—	3.017	2.929
10,000	3.035	—	3.130	3.044
20,000	3.147	—	3.239	3.155
50,000	3.289	—	3.377	3.296
100,000	3.393	—	3.478	3.400

highest one-tenth of the waves, $p = 0.333$ refers to the average of the highest one-third of the waves, $p = 1.0$ refers to the mean of all of the waves, etc. The ratio $H_{1/10}/H_{1/3}$ can be obtained from Table 9.3 as $1.800/1.416 = 1.28$; and the ratio $H_{1/3}/H_{\text{mean}}$ as $1.416/0.886 = 1.60$. It is evident from a comparison of these values with those shown in Table 9.1 that the wave height distribution function of Longuet-Higgins is of practical importance.

The wave height distribution function given in Tables 9.2 and 9.3 is for the case of a narrow wave frequency spectrum, and is the Rayleigh distribution. Other distributions are useful for wider spectra. Putz (1954) found that the parameter of greatest importance in choosing the proper distribution was the ratio of zero crossings to the number of wave maximums and minimums (see also Cartwright and Longuet-Higgins, 1956; Williams and Cartwright, 1957). When this ratio is 1, the distribution is a Rayleigh distribution. Putz found that this ratio and the root mean square value of the ordinates of a wave record were sufficient to describe the wave height distribution. He also found that the theoretical relationship between the average wave height H_m and the root mean

Table 9.3. REPRESENTATIVE VALUES OF $a^{(p)}/\bar{a}$ IN THE CASE OF A NARROW WAVE SPECTRUM (After Longuet-Higgins, 1952)

p	$a^{(p)}/\bar{a}$	p	$a^{(p)}/\bar{a}$
0.01	2.359	0.4	1.347
0.05	1.986	0.5	1.256
0.1	1.800	0.6	1.176
0.2	1.591	0.7	1.102
0.25	1.517	0.8	1.031
0.3	1.454	0.9	0.961
0.333	1.416	1.0	0.886

square value of the wave ordinates H_{rms} was verified by an analysis of the wave records. This relationship is given by

$$H_m = \frac{1}{2} H_{\text{rms}} \sqrt{2\pi} \frac{N_0}{N_1}, \qquad (9.1)$$

where N_0 is the number of zero crossings (the number of times the wave record goes through the mean abscissa) and N_1 is the sum of the number of wave maximums and minimums (wave crests and troughs).

Another important set of statistical data deals with the wave period (or frequency) spectrum. Putz (1952) made measurements of 25 wave records (bottom-pressure type), each record of approximately 20-min duration. He found the relationship between the significant wave period to the mean wave period shown in Fig. 9.5. The complete period distribution function of Putz is shown in Fig. 9.6. As an example of its use, suppose the significant period was 11.5 sec, then from Fig. 9.5 it would be found that the mean wave period would be 11.0 sec. From Fig. 9.6 it would be found that 99.5 per cent of the wave periods would be less than 18 sec and that only 0.5 per cent would be less than 4.5 sec. Of more importance in the consideration of waves in the generating area is the work of Darbyshire (1959), the data on wave frequency distribution being given in Fig. 9.12.

Darlington (1954) has made a similar study of wave records obtained with the NIO shipborne wave recorder (Tucker, 1956) in the North Atlantic. Most of the recordings reported were made with the ship stopped. As can be seen in Table 9.1, his results on wave height distribution agree well with those of Putz, and the data extended the results to a mean wave height of up to 28 ft with a maximum wave height of 42 ft. The results of the relationship between the period of the highest one-third of the waves and the mean period also are shown in Fig. 9.5. The results are not the same as those obtained by Putz, but are not too different. Part of the difference must be attributed to Putz's records being obtained from a bottom-pressure recorder. In addition, Darlington's measurements were made in storm areas (with background swell in many cases), whereas Putz's measurements were mostly of swell. It is interesting to note that both results show that for lower-period waves T_m should be greater than $T_{H_{1/3}}$. Darlington found that, by interpolating the best fit straight line, $T_m > T_{H_{1/3}}$ for $T_m < 6$ sec, and Putz found that $T_m > T_{H_{1/3}}$ for $T_m < 9$ sec. No actual measurements of this condition were made, however. Both of these findings are in conflict with Sibul's (1955) measurements of short-period waves generated in a wind-wave tunnel, Fig. 9.7. It would appear then that the relationship between $T_{H_{1/3}}$ and T_m is nonlinear. Part of the difference may be that both Putz's and Darlington's wave measurements are either wholly or partially dependent upon the prediction of surface waves from subsurface pressure measurements with some inher-

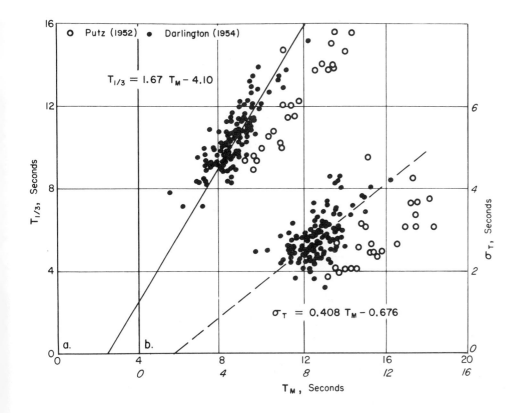

Fig. 9.5. (a) Mean period of the highest third of the waves versus mean period; (b) standard deviation of wave periods versus mean wave period (abscissa shifted to right) (open-circle data, after Putz, 1952; closed-circle data, after Darlington, 1954)

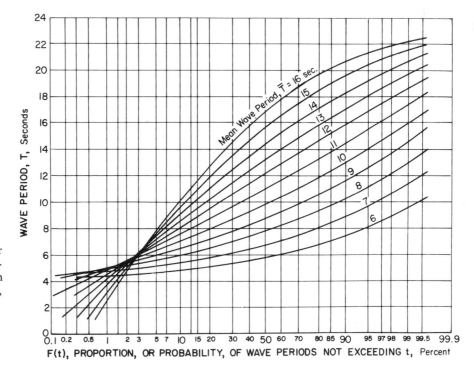

Fig. 9.6. Prediction curves for wave period distributions for various mean wave periods (after Putz, 1952)

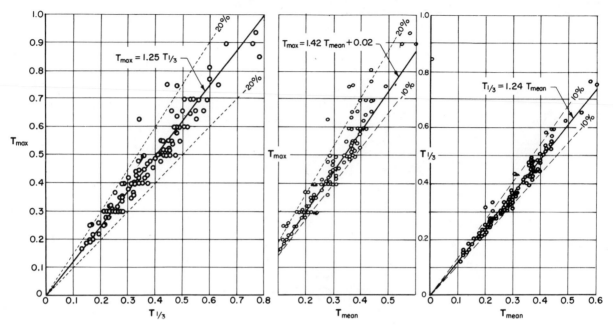

Relationship between mean, significant and maximum wave periods

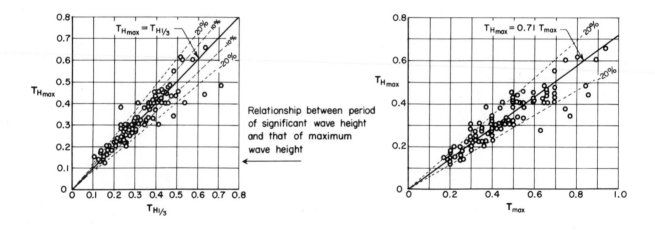

Relationship between period of significant wave height and that of maximum wave height ←

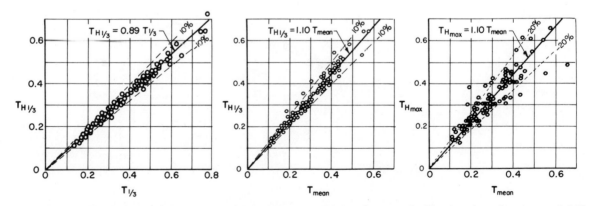

Relationship between individual wave periods and wave periods of mean, significant and maximum wave heights

Fig. 9.7. (after Sibul, 1955)

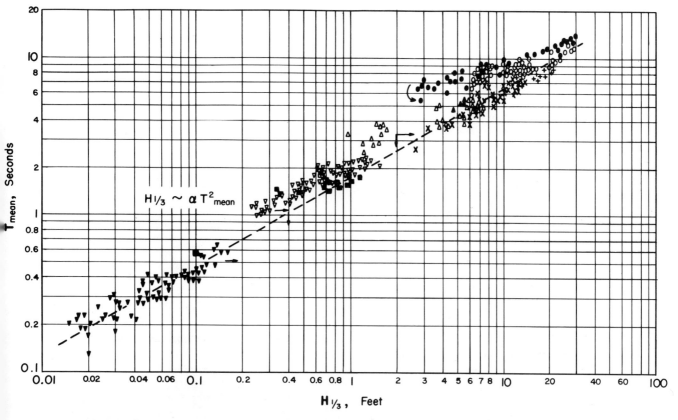

$H_{1/3} \sim \alpha T^2_{\text{mean}}$

T_{mean}, Seconds

$H_{1/3}$, Feet

o **North Atlantic, NIO** shipboard wave recorder (Darlington, 1954)

• **North Atlantic, NIO** shipboard wave recorder (Darbyshire, 1959)

x **Gulf of Mexico (California**
+ **Research Corp., 1960)**
Step resistance gage

Δ **Gulf of Mexico, WHOI** pressure recorder, ¼ the water depth below SWL (Bretschneider, 1954)

▲ **Woods Hole, WHOI** capacitance wave pole (Marks, 1954)

▽ **Abbots Lagoon,** subsurface pressure recorder, just under the surface (Johnson, 1950)

▼ **UCB Wind Wave Tunnel,** parallel wire resistance gage (Sibul, 1955)

■ **Roll, 1949**

Fig. 9.8. Relationship between $H_{1/3}$ and T_{mean} (after Wiegel, 1961)

ent difficulties in measuring the true mean and significant wave periods.

What is the relationship between the heights and periods of the wind waves? In Fig. 9.8 is shown the relationship between T_{mean} (the length of the wave record divided by the number of waves in the record) and $H_{1/3}$. These two parameters were chosen because they were the most generally available. The data shown are for wind waves rather than swell, although there may be some swell present in the records of Bretschneider (1954) and Darlington (1954). A line was drawn through the wind-wave tunnel data of Sibul (1955), neglecting the smallest waves, as these probably were affected considerably by surface tension. This line was extended through several cycles on the log-log paper. This line is expressed by

$$H_{1/3} = 0.45 T^2_{\text{mean}}. \qquad (9.2)$$

If it is assumed that, for nonperiodic waves, $L = gT^2/2\pi$

$$\frac{H_{1/3}}{(L_{\text{mean } o})} = \frac{1}{11.4}. \qquad (9.3)$$

If it is assumed that the relationship derived by Pierson, Neumann, and James (1955) for a Gaussian sea surface is correct, that is, the apparent wave length L^* is about two-thirds the value of a periodic wave, then

$$\frac{H}{(L^*_{\text{mean } o})} = \frac{1}{11.4} \times \frac{3}{2} = \frac{1}{7.6},$$

which is very close to the theoretical maximum wave steepness, and nearly identical to the maximum steepness of mechanically generated waves, as can be seen in Fig. 2.10. This appears to be the physical reason for the limit of the maximum wave heights; for a given mean wave period this is the maximum that can exist without breaking.

The California Research Corporation (1960) has measured waves from several oil platforms in the Gulf of Mexico using a step resistance gauge that is essentially self-calibrating in that, as each electrode is shorted by the sea water passing over it, a "step" occurs on the record. The data shown in Fig. 9.8 are for several hurricanes in the Gulf of Mexico, with the measurements being made in water 30 ft deep. Each point shown was obtained from a sample of between 100 and 200 waves. The fact that the water was not deep should be considered, as the maximum wave steepness in transitional water is not so great as it is in deep water (see Fig. 2.10). On the other hand, some of the data were for essentially deep water waves, and these data show the same trend with respect to the dashed line as do transitional waves.

The extension of the line through Sibul's data does not go through all the data taken in the field. Many of the field data were obtained with instruments either entirely or partially dependent upon subsurface pressure measurements. It is known that these pressure measurements

projected to surface values by use of the linear theory relationship between wave height and subsurface pressures, when using an average pressure factor associated with a mean period, underpredict the surface waves by an average of about 25 per cent (Folsom, 1959; Gerhardt, John, and Katz, 1955). This same technique utilizes a subjective method of determining the number of waves in a record which, when combined with rapid attenuation with distance below the water surface of the short-period wave components, results in a mean wave period that is longer than would be the case in analyzing a surface wave record. The field data, then, should be shifted as indicated by the arrows in Fig. 9.8. The

Fig. 9.9. Joint frequency distribution of wave period and wave height ($U = 42.7$ ft/sec) (after Johnson and Rice, 1952)

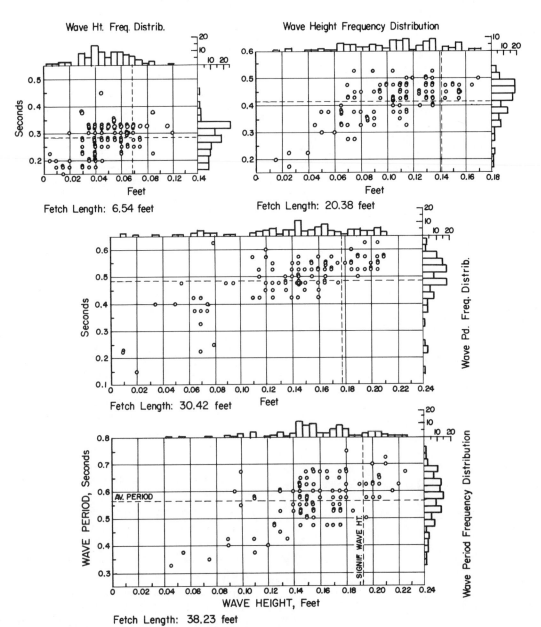

data of Darlington (1954) and Darbyshire (1959) should probably be rotated in the manner shown in the figure, as the recorder used in obtaining the data is very insensitive to waves with periods less than about 6 sec (Marks, 1955, Marks, 1956; Williams and Cartwright, 1957). In addition, the NIO shipborne recorder utilized a vertical acceleration sensing device, the output of which must be integrated twice to give a reference for the pressure cells. The reliability of the results of such a technique is not yet fully understood. This would tend to make them conform to the other data.

The data shown in Fig. 9.8 due to Sibul (1955), Johnson (1950), and the California Research Corporation (1960) are for only wind waves; those of Bretschneider (1954) and Darlington (1954) probably include swell as well as wind waves. Studies made of data largely of swell do not show this trend (Putz, 1951).

Is there any clear relationship between the heights and periods of individual waves? For wind waves generated in the laboratory the answer is a qualified yes. The results for this case can be seen in Fig. 9.9 (Johnson and Rice, 1952). In general, the longer the wave, the higher the wave. For swell, or a mixture of swell and wind waves, the joint frequency distributions are shown in Fig. 9.10 (Putz, 1951, 1952). There is no evidence of a relationship between these quantities although there is a tendency for highest wave heights to occur with near average periods.

Ocean waves can be considered to be the combination of a series of components of different periods, or frequencies, with a certain amount of power being transmitted by each component. In describing waves using this concept the term *wave spectrum* is used. The spectrum describes in some manner the distribution of the energy density present in a wave system with respect to the wave period, or frequency. An example of one type of wave spectrum is shown in Fig. 9.2.

Fig. 9.10. Wave height versus wave period for individual waves (after Putz, 1951)

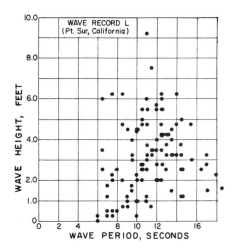

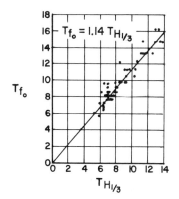

Fig. 9.11. Relationship between T_{f_0} and $T_{H1/3}$ (after Darbyshire, 1959)

Darbyshire (1959) has obtained the wave spectra for recorded waves of a large number of storms in the North Atlantic using the NIO shipborne wave recorder. The records were analyzed first to obtain values of $T_{H_{1/3}}$, $H_{1/3}$, etc., using a wave recorder calibration curve based upon the wave component period associated with f_0. The records were also analyzed by an approximate Fourier method to obtain certain information on the wave spectra, namely, the square of the wave height components, H_f, within each frequency interval $\Delta f = 0.007$ sec^{-1}; f_0 was defined as the frequency of the class having the largest value of H_f^2 in the spectrum. The heights H_f were corrected, using a calibration curve based upon the component period associated with each frequency f. The wave period associated with this, T_{f_0}, was found to be closely related to the significant wave period ($T_{f_0} = 1.14\,T_{H_{1/3}}$) as is shown in Fig. 9.11. It appears from this and from the relationships shown by Putz and Sibul, that the choice of the significant wave to describe waves was a good one.

Darbyshire (1959) has found a consistent relationship between H_f^2/H^2 and $f - f_0$ (Fig. 9.12), where H is defined as the height of a hypothetical single sine wave train which has the same energy density as the actual wave system. $H^2 = \sum H_f^2$. The close relationship between this hypothetical wave and other surface wave characteristics can be seen in Fig. 9.13, where $H_{\max} = 2.40H$ and $H = 0.604\,H_{1/3}$ with very little scatter of the data.

The scatter of data of H_f^2/H^2 versus $f - f_0$ is considerable when compared with the scatter of data for $H_{1/3}$ versus H_m, and $T_{H_{1/3}}$ versus T_m. For example, if H were 10 ft, H_f could be between 41 and 56 ft at its maximum.

As pointed out by Darbyshire, this empirically determined spectrum is inconsistent with the previous spectrum of Darbyshire (1952; 1955), the spectrum of Neumann (1953), and the spectrum of Roll and Fisher (1956).

Burling (1955; 1959) computed the spectra associated with various fetches and meteorological conditions on a reservoir (fetches from 1200 to 4000 ft and wind

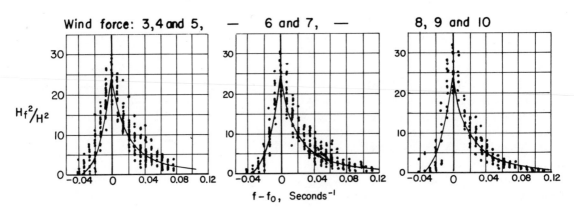

Fig. 9.12. Relationship between H_f^2/H^2 and $f - f_0$

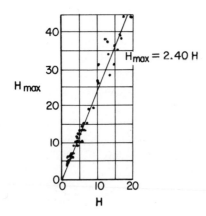

Fig. 9.13. Relationship between H_{max} and H (after Darbyshire, 1959)

speeds from about 15 to 25 ft/sec). The results are shown in Fig. 9.14(a), where σ is the circular wave frequency component $(2\pi/T)$ and

$$\Phi(\sigma) = \frac{1}{2\pi} \int_{-\infty}^{\infty} y_s(x, t) y_s(x, t + \tau) e^{-i\sigma\tau} \, d\tau. \quad (9.4)$$

Equation 9.4 is given in the form of Phillips (1958b) where $y_s(x, t)$ is the surface displacement at a fixed point, and τ is a time displacement. $\Phi(\sigma)$ has the dimensions of ft²-sec, (or cm²-sec) and the area under the curve $\Phi(\sigma)$ versus σ is related to the energy per unit area being transmitted by the wave system. It is important to note that Burling's data seem to lie along a single curve for circular frequencies greater than about 6 radians/sec (wave periods smaller than about 1 sec). The curve in

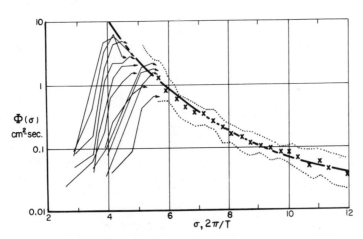

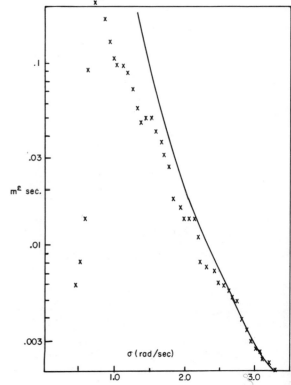

Fig. 9.14. (a) Spectra of wind-generated waves measured by Burling (1955). The cluster of lines on the left are representative of the spectra at low frequencies for which equilibrium has not been attained. On the right the curves merge over the equilibrium range, and the dotted lines indicate the extreme measured values of $\Phi(\sigma)$ at each frequency σ. The x's represent the mean observed value at each σ, and the heavy line the relation $\Phi(\sigma) = \alpha g^2 \sigma^{-5}$ with $\alpha = 7.4 \times 10^{-3}$ (after Phillips, 1958a) (b) the frequency spectrum $\Phi(\sigma)$ from the S.W.O.P. wave-pole data. The solid curve represents the equilibrium range spectrum $\Phi(\sigma) \sim 7.4 \times 10^{-3} g^2 \sigma^{-5}$ (after Phillips, 1958b)

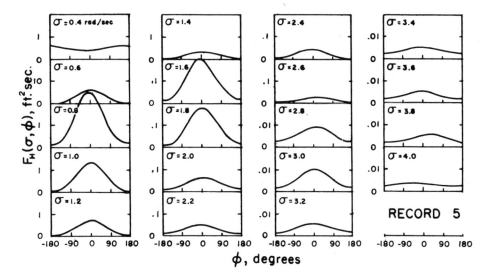

Fig. 9.15. Example of two-dimensional smoothed power spectrum (from Longuet-Higgins, *et al,* 1961)

this region has been developed theoretically by Phillips (1958c).

$$\Phi(\sigma) \approx \alpha g^2 \sigma^{-5} = \text{constant}/\sigma^5. \qquad (9.5)$$

The solid curves in Fig. 9.14 are $\Phi(\sigma) = 7.4 \times 10^{-3} g^2 \sigma^{-5}$, in cgs units. In Fig. 9.14(b) the results of Project SWOP (Chase *et al.*, 1957) are shown compared with Eq. 9.5 (Phillips, 1958c). This extends the relationship shown in Fig. 9.14(a) to ocean conditions, and it appears that the "fully developed sea" extends only to wave components down to $\sigma \approx 2$; That is, wave period components up to about 3 sec for the particular set of waves which were measured. The physical significance of this limiting curve is that a wave of a given frequency can increase in energy up to only a certain limit ($H/L = 1/7$ in deep water for a uniform periodic wave). Hence, if the fetch and duration are long enough, $\Phi(\sigma)$ must have a unique value which depends only upon the frequency.

Some of the differences between the spectra of different investigators have been explained by Pierson (1959a) based upon a random nonlinear model of waves (Tick, 1958).

Directional spectra have been measured by Chase *et al.* (1957), but their results are difficult to interpret. Phillips (1958c) and Cox (1958b) offer conflicting interpretations of the results. The major difficulty, aside from experimental errors, is that the measurements were made of ocean waves, and it is not possible to be sure of the level of wave action with respect to the local winds.

Two-dimensional spectra have also been measured by Longuet-Higgins *et al.* (1961), and the results of one set of measurements are shown in Fig. 9.15. In this figure, σ is the circular frequency of the waves (radians/second), ϕ is the azimuth measured from the direction in which the mean surface wind is moving, and $F_3(\sigma,\phi)$ is a smoothed power spectrum. The illustration shown is for the case of waves being generated by nearly steady waves, with little swell present. The shape of the unsmoothed

power spectrum $F(\sigma,\phi)$ was found to be approximately

$$F(\sigma, \phi) \propto (1 + \cos \theta)^s \propto \cos^{2s}(\tfrac{1}{2} \phi), \qquad (9.6)$$

where s was a function of the ratio of the wind speed to the phase speed of the component wave of circular frequency σ. The values of s were determined empirically, and varied between 7 for wind speeds approximately equal to the wave phase speed to about 0.25 for winds 5 times the component wave phase speed. There was considerable uncertainty for the higher values of s.

Waves are short crested; i.e., they have a dimension in the horizontal direction normal to the direction of wave advance (Fig. 2.36). This dimension has been termed the *crest length* of a wave. There are very few measurements of this wave dimension. The average ratio of crest length to wave length (L'/L) of the waves in Fig. 2.36 is from 2 to 3. Johnson (1948) made some measurements on waves in a lake and found (L'/L) = 3. Yampol'ski (1955) found that the distribution curves for wave lengths and crest lengths were nearly identical in appearance, the measurements being made from aerial photographs of the sea surface.

A three-dimensional laboratory study was made by Ralls and Wiegel (1956) of wind-generated waves. The results of measurements of crest lengths and wave lengths are shown in Fig. 9.16. In deep water, the ratio is approximately 3.

2. RELATIONSHIPS AMONG WAVE DIMENSIONS, WINDS, AND FETCHES

The relationships among the wave dimensions, wind speed and direction, atmospheric stability, fetch length, and wind and fetch variability are not well established. It is not surprising, considering the extreme difficulty in obtaining reliable measurements in the open ocean combined with the near impossibility of obtaining sea conditions consisting of only wind waves, with no swell

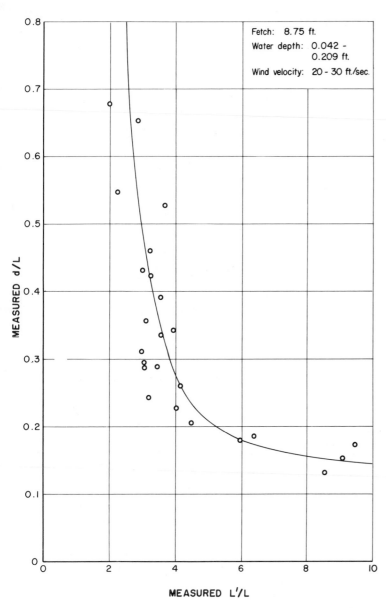

Fetch: 8.75 ft.

Water depth: 0.042 – 0.209 ft.

Wind velocity: 20 – 30 ft./sec.

Fig. 9.16. Measured L'/L (data from Ralls and Wiegel, 1956)

present. To these difficulties must be added the problems of determining the wind fetches and durations. Because of this, the data obtained will be presented, but ideas from studying data obtained in the laboratory and in lakes will be used to interpret the ocean data.

First, let us consider waves from the standpoint of wave spectra. The data of Burling (1955) have already been covered. The remaining data will pertain to ocean waves.

In deep water, the average power, averaged over a complete cycle, transmitted per unit area of surface (Eq. 2.67) is proportional to the square of the wave height and inversely proportional to the wave period. The energy per unit area, the energy density, is proportional to the square of the wave height. Darbyshire (1952; 1955) used the equation

$$E_T = \tfrac{1}{8}\rho g \sum_{T-1/2}^{T+1/2} h_n^2 = \tfrac{1}{8}\rho g H_T^2 \qquad (9.7)$$

where E_T is the energy per unit area of horizontal sea surface in a wave-period interval from $T - \tfrac{1}{2}$ to $T + \tfrac{1}{2}$ sec, h_n is the height of the nth peak in the spectrum, and H_T is defined as the equivalent wave height for waves of period between $T - \tfrac{1}{2}$ and $T + \tfrac{1}{2}$ (that is, in a 1-sec interval) as obtained by a Fourier analysis and taking the square root of the sum of the squares of the peaks within each 1-sec interval of wave period. Examples of the results, plotted in terms of H_T/T versus T/U, where U is the gradient wind speed, are shown in Fig. 9.17. The solid line curve shown in the figures is the Gaussian function that Darbyshire states expresses the relationship between H_T/T and T/U. The scatter is extreme, and the relationship between the data and the curves practically nonexistent.

Darbyshire has found a much more reliable spectrum, leaving out the relationship of the spectrum to the wind (Fig. 9.12).

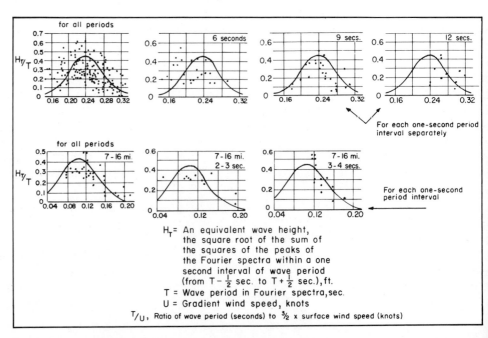

Fig. 9.17. Examples of the relationship between H_T/T and T/U (after Darbyshire, 1952)

H_T = An equivalent wave height, the square root of the sum of the squares of the peaks of the Fourier spectra within a one second interval of wave period (from $T - \tfrac{1}{2}$ sec. to $T + \tfrac{1}{2}$ sec.), ft.

T = Wave period in Fourier spectra, sec.

U = Gradient wind speed, knots

T/U, Ratio of wave period (seconds) to $\tfrac{3}{2}$ × surface wind speed (knots)

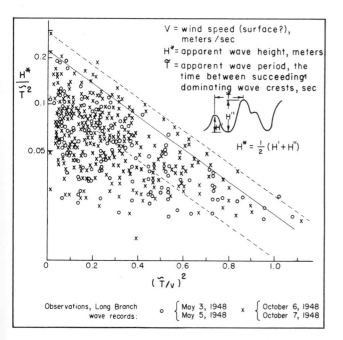

Fig. 9.18. Ratio of the wave heights to the square of the apparent wave periods plotted against the square of the ratio of the apparent wave period to the wind velocity (after Neumann, 1953)

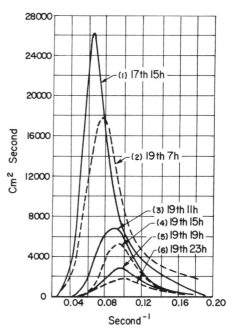

Fig. 9.19. Time changes of wave energy spectrum in the decrease of swell of a typhoon (after Ijima, 1957)

Other data for which a spectrum equation has been developed are shown in Fig. 9.18 (Neumann, 1953). The relationship between the ratio of wave height to wave period and the ratio of wave period to wind speed, as given by Neumann, is stated to be an exponential curve, and this curve forms the upper envelope of the data in Fig. 9.18. The values of H and T used in this figure refer to the height and period of surface waves, many of them visual observations from a ship. They do not refer to data obtained by a Fourier analysis or similar method. This upper envelope is considered by Neumann to represent the "fully arisen sea," where the term refers to the case where no more energy can be added to the wave system regardless of the length of time the wind blows or the length of the fetch (Pierson, Neumann, and James, 1955). This means that the waves, at all frequencies, are dissipating energy and radiating it from the storm area at the same rate that wind is transferring wave energy to the sea surface. It is doubtful that this condition exists in the open ocean for all frequencies although it does exist for the higher-frequency components (Fig. 9.14). It is, in fact, contradicted by the tendency for wave steepness to decrease with increasing values of T/U.

Some analyses of ocean waves to obtain the energy spectra have been made by Ijima (1957), but these are for cases so complicated that it is difficult to compare anything but their gross characteristics with the simple expressions of wave spectra (Fig. 9.19). The spectrum for a relatively simple case is shown in Fig. 9.14b (Chase *et al.*, 1957).

It is difficult to obtain certain data for wind waves in the ocean, such as the wave height and period as a function of wind speed for a constant fetch, or the wave height and period as a function of fetch for a constant wind speed. In fact it is difficult to obtain data even for either a constant wind speed or fetch, or for a stationary storm. It is possible to obtain these data in the laboratory or in relatively small lakes or reservoirs. These data are presented in Figs. 9.20 and 9.21, together with some ocean data where the fetches were not constant, nor were the fetch limits even stated. The wind velocities were often measured at different elevations above the water surface. The elevation where the winds are measured, however, as long as it is fairly close to the surface, should not affect the power relationship between H and U, and T and U, although it will affect the constant of proportionality. That this is true can be seen from a study of the wind-speed data of Thornthwaite and Kaser (1943) for winds at 1/2 and at 15 ft above the ground, the data of Deacon (1949) for winds at 1/2 and 4 meters above the ground, and the data of Deacon, Sheppard, and Webb (1956) for winds at 6.4 and 13 meters above the sea surface. These data show that the winds are linearly related (Fig. 9.22), the slope of the line on the log-log plot being unity, and that the constant of proportionality is related to Richardson's number.

There is scatter in the data. Some data were taken for neutral stability winds (all of Sibul's laboratory data, for example), some of the data for unstable winds and some of the data for stable winds. A good deal of the scatter is probably due to the different wind types. That this is

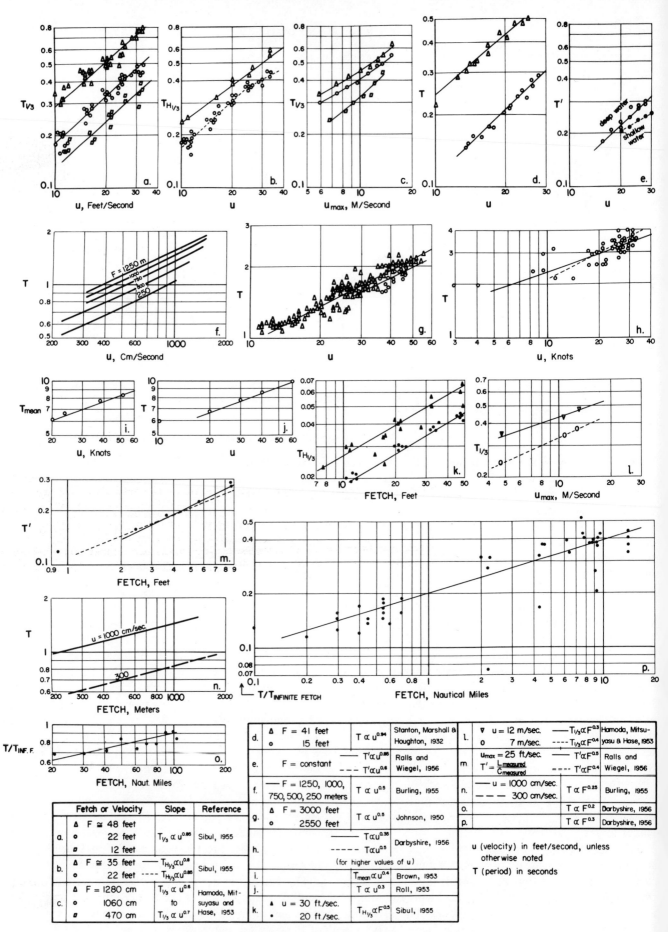

Fig. 9.20. Wave period as a function of wind speed and fetch

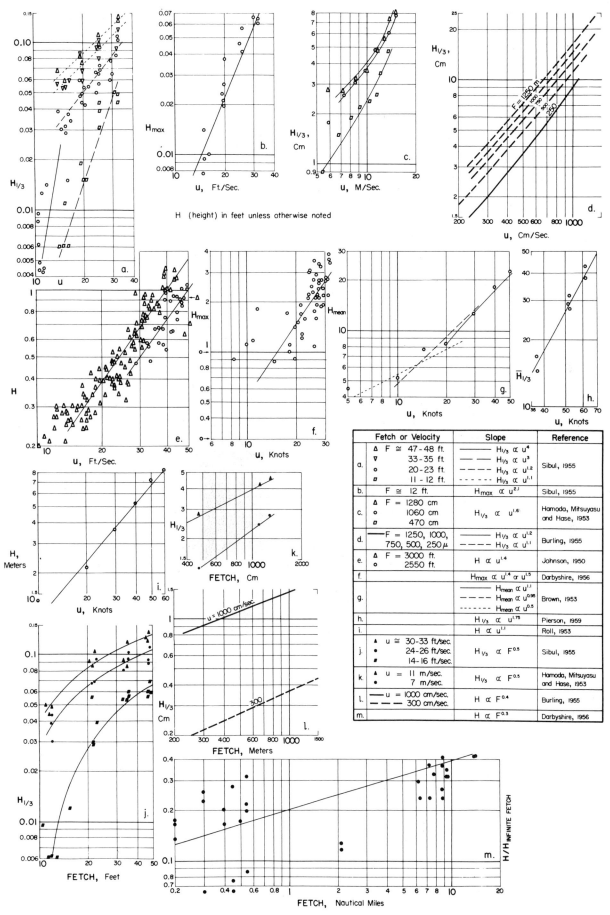

H (height) in feet unless otherwise noted

	Fetch or Velocity		Slope	Reference
a.	Δ ∇ ○ ◻	F ≅ 47 - 48 ft. 33 - 35 ft. 20 - 23 ft. 11 - 12 ft.	$H_{1/3} \propto u^4$ $H_{1/3} \propto u^3$ $H_{1/3} \propto u^{1.2}$ $H_{1/3} \propto u^{1.1}$	Sibul, 1955
b.	◻	F ≅ 12 ft.	$H_{max} \propto u^{2.1}$	Sibul, 1955
c.	Δ ○ ◻	F = 1280 cm 1060 cm 470 cm	$H_{1/3} \propto u^{1.6}$	Hamoda, Mitsuyasu and Hase, 1953
d.	F = 1250, 1000, 750, 500, 250 μ		$H_{1/3} \propto u^{1.2}$ $H_{1/3} \propto u^{1.1}$	Burling, 1955
e.	Δ ○	F = 3000 ft. 2550 ft.	$H \propto u^{1.4}$	Johnson, 1950
f.			$H_{max} \propto u^{1.4}$ or $u^{1.5}$	Darbyshire, 1956
g.			$H_{mean} \propto u^{1.1}$ $H_{mean} \propto u^{0.95}$ $H_{mean} \propto u^{0.5}$	Brown, 1953
h.			$H_{1/3} \propto u^{1.75}$	Pierson, 1959
i.			$H \propto u^{1.1}$	Roll, 1953
j.	▲ ● ◻	u ≅ 30-33 ft./sec. 24-26 ft./sec. 14-16 ft./sec.	$H_{1/3} \propto F^{0.5}$	Sibul, 1955
k.	▲ ●	u = 11 m/sec. 7 m/sec.	$H_{1/3} \propto F^{0.5}$	Hamoda, Mitsuyasu and Hase, 1953
l.	—— - - -	u = 1000 cm/sec. 300 cm/sec.	$H \propto F^{0.4}$	Burling, 1955
m.			$H \propto F^{0.3}$	Darbyshire, 1956

Fig. 9.21. Wave height as a function of fetch and wind speed

213

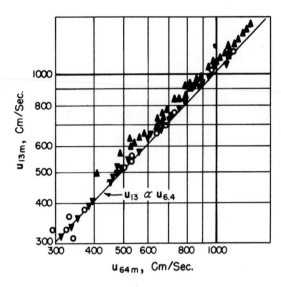

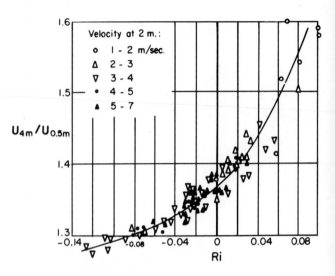

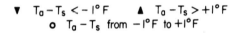

$\blacktriangledown$ $T_a - T_s < -1°F$ $\blacktriangle$ $T_a - T_s > +1°F$

$\circ$ $T_a - T_s$ from $-1°F$ to $+1°F$

Fig. 9.22. (left) Winds over the sea (data from Deacon, Sheppard, and Webb, 1956); (right) 4:0.5m wind ratio related to Richardson number; short grass surface (after Deacon, 1949)

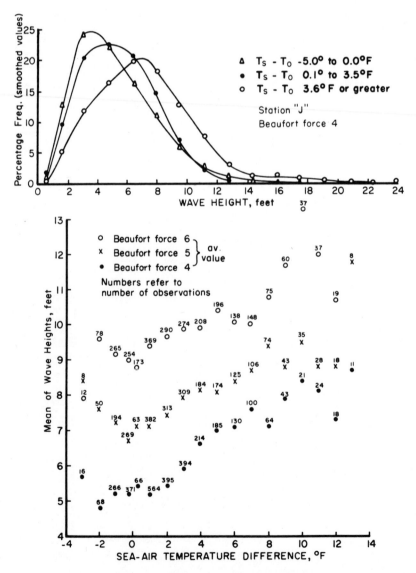

Fig. 9.23. (after Brown, 1953)

so has been shown by Roll (1952), Burling (1955), and Brown (1953). Brown's data are presented in Fig. 9.23. It can be seen that, for a given wind speed, the wave heights are higher for both stable and unstable winds than for neutral winds (essentially $T_s - T_a = 0$, where T_s is the temperature of the sea surface and T_a is the air temperature). Roll's data (1952) showed about the same variation, and also showed an increase in wave length with increasing air-sea temperature difference, with about a 15 per cent increase in length when the sea was about 10°F warmer than the air. Measurements of the three types of wind profiles over the ocean have been made by Deacon, Sheppard, and Webb (1956).

Laboratory studies by the author using water 20° to 30°F colder and warmer than the wind blowing over it have shown some effect of $T_s - T_a$. These studies have, however, indicated that the main effect of the $T_s - T_a$ difference in the ocean is to modify the wind velocity profile. This will cause the curvature of the wind profile for stable and unstable winds to be different for a given wind speed at a given elevation above the sea surface than is the case for neutral winds. Considering this in the light of Miles's (1960) theory of the generation of waves by turbulent shear flows, and the effect of the curvature of the velocity profile at the elevation where the wind speed is the same as the speed of the wave component in question upon the rate of energy transfer from the wind to the waves, would lead one to believe that the type of wind profile is in turn responsible for the different wave heights observed in the ocean for a given duration, fetch, and wind speed at the normal height of an anemometer on a ship.

Figures 9.20 and 9.21 show that in the laboratory $T_{1/3} \propto U^{0.8}$ and $H_{1/3} \propto U^{1.1}$, approximately (Stanton, Marshall, and Houghton, 1932; Hamada, Mitsuyasu, and Hase, 1953; Flinsch, 1946; Sibul, 1955; Ralls and Wiegel, 1956). Studies in a lagoon and in a reservoir show that $T_{mean} \propto U^{0.5}$, $T_{1/3} \propto U_{0.4}$ to $U^{0.5}$, and $H_{1/3} \propto U^{1.1}$ to $U^{1.4}$ (Johnson, 1950; Burling, 1955). Studies in a lake show that $T \propto U^{0.35}$ to $U^{0.5}$ and $H_{max} \propto U^{1.2}$ to $U^{1.5}$ (Darbyshire, 1956). Most of the data available for ocean waves do not include information even as to the range of fetches for which the measurements were made. The relationships among T, H, and U are for a variety of fetches. The ocean data show $T \propto U^{0.3}$ to $U^{0.4}$ and $H \propto U^{0.5}$ to $U^{1.1}$ (Brown, 1953; Roll, 1953).

The relationships between wave period and fetch and wave height and fetch are not as clear as the relationships among H, T, and U. The laboratory data for constant wind velocities show approximately that $T_{1/3} \propto F^{0.3}$ to $F^{0.5}$, $T' \propto F^{0.5}$, and $H_{1/3} \propto F^{0.5}$ (Hamada, Mitsuyasu, and Hase, 1953; Sibul, 1955; Ralls and Wiegel, 1956). T' refers to a wave period calculated from measured values of wave length and speed. The reservoir measurements, for constant wind velocity (Burling, 1955) show $T_{1/3} \propto F^{0.25}$ and $H_{1/3} \propto F^{0.4}$. Darbyshire's (1956)

measurements on a lough give $T/T_{infinity} \propto F^{0.3}$ and $H_{max}/H_{infinity} \propto F^{0.3}$, for varying wind speeds, where $T_{infinity}$ and $H_{infinity}$ refer to values of T and H for infinite fetches. It is interesting to note that Stevenson (1852) found the wave height to be proportional to the square root of the fetch. Because the wave height is proportional to $F^{0.3}$ to $F^{0.5}$, an increase in fetch from 400 nautical miles, say, to 600 nautical miles would cause an increase in wave height of from $12\frac{1}{2}$ to $22\frac{1}{2}$ per cent which might not be noticed in the scatter of the data.

The physical reason for the increase in height and period with wind velocity and fetch is clear, considering the way wave records are analyzed. Waves in deep water can have a maximum steepness (H/L) of 1/7. Thus at the end of a short fetch, enough energy has been transferred from the wind to water to make only the shortest waves of any significant height. As the fetch is increased, more energy can be added only to the longer waves as the short ones have reached their maximum steepness; the longer waves then dominate the wave system, as far as the eye is concerned.

If the wave height and wave period depend upon wind velocity, fetch, duration of wind, air temperature, and sea surface temperature, then by dimensional analysis

$$\frac{gT}{U} = f_1\left(\frac{gF}{U}, \frac{gt}{U}, \frac{T_a}{T_s}\right), \tag{9.8a}$$

$$\frac{gH}{U^2} = f_2\left(\frac{gF}{U^2}, \frac{gt}{U}, \frac{T_a}{T_s}\right). \tag{9.8b}$$

The functions f_1 and f_2 cannot be determined by dimensional analysis; they must be determined either empirically or theoretically. Effectively then, after quasi-steady-state conditions are reached (gt/U large), the wave height and period depend upon the stability of the wind blowing over the water surface and a Froude number based upon the linear dimension of the storm, as long as the waves are in "deep water." The wave height and period should depend critically upon the turbulence of the air flow, and a logical criterion for this would be a modified Froude number based upon the mean horizontal length of the eddies in the air. Such a criterion would be useful in studying coupled wave effects. A more complete analysis would show that, in deep water, the wave height and period would also depend upon Reynolds number (damping), Weber number (surface tension effects), Richardson number rather than simply T_a/T_s, the ratio of fetch length to fetch width, and a boundary roughness parameter H/L (which of course would be a function of the wind speed, etc., again). The air-sea boundary process is one of fluid flow and thus the dimensionless numbers controlling all other fluid flow processes should be the ones used in describing it, neglecting those which obviously are not important in this, such as Mach number and the cavitation number. In Fig. 9.24 are shown the

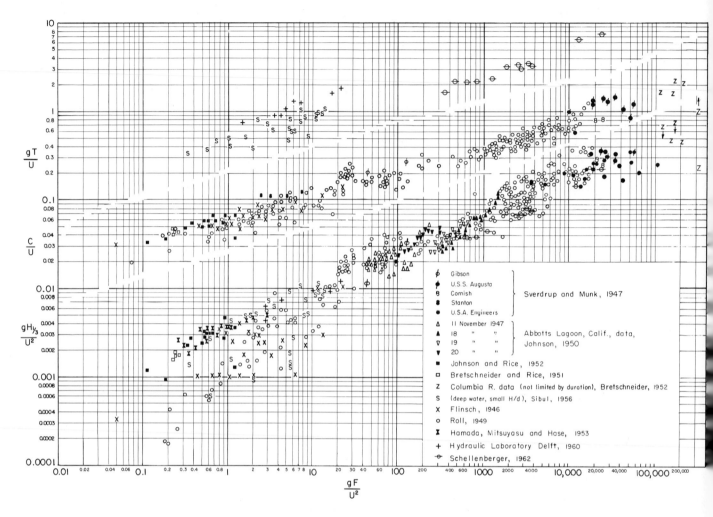

Fig. 9.24. Relationships among fetch, wind velocity and wave height, period, and velocity (after Wiegel, 1961)

empirical relationships between gT/U and gF/U^2, and gH/U^2 and gF/U^2 in the manner suggested by O'Brien (1943) with the effect of T_a/T_s being lost in the scatter of data. The empirical relationship between gT/U and gt/U, and gH/U^2 and gt/U is shown in Fig. 9.26. In some references, curves have been drawn through the data in such a manner that they leveled off in the region of gt/U and gF/U^2 greater than about 10^5. This occurs because the lines are drawn through the averages of all the data. A close inspection of the data shows that this leveling off did not occur for any of the individual sets of measurements, with the exception of Darbyshire's (1959) data for $gH_{1/3}/U^2$. The best curves drawn through these individual sets of measurements give the approximate relationships; $gT/2\pi U \propto (gF/U^2)^{0.3}$ to $(gF/U^2)^{0.4}$, or $T \propto U^{0.2}$ to $U^{0.4}$, and $T \propto F^{0.3}$ to $F^{0.4}$; and $gH/U^2 \propto (gF/U^2)^{0.25}$ to $(gF/U^2)^{0.35}$, or $H \propto U^{1.3}$ to $U^{1.5}$, and $H \propto F^{0.25}$ to $F^{0.35}$. These data seem to indicate that the "fully developed sea" does not occur in the open ocean for winds of any importance. Figure 9.28 can be treated in a similar manner. Here, $gT/2\pi U \propto (gt/U)^{0.3}$, or $T \propto U^{0.7}$,

and $T \propto t^{0.3}$; and $gH/U^2 \propto (gt/U)^{0.4}$ to $(gt/U)^{0.5}$, or $H \propto U^{1.5}$ to $U^{1.6}$, and $H \propto t^{0.4}$ to $t^{0.5}$.

It is clear that these data are in conflict with the original empirical spectrum of Neumann, where $H \propto U^{2.5}$ (Neumann, 1952; 1953; Neumann and Pierson, 1957). In order for $H \propto U^{2.5}$, the curve relating gH/U^2 and gF/U^2 would have to have a negative slope, which would also mean that $H \propto F^{-0.25}$, which is in obvious conflict with the physical situation. The original spectrum, however, is apparently in the process of being modified to include recent findings (Pierson, 1959b).

There is evidently still a pressing need for a large number of reliable measurements of waves in the open oceans.

The data shown in Fig. 9.24 are for the case described by Sverdrup and Munk (1947) as *fetch-limited*, the other condition being *duration-limited* (Fig. 9.25). Phillips (1957; 1958) developed a mathematical model that predicted that the mean square wave height is proportional to the square root of time for the duration-limited case and to the square root of fetch for the fetch-limited case.

What is meant by these two terms? Consider an infinite fetch, with the wind suddenly starting with a given velocity and then remaining at this velocity. At any point in the fetch, the significant wave height and period will increase with time. The significant waves moving past this point will have traveled a distance equal to the product of the group velocity of the significant waves and the length of time the wind has been blowing. Power will have been added to the wave system by the wind during the entire time. Now, real fetches are finite, so eventually the significant waves reaching a given point will be associated with the component wave periods that originated at the beginning of the fetch. The time for this to occur will be the distance divided by the group velocity (Sverdrup and Munk, 1947; Phillips, 1958a). After this duration has been reached, the waves are said to be fetch-limited.

It is possible for a third case to exist, the fully developed sea. This case would require a storm duration and fetch both long enough so that energy is being dissipated and radiated at the same rate that it is being transferred from the wind to the water in the form of waves.

Some measurements in the laboratory of Sibul (1955) show the effect of duration-limited and fetch-limited conditions on the wave height, as can be seen in Fig. 9.26.

In Fig. 9.24 are shown the data for C/U versus gF/U^2. There are a considerable number of measurements that show C/U in excess of unity. This is not surprising, nor is it a paradox. The flow of air over water is a boundary layer phenomenon and the phase speeds of the waves should be related to the free stream wind speed (the geostrophic wind speed) which is in considerable excess of the wind speed at anemometer height, which is the wind speed given for the ocean data of Figs. 9.24 and 9.25.

3. GENERATION OF WAVES

Although there is a considerable body of data on the relationships among fetch, wind velocity, duration of the wind, and the statistical values of the height and period of the waves so generated, little is known of the detailed mechanics of wave generation by wind. It is generally considered that below a certain wind speed (from about 4 to 10 ft/sec, the exact speed apparently depending upon the stability of the air immediately adjacent to the water surface and the turbulence in the air stream) no waves are formed. Ursell (1956) presents data and arguments that indicate that waves will form for practically zero wind velocity provided the fetch is long enough, the main reason being that, from the flat plate analogy of boundary layer growth, the air flow over the earth must be turbulent since the distance over which it blows is always large. Phillips (1958d) also discusses this in explaining some of the laboratory measurements of Cox (1958). When very low speed winds are blowing over a water surface, the appearance of the surface is glassy. Occasionally a disturbance is noted. It may be either a series of small ring waves moving out from a point or a series of small periodic waves similar to the waves caused by a ship moving through the water (Fig. 9.27). The lengths of these waves are very small, only an inch or so. They are caused by small eddies (or atmospheric waves) in the turbulent wind blowing over the water surface. When the wind speed increases a little, the surface looks more like Fig. 9.28. (Note the surface tension waves in the lower left corner, in addition to the gravity waves.)

Waves caused by ships are both divergent and transverse. The divergent waves do not make as large an angle with the transverse waves as appears. This is because the eye sees the superposition of the transverse and divergent waves as the "ship wave." in reality, the two wave systems are as shown in Figs. 9.29 and 9.30. The transverse waves

Fig. 9.25. Relationships among duration, wind velocity and wave height, period, and velocity (after Wiegel, 1961)

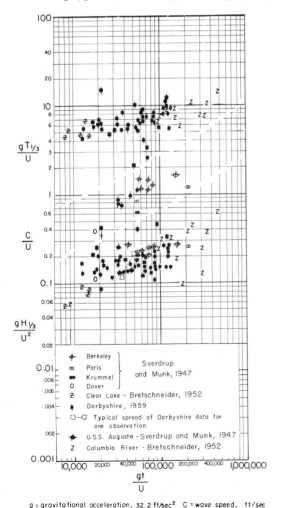

g = gravitational acceleration, 32.2 ft/sec²   C = wave speed, ft/sec.
T₁/₃ significant wave period, sec.   H₁/₃ significant wave height, ft.
U = surface wind speed (at various   t = duration, sec
    elevations), ft/sec   F = fetch, ft

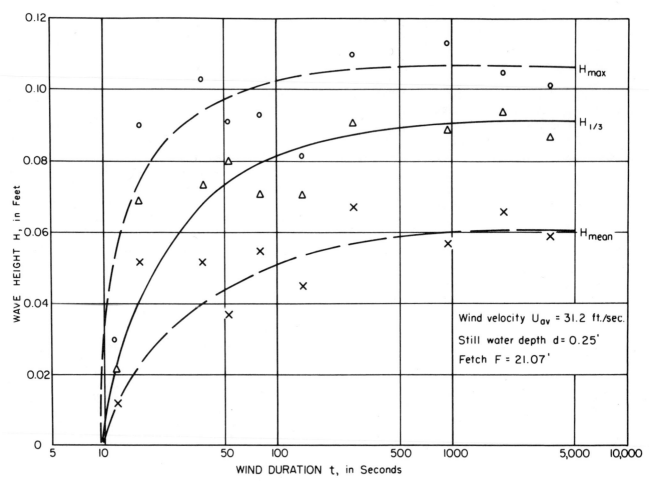

Fig. 9.26. Wave height as a function of wind duration (after Sibul, 1955)

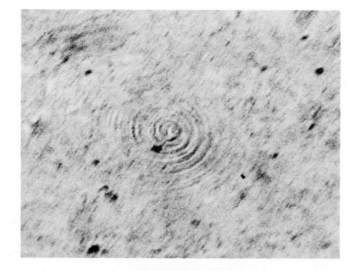

Fig. 9.27. (left) Waves generated by wind gust striking water surface; (right) waves generated by small wind gust moving over water surface

Fig. 9.28. Pattern of waves for winds just able to generate waves

Fig. 9.29. Vertical aerial photo showing ship waves

that exist can be seen in the aerial photograph of Fig. 9.29.

It has been stated that surface tension plays an important role in the generation of wind waves (Keulegan, 1951 and Van Dorn, 1952) in winds of low speeds. These investigators placed detergents in the water to decrease the surface tension. They found that, apparently, the lower the surface tension, the higher was the wind speed necessary to generate waves and attributed this effect to the lessening of the surface tension. Ursell (1956), however, points out the importance of the influence of surface film on the phenomenon, and states that it might be the surface film effect of the detergent rather than the surface tension; anyone who has worked in the laboratory on water-air phenomena will recognize the importance of this. In a laboratory study performed by R. H. Cross and the author, the water was heated by steam, and under some conditions it was observed that a group of thirty or so waves would form, followed by an interval of relative calm, which was followed by a group of waves, followed by relative calm, and so on.

With higher winds, it appears that gravity waves form directly, although both gravity waves and surface tension waves will be present together. Schooley (1958) took motion pictures of the waves formed at low wind speeds in a wind-wave tunnel. He found that the surface tension waves and the gravity waves traveled at the same speed, with their lengths being approximately the lengths predicted by theory. The greater the wind speed, the greater the speed of both the surface tension and gravity waves, hence the longer the gravity wave and the shorter the surface tension wave. He also found a wave profile that had a double peak downward (Schooley, 1960), as predicted theoretically by Wilton (1915) and Crapper (1957).

Eckart (1953) has advanced a theory that the waves which are relatively long and high (as opposed to those that might be termed *ripples*, which were just discussed) are created by large-scale eddies of low or high pressure swooping down and moving over the water surface for a minute or two with speeds with a statistical distribution about the mean wind velocity. Eckart states that these would cause coupled waves moving at small angles with the direction of motion of the gust. Waves of this sort have been observed by Van Dorn (1952).

Measurements of air pressure micro-oscillations have been made by Roschke (1954) using three microbaro-

Fig. 9.30. First few crests of wave patterns due to discontinuity of source distribution on a vertical plane (after Jinnaka, 1957)

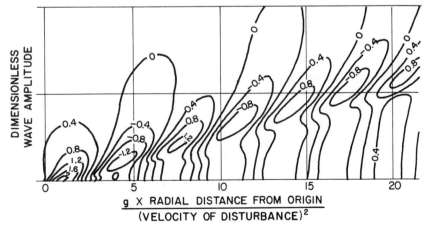

DIMENSIONLESS WAVE AMPLITUDE

g X RADIAL DISTANCE FROM ORIGIN
(VELOCITY OF DISTURBANCE)²

graphs located a few feet above the ground at the apexes of a horizontal triangle about 60 feet apart. It was found that gusts (air pressure "waves") do exist and that many of them moved at ground level long enough to be recorded on all three recorders. Furthermore, the average measured velocities (magnitude and direction) were approximately the same as the average wind velocity. These gusts that were studied had periods in the range of 10 to 25 sec, although there were many shorter periods as well as longer periods.

Wiegel, Snyder, and Williams (1957) have generated two-dimensional waves by moving a low-pressure area over the water surface in a towing tank, and Abraham (1959) has done the same thing for the three-dimensional case. In the two-dimensional case, it was found that trains of waves were generated with periods associated with the wave speed identical to the forward speed of the low-pressure area. In the three-dimensional case, the wave periods of the transverse wave were associated with the wave speed identical to the forward speed of the low-pressure area and the wave periods of the diverging waves were associated with a forward component the same as the speed of the low-pressure area. This gust mechanism would generate a series of groups of nearly periodic waves, the interaction of the groups resulting in a wave system with many characteristics of a Gaussian distribution.

Wind gusts are a large-scale turbulence phenomenon. Their velocities and pressures have some sort of a partial random distribution about some mean values. Consequently waves generated by this mechanism must have similar distributions of dimensions, hence the formation of short-crested waves. The degree of randomness is modified, however. Since the gust moves over the water surface for a finite length of time and since wave energy is transmitted with the group velocity, which in deep water is one-half the phase velocity, a group of waves are formed that trail the gust area, and are close to being uniform in height and period. This group may consist of as many as five to ten waves. The existence of groups (see Fig. 9.3) is an observed phenomenon which has apparently been lost sight of in the recent advance in studying waves by statistical means. One of the wave records shown in Fig. 9.3 was obtained in a small body of water (Johnson, 1950) where there was no swell present to complicate the pattern. Only the effect of the small gusts moving in the reduced speed boundary close to the water would be noticeable. The use of statistics is a necessary and useful tool in dealing with ocean waves. Unfortunately, one fundamental property has been thrown away, and that is the phase relationship which exists in any time series.

The theory of Eckhart (1953) was not able to predict the observed wave heights, the predicted values being an order of magnitude too low. Phillips (1957, 1958a, 1958b, 1958c), using a concept of resonance, was able to develop a theory which predicted wave heights which were originally believed to be of the same order of magnitude as those observed in the ocean. Measurements made in the ocean, however, have shown that the mean-square pressure fluctuations assumed by Phillips were too high by a factor of the order of 100 (Longuet-Higgins, Cartwright, and Smith, 1961). In addition, the gusts did not have to move down into the lower boundary, but were already there. The wind pressure records of Roschke (1954) show that the gusts close to the surface are "periodic" in a manner somewhat similar to ocean waves; the pressure rises and falls in a time series which has some sort of an average "period" (hence, "length" and "height") with a Gaussian distribution about these averages. Now a single gust moving over the water surface for an interval of time generates a series of periodic waves. If the length of these waves is the same as the length of the pressure fluctuations (distance between centers of succeeding gusts), then the following gust will add energy to the existing waves, and the next gust will add more energy, etc., this being the resonance phenomenon considered by Phillips. In order for this resonance mechanism to exist, the gust "period" must be $2\pi/g$ times the speed of the gust. Because both divergent and transverse waves are generated, a two-dimensional wave spectrum is developed, with the shorter the wave length, the greater the relative importance of angular spreading (Phillips, 1958c).

There is almost no information on spatial distribution of pressure and velocity. What is the relationship between the length of the gusts in the direction of their advance, normal to this direction, and perpendicular to the ground? Some work on the medium-scale fluctuating velocity components higher in the atmosphere by Charnock, Francis, and Sheppard (1955) has shown that the vertical component of the velocity fluctuation was from one to two orders of magnitude smaller than the two components in the plane parallel to the ground. These conclusions appear to be at variance with the measurements of Sherlock and Stout (1937), as can be seen in Fig. 9.31.

If the vertical dimensions of the gusts were an order of magnitude smaller than horizontal dimensions, and if one considers that these gusts move with a speed associated with the mean wind speed at an elevation of their mid-heights, then a 10-sec period water wave would be moving with the speed of the wind measured about 25 ft above the water surface. A 1-sec wave would be associated with the mean wind speed measured a few inches above the water surface and a 20-sec wave would be associated with the mean wind speed 100 ft above the water. Now, wind blowing over the water surface is a boundary layer phenomenon; hence, the closer to the ground, the lower will be the average wind speed for a given geostrophic wind speed. Thus, we have the condition that, on the average, the smaller the eddy spacing, the lower will be the forward speed of the eddy; and this is the relationship needed for this particular resonance mechanism to occur often. It is evident that a considerable body of information must be obtained on the details of wind structure before further advances can be made in the understanding of the initial mechanism of wave generation.

What will be the periods of the waves? Our answer will depend upon how we measure the waves and how we analyze the measurements. In certain types of records, such as aerial photographs, the steepness of the waves is the feature most in evidence. In analyzing surface time histories at a point or watching waves along the shore, the actual height is the feature most in evidence. In either case, the wave height is important in analyzing waves. There is a tendency to consider the higher waves and to neglect the lower waves. From this tendency arose the concept of significant wave height, the significant wave height being the average of the highest one-third of the waves. The period that is observed is probably related to these significant waves; in fact, one of the simplest methods of designating the wave period

is by measuring the periods of the high groups of waves, this being the so-called significant wave period. The power transmitted by a train of uniform periodic waves is proportional to the wave period and the square of the wave height. Thus, it takes a greater transfer of power from the wind to the waves to double the height of a long-period wave than it does to double the height of a short-period wave. It takes time and fetch to do this. Thus, for a given wind velocity, the greater the time the wind has blown (up to some limit) over a given fetch, the higher will be the waves and the longer will be the period. The longer the fetch (up to a limit?), the higher will be the waves and the longer the periods. For a given fetch and duration, the higher the wind velocity, the higher will be the significant wave and the longer will be the significant period.

Using this concept of wave period and wave height, Phillips (1957) showed, for the case where the waves were still growing with respect to time at a given place, that the wave height was proportional to the wind speed and proportional to the square root of the length of time the wind had been blowing (the wind duration). The

Fig. 9.31. Wind gust shown by iso-velocity contours: (*top*) horizontal section of the wind 50 ft above the ground; (*bottom*) vertical section of the wind at the tower (after Sherlock and Stout, 1937)

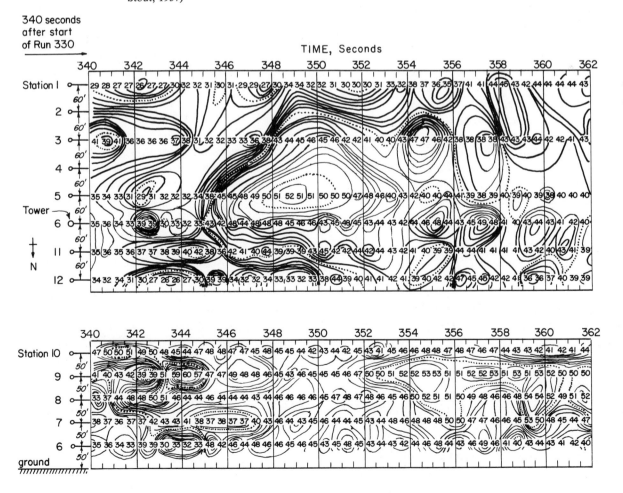

data shown in Fig. 9.25 appear to confirm this result. In addition, Phillips' work appears to predict the observed constant of proportionality.

For the case of the fetch limiting the wave dimensions rather than the wind duration, Phillips (1958a) has shown that the average wave height should be proportional to the square root of the fetch. This has been approximately verified, as can be seen in Fig. 9.24.

Scientists often use the term *fully developed sea*. What is meant by it? It is not the same as the quasi–steady-state condition, which refers to the case where the statistical values of wave height and wave period do not vary with time at a given position. Rather, it refers to the condition of no change in the statistical values of wave height and period regardless of the duration or fetch. This problem has been studied theoretically by Phillips (1958b), and his conclusions have been presented previously (Eq. 9.5). The measurements shown in Fig. 9.14 indicate that a true fully developed sea does not exist in the ocean, although the wave components of the shorter periods are fully developed.

An idealized mechanism by which some insight might be gained of the generation of waves by winds is the Kelvin-Helmholtz instability (Kelvin, 1871; Lamb, 1945, pp. 232, 268; Ursell, 1956). This mathematical model consists of two fluids with a common surface, the surface being a horizontal plane when undisturbed. The upper fluid of density ρ' has a horizontal velocity U'; the lower fluid of density ρ has a horizontal velocity U. Thus, there is a flow discontinuity at the common surface and no viscosity is admitted; it is an "ideal fluid" model. The common surface is now considered to be disturbed by some unspecified mechanism in such a manner that a linear harmonic surface condition is created. The two flows, U' and U, and the pressures in the fluids will now be modified. There are two solutions to the problem, one is a stable boundary solution and the other is an unstable boundary solution. The equation which must be examined to determine which of the two solutions will exist is

$$C = \frac{\rho U + \rho' U'}{\rho + \rho'} \pm \left\{ \frac{g}{k} \frac{\rho - \rho'}{\rho + \rho'} - \frac{\rho \rho'}{(\rho + \rho')^2} (U - U')^2 \right\}^{1/2};$$

$$(9.9)$$

$(\rho U + \rho' U')/(\rho + \rho')$ can be considered to be a mean current. The waves travel relative to this current with velocity C

$$C^2 = C_o^2 - \frac{\rho \rho'}{(\rho + \rho')^2} (U - U')^2 \qquad (9.10)$$

where C_o is the wave velocity in the absence of currents. If C^2 is positive, the boundary will be stable, but if C^2 is negative, the circular frequency $2\pi/T$ will be complex and the boundary will be unstable. Now $C < O$ if

$$|U - U'| > \sqrt{\frac{g}{k} \frac{\rho^2 - \rho'^2}{\rho \rho'}} \qquad (9.11a)$$

or, if $\rho' \ll \rho$

$$|U - U'| > \sqrt{\frac{g}{k} \frac{\rho}{\rho'}} \approx 65.5 \sqrt{L}. \qquad (9.11b)$$

As $U \ll U'$, a wave only 1 ft long would require a wind speed of at least 65.5 ft/sec (39 knots) in order for the boundary to be unstable so that the wave would grow; for a wave only 10 ft long, the necessary wind speed would be in excess of 207 ft/sec (123 knots). These values of required wind speed show that the mechanism, in addition to being unrealistic in respect to describing the actual air flow, cannot account for any but the shortest waves at any reasonable wind speed. Unless some mechanism can be described for the efficient transfer of energy from very short waves to long waves, it is difficult to see how the Kelvin-Helmholtz instability can account for the abrupt changes of air-sea processes described by Munk (1947).

If the effect of surface tension is considered, it can be shown that there is a minimum wind velocity that can cause an instability and that this minimum is associated with the minimum wave velocity of a surface tension-gravity wave. If this minimum wave velocity is substituted for C_o in Eq. 9.9, it is found that $U'_{\min}$ is about 21.3 ft/sec for air blowing over water, which is considerably in excess of the observed case. However, a laboratory study (Francis, 1954) of air blowing over oil appears to confirm an instability of a type related to the Kelvin-Helmholtz instability. Miles (1959b) has predicted theoretically the values measured by Francis for air flowing over oil, assuming shear flow.

Wuest (1949) and Lock (1954) have treated the formation of waves as a boundary layer stability problem for two viscous fluids for which the free stream flow is nonturbulent. The mathematics is difficult to follow and experimental studies of a less difficult system, that of the flow over a flat plate, show that flow instability, as indicated by the occurrence of a turbulent boundary layer, is much more dependent upon the turbulence level of the incident stream than it is upon the theory for viscous nonturbulent flow (Goldstein, 1938, p. 199).

An instability of this type does occur, however, at a much lower minimum wind speed than required by the Kelvin-Helmholtz instability, and the instability is dependent upon the fetch as well as the wind speed. The stability is of a nonuniform type in that, for a particular wind speed, the interface may be stable for a range of given wave lengths from zero fetch up to some small fetch, then unstable for a distance, and then stable after this as the fetch increases. What is the physical significance of this new stability? Perhaps it is associated with the fact that a wave of a given length can build up to only a certain limiting height before it breaks and the flow becomes turbulent. The wave height then decreases, gradually builds up a little, breaks again, and so on, with the excess energy being transferred from the water going into the nondirected energy of turbulence.

Miles (1957; 1959a) developed a shear flow model. The distribution of the wind was assumed to be some function $U(y)$, with there being no mean water motion due to the flow of air over the surface. The flows were considered to be inviscid. Under these conditions, the energy transfer was by normal pressures. His general conclusions were that the rate at which energy would be transferred from the wind to a wave with a speed C was proportional to the curvature of the wind profile at the elevation where $U(y) = C$; hence, the longer the wave component (the greater C), the slower the rate of energy transfer, as the curvature $-U''(y)$ decreases with increasing elevation, increasing $U(y)$. This would, of course, mean that, at first, the short waves would predominate. It also predicts wave components with speeds as great as the free stream speed (the geostrophic wind speed). It is interesting to note that if the curvature $-U''(y)$ was the reverse of the curvature actually encountered in fluid flow, the boundary would never be unstable.

One other general conclusion due to Miles was that measurements of the pressure distribution on a stationary wave model with wind blowing over it would not necessarily result in sheltering coefficients that would be similar to those obtained if the wave surface were moving at some speed C. The reason for this is that $U(y) = 0$ at the stationary boundary.

At this stage, Miles introduced a velocity distribution $U(y)$ which is the generally accepted distribution for the neutral condition in wind flow. In addition, it was postulated that the water wave energy was damped by the viscosity of the water. Under these conditions, it was found that the minimum wind speed for the initiation of gravity (not surface tension) waves would be from 2.6 to 3.3 ft/sec (80 to 100 cm/sec) which is about the generally accepted observed minimum obtained from observations. The wind speed given by Miles is at a level of about 6 ft above the undisturbed water surface. Using this same wind velocity profile, Miles obtains a theoretical coefficient which should be analogous to Jeffreys' sheltering coefficient; he finds this to be 0.3 This is remarkably close to the value of 0.32 Jeffreys found necessary to cause waves to grow with a minimum wind speed of 3.3 ft/sec (100 cm/sec).

The essential feature of Miles's theory is that the air pressure gets out of phase with the water surface. The water surface (linear theory) is

$$y_s = \text{Real Part } \{a \, e^{i(kx - \sigma t)}\} \qquad (9.12)$$

and the variation in pressure (p_a) from the static value along the water surface is

$$p_a = \text{Real Part } \{\rho_a(\alpha + i\beta)u_1^2 ka \, e^{i(kx - \sigma t)}\} \qquad (9.13)$$

where a is the wave amplitude, k is $2\pi/L$, L is the wave length, x is the horizontal coordinate, σ is $2\pi/T$, T is the wave period, t is time, ρ_a is the mass density of air, and α and β are dimensionless functions of the wave and wind characteristics which are predicted from Miles's theory u_1 given by

$$u_1 = u_*/k_0 \qquad (9.14)$$

in which u_* is the "friction velocity" and k_0 is von Karman's constant, usually taken as 0.4.

The power per wave length (P_{aw}) transferred from the air to the water is obtained by integrating the product of p_a and the component of water surface velocity normal to the surface. This gives

$$P_{aw} = \pi \rho_a g a^2 C \left(\frac{u_1}{C}\right)^2 \sqrt{\alpha^2 + \beta^2} \sin \phi \qquad (9.15)$$

where ϕ is the angle by which the maximum air pressure leads the wave trough, and is

$$\tan \phi = -\beta/\alpha \qquad (9.16)$$

Thus, the maximum air pressure occurs a little before the wave trough, and the minimum pressure occurs a little before the wave crest, rather than these occurring at the trough and crest as would be the case of an inviscid air flow. Theoretical work by Benjamin (1959) also shows this to be the case. Measurements made at sea by Longuet-Higgins, Cartwright, and Smith (1961) have been inconclusive in showing whether or not this does occur. Measurements made in the laboratory by R. H. Cross, J. D. Cumming and the author have shown that the phase shift does occur, and that the value of ϕ predicted by Miles's theory is correct.

By setting the time average power being transferred by the wave across a vertical plane (normal to the direction of wave advance) at some value of $x + dx$ equal to the sum of time average power being transmitted across a vertical plane at x and the time average power being transmitted from the air to the water per differential increment of surface, the rate of growth of wave amplitude can be found. It is

$$a = a_0 \exp \left[\pi \frac{\rho_a}{\rho_w} \frac{C}{C_G} \left(\frac{u_1}{C}\right)^2 \beta \frac{x}{L} \right] \qquad (9.17)$$

where ρ_w is the mass density of the water and C_G is the group velocity of the waves, given by Eq. 2.68. If a_0 can be expressed as a function of frequency at the start of a fetch, then the one dimensional distribution of a as a function of fetch can be calculated for any fetch provided the air shear flow is known. Calculations of this type show that a rises very slowly with increasing σ, until it reaches a critical value, then it rises extremely rapidly. The upper limit of a for any value of σ is determined by Eq. 2.144 (see also Eq. 9.5).

The mechanisms of Miles (1957; 1959a) and Phillips (1957) probably occur simultaneously, one mechanism

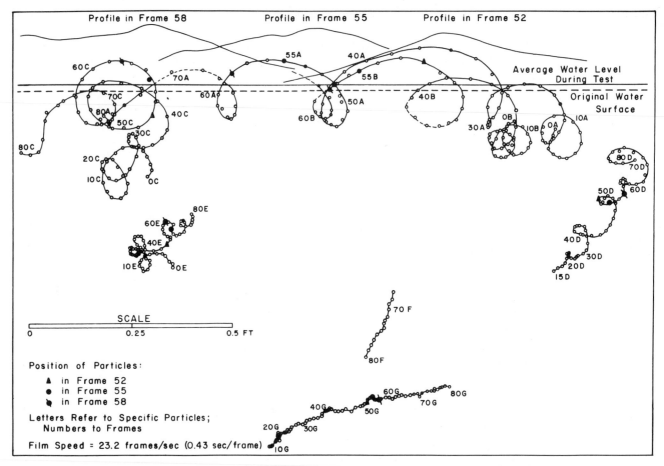

Fig. 9.32. Particle paths in wind waves. Note that particles near the water surface tend to follow an orbital path; however, when a particle is at the crest of a wave the motion due to surface drag of the wind is several times the orbital diameter (from Johnson and Rice, 1952)

being associated with the mean air flow and the other mechanism being associated with the time and spatial variation (large-scale turbulence in this case) from the mean air flow. The two mechanisms cannot be treated independently with the results being simply added if only because the turbulence energy in the wind must come from the mean flow energy in the wind. Both models rely upon normal forces (pressure) to transmit energy from the wind to the waves. It is probable that the mechanism assumed by Phillips is responsible for the initiation of the waves, and that the mechanism in Miles's theory is responsible for their growth to the large waves observed in the ocean. Miles (1960) has considered the two mechanisms together but from a slightly different viewpoint.

Jeffreys (1925; 1926) developed a model consisting of two parts. One part consisted of a net force in the direction of wave motion due to pressures acting on the wave surface with the pressure on the upwind side of the wave being greater than the pressures on the downwind side of the wave because of the separation just past the crest of the wind flowing over the waves. This is often called *form drag*. The other part consisted of a net force in the direction of wave advance due

to the shear stress exerted by the wind on the water surface. The second part is difficult to formulate, as a portion of the power transferred from the air to the water is transmitted by a water current whereas another portion is transmitted by water waves. This dual effect is illustrated in Fig. 9.32 which shows the motion of some water particles in a wind-wave tunnel.

Details of the formulation of the problem will not be presented here, only the conclusions. The power per unit area, P_N, transferred from the wind to the water waves by normal forces (using the notation of Sverdrup and Munk, 1947) is

$$P_N = \tfrac{1}{2} s \rho_a (U - C)^2 k^2 a^2 C, \quad \text{for } C < U \quad (9.18)$$

where s is the dimensionless sheltering coefficient, ρ_a the density of air and U is the wind velocity over the crests. If this is considered to be the equivalent of the form drag on a body, then U should be the free stream wind speed (geostrophic wind speed); hence, C can be considerably greater than the wind speed measured at normal anemometer height and still permit the transfer of energy from the wind to the waves.

Jeffreys was interested primarily in determining the criterion for the initial formation of waves. He equated

Eq. 9.10 to the equation expressing the rate at which wave energy is dissipated by viscosity per unit area, P_μ

$$P_\mu = -2\mu_w k^3 a^2 C^2 \qquad (9.19)$$

where μ_w is the absolute viscosity of the water. This leads to the inequality

$$(U - C)^2 C > \frac{4\mu_w g}{s\rho_a} \qquad (9.20)$$

which must exist for waves to grow. By differentiating Eq. 9.20, Jeffreys found that

$$U_{min} = 3C_{min} \geq 3\left(\frac{\mu_w g}{s\rho_a}\right)^{1/3} = \frac{73}{s^{1/3}}. \qquad (9.21)$$

For an observed wind speed of 3.6 ft/sec (110 cm/sec), this would lead to a value of $s = 0.29$. If the wind speed were doubled to 220 cm/sec (7.2 ft/sec), s would be 0.036. It is evident that the minimum wind speed (and hence, wave length) is not critically dependent upon the value of the sheltering coefficient s.

Several laboratory studies have been made to determine the value of s. Stanton, Marshall, and Houghton (1932) constructed a series of 27 waves made of wood and placed these in a wind tunnel. The waves were all of the same steepness ($H/L = 0.206$), but the height and length gradually increased (from $L = 5.1$ cm to $L = 21.6$ cm) from the first to the twenty-seventh wave. Pressure taps were placed at intervals along the tenth and the twenty-seventh waves, and pressure measurements made. Sverdrup and Munk used these data to compute the sheltering coefficient s; it was found to vary from 0.036 to 0.90, depending upon the height of the wave and the wind speed. The investigators thought that a tunnel blockage effect might have existed, as the wind tunnel was only 12 in. high, so a second test was made in a 36-in. wind tunnel. Two sets of wood models of simple harmonic waves of uniform height were constructed ($H/L = 0.4$), one set having a wave length of 7.62 cm and the other having a wave length of 2.6 cm. Both series consisted of many waves. For this case, Sverdrup and Munk give a value of $s = 0.006$.

Mostzfeld (1937) used four sets of rigid corrugated metal waves, covered with plaster of paris and varnish, in a wind tunnel, two of these sets being sine waves, the third being nearly trochoidal in shape, and the fourth

set consisting of a series of arcs with the crest having a total angle of 120 degrees. In the first two sets of waves (three each), the wave lengths were 30 cm, one being 1.5 cm high and the second being 3.0 cm high. In the other two sets (six waves each), the wave lengths were 15 cm long, with the trochoidal wave being 1.45 cm high and the sharp-crested wave being 2.0 cm high. The Reynolds numbers ranged from 63,000 to 162,000, based upon the wave amplitude. The sheltering coefficients (after Ursell, 1956) were found to be 0.034, 0.024, 0.028, and 0.11. It appears that the sharp-crested wave leads to greater separation with a higher sheltering coefficient than is the case for the sine or trochoidal wave.

A series of tests in a wind tunnel were made by Bonchkovskaya (1955). It was found that a long series of waves had to be used as a stable flow did not develop rapidly. The results shown in Figs. 9.33–9.36 consist of data taken far enough downstream for stable conditions to exist. It was found that flow over a nonsymmetrical model led to greater separation than flow over a symmetrical model (Figs. 9.33 and 9.35). The wind velocity profiles used to construct Fig. 9.35 are shown in Fig. 9.34.

It can be seen in Fig. 9.1 that many of the waves in the generation area are nonsymmetrical and many are relatively sharp-crested. It appears that, once the waves have attained appreciable steepness and nonuniformity, flow separation plays an important part in the transfer of power from wind to waves. Photographs taken by Bonchkovoskaya of the air flow over the waves, using a very fine powder in suspension in the air stream, showed this to be true.

The measurements of the pressures over the wave surfaces for a series of different wave steepnesses are presented in Fig. 9.36. Flow separation is small for waves of low steepness. It does not appear that Jeffreys' model can predict the growth of the first waves, as these must start with minimum steepness. The argument that measured values of s are much smaller than the value predicted by Jeffreys for the initial growth of waves is academic, as s must go to zero at the limit of waves of zero height.

The second mechanism of energy transfer from wind to waves considered by Jeffreys was that of *skin friction*, the transfer of energy by tangential (shear) stress. Using a reasonable value of the skin friction coefficient, s', of

Fig. 9.33. Flow over symmetrical and non-symmetrical waves (after Bonch-kovskaya, 1955)

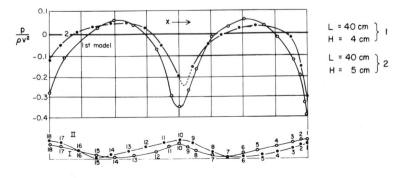

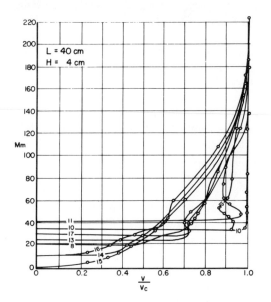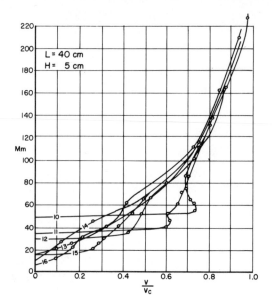

Fig. 9.34. Air velocity profiles (after Bonchkovskaya, 1955)

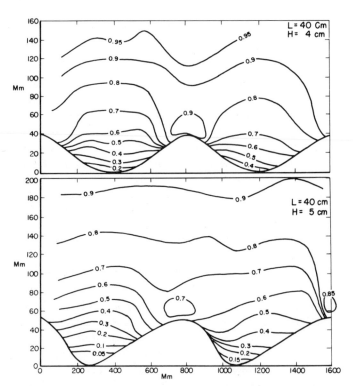

Fig. 9.35. Air velocity contours (after Bonchkovskaya, 1955)

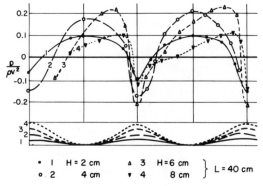

| • 1 | H = 2 cm | ▲ 3 | H = 6 cm | } L = 40 cm |
| ○ 2 | 4 cm | ▼ 4 | 8 cm | |

0.002, Jeffreys found that the lowest wind speed needed to transfer energy at a rate faster than the rate dissipated by viscosity in the water was 480 cm/sec, with a minimum wave length of 140 cm. This is higher than the observed value of minimum wind speed necessary to generate waves that will grow; it does not, however, detract from the importance of the mechanism in adding energy of waves when the winds exceed 480 cm/sec.

Sverdrup and Munk (1947) also considered the effect of wind stress as a wave-generating mechanism to the second order. Schaaf and Sauer (1950) reanalyzed the problem and found that the Sverdrup and Munk analysis was not consistent to the second order, and as a consequence, some of the conclusions were in error, particularly the conclusion that this mechanism could transfer energy from the wind to the waves even when $C > U$.

Sverdrup and Munk (1947) developed a mathematical model for the prediction of the increase in wave height and length as a function of wind speed, fetch, and time, considering both normal and tangential forces acting on the water surface. Each of these mechanisms was treated as if the other mechanism did not operate, and then predicted power transferred by each mechanism independently was added linearly to give a total power transferred between the two fluids. Two difficulties are readily apparent. First, if the sheltering mechanism occurs, a considerable modification of the tangential air flow over the water must result. Secondly, some power transferred by tangential stresses must set up a water current.

Perhaps the simplest manner of viewing the problem of the transfer of power from wind to waves is to view the waves as a solid undulation of height H, a length L,

Fig. 9.36. Flow over waves of increasing steepness (after Bonchkovskaya, 1955)

226

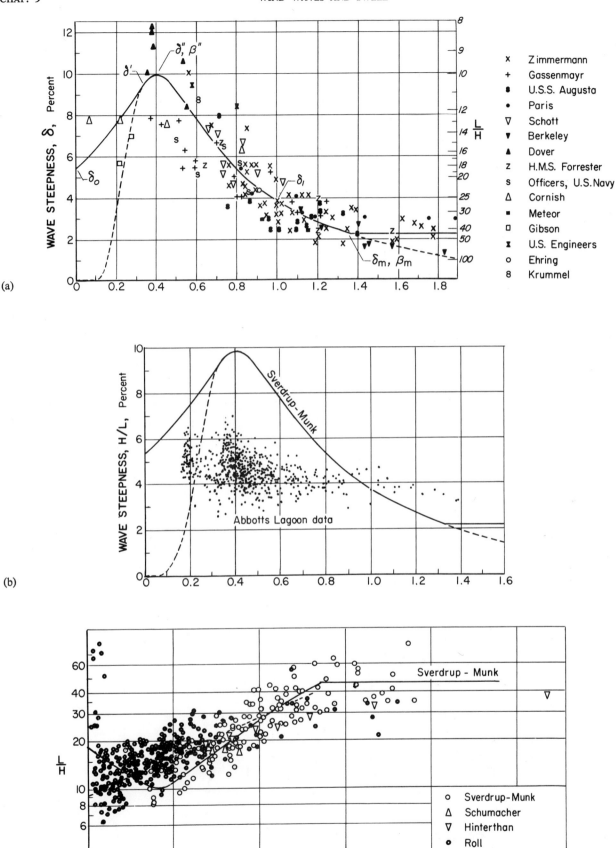

(a)

(b)

(c)

Fig. 9.37. (a) Wave age, β (after Sverdrup and Munk, 1946) (b) Wave age, C/U (after Johnson, 1950)
(c) (after Roll, 1949)

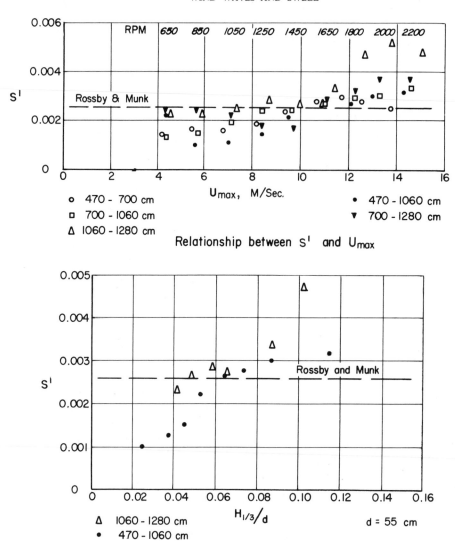

Relationship between s' and U$_{max}$

Relationship between s' and H$_{1/3}$/d

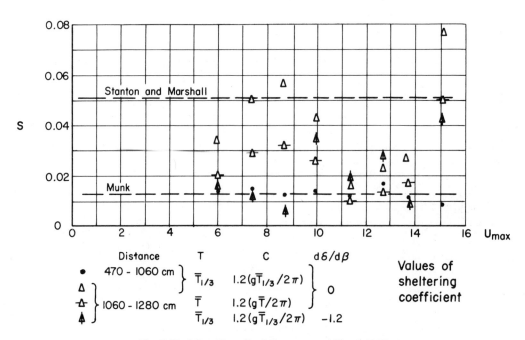

Values of
sheltering
coefficient

Fig. 9.38. (after Hamada, Mitsuyasu and Hase, 1953)

with a forward velocity C. If the free stream wind velocity is U_o, ρ_a the density of the air, then the force in direction of wave advance acting on a wave per unit width of wave is

$$F_o = \tfrac{1}{2} C_D \rho_a H (U_o - C)^2,$$

and power, P_{aw}, transferred per wave is $F_D C$

$$P_{aw} = \tfrac{1}{2} C_D \rho_a H (U_o - C)^2 C \qquad (9.22)$$

where C_D is a function of a dimensionless boundary roughness (wave height and length), a Reynolds number (based upon wave height, wind velocity, and the viscosity of air), the stability of the air, and the free stream turbulence. C_D includes both form and shear drag effects. Kapitsa (1949) showed by a simplified theoretical analysis that this could be expressed as

$$\bar{P} = \frac{K}{\pi} \left(\frac{\pi H}{L} \right)^2 \rho_a (U_o - C)^2 C \qquad (9.23a)$$

or

$$P_{aw} = K \left(\frac{\pi H}{L} \right)^2 H \rho_a (U_o - C)^2 C, \qquad (9.23b)$$

where $\bar{P}$ is the power transferred ($P_{aw} = L\bar{P}$) per unit horizontal surface area with the dimensionless coefficient K being less than unity. This is of the same form as the Sverdrup and Munk equation. It is interesting to note that, assuming K to be unity and assuming that the only dissipative force with respect to the wave energy is the viscosity of the water, Eq. 9.23 leads to the prediction that the minimum wind velocity that will result in the growth of water waves is 85 cm/sec, and the wave length associated with this will be 5 cm. A value of K of unity is associated with the physical phenomenon of separation of the air flow occurring at the wave crest.

This combined form and shear drag mechanism allows the computation of the power transfer from wind to waves for a system of uniform periodic waves, of a given wave number. It does not predict the frequency spectrum nor any of the nonlinear mechanisms of power transfer between waves of different frequencies. Sverdrup and Munk got around this from a practical standpoint by considering that the wave system at any place and time could be represented by a significant wave train. It was then assumed that the significant waves increased in both height and length as a function of time and fetch. The partition of energy between the growth of wave height and the growth of wave length was determined from the empirical relationship between wave steepness (H/L) and wave age (C/U). This particular representation is an approximate energy spectrum (Ichiye, 1952). The data used by Sverdrup and Munk are shown in Fig. 9.37, together with data from other sources. It is evident that the various sets of data are in conflict. Furthermore, there is the usual considerable scatter of data from any one source. For example, Roll's values of L/H vary between about 10 and 20 at a value of C/U of 0.5. Considering a given value of U and C (hence, L), the wave

steepness varies by a factor of 2. This fact is not brought up as a criticism of the method of Sverdrup and Munk, rather it is brought up to emphasize that we are dealing with a phenomenon that has not been described in precise terms.

The procedure of Sverdrup and Munk also requires two empirically determined coefficients, the sheltering coefficient, s, and the shear stress (skin friction) coefficient s'. Indirect measurements of these values were made by Hamada, Mitsuyasu, and Hase (1953) from observations of waves generated by winds in a wind-wave tunnel. These data are shown in Fig. 9.38. They are higher than the values used by Sverdrup and Munk.

All the foregoing theories require empirical data. These data may consist of sheltering and shear stress coefficients, wind profiles, or wind eddy magnitudes and distributions. The determination of these data is difficult in the ocean, and yet it is in the ocean that they must ultimately be obtained. One of the most fruitful avenues of research will be the continuation of measuring ocean waves in generating areas, calculating frequency spectra from them, and then relating these spectra to meteorological and oceanographical criteria. It is of primary importance to measure any change in wind profiles as a function of fetch.

4. WIND WAVES IN SHALLOW WATER

Measurements of waves in the North Sea, in the Gulf of Mexico, and in Lake Okeechobee have shown that wave height and period differ from the values in deep water, for the same wind speed, fetch duration, and atmospheric stability. That this should be so is easily understood from the fact that the wave speed is less in shallow water than in deep water, and the theoretical limiting wave steepness in shallow water is less than it is in deep water (Miche, 1944), and is $H/L = (1/7) \tanh 2\pi d/L$. In addition to this, energy is lost because of bottom friction, which will also modify the wave height and period, everything else being the same.

Sibul (1955) studied the phenomenon in a wind-wave tunnel. He plotted $H_{1/3}/H_{o 1/3}$ versus $d/H_{o 1/3}$ and $T_{1/3}/T_{o 1/3}$ versus $d/L_{o 1/3}$ where $H_{1/3}$ refers to the significant wave height in water of depth d, $T_{1/3}$ refers to the average of the highest one-third of the wave periods, and $H_{o 1/3}$ and $T_{o 1/3}$ refer to the wave height and period in deep water for the same fetch and wind speed. Samples of Sibul's data are shown in Fig. 9.39. The graphs of $H_{\max}$ versus H_{mean}, etc., showed no significant trends with respect to the shallow or deep water.

Bretschneider (1954) analyzed data obtained in the Gulf of Mexico and Lake Okeechobee. His results are shown in Fig. 9.40. In addition to these data, he found that the ratio of maximum wave height to significant wave height was a function of water depth, in conflict with the findings of Sibul.

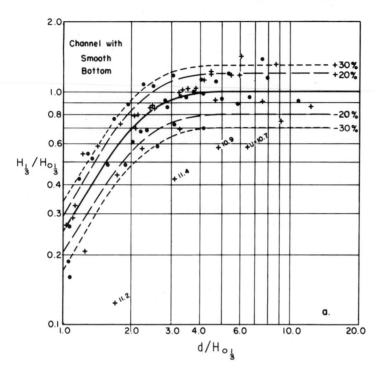

Fig. 9.39. Height of wind generated waves as a function of water depth; wave period as a function of water depth (after Sibul, 1955)

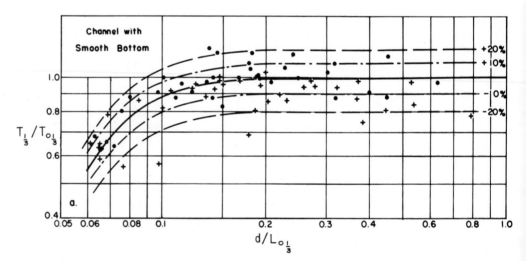

5. LIMITED WIDTH WIND FIELD

Arthur (1949) has found that waves within the wind field will grow even though they are moving at angles of as much as plus or minus 45 degrees with the mean wind direction, and that the height of these waves will be at least 50 per cent of the height of the waves moving in the mean wind direction.

If the width of the wind field is limited (not the usual case in the ocean), as is often the case in a reservoir or a bay then the significant wave height will be different for a given fetch, all other conditions being equal.

Saville (1954) developed a physical model for handling the effect of limited width of the wind field. This model was for a rectangular wind field of width W and fetch length F. Several wind distributions were used, the one most likely for the practical case being a wind effectiveness that varied with the cosine of the angle measured from the mean wind direction up to plus or minus 45 degrees. The study of Arthur and unpublished measurements by the U.S. Army Corps of Engineers on reservoirs appear to substantiate this particular model. The effective fetch F_e to be used in conjunction with Figs. 9.24 and 9.25 is given in Table 9.4.

Table 9.4. EFFECT OF WIND FIELD WIDTH ON WAVE HEIGHT

W/F	0.1	0.2	0.3	0.4	0.5	0.6	0.7	0.8	0.9	1.0	1.1	1.2	1.3	1.4	1.5	1.75	2.0
F_e/F	0.23	0.38	0.49	0.58	0.65	0.72	0.77	0.81	0.85	0.88	0.91	0.93	0.94	0.96	0.98	0.99	1.00

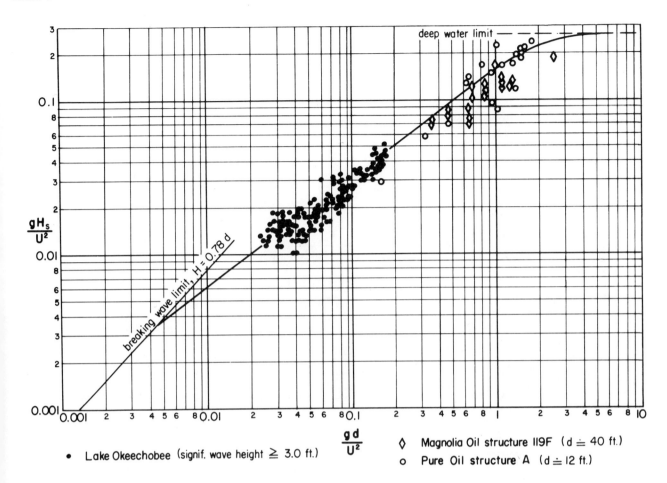

- Lake Okeechobee (signif. wave height ≧ 3.0 ft.)

◊   Magnolia Oil structure 119F (d ≐ 40 ft.)

o   Pure Oil structure A (d ≐ 12 ft.)

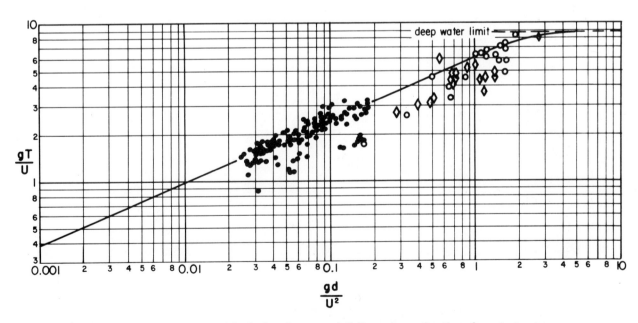

Fig. 9.40. Significant wave height (*top*) and wave period (*bottom*) as a function of constant water depth and wind speed using non-dimensional parameters; based on steady state conditions and bottom friction factor f = 0.01 (after Bretschneider, 1954)

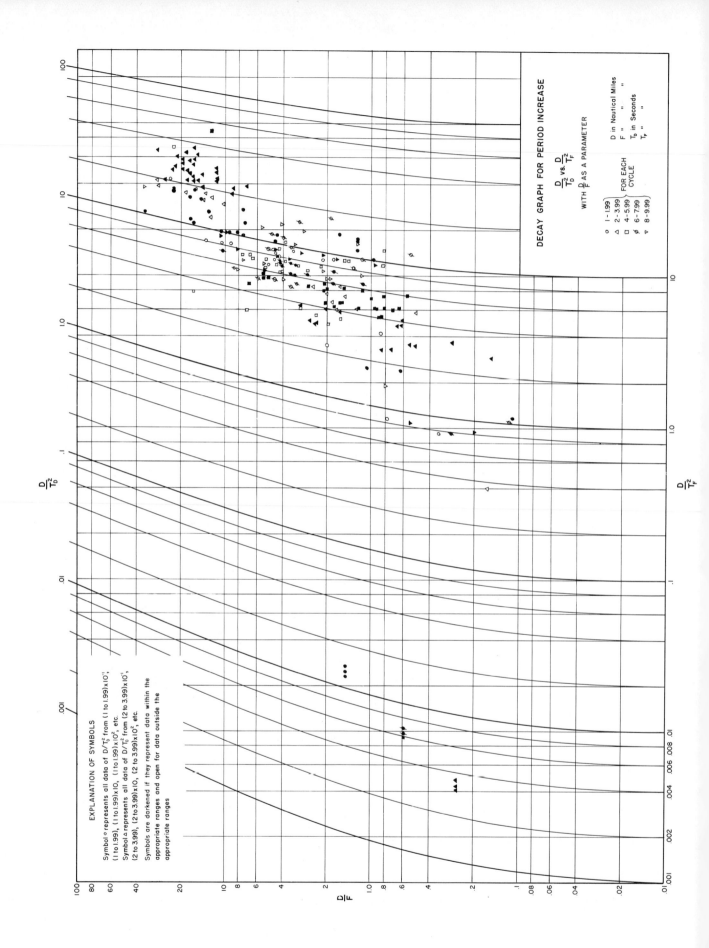

DECAY GRAPH FOR PERIOD INCREASE

$\frac{D}{T_D^2}$ vs. $\frac{D}{T_F^2}$

WITH $\frac{D}{F}$ AS A PARAMETER

			D in Nautical Miles
			F " " "
	FOR EACH		T_D " in Seconds
	CYCLE		T_F " " "

○ 1 - 1.99
△ 2 - 3.99
□ 4 - 5.99
ø 6 - 7.99
▽ 8 - 9.99

$\frac{D}{T_D^2}$

$\frac{D}{T_F^2}$

EXPLANATION OF SYMBOLS

Symbol ○ represents all data of D/T_D^2 from (1 to 1.99) from (1 to 1.99)x10⁻¹,
(1 to 1.99), (1 to 1.99)x10, (1 to 1.99)x10², etc.
Symbol △ represents all data of D/T_D^2 from (2 to 3.99)x10⁻¹,
(2 to 3.99), (2 to 3.99)x10, (2 to 3.99)x10², etc.

Symbols are darkened if they represent data within the
appropriate ranges and open for data outside the
appropriate ranges

$\frac{D}{F}$

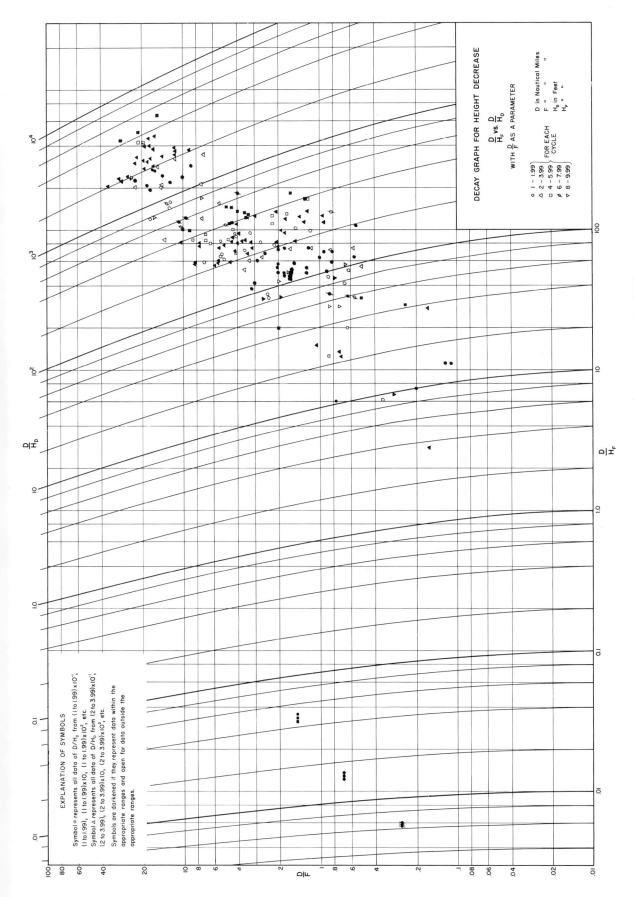

Fig. 9.42. Decay graph for height decrease (from Bretschneider, 1952)

6. DECAY OF SWELL

After leaving the generating area, waves change. Energy is dispersed, spread, dissipated, perhaps transferred between wave components of different frequencies, and scattered. The effect of the combination of all these factors is known as the *decay of swell*. Furthermore, the swell may travel through regions of following, cross, or opposing winds. If the decay of swell were due only to dispersion, it would be relatively simple to predict the wave spectrum at any given distance from the generating area. The theory of the group velocity of waves could be applied to individual frequency components of the wave spectrum to predict their characteristics as a function of time at the desired location, the individual components recombined, and the spectrum of the resulting time series obtained as a function of time. The techniques for handling the dispersion problem have been given by Pierson, Neumann, and James (1955). There are very few data on the effect of travel distance on the spectrum of swell (Bretschneider, 1952; Darbyshire, 1957). One approach, based upon field measurements, was given by Bretschneider (1952). The primary factors were considered to be the ratio of the decay distance to the fetch length. The results are presented in Figs. 9.41 and 9.42. The relationships for period increase can be made dimensionless by multiplying D/T_F^2 and D/T_D^2 by $1/g$, where g is the acceleration of gravity, and by using consistent units for D, T, H, and g.

REFERENCES

Abraham, G., *Model study of water gravity waves generated by moving circular low pressure area*, Univ. Calif., Inst. of Eng. Res., Tech. Rept. No. 99–5 May, 1959. (Unpublished.)

Arthur, Robert S., Variability in direction of wave travel, *Annals New York Acad. Sci.*, **51**, Art. 3 (May, 1949), 511–22.

Barber, N. F. and F. Ursell, The generation and propagation of ocean waves and swell; I: Wave periods and velocities, *Trans. Roy. Soc.* (London), ser. A, **240**, 824 (February, 1948), 527–60.

Benjamin, T. Brooke, Shearing flow over a wavy boundary, *J. Fluid Mech.*, **6**, Part 2 (August, 1959), 161–205.

Bonchkovskaya, T. V., Wind flow over solid wave models, *Akademia Nauk SSSR*, Morskot Gidrofizicheskii Institut, **6** (1955), 98–106.

Bretschneider, C. L., The generation and decay of wind waves in deep water, *Trans. Amer. Geophys. Union*, **33**, 3 (June, 1952), 381–89.

———, *Field investigation of wave energy loss of shallow water ocean waves*, U.S. Army, Corps of Engineers, Beach Erosion Board, Tech. Memo. No. 46, Sept. 1954.

———, *Wave variability and wave spectra for wind generated gravity waves*, U.S. Army Corps of Engineers, Beach Erosion Board, Tech. Memo. No. 118 August, 1959.

——— and E. K. Rice, *The generation and decay of wind waves in a sixty-foot channel*, Univ. Calif., Inst. of Eng. Res., Tech. Rept. 3–327 July, 1951. (Unpublished.)

Brown, P. R., Wave data for the Eastern North Atlantic, *Marine Observer*, **23**, 160 (April, 1953), 94–98.

Burling, R. W., Surface waves on enclosed bodies of water, *Proc. Fifth Conf. Coastal Eng.* Berkeley, Calif.: The Engineering Foundation Council on Wave Research 1955a, pp. 1–10.

———, Wind generation of waves on water, Ph.D. dissertation, Imperial College, Univ. London, 1955b.

———, The spectrum of waves at short fetches, *Deut. Hydrograph. Ze.*, **12**, 2 No. 3 (1959), 45–64; 96–117.

Caldwell, Joseph M., The step-resistance wave gage, *Proc. First Conf. Coastal Eng. Instr.*, Berkeley, Calif.: The Engineering Foundation Council on Wave Research, 1956, pp. 44–60.

California Research Corporation, Table of significant wave heights and mean periods for four hurricanes in the Gulf of Mexico in 30 feet of water. (Private correspondence, May, 1960.)

Cartwright, D. E. and M. S. Longuet-Higgins, The statistical distribution of the maxima of a random function, *Proc. Roy. Soc.* (London), ser. A, **230** (1956), 212–32.

Charnock, H., J. R. D. Francis, and P. A. Sheppard, Medium-scale turbulence in the trade winds: summary, *Quart. J., Roy. Met. Soc.* (London), **81**, 350 (October, 1955), 634.

Chase, Joseph, *et al.*, *The directional spectrum of a wind generated sea as determined from data obtained by the Stereo Wave Observation Project*, New York Univ. College of Engineering, Research Division, Contract Nonr-285(03) July, 1957.

Cox, Charles S., Measurements of slopes of high-frequency wind waves, *J. Mar. Res.*, **16**, 3 (October, 1958a), 199–225.

———, Comments on Dr. Phillip's paper, *J. Mar. Res.*, **16**, 3 (October, 1958b), 241–45.

Crapper, G. D., An exact solution for progressive capillary waves at arbitrary amplitude, *J. Fluid Mech.*, **2** (1957), 532–39.

Darbyshire, J., The generation of waves by wind, *Proc. Roy. Soc.*, ser. A 215 (1952), 299–328.

———, An investigation of storm waves in the North Atlantic Ocean, *Proc. Roy. Soc.* (London), ser. A, **230** (1955), 560–69.

———, The distribution of wave heights, a statistical method based upon observations, *Dock and Harbour Authority*, **37**, 427 (May, 1956), 31–32.

———, Attenuation of swell in the North Atlantic Ocean, *Quar. J. Roy. Met. Soc.*, **83**, 357 (July, 1957), 351–59.

———, A further investigation of wind generated waves, *Deut. Hydrograph. Z.* **12**, 1 (1959), 1–13.

Darlington, C. R., The distribution of wave heights and periods in ocean waves, *Quar. J. Roy. Met. Soc.*, **80**, (1954), 619–26.

Deacon, E. L., Vertical diffusion in the lowest layer of the atmosphere, *Quart. J. Roy. Met. Soc.*, **75**, 323 (January, 1949), 89–103.

———, P. A. Sheppard and E. K. Webb, Wind profiles over the sea and the drag at the sea surface, *Australian J. Physics*, **9**, 4 (December 1956), 511–41.

Donn, William L., and William T. McGuinness, *Some results of the IGY island observatory program in the Atlantic*, National Academy of Sciences IGY Bull., No. 22 (April, 1959), pp. 9–12.

Eckart, Carl., The generation of wind waves on a water surface, *J. Applied Physics*, **24**, 12 (December, 1953), 1485–94.

Ehring, H., Kennzeichnung des gemessener Seegangs auf Grund der Haufigkeitsverteilung von Wellenhohe, Wellenlange, und Steilheit, *Techn. Ber.* **4** (1940), 152–55; translation by Scripps Inst. Oceanog. Rep. 54. Contract NObs 2490, 1944.

Farmer, H. G., *Some recent observations of sea surface elevation and slope*, Woods Hole Oceanog. Inst. Ref. 56–37 June, 1956. (Unpublished manuscript.)

Flinsch, H. V. N., An experimental investigation of wind-generated surface waves, Ph.D. thesis, Univ. Minn., June, 1946. (Unpublished.)

Folsom, R. G., Measurement of ocean waves, *Trans. Amer. Geophys. Union*, **30** (1949), 691–99.

Francis, J. R. D., Wave motions and the aerodynamic drag on a free oil surface, *Phil. Mag.*, ser. 7, **45** (July, 1954), 695–702.

———, Wind action on a water surface, *Proc. Inst. Civ. Engrs.* (England), **12** (February, 1959), 197–216.

Fuchs, R. A., On the theory of irregular waves, *Proc. First Conf. Ships and Waves*, Berkeley, Calif.: Council on Wave Research and the Society of Naval Architects and Marine Engineers 1955, 1–10.

Gerhardt, J. R., K. H. Jehn, and I. Katz, A comparison of step-pressure, and continuous-wire-gage wave recordings in the Golden Gate Channel, *Trans. Amer. Geophys. Union*, **36** (1955), 235–50.

Gibson, George E., *Investigation of the calming of waves by means of oil films*, Woods Hole Oceanog. Inst., Apr. 20, 1944. (Unpublished.)

Goldstein, S., ed., *Modern developments in fluid dynamics.* Oxford; Oxford University Press (Clarendon Press), 1938.

Hamada, Tokuichi, Hisashi Mitsuyasu, and Maoki Hase, *An experimental study of wind effect upon water surface*, Report of Transportation Technical Research Institute (Tokyo), June, 1953.

Harney, L. A., J. F. T. Saur, Jr. and A. R. Robinson, *A statistical study of wave conditions at four open-sea localities in the North Pacific Ocean*, N.A.C.A. Tech. Note 1493, 1949.

Hydraulics Laboratory Delft, *Golfaanval haringvlietsluizen. Deel II, systematisch onderzoek golfaangroei en dynamische golfbelasting*, Code 32.79, M 399–II March, 1960.

Ichiye, Takashi, A short note on energy transfer from wind to waves and current, *Oceanogr. Mag.*, **4**, 3 (December, 1952), 89–93.

Ijima, Takeshi, *The properties of ocean waves on the Pacific Coast and the Japan Sea coast of Japan*, Transportation Technical Research Institute, Tokyo, Report No. 25, June, 1957.

Jeffreys, Harold, On the formation of water waves by wind, *Proc. Roy. Soc.*, ser. A, **107** (1925), 189–206.

———, On the formation of water waves by winds: (second paper), *Proc. Roy. Soc.*, ser. A, **110** (1926), 241–47.

Jinnaka, Tatsuo, *Wave patterns and ship waves*, The Society of Naval Architects of Japan 60th Anniversary Series, **2** (1957), 83–94.

Johnson, J. W., The characteristics of wind waves on lakes and protected bays, *Trans. Amer. Geophys. Union*, **29**, 5, (October, 1948), 671–81.

———, Relationship between wind and waves, Abbots Lagoon, California, *Trans. Amer. Geophys. Union*, **29**, 3 (June, 1950), 671–81.

——— and E. K. Rice, A laboratory investigation of wind-generated waves, *Trans. Amer. Geophys. Union*, **33**, 6 (December, 1952), 845–54.

Kapitsa, L. P., The formation of sea waves by the wind, *Doklady Akademii Nauk SSSR*, **64**, 4 (1949), 513–16.

Keulegan, Garbis H., Wind tides in small closed channels, *J. Research National Bur. Standards*, **46**, 5 (1951), 358–81.

Lock, R. C., Hydrodynamic stability of the flow in the laminar boundary layer between parallel streams, *Proc. Camb. Phil. Soc.*, **50** (1954), 105–24.

Longuet-Higgins, M. S., On the statistical distribution of the heights of sea waves, *J. Mar. Res.*, **11**, 13 (December, 1952), 245–66.

———, D. E. Cartwright, and N. D. Smith, Observations of the directional spectrum of sea waves using the motions of a floating buoy, *Proc. Conf. Ocean Wave Spectra*, Easton, Md., May 1–4, 1961, Nat. Acad. Sci.–Nat. Res. Council. Englewood Cliffs, N. J.: Prentice-Hall, Inc. (1963), 111–31.

Marks, Wilbur, *Analysis of the performance of the NIO ship-borne wave recorder installed in the R. V. Atlantis*, Woods Hole Oceanog. Inst., Ref. No. 55–64, November 1955. (Unpublished manuscript.)

———, *Woods Hole Oceanographic Institution participation in the stereo-wave observation program, October 1954*, Woods Hole Oceanog. Inst., Ref. No. 56–44, July, 1956. (Unpublished manuscript.)

Miche, R., Mouvements ondulatoires des mers en profondeur constante au décroissante, *Annales des Ponts et Chaussées*, (1944), 25–78; 131–64; 279–92; 309–406.

Miles, John W., On the generation of surface waves by shear flows, *J. Fluid Mech.*, **3**, Part 3 (November, 1957), 185–204.

————, On the generation of surface waves by shear flows, Part 2, *J. Fluid* Mech. **6**, Part 4 (1959a), 568–82.

————, On the generation of surface waves by shear flows, Part 3, Kelvin-Helmholtz instability, *J. Fluid Mech.* **6**, Part 4 (1959b), 583–98.

————, On the generation of surface waves by turbulent shear flows, *J. Fluid Mech.*, **7**, Part 3 (March, 1960), 469–78.

Motzfeld, Heinz, Die turbulente Strömung auf welligen Wänden, *Z. angewandte Math. Mech.*, **17** (August, 1937), 193–212.

Munk, Walter H., A critical wind speed for air-sea boundary processes, *J. Mar. Res.*, **6**, 3 (1947), 203–18.

———— and R. S. Arthur, "Forecasting ocean waves," *Compendium of Meteorology*, Amer. Met. Soc. (1951), 1082–89.

Neumann, Gerhard, Über die komplexe Natur des Seeganges, *Deut. Hydrograph. Z.* **5**, 2/3 (1952), 95–110.

————, *On wind-generated ocean waves with special reference to the problem of wave forecasting*, New York Univ. College of Engineering, Department of Meteorology, Contract Nonr-285(05), May, 1952. (Unpublished.)

————, *On ocean wave spectra and a new method of forecasting wind-generated sea*, U.S. Army Corps of Engineers, Beach Erosion Board, Tech. Memo. No. 43, December, 1953.

————, On wind-generated wave motion at subsurface levels, *Trans. Amer. Geophys. Union*, **36**, 6 (December, 1955), 985–92.

————, and Willard J. Pierson, A detailed comparison of theoretical wave spectra and wave forecasting methods, *Deutsche Hydrograph. Z.* **10**, 3, 4 (1957), 73–92, 134–46.

O'Brien, M. P., Letter, December 31, 1943. (Unpublished.)

Phillips, O. M., On the generation of waves by turbulent winds, *J. Fluid Mech.* **2**, Part 5 (July, 1957), 417–45.

————, Wave generation by turbulent wind over a finite fetch, *Proc. Third U.S. National Congress of Applied Mechanics*, *ASME* (June, 1958a), 785–89.

————, The equilibrium range in the spectrum of wind-generated waves, *J. Fluid Mech.*, **4**, Part 4 (August, 1958b), 426–34.

————, On some properties of the spectrum of wind-generated ocean waves, *J. Mar. Res.*, **16**, 3 (October, 1958c), 231–40.

————, Comments on Dr. Cox's Paper, *J. Mar. Res.*, **16**, 3 (October, 1958d), 226–30.

Pierson, Willard J., Jr., *A note on the growth of the spectrum of wind-generated gravity waves as determined by nonlinear considerations*, New York University, College of Engineering, Research Division, Contract Nonr-285(03), February, 1959(a). (Unpublished.)

————, *A study of wave forecasting methods and the height of a fully developed sea on the basis of some wave records obtained by the O. W. S. Weather Explorer during a storm at sea*, New York Univ., College of Engineering, Research

Div., Contract Nonr–285(03), 33 pp, June, 1959 (b). (Unpublished.)

————, and Wilbur Marks, The power spectrum analysis of ocean-wave records, *Trans. Amer. Geophys. Union*, **33** (1952), 834–44.

————, Gerhard Neumann, and Richard W. James, *Practical methods for observing and forecasting ocean waves*, U.S. Navy Hydrographic Office, Pub. 603, 1955.

Putz, R. R., *Joint variation of wave height and wave period for ocean swell*, Univ. Calif., Inst. of Eng. Res., Tech. Rept. 3–328, October, 1951. (Unpublished.)

————, Statistical distributions for ocean waves, *Trans. Amer. Geophys. Union*, **33**, 5 (October, 1952), 685–92.

————, Statistical analysis of wave records, *Proc. Fourth Conf. Coastal Eng.*, Berkeley, Calif.,: The Engineering Foundation Council on Wave Research, 1954, pp. 13–24.

Ralls, G. C., Jr. and R. L. Wiegel, *A laboratory study of short-crested wind waves*, U.S. Army, Corps of Engineers, Beach Erosion Board, Tech. Memo. No. 81, June, 1956.

Roll, Hans Ulrich, Über die Ausbreitung der Meereswellen unter der Wirkung des Windes (auf Grund von Messungen im Wattenmeer), *Deut. Hydrograph. Z.* **2**, 6 (1949), 268–80.

————, Über Grössenunterschiede der Meereswellen bei Warm-und-Kaltluft, *Deut. Hydrograph. Z.* **5**, 213 (1952), 111–14.

————, *Height, length and steepness of seawaves in the North Atlantic and dimensions of seawaves as functions of wind force*, Special Publication No. 1 of the Office of Sea-Weather, German Weather Service, 1953, Manley St. Denis, trans. Tech. Res Bull. No. 1–19, The Society of Naval Architects and Marine Engineers (December, 1958).

————, and G. Fischer, Eine kritische Bemerkung zum Neumann-Spektrom des Seeganges, *Deut. Hydrograph. Z.* **9**, 1 (1956), p. 9.

Roschke, W. H., Jr., The propagation of short period air pressure microoscillations, *Bull. Amer. Meteor. Soc.*, **35**, 1 (January, 1954), 20–25.

Rudnick, P., Correlograms for Pacific Ocean waves, *Proc. Second Berkeley Symposium on Mathematical Statistics and Probability* (1951), 627–38.

Saville, Thorndike, Jr., *The effect of fetch width on wave generation*, U.S. Army, Corps of Engineers, Beach Erosion Board, Tech. Memo. No. 70 (December, 1954).

Schaaf, S. A. and F. M. Sauer, A note on the tangential transfer of energy between wind and waves, *Trans. Amer. Geophys. Union*, **31**, 6 (December, 1950), 867–69.

Schellenberger, G., Untersuchungen über Windwellen auf einem Binnensee, *Gerlands Beiträge zur Geophysik*, **71**, (1962), 67–91.

Schooley, Allen H., Profiles of wind-created water waves in the capillary-gravity transition region, *J. Mar. Res.* **16**, 2 (August, 1958), 100–108.

————, Double, triple, and higher-order dimples in the profiles of wind-generated water waves in the capillary-gravity transition region, *J. Geophys. Res.*, **65**, 12 (December, 1960), 4075–79.

Scripps Institution of Oceanography, *Proposed uniform procedure for observing waves and interpreting instrument records*, Wave Project Report No. 26, 1944.

Seiwell, H. R., Results of research on surface waves of the western North Atlantic, *Paper Phys. Oceanog. Met.*, **10**, 4, 1948.

Sherlock, R. H. and M. B. Stout, Wind structure in winter storms, *J. Aeronautical Sciences*, **5**, 2 (December, 1937), 53–61.

Sibul, Oswald, *Laboratory study of the generation of wind waves in shallow water*, U.S. Army, Corps of Engineers, Beach Erosion Board, Tech. Memo. No. 72, March, 1955.

Snodgrass, F. E., Wave recorders, *Proc. First Conf. Coastal Eng.*, Berkeley, Calif.: The Engineering Foundation Council on Wave Research, 1951, 69–81.

——, Mark IX shore wave recorder, *Proc. First Conf. Coastal Eng. Instr.*, Berkeley, Calif.: The Engineering Foundation Council on Wave Research (1956), 61–100.

Stanton, T., D. Marshall and R. Houghton, The growth of waves on water due to action of the wind, *Proc. Roy. Soc.*, ser. A, **137** (1932), 283–93.

Stevenson, Thomas, *Generation of waves*, 3rd ed., chap. 3, "The design and construction of harbours," Edinburgh: A. & C. Black, 1886, pp. 26–40.

Sverdrup, H. V. and W. Munk, *Wind, sea and swell. Theory of relations for forecasting*, U.S. Navy Hydrographic Office Pub. No. 601, 1947.

Thornthwaite, C. W. and Paul Kaser, Wind-gradient observations, *Trans. Amer. Geophys. Union*, Part I (October, 1943), 166–82.

Tick, Leo J., *A non-linear random model of gravity waves*, New York Univ. College of Engineering, Research Division, Scientific Paper No. 11, October, 1958. (Unpublished.)

Tucker, M. J., A ship-borne wave recorder, *Proc. First Conf.*

Coastal Eng. Inst., Berkeley, Calif.: The Engineering Foundation Council on Wave Research 1956., 112–18.

Ursell, F., "Wave generation by wind," *Surveys in Mechanics*, ed. G. K. Batchelor and R. M. Davies. London: Cambridge University Press, 1956, pp. 216–49.

Van Dorn, William G., *Wind stress over water*, Scripps Inst. Oceanog., Ref. No. 52–60, 1952.

Watters, Jessie K. A., Distribution of height in ocean waves, *New Zealand J. Sci. Tech.*, B, *34* (1953), 408–22.

Wiegel, R. L., An analysis of data form wave recorders on the Pacific Coast of the United States, *Trans. Amer. Geophys. Union* 30, (1949), 700–4.

——, *Waves, tides, currents, and beaches; glossary of terms and* list of standard symbols. Berkeley, Calif.: The Engineering Foundation Council on Wave Research, 1953.

——, and R. A. Fuchs, Wave transformation in shoaling water, *Trans. Amer. Geophys. Union*, **36** (1955), 975–84.

——, and J. Kukk, Wave measurements along the California coast, *Trans. Amer. Geophys. Union*, **38**, 5 (October, 1957), 667–74.

——, C. M. Snyder, and J. B. Williams, Water gravity waves generated by a moving low pressure area, *Trans. Amer. Geophys. Union*, **39**, 2 (April, 1958), 224–35.

Williams, A. J. and D. E. Cartwright, A note on the spectra of wind waves, *Trans. Amer. Geophys. Union*, **38**, 6 (December, 1957), 864–66.

Wilton, J. R., On ripples, *Phil. Mag.*, *29* (1915), 688–700.

Wuest, W., Beitrag zur Entstehung von Wasserwellen durch Wind, *Z. Angew. Math. Mech.*, *29*, 7–8 (July–August, 1949), 239–52.

Yampol'ski, A. D., On a certain characteristic of three dimensional wave motion, *Priroda*, *5* (1955), 80–81.

Yoshida, K., K. Kajiura, and K. Hidaka, *Preliminary report of the observation of ocean waves at Hachijo Island*, Rec. Oceanog. Works Japan ser., 1 (March, 1953), 81–87.

CHAPTER TEN

Wave Prediction

1. INTRODUCTION

The prediction of waves is of considerable importance from both the design and operational standpoints. In order to obtain design data, it is possible to make use of files of old weather maps, the calculation of the wave characteristics from the meteorological information on these maps being known as *hindcasting*. In using these maps, care must be exercised to be sure of the millibar spacing being used, and whether wind speeds are in Beaufort scale or in knots. Operational use of wave prediction techniques may consist of hindcasting for the purpose of obtaining statistical information on the possible wave conditions to be expected at a given location, using this information as a basis for choosing the equipment to be used, or for choosing the best time of the year for operations. Wave forecasting may be used to predict the wave dimensions at a given location on a day-to-day basis, and using this information to decide a day or so in advance whether a dredge can operate,

a mobile offshore oil platform can be moved, a barge can be towed, etc. Of increasing importance are applications to marine navigation (James, 1957). The value of these day-to-day forecasts has been subject to criticism by marine operators due to their inconsistencies, and they must be used with caution at the present time.

Formulas for wave forecasting date back at least as far as Stevenson (1886). A major advance in wave forecasting was made during World War II in connection with amphibious operations. A technique was developed by Sverdrup and Munk (U.S. Navy Hydrographic Office, 1944a, 1944b; Sverdrup and Munk, 1946, 1947) who advanced the concept of duration-limited and fetch-limited cases. In the duration-limited case, the height and period of the significant waves depend upon the length of time the wind blows; in the fetch-limited case, which is the quasi–steady-state case, the wave dimensions depend upon the fetch. In forecasting or hindcasting, it is necessary to distinguish between these two cases and to proceed accordingly. A third case might exist, the fully developed sea, which might be termed *wind-speed limited*. This concept would be valid if the gH/U^2 versus gF/U^2 and gt/U curves became horizontal. It is difficult to draw definite conclusions in regard to this owing to the lack of reliable data. If this case were to occur, however, the curve of C/U versus gF/U^2 should also become horizontal, as it is difficult to see how $H_{1/3}$ can cease increasing unless the wave length, and hence C, ceases to increase, as the limiting value of $H_{1/3}$ must be associated with the limiting wave steepness. This limiting wave steepness could not occur unless the wave stopped increasing in speed and length.

Sverdrup and Munk plotted the measured wave and wind data as C/U and gH/U^2 versus gF/U^2, and C/U and gH/U^2 versus gt/U together with a curve of $t_{\min}U/F$ versus gt/U. The curve $t_{\min} U/F$ versus gt/U was obtained using the theoretical value of the group velocity of the significant wave, together with some data which are in disagreement with recent measurements. In order to determine the real relationship, data must be obtained for ocean waves of the type given in Fig. 9.26, that is, as a function of time. Until the mechanisms of wave generation by winds are thoroughly understood, a semi-theoretical determination of $t_{\min}$ must be used with caution. An approximation of $t_{\min}$ can be computed by obtaining C from Fig. 9.24, and dividing F by $\frac{1}{2} C$ as $\frac{1}{2} C$ is about the group velocity of the significant waves. This approximation is based upon the belief that wave components of the period associated with the speed in question were generated at the beginning of the fetch as well as elsewhere in the fetch. In practice, the lower set of values of C and H obtained from Figs. 9.24 and 9.25 should be used.

A considerable body of additional data has been obtained since Sverdrup and Munk published their theory and those data are given in Figs. 9.24 and 9.25. Few of them are for the open ocean, however, so that it is still difficult to make reliable predictions for large fetches. The scatter of data is large and it is difficult to see how predictions can be made more reliable than about plus or minus 50 per cent of the mean value. Additional information on the effect of the air-sea temperature difference (Fig. 9.23) will improve the forecasts in the future; this refinement, however, has not yet been incorporated within the body of forecasting curves. In any case, the laboratory studies, for which the difference between the air and water temperatures cannot have any large effect, also show considerable scatter. The relationships among winds and sea are not precise, especially when considering the case in the ocean of moving storm areas and varying wind fields: the numerical effects of these factors must be determined and incorporated into the body of wave forecasting data.

There is a large body of writings on wave forecasting, only some of which is referred to here, although many of the papers are cited in Chapter 9.

2. FORECASTING PROCEDURES AND THEIR RELIABILITY

In general there are three classes of forecasting procedures. One class might be termed the class for a specific case, such as given by Stevenson (1886). The second class is the spectrum method (Neumann, 1953; Pierson *et al.*, 1955). The third class is the significant wave method (Sverdrup and Munk, 1947; Bretschneider, 1952, 1950; Darbyshire, 1952). Comparisons of the results obtained using these various classes, together with variants of them, have been made (Kaplan and Saville, 1954; Saville, 1955; Rattray and Burt, 1956; Bretschneider, 1957; Ijima, 1957; Neuman and Pierson, 1957). One thing is apparent from the few comparisons of forecasts with observed waves: no major procedure is appreciably better than the others in use today. It is difficult to compare the results of the several methods of wave forecasting, as each set of forecasting curves is based largely upon its own set of original data, none of which can be considered to be adequate. There is no theoretical set of forecasting curves; all the "theoretical," semi-empirical, and empirical procedures rely on physical measurements, and these measurements in the open ocean are few, largely unreliable for a number of reasons, usually for varying winds and fetches, and often for moving fetches. Programs now under way should lead to improvements in the state of the art in the near future. Because of this, no detailed set of forecasting curves will be presented here, as such curves tend to be kept in use long after they should be retired. It is preferable to use Figs. 9.24 and 9.25, together with Figs. 9.41 and 9.42, and to modify them as new data become available.

There is not as much difference between the spectrum method and the significant wave method as numerous papers on the subject would lead one to believe. In practice, the spectrum method introduces the concept that the total energy density of waves generated by wind can be defined by a quantity designated as E. This quantity E apparently was originally obtained as a function of wind speeds for what was termed a *fully developed sea*. It was obtained in the following manner: the ordinates of a wave record of the desired length were measured at equal intervals throughout the record, using an arbitrary zero for the ordinates, then an average value of the ordinates was computed, and this average subtracted from the measured ordinates. The resulting numbers were then squared and the average of the sum of the squares computed. E was defined as twice this number, with the dimensions of foot^2 in the English system. According to Pierson, Neumann, and James (1955), this value of E is the same as the area under the wave spectrum curve, with the wave spectrum curve being the product of the square of the amplitude of each wave component and the period of the same wave component versus the frequency (the reciprocal of the wave period, rather than the circular frequency) of the same wave component. Use is then made of the distribution function of Longuet-Higgins (1952) to obtain the significant wave height, the mean wave height, etc., from E. For example, the most frequent wave height is $1.41\sqrt{E}$, $H_{\mathrm{avg}} = 1.77\sqrt{E}$, $H_{1/3} = 2.83\sqrt{E}$, and $H_{1/10} = 3.60\sqrt{E}$ ft.

The main forecasting curves of Pierson, Neumann, and James present E as a function of wind speed for the fully developed sea, with additional sets of curves for seas that are limited by the duration of the storm or the length of the fetch. In addition, the average wave period is also given, together with a range of periods to be expected. This range of periods is taken as between T_U and T_L, where T_U is defined as the period at $0.95\,E$ and T_L is defined as the period at $0.03\,E$ on the co-cumulative spectrum. The relationships between periods and heights are based upon the spectrum presented by Pierson, Neumann, and James, and recent findings have shown it to be in need of modification (Darbyshire, 1950; Pierson, 1959).

As stated previously, the distribution of wave amplitudes in the procedure just described was taken from Longuet-Higgins (1952), and this is essentially the same as that obtained by Putz (1952) from the measurements of ocean waves (see, for example, the comparisons made by Bretschneider, 1959).

The significant wave procedure as originally developed by Sverdrup and Munk (1946) had a serious drawback in that it did not predict a distribution of wave heights and periods. This was known at the time and several programs were initiated to discover these distributions by measuring and analyzing ocean waves. The first results (Seiwell, 1948; Wiegel, 1949) were developed more thoroughly (Putz, 1952; Longuet-Higgins, 1952). If the wind speed, fetch, and duration are reliably known, Figs. 9.24 and 9.25 can be used to obtain the significant wave height and period at the end of the fetch. Then Table 9.3 and Fig. 9.6(c) can be used to obtain the distribution of heights and periods. The relationship between the components of H and T (the spectrum) can be obtained from Fig. 9.12.

How can T be obtained from C? It must be remembered that we are not dealing with truly periodic waves, but with a wave system comprised of wave components of a spectrum of periods. Pierson, Neumann, and James (1955) have shown that, for their model, the average wave length is not equal to $g/2\pi$ times the square of the average wave period, but about two-thirds of this value. They were not able, however, to determine how to compute the average wave speed. Until there is evidence to the contrary, it might as well be assumed that $T = C(2\pi/g)$.

How these results are applied depends upon the nature of the problem. The movement of sand by littoral pro-

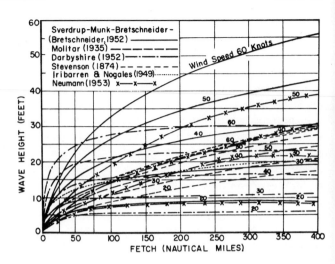

Fig. 10.1. Comparison of forecasting procedures (after Saville, 1955)

cesses does not require quite the same data as are required in the design of a sewer outfall, for example. For other problems, such as a moored ship or a ship under way, the periodicity of the large groups of waves (Fig. 9.3) is of considerable importance. For detailed analyses, the complete spectrum may be necessary.

Other investigators have developed wave forecasting formulas of their own. Saville (1955) has compared the significant wave height predicted by many different formulas as a function of fetch and wind speed for storms of unlimited duration. The results are shown in Fig. 10.1. It is not possible to decide which formula is the most desirable, as no measured values are presented for the idealized case considered.

Once the waves leave the generating area, they change in appearance, becoming smoother. They lose energy through internal resistance, damping, and relatively large-scale scattering in traveling through other wave systems (Phillips, 1959). Because the waves in the generating area are going in various directions, the energy density decreases in the decay area owing to the angular spreading of the wave system. In addition, they are dispersive. This latter characteristic was not considered in the original work of Sverdrup and Munk; the defect was soon apparent, however, and a method of handling this characteristic was proposed by Bretschneider (1952). The relevant parameters were the ratio of the decay distance, D (distance from the end of the fetch to the place of interest) to the fetch, F, and the ratio of the wave height at the end of the fetch to the decay distance (hence, fetch). Values of D/H_F were plotted as a function of D/F and D/H_D, and values of D/T_F^2 were plotted as a function of D/F and D/T_D^2, where the subscripts F and D refer to the values at the end of the fetch and the decay distance, respectively, (Figs. 9.41 and 9.42). The curves, however, are based on very few data. The distribution of wave heights and periods about H_D and T_D

(which are approximately the significant wave height and period) are the same as at the end of the fetch. What are the relative advantages of this representation and of the method presented by Pierson, Neumann, and James (1955)? The latter permits the handling of wave dispersion and angular spreading of wave systems by a linear theory, with the flexibility inherent in a linear theory. On the other hand, the empirical method includes nonlinear effects that may exist, although these effects are not distinguished in the procedure; this needs to be done. Not enough reliable measurements are available at the present time to determine which is the more useful of the two procedures. This appears to be an area in which a considerable amount of effort must be spent in obtaining measurements.

One problem of considerable importance is the effect of following, opposing, or cross winds on an existing wave system. There are almost no data on this. Some limited laboratory measurements (Morison and Todd, 1953) showed that winds added energy to existing waves under some circumstances, but primarily generated a new wave system that was superimposed on the old wave system under other conditions.

The travel time t_D of the swell from the end of the fetch is obtained by dividing the decay distance by the group velocity C_G associated with the wave period at the end of the decay distance D. C_G is given by Eq. 2.68 and is given graphically in Fig. 10.2.

Fig. 10.2. Travel time of swell based on $t_D = D/C_G$

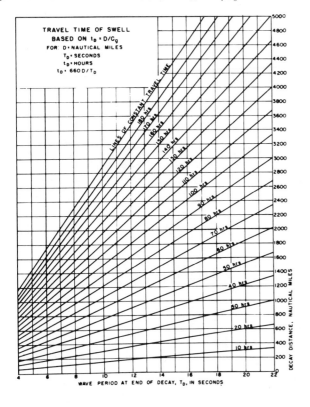

One difficulty in predicting wave characteristics from weather data is that each meteorologist would construct a slightly different weather chart from another meteorologist, and each person would interpret a given weather chart differently in regard to fetch, wind speed, and decay distance. In addition, due to the interval between weather charts, the wind duration and the time of the start of the winds are difficult to define. A study has been made to determine the effect of different weather charts analyzed by one person, and the effect of three meteorologists experienced in wave forecasting analyzing the same set of weather maps, all using the same wave forecasting theory (Bretschneider, Todd, and Kimberley, 1950). The three sets of weather maps were U.S. Navy (12-hourly surface maps and 3-millibar isobars), U.S. Air Force (6-hourly surface maps and 3-millibar isobars), and U.S. Weather Bureau (12-hourly surface maps and 5-millibar isobars). The comparisons were made for an interval of several months, and included the problem of wave decay.

An understanding of the reliability of forecasts, from the standpoints of the relationship between forecasts from several sets of weather maps and the relationship between forecasts by several forecasters using the same set of weather maps, can be obtained by comparing the forecast values of wave height and period. This has been done and presented in Figs. 10.3 and 10.4. These values were calculated by Bretschneider, Todd, and Kimberley (1950) but were not included in their report.

Considering the heights predicted by the same person from the three sets of weather charts, 33 per cent of the time forecasts from C_2 (set of charts No. 2) was equal to, or higher than, forecasts from C_1, 66 per cent of the forecasts from C_3 was equal to, or higher than, forecast C_1, and 80 per cent of the time forecast from C_3 was equal to or higher than forecasts from C_2. Considering the wave periods predicted from the three sets of weather charts, 62 per cent of the time forecast from C_2 was equal to, or longer than, forecasts from C_1, 60 per cent of the time forecast from C_3 was equal to, or higher than, forecasts from C_2, and 56 per cent of the time forecast from C_3 was equal to, or longer than, forecasts from C_2.

Considering the heights predicted by three meteorologists, using the same set of weather maps, more than 90 per cent of the time, forecaster B predicted wave heights equal to, or greater than, either forecaster A or C; 63 per cent of the time forecaster A predicted waves equal to, or higher than, forecaster C. Considering the predicted periods, 50 per cent of the time, forecaster A predicted periods equal to, or longer than, forecaster B; 40 per cent of the time forecaster B predicted periods equal to, or longer than, forecaster C; and 46 per cent of the time forecaster A predicted periods equal to, or longer than, forecaster C.

In general, the disagreement between forecasts by one

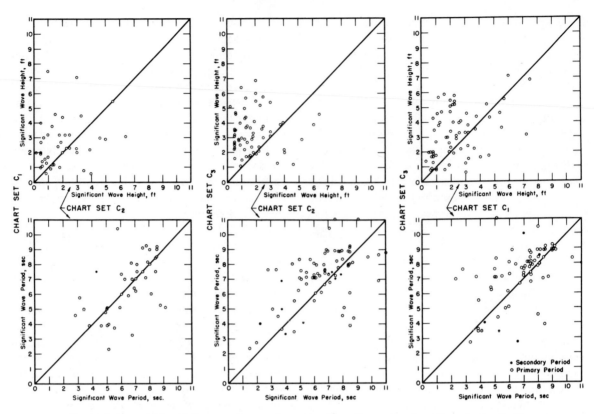

Fig. 10.3. Comparison of forecasts by one forecaster from three sets of weather charts, C_1, C_2, and C_3 (from Wiegel, 1961)

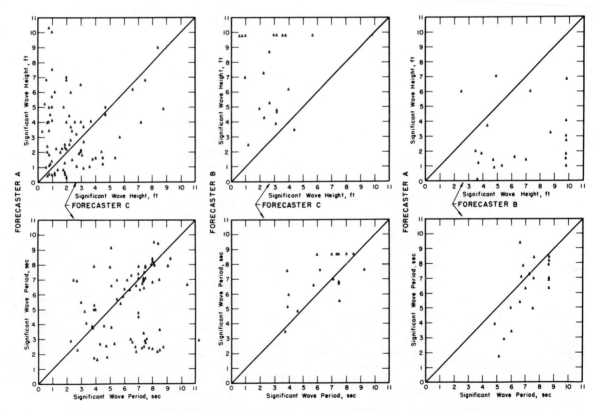

Fig. 10.4. Comparison of forecasts by three forecasters, A, B, and C, from one set of weather charts (from Wiegel, 1961)

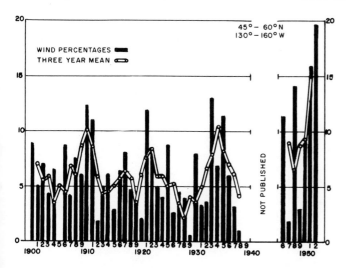

Fig. 10.5. Percentage of winds of force 8 or greater which were reported by ships in the Gulf of Alaska during January, 1900–1952 (after Danielsen, Burt, and Rattray, Jr., 1957)

forecaster using three sets of weather charts was less than the disagreement between three different forecasters using the same set of weather charts.

There have been few published studies of the comparison of wave forecasts with measured waves, except for a few specific storms. One such study (Isaacs and Saville, 1949) was based upon the original data of Sverdrup and Munk (1947), and so is no longer completely applicable in detail. They made 271 forecasts for one region (a nine-month interval) and 201 forecasts for another region (an eight-month interval). One of their findings is still valid, and of great importance in utilizing wave forecasting techniques for the day-to-day planning of certain operations, such as drilling from a moored ship in the open ocean. It was found that 97 per cent of the recorded significant increases in wave height were forecast, but 23 per cent of the forecast wave trains failed to arrive. The rather large proportion of nonarrivals apparently resulted from the erroneous selection of fetches, frequently because of difficulty in determining the limits of the effective angles of the winds with respect to the point at which the waves were recorded.

In hindcasting waves in order to determine the "wave climate" of a region, care must be given in choosing the number of years, and the particular years, for which the information on the weather maps will be reduced to wave characteristics. Danielson, Burt, and Rattray (1957) examined weather maps for the interval of 1900–1952 in order to determine the percentage of times extremely severe storms occurred in the Gulf of Alaska (high wind, great fetch, and long duration—at least 24 hours). These studies showed that most of the severe storms occurred in groups of a few years with relatively less stormy groups of years between. A simplified visual description of this

can be seen in Fig. 10.5 where the percentage of times extremely high winds were encountered by ships in the Gulf of Alaska in January is given for the years 1900–1952.

Forecasting waves generated by moving fetches is detailed and the reader is referred to the original papers for the necessary techniques (Kaplan, 1953; Wilson, 1955). The reader is also referred to the original papers for forecasting waves generated by hurricanes (Unoki and Nakano, 1955; Bretschneider, 1957b).

3. SURFACE WIND VELOCITY AND FETCH DETERMINATION

The geostrophic wind occurs only with straight parallel isobars above the layer of air in the boundary layer. This wind speed, V_g, is given by

$$2V_g \Omega \sin \phi - \frac{1}{\rho_a} |\Delta p / \Delta n| = 0 \qquad (10.1)$$

where Ω is the angular velocity of the earth (0.729×10^{-4} rad/sec), ϕ is the latitude in degrees, ρ_a is the density of the air, and $\Delta p / \Delta n$ is the absolute value of the pressure gradient. The direction of the flow is determined by Buys Ballot's law: with your back to the wind the low pressure is to your left and the high pressure is to your right in the Northern Hemisphere; the low pressure to the right and the high pressure to the left in the Southern Hemisphere. The solutions for Eq. 10.1 for 3-millibar and 5-millibar isobar spacings are given in Fig. 10.6. For 1-millibar isobar spacing use one-fifth of the values given by the 5-millibar isobar spacing.

Equation 10.1 is for straight parallel isobars, and these seldom occur. It is necessary to consider the curvature of the air flow, which is not necessarily the same as the isobar curvature. The curvature of the flow is about the same as the curvature of the isobars for stationary or slowly advancing pressure systems, whereas for systems that are advancing rapidly, the curvature is greatest in the forward part of the storm, least on the right-hand side of the path of low centers and left-hand side of high centers. In addition, the centrifugal force due to the curvature of the air flow must be considered. For high-pressure centers, the centrifugal force acts in the same direction as the pressure gradient force, and in low-pressure centers it acts against the pressure gradient. The wind velocity is such that a balance exists among these forces for equilibrium conditions above the boundary layer (frictional air layer).

Most wave observations have been correlated with surface winds, and these winds, being well down in the boundary layer, are considerably less than the speed obtained from the geostrophic wind speed of Eq. 10.1, and there is a change in direction, with the winds in the Northern Hemisphere blowing between 15 degrees and 30 degrees to the left of the isobars. In addition, the rela-

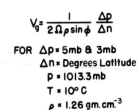

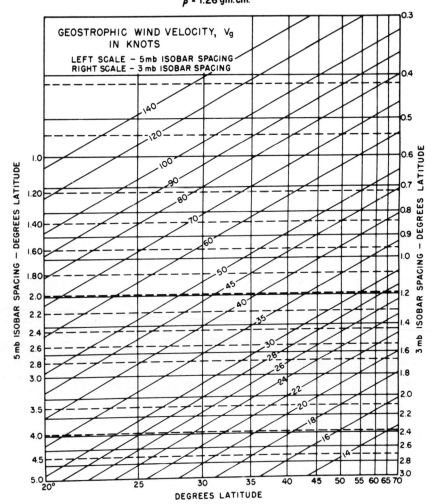

Fig. 10.6. Geostrophic wind scale (after University of California, 1951)

tionship between the wind speed at the geostrophic level and near the surface depends critically upon the stability of the air, or the lapse rate. This effect is dependent upon the difference in temperature between the surface layer of the ocean and the air above it. Sea surface temperatures are available from the U.S. Navy Hydrographic Office (1944c) and air temperatures can be obtained from ship reports.

The relationship between surface wind speeds (at about shipboard height above the water surface) and the geostrophic wind speed of Eq. 10.1, has been given in Fig. 10.7, the correction for the difference in temperature between the sea and the air being obtained from the data in Fig. 10.8 (Scripps Institution of Oceanography, 1945a, 1945b; Arthur, 1948). Surface wind speeds should be checked with values reported by ships in the area and

which are plotted on the weather charts. In using these surface wind speeds, it must be remembered that reports from most nonmilitary ships are merely estimates by ship personnel based upon the state of the sea, flapping of a flag, etc. (U.S. Dept. of Commerce, Weather Bureau, 1960).

The fetch is the horizontal length of the wave-generating area in the direction of the wind. The fetch boundaries are delineated by (a) the coastline, (b) meteorological fronts, (c) curvature of the isobars, (d) spreading of the isobars, as shown in Fig. 10.9. The determination of the fetch is probably the most subjective factor in the entire process of wave forecasting. Considerable experience is needed to do a proper job. There may be two or more wave-generating areas that must be considered, as can be seen in Fig. 10.10, where hindcasts of waves were to

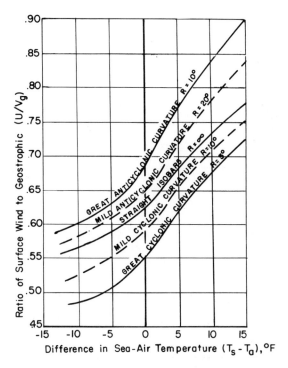

Fig. 10.7. Surface wind scales (after University of California, 1951)

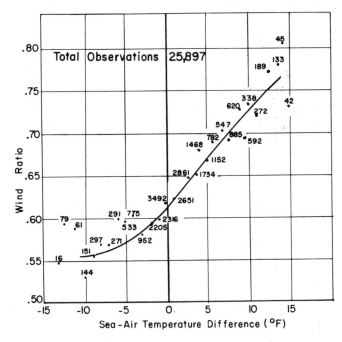

Fig. 10.8. Ratio between average observed and average computed wind as a function of sea-air temperature difference (after Scripps Institute of Oceanography, 1945)

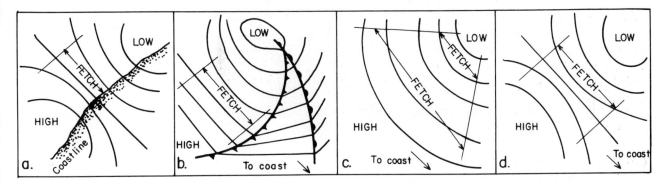

Fig. 10.9. Boundaries of the fetch for different types of isobars (after U. S. Navy Hydrographic Office, 1951)

be made for Point Arguello, California. In choosing the generating area with respect to an area for which wave forecasts or hindcasts are to be made, consideration must be given to the observation that waves decaying from a generating area where the wind direction is uniform reach points not only in line with the wind, but also points at angles up to between 30 and 45 degrees (Arthur, 1949). This is because of the variability in the direction of waves generated by winds.

It has been recommended that the following rules of thumb be used in delineating the generating area for relatively simple meteorological conditions (U.S. Army, Corps of Engineers, Beach Erosion Board, 1961): (a) when there is no spreading of the isobars, the front of the generating area can be located by rotating a template cut to the proper great circle curvature about the area

for which the forecast is to be made, until it intersects the isobars at a storm front at about 10 to 15 degrees, and the rear of the generating area is delineated by rotating the template and marking the points where it intersects the isobars at the rear of the storm at about 45 to 60 degrees; (b) if the isobars spread apart at either the front or the rear of a storm area, the front and rear of the generating area are located where the spreading occurs.

4. EXAMPLE

As an example, consider a storm that is nearly stationary, and an examination of six-hourly weather maps show that its duration is about fifteen hours. The center of the wave-generating area is at 35° N latitude, the 3-

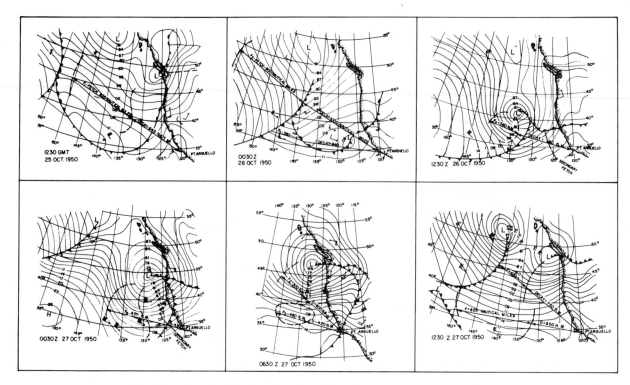

Fig. 10.10. Example of a determination of fetch area decay for Pt. Arguello, California

millibar isobar spacings are approximately 1.0 degrees, the curvature is great cyclonic with $R = 5°$ latitude, the difference between the sea and air temperature is $+ 5°F$, the fetch is 200 nautical miles, and the decay distance is 1000 nautical miles. From Fig. 10.6, V_G is 48 knots; from Fig. 10.7, $U/V_G = 0.62$, so $U = 30$ knots, $gF/U^2 = 32.3 \times 200 \times 6080/(30 \times 6080/3600)^2 = 15,200$, and from Fig. 9.24, $gH_{1/3}/U^2 = 0.15$ to 0.40 and $C/U = 0.75$ to 1.3. Now, $gt/U = 32.2 \times 15 \times 3600/(30 \times 6080/3600) = 34,200$, and from Fig. 9.25, $gH_{1/3}/U^2 = 0.13$ to 0.28, $C/U = 0.7$ to 0.9, and $gT_{H_{1/3}}/U = 5$ to 8. As the latter values are smaller than the former values, the waves are in the duration-limited range, and $H_{1/3} = 0.13 \times (30 \times 6080/3600)^2/32.2$ to $0.28 \times (30 \times 6080/3600)^2/32.2 = 10.4$ ft to 22.4 ft; $T_{H_{1/3}} = 5 \times 30 \times 6080/32.2 \times 3600$ to $8 \times 30 \times 6080/32.2 \times 3600 = 7.9$ to 12.6 sec. Looking at Fig. 9.23, it would be expected that the most likely value of $H_{1/3}$ would be in the vicinity of about 16 ft.

The spectrum can be obtained from Fig. 9.12 for each pair of $H_{1/3}$ and $T_{H_{1/3}}$ as T_{f_o} can be obtained from Fig. 9.11. For example, if we choose $H_{1/3} = 16$ ft and $T_{H_{1/3}} = 10$ sec., $T_{f_o} = 1.14 \times 10 = 11.4$ sec., and $f_o = 1/11.4 = 0.088$.

Now, consider the wave characteristics at the end of the decay. $D/F = 1000/200 = 5$, and $D/H_F = 1000/10.4$ to $1000/22.4 = 96.2$ to 45.1, so from Fig. 9.42, $D/H_D = 360$ to 150, and $H_D = 1000/360$ to $1000/150 = 2.8$ to 6.7 ft; with the most likely significant height being 4.2 ft; $D/T_F^2 = 1000/7.9^2$ to $1000/12.6^2 = 16$ to 6.3, so from Fig. 9.41, $D/T_D^2 = 8.5$ to 3.5, and $T_D = (1000/8.5)^{1/2}$ to $(1000/3.5)^{1/2} = 10\frac{3}{4}$ to 17 sec. From Fig. 10.2, the travel time from the generating area to the forecast area is from 38 to 60 hr. The foregoing ranges of values are for the significant wave height and period of the waves, and it can be seen that there is approximately a 2:1 ratio between the upper and lower limits of the two sets of values. The variation in heights and periods from the limits of the range can be obtained from Table 9.3 and Fig. 9.6(c).

REFERENCES

Arthur, Robert S., *Revised wave forecasting graphs and procedures*, Scripps Inst. Oceanog. Wave Report No. 73, March, 1948. (Unpublished.)

———, Variability in direction of wave travel, *Ann. New York Acad. Sci.* **51**, Art. 3 (May, 1949), 511–22.

Bretschneider, C. L., The generation and decay of wind waves in deep water, *Trans. Amer. Geophys. Union*, **33**, 3 (June, 1952), 381–89.

———, Review of "Practical methods for observing and forecasting ocean waves by means of wave spectra and statistics," *Trans. Amer. Geophys. Union*, **38**, 2 (April, 1957a), 264–66.

———, Hurricane design wave practice, *J. Waterways Harbors Div.*, ASCE, **83**, WW2, Paper No. 1238 (May, 1957b).

———, *Wave variability and wave spectra for wind-generated*

gravity waves, U.S. Army, Corps of Engineers, Beach Erosion Board, Tech. Memo. No. 118, August, 1959.

——, D. K. Todd, and H. L. Kimberley, *Comparisons of wave forecasts*, Univ. Calif. IER, Tech. Rept. 29–39, December, 1950. (Unpublished.)

Danielson, E. F., W. V. Burt, and M. Rattray, Jr., Intensity and frequency of severe storms in the Gulf of Alaska, *Trans. Amer. Geophys. Union*, **38**, 1 (1957), 44–49.

Darbyshire, J., The generation of waves by wind, *Proc. Roy. Soc.*, ser. A, **215**, 1122 (1952), 299–328.

——, A further investigation of wind-generated waves, *Deut. Hydrograph. Z.* **12**, 1 (1959), 1–13.

Ijima, T., *The properties of ocean waves on the Pacific coast and the Japan Sea coast of Japan*, Transportation Techn. Res. Inst. (Japan), Rept. No. 25 June, 1957.

Isaacs, John D., and Thorndike Saville, Jr., A comparison between recorded and forecast waves on the Pacific coast, *Ann. New York Acad. Sci.* **51**, Art. 3 (May, 1949), 502–10.

James, Richard W., *Application of wave forecasts to marine navigation*, U.S. Navy Hydrographic Office, Pub. No. SP-1, July, 1957.

Kaplan, K., *Analysis of moving fetches for wave forecasting*, U.S. Army, Corps of Engineers, Beach Erosion Board, Tech. Memo. No. 35, 1953.

—— and Thorndike Saville, Jr., Comparison of hindcast and observed waves along the northern New Jersey coast for the storm of Nov. 6–7, 1953, U.S. Army, Corps of Engineers, *Beach Erosion Board Bulletin*, vol. **8**, 3 (July, 1954), 13–17.

Longuet-Higgins, M. S., On the statistical distribution of the heights of sea waves, *J. Mar. Res.*, **11**, 3 (December, 1952), 245–66.

Morison, Jack R. and David K. Todd, *The effect of onshore and offshore winds upon wave velocities on a sloping beach*, Univ. Calif., IER, Tech. Rept. 74–1, September, 1953. (Unpublished.)

Neumann, G., *On ocean wave spectra and a new method of forecasting windgenerated sea*, U.S. Army, Corps of Engineers, Beach Erosion Board, Tech. Memo. No. 43, December, 1953.

——, and Willard J. Pierson, Jr., A detailed comparison of theoretical wave spectra and wave forecasting methods, *Deut. Hydrograph. Z.* **10**, 3, 73–92; **10**, 4 (1957), 134–46.

Phillips, O. M., The scattering of gravity waves by turbulence, *J. Fluid Mech.* **5**, Part 2 (February, 1959), 177–92.

Pierson, Willard J., Jr. *A study of wave forecasting methods and of the height of a fully developed sea on the basis of some wave records obtained by the O. W. S. Weather Explorer during a storm at sea*, New York Univ., College of Engineering, Research Div., Dept. of Meteor. and Ocean., June, 1959. (Unpublished.)

——, Gerhard Neumann, and Richard W. James, *Practical methods for observing and forecasting ocean waves*, U.S. Navy Hydrographic Office, Pub. No. 603, 1955.

Putz, R. R., Statistical distributions of ocean waves, *Trans. Amer. Geophys. Union*, **33**, 5 (October, 1952), 685–92.

Rattray, J., Jr., and W. V. Burt, A comparison of methods for forecasting wave generation, *Deep Sea Res.* **3**, 2 (1956), 140–44.

Saville, Thorndike, Jr., Wave forecasting, *Proc. First Conf. Ships and Waves*, Berkeley, Calif.: The Engineering Foundation Council on Wave Research, 1955, 78–91.

Scripps Institution of Oceanography, *Procedure for computing surface wind velocities*, Wave Report No. 35, February, 1945a. (Unpublished.)

——, *Surface wind velocity and sea-air temperature differences*, SIO Wave Project Rept. No. 28, 1945b. (Unpublished.)

Seiwell, H. R., Results of research on surface waves of the western North Atlantic, *Papers Phys. Oceanog. Met.*, **10**, 4, 1948.

Stevenson, Thomas, *The design and construction of harbours, a treatise on maritime engineering*, 3rd ed. Edinburgh: A. C. Black, Ltd. 1886.

Sverdrup, H. V., and W. H. Munk, Empirical and theoretical relations between wind, sea, and swell, *Trans. Amer. Geophys. Union*, **27**, 6 (December, 1946), 823–27.

——, and W. H. Munk, *Wind, sea, and swell; theory of relations for forecasting*, U.S. Navy Hydrographic Office, H. O. Pub. No. 601, March, 1947.

U.S. Army, Corps of Engineers, Beach Erosion Board, *Shore protection, planning and design*. Tech. Rept. No. 4, 1961.

U.S. Dept. of Commerce, Weather Bureau, *Manual of marine meteorological observations*, Circular M, 10th ed., 1960.

U. S. Navy, Hydrographic Office, *World Atlas of sea surface temperatures*, H.O. Pub. No. 225, 2nd ed., 1944.

——, *Wind waves and swell; principles in forecasting*, H.O. Misc. 11, 275, 1944a.

——, *Breakers and surf; principles in forecasting*, H.O. No. 234, 1944b.

——, *Techniques for forecasting wind waves and swell*, H.O. Pub. No. 604, 1951.

University of California, *Manual on Amphibious Oceanography*, R. L. Wiegel, ed., IER, Contract N7onr–29535, Washington, D.C., Pentagon Press, 1951.

Unoki, Sanae, and Masito Nakano, A note on forecasting ocean waves caused by typhoons, *Rec. Oceanog. Works in Japan*, **2**, 1 (March, 1955), 151–61.

Wiegel, R. L., An analysis of data from wave recorders on the Pacific Coast of the United States, *Trans. Amer. Geophys. Union*, **30**, 5 (1949), 700–4.

——, Some engineering aspects of wave spectra, *Conf. Ocean Wave Spectra: Proceedings of a Conference*, National Acad. Sci.–National Res. Council, Prentice-Hall, Inc., 1963, 309–21.

Wilson, Basil, *Graphical approach to the forecasting of waves in moving fetches*, U.S. Army, Corps of Engineers, Beach Erosion Board, Tech. Memo, No: 73, April, 1955.

CHAPTER ELEVEN

Wave Forces

1. INTRODUCTION

Structural design is the design of structures so as to insure that they successfully withstand the forces to which they are subjected. In addition to withstanding the hydraulic forces, it is necessary that marine structures withstand the abrasive action of suspended sediments, marine borers, corrosion, etc. Furthermore, the foundation must be designed with special consideration.

This chapter deals with those phases of structural design which have to do with the forces peculiar to oceans and other large bodies of water, such as forces induced by wave action.

The determination of the forces exerted by waves on structures is difficult. In most cases, semiempirical methods have been taken from the treatment of more standard hydraulic problems and adapted to the special conditions associated with wave motion.

2. FORCES ON RIGID SUBMERGED BODIES IN UNSTEADY FLOW

We shall consider the forces acting on a body which is rigidly mounted within the water. Two types of solutions can be used to determine forces on a body. One is solving the boundary value problem for a potential flow. Such a solution is for a perfect fluid, however, and to this solution must be added a semiempirical equation for the drag forces. The second type of solution uses the semiempirical type of equations which have been employed successfully in unidirectional flow.

In regard to the semiempirical approach, two methods have been advanced to date for predicting the wave-induced forces. The first method, described by Morison, O'Brien, Johnson, and Schaaf (1950), is based upon the assumption that the force consists of two parts (a drag force due to the water particle velocity and an inertia force due to the water particle acceleration) and that the solution can be obtained by treating each component separately and adding the solutions linearly to obtain the total force. The equation contains two coefficients which must be determined empirically. The second method, described by Crooke (1955), is based upon the study by Iversen and Balent (1951) of the forces exerted upon a body in accelerated motion through a fluid. This method assumed that there was a linear dependence of velocity upon acceleration so that the force could be expressed as the product of one coefficient, and fluid density, the projected body area, and the square of the particle velocity. It was found that the coefficient was related primarily to Iversen's modulus (the product of the particle acceleration and the body diameter divided by the square of the water particle velocity) and Reynolds number. Published data indicate that improvements in both methods are necessary. A review of work in similar fields and of the limited data on wave forces shows that it will be necessary to determine the relationships among the coefficients in the force equations and other factors, such as the degree of "upstream" turbulence, roughness of the object (Fage and Warsap, 1929), and past history of the flow indicated by the differences between the data of Iversen and Balent (1951) and Luneau (1948).

The last factor deserves considerable attention, as we are dealing with a flow which reverses periodically. Thus, the eddies which are formed behind a body move against the body when the direction of water particle motion reverses, and the "wake" is then at the leading edge. The effect of such motion is almost unknown although some studies which have been made emphasize that large variation in force on the body can be expected, especially if cross currents are present. Some work by Laird, Johnson, and Walker (1959) on moving cylinder pairs shows that under certain conditions the effect of the eddies can be considerable, Fig. 11.5(a). That eddies can form behind piles is shown in the photograph of Fig. 11.1, at least for flow at subcritical Reynolds numbers.

Both methods of force prediction depend upon a knowledge of water particle motion and empirically determined coefficients. The coefficients are combined in the force formulas with theoretical equations for water particle velocity and acceleration. The values of the coefficients which have been obtained to date have been determined by the same means (that is, relying upon the theoretical values of fluid motion rather than measured values); this fact leads to the difficulty that, if the equations do not correctly predict the particle motion, then the coefficients can be used only under similar conditions. Theoretically, the equations are valid only for limited conditions: for example, the Stokes' wave theory is valid for waves, either of low or finite height in deep water, and for waves of infinitesimal height (compared with both wave length and water

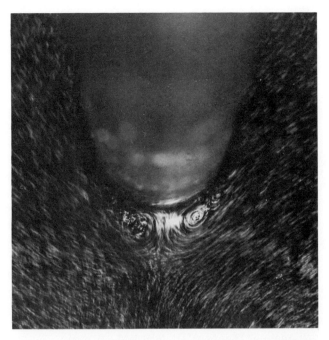

Fig. 11.1. Eddies formed behind a one-inch diameter circular pile in a laboratory channel with water gravity waves running by the pile

depth) in relatively shallow water (Miche, 1954; Reid and Bretschneider, 1954). Few data are available with which to determine the degree of accuracy of these equations even under the most ideal conditions (Elliott, 1953; Morison and Crooke, 1953; and Suquet and Wallet, 1953). The cnoidal wave theory is valid only in relatively shallow water and then strictly only for waves of relatively small height, although there is some evidence that it is valid for waves of appreciable height from a practical standpoint.

The problem of determining the forces exerted on piles by waves which are breaking, or nearly breaking, due to depth limitations, is extremely difficult to handle. There is no satisfactory theory which expresses the water particle velocities and accelerations associated with the motion of period waves in the region in which the wave breaks.

One method which has been used to treat the problem of the forces exerted on a submerged object depends upon the object being at a sufficient distance below the free surface so that it can be considered to be in a fluid of infinite extent. Furthermore, the flow has been treated as if it were unidirectional.

In a frictionless, incompressible fluid (and including the limitations just stated) the force exerted on a submerged body may be expressed as (Lamb, 1945, p.93):

$$F_I = (M_o + M_v)f, \qquad (11.1a)$$

where F_I is the inertia force, M_o is the mass of the displaced fluid, M_v is the so-called added mass which is dependent upon the shape characteristics of the body and the flow characteristics around the body, and f is the acceleration of the undisturbed fluid at the center of the body were the body not there. The sum $M_o + M_v$ is known as the *virtual mass*.

What is the physical model of the inertial force? The understanding of part of it is simple. There must be a force field to cause the fluid to accelerate with time at a point. This force is expressed in terms of the fluid mass. This same force field will try to accelerate any body held in the fluid, and this force is expressed in terms of the mass of fluid displaced by the body. In addition the fluid must move around the body. If the flow were steady (and inviscid), the pressure distribution around the body would be symmetrical fore and aft and there would be no net force; as the flow is accelerating, however, the reaction of the forward portion of the body to the change in one component of momentum of the fluid that must flow around it must be of a different magnitude than the reaction of the rear portion of the body on that same fluid.

Equation 11.1a can be rewritten

$$F_I = C_M \rho V f, \qquad (11.1b)$$

where C_M is the coefficient of mass (also called the *coefficient of inertia*), ρ is the mass density of the fluid, and V is the volume of the fluid displaced by the body. Lamb showed theoretically that for a right circular cylinder $M_v = M_o = \pi D^2/4$ per foot of cylinder length, where D is the diameter of the cylinder. For the right circular cylinder, then,

$$F_I = 2\rho V f. \qquad (11.1c)$$

Lamb (1945, p. 124) has shown that for a sphere the total force on the sphere is

$$F_I = \tfrac{3}{2} \rho V f, \qquad (11.1d)$$

where V is the volume of fluid displaced by the sphere. Taylor (1928) studied theoretically the effect of convective accelerations in a curved or converging stream on ellipsoids of revolution. The major axis was directed into the flow. His calculations were in terms of the pressure gradient $\partial p/\partial x$, but as $-\partial p/\partial x = \rho u \partial u/\partial x$, the results are applicable to the problem being considered here, as the force is proportional to each of the acceleration terms through the coefficient of mass. For ratios of major to minor axes of 1.00 (a sphere), 1.34, 1.81, 2.5, 3.64, 5.08, and 7.12 he found C_M to be 1.50, 1.35, 1.24, 1.16, 1.093, 1.057, and 1.035, respectively.

In any real fluid there is an additional force due to viscosity, which depends upon Reynolds number, shape, roughness, turbulence of the flow, Mach number, etc. The equation for this force is generally given as

$$F_D = \tfrac{1}{2} C_D \rho A u^2, \qquad (11.2)$$

where F_D is the drag force which consists of both fric-

tional drag and form drag, A is the projected area perpendicular to the stream velocity u, C_D is the coefficient of drag, and the factor $\tfrac{1}{2}$ is due to the American practice of reporting measured values of C_D.

It is assumed by Morison, O'Brien, Johnson, and Schaaf (1953) that the total force can be obtained by adding the two terms linearly to get

$$F = F_I + F_D = C_M \rho V \frac{\partial u}{\partial t} + \frac{1}{2} C_D \rho A |u| u, \qquad (11.3)$$

where $|u|u$ has been introduced in place of u^2 in order to maintain the proper sign for the drag force. It is evident that this step is open to question, especially since admitting Eq. 11.2 violates the assumptions of Eq. 11.1a. This is especially true in wave motion as the flow reverses periodically so that the wake that forms behind a body becomes the fluid at the leading edge of the body at the start of the next step in the flow cycle. From an engineering standpoint, however, such steps are admissible provided that useful results are obtained. $\partial u/\partial t$ is the local acceleration of the fluid and it is valid to use this approximation only if $u(\partial u/\partial x)/(\partial u/\partial t) \ll 1$ and $v(\partial u/\partial y)/(\partial u/\partial t) \ll 1$ (Schlichting, 1955); that is, if the convective acceleration is much less than the local acceleration. Using the linear theory for waves, we find that for one of the convection terms

$$\frac{u(\partial u/\partial x)}{\partial u/\partial t} = -\frac{\pi H}{L} \frac{\cosh [2\pi(y+d)/L]}{\sinh 2\pi d/L} \cos 2\pi \left(\frac{x}{L} - \frac{t}{T} \right). \qquad (11.4)$$

Thus the use of $\partial u/\partial t$ in place of du/dt is most nearly valid for waves of low steepness in relatively deep water. In addition, as the phase of the ratio is described by the cosine function, whereas the phase of $\partial u/\partial t$ is described by the sine function, Eq. 11.4 is a minimum when the inertial force is a maximum which tends to minimize the deleterious effects of this approximation.

The effect of a free surface on the inertial force has been studied because of its importance in the physics of ship motion. Murtha (1954) measured the forces necessary to vibrate a circular cylinder vertically with very small motions at a series of distances below the free surface. The aspect ratios of the cylinders were small, the largest being only 4. These data show that the value of the added mass constant rises from a minimum value of 0.5 to 0.6 (which would be 0.6 to 0.7 for a long cylinder) near the surface to 1.0 (corrected to the long cylinder value) when only two diameters below the surface.

The second method, described by Crooke (1955), is based upon the study of Iversen and Balent (1951) of the forces exerted by a flat disk in accelerated motion. Iversen and Balent showed that, if there is a linear dependency of velocity upon acceleration, the equation for the total force may be given by

$$F = \tfrac{1}{2} C A \rho |u| u, \qquad (11.5)$$

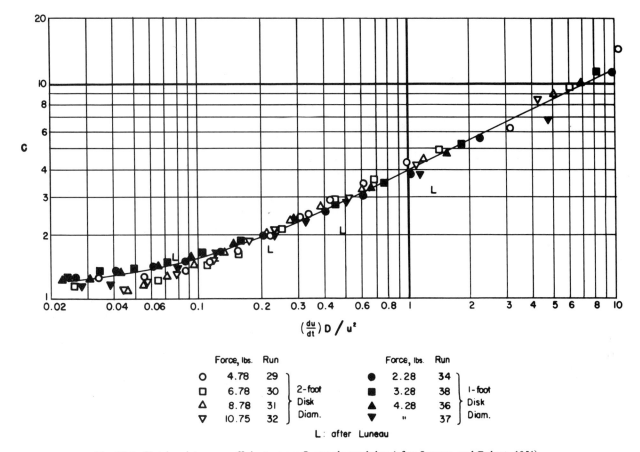

Force, lbs.	Run			Force, lbs.	Run		
○	4.78	29		●	2.28	34	
□	6.78	30	2-foot	■	3.28	38	1-foot
△	8.78	31	Disk	▲	4.28	36	Disk
▽	10.75	32	Diam.	▼	"	37	Diam.

L: after Luneau

Fig. 11.2. Total resistance coefficient versus Iversen's modulus (after Iversen and Balent, 1951)

where C is the coefficient which depends not only upon the shape of the body, Froude modulus, Reynolds number, friction, and Mach number, but also upon the modulus $(du/dt) D/u^2$.

The data obtained by Iversen and Balent are shown in Fig. 11.2 for disks 1 and 2 ft in diameter. The data seem to verify Iversen's and Balent's reasoning for the case of a disk being accelerated by an almost constant force. Iversen and Balent then proceeded to compute coefficients of mass, C_M, after assuming that the coefficient of drag, C_D, was independent of this modulus. It should be stated here that a disk was used in this study because, for disks, C_D is nearly independent of the Reynolds number for uniform velocities greater than 0.005 ft per second for the 2-ft diameter disk and 0.01 fps for the 1-ft diameter disk. A constant value of 1.12 was used for C_D, and all the measured values put into the following equation:

$$ F - M_e \frac{du}{dt} = \frac{1}{2} C_D \rho A u^2 - K \rho V \frac{du}{dt}, \quad (11.6) $$

where F is the net driving force, M_e is the mass of the solid, A is the projected area, and V is the displaced volume of the fluid (for a disk this is taken as the volume of a sphere, with the same diameter as the disk). The value of K, the added mass constant, as computed by

Iversen and Balent, is shown in Fig. 11.3. It can be seen that the value approaches the theoretical value of $2/\pi$ for a disk for high values of $(du/dt) D/u^2$, that is, where the acceleration is high compared with the velocity. The investigators stated that the scatter for small values of the modulus occurred because the solution depends upon taking differences, and as the differences become very small in this region, small experimental errors lead to large errors in the computed values of K. One important point is that K was found to be as much as 5 times as great as the theoretical value, provided that C_D is independent of the modulus. Actually, both coefficients must be dependent upon the modulus. The total error in calculating total forces using steady state values of C_D and theoretical values of K in Eq. 11.6 would be large.

In addition to this study of the force on a disk being accelerated through a fluid, some work has been performed by Luneau (1948) for the case of constant acceleration, rather than constant driving force. The data are presented in Fig. 11.3 in addition to those of Iversen and Balent. It can be seen that there is quite a difference. It is not known whether this is present because the past history of the motion must be taken into consideration in addition to Iversen's modulus, or whether the difference reflects difficulties with experimental techniques.

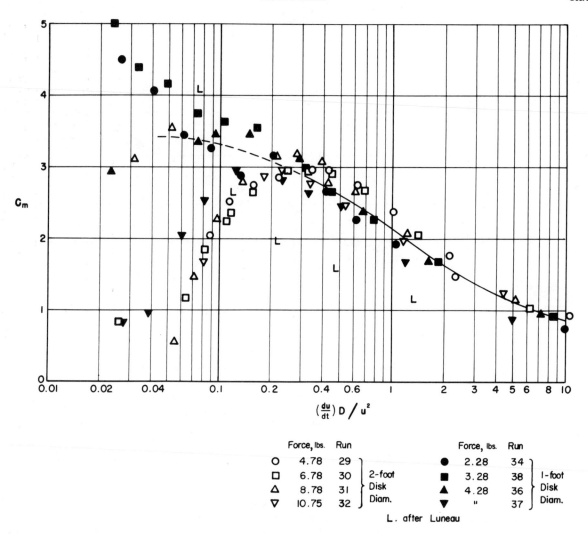

Fig. 11.3. C_m versus Iversen's modulus (after Iversen and Balent, 1951)

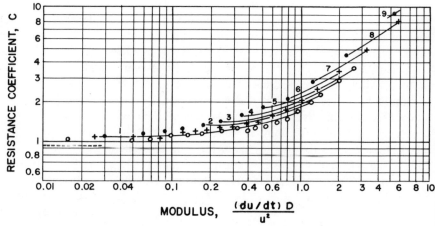

Fig. 11.4. Acceleration modulus versus co-efficient of resistance, for constant instantaneous Reynolds number, N_R (after Keim, 1956)

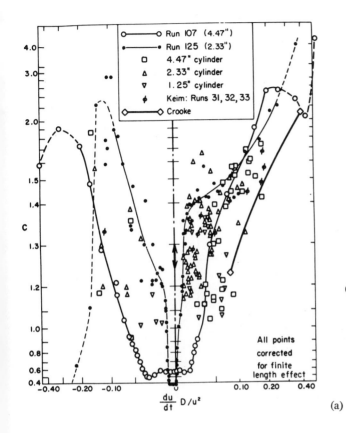

(a)

(b)

Fig. 11.5. (a) Infinite cylinder resistance coefficient, C, versus Iversen's modulus, $(du/dt)D/u^2$; (b) Uniform motion drag forces for cylinder alone and with neighbor placed as stated (from Laird, Johnson, and Walker, 1959)

Keim (1956) studied a series of cylinders accelerating through water using essentially the equipment of Iversen and Balent. Keim's data are shown in Fig. 11.4 in a manner to show the effect of Reynolds number.

Laird, Johnson, and Walker (1959) made measurements of the forces necessary to move a right circular cylinder at constant velocity and at constant acceleration in a towing tank. Some of the results are shown in Fig. 11.5.

The same approach has been used to study the frictional drag in a tube and the headloss of an orifice in accelerated flow (Daily and Hankey, 1953). The effect of acceleration on frictional drag in the tube was found to be very small; in the orifice, however, the dependency of the headloss upon the modulus was of significant magnitude.

3. WAVE-INDUCED FORCES

The equations for the water particle velocity and acceleration must be substituted into the general equations of forces exerted upon submerged bodies in unsteady flow. The linear theory horizontal components of velocity and local acceleration (Stokes' wave) are

$$u = \frac{\pi H}{T} \frac{\cosh\left[2\pi(y + d)/L\right]}{\sinh 2\pi d/L} \cos 2\pi \left(\frac{x}{L} - \frac{t}{T}\right), \quad (11.7a)$$

$$\frac{\partial u}{\partial t} = \frac{2\pi^2 H}{T^2} \frac{\cosh\left[2\pi(y + d)/L\right]}{\sinh 2\pi d/L} \sin 2\pi \left(\frac{x}{L} - \frac{t}{T}\right), \quad (11.7b)$$

where y is the vertical distance from the still water surface to the water particle (measured negative downwards), d is the water depth, L is the wave length, H is the wave height, T is the wave period, x is the horizontal distance measured in the direction of wave advance from the wave crest, and t is the time measured as positive from the time the crest of the wave passes the center of the object. The vertical components of water particle velocity and local acceleration are

$$v = \frac{\pi H}{T} \frac{\sinh\left[2\pi(y + d)/L\right]}{\sinh 2\pi d/L} \sin 2\pi \left(\frac{x}{L} - \frac{t}{T}\right), \quad (11.7c)$$

$$\frac{\partial v}{\partial t} = -\frac{2\pi^2 H}{T^2} \frac{\sinh\left[2\pi(y + d)/L\right]}{\sinh 2\pi d/L} \cos 2\pi \left(\frac{x}{L} - \frac{t}{T}\right). \quad (11.7d)$$

In dealing with practical problems Eqs. 11.7a, b, c, and d are often used considering either the velocity and acceleration fields at a fixed time or the variation with time of velocities and accelerations at a point. The sign conventions have been shown in Fig. 11.6.

If one considers a submerged body, the diameter of which is small compared with the wave length and water depth, fixed in space, the horizontal component

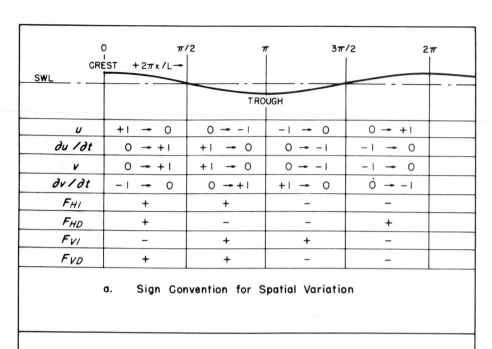

	0	π/2	π	3π/2	2π
	CREST +2πx/L→				
SWL			TROUGH		
u	+1 → 0	0 → -1	-1 → 0	0 → +1	
$\partial u/\partial t$	0 → +1	+1 → 0	0 → -1	-1 → 0	
v	0 → +1	+1 → 0	0 → -1	-1 → 0	
$\partial v/\partial t$	-1 → 0	0 → +1	+1 → 0	0 → -1	
F_{HI}	+	+	-	-	
F_{HD}	+	-	-	+	
F_{VI}	-	+	+	-	
F_{VD}	+	+	-	-	

a. Sign Convention for Spatial Variation

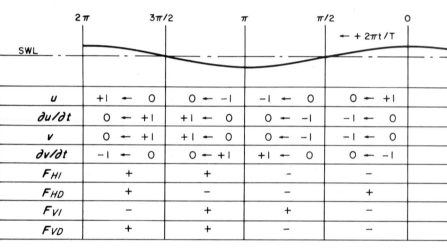

	2π	3π/2	π	π/2	0
				← +2πt/T	
SWL					
u	+1 ← 0	0 ← -1	-1 ← 0	0 ← +1	
$\partial u/\partial t$	0 ← +1	+1 ← 0	0 ← -1	-1 ← 0	
v	0 ← +1	+1 ← 0	0 ← -1	-1 ← 0	
$\partial v/\partial t$	-1 ← 0	0 ← +1	+1 ← 0	0 ← -1	
F_{HI}	+	+	-	-	
F_{HD}	+	-	-	+	
F_{VI}	-	+	+	-	
F_{VD}	+	+	-	-	

b. Sign Convention for Temporal Variation

Fig. 11.6. Linear theory sign convention: (a) for spatial variation; (b) for temporal variation

of force expressed as a function of time is

$$F_h = -2\pi^2 C_M \rho V \frac{H}{T^2} \frac{\cosh\,[2\pi(y+d)/L]}{\sinh 2\pi d/L} \sin \frac{2\pi t}{T}$$

$$+ \frac{1}{2} C_D \rho A \pi^2 \frac{H^2}{T^2} \left(\frac{\cosh\,[2\pi(y+d)/L]}{\sinh 2\pi d/L}\right)^2 \quad (11.8a)$$

$$\cdot \left|\cos \frac{2\pi t}{T}\right| \cos \frac{2\pi t}{T},$$

where y is the distance below the still water level of the small body of volume V. The minus sign is in front of the inertia force term as $\sin - 2\pi t/T = -\sin 2\pi t/T$. It can be seen that the maximum horizontal component of force does not occur when the crest (or trough) of a wave passes the object, but rather it leads the wave crest (or trough). The phase angle, β_{hf}, at which the maximum horizontal component of force occurs can

be found by differentiating Eq. 11.8a with respect to $2\pi t/T$ and setting it equal to zero.

$$0 = \frac{dF}{d(2\pi t/T)}$$

$$= -2\pi^2 C_M \rho V \frac{H}{T^2} \frac{\cosh\,[2\pi(y+d)/L]}{\sinh 2\pi d/L} \cos \frac{2\pi t}{T}$$

$$- C_D \rho A \pi^2 \frac{H^2}{T^2} \left(\frac{\cosh\,[2\pi(y+d)/L]}{\sinh 2\pi d/L}\right)^2$$

$$\cdot \left|\cos \frac{2\pi t}{T}\right| \sin \frac{2\pi t}{T}.$$

The solution is either

$$\sin \beta_{hf} = \pm \frac{2 C_M V}{C_D A H} \cdot \frac{\sinh 2\pi d/L}{\cosh\,[2\pi(y+d)/L]}, \quad (11.9a)$$

(where β_{hf} is in the second and fourth temporal quadrant; i. e., when the trough or crest is approaching

254

the object), or, if

$$\frac{2C_M V \sinh (2\pi d/L)}{C_D A H \cosh [2\pi (y + d)/L]} > 1,$$

$$\cos \beta_{hf} = 0, \qquad (11.9b)$$

(that is, the maximum force occurs as one of the "still water levels" of the wave passes the object). If, in computing forces, the pile is "moved" while the wave is considered stationary, then Eq. 11.9a applies to the first and third quadrants.

It is difficult to know the proper way to determine the forces when both currents and waves are present. It has been suggested (Reid, 1956) that the drag portion of Eq. 11.8a be modified and that in place of $|u|u$, $|U + u|(U + u)$ be used, where U is the current velocity (which varies with depth) and u is the horizontal component of water particle velocity due to wave motion. One is then faced with the problem of determining whether the coefficient of drag obtained for oscillatory flow is satisfactory for the steady component, U, as well.

The equation for the vertical component of force is

$$F_v = -2\pi^2 \rho C_M V \frac{H}{T^2} \frac{\sinh [2\pi (y + d)/L]}{\sinh 2\pi d/L} \cos \frac{2\pi t}{T}$$

$$- \frac{1}{2} \rho C_D A \frac{\pi^2 H^2}{T^2} \left\{ \frac{\sinh [2\pi (y + d)/L]}{\sinh 2\pi d/L} \right\}^2 \quad (11.8b)$$

$$\cdot \left| \sin \frac{2\pi t}{T} \right| \sin \frac{2\pi t}{T}.$$

The phase angle, β_{vf}, at which the maximum vertical component of force occurs is either

$$\cos \beta_{vf} = \pm \frac{2C_M V}{C_D A H} \cdot \frac{\sinh 2\pi d/L}{\sinh [2\pi (y + d)/L]}, \quad (11.8c)$$

(where β_{vf} is in the second and fourth spatial quadrants, first and third temporal) for Eq. 11.8b or, if $[(2C_M V \sinh 2\pi d/L)]/[(C_D AH \sinh 2\pi (y + d)/L)] > 1$,

$$\sin \beta_{vf} = 0. \qquad (11.9d)$$

The preceding equations could be refined by substituting higher-order solutions for u and v, and using du/dt and dv/dt for $\partial u/\partial t$, and $\partial v/\partial t$ for the Stokes wave in relatively deep water and solutions for these quantities for cnoidal waves in relatively shallow water.

Two examples of applications will be given.

1. A sphere 1.50 ft in diameter is held rigidly 6.00 ft above the bottom in water 40.00 ft deep. Waves 6.00 ft high with a period of 6.00 sec are present. Compute the maximum horizontal force on the sphere. Use $\rho = 2.0$ slugs/cubic ft for the density of sea water.

$$L_0 = \frac{g}{2\pi} T^2 = \frac{32.2}{2\pi} \times 6.00^2 = 185 \text{ ft};$$

$$\frac{d}{L_0} = \frac{40.00}{185} = 0.216.$$

From Appendix 1,

$$\frac{d}{L} = 0.239 \quad L = \frac{40.00}{0.239} = 167 \text{ ft};$$

$$\frac{y + d}{L} = \frac{6.00}{167} = 0.036.$$

From Appendix 1,

$$\cosh \frac{2\pi (y + d)}{L} = 1.02; \quad \sinh \frac{2\pi d}{L} = 2.13;$$

$$u_{max} = \frac{\pi \times 6.00}{6.00} \times \frac{1.02}{2.13} = 1.50 \text{ ft/sec};$$

$$N_R = \frac{u_{max} D}{\nu} = \frac{1.50 \times 1.50}{1.0 \times 10^{-5}} = 2.25 \times 10^5.$$

From published curves of steady flow past a sphere for this Reynolds number, $C_D = 0.5$. Use $C_M = 1.5$ (the theoretical value for a sphere)

$$V = \frac{\pi D^3}{6} = \frac{\pi \times 1.5^3}{6} = 1.77 \text{ ft}^3;$$

$$A = \frac{\pi D^2}{4} = 1.77 \text{ ft}^2;$$

$$\sin \beta_{hf} = \frac{2 \times 1.5 \times 1.77 \times 2.13}{1.5 \times 1.77 \times 6.00 \times 1.02} = 2.09$$

and as it is greater than 1, the solution must be $\cos \beta_{hf} = 0$; that is, the maximum force leads the wave crest (or trough) by $\pi/2$. Hence the maximum force is entirely an inertial force (see Fig. 11.6 for sign convention)

$$F_h = +2\pi^2 \times 1.5 \times 2.0 \times 1.77 \times \frac{6.00}{6.00^2} \times \frac{1.02}{2.13}$$

$$\times \pm 1 + \frac{1}{2} \times 0.5 \times 2.0 \times 1.77 \times \pi^2$$

$$\times \frac{6.00^2}{6.00^2} \left(\frac{1.02}{2.13} \right)^2 |0| \times \pm 0$$

$$= \pm 8.3 \pm 0 = \pm 8.3 \text{ lb}.$$

The force is 8.3 lb in the direction of wave advance when t/T leads the wave crest by 1/4 and the force is 8.3 lb in the opposite direction when t/T leads the crest by 3/4.

2. A long circular cylinder 0.50 ft in diameter is mounted parallel to the bottom, 34.0 ft above the bottom, in water 40.0 ft deep. Waves 6.00 ft high with a period of 6.00 sec are present. Compute maximum vertical component of force per foot of cylinder.

d/L_0, d/L, and L same as previous problem;

$$\frac{d + y}{L} = \frac{34}{167} = 0.204 \quad \sinh \frac{2\pi (d + y)}{L} = 1.66;$$

$$\sinh \frac{2\pi d}{L} = 2.13 \text{ (from the appendix)}.$$

$$v_{max} = \frac{\pi \times 6.00}{6.00} \times \frac{1.66}{2.13} = 2.45 \text{ ft/sec};$$

$$N_R = \frac{2.45 \times 0.50}{1 \times 10^{-5}} = 1.2 \times 10^5;$$

from the data of Fig. 11.8, $C_D = 1.6$ Use the theoretical value of $C_M = 2.0$.

$$V = \frac{\pi D^2}{4} \times 1 = \frac{\pi \times 0.50^2 \times 1}{4}$$

$$= 0.197 \text{ ft}^3 \text{ per foot of length};$$

$$A = D \times 1 = 0.50 \times 1$$

$$= 0.50 \text{ ft}^2 \text{ per foot of length};$$

$$\cos \beta_{vf} = \pm 0.210$$

and β_{vf} follows the wave crest or trough by 1.359 radian,

$$\frac{2\pi t}{T} = 1.359; \qquad \beta_{vf} = 77.8 \text{ or } 257.8 \text{ degrees}$$

$$\sin \beta_{vf} = \pm 0.978.$$

$$F_v = -2\pi^2 \times 2.0 \times 2.0 \times 0.197 \times \frac{6.00}{6.00^2} \times \frac{1.66}{2.13}$$

$$\times \pm 0.210 - \frac{1}{2} \times 2.0 \times 1.6 \times 0.50 \times \pi^2$$

$$\times \frac{6.00^2}{6.00^2} \left(\frac{1.66}{2.13}\right)^2 |0.978| \times \pm 0.978$$

$$= \pm 4.6 \pm 0.4 = \pm 5.0 \text{ lb per foot of cylinder},$$

with the force being negative (down) in the first temporal quadrant and positive in the third quadrant.

MacCamy and Fuchs (1952) have treated the case of a right circular cylinder extending through the water surface to the ocean bottom as a boundary value problem in diffraction theory in the manner of Morse (1936): the solution is for a perfect fluid in that no drag forces are admitted (that is, neglecting viscosity). For the case of a cylinder with a diameter which was small compared with the wave length, the force per unit length of pile is

$$\frac{F_h(y)}{\Delta y} = -\frac{1}{2} \pi^2 \rho g H D^2 \frac{\cosh [2\pi (y + d)/L]}{\cosh 2\pi d/L} \sin \frac{2\pi t}{T}$$

$$(11.10a)$$

(in our coordinate system). Here $Fh(y)/\Delta y$ is the horizontal force per unit length of cylinder, ρ is the density of the water, g is the acceleration of gravity, D is the pile diameter, y is the distance from the still water level to the level at which F_h is computed (measured negatively downward), L is the wave length in water of depth d, t is time, and T is wave period. Using the relationships $C = (g/2\pi) T \tanh 2\pi d/L$, $L = CT$, and the volume per unit length, $V = \pi D^2/4$, it can be shown that Eq. 11.10a can be written

$$\frac{F_h(y)}{\Delta y} = -4\pi^2 \rho V \frac{H}{T^2} \frac{\cosh [2\pi (y + d)/L]}{\sinh 2\pi d/L} \sin \frac{2\pi t}{T},$$

$$(11.10b)$$

which is the same as the inertia force term in Eq. 11.8a, provided that the theoretical value of $C_M = 2.0$ for a circular cylindrical pile is used.

4. WAVE FORCES ON CIRCULAR CYLINDRICAL PILES

The forces exerted by waves on circular cylindrical piles have been measured in the laboratory by Morison (1951) and Morison, Johnson, and O'Brien (1954) in order to determine the general characteristics and also to determine the coefficients of drag and mass. These data showed considerable scatter even though the waves were nearly uniform.

Measurements of the forces exerted on piles by ocean waves have been made by Reid (1956) and Wiegel, Beebe, and Moon (1957). The latter measurements were made of forces resulting from waves as high as 20 ft with periods from 10 to 20 sec in water about 50 ft in depth. These measurements were made with a test section 1 ft high mounted on a supporting pile. The test sections were $6\frac{5}{8}$ in., $12\frac{3}{4}$ in. and 2 ft in diameter. Dummy sections of the same diameters were mounted below and above the test sections to reduce end effects. Examples of the data are shown in Fig. 11.7. These data are for a unit length of pile; that is,

$$F_h(y) = \left[C_M \rho \left(\frac{\pi D^2}{4}\right) \frac{\partial u}{\partial t} + \frac{1}{2} C_D \rho D |u| u \right] \Delta y. \quad (11.11)$$

$\partial u/\partial t$ and u are given by Eqs. 11.7b and 11.7a, respectively, for the distance y below the still water level of the pile test section location, Δy is the unit length of pile, and D is the pile diameter. As the waves and forces were recorded continuously, it was possible to solve for C_D when $\cos 2\pi t/T = \pm 1$ (when the crest or trough of the wave passed the pile); and it was possible to solve for C_M by measuring the force record when the wave passed through the still water level (when $\sin 2\pi t/T = \pm 1$ for linear wave theory).

In Fig. 11.8, the values of C_D are plotted with respect to Reynolds number; included in this figure are the field data of Reid (1956) and Wilson and Reid (1963), and the laboratory data of Morision (1951). Because the engineer is interested in the effect of the larger waves, a graph of C_D versus N_R was prepared for waves greater than 10 ft in height. The data showed no different trend from that for all the data presented in Fig. 11.8.

In Fig. 11.8 the commonly accepted curve of C_D versus N_R for steady flow is given. A similar curve for the oscillatory flow associated with wave motion does not seem to exist. Bacon and Reid (1923) state that the "grain" of the turbulence is extremely important in connection with drag forces. In some work on spheres, they found that if the scale of the turbulence was small compared with the diameter of the sphere, the Reynolds number was a good criterion, but if the grain were coarse, then the Reynolds number no longer served as even an indicator. In this regard, it should be emphasized that the scale of turbulence in the vicinity of a pile structure in the ocean is often large (the eddies forming at, and moving away from the piles). Keulegan and

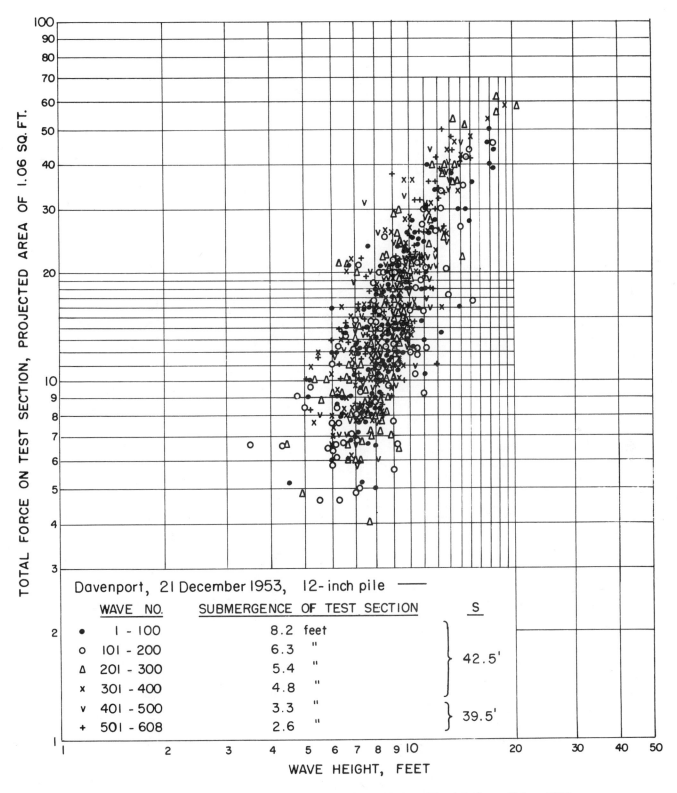

Fig. 11.7. Wave force as a function of wave height (1-foot pile) (from Wiegel, Beebe, and Moon, 1957)

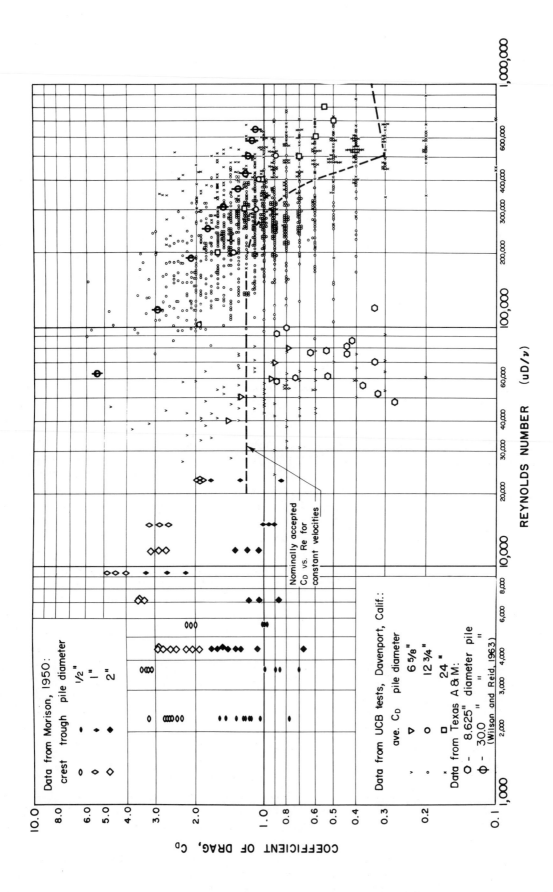

Fig. 11.8. Coefficient of drag for circular cylindrical piles of various diameters (from Wiegel, Beebe, and Moon, 1957)

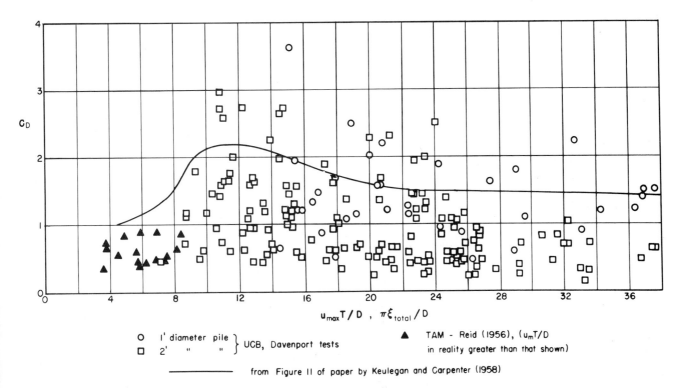

Fig. 11.9. Comparison of C_D with the Keulegan-Carpenter modulus

Carpenter (1958) studied the forces exerted on a circular pile and a flat plate subjected to standing waves in the laboratory and found that their values of C_D could not be related to Reynolds number, but appeared to be a function of the parameter $u_{max} T/D$. In the analysis of Keulegan and Carpenter, the values of C_D and C_M were not obtained for single points as were those of Wiegel, Beebe, and Moon. Rather, they were related to the two primary coefficients of a Fourier series and were obtained by a Fourier analysis of an entire wave cycle. Many of the C_D data of Wiegel, Beebe, and Moon (1957) are shown in relationship to this parameter in Fig. 11.9. The empirical curve of Keulegan and Carpenter forms an approximate upper envelope for these data.

The parameter $u_{max} T/D$ is proportional to the ratio of the horizontal motion of the water particle to the pile diameter. This can be shown easily. For a progressive wave, the total horizontal motion is Eq. 2.55

$$\xi_{total} = H \frac{\cosh [2\pi(y_0 + d)/L]}{\sinh 2\pi d/L},$$

and $u_{max} T$ is (from Eq. 2.45),

$$\frac{\pi H}{T} \frac{\cosh [2\pi(y + d)/L]}{\sinh 2\pi d/L} \cdot T = \pi H \frac{\cosh [2\pi(y + d)/L]}{\sinh 2\pi d/L},$$

so that
$$\frac{u_{max} T}{D} = \frac{\pi \xi_{total}}{D}. \qquad (11.12)$$

By putting dye streaks into the water, Keulegan and Carpenter (1958) were able to determine that eddy formation was associated with the parameter. When the

Keulegan-Carpenter parameter was small, separation of flow from the cylinder did not occur; for the values of the parameter in the vicinity of 15, a single eddy formed; and for relatively high values of the parameter, numerous eddies formed in a Karman vortex street. Thus the water particles must travel several times the pile diameter at the higher portion of the particle speeds for an eddy, or eddies, to form. Another way of looking at it is to consider the length of time for an eddy to form in steady flow, based upon the Strouhal relationship, and to compare this time with the wave period. For relatively high Reynolds numbers (but below critical), the Strouhal relationship is $1/T_e = 0.2\, u/D$, or $u T_e/D = 5$, where u is a constant velocity. The similarity is evident. The wake is thus an important part of the problem. This has been treated theoretically to a certain extent by McNown (1957) and McNown and Keulegan (1959) using the method of Riabouchinski (1920). In essence, it was shown that the larger the wake, the greater the pressure within it and the smaller the coefficient of drag; and the larger the wake, the greater the disturbance in the surrounding flow, and the larger the coefficient of mass. The relationship between the drag coefficient and mass coefficient for a flat plate was computed. It would be expected that a similar relationship should exist for bodies of other shapes.

The laboratory studies of Keulegan and Carpenter were performed with standing waves with the cylinder being at the node to insure one dimensional flow. In progressive waves, the water particles undergo orbital motion and as such there is always flow adjacent to the cylinder. Just how eddies form under these circumstances is uncertain.

There is considerable scatter of the data presented in Figs. 11.8 and 11.9, but this is to be expected considering the variation in degree of turbulence that exists in the vicinity of a pier (both that inherent in the flow and that associated with the wakes of neighboring piles), the variation of the waves from those described by theory, and the fact that locally generated wind waves often exist in addition to the large storm waves, and finally, the fact that currents exist in addition to the waves. When considering the amount of scatter, one should keep in mind the difficulty encountered in correlating aerodynamic data from several wind tunnels, even though the control is far superior to that which can possibly be obtained in nature. Dryden and Abbott (1948) state that even early studies on spheres showed that turbulence of the air stream could have effects on aerodynamics measurements comparable with the effects of Reynolds number. They further state that studies in various wind tunnels have shown that the drag on a sphere may vary by a factor of as large as 4, the minimum drag of an airfoil model by a factor of at least 2, and the maximum lift of an airfoil by the factor of as much as 1.3 at the same Reynolds and Mach numbers.

A word of caution should be given in regard to the trend of the steady flow coefficient of drag versus Reynolds number. Fage and Warsap (1929) have shown that the effect of surface roughness is to cause the coefficient of drag to remain much higher at Reynolds numbers in the critical and supercritical region ($N_R > 2 \times 10^5$). Delany and Sorensen (1953) found in wind-tunnel tests on circular cylinders that the coefficient of drag increased with increasing Reynolds number for $N_R > 4 \times 10^5$. Blumberg and Rigg (1961) had circular cylinders dragged through water at supercritical Reynolds numbers ($> 10^6$ for the highest values) and found that C_D depended upon the surface roughness. For a very rough cylinder, the equivalent of a pile covered with barnacles, C_D was nearly unity in the region of Reynolds number greater than 10^6.

Keulegan and Carpenter developed a technique of obtaining instantaneous values of C_D and C_M through the complete wave cycle. C_D was found to remain at its mean value except for a small interval about the zero velocity phases (maximum acceleration) when they would rise to several times (or more) their mean values (Fig. 11.10).

After considering all of the foregoing observations it would appear that the use of a small value of C_D is not appropriate from an engineering standpoint except perhaps for very small values of the ratio of water particle motion to pile diameter.

The greatest amount of published data on the coef-

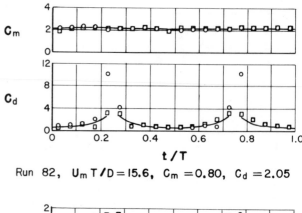

Run 82, $U_m T/D = 15.6$, $C_m = 0.80$, $C_d = 2.05$

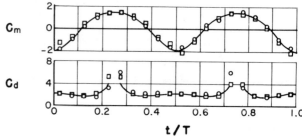

Run 9, $U_m T/D = 3.0$, $C_m = 2.14$, $C_d = 0.70$

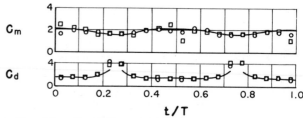

Run 93, $U_m T/D = 44.7$, $C_m = 1.76$, $C_d = 1.54$

Fig. 11.10. Example of variation of inertia and drag coefficients of a cylinder during a wave cycle (after Keulegan and Carpenter, 1958)

ficient of mass C_M is that of Wiegel, Beebe, and Moon (1957). Some values of C_M, together with associated wave parameters, are shown in Table 11.1. In Fig. 11.11, these data have been plotted on statistical graph paper as a function of the cumulative percentage of values lying below the value in question. It can be seen that the points form a straight line in the major range of interest, indicating that the scatter of data about the mean is a normal Gaussian distribution. The mean value is 2.5, the standard deviation 1.2, and the skewness approximately zero. The mean value is somewhat greater than the theoretical value of 2.0 (Lamb, 1945, p. 93; Fuchs and MacCamy, 1954, Keulegan and Carpenter, 1958). Plots of these values of C_M versus Reynolds number and water particle acceleration showed no apparent correlation; a plot of C_M versus wave period showed a slight tendency for C_M to increase with increasing wave period.

Keulegan and Carpenter (1956) found a relationship between C_M and $u_{max} T/D$. The curve through their data

Table 11.1. COEFFICIENT OF MASS
(from Wiegel, Beebe, and Moon, 1957).[a]

Wave	F_{SWL} (lb)	H (ft)	T (sec)	C_M	Wave	F_{SWL} (lb)	H (ft)	T (sec)	C_M	Wave	F_{SWL} (lb)	H (ft)	T (sec)	C_M
46	—	8.1	15.4	—	96	40.6	9.5	12.3	3.3	146	27.8	7.6	14.3	3.3
47	11.6	8.0	13.2	1.2	97	29.0	9.1	15.2	3.0	147	13.9	10.0	12.4	1.1
48	23.2	8.5	12.0	2.1	98	29.0	10.6	14.8	2.5	148	48.1	9.3	10.1	3.2
49	23.2	8.5	13.5	2.3	99	17.4	7.3	14.8	2.2	149	29.0	8.2	14.8	3.2
50	29.0	11.5	14.5	2.3	100	52.3	11.0	14.1	4.1	150	34.8	9.0	12.5	3.2
51	23.2	7.5	15.3	3.0	101	17.4	9.1	13.6	1.6	151	58.0	11.0	12.9	4.3
52	17.4	10.2	12.2	1.3	102	34.8	8.2	14.0	3.6	152	46.4	10.0	14.1	4.0
53	52.2	10.0	13.0	4.3	103	23.2	8.0	13.5	2.4	153	48.7	9.5	14.6	4.6
54	11.6	6.0	12.0	1.5	104	23.2	10.5	15.2	2.1	154	81.2	10.6	14.0	8.6
55	46.4	11.5	14.2	3.5	105	23.2	9.2	11.7	1.8	155	30.7	8.0	12.2	3.0
56	34.8	11.0	14.4	2.9	106	31.9	10.0	15.0	2.9	156	40.6	8.6	12.2	3.9
57	52.2	12.5	14.5	3.7	107	29.0	7.5	14.2	3.4	157	44.1	11.3	13.4	3.3
58	43.5	11.5	13.5	3.2	108	40.6	10.3	14.5	3.5	158	13.3	8.7	15.2	1.4
59	40.6	12.2	15.0	3.0	109	17.4	12.0	16.9	1.5	159	31.9	10.2	13.9	2.7
60	11.6	8.0	14.4	1.6	110	23.2	9.3	17.0	2.5	160	26.1	7.0	13.0	3.0
61	26.1	8.0	12.0	2.3	111	11.6	8.5	13.7	1.4	161	31.3	10.4	13.0	2.4
62	11.6	5.2	10.0	1.4	112	23.2	7.0	9.8	2.1	162	39.5	7.0	12.8	4.5
63	26.1	9.0	12.2	2.2	113	29.0	7.3	12.7	3.2	163	27.3	10.2	13.4	2.2
64	29.0	8.5	14.3	3.0	114	34.8	9.0	15.2	3.6	164	26.7	6.0	14.0	3.8
65	29.6	6.2	15.7	4.7	115	13.4	9.0	15.1	1.4	165	34.8	7.6	14.7	4.1
66	37.7	9.5	11.0	2.8	116	31.9	8.3	13.1	3.1	166	26.7	10.0	14.8	2.4
67	17.4	8.0	11.0	1.5	117	12.8	10.8	14.1	1.0	167	23.2	10.8	15.4	2.0
68	31.9	8.5	11.8	2.8	118	45.3	11.5	11.0	2.7	168	34.5	9.0	15.2	3.6
69	13.3	6.0	12.5	1.8	119	11.6	7.5	11.4	1.1	169	24.9	9.0	15.8	2.7
70	34.8	9.0	13.9	3.3	120	41.7	11.0	12.0	2.8	170	26.7	7.0	13.8	3.3
71	29.0	8.5	16.2	3.4	121	26.1	7.0	13.7	3.2	171	34.7	8.5	14.5	3.7
72	17.4	9.2	15.0	1.8	122	33.7	9.6	12.8	2.9	172	36.6	10.5	14.6	3.1
73	29.0	9.5	13.2	2.5	123	36.0	7.5	14.8	4.3	173	34.7	10.6	14.9	3.0
74	40.6	10.0	15.4	3.9	124	45.3	8.1	13.6	4.8	174	53.9	9.2	15.5	5.6
75	11.6	9.0	12.3	1.0	125	29.0	10.0	15.3	2.7	175	17.4	6.0	12.3	2.2
76	31.9	8.2	13.9	3.3	126	30.2	11.0	14.5	2.5	176	23.2	5.6	14.3	3.7
77	29.0	6.6	11.9	3.3	127	36.0	10.0	15.3	3.4	177	30.2	8.0	13.0	3.1
78	37.7	8.0	12.0	3.8	128	37.7	8.8	15.7	4.1	178	31.9	7.1	11.3	3.2
79	27.8	9.2	13.2	2.5	129	16.6	10.0	12.8	1.5	179	40.6	8.0	13.3	4.2
80	37.7	10.3	13.5	3.1	130	29.0	10.0	13.5	2.5	180	38.9	10.0	12.4	3.0
81	29.0	10.6	11.8	2.0	131	17.4	8.0	13.3	1.8	181	49.3	9.5	12.2	3.6
82	11.6	8.0	11.2	1.0	132	30.1	11.5	13.0	2.1	182	17.4	9.6	13.7	1.5
83	23.2	6.2	10.4	2.4	133	40.6	10.2	13.7	3.7	183	39.4	10.0	10.1	2.4
84	29.0	7.5	12.6	3.1	134	36.5	9.3	11.7	2.8	184	27.9	5.1	8.7	2.9

[a] Feb. 13, 1954; roll 51; 24″ test pile; $S = 33′$; d varied from 48′ to 46′.

is shown in Fig. 11.12. In Fig. 11.13 are also shown the data of Wiegel, Beebe, and Moon. There is a definite trend in that the mean values of the data increase with increasing $u_{max} T/D$. It can be seen that the curve of Keulegan and Carpenter forms the lower envelope to the data obtained in the ocean. This is especially interesting in view of the fact that the curve of C_D versus $u_{max} T/D$ of Keulegan and Carpenter formed an approximate upper envelope to the data obtained in the ocean. In comparing the two sets of coefficients, it should be kept in mind that the Keulegan and Carpenter measurements were made in standing waves at the node where the water moved in a horizontal direction only, whereas the other set of measurements were made in progressive waves so that the particles moved in elliptical paths. The mean values of C_M and also C_d obtained by Reid

(1956) are rather difficult to compare with the data presented here as they were obtained from the total force on a long pile suspended in waves. In a survey of this, and additional work, Wilson and Reid (1963) decided that a value of $C_M = 1.5$ was a good average. The mean values of C_M varied between 0.41 and 1.92. The value of a parameter similar to $u_{max} T/D$ showed that these values were in the vicinity of the critical region of Keulegan and Carpenter. The parameter based upon a mean value of the average with respect to both time and distance below the water surface ranged from about $3\frac{1}{2}$ to $8\frac{1}{2}$. This would probably correspond to a value from $1\frac{1}{2}$ to 2 times as great if it were in terms of $u_{max}/T/D$.

Keulegan and Carpenter also presented instantaneous values of C_M as a function of time over an entire wave period (Fig. 11.10). It was found for small values of

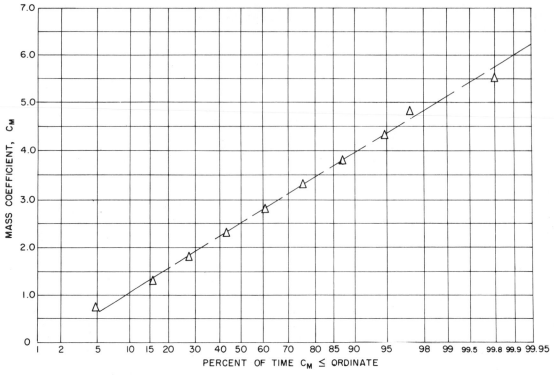

Values from Rolls 41, 43, 51 & 53, 400 waves

Fig. 11.11. Distribution of values of C_M (from Wiegel, Beebe, and Moon, 1957)

Fig. 11.12. Comparison of C_M with the Keulegan-Carpenter modulus

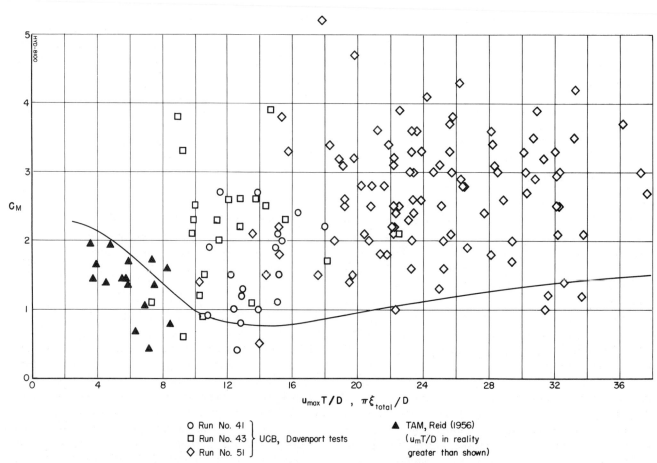

O Run No. 41 ⎫
□ Run No. 43 ⎬ UCB, Davenport tests
◇ Run No. 51 ⎭

▲ TAM, Reid (1956)
(u$_m$T/D in reality
greater than shown)

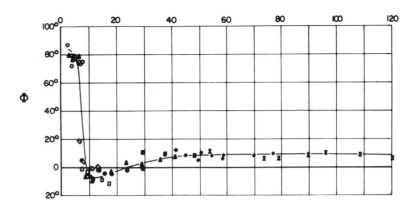

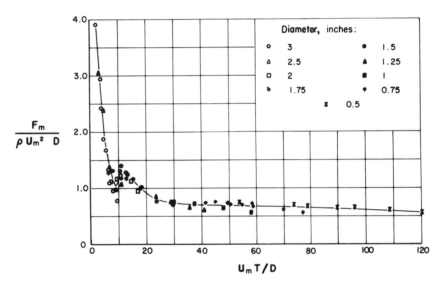

Fig. 11.13. Variations of the magnitude and phase of the maximum force on cylinders (after Keulegan and Carpenter, 1958)

Fig. 11.14. Maximum horizontal forces on (a) a $12\frac{3}{4}$-inch pile and (b) a 2-foot pile (after Wiegel, Beebe, and Moon, 1957)

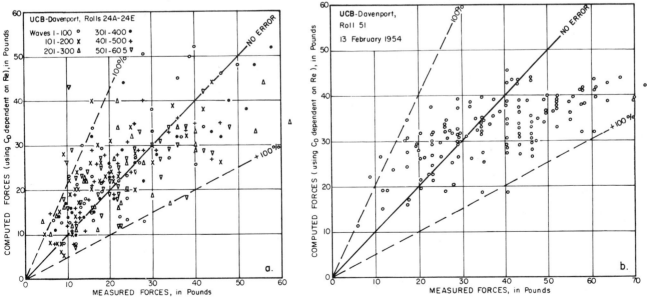

$u_{\max}T/D$ that the instantaneous values of C_M varied hardly at all from the mean value, whereas for values of the parameter in the vicinity of 15, C_M varied considerably, dropping to as low as -2; for relatively large values of the parameter, C_M varied a little above the mean value.

From the relationship between eddy formation and $u_{\max}T/D$, it appears that the low values of C_M in the vicinity of $u_{\max}T/D = 15$ are associated in some way with eddies, perhaps in relationship to the phase at which the single eddy breaks away from the pile—or, perhaps this single eddy doesn't break away at all. Certainly there is something peculiar about the behavior at $u_{\max}T/D \cong 15$, as it is not at all clear how C_M can drop below 1.0 in any circumstance. The single eddy must be responsible for the formation of a back pressure in some way so as to reduce the apparent inertial force. The fact that C_D is high for critical regions of $u_{\max}T/D$ also seems to lead to this conclusion.

One additional type of information that should be considered in regard to using the relationships among $u_{\max}T/D$, C_D and C_M is the relationship between the maximum combined force and the phase angle at which it occurs. In Fig. 11.13, $\beta_{hf} = 90°$ would mean that the total maximum force occurred for $u = 0$ and $\partial u/\partial t =$ maximum, whereas $\beta_{hf} = 0°$ would mean that the total maximum force occurred for $u =$ maximum and $\partial u/\partial t = 0$.

In order to check partially the validity of some of the assumptions made in the wave force theory and the analysis of the data, the average values of C_D versus N_R and the mean value of C_M (2.5) were used to predict the maximum force on the $12\frac{3}{4}$ in. and the 2-ft diameter piles. These values were then compared with the maximum measured forces (Fig.11.14). The scatter of the predictions with respect to the "no error" line is of the order of plus or minus 100 per cent for both piles. The predicted forces, however, are higher than the measured forces for low measured forces and lower than the measured forces for the high measured forces; this is especially true in the case of the 2-ft diameter pile. An examination of the records from the standpoint of vibrations indicates that this difficulty is not due to vibrations, per se, although it might be due to a somewhat similar problem. This type of variation cannot be explained by currents, mass transport, or a shift in the "zero force line," as these would cause a translation of the points in Fig. 11.14 in one direction or the other, rather than a rotation.

A typical example of a measured force that was approximately twice as great as the computed force is shown in Fig. 11.15(b), wave no. 146. It can be seen that the force rises rapidly and then drops very rapidly. This type of force often caused the pile to vibrate severely, although in the example shown the vibrations are quite small. A force trace of this type could be caused by

the superposition of wind waves—which were present, see Fig. 11.15(b)—on the swell with a critical phase relationship; it could have been caused by the shifting of the phase of the water particle accelerations with respect to wave period; it could be a transient phenomenon due to the higher harmonics which are apparent in the wave records; it could be caused by an eddy hitting the pile in such a manner that the eddy and stream velocities were additive; or it could be caused by a combination of any or all of the aforementioned phenomena.

Keulegan and Carpenter found that the peak forces computed using their measured values of C_M and C_D were generally lower than the measured peak values, although the differences seldom exceeded 10 per cent, and in addition the difference between the measured and computed phases of the peak forces varied by as much as 15 degrees.

C_D and C_M were computed from the measured data of Wiegel, Beebe, and Moon using the linear wave theory. The main disadvantage of the linear theory is in regard to the prediction of the water particle acceleration. In addition to the magnitude changing, the phase angle at which the maximum acceleration occurs shifts toward the wave crest. This would not introduce much error in the determination of the C_M, as the actual still water level position was used as the reference in measuring the data. It would, however, introduce an error in predicting the maximum total force that would occur when the inertia force is a significant part of the maximum total force, and it apparently would result in an under-prediction of the maximum total force, and a shift in the phase angle of this maximum.

Some wave force data have been analyzed from the standpoint of Iversen and Balent (Eq. 11.5). The first of these studies was due to Crooke (1955), Fig. 11.16. Some ocean wave data are also shown in Fig. 11.16, including values of C for negative values of Iversen's modulus.

It is interesting to note that Iversen's modulus and the Keulegan-Carpenter parameter are closely related to each other for waves. Using $\partial u/\partial t$ given by Eq. 2.50 and u given by Eq. 2.45, it is found that

$$\frac{(\partial u/\partial t)D}{u^2} = \frac{2\pi D}{uT} \frac{\sin 2\pi[(x/L) - (t/T)]}{\cos 2\pi[(x/L) - (t/T)]}.$$

If the maximum values of $\partial u/\partial t$ and u are used, then

$$\frac{(\partial u/\partial t)D}{u^2} = \frac{2\pi D}{u_{\max}T} = \frac{2\pi}{\text{Keulegan-Carpenter parameter}}. \tag{11.13}$$

The horizontal unit component of differential force, $F_h(S)$ on a differential length of a vertical circular cylindrical pile dS is

$$F_h(s) = \left[C_M \rho \frac{\pi D^2}{4} \frac{\partial u}{\partial t} + \frac{1}{2} C_D \rho D|u|u \right] dS, \tag{11.14}$$

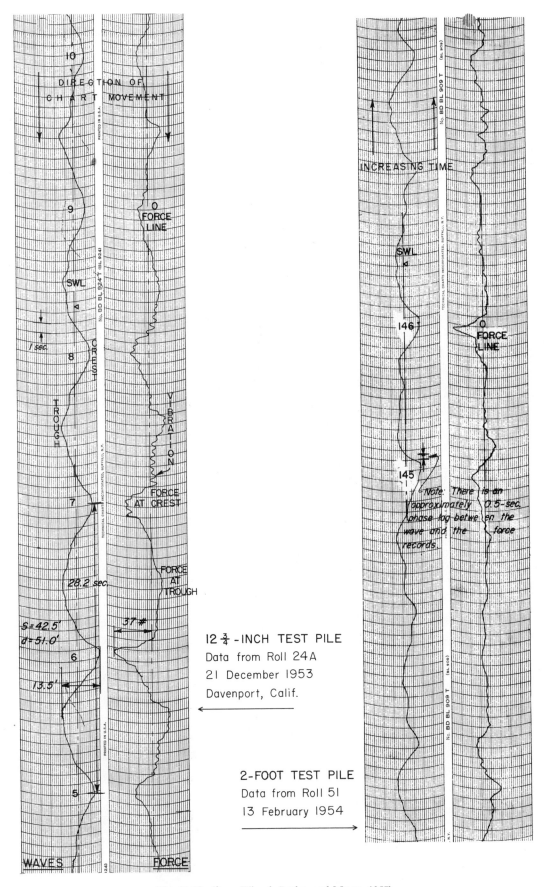

Fig. 11.15. (from Wiegel, Beebe, and Moon, 1957)

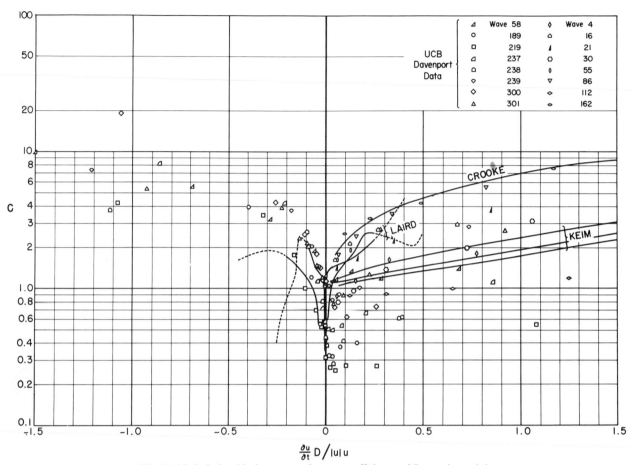

Fig. 11.16. Relationship between resistance coefficient and Iversen's modulus

where $\partial u/\partial t$ and u may be given by Eq. 11.7b and 11.7a, or other higher-order expressions; S is $y + d$ and is positive measured up from the ocean bottom (Fig. 11.17). If C_D can be taken as a constant throughout the entire length of pile, the total force on a section of pile between S_1 and S_2 is given by

$$F_h \int_{S_1}^{S_2} F_h(s) = \pi \rho D \frac{H^2 L}{T^2} \left[\frac{\pi D}{4H} C_M K_2 \sin 2\pi \left(\frac{x}{L} - \frac{t}{T} \right) \right.$$
$$\left. + C_D K_1 \left| \cos 2\pi \left(\frac{x}{L} - \frac{t}{T} \right) \right| \cos 2\pi \left(\frac{x}{L} - \frac{t}{T} \right) \right],$$
(11.15)

where S_1 and S_2 must be within the body of the fluid at the particular x and t in question, and where

$$K_1 = \frac{\dfrac{4\pi S_2}{L} - \dfrac{4\pi S_1}{L} + \sinh \dfrac{4\pi S_2}{L} - \sinh \dfrac{4\pi S_1}{L}}{16 \left(\sinh \dfrac{2\pi d}{L} \right)^2},$$
(11.16a)

$$K_2 = \frac{\sinh (2\pi S_2/L) - \sinh (2\pi S_1/L)}{\sinh 2\pi d/L}.$$
(11.16b)

The maximum total horizontal force will be developed at a phase angle β_{hf} given by

$$\sin \beta_{hf} = \pm \frac{\pi D C_M K_2}{8 H C_D K_1},$$
(11.17a)

where the sign of $\sin \beta_{hf}$ is the opposite to that of $\cos \beta_{hf}$ (that is, in the second and fourth temporal quadrants), or, if $\sin \beta_{hf} > 1$,

$$\cos \beta_{hf} = 0,$$
(11.17b)

that is, 90 degrees in advance of the crest (or trough) of the wave.

Fig. 11.17. Wave force schematic diagram

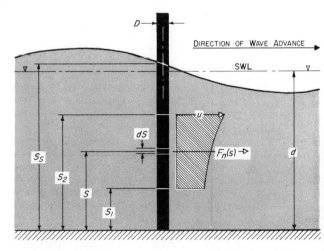

The total horizontal force on a vertical circular pile extending from the bottom through the surface is obtained by substituting S_s for S_2 and putting $S_1 = 0$. Then

$$F_h = \pi \rho D \frac{H^2 L}{T^2} \left[\frac{\pi D}{4H} C_M K_2' \sin 2\pi \left(\frac{x}{L} - \frac{t}{T} \right) \right.$$
$$\left. + C_D K_1' \left| \cos 2\pi \left(\frac{x}{L} - \frac{t}{T} \right) \right| \cos 2\pi \left(\frac{x}{L} - \frac{t}{T} \right) \right],$$
(11.18)

where
$$K_1' = \frac{(4\pi S_s/L) + \sinh (4\pi S_s/L)}{16 \left[\sinh (2\pi d/L) \right]^2}, \quad (11.19a)$$

$$K_2' = \frac{\sinh 2\pi S_s/L}{\sinh 2\pi d/L}. \quad (11.19b)$$

The phase angle for maximum total horizontal thrust is obtained by substituting K_1' and K_2' for K_1 and K_2 in Eqs. 11.17a and 11.17b, respectively. This must be obtained by successive approximations, as $\sin \beta_{hf}$ is a function of K_1' and K_2', both of which incorporate hyperbolic functions of S_s/L, S_s being the elevation of the water surface at the phase angle, which, in this case, must be β_{hf}. It has been found that only two or three attempts are needed to solve it for a particular case. When the wave height is quite small compared with the water depth and wave length, $\bar{K}_1$ and $\bar{K}_2$ can be used in place of K_1' and K_2',

$$K_1' \to \bar{K}_1 = \frac{(4\pi d/L) + \sinh (4\pi d/L)}{16 \left[\sinh (2\pi d/L) \right]^2}, \quad K_2' \to \bar{K}_2 = 1.$$

These approximations also can be used to obtain a value of β_{hf} to start with.

The total moment above any point S_1 on a vertical circular cylindrical pile due to the horizontal component of force is

$$M_{S_1} = \int_{S_1}^{S_2} (S - S_1) \, dF_h$$

$$= \rho D \left(\frac{HL}{T} \right)^2 \left\{ \frac{\pi D}{4H} C_M K_4 \sin 2\pi \left(\frac{x}{L} - \frac{t}{L} \right) \right.$$

$$+ C_D K_3 \left| \cos 2\pi \left(\frac{x}{L} - \frac{t}{T} \right) \right| \cos 2\pi \left(\frac{x}{L} - \frac{t}{T} \right)$$

$$- \frac{2\pi S_1}{L} \left[\frac{\pi D}{8H} C_M K_2 \sin 2\pi \left(\frac{x}{L} - \frac{t}{T} \right) \right.$$

$$+ \frac{1}{2} C_D K_1 \left| \cos 2\pi \left(\frac{x}{L} - \frac{t}{T} \right) \right|$$

$$\left. \left. \cdot \cos 2\pi \left(\frac{x}{L} - \frac{t}{T} \right) \right] \right\}, \quad (11.20)$$

where K_1 and K_2 are given by Eqs. 11.16a, 11.16b, and

$$K_3 = \frac{1}{64 \left[\sinh (2\pi d/L) \right]^2} \left\{ \frac{1}{2} (4\pi S_2/L)^2 \right.$$

$$- \frac{1}{2} (4\pi S_1/L)^2 + (4\pi S_2/L) \sinh (4\pi S_2/L)$$

$$- (4\pi S_1/L) \sinh (4\pi S_1/L)$$

$$- \cosh (4\pi S_2/L) + \cosh (4\pi S_1/L) \}, \quad (11.21a)$$

$$K_4 = \frac{1}{2 \sinh (2\pi d/L)} \{ (2\pi S_2/L) \sinh (2\pi S_2/L)$$

$$- (2\pi S_1/L) \sinh(2\pi S_1/L) \quad (11.21b)$$

$$- \cosh (2\pi S_2/L) + \cosh (2\pi S_1/L) \}.$$

If S_2 is less than S_s, the phase angle of the maximum moment about S is

$$\sin \beta_{hm} = \pm \frac{\pi D C_M [K_4 - (2\pi S_1/L)(K_2/2)]}{8H C_D [K_3 - (2\pi S_1/L)(K_1/2)]}, \quad (11.22a)$$

where the sign is the same as for $\sin \beta_{hf}$; or, if $\sin \beta_{hm} > 1$,

$$\cos \beta_{hm} = 0; \quad (11.22b)$$

that is, β_{hm} leads the wave crest (or trough) by $\pi/2$. Care must be taken that S_1 and S_2 are within the body of the fluid. If $S_2 = S_s$, an explicit solution for β_{hm} does not exist, and the position of maximum moment must be found graphically.

When the wave height is quite small compared with the water depth and wave length, the total moment about the ocean bottom of a pile extending from the bottom through the water surface may be expressed as

$$\bar{M}_{S_0} = \rho D \left(\frac{HL}{T} \right)^2 \left[\frac{\pi D}{4H} C_M \bar{K}_4 \sin 2\pi \left(\frac{x}{L} - \frac{t}{T} \right) \right.$$

$$\left. + C_D \bar{K}_3 \left| \cos 2\pi \left(\frac{x}{L} - \frac{t}{T} \right) \right| \cos 2\pi \left(\frac{x}{L} - \frac{t}{T} \right) \right],$$
(11.23)

where

$$\bar{K}_3 = \frac{1 + \frac{1}{2} \left(\frac{4\pi d}{L} \right)^2 + \frac{4\pi d}{L} \sinh \frac{4\pi d}{L} - \cosh \frac{4\pi d}{L}}{64 \left[\sinh (2\pi d/L) \right]^2},$$
(11.24a)

$$\bar{K}_4 = \frac{1 + (2\pi d/L) \sinh (2\pi d/L) - \cosh (2\pi d/L)}{2 \sinh (2\pi d/L)}.$$
(11.24b)

The maximum total moment will occur at the phase angle

$$\sin \bar{\beta}_{hm} = \pm \frac{\pi D C_M \bar{K}_4}{8H C_D \bar{K}_3}, \quad (11.25a)$$

with the sign of $\sin \beta_{hm}$ being determined as before, or if Eq. 11.26a > 1

$$\cos \bar{\beta}_{hm} = 0. \quad (11.25b)$$

The shorter the wave period, the lower the wave height and the deeper the water and the greater the diameter, the more important is the effect of the inertia term with respect to the drag term.

The location of the application of the resultant horizontal component of force is given by

$$l_1 = \frac{M_{S_1}}{F_h}, \quad (11.26a)$$

where l_1 is the distance above S_1 or, for the case of

a pile extending from the bottom through the water surface, or

$$\bar{l}_0 = \frac{\bar{M}_{S_0}}{\bar{F}_h}, \tag{11.26b}$$

where $\bar{l}_0$ is the distance above the bottom, and $\bar{F}_h$ is obtained from Eq. 11.18 using $\bar{K}_1$ and $\bar{K}_2$.

As an example consider a pile 1.06 ft in diameter extending through the water surface in water 37.0 ft deep subjected to a wave of 5.70 sec period and a height of 1.40 ft. Now C_D would normally vary with depth; however, as the total moment about the bottom is largely due to forces near the surface where the velocities are the largest, it is assumed that the use of a C_D associated with the surface velocity will be adequate, especially in consideration of the scatter of C_D in Figs. 11.8 and 11.9

$$L_0 = \frac{g}{2\pi} \times 5.7^2 = 166 \text{ ft}; \qquad \frac{d}{L_0} = \frac{37.0}{166} = 0.221.$$

From Appendix 1,

$$\frac{d}{L} = 0.243; \quad L = \frac{37.0}{0.243} = 152 \text{ ft}; \quad \cosh\frac{2\pi d}{L} = 2.409;$$

$$\sinh\frac{2\pi d}{L} = 2.192; \quad \frac{2\pi d}{L} = 1.526; \quad \frac{4\pi d}{L} = 3.052;$$

$$\cosh\frac{4\pi d}{L} = 10.61; \quad \sinh\frac{4\pi d}{L} = 10.56;$$

$$u_{\max} = \frac{\pi \times 1.40}{5.70} \times \frac{2.409}{2.192} \times \pm1 = 0.848 \text{ ft/sec};$$

$$N_R = \frac{1.06 \times 0.848}{1 \times 10^{-5}} = 9.0 \times 10^4;$$

$$\frac{u_{\max} T}{D} = \frac{0.848 \times 5.70}{1.06} = 4.83.$$

From Figs. 11.8 and 11.09 it appears that $C_D = 1.6$ and $C_M = 2.0$ should be fairly reliable, provided that an adequate safety factor is used in the final design:

$$\bar{K}_1 = \frac{3.052 \times 10.56}{16(2.192)^2} = 0.190,$$

$$\bar{K}_3 = 1,$$

$$\sin\bar{\beta}_{hf} = \frac{\pi \times 1.06 \times 2.0 \times 1}{8 \times 1.4 \times 1.6 \times 0.190} = 1.96,$$

and as this is > 1, the solution is $\cos\bar{\beta}_{hf} = 0$.

$$\bar{F}_h = \pi \times 2.0 \times 1.06 \frac{1.4^2 \times 152}{5.70^2}\left[\frac{\pi \times 1.06}{4 \times 1.4} \times 2.0\right.$$
$$\left. \times 1 \times \pm1 + 1.6 \times 0.190|0| \times \pm 0\right]$$
$$= 60.0\,[\pm1.19 \pm 0] = \pm71.4 \text{ lb}.$$

There is a maximum force of 71.4 lb in the direction of wave advance at $t/T = 1/4$ before the wave crest passes the pile and a maximum force of 71.4 lb in the opposite direction at $t/T = 1/4$ after the wave crest passes the pile.

$$\bar{K}_3 = \frac{1 + \frac{1}{2}(3.052)^2 + 3.052 \times 10.56 - 10.61}{64(2.192)^2}$$
$$= 0.092,$$

$$\bar{K}_4 = \frac{1 + 1.526 \times 2.192 - 2.409}{2 \times 2.192} = 0.442,$$

$$\sin\bar{\beta}_{hm} = \pm\frac{\pi \times 1.06 \times 2.0 \times 0.442}{8 \times 1.4 \times 1.6 \times 0.092} = \pm1.78$$

and as $\sin\bar{\beta}_{hm} > |1|$, the solution is $\cos\bar{\beta}_{hm} = 0$.

$$\bar{M}_{S_0} = 2.0 \times 1.06\left(\frac{1.40 \times 152}{5.70}\right)^2\left[\frac{\pi \times 1.06}{4 \times 1.40} \times 2.0\right.$$
$$\left. \times 0.442 \times \pm1 + 1.60 \times 0.092|0| \times \pm0\right]$$
$$= 2950\,[\pm0.523 \pm 0] = \pm1550 \text{ ft-lb},$$

with the plus and minus values occurring when $t/T = 1/4$ before the wave crest passes and after the wave crest passes the pile, respectively.

$$\bar{l}_0 = \frac{\bar{M}_{S_0}}{\bar{F}_h} = \frac{1550}{71.4} = 21.7 \text{ ft above the bottom}.$$

For many purposes, the problem of wave forces acting on a pile can be handled by plotting the force distribution against depth for various phase angles.

Linear wave theory has been used in this section, as it is easier to see the trends when this theory is applied than when nonlinear wave theories are used. Linear theory can also be used to represent irregular wave systems by the principle of linear superposition. The author has used this principle to obtain the spectral analysis of an irregular wave system and the spectral analysis of the resulting force on a section of a pile. The power spectra were similar, the linear coherence was between 0.6 and 0.92 in the range of frequencies of interest, and the phase angles were about as to be expected. In relatively shallow water, however, Stokes' third-order or cnoidal wave theory should probably be used.

5. VIBRATIONS

In the previous section, the relationship between $\pi\xi_{\text{total}}/D$ and the formation of eddies was described. Not much is known of the frequency of these eddies for wave motion. They have been studied for uniform rectilinear flow, and the frequency of the eddy pair formation, f_e, has been found for sub-critical Reynolds number to be

$$f_e = S_e\left(1 - \frac{19.7}{N_R}\right)\frac{u}{D} \tag{11.27}$$

where u is the flow velocity, D is the diameter of the structural member, N_R is the Reynolds number and S_e is the Strouhal number. For circular cylinders, a commonly reported value is $S_e \approx 0.2$. Values of S_e for other shapes have been tabulated by Delany and Sorensen (1953), and the American Society of Civil Engineers (1961). For various H beams, channels, and plates it varies from 0.12 to 0.15.

For the high Reynolds numbers associated with structures in ocean waves, $f_e = S_e u/D$. However, no values of S_e have been obtained for wave motion. In addition, the phenomenon for waves must be complicated by the fact that the horizontal component of velocity varies with distance beneath the water surface, so that it is not clear how to calculate f_e from Eq. 11.27. Also, in rectilinear flow for relatively high Reynolds numbers ($N_R > 2 \times 10^5$), f_e is not a constant for a given u and D, but is the mean of a spectrum of eddy-shedding frequencies (Delany and Sorensen, 1953; Fung, 1960; Humphreys, 1960). The average value of S_e has been found to rise from 0.2 for $N_R \approx 2 \times 10^5$ to 0.4 for $N_R = 2 \times 10^6$ for circular cylinders in rectilinear flow by Delany and Sorensen, but other investigators found other results; for example, the spectral analyses by Fung indicated peak power density for a value of S_e of about 0.1. Both Humphreys and Fung feel that for supercritical Reynolds numbers the wake consists of three dimensional turbulent flow, with no true two dimensional vortices being shed.

Thus, not only do ocean waves cause a "periodic" forcing function along an axis in the direction of wave advance, but they cause a "periodic" forcing function along the transverse axis to this owing to the shedding of the eddies. Little is known of the force coefficients associated with these transverse eddies, but they apparently are of the same order of magnitude as the forces in the direction of wave advance (Laird, 1962). This is consistent with the results of experiments in unidirectional rectilinear flow (Humphreys, 1960). In addition, there is a component of force in the direction of wave advance associated with the shedding of eddies. As this component goes through a complete cycle for each eddy that is shed, the frequency is twice the frequency of the transverse force. Other, higher, frequencies are also observed.

Few structures in the ocean are really rigid. The author has observed structural members of piers oscillating with periods of several seconds, and has felt entire structures oscillate when subject to relatively big waves (15 to 20 ft in height). Texas Tower No. 4 was found to have a natural period of about 3.2 sec (U.S. Senate, 1961, p. 195). If Eq. 11.27 is applicable, then a wave 40 ft high, with a 12 sec period in water 185 ft deep would cause eddies to shed from a 12.5-ft diameter pile with a mean period of between 2.3 and 4.5 sec, depending upon the value of S_e chosen. Considering the fact that the eddy-shedding frequency is probably best represented by a spectrum of frequencies indicates the importance of these to the vibrations of a structure. If the results of Fung are similar to what occurs in the ocean, then peak eddy induced vibrations in the 2.3 to 4.5 sec range of periods would result from waves of less than half the height given above.

A considerable insight on this problem may be obtained from a review of the work of Laird (1962) on circular cylinders oscillating in water. The cylinder was designed so that its flexibility could be changed to give it a range of natural vibration frequencies from 1.5 to 10.8 cycles per second. Oscillation frequencies were varied from 0.17 to 0.34 cps, and the eddy-shedding frequencies varied from 1.17 to 1.81 cps. When the natural vibration frequencies were approximately the same as the frequency of eddy shedding, it was found that both the drag and transverse forces were as much as $4\frac{1}{2}$ times as great as the drag force that would be experienced by a stiff circular cylinder in normal uniform rectilinear flow. The experiments from which these measurements were obtained were set up in such a manner as to minimize the effect of inertia forces.

It appears that the dynamical design of ocean structures is necessary under certain conditions.

When a cylinder is very flexible, a coupling between the cylinder and the fluid occurs so that the vibration frequency lies between the natural frequency of the cylinder and the frequency predicted using the Strouhal number for fluid velocities greater than the velocity necessary to cause a match of the natural frequency and the "Strouhal frequency" (Price, 1956; Laird, 1962).

6. WAVE FORCES ON SPHERES

The horizontal and vertical components of force on a sphere that is small compared with the wave length and water depth are given by Eqs. 11.8a and 11.8b.

O'Brien and Morison (1952) have made some laboratory studies in which one set of data was obtained by measuring the time history of the horizontal component of force on a sphere suspended at different distances below the water surface. By measuring the force record at the wave crest, trough, and when the two still water levels passed the center of the sphere, values of C_D and C_M were computed, using Eqs. 11.8a and 11.8b. C_D varied from about 0.8 to 3.0 and C_M varied from about 0.7 to 1.7. The other set of data was obtained by setting a sphere on a razor edge tee at the bottom of the wave tank, holding the wave period constant, and varying the wave height very slowly until the sphere just moved off the tee. The force at this instant was considered to be the maximum force and to be equal to the weight of the sphere in water, the phase at which the sphere moved was measured. Using the appropriate theory with the values of C_D obtained in the first series of experiments, values of C_M were computed. The authors found values of C_M as high as 3.8. The average value of C_M for both sets of tests was found to be 1.59, which compares favorably with the theoretical value of 1.5. Using the average values of C_M, values of C_D versus N_R were computed, using the data obtained from the second series of tests. All values of C_D versus N_R are shown in Fig. 11.18. There seems to be a correlation between C_D and N_R, with the experimental values of C_D approaching the commonly accepted curve for N_R something in excess of 10^4.

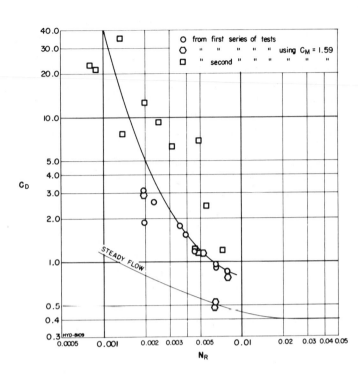

Fig. 11.18. Coefficient of drag for a sphere in oscillatory flow
(after O'Brien and Morison, 1952)

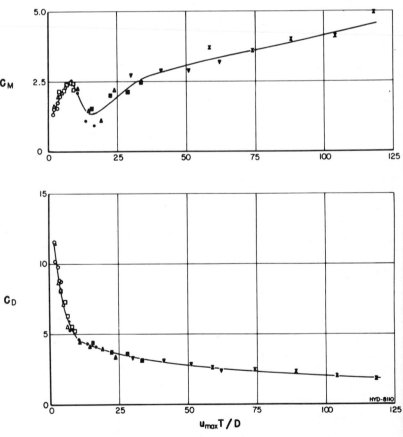

Fig. 11.19. Variation of inertia (top) and drag
(bottom) coefficients of flat plates
(after Keulegan and Carpenter,
1958)

○ 3″ plate diameter
▲ 2.5″
□ 2″
• 1.5″

▲ 1.25″
■ 1″
▼ 0.75″
✗ 0.5″

7. WAVE FORCES ON SUBMERGED VERTICAL PLATES

Laboratory studies of wave forces on a flat plate have been made for standing waves by Keulegan and Carpenter (1956) and Brater, McNown, and Stair (1958).

The results of Keulegan and Carpenter (1958) are shown in Fig. 11.19. For flat plates, C_M is associated with the displaced volume of a circular cylinder having the same diameter as the width of the plate, provided that the width of the plate is normal to the direction of wave advance. As the plate extended from one wall of the wave channel to the other, the results should be nearly those of a flat plate infinitely long. The plate was mounted horizontally.

Brater, McNown, and Stair (1958) determined values of C_D and C_M for a flat plate 30 in. long, $1\frac{7}{8}$ in. high and $\frac{1}{4}$ in. thick with the 30-in. $\times$ $1\frac{7}{8}$-in. face normal to the direction of flow, for waves of the order of 5 ft long in water 1 ft deep. Measurements were obtained with the center of the model $\frac{1}{4}$, $\frac{1}{2}$, and $\frac{3}{4}$ of the water depth below the surface. They determined a type of average C_M and C_D by varying the values until a force-time-history curve was obtained that nearly matched the measured time history curve, with C_M based upon the volume of a circular cylinder $1\frac{7}{8}$ in. in diameter and 30 in. long. It was found that the average C_M was about 1.75 and the average C_D was about 3.5. The values of $u_{max}T/D$ varied from 0.9 to 5.3, the average being 3.1. C_M compares favorably with the curve of Keulegan and Carpenter whereas C_D is lower; however, the C_D obtained by Keulegan and Carpenter varies rapidly in this region.

8. WAVE FORCES ON SUBMERGED BARGELIKE STRUCTURES

Laboratory studies of wave forces on a series of submerged bargelike structures were made by Brater, Mc-Nown, and Stair (1958). The inertial forces predominated owing to the size of the structures relative to the wave dimensions. As the theory used by these investigators in their studies neglected the convective acceleration terms, the horizontal component of wave force would be given by Eqs. 11.8a and 11.8b, neglecting the portions due to the drag forces. Instead of using it in this form, they chose to use the equations for pressure at the upwave and downwave ends of the barges (which were mounted normal to the direction of wave advance).

In the linear theory,

$$\frac{p}{\rho} = -gy - \frac{\partial \phi}{\partial t}$$

$$= -gy + ga\frac{\cosh k(y+d)}{\cosh kd}\cos(kx - \sigma t),$$

and as $a = H/2$, $k = 2\pi/L$, and $\sigma = 2\pi/T$

$$p = \rho gy + \frac{1}{2}\rho gH\frac{\cosh[2\pi(d+y)/L]}{\cosh 2\pi d/L}\cos(kx - \sigma t).$$

$$(11.28)$$

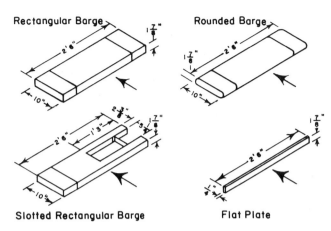

Rectangular Barge   Rounded Barge

Slotted Rectangular Barge   Flat Plate

Fig. 11.20. Models

It was shown to this order of approximation when the center of the box was at $x/L = 1/4$ or $3/4$ that difference in pressure between the upwave face and the downwave face was

$$p_1 - p_2 = \rho g\,\Delta H\frac{\cosh[2\pi(y+d)/L]}{\cosh 2\pi d/L}, \quad (11.29)$$

where ΔH is the difference in elevation of the water surface at the upwave and downwave position above the barge. ΔH is given by

$$\Delta H = H\cos\frac{2\pi x_1}{L}, \quad (11.30)$$

where x_1 is the horizontal distance between the wave crest and the upwave edge of the barge. The inertia coefficient associated with the horizontal component of wave-induced forces was obtained using the relationship

$$(p_1 - p_2)A = \rho V\frac{\partial u}{\partial t}, \quad (11.31)$$

where A is the projected area of the barge normal to the direction of wave advance and V is the displaced volume of the barge.

The various bargelike bodies studied are shown in Fig. 11.20. The results of the values of C_M obtained by measuring the horizontal component of force are shown in Fig. 11.21. There is a definite trend of C_M with respect to increasing the wave height. C_M does not appear to be associated with the wave steepness as is evident Fig. 11.21(b) where a separate curve of C_M versus H/L is needed for each wave period. This same effect was found by Keulegan and Carpenter (1956) if one considers another form of their modulus,

$$\frac{u_{max}T}{D} = \frac{\pi H}{T}\frac{\cosh[2\pi(y+d)/L]}{\sinh 2\pi d/L} \times \frac{T}{D}$$

$$= \frac{\pi H}{D}\frac{\cosh[2\pi(y+d)/L]}{\sinh 2\pi d/L}, \quad (11.32)$$

so that for a given wave length, water depth, and body submergence, C_M should vary with H/D.

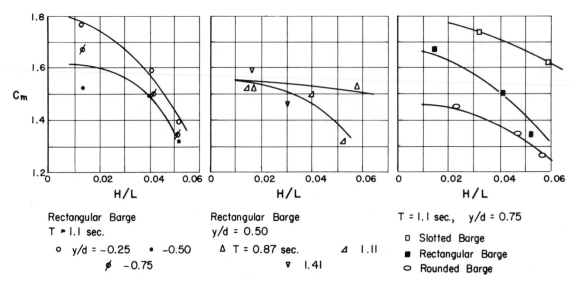

Fig. 11.21. Coefficients of inertial resistance for horizontal forces on barge models

The theoretical results of Riabouchinski (1920) for the coefficients of infinitely long rectangular bodies of different values of a/b are shown in Fig. 11.22. For the bodies tested, $a/b = 0.53$, and the theoretical value of C_M is 1.33 which compares favorably with the measured values of 1.31 to 1.76.

Measurements were also made of the vertical forces on these barges. Within narrow limits, the maximum forces were found to occur when the wave crest or wave trough was over the center of the barge (that is, inertia forces). Experimental values of C_M were determined in this case using the inertial force portion of Eq. 11.8b. The displaced volume of each model was used in the equation. Measured values of C_M vary between 3.4 and 10.0 for the rectangular barge, and between 4.8 and 7.6 (averaging 5.9) for the rounded barge. The value of $a/b = 0.19$ for the rectangular and rounded barges shows $C_M = 6.1$ in Fig. 11.22. Brater, McNown, and Stair state that, for two-dimensional flow around an elliptical section with the same proportions, the theor-

etical value is 6.33. It appears that the theory compares favorably with the measurements even though the models were of finite length. For the slotted barge, C_M varies between 2.9 and 4.7, averaging 3.8. The weighted average of the theoretical values of the unslotted portion and the legs is 4.9.

No measurements were made with the barge on the bottom.

9. WAVE FORCES ON LARGE CIRCULAR CYLINDERS

In dealing with the forces on a large cylinder (or other large body), it is useful to understand the difference between $\partial u/\partial t$ and du/dt.

$$\frac{du}{dt} = \frac{\partial u}{\partial t} - gak \frac{ak}{\sinh 2kd} \sin 2(kx - \sigma t),$$

$$\frac{du/dt}{gak} = \frac{\cosh k(y + d)}{\cosh kd} \sin (kx - \sigma t) \qquad (11.33)$$

$$- \frac{ak}{\sinh 2kd} \sin 2(kx - \sigma t),$$

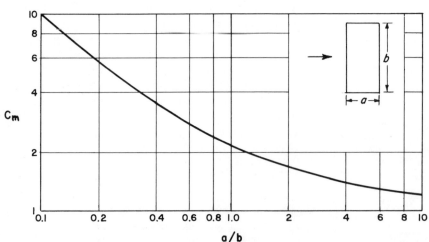

Fig. 11.22. C_M for two-dimensional flow past rectangular bodies (after Brater, McNown, and Stair, 1958)

so that the convective portion becomes more important for steeper waves and shallower water. Furthermore, the convective acceleration has a frequency twice that of the local acceleration and is probably responsible for the type of motion observed in large moored submerged bodies in steep waves in shallow water.

MacCamy and Fuchs (1959) presented a theory for predicting the force exerted by waves on a circular cylindrical caisson which extends from the ocean bottom through the surface. The problem was solved for the case of linearized waves of small steepness. This neglected the forces due to fluid viscosity. When the cylinder was large compared with the wave length, the force due to diffraction (which includes inertial forces) becomes large compared with drag forces. It was shown that the problem of scattering of plane waves by a circular cylinder in acoustic theory could also be applied to ocean waves. The acoustic solution has been presented by Morse (1936) together with tabulations of the necessary functions.

MacCamy and Fuchs found that the total force in the horizontal direction at a depth below the surface, y, per unit length of caisson is

$$F_h(y) = \frac{\rho g H L}{\pi} \frac{\cosh[2\pi(y+d)/L]}{\cosh 2\pi d/L} f_A \cos\left(\frac{2\pi t}{T} - \alpha\right),$$

$$(11.34)$$

where

$$\tan\alpha = \frac{J_1'(\pi D/L)}{Y_1'(\pi D/L)}$$

and

$$f_A = \frac{1}{\sqrt{[J_1'(\pi D/L)]^2 + [Y_1'(\pi D/L)]^2}},$$

and

$$y_S = -\frac{1}{2}H\sin 2\pi\left(\frac{x}{L} - \frac{t}{T}\right)$$

where J_1 and Y_1 are Bessel functions of the first and second kinds, respectively. The prime indicates differentation and α is the angle of phase lag. It should be noted that the location of $t/T = 0$ and $x/L = 0$ for Eq. 11.34 is when the wave goes through its still water level in advance of the crest (see sketch in Fig. 11.23). The functions α and f_A are plotted in Fig. 11.24.

When $D/L \to 0$, $f_A \to (\pi/2)(\pi D/L)^2$ and $\alpha \to (\pi/4)(\pi D/L)^2$, and Eq. 11.34 reduces to the equation for the inertia force on a pile provided that $C_M = 2$.

The total moment about the bottom of a caisson extending from the ocean bottom through the water surface is (integrating from $y = -d$ to $y = 0$):

$$\overline{M}_{S_0} = \frac{2\rho g H L^3}{8\pi^3} f_A \cdot f_B \cos(\sigma t - \alpha), \quad (11.35)$$

where

$$f_B = \frac{1 - \cosh(2\pi d/L) + (2\pi d/L)\sinh(2\pi d/L)}{\cosh 2\pi d/L}.$$

Fig. 11.23. Coordinate system, MacCamy-Fuchs Large Caisson Theory

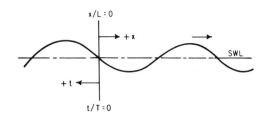

Fig. 11.24. Functions for MacCamy-Fuchs Large Cylinder Equations

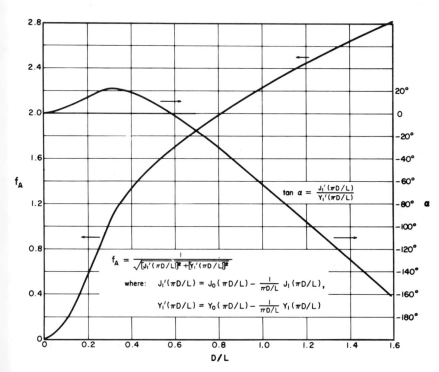

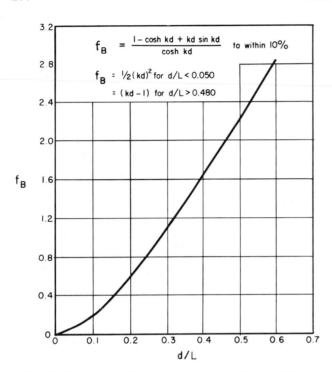

$$f_B = \frac{1 - \cosh kd + kd \sin kd}{\cosh kd} \quad \text{to within 10\%}$$

$$f_B = \tfrac{1}{2}(kd)^2 \text{ for } d/L < 0.050$$

$$= (kd - 1) \text{ for } d/L > 0.480$$

Fig. 11.25. f_B versus d/L (after MacCamy and Fuchs, 1954)

The function f_B has been plotted in Fig. 11.25. The total moment about the bottom of a pile extending through the water surface is (integrating from $y = -d$ to $y = y_s$)

$$M_{S_0} = \overline{M}_{S_0}\left[1 + \frac{2\pi^2 Hd}{L^2 f_B}\sin\frac{2\pi t}{T}\right]. \quad (11.36)$$

The maximum moment occurs for

$$\sin\left(\frac{2\pi t}{T}\right)_{\max} = \frac{1 - \sqrt{1 + 2\left[\dfrac{4\pi^2 Hd}{L^2 f_B}\right]^2}}{2\dfrac{4\pi^2 Hd}{L^2 f_B}}. \quad (11.37)$$

The total horizontal force in the case of small waves is given by (integrating from $y = -d$ to $y = 0$)

$$\overline{F}_h = \frac{\rho g H L^2}{2\pi^2} f_A \tanh\frac{2\pi d}{L}\cos\left(\frac{2\pi t}{T} - \alpha\right). \quad (11.38)$$

The distance of application of the resultant horizontal force above the bottom is

$$\bar{l}_0 = \frac{\overline{M}_0}{\overline{F}_h} = \frac{L f_B}{2\pi \tanh 2\pi d/L}. \quad (11.39)$$

There are almost no data with which to check the validity of the foregoing equations. Laird (1955) per-

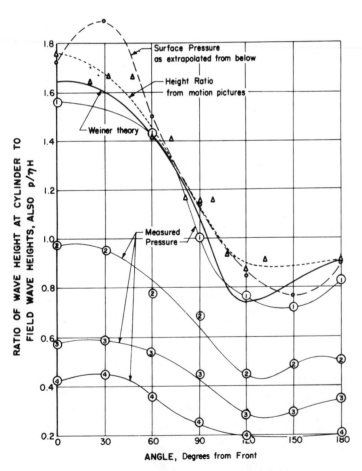

Fig. 11.26. Ratios of maximum pressures and heights to field wave heights as a function of angular and vertical position; pressure readings 1, 2, 3, and 4 at depth-to-water-depth ratios of 0.050, 0.286, 0.523, and 0.762 respectively

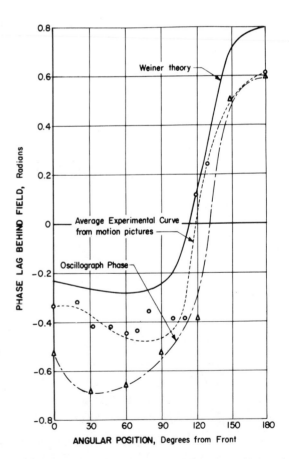

Fig. 11.27. Variation of phase, relative to field waves, with angular position around the cylinder (after Laird, 1955)

Table 11.2. H_c/H on the surface of a Rigid Circular Cylinder of Infinite Length
(after Wiener, 1947)

θ (degrees)	$\frac{\pi D}{L}$: $\frac{1}{2}$	1	2	3	4	5	6	7	8	9	10
0	1.4316	1.7071	1.8586	1.9179	1.9467	1.9624	1.9717	1.9765	1.9807	1.9823	1.985
10	1.4239	1.7038	1.8544	1.9125	1.9401	1.9550	1.9639	1.9698	1.9750	1.9791	1.983
20	1.4012	1.6931	1.8394	1.8934	1.9180	1.9329	1.9448	1.9561	1.9658	1.9737	1.978
30	1.3641	1.6734	1.8079	1.8535	1.8784	1.9020	1.9250	1.9425	1.9523	1.9563	1.959
40	1.3127	1.6416	1.7526	1.7909	1.8308	1.8723	1.8992	1.9077	1.9135	1.9246	1.937
50	1.2528	1.5935	1.6692	1.7168	1.7864	1.8267	1.8338	1.8452	1.8694	1.8832	1.886
60	1.1837	1.5248	1.5628	1.6494	1.7262	1.7290	1.7454	1.7832	1.7902	1.7959	1.818
70	1.1112	1.4318	1.4513	1.5872	1.6076	1.6084	1.6628	1.6594	1.6737	1.6997	1.696
80	1.0407	1.3131	1.3591	1.4925	1.4442	1.5134	1.5062	1.5192	1.5431	1.5321	1.561
90	0.9783	1.1713	1.2965	1.3283	1.3209	1.3600	1.3356	1.3713	1.3463	1.3758	1.354
100	0.9305	1.0153	1.2422	1.1227	1.2210	1.1397	1.2100	1.1466	1.1884	1.1510	1.168
110	0.9017	0.8622	1.1529	0.9777	1.0478	1.0270	0.9689	1.0218	0.9486	0.9753	0.957
120	0.8933	0.7390	0.9944	0.9397	0.8007	0.9029	0.8348	0.7732	0.8281	0.7705	0.738
130	0.9026	0.6760	0.7670	0.8886	0.7187	0.6361	0.7250	0.6866	0.6002	0.6079	0.632
140	0.9237	0.6853	0.5267	0.7046	0.7264	0.5764	0.4613	0.5134	0.5585	0.5016	0.413
150	0.9493	0.7443	0.4201	0.4131	0.5416	0.5843	0.5187	0.3949	0.3060	0.3228	0.379
160	0.9730	0.8152	0.5289	0.3119	0.2326	0.2927	0.3704	0.4091	0.4067	0.3675	0.302
170	0.9893	0.8687	0.6736	0.5195	0.3919	0.2855	0.1968	0.1295	0.0939	0.1007	0.129
180	0.9951	0.8882	0.7319	0.6235	0.5432	0.4822	0.4330	0.3929	0.3592	0.3314	0.306

formed some small-scale experiments in which water surface and pressure distribution about a circular cylindrical caisson were measured. These data could not be compared directly with the preceding equations, but they were compared with tabulations of Wiener (1947) for acoustic waves. The results (Figs. 11.26 and 11.27) show that the theory predicts the gross effects of water surface profile and phase lag at the surface of the caisson.

The theoretical values calculated by Wiener (1947) are given in Tables 11.2 and 11.3. Table 11.2 gives values of H_c/H, the ratio of the wave height at the cylinder wall to the field wave height, as a function of $\pi D/L$, where D is the cylinder diameter, L is the field wave length, and θ is the angle around the cylinder with $\theta = 0$ degrees coincident with the cylinder radius heading into the waves. Table 11.3 gives values of phase lag, in radians, as a function of $\pi D/L$, with negative values of the phase lag indicating the wave at the cylinder leads the field wave. This theory is applicable for "linear waves." Some laboratory measurements of the motion of water waves around a vertical circular cylinder by Hellstrom and Rundgren (1954) show that, although low amplitude waves behave in general as predicted by acoustic theory, the higher-amplitude waves moving around the cylinder met in the vicinity of $\theta = 180$ degrees and sometimes caused vertical jets of water to occur.

Table 11.3. Phase of H_c with Respect to H, on the Surface of a Rigid Circular Cylinder of Infinite Length
(after Wiener, 1947)

θ (degrees)	$\frac{\pi D}{L}$: $\frac{1}{2}$	1	2	3	4	5	6	7	8	9	10
1	−0.394	−0.207	−0.152	−0.120	−0.100	−0.085	−0.075	−0.066	−0.059	−0.054	−0.049
10	−0.393	−0.210	−0.155	−0.124	−0.104	−0.089	−0.077	−0.068	−0.060	−0.054	−0.049
20	−0.389	−0.213	−0.164	−0.134	−0.113	−0.095	−0.081	−0.071	−0.063	−0.058	−0.053
30	−0.383	−0.221	−0.178	−0.147	−0.122	−0.100	−0.086	−0.079	−0.073	−0.068	−0.063
40	−0.371	−0.232	−0.194	−0.155	−0.123	−0.106	−0.100	−0.095	−0.086	−0.077	−0.072
50	−0.352	−0.245	−0.204	−0.151	−0.124	−0.123	−0.117	−0.103	−0.094	−0.093	−0.089
60	−0.323	−0.260	−0.199	−0.138	−0.138	−0.138	−0.116	−0.111	−0.114	−0.103	−0.098
70	−0.280	−0.273	−0.170	−0.130	−0.152	−0.121	−0.117	−0.123	−0.106	−0.110	−0.108
80	−0.220	−0.278	−0.116	−0.136	−0.125	−0.097	−0.117	−0.089	−0.103	−0.091	−0.092
90	−0.142	−0.265	−0.056	−0.132	−0.046	−0.087	−0.043	−0.064	−0.040	−0.051	−0.037
100	−0.045	−0.218	−0.015	−0.056	0.003	0.011	0.027	0.049	0.041	0.075	0.056
110	0.062	−0.115	0.002	0.128	0.057	0.196	0.148	0.207	0.241	0.225	0.292
120	0.171	0.065	0.018	0.316	0.260	0.305	0.466	0.422	0.482	0.591	0.563
130	0.272	0.318	0.089	0.409	0.658	0.572	0.659	0.887	0.939	0.916	1.076
140	0.356	0.578	0.350	0.475	0.881	1.187	1.176	1.178	1.431	1.710	1.807
150	0.420	0.776	0.975	0.759	0.996	1.443	1.892	2.212	2.264	2.251	2.493
160	0.465	0.898	1.514	1.832	1.712	1.744	2.071	2.598	3.116	3.628	4.092
170	0.491	0.961	1.718	2.390	3.011	3.578	4.062	4.388	4.440	4.430	4.678
180	0.499	0.980	1.766	2.480	3.162	3.824	4.473	5.111	5.744	6.371	6.993

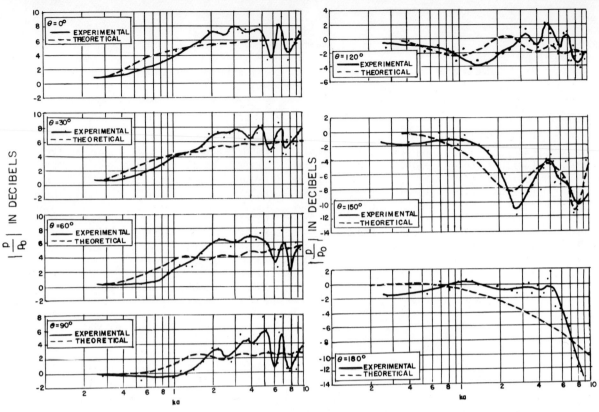

The pressure distribution on the surface of a rigid circular cylinder for points lying in the median plane perpendicular to the axis. The wave front of the incident plane wave and the axis of the cylinder are parallel. The cylinder has a length equal to its diameter D(15 cm.). Note low pass filter characteristic for $\theta = 180°$. For comparison, the theoretical values for a cylinder of infinite length are also shown (after Wiener, 1947)

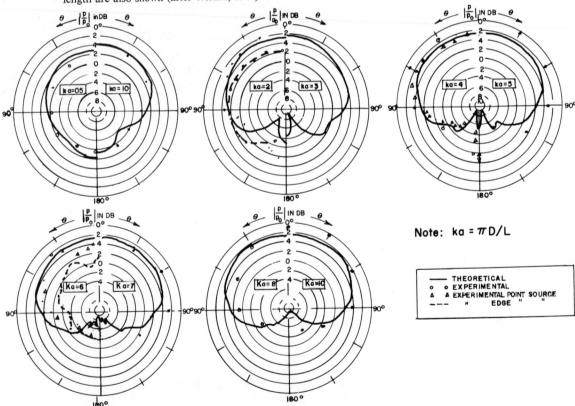

Note: $ka = \pi D/L$

Some of the above data replotted in polar form with data obtained with a point source of sound added. Circles and triangles denote the data for the median plane while squares show the pressures near the edge of the cylinder (after Wiener, 1947)

Fig. 11.28.

276

Laird (1955) found that the dynamic pressure decreased much more rapidly with depth than predicted by linear theory. This also has been noted by engineers working with subsurface pressure type wave recorders (Folsom, 1949; Gerhardt, Jehn, and Katz, 1955).

The theory just described is for a caisson extending from the bottom through the water surface. Wiener (1947) made study of the effect of acoustic waves on a truncated caisson the length of which was the same as the diameter. Measurements showed that the pressure distribution for an infinitely long circular cylinder compared favorably even near the ends, except in the vicinity of $\theta = 180$ degrees (Fig. 11.28). Here the ratio $|p/p_0|$ is in decibels. These results should be valid for a truncated caisson in ocean waves.

10. WAVE FORCES ON VERTICAL WALLS

If a vertical wall breakwater is founded in water deep enough, it will completely reflect incident waves, provided that the horizontal angle between the breakwater and the waves is less than some limiting angle. The assumption is made that waves reflecting from a vertical wall form a standing wave pattern (commonly called a clapotis). The validity of this assumption has not been proved for periodic incident waves and if one considers the irregularity of ocean waves the assumption is open to considerable doubt. Measurements of pressure distribution on model and prototype breakwaters, however, show that use of this assumption leads to reliable results.

The pressure (second-order theory, Miche, 1944) in a standing wave is given by Eq. 2.90 where H is the height of the standing wave and the origin of x is at $L/4$. The validity of this equation for predicting the dynamic pressure distribution in a model is shown in Fig. 2.15.

The theory usually used in the design of vertical wall breakwaters is due to Sainflou (1948), although the theory of Miche appears to be better. In considering the difference between the waves in the ocean and the theoretical wave, however, the differences between the two theories would appear to be of less importance. At the wall, both theories reduce to the same pressure at the bottom.

$$\frac{p_b}{\rho g} - d = \pm \frac{\overline{H}}{\cosh 2\pi d/L} \qquad (11.40)$$

where $\overline{H}$ refers to the wave height that would exist at the wall were the wall not there and p_b refers to the maximum and minimum pressure during the wave cycle (that is, when the crest and the trough are at the wall). This definition of $\overline{H}$ must be kept in mind when comparing Eq. 11.41 with Eq. 2.89a. It has been found to be adequate from a design standpoint to use a linear pressure distribution taken as zero at the maximum or

minimum wave elevation at the wall to the value given by Eq. 11.40 at the bottom. In order to do this, it is necessary to compute the maximum and minimum water elevations at the wall. The mean level of a standing wave (second-order) above the still water level, Δh is (Miche, 1944)

$$\Delta h = \frac{\pi \overline{H}^2}{L} \left[1 + \frac{3}{4 \sinh^2 (2\pi d/L)} - \frac{1}{4 \cosh^2 (2\pi d/L)} \right] \coth \frac{2\pi d}{L}. \qquad (11.41)$$

Since $\cosh 2\pi d/L > \sinh 2\pi d/L$, $\Delta h > (\pi \overline{H}/L) \cosh 2\pi d/L$ which is the commonly used expression for Δh. Comparison of the two expressions with some measurements have been given in Table 2.3. The maximum water elevation at the wall above the still water level (SWL) is $\overline{H} + \Delta h$ and the minimum elevation below SWL is $\Delta h - \overline{H}$. In Fig. 11.29, the solid lines represent the linear distribution and the dotted lines represent the theoretical pressure distribution.

(1) is the maximum elevation of the wave at the wall and (10) is the minimum elevation. The mean wave level above SWL is Δh and the distance (1) (2) is $\overline{H} + \Delta h$ while (2) (10) is $\Delta h - \overline{H}$. The hydrostatic head, d, at the bottom is scaled out from (12) and plotted as (3), or (7). The triangle formed by (12) (2) (3) is the hydrostatic pressure distribution against the wall due to the water at SWL. The dynamic pressure is obtained by plotting $(p_b/\rho g) - d$ in both plus and minus directions from (3); that is, (4) and (11). These are the maximum and minimum pressures at the bottom. The approximate total forces per foot of length of wall are the triangles (4) (12) (1) and (11) (12) (10).

In the case where there is water of the same depth on both sides of a vertical wall with waves on only one side, such as the wall between the ocean and a harbor, there is a seaward pressure distribution which is given by (3) (2), or (7) (2) in Fig. 11.29. The resultant pressure distribution on the vertical wall is given by (6) (5) (1) in Fig. 11.29, when the water is at its maximum elevation, and by (9) (8) (2) when the water is at its minimum elevation.

For a unit length of vertical wall, the resultant force R, the moment about the bottom M, and the point of application l measured from the bottom are given by the following equations, where the subscript $_c$ refers to the maximum wave elevation at the wall and subscript $_t$ refers to the minimum wave elevation at the wall:

$$R_c = \frac{1}{2}(d + \Delta h + \overline{H})\left[d + \frac{\overline{H}}{\cosh(2\pi d/L)}\right] - \frac{d^2}{2}, \qquad (11.42a)$$

$$M_c = \frac{1}{6}(d + \Delta h + \overline{H})^2\left[d + \frac{\overline{H}}{\cosh(2\pi d/L)}\right] - \frac{d^3}{6}, \qquad (11.43a)$$

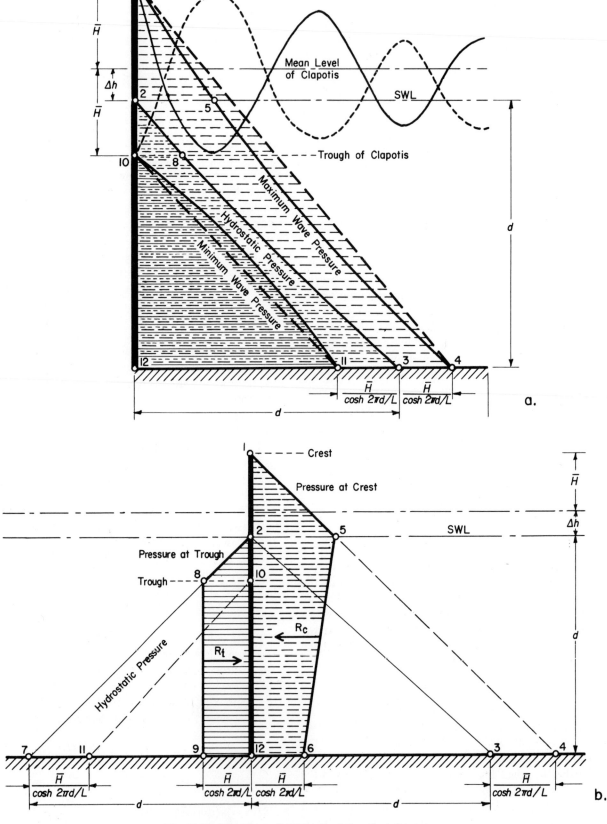

Fig. 11.29. Wave force diagrams for fully reflected waves

$$l_c = \frac{M_c}{R_c}, \qquad\qquad (11.44a)$$

$$R_t = \frac{1}{2}(d + \Delta h - \bar{H})\left[d - \frac{\bar{H}}{\cosh{(2\pi d/L)}}\right] - \frac{d^2}{2}, \qquad (11.42b)$$

$$M_t = \frac{1}{6}(d + \Delta h - \bar{H})^2\left[d - \frac{\bar{H}}{\cosh{(2\pi d/L)}}\right] - \frac{d^3}{6}, \qquad (11.43b)$$

$$l_t = \frac{M_t}{R_t}. \qquad\qquad (11.44b)$$

The preceding equations are for the case of waves on the sea side of the wall and still water on the harbor side. The forces and moments for the case of water and waves on one side only would be the same as the foregoing equation except that the $d^2/2$ and $d^3/6$ terms would be dropped.

Example The approaching waves 6.00 ft high and 100.0 ft long (in this depth of water) are fully reflected from a vertical impervious wall standing in 20.0 ft of water. Still water of the same depth backs up the sea wall.

$$\frac{d}{L} = \frac{20.0}{100.0} = 0.200.$$

From the appendix,

$$\sinh{\frac{2\pi d}{L}} = 1.614; \quad \cosh{\frac{2\pi d}{L}} = 1.899;$$

$$\tanh{\frac{2\pi d}{L}} = 0.8501;$$

$$\Delta h = \frac{\pi \times 6.00^2}{100}\left(1 + \frac{3}{4 \times 1.614^2}\right.$$

$$\left. - \frac{1}{4 \times 1.899^2}\right)\frac{1}{0.8501} = 1.62\ \text{ft};$$

$$\frac{\bar{H}}{\cosh{(2\pi d/L)}} = \pm\frac{6.00}{1.899} = \pm 3.15\ \text{ft};$$

$R_c = \frac{1}{2}(20.0 + 1.62 + 6.00)(20.0 + 3.15) - 20.0^2/2 = 119$ ft of water per linear foot of wall, or, taking the specific weight of sea water as 64.0 lb/cu ft, $R_c = 7620$ lb per linear foot of wall.

$$M_c = \frac{1}{6}(20.0 + 1.62 + 6.00)^2(20.0 + 3.15) - \frac{20.0^3}{6}$$

$$= 1590\ \text{ft-ft/ft}$$

$$= 102{,}000\ \text{ft-lb per linear ft of wall};$$

$$l_c = \frac{102{,}000}{7620} = 13.4\ \text{ft above bottom};$$

$$R_t = \frac{1}{2}(20.0 + 1.62 - 6.00)(20.0 - 3.15) - \frac{20.0^2}{2}$$

$$= -68.5\ \text{ft/ft}$$

$$= -4380\ \text{lb per linear ft of wall};$$

$$M_t = \frac{1}{6}(20.0 + 1.62 - 6.00)^2(20.0 - 3.15) - \frac{20.0^3}{6}$$

$$= -650\ \text{ft-ft/ft}$$

$$= -41{,}500\ \text{ft-lb per linear ft of wall};$$

$$l_t = \frac{41{,}500}{4380} = 9.5\ \text{ft above the bottom.}$$

The minus signs indicate that R_t and M_t are toward the sea.

For deep water waves ($L/d > 2$), a second-order effect becomes important. The amplitude and phase of the second harmonic are such that both the maximum resultant force and the maximum moment occur when the wave trough is at the wall, and the direction of the force and moment is seaward; this has been shown both theoretically and experimentally (Carr, 1953). This might explain some of the observations of vertical wall breakwaters overturning seaward (Minikin, 1950).

11. SLOPING, CURVED, AND STEPPED BARRIERS

Because of space limitations, it is not possible to discuss the forces exerted by waves on the sloping, curved, and stepped type of barriers. Carr (1953; 1954a) has made some studies which are of use to the design engineer.

12. BREAKING WAVE FORCES

There have been very few attempts to correlate with measured wave dimensions the high pressure known to exist at times when waves break against certain bodies. Molitor (1934) published a suggested solution of this problem, using a semitheoretical approach and making use of the observations of Gaillard (1904) taken on Lake Superior and of his own measurements taken with a spring dynamometer at Toronto in 1915. These measurements were made with devices which were not able to record faithfully the impact forces. They would record the long period pressure of relatively low intensity associated with the reflection of the bulk of the wave energy but not the impact pressure of great intensity but of very short time intervals of the order of 0.001 to 0.01 sec for the more intense shocks.

The first work done using modern recording equipment was by Bagnold (1939) who studied the basic physics of the problem using waves breaking against a vertical wall in a laboratory tank. It was found that impact forces rarely occurred unless the timing of the wave was exact and that, even when waves broke right at the wall, impact occurred only under certain conditions. This finding, verified by other investigators (Morison, 1948; Denny, 1951), is one of great importance. It explains the rarity of the impact loadings. In nature, waves vary considerably and Bagnold states

Fig. 11.30. Wave breaks before reaching wall (after Bagnold, 1939)

that some French measurements at Dieppe (Rouvill, Besson and Petry, 1938) showed that only 2 per cent of the waves that struck the measuring device produced appreciable pressure. The maximum value recorded was 100 lb/sq in. In considering small percentages of occurrences in regard to ocean waves, however, it must be remembered that several million waves act on a structure each year.

Bagnold's experiments were conducted in a wave tank with several sections of glass through which the action of waves breaking against a vertical concrete wall could be observed. Solitary waves were created by a complex wave generator which was a combination of the piston type and the rigid flap type. The waves traveled over a horizontal bottom until they neared the concrete wall and then they traveled over a sloping bottom which was designed to cause the waves to peak up and to break against the vertical wall.

Bagnold observed that there were three conditions of breaking waves and that impact forces occurred for only one of these conditions. The three conditions were (1) the wave broke well before reaching the wall (Fig. 11.30). (2) The wave broke just before reaching the wall in which case the crest jet curled down a little before striking the wall and entrapped a large pocket of air (Fig. 11.31) which was compressed, cushioning the blow, and then escaped vertically, dragging water with it and forming the plume that is often observed in waves breaking on certain seacoast structures. (3) The wave broke just at the wall, trapping a thin lens of air (Fig. 11.32). Impact occurred when the wave broke as described in type 3. It can be seen in Fig. 11.31 that the water level at the wall when the crest jet hits the wall is below the general water surface in front of the crests and behind

the crest just before the crest jet hits, whereas in the type 3 break, the water level is above the general level. The rate of rise of the water level at the wall is very rapid and if the "break" occurs a fraction of a second too late, the water level has reached the crest level and no break really occurs, rather the wave is completely reflected. It was found that the wave-induced pressures were negligible, from the standpoint of impact, when the thickness of the air pocket exceeded half its height and that the forces increased in intensity with decreasing thickness of the air pocket. Bagnold concluded that the phenomenon was not a water hammer effect and that the energy of the impact was apparently stored in the compression of the air cushion rather than in the compression of the water.

After considerable experimental work, Bagnold was able to generate a type of wave such that an impact load would occur every time a wave hit the wall. The maximum pressure varied considerably, however. The frequency of pressures of different intensities was studied in more detail by Denny (1951) and his results are shown in Fig. 11.33. Bagnold and Denny both felt that these variations were due to small irregularities ("parasitic wavelets") on the primary waves which caused variations in the shape of the air picket. The effect of waves in disturbed water can be seen clearly in Fig. 11.33, where the most frequent ratio of shock pressure to wave height was about twenty-four for waves 7 and 14 in. high in disturbed water, but was about forty-three for 12-in. waves in calm water; this being a factor of nearly 2. The irregularities were of the order of one-tenth the main wave height. If this is the case, it is easy to understand the low number of impact pressures measured by the French at Dieppe, as ocean waves contain many irregularities.

Ross (1954) measured the forces due to periodic waves. In his experiment, the wave machine was started at a precise phase to cause the first wave to break on a vertical wall mounted on a sloping beach. The water depth was adjusted so that the next several waves would all break at the wall. By the time some energy had reflected from the wall to the wave generator and back again, the waves were somewhat irregular and it was

Fig. 11.31. Wave breaks at wall, but a little too early, and crest jet curls down (after Denny, 1951)

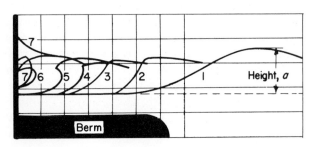

Fig. 11.32. Wave breaks fully against vertical wall (after Denny, 1951)

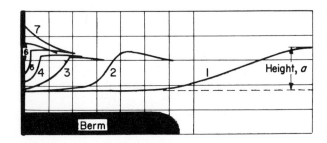

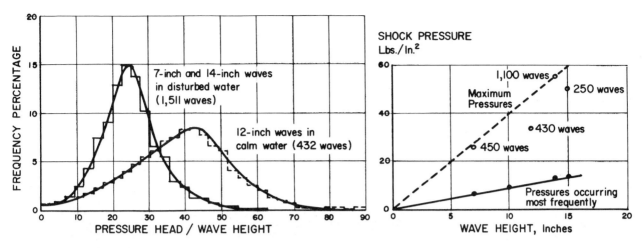

Fig. 11.33. (a) Frequency of impact pressures (b) Variation of shock pressure with wave height for similar waves in disturbed water (after Denny, 1951)

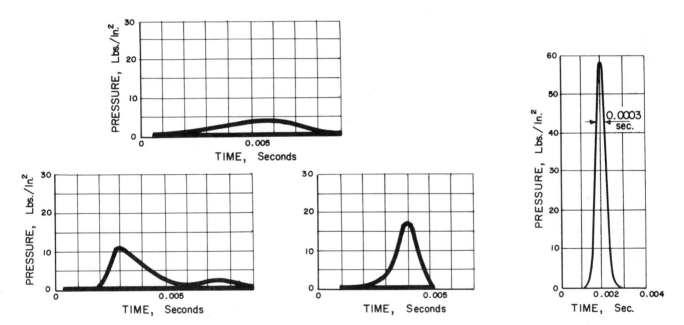

Fig. 11.34. Sample model wave pressure impulse records (after Bagnold, 1939)

found that most of the waves broke too early or too late to cause much pressure. Sometimes, however, very high pressures were recorded.

Both Bagnold and Denny found that, for the same waves, the lower the peak load, the longer the duration of the impact load (Fig. 11.34). The area under the pressure-time curves tended to approach but never exceed a definite maximum value. This was well illustrated by one of Denny's figures which showed a definite skewed distribution when the ratio of shock impulse per unit area to a measure of wave momentum was plotted with respect to frequency of occurrence (Fig. 11.35).

Bagnold and Denny both made pressure pick-up traverses of the face of the wall to determine the shape of

the pressure curve. This was found to vary from wave to wave. Certain statistical distributions are shown in Fig. 11.36. It can be seen that the impact force occurs only within the limits of the location of the flat air pocket.

In another set of experiments, Denny was able to measure directly the total impulse exerted by waves on the wall by use of a massive concrete wall horizontal spring system. Some sample records are shown in Fig. 11.37. It can be seen that, when a wave doesn't break but is merely reflected, there results a gradual movement of the wall. A wave that breaks against the wall without causing shock results in a force trace that has small oscillations with a mean line through them similar to the dashed line in Fig. 11.37. When a light shock load

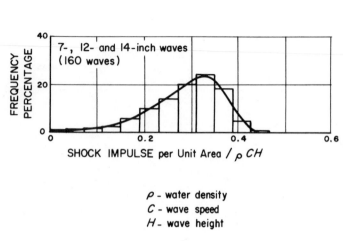

Fig. 11.35. Frequency of shock impulse (after Denny, 1951)

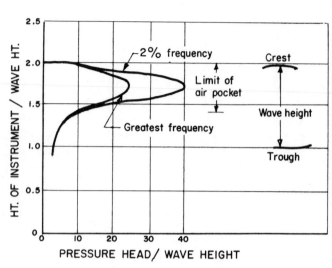

Fig. 11.36. Pressure distribution on wall with respect to wave crest and trough for wave 14 inches high in disturbed water (from Denny, 1951)

Fig. 11.37. Movement of heavy sprung wall caused by a wave 12 inches high in calm water (from Denny, 1951)

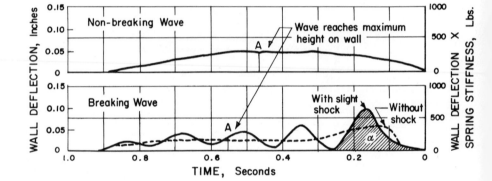

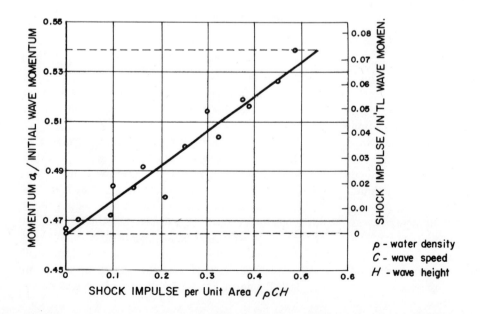

Fig. 11.38. Relationship between momentum and time interval of shock pressure (after Denny, 1951)

occurs, the curve consists of a series of large oscillations (solid line in Fig. 11.37). The shaded area, α, of the first impulse represents the momentum transferred to the wall from the water over that period of time. It was found that the area α increased linearly with increase in the time integral of shock pressure measured on the pressure recorder (Fig. 11.38). For waves that broke without shock, the momentum transferred to the walls was a constant 0.466 of the initial wave momentum (the lower dotted line in Fig. 11.38).

Some calculations of momentum transfer for reflected and breaking waves were made by Denny. The forward momentum per foot of wave crest of the waves tested was approximately $3.2\,\rho H^2 C$, where ρ is the water density, H is the height of the wave above the still water level, and C is the wave speed (solitary wave).

Denny concluded that for breaking waves (considering them to be represented by the highest solitary wave) the forward momentum per foot of wave crest, U_b, is $4.5\,\rho H^2 C$, where $C = \sqrt{gH(1 + 1/0.78)}$, and that the shock impulse, I, corresponding to 0.07 of the wave momentum is given by $\frac{1}{2}\rho g^{1/2} H^{5/2}$ per foot of wave crest. This latter figure refers to the maximum value obtained during the tests (Fig.11.38). Carr (1954) found a maximum ratio of shock impulse to initial wave momentum of about the same order (maximum of about

0.11) for waves breaking against a vertical wall. For waves breaking on a 30-degree sloping barrier, Carr found shock impulses more infrequent and with a maximum I/U_b of about 0.02.

Carr (1954b) made laboratory studies and presented data in a form that permits the calculation of the forces by semiempirical means. Data were obtained for periodic waves with a large range of heights and periods for vertical bulkheads on beaches of 1:3, 1:10, and 1:30 slopes, and a 30-degree bulkhead on a beach with a 1:10 slope. An example of the type of data obtained can be seen in Fig. 11.39 where the solid lines form the approximate envelope of the data. The experimental records varied considerably: some appeared to be "typical"; others appeared to be "exceptional." Both types are represented in the figure.

Using the concept developed by Bagnold, together with the concept that a breaker can be represented with a fair degree of accuracy by the theory of a solitary wave of maximum steepness, Carr developed the following relationship for the momentum, U_b, of a breaking solitary wave.

$$U_b = 55.5 H_b^{5/2}, \qquad (11.45)$$

where H_b is the breaker height in feet. For the relationship between H_b and the unrefracted deep water wave

Fig. 11.39. Wave force data, vertical barrier, 1: 30 slope (after Carr, 1954b)

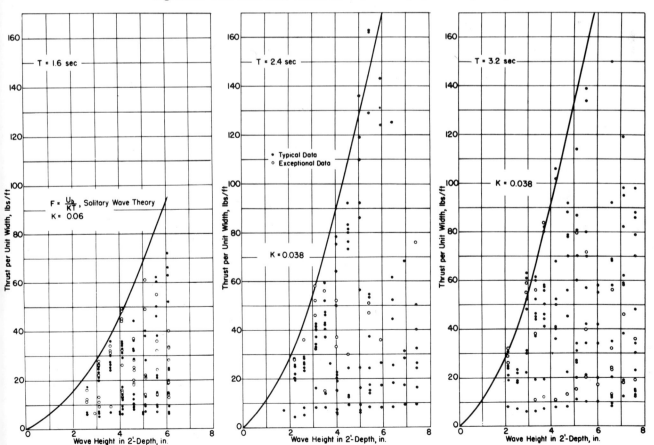

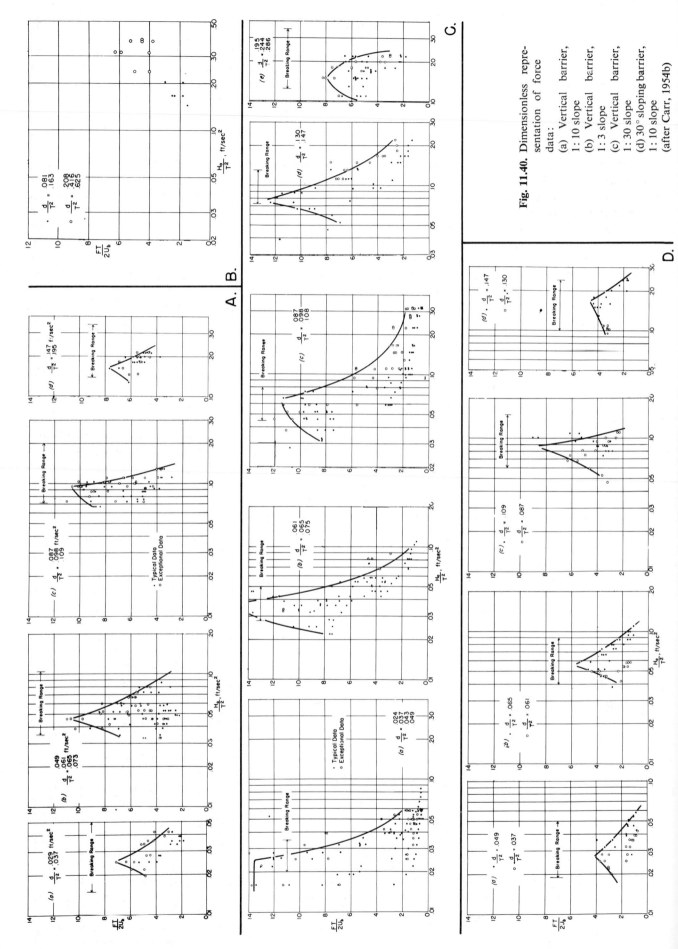

Fig. 11.40. Dimensionless representation of force data:

(a) Vertical barrier, 1:10 slope

(b) Vertical barrier, 1:3 slope

(c) Vertical barrier, 1:30 slope

(d) 30° sloping barrier, 1:10 slope

(after Carr, 1954b)

284

height H_o, Carr used

$$\frac{H_b}{H_o} = \frac{1}{33.3\sqrt[3]{H_o/L_o}}. \qquad (11.46)$$

Carr then plotted the empirical results as $FT/2U_b$ versus H_o/T^2 with d/T^2 as a parameter, where F is the maximum effective force (which would be the impact force if it occurred) and d is the water depth (below Still Water Level) at the toe of the bulkhead. These data and their envelopes are shown in Fig. 11.40.

Carr also measured the total moments on the bulkheads. Rather than present the results as moments, the ratios of the elevation of the center of the resultant force above the toe of bulkhead to the deep water wave height, c.p., were given as functions of H_o/T^2 and d/T^2 (Fig. 11.41).

As an example of the use of these curves consider a vertical bulkhead located in 10.0 ft of water on a 1:10 sloping beach subject to waves 8.0 ft high in deep water and 12.0 sec period.

$$\frac{H_0}{L_0} = \frac{8.0}{738} = 0.0108;$$

$$H_b = 8.0\frac{1}{3.3\sqrt[3]{0.0108}} = 11.0 \text{ ft};$$

$$U_b = 55.5 \times 11.0^{5/2} = 22,200 \text{ lb-sec/ft (fresh water)};$$

$$\frac{d}{T^2} = \frac{10.0}{12.0^2} = 0.0694; \quad \frac{H_0}{T^2} = \frac{8.0}{12.0^2} = 0.0556.$$

From Fig. 11.40,

$$\frac{FT}{2U_b} = 8.$$

$$F = 2 \times 22,200 \times \frac{8}{12.0} = 29,500 \text{ lb/ft (fresh water)}.$$

From Fig. 11.41,

$$\frac{\text{c.p.}}{H_0} = 1.6;$$

c.p. $= 1.6 \times 8.0 = 12.8$ ft above the toe of the breakwater;

$$M = F \times \text{c.p.} = 29,500 \times 12.8 = 318,000 \text{ ft-lb/ft of}$$ crest (fresh water).

In order to design for the general case, much additional data of this type must be obtained.

Considerable detail has been given on the impact (shock) forces exerted by breaking waves on plane walls because of their great importance if they were to occur against a structure. It is very unlikely that a circular cylindrical pile would be subject to wave-induced impact forces. In the first place, the waves in the open sea are very irregular and are covered with parasitic wavelets which would decrease the likelihood of impact. Secondly, it has been shown definitely that an air cushion of a particular shape must be formed. This air cushion can be formed by a wave moving over a sloping bottom, but it would be nearly impossible to form it in the open ocean. Thirdly, if such an air pocket did tend to form, the circular cylindrical shape of the pile would tend to cause the air to move around it rather than act as a cushion. In support of this, measurements of wave forces against a circular pile 2 ft in diameter plus observations of waves acting against a circular pile 5 ft in diameter (Wiegel, Beebe, and Moon, 1957) showed that this type of impact did not occur. During one series of tests, the test section was placed so that it would be in the air when the wave trough passed and submerged when the wave crest passed. Waves up to 15 ft in height were observed to cause only normal forces.

Bagnold (1939) and Morison (1948) comment that when impact occurs a distinct sound is heard, a sharp crack. Wiegel, Beebe, and Moon (1957) mention that this was heard when the waves acted against I beams in a pier where air was trapped in the web, but that it was not heard when waves acted against a circular pile.

Studies have been made of the forces exerted on vertical circular piles subjected to breaking waves (Morison, Johnson, and O'Brien, 1954; Hall, 1958). Hall (1958) developed laboratory equipment to study the force exerted by waves breaking against a circular pile, the wave being forced to break by a beach with a 1:10 slope. The model pile extended from the bottom through the surface and was placed at a series of positions with respect to the location where the wave broke. It was found that the maximum force and maximum moment about the bottom occurred very close to the wave crest, within about one frame of a motion picture exposed at 32 frames per second (Wiegel and Skjei, 1958). Sample records of the force at the top and bottom reaction points of the pile are shown in Fig. 11.42 for waves breaking seaward, at, and shoreward of the pile. The data are presented in Fig. 11.43 in terms of $F/DH_B^2\gamma$ versus d_p/H_b, where F is the maximum total force, D is the pile diameter, H_B is the breaker height, γ is the specific weight of water and d_p is the still water depth at the pile. An example of the data obtained for one wave condition is shown in Fig. 11.42(a) to illustrate the scatter of the data. In Fig. 11.42(b) are shown the mean curves for all of the data. It was also found that the force and moment were functions of the wave steepness, with the steeper the wave, the lower the force and the moment.

The prototype tests were made with a pile $3\frac{1}{2}$ in. in diameter (Morison, Johnson, and O'Brien, 1954). Some of the results of these tests are given in Fig. 11.44.

The prediction of forces exerted by breaking waves should include the theoretical velocity and acceleration fields of a breaking wave. These are not known. Rather, attempts have been made to "force fit" the solitary wave theory to the problem, although use of the cnoidal wave theory would be a better approximation. The

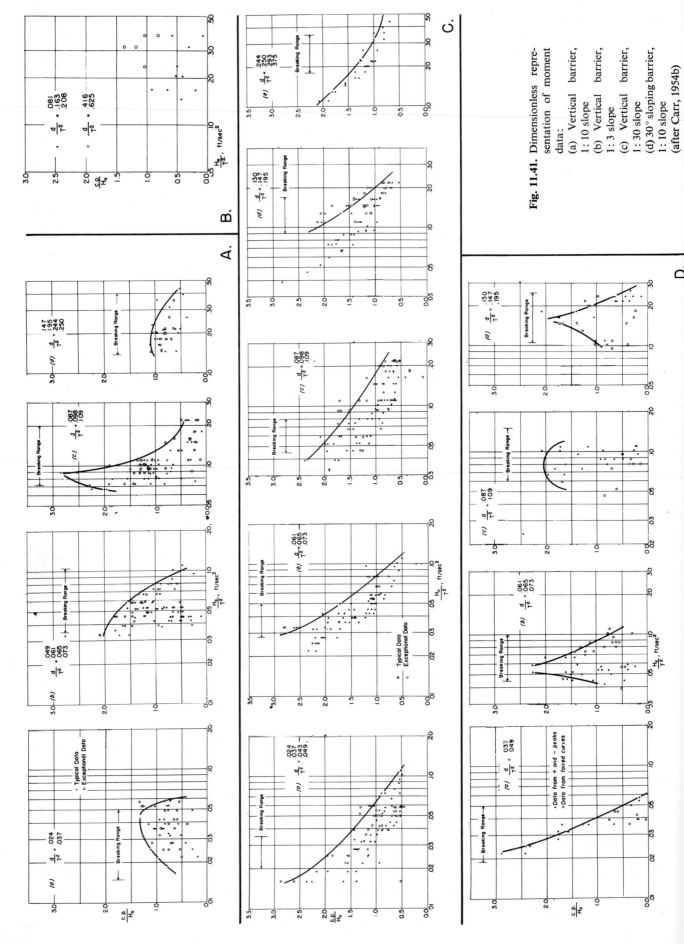

Fig. 11.41. Dimensionless representation of moment data:

(a) Vertical barrier, 1:10 slope
(b) Vertical barrier, 1:3 slope
(c) Vertical barrier, 1:30 slope
(d) 30° sloping barrier, 1:10 slope

(after Carr, 1954b)

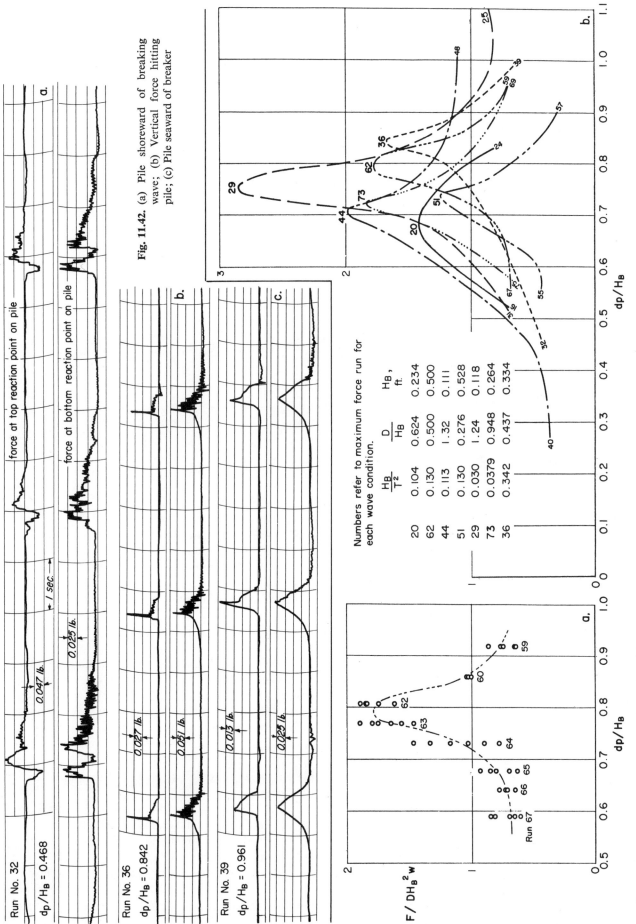

force at top reaction point on pile

force at bottom reaction point on pile

← 1 sec. →

0.047 lb.

0.025 lb.

Run No. 32
$d_p/H_B = 0.468$

a.

Run No. 36
$d_p/H_B = 0.842$

0.027 lb.

0.051 lb.

b.

Run No. 39
$d_p/H_B = 0.961$

0.013 lb.

0.025 lb.

c.

Fig. 11.42. (a) Pile shoreward of breaking wave; (b) Vertical force hitting pile; (c) Pile seaward of breaker

Numbers refer to maximum force run for each wave condition.

	$\dfrac{H_B}{T^2}$	$\dfrac{D}{H_B}$	H_B, ft.
20	0.104	0.624	0.234
62	0.130	0.500	0.500
44	0.113	1.32	0.111
51	0.130	0.276	0.528
29	0.030	1.24	0.118
73	0.0379	0.948	0.264
36	0.342	0.437	0.334

b.

d_p/H_B

a.

$F/DH_B{}^2 w$

Run 67

d_p/H_B

Fig. 11.43. (a) Maximum force, vertical pile position; (b) Summary of forces for different wave conditions versus pile position (after Hall, 1958)

287

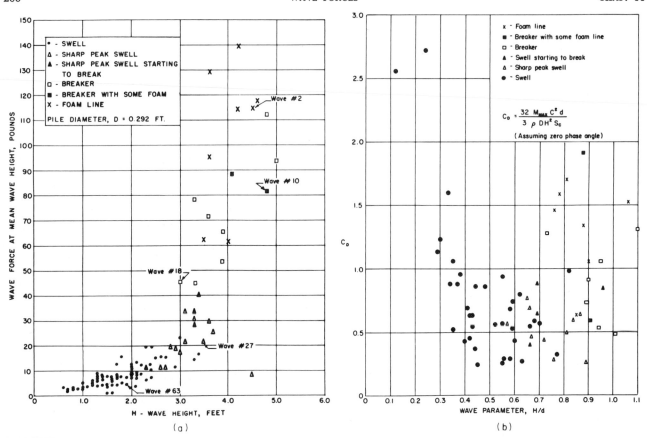

Fig. 11.44. (a) Wave force at mean wave height as a function of wave height; (b) Coefficient of drag computed from field tests on a circular pile (after Morison, O'Brien, and Johnson, 1954)

reliability of the solitary wave theory as an approximation to a breaker has been discussed in detail in a previous chapter. From the standpoint of wave forces, the important conclusions are that the solitary wave theory predicts the velocity field in the vicinity of the breaker crest with an acceptable accuracy, but that it does not predict the accelerations with an acceptable accuracy (Wiegel and Beebe, 1956; Wiegel and Skjei, 1958). As a result, it appeared that a modified solitary wave theory should be useful to the design engineer, provided that the maximum total moment on a pile occurred when a breaker crest passed a pile because of the small effect of water particle accelerations under the crest. Hall's work (1958) showed this to a certain extent, and a study by Wiegel and Skjei (1958) using the kinematics of breakers obtained in a laboratory study (Iversen, 1952) also showed it. In this study, the force fields on circular piles of different diameters were computed from the measured breaker kinematics and presented as shown in Fig. 11.45. The total moments were then computed about the bottom for the pile in various positions with respect to the wave crest. The total moments were also computed using the modified solitary wave theory. The results were quite close except for one run, which was a special case. In this run a "foam line" hit the pile, thus the actual breaker height was

quite a bit larger than the height of the "foam line" used in the calculation. The field measurements of Morison, O'Brien, and Johnson (1954) showed the same thing, i.e., the forces exerted on a pile by a "foam line" of a certain height were considerably higher than the forces for a breaking wave of the same height.

13. RUBBLE MOUND BREAKWATER

The bodies described so far in this chapter have been impervious, but this is not true of one of the most important classes of structures, the rubble mound breakwater (Fig. 11.46). This class is even more important than might appear at first because many so-called vertical wall breakwaters are really composite in that they are founded on rubble mounds that have been placed on a bottom that is poor from a foundation standpoint.

The rubble mound breakwater has one very important advantage in that failures are usually partial and can be repaired by simply dumping more rock during periods of low waves (Gesler, Eaton, and Hall, 1953). A study of the various engineering journals leads to the conclusion that, historically, rubble mound breakwaters have been underdesigned. Time after time, breakwaters have failed when subjected to particularly severe

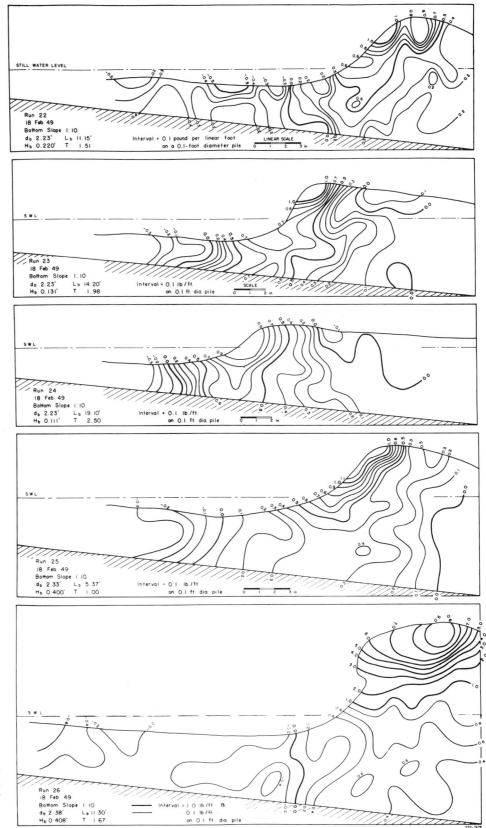

Fig. 11.45. Force contours, 0.10-ft diameter pile in breaking waves (after Wiegel and Skjei, 1958)

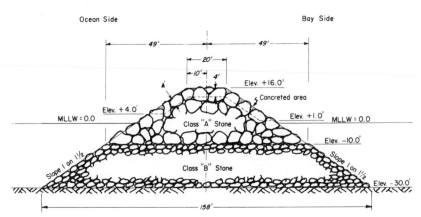

Fig. 11.46. Example of rubble mound breakwater. Morro Bay, California (after Beach Erosion Board, 1953)

wave conditions. Sometimes the underdesign occurs because heavy enough rock was not available (Johnson and Weymouth, 1956); in most instances the breakwater was subject to waves more severe than was judged likely to occur within a reasonable time; in other words, it was judged better engineering practice to build the structure less strong at lower initial expense and to meet the maintenance expense when it should occur. Because of the change of costs since World War II, design criteria in the United States now use no-damage data. The cost of either keeping a full-time maintenance crew or mobilizing plant for breakwater repairs is too high.

The flow phenomenon associated with rubble mound breakwaters is extremely complicated. Because of the slope of the breakwaters and their locations, waves may break before hitting the structure, break on the structure, or reflect from the structure without breaking, depending upon the wave steepness, the tide stage, breakwater slope, and the depth of water in which the structure is founded. The problem is not amenable to rigorous analysis at the present time. Several formulas are used, some of which are backed up by a considerable body of experimental studies. Iribarren (1948) studied the failure of several prototype breakwaters and developed a formula to relate the weight of cap rock to breakwater slope, wave height, specific weight of the rock, and the coefficient of sliding friction of the rock. His analysis was based upon the concept of a wave breaking on a breakwater and acting as a jet, hitting the breakwater surface in a normal direction. It is interesting to note that Hedar (1953) arrived at the same formula by considering that the water motion was parallel to the face of the breakwater. Svee's (1962) theoretical and laboratory work confirms Hedar's approach. Hudson (1953) used the reasoning of Iribarren to present the formula in more general terms. The equation is

$$W = \frac{K' \gamma_f^3 \gamma_r H^3 \epsilon^3}{(\epsilon \cos \alpha - \sin \alpha)^3 (\gamma_r - \gamma_f)^3}, \qquad (11.47)$$

where W is the weight of the cap stone (pounds); K' is an empirically determined dimensionless coefficient;

γ_f is the specific weight of the fluid, (lb/cu ft); γ_r is the specific weight of the stone (lb/cu ft), ϵ is the effective coefficient of friction, rock on rock (ϵ is 1.05 for quarry rock); H is the height of the wave at the structure location were no structure present (feet), and α is the slope the rubble mound makes with the horizontal (degrees). Hudson showed that this was also the same equation proposed by Epstein and Tyrell (1949) provided that their R_t be taken as $K' \epsilon^3 \gamma_f / \cos^3 \alpha$.

The coefficient K', or its equivalent, was obtained for two prototype structures by Iribarren by measuring the final slope of the structures after the rocks had been rearranged by large waves. The coefficients obtained by Iribarren in terms of Eq. 11.47 were $K' = 0.015$ and 0.019 (Hudson, 1953). Laboratory tests have been made to determine the relationships among the variables for incipient failure (where the first rocks moved out of their positions and down the slope) and K' was found to be a function of breakwater slope and the relative water depth, d/L (U.S. Army, Corps of Engineers, Waterways Experiment Station, 1953). Some of these data are shown in Figs. 11.47 and 11.48. These tests were for nonbreaking waves.

In this same report, results of the effect of water of several specific gravities were presented (the different specific gravities being obtained by adding calcium chloride to the water), the values being 1.08, 1.17, and 1.27. These tests showed the general validity of the form in which the specific gravity enters the equation. It can be seen in Eq. 11.47 that the rock and water densities enter as $\gamma_f^3 \gamma_r / (\gamma_r - \gamma_f)^3 = \gamma_r / (S_r - 1)^3$ where S_r is the specific gravity of the rock relative to the fluid in which it is immersed. This criterion was used in laboratory tests by Barbe and Beaudevin (1953) and found to be approximately correct, but not completely so.

In studying the way rocks were moved by waves in the laboratory models of rubble mound breakwaters, Barbe and Beaudevin observed that they were lifted out of the surface layer of rocks by the wave motion and then they rolled, rather than slid, down the slope.

Another observation of a model study of great impor-

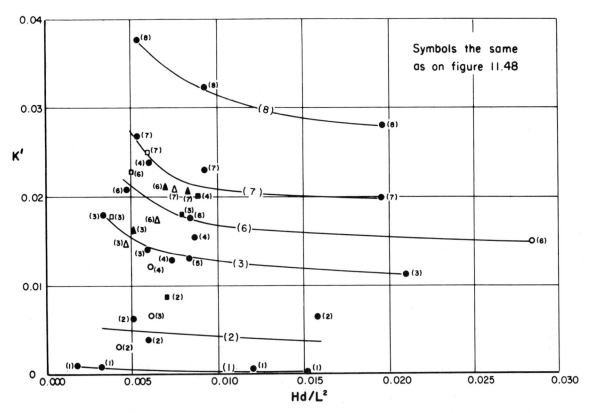

Fig. 11.47. Variations of K' with α and Hd/L^2. Design waves for no-damage criterion were used for computing K'; angle of incidence of wave attack = 90 degrees (after Waterways Experiment Station, 1953)

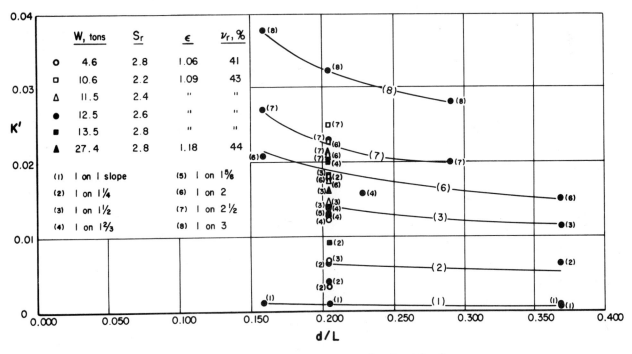

Iribarren's formula: $\quad W = \dfrac{K' \, \gamma_f^3 \, \gamma_r \, H^3 \, \epsilon^3}{(\epsilon \cos \alpha - \sin \alpha)^3 \, (\gamma_r - \gamma_f)^3}$

Fig. 11.48. Variations of K' with α and d/L. Design waves for no-damage criterion were used for computing K'; angle of incidence of wave attack = 90 degrees (after Waterways Experiment Station, 1953)

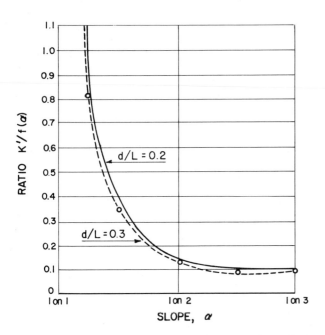

Fig. 11.49. $K'/f(\alpha)$ versus slope (cot α) for two values of d/L (after Kaplan, 1953)

where K is a constant for any one condition of rock and fluid and $f(\alpha) = (\epsilon \cos \alpha - \sin \alpha)^3$. Using the data of Hudson (1953), Kaplan plotted $K'/f(\alpha)$ versus slope (α) for two values of d/L, Fig. 11.49. The most noticeable characteristic is that $K'/f(\alpha)$ approaches a limiting value at about a slope of 1:2.5. This would indicate that, for a given design, wave height, and stone weight, a decrease in slope would not add to the stability of the breakwater.

There are circumstances where it has not proved economical to use natural rock either because it was impossible to obtain rock of sufficient weight or because the price was not acceptable. One way to overcome these problems is to use concrete castings in various shapes, such as rectangular blocks, tetrahedrons (Johnson and Weymouth, 1956), tetrapods (Danel, 1954), tribars (Engireering News-Record, 1958), and hollow tetrahedrons (Nagai, 1961). The advantage of shapes such as tetrapods, tribars, or hollow tetrahedrons is that they interlock when properly placed (Fig. 11.50). Because of this, a tetrapod of a given weight has been found to be more effective than concrete blocks or rocks of the same weight. A comparison of the results of several sets of laboratory experiments shows that concrete cubes with sharp edges are as effective as quarry stone 3 times as heavy, concrete cubes with rounded edges are only 2 times as effective, and tribars, 2.6 times as effective (Carvalho and Vera-Cruz, 1961). The value in prototype is undetermined, as how much breakage might occur over the year is unknown. In the use of Eq. 11.47, W can be plotted versus H with the breakwater slope as a parameter. This method has been used to show a comparison between natural stone and tetrapod, as obtained from model tests (Fig. 11.51).

Hudson (1959) has reanalyzed the problem of rubble mound breakwater design and used a modified version of Eq. 11.47.

$$\frac{\gamma_r^{1/3} H}{[(\gamma_r/\gamma_f) - 1] W^{1/3}} = f\left(\alpha, \frac{d}{L}, \frac{H}{L}, \Delta\right). \quad (11.49)$$

where Δ is the damage parameter, defined as the percentage of armor units displaced from the cover layer by wave action. As a result of a series of laboratory tests,

tance was made by Stucky and Bonnard (1937). The first models had armor stone which consisted of 54-ton (prototype) blocks, together with rocks of 0.5 to 2 tons (prototype). Waves lifted the smaller rocks out, causing the larger blocks to become unstable. In later experiments, the small rocks were eliminated and the structure was more stable. From this, it would appear unwise to allow any smaller rocks in the cap or armor section of a breakwater.

It is evident that, if Eq. 11.47 is correct, the smallest wave will move the largest rock when α is slightly greater than 45 degrees ($\epsilon \cong 1$, or a little greater). Model studies (Barbe and Beaudevin, 1953) as well as prototype experience show that this is not so. Another interesting characteristic of the equation has been pointed out by Kaplan (1953). Equation 11.47 can be put in the form

$$\frac{W}{KH^3} = \frac{K'}{f(\alpha)}, \quad (11.48)$$

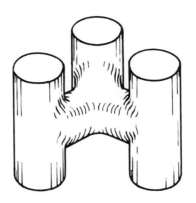 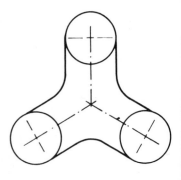

Fig. 11.50. The tri-bar, an interlocking breakwater structural material

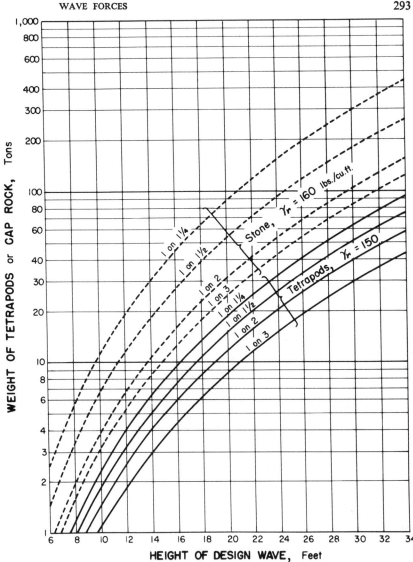

Fig. 11.51. Stone and tetrapod weight versus wave height, average K' values (after Johnson and Weymouth, 1956)

Iribarren's formula:

$$W = \frac{K' \, w \, \gamma_r^{\,3} \, \gamma_r \, \epsilon^{\,3} \, H^3}{(\mu \cos \alpha - \sin \alpha)^3 \, (\gamma_r - \gamma_f)^3}$$

he found that this could be expressed as

$$\frac{\gamma_r^{1/3} H}{[(\gamma_r/\gamma_f) - 1] W^{1/3}} = N_s = a \, (\cot \alpha)^{1/3}$$

or, letting $a^3 = \overline{K}_\Delta = N_s^3/\cot \alpha$.

$$W = \frac{\gamma_r H^3}{\overline{K}_\Delta [(\gamma_r/\gamma_f) - 1]^3 \cot \alpha}. \qquad (11.50)$$

Values of N_s for two materials are shown in Fig. 11.52. These data are for no wave damage to the breakwater, and when the waves neither overtop the structure nor break against it.

In designing a breakwater, it may be permissible to allow for possible minor failure. The relationship between partial failure and $\overline{K}_\Delta$ for quarry stone (model tests), as obtained by Hudson (1959), can be obtained from Fig. 11.53, using the relationship $\overline{K}_\Delta = N_s^3/\cot \alpha$.

A simplified method of obtaining the size of armor stone, tribars, quadripods, and tetrapods as a function

of wave height, specific weight of material, and slope of the face of the breakwater has been developed by Palmer (1962) and is given in Fig. 11.54. The values of $\overline{K}_\Delta$ shown are based upon both laboratory and prototype observations.

For supplementary design details, such as the effects of the number of layers of armor stone, method of placement of stone, porosity, etc., the reader is referred to the paper by Hudson (1959).

It has been generally considered that the action of waves breaking on a rubble mound breakwater is most severe in the region above the wave trough. For the design of rock size and breakwater slope below the trough of the design wave, it has been recommended that H in Eq. 11.47 be replaced by H', where (Iribarren and Nogales, 1951; U.S. Army, 1961),

$$H' = \frac{\pi H^2}{L_0 \sinh^2 2\pi d/L}. \qquad (11.51)$$

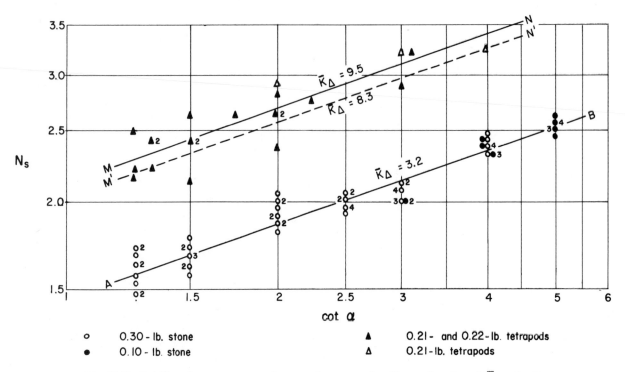

Fig. 11.52. Stability of quarry-stone and tetrapod armor units: N_s as a function of $\overline{K}_\Delta$ and α for the no-damage and no-overtopping criteria. Numbers beside data points indicate the number of tests when greater than one. (after Hudson, 1959)

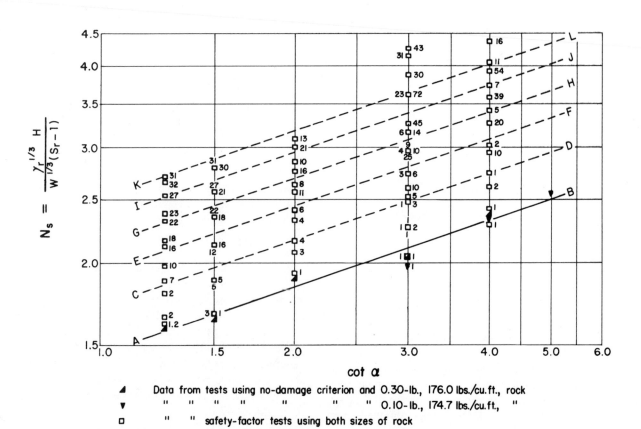

Fig. 11.53. Stability of quarry-stone cover layers: N_s as a function of Δ and α. Numbers beside data points are experimentally determined percentage values of displacement (Δ) of armor stone for corresponding values of N_s and cot α. (after Hudson, 1959)

γ_r = pounds per cubic foot

$\bar{K}$	144	146	148	150	152	154	156	158	160	162	164	166	168	170	172	176	180
1.5	119	112	106	100	95	90	85	81	77	73	70	67	64	61	58	53	49
2.0	89	84	80	75	71	67	64	61	58	55	52	50	47	45	43	40	37
2.5	71	67	64	60	57	54	51	48	46	44	42	40	38	36	35	32	29
3.0	60	56	53	50	47	45	42	40	38	37	35	33	31	30	29	27	25
3.5	51	48	46	43	40	38	36	34	33	31	30	28	27	26	25	23	21

Percentage of W_{150}

Conditions:

1. Trunk in deep water subject to nonbreaking waves.
2. Trunk in shallow water subject to breaking waves.
3. Head conical shaped subject to nonbreaking waves.
4. Head conical shaped subject to breaking waves.

$\bar{K}$ VALUES

Armor	Condition			
	1	2	3	4
Stone, rounded, 2 layers pell-mell	2.6	2.5	2.4	2.0
Stone, angular, 2 layers pell-mell	3.5	3.0	2.9	2.5
Stone, angular, 2 layers placed	5.3	3.0	2.9	2.5
Stone, keyes and fitted, 1 layer	7.8	6.7	5.5	4.4
Tetrapods, 2 layers pell-mell	8.3	8.0	6.5	4.5
Quadripods, 2 layers pell-mell	8.3	8.0	6.5	4.5
Tribars, 2 layers pell-mell	10.0	9.5	7.5	6.0
Tribars, 1 layer uniform	15.0	12.0	9.5	7.5

Notes: Design-factor relationships are from the Waterways Experiment Station formula

$$W = \frac{\gamma_r H^3}{\bar{K}(S_r - 1)^3 \cot \alpha}$$

where $\cot \alpha = 1.5$, $\gamma_r = 150$ lb/cu ft, unit weight of water = 64 lb/cu ft. For fresh water, unit weight = 62.4; a factor of 0.9 must also be applied. Subscript 150 indicates γ_r value of 150 lb/cu ft.

Problem: For the purpose of illustration the following values have been assumed: $H = 9'$, $\bar{K} = 2.9$, γ_r 160, and $\cot \alpha = 2$. The armor weight is desired.

Solution: Enter graph from left side at $\bar{K} = 2.9$, follow horizontal dashed line to point of intersection with diagonal line $H = 9$, drop vertically, read $W_{150} = 5.3$ tons. The value of 58 per cent is obtained from the upper right-hand table for $\cot \alpha = 2$ and $\gamma_r = 160$. *5.3 tons × 0.58 = 3.1 tons. If in fresh water 0.9 × 3.1 = 2.8 tons*

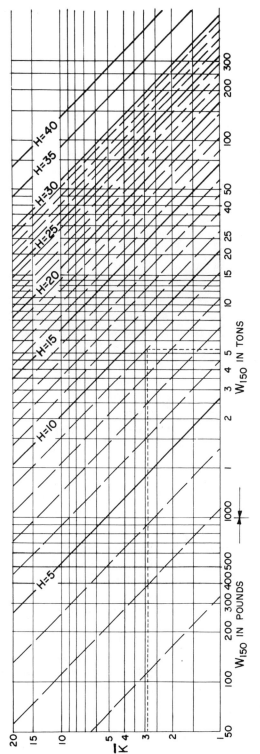

Fig. 11.54. Breakwater armor design-factor relationships for various conditions (from Palmer, 1962)

295

Palmer (1962) recommends that armor units be placed to a distance of approximately one wave height below the low water level. Detailed laboratory studies by Sigurdsson (1962) showed that under certain conditions the maximum wave force acting normal to the breakwater occurred more than one wave height below the still water level. Because of these two sets of observations, it appears that the armor stone should be used to at least one design wave height below the low water level, and preferably to almost one and a half design wave heights below the low water level.

The studies of Sigurdsson revealed three types of breakwater failures: sliding failure, lifting failure, and impact failure. It was stated that impact failure would occur if the breakwater was not able to provide sufficient reaction agaitns the high parallel forces associated with the impact. These forces would push or roll the stones over the top of the breakwater. This can be prevented by having a breakwater crown which has adequate width and height, or by capping the breakwater with concrete or asphalt.

REFERENCES

American Society of Civil Engineers, Wind forces on structures: final report of the Task Committee on Wind Forces of the Committee on Loads and Stress of the Structural Division, *Trans. ASCE*, **126**, Part II (1961), 1124–98.

Bacon, David L., and Elliott G. Reid, The resistance of spheres in wind tunnels and in air, U.S. NACA, *Ninth Annual Report*, Rept. No. 479 (1923), 471–87.

Bagnold, R. A., Interim report on wave-pressure research, *J. Inst. Civil Engrs* (June, 1939), 202–26.

Barbe, R., and C. Beaudevin, Experimental research on the stability of a rubble mound breakwater subjected to wave action, *La Houille Blanche*, **8**, 3 (1953), 346–59.

Blumberg, R., and A. M. Rigg, Hydrodynamic drag at super-critical Reynolds numbers, Paper presented at ASME meeting (Pet. Session), Los Angeles, June 14, 1961.

Brater, Ernest F., John S. McNown, and Leslie D. Stair, Wave forces on submerged structures, *J. Hyd. Div. ASCE*, Paper 1833, **84**, HY6, November, 1958.

Carr, John H., *Wave forces on plane barriers*, Calif. Inst. of Tech., Hydro. Lab., Contract NOy-12561, Rept. No. E-11.1, October, 1953. (Unpublished.)

————, *Wave forces on curved and stepped barriers*, Calif. Inst. of Tech., Hydro. Lab. Contract NOy-12561, (June, 1954a). (Unpublished.)

————, *Breaking wave forces on plane barriers*, Calif. Inst. of Tech., Hydro. Lab., Contract NOy-12561, Rept. No. E-11.3, November, 1954, (b). (Unpublished.)

Carvalho, Jose Joaquin Reis de, and Daniel Vera-Cruz, On the stability of rubble-mound breakwaters, *Proc. Seventh Conf. Coastal Eng.*, Berkeley, Calif.: The Engineering Foundation Council on Wave Research, 1961, pp. 633–58.

Crooke, R. C., *Re-analysis of existing wave force data on model piles*, U.S. Army, Corps of Engineers, Beach Erosion Board, Tech. Memo. No. 71, April, 1955.

Daily, James W., and Wilbur L. Hankey, Jr., *Resistance coefficients for accelerated flow through orifices*, Mass. Inst. Tech., Hydro. Lab., Tech. Rept. No. 10, October, 1953.

Danel, Pierre, Tetrapods, *Proc. Fourth Conf. Coastal Engineering*, Berkeley, Calif.: The Engineering Foundation Council on Wave Research, 1954, pp. 390–98.

Delany, N. K., and N. E. Sorensen, *Low speed drag of cylinders of various shapes*, U.S. NACA, TN3038, November 22, 1953.

Denny, D. F., Further experiments on wave pressures, *J. Inst. Civil Engrs.*, February, 1951, 330–45.

Dryden, H. L. and I. H. Abbott, *The design of low-turbulence wind tunnels*, NACA Tech. Note No. 1755, November, 1948.

Elliott, John H., *Interim report*, Calif. Inst. Tech., Hydro. Lab., Contract NOy-12561, July, 1953. (Unpublished.)

Engineering News-Record, Tetrapods challenged by new tribar shape, (July, 3, 1958), 36.

Epstein, Harris and F. C. Tyrell, Design of rubble-mound breakwaters, Section 2, Communication 4, *XVIIth International Cong. Navigation*, Lisbon, Portugal, 1949.

Fage, A. and J. H. Warsap, *The effects of turbulence and surface roughness on the drag of a circular cylinder*, Aeronautical Research Committee (Air Ministry, Great Britain), R and M No. 1283, October, 1929.

Folsom, R. G., Measurement of ocean waves, *Trans. Amer. Geophys. Union*, **30**, 5 (1949), 691–99.

Fung, Y. C., Fluctuating lift and drag acting on a cylinder in a flow at supercritical Reynolds numbers, *J. Aerospace Sciences*, **27**, 11, November, 1960, 801–14.

Gaillard, D. D., *Wave action in relation to engineering structures*, reprinted 1935, Ft. Belvoir, Va.: U.S. Army, Corps of Engineers, The Engineer School, 1904.

Gerhardt, J. R., K. H. Jehn, and I. Katz, A comparison of step-pressure, and continuous-wire-gage wave recordings in the Golden Gate Channel, *Trans. Amer. Geophys. Union*, **36**, 2 (April, 1955), 235–50.

Gesler, Earl E., Richard O. Eaton, and Jay V. Hall, Jr., Breakwater design, construction and maintenance in the United States of America, *XVIII Intern. Navigation Congress*, Rome, 1953, Section II, Question I, 1953, 67–90.

Hall, M. A., *Laboratory study of breaking wave forces on piles*, Tech. Memo. No. 106, Beach Erosion Board, Corps of Engineers, U.S. Army, August, 1958.

Hedar, Per Anders, Design of rock-fill breakwaters, *Proc. Minnesota Inter. Hydraulics Conv.*, IAHR and ASCE, 1953, pp. 241–60.

Hellstrom, B., and L. Rundgren, *Model tests on Olands Sodra Grund Lighthouse*, R. Inst. Tech., Stockholm, Inst. Hyd. Bull. No. 39, 1954.

Hudson, Robert Y., Wave forces on breakwaters, Engineering Aspects of Water Waves: A Symposium, *Trans., ASCE*, **118** (1953), 653–74.

———, Laboratory investigation of rubble-mound breakwaters, *J. Waterways Harbors Div., ASCE*, **85**, WW3, Paper No. 2171 (September, 1959) 93–121.

Humphreys, John S., On a circular cylinder in a steady wind at transition Reynolds numbers, *J. Fluid Mechanics*, **9**, 4 (1960), 603–12.

Iribarren, Cavanilles, R., *A formula for the calculation of rock-fill dikes*, D. Heinrich, trans., Univ. Calif., Berkeley, Inst. Eng. Res. Tech. Rept. 3-295, 1948. (Original booklet published in July, 1938.)

———, and Casto Nogales y Olano, Generalization of the formula for calculation of rock-fill dikes and verification of its coefficients, trans. *Bulletin*, U.S. Army Corps of Engineers, Beach Erosion Board, **5**, 1 (January, 1951), 4–24.

Iversen, H. W., and R. Balent, A correlating modulus for fluid resistance in accelerated motion, *J. Applied Physics*, **22**, 3 (March, 1951), 324–28.

———, *Laboratory study of breakers*, U.S. National Bureau of Standards, Circular No. 521, 1952, pp. 9–31.

Johnson, Reuben J., and Olin F. Weymouth, Alternatives to stone in breakwater construction, *J. Waterways Harbors Div., ASCE*, **82**, WW4, Paper 1059 (September, 1956).

Kaplan, Kenneth, Discussion of wave forces on breakwaters; Engineering aspects of water waves: A Symposium, *Trans. ASCE*, **118** (1953), 675–76.

Keim, S. R., Fluid resistance to cylinders in accelerated motion, *J. Hyd. Div. ASCE*, **72**, HY6, Paper No. 1113 (December, 1956).

Keulegan, Garbis H., and Lloyd H. Carpenter, Forces on cylinders and plates in an oscillating fluid, *J. Res. National Bur. Standards*, **60**, 5 (May, 1958), 423–40.

Laird, A. D. K., A model study of wave action on a cylindrical island, *Trans. Amer. Geophys. Union*, **36**, 2 (April, 1955), 279–85.

———, Water forces on flexible oscillating cylinders, *J. Waterways Harbors Div., Proc. ASCE*, **88**, WW3 (August, 1962), 125–37.

———, C. A. Johnson, and R. W. Walker, Water forces on accelerated cylinders, *J. Waterways Harbors Div., ASCE*, **85**, WW1, Paper 1982 (March, 1959), 99–119.

Lamb, Sir Horace, *Hydrodynamics*, 6th ed. New York: Dover Publications, Inc., 1945, p. 93.

Luneau, Jean, Sur l'effet d'inertie des sillages de corps se déplaçant dans un fluide d'un mouvement uniformement accéléré, *Comptes Rendus Académie des Sciences* (Paris) **227**, 17 (October 27, 1948), 823–24.

MacCamy, R. C., and R. A. Fuchs, *Wave forces on piles: a diffraction theory*, U.S. Army, Corps of Engineers, Beach Erosion Board, Tech. Memo. No. 69, December, 1954.

McNown, John S., Drag in unsteady flows, *IX Congrés International de Mécanique Appliquée, Actes*, Tome III (1957) 124–34.

———, and G. H. Keulegan, Vortex formation and resistance in periodic motion, *J. Eng. Mech. Div., ASCE*, **85**, EM1, Part 1, Paper 1894 (January, 1959), 1–6.

Miche, Robert, Mouvements ondulatoires des mers en profondeur constante ou décroissante, *Annales des Ponts et Chausseés* (1944), 25–78; 131–64; 270–92; 369–406.

Minikin, R. R., *Winds, waves and maritime structures*. London: Chas. Griffin and Co., Ltd., 1950.

Molitor, D. A., Wave pressures on seawalls and breakwaters, *Trans. ASCE*, **100** (1935), 984–1002.

Morison, J. R., *Wave pressures on a vertical wall*, Univ. Calif. IER, Tech. Rept. 3–298, 1948. (Unpublished.)

———, The design of piling, *Proc. First Conf. Coastal Engineering*, The Engineering Foundation Council on Wave Research, Berkeley, Calif.: 1951, pp. 254–58.

———, and R. C. Crooke, *The mechanics of deep water, shallow water and breaking waves*, Tech. Memo. No. 40, Beach Erosion Board, U.S. Army, Corps of Engineers March, 1953.

———, J. W. Johnson, and M. P. O'Brien, Experimental studies of forces on piles, *Proc. Fourth Conf. Coastal Council on Wave Research*, Berkeley, Calif.: The Engineering Foundation Council on Wave Research, 1954, pp. 340–70.

———, M. P. O'Brien, J. W. Johnson and S. A. Schaaf, The force exerted by surface waves on piles, *Petroleum Trans.*, **189**, TP 2846 (1950) 149–54.

Morse, Philip M., *Vibration and sound*. New York: McGraw-Hill, Inc., 1936.

Murtha, Joseph P., Virtual mass of partially submerged bodies, M. S. Thesis, Carnegie Inst. Tech., Dept. of Civil Eng., 1954.

Nagai, Shoshichiro, Experimental studies of specially shaped concrete blocks for absorbing wave energy, *Proc. Seventh Conf. Coastal Eng.*, Berkeley, Calif.: The Engineering Foundation Council on Wave Research, 1961, pp. 659–73.

O'Brien, M. P., and J. R. Morison, The forces exerted by waves on objects, *Trans. Amer. Geophys. Union*, **33**, 1 (February, 1952), 32–38.

Palmer, Robert Q., Discussion of Breakwaters in the Hawaiian Islands, *J. Waterways Harbors Div., Proc. ASCE*, **88**, WW3 (August, 1962), 161–69.

Price, Peter, Suppression of the fluid-induced vibration of circular cylinders, *J. Eng. Mech. Div., Proc. ASCE*, **82**, EM3, July 1956.

Reid, R. O., *Analysis of wave force experiments at Caplen, Texas*, Agricultural and Mechanical College of Texas, Dept. of Oceanography, Tech. Rept. No. 38–4, graphs, tables, January, 1956. (Unpublished.)

———, and C. L. Bretschneider, *The design wave in deep water or shallow water, storm tide, and forces on vertical piling and large submerged objects*, Agricultural and Mechanical College of Texas, Dept. of Oceanography, Tech. Rept. on Contract NOy–27474, DA-49-005-eng 18, and N 7onr-48704, February, 1954. (Unpublished.)

Riabouchinski, P., Sur la résistance des fluides, *Int. Congr. Math.*, Strasbourg, 1920, pp. 568–85.

Ross, Culbertson W., Shock pressure of breaking waves, *Proc. Fourth Conf. Coastal Engineering*, Berkeley, Calif.: The Engineering Foundation Council on Wave Research, 1954, pp. 323–32.

Rouville, M. A., P. Besson, and P. Pétry, État actual des études internationales sur les efforts dus aux lames, *Annales Ponts et Chaussées* (France) **108** (II), 1938, 5–113.

Schlichting, Hermann, *Boundary layer theory*, J. Kestin, Trans. London: Pergamon Press, 535 pp, 1955.

Sigurdsson, Gunnar, Wave forces on breakwater capstones, *J. Waterways Harbors Div., Proc. ASCE*, **88**, WW3 (August, 1962), 27–60.

Stucky, A., and D. Bonnard, Contribution to the experimental study of marine rock fill dikes, *Bull. Technique de la Suisse Romande*, August, 28, 1937.

Suquet, F. and A. Wallet, Basic experimental wave research, *Proc. Minn. Intern. Hydraulics Conv., IAHR ASCE* (August, 1953), 173–91.

Svee, Roald, Formulas for design of rubble-mound breakwaters, *J. Waterways Harbors Div., Proc. ASCE*, **88**, WW2 (May, 1962), 11–22.

Taylor, G. I., *The force acting on a body placed in a curved and converging stream of fluid*, Aero. Res. Comm. (Gt. Britain), R and M No. 1166, April, 1928.

U.S. Army, Corps of Engineers, Beach Erosion Board, *Shore protection planning and design*, Tech. Rept. No. 4, 1961.

———, Waterways Experiment Station, *Stability of rubble-mound breakwaters*, Tech. Memo. No. 2–365, June, 1953.

U.S. Senate, *Inquiry into the collapse of Texas Tower No. 4*, Hearings before the Preparedness Investigating Committee on Armed Services, United States Senate, 87th Cong. 1st Sess., 1961.

Wiegel, R. L. and K. E. Beebe, The design wave in shallow water, *J. Waterways Div., ASCE*, **82**, WW1, Paper 910, March, 1956.

———, K. E. Beebe, and James Moon, Ocean wave forces on circular cylindrical piles, *J. Hyd. Div., ASCE*, **83**, HY2, Paper 1199, April, 1957.

———, and R. E. Skjei, Breaking wave force predictions, *J. Waterways Harbor Div., ASCE*, **84**, WW2, Paper 1573, March, 1958.

Wiener, F. M., Sound diffraction by rigid spheres and circular cylinders, *J. Acoust. Soc. Amer.*, **19**, 3 (May, 1947), 444–51.

Wilson, Basil W., and Robert O. Reid, Discussion of wave force coefficients for offshore pipelines, *J. Waterways Harbors Div., Proc. ASCE*, **89**, WW1 (February, 1963), 61–65.

CHAPTER TWELVE

Tides and Sea Level Changes

1. INTRODUCTION

Tides have probably attracted man's attention since time immemorial. Those living along the shores of oceans made use of the low tides to collect shellfish to eat. The tides have always been of considerable importance in the navigation of ships, with tidal currents affecting the movement of ships to and from river ports, with ships moving upstream in the flood current and downstream with the ebb current, and the stage of the tide affecting the movement of ships over bars, as well as the unloading of ships alongside docks. Much is known empirically about the tides along the coasts; little is known empirically about the tides in the open ocean because of the lack of an adequate reference from which to measure the changing elevation of the water surface.

The word *tide* usually refers to the periodic rising and falling of the water that results from the gravitational attraction of the moon and sun acting on the rotating earth. Although the accompanying horizontal movement

of the water resulting from the same cause is sometimes called the tide also, this is usually designated as a *tidal current*. The elevations of tides will be considered in this chapter; the tidal currents will be covered briefly in Chapter 13.

In addition to the tides caused by the forces just mentioned there are changes in the water level due to meteorological conditions which are called meteorological tides, storm surges, storm floods, etc. Part of the change in water level is due to a coupled wave system caused by a low- or high-pressure region moving over relatively shallow water as described in Chapter 4; part of it is due to the piling up of water along the coast because of the stress of the wind on the water surface. Some results of the combination of both of these mechanisms have been given in Chapter 5.

There are seasonal and longer changes which occur in the level of the oceans with respect to land. Sometimes the cause is known, such as the land subsidence in the Long Beach Harbor area in California (Gilluly and

Grant, 1949), but usually it is not known whether the land is subsiding or the sea level rising (Marmer, 1952).

An extensive bibliography on tides, much of it annotated, has been prepared by the U.S. Army, Corps of Engineers (1954). This bibliography includes titles and abstracts of papers on estuarial problems as well.

2. ASTRONOMICAL TIDES

The forces which are of primary importance in the tides of the oceans are the gravitational forces of the moon and the sun, the centrifugal force due to the movement of the earth in its orbit, the Coriolis force due to the earth rotating about its axis, and the frictional force due to the movement of the water with respect to its boundaries. The gravitational forces of the other planets have been shown to be negligible. Although the moon's mass is much less than the mass of the sun, the moon is so much closer than the sun that its gravitational field is a greater tide-generating force than is the sun.

Consider the relative effect of two forces, the gravitational pull of the moon revolving around the earth (at the equator, for simplicity), and the centrifugal force of the earth moving in its orbit. The two sets of forces are shown in Fig. 12.1. The resultant force is normal to the earth along the line extending from the moon through the earth, and along a great circle through the poles. At all other points on the earth's surface there is a component tangential to the surface.

As in most geophysical phenomena, it is necessary to make a simplification in order to understand the phenomenon qualitatively. Consider a narrow channel with vertical walls extending around the equator, with the moon revolving around the earth above the equator. This system is similar to the diffused line pressure source described in Chapter 4, except that there is now a horizontal component, and the Coriolis force which causes

a transverse slope of the water in the channel. The controlling parameter is a Froude number based upon the speed at which the projection of the moon on the earth moves over the earth's surface, and $\sqrt{gd}$, where d is the water depth. A forced wave system will result, moving with the speed of the moon, but its amplitude and phase depend upon the Froude number.

If the Froude number were less than unity, a dispersive wave would result; if the Froude number were unity, a build-up of the wave to a large amplitude would result; and if the Froude number were greater than unity, a much smaller wave would result (Fig. 4.9). For Froude numbers less than unity, the wave is essentially in phase with the disturbance, which in this case would mean that the maximum tide level would occur when the moon was overhead or 12 lunar hours later when the moon was over the opposite side of the earth. When the Froude number is greater than unity, the phase goes through a 180-degree shift, and the forced wave would lag the moon by 180 degrees (6 hr and 18 hr, respectively). The phase shift occurs over a range of Froude numbers in the vicinity of unity, so that other phases between zero and 180 degrees occur. Inui (1936) has shown for Froude numbers greater than unity that not only does a phase shift occur, but that negative force (or negative pressure area, which is large with respect to the water depth) generates a decrease in the water level first, followed by an increase in the water level, rather than just an increase in the water level.

Suppose the channel under consideration to be filled with water 20,000 ft deep, so that $\sqrt{gd}$ is about 802 ft/sec. At the equator, the moon's projection moves at a speed of about 1470 ft/sec; hence $V/\sqrt{gd} = 1470/802 = 1.83$, and the tide wave is not in phase with the moon's transit. If the channel were at 60 degrees latitude, the speed would be about 735 ft/sec, and $V/\sqrt{gd} = 735/802 = 0.916$, and the tide wave would be in phase.

If a series of parallel tanks were considered, and their walls removed, flow would have to take place from regions of high to regions of low waves, with the resulting complications. The problem is further complicated by variations in the depths of the oceans and by the Coriolis force. The Coriolis force will be described in more detail in Chapter 13. The Coriolis force deflects fluid motion to the right in the Northern Hemisphere, and to the left in the Southern Hemisphere. Because of this, tides which are rotational in character must occur in the ocean. An example of such a large-scale system can be seen in Fig. 12.2, where the semidiurnal component (M_2) is shown for the oceans. Rotation occurs about the amphidromic point at about 45° W longitude and 45° N latitude. Lines of equal phase and lines of equal amplitude are given.

Another complication is due to the lateral boundaries of the oceans, the continents. The tide wave is reflected by the land mass, so that the simple model considered

Fig. 12.1. The magnitude and direction of the tide-generating force as the difference between attractive and centifugal forces at given points on the surface of the earth. Plain arrows: attractive force; striped arrows: centrifugal force; black arrows: tide-generating force (from Defant, 1953)

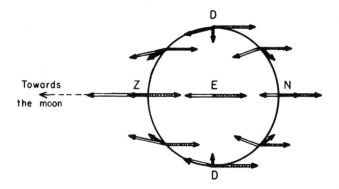

Towards the moon

GREENWICH TIMES

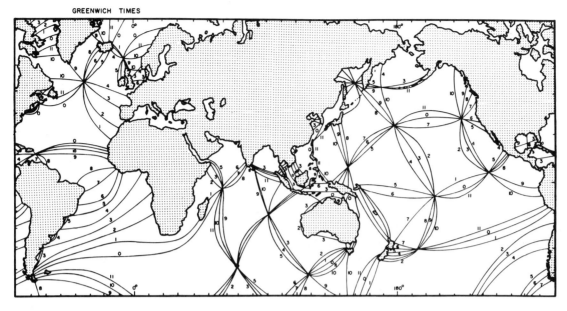

Fig. 12.2. Cotidal lines of the semidiurnal tide in the oceans (from Sterneck, 1920)

before would result in incorrect predictions. A study of some of the types of resonant problems described in Chapter 5 will suffice to show how complicated the tide motions must be in coastal regions. In addition, little is known on the dissipation of energy of the tides, and this information is needed if theoretical predictions are to be made.

Consider another case, that of a tide wave entering a rectangular channel from the open ocean (the North Sea can be considered to be an approximation to this). Because of the Coriolis force, the wave is higher to the right of direction advance in the Northern Hemisphere. For the case of the North Sea, the tide wave moving southward would be higher along the coast of Scotland and England than it would be along Scandinavia. Upon reflection from the Netherlands and Germany, it would travel northward and be higher along Scandinavia than along the British Isles. Because of this, it would have

the appearance of a wave traveling southward along the British Isles, swinging eastward, and then moving northward along Scandinavia. Such a wave is known as a *Kelvin wave*, and has been found to be important for storm surges as well as tides (see Chapter 5).

In addition to all this, the moon does not just simply revolve about the earth's equator, it moves in an ellipse; it has minor variations in its motion, and the effect of the sun's gravitational field must be considered. The dynamical theory of the tides was formulated by Laplace (1775), and special solutions have been obtained (see, for example, Proudman and Doodson, 1936; 1938). In 1960 at the Helsinki meeting of the International Association of Geodesy and Geophysics, Pekeris and Dishon reported considerable headway in solving the equations numerically for a real ocean.

Because of the enormous complications associated with the theory of tides, most of the useful information

Table 12.1. PRINCIPLE TIDAL COMPONENTS
(from Defant, 1953)

Component	Semidiurnal components				Diurnal components			Long period
	M_2	S_2	N_2	K_2	K_1	O_1	P_1	M_f
Period, solar hours	12.42	12.00	12.66	11.97	29.93	25.82	24.07	327.86
Amplitude relative to M_2	1.000	0.466	0.191	0.127	0.584	0.415	0.193	0.172
Cause	Main lunar	Main solar	Monthly variation in moon's distance	Changes in declination of sun and moon during orbital cycle	Solitary lunar	Main lunar diurnal	Main solar diurnal	Moon, fortnightly

has been obtained from measurements, using the simplified theories of tides as guides in the analysis of the measurements (Schureman, 1950; Defant, 1958; Doodson, 1958). The vertical movement of the tides at a point may be considered to be the linear superpositions of a series of harmonic terms, the periods of the components depending upon the periods of motion of the moon and the sun. These are known. The amplitude and phases of the components have been determined empirically for many coastal regions. The periods and relative importance of the principal components are given in Table 12.1. There are many more constituents of varying importance. Different agencies use different numbers of constituents in their analysis. The U.S. Coast and Geodetic Survey uses M_2, S_2, K_1, O_1, N_2,

M_4, and 31 other constituents in its tide prediction machine (Schureman, 1940).

Tides have been recorded in many areas for a great number of years and they have been analyzed to obtain the amplitude and phase of the tidal components. This information is then used to predict the astronomical tides in advance, and these are available in such publications as the tide tables published by the U.S. Coast and Geodetic Survey.

The oceans' tides have been classified as semidiurnal, mixed, and diurnal. Examples of the types are shown in Figs. 12.3 and 12.4. The type that occurs at a particular place depends upon the ratio $(K_1 + O_1)/(M_2 + S_2)$. When the ratio is of the order of 0.1 the tides are semidiurnal, when it is about unity they are mixed, but predominantly semidiurnal; when it is about 2, they are mixed but predominantly diurnal; and when it is 15 or so, they are diurnal (Defant, 1958). The semidiurnal tide has a cycle of approximately one-half a tidal day, whereas the diurnal tide has only one high water and one low water per day. Mixed tides are conspicuous by the presence of a large inequality in either the high or low water heights, with two high waters and two low waters usually occurring each tidal day. Most tides are of this type, and are closer to the semidiurnal end of the spectrum. The difference in height of the two high waters or of the two low waters of each day is called the *diurnal inequality*. Some other definitions connected with tides are shown in Fig. 12.4.

The high waters are much higher and the low waters much lower than usual for several days every 14.3 days and these are called *spring tides*, these being associated with full or new moons, i.e., when the sun, moon, and earth are either in opposition or conjunction. At the other extreme, each 14.3 days there are several days when the high waters are lower than usual, and the low waters are higher than usual, these being called *neap tides*. Twice a year there are extra-high spring tides, called *equinoctial* spring tides; these occur at the times of vernal and autumnal equinoxes.

In referring to tide heights, some datum is chosen, which is usually the long-time average of some property of the tides, such as the mean lower low water (MLLW), which is the average of the lower low water over a 19-year interval (Marmer, 1951). In the United States, the mean sea level (MSL) is the average height of the surface of the sea for all stages of the tide for a 19-year interval. The U.S. Coast and Geodetic Survey level net is based upon the MSL, and is known as the Sea Level Datum of 1929.

3. SEA LEVEL CHANGES

Sea levels change with the seasons, and in addition there are longer-term changes. Pattullo *et al.* (1955) have analyzed tide gauge records from 419 stations throughout

Fig. 12.3. Typical tide curves (from U.S.C. & G.S. Tide Tables)

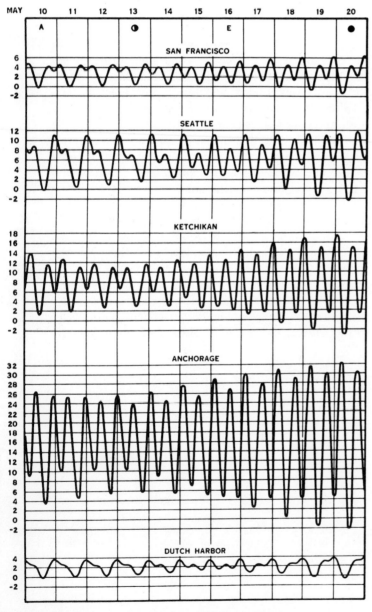

A   Apogee   E   Moon on equator

Note the diurnal tide at Dutch Harbor

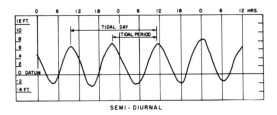

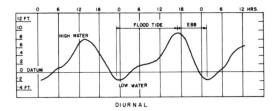

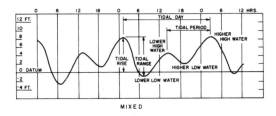

Fig. 12.4. Types of tides

the world, many of these data being obtained from a compilation by the International Union of Geodesy and Geophysics (1940; 1950), from the U.S. Coast and Geodetic Survey, and from the Geophysical Survey Institute of Japan. The data were analyzed from the standpoint of seasonal oscillations of sea level compared with the annual means. Most of the fluctuations were found to be annual rather than semiannual in character. The variations ranged from a few centimeters in the tropics to the order of 20 centimeters and more in higher latitudes, and exceeding 1 meter in the Bay of Bengal.

This is in general agreement with the findings of Marmer (1952) for United States tide gauges. In addition, Marmer studied the variations along coasts from station to station and showed how the seasonal pattern varied gradually from one side of the Gulf of Mexico to the other (Fig. 12.5), from one end of the Atlantic Coast to the other, and from one end of the Pacific Coast to the other.

Marmer has also shown that there are long-term changes in the sea level, Fig. 12.6. Each point in this figure is an average of 9000 consecutive hourly heights. It can be seen that the sea level has been gradually rising with respect to the land since the beginning of the records. Similar observations were made along the Gulf of Mexico and along the Pacific Coast. Records for

Fig. 12.5. Seasonal variation in sea level, Gulf Coast of United States (from Marmer, 1952)

Fig. 12.6. Yearly sea level, Atlantic Coast of United States (from Marmer, 1952)

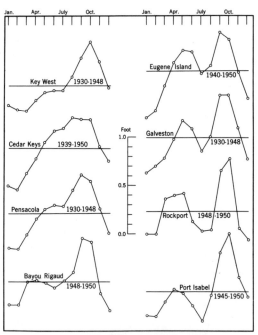

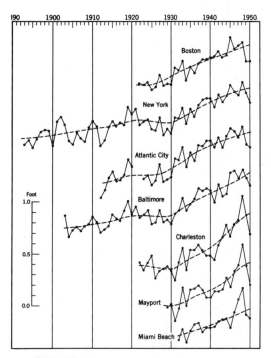

several Alaskan stations have shown a substantial decrease in the tide level, but these records were not for so long a time as for those shown in Fig. 12.6, except for Ketchekan, and in this area little change has been observed.

4. METEOROLOGICAL TIDES

Some changes in sea level are caused by meteorological conditions. Part of the rise or fall of the sea level is due to the changing barometric pressure, but this is small and usually of little importance except when considering annual variations of mean sea level. The greatest effects occur when there are violent storms, such as hurricanes, which cause coupled long waves and wind setup. Details on coupled wave systems of this type have been given in Chapter 4 and information on the combined phenomena has been presented in Chapter 5. A few additional data are given in Table 12.2; these data include the average yearly highest tides, average yearly lowest

Table 12.2. MEAN RANGE AND HIGHEST AND LOWEST TIDES—ATLANTIC, GULF, AND PACIFIC COASTS OF THE UNITED STATES (from Disney, 1955; Harris and Lindsay, 1957; U. S. Coast and Geodetic Survey, 1962)

Place	Period of observation	Mean range (ft)	Highest tides above mean high water			Lowest tides below mean low water		
			Average yearly highest (ft)	Extreme high (ft)	(date)	Average yearly lowest (ft)	Extreme low (ft)	(date)
Atlantic Coast								
Eastport, Me.	1930–61	18.1	4.1	5.0	Nov. 20, 1945	3.6	4.2	Jan. 7, 1943 May 23, 1959
Portland, Me.	1912–61	9.0	3.1	4.3	Nov. 30, 1944 Nov. 20, 1945	2.6	3.5	Nov. 30, 1955
Portsmouth, N. H.	1927–61	8.1	3.0	3.9	Nov. 30, 1944 Dec. 29, 1959	2.5	3.2	Nov. 30, 1955
Boston, Mass.	1922–61	9.5	3.3	4.7	Dec. 29, 1959	2.8	3.5	Jan. 25, 1928 Mar. 24, 1940
Woods Hole, Mass.	1933–61	1.8	3.0	9.4	Sept. 21, 1938	1.8	2.5	Jan. 24, 1936
Newport, R. I.	1931–61	3.5	3.0	10.3	Sept. 21, 1938	1.8	2.6	Jan. 25, 1936
Providence, R. I.	1938–47, 1957–61	4.6	3.3	15.6	Sept. 21, 1938	2.4	3.2	May 5, 1959
New London, Conn.	1938–61	2.6	3.4	8.5	Sept. 21, 1938	1.8	3.0	Dec. 11, 1943
Willets Point, N. Y.	1932–61	7.1	4.1	9.9	Sept. 21, 1938	2.9	3.8	Mar. 24, 1940
Fort Hamilton, N. Y.	1893–1932	4.7	2.8	4.0	Feb. 5, 1920 Nov. 10, 1932	3.0	4.1	Feb. 2, 1908
New York (Battery), N. Y.	1920–61	4.5	3.2	6.0	Sept. 12, 1960	2.8	3.8	Mar. 8, 1932
Sandy Hook, N. J.	1933–61	4.6	3.3	6.0	Sept. 12, 1960	2.6	3.7	Jan. 24, 1936
Atlantic City, N. J.	1912–20, 1923–61	4.1	2.9	5.4	Sept. 14, 1944	2.6	3.5	Mar. 8, 1932
Lewes, Del. (Ft. Miles)	1936–39, 1947–50, 1952–57	4.2		3.6	Nov. 23, 1953		2.8	Mar. 28, 1955
Philadelphia, Pa.	1900–20, 1922–61	5.9	2.3	4.8	Nov. 25, 1950	3.1	5.1	Jan. 25, 1945
Baltimore, Md.	1902–61	1.1	2.8	7.2	Aug. 23, 1933	2.8	4.5	Jan. 24, 1908
Annapolis, Md.	1928–57	0.9		5.9	Aug. 23, 1933		2.9	Sept. 18, 1936 Jan. 8, 1929
Solomons, Md.	1907–08, 1937–57	1.2		2.9	Oct. 15, 1954		2.2	Mar. 23, 1928 Mar. 26, 1928
Washington, D. C.	1931–61	2.9	3.0	8.5	Oct. 17, 1942	2.6	3.4	Feb. 15, 1940 Mar. 1, 1940
Morehead City, N. C.	1953–57	2.8		4.2	Oct. 15, 1954 Sept. 19, 1955		1.7	Dec. 11, 1954
Norfolk (Sewall Pt.), Va.	1928–61	2.5	3.1	6.3	Aug. 23, 1933	1.8	2.7	Jan. 23, 1928 Jan. 26, 1928
Wilmington, N. C.	1935–57	3.7		4.6	Oct. 15, 1954		1.6	Feb. 3. 1940
Southport, N. C.	1933–61	4.1	2.4	3.4	Nov. 2, 1947	1.2	1.9	June 28, 1934
Charleston, S. C.	1922–61	5.2	2.6	5.6	Aug. 11, 1940	2.1	2.8	Feb. 15, 1953
Fort Pulaski, Ga.	1936–61	6.9	2.8	4.5	Oct. 15, 1947	2.7	4.1	Mar. 20, 1936
Fernandina, Fla.	1897–1924, 1939–61	6.0	2.7	7.8	Oct. 2, 1898	2.4	3.7	Jan. 24, 1940
Miami Beach, Fla.	1931–51 1955–61	2.5	1.8	3.9	Oct. 18, 1950	1.1	1.4	Mar. 24, 1936

Table 12.2. (continued)

Place	Period of observation	Mean range (ft)	Highest tides above mean high water			Lowest tides below mean low water		
			Average yearly highest (ft)	Extreme high (ft)	(date)	Average yearly lowest (ft)	Extreme low (ft)	(date)
Gulf Coast								
Key West, Fla.	1926–61	1.3	1.5	2.6	Oct. 18, 1944	1.0	1.4	Feb. 19, 1928
Cedar Key, Fla.	1914–25, 1939–61	2.6	2.6	3.5	Feb. 15, 1953	2.7	3.3	Aug. 27, 1949 Oct. 21, 1952
Pensacola, Fla.	1923–61	1.2	1.8	7.8	Sept. 20, 1926	1.2	2.0	Jan. 6, 1924
Bayo Rigaud, La. (Barataria Pass)	1947–58	1.0ᵃ 1.9ᵃ		3.5	Sept. 19, 1947		1.1	Feb. 3, 1951
Eugene Island, La. (Atchafalaya Bay)	1939–58	1.1		5.2	June 27, 1957		2.3	Jan. 25, 1940
Galveston, Texas	1908–61	1.0	0.8	10.1	Aug. 16, 1915 Aug. 17, 1915	2.2	4.9	Jan. 11, 1908
Port Isabel, Texas	1944–57	1.2ᵃ 0.9		2.4	Aug. 26, 1945		1.8	Jan. 17, 1946
Pacific Coast								
San Diego, Calif.	1906–61	5.7	2.0	2.5	Dec. 17, 1914 Dec. 18, 1914	1.9	2.6	Dec. 17, 1933 Dec. 17, 1937
La Jolla, Calif.	1925–53, 1956–61	5.2	1.9	2.3	Dec. 17, 1952 Dec. 29, 1959	1.8	2.5	Dec. 17, 1933
Los Angeles, Calif.	1924–61	5.4	1.9	2.2	Jan. 25, 1948 Jan. 26, 1948	1.8	2.6	Dec. 26, 1932 Dec. 17, 1933
Santa Monica, Calif.	1933–61	5.4	2.0	2.3	Dec. 27, 1936 July 17, 1951 July 18, 1951 Feb. 4, 1958	1.7	2.5	Dec. 17, 1933
San Francisco, Calif.	1898–1961	5.7	1.7	2.5	Dec. 24, 1940	1.9	2.5	Dec. 26, 1932 Dec. 17, 1933
Crescent City, Calif.	1933–46, 1950–61	6.9	2.6	3.3	Feb. 4, 1958	2.2	2.7	May 22, 1955
Astoria, Oregon	1925–61	8.3	2.7	3.9	Dec. 17, 1933	1.9	2.8	Jan. 16, 1930
Neah Bay, Wash.	1935–61	7.9	2.8	4.0	Nov. 30, 1951	3.0	3.6	Nov. 29, 1936
Seattle, Wash.	1899–1961	11.3	2.3	3.4	Feb. 6, 1904	3.8	4.6	Jan. 4, 1916
Friday Harbor, Wash.	1934–61	7.7	2.4	3.3	Dec. 30, 1952	3.0	3.9	Jan. 7, 1947
Ketchikan, Alaska	1918–61	15.4	4.3	5.4	Nov. 2, 1948	4.5	5.2	Dec. 8, 1919 Jan. 16, 1957 Dec. 30, 1959
Juneau, Alaska	1936–41, 1944–61	16.4	3.9	5.2	Nov. 2, 1948	5.4	6.6	Jan. 16, 1957
Skagway, Alaska	1945–61	16.7	4.0	5.8	Oct. 22, 1945	5.7	6.7	Jan. 16, 1957
Sitka, Alaska	1938–61	9.9	3.2	4.5	Nov. 2, 1948	3.4	4.0	June 19, 1951 Jan. 16, 1957
Yukatut, Alaska	1940–61	10.1	3.4	4.5	Nov. 2, 1948	3.5	4.3	Dec. 29, 1951 Jan. 16, 1957
Seward, Alaska	1925–38	10.5	3.4	4.1	Oct. 18, 1927	3.6	4.3	Jan. 14, 1930
Anchorage, Alaska	1918 1922–25	29.6	4.5	5.8	Oct. 12, 1923	4.3	4.9	June 19, 1924
Dutch Harbor, Unalaska I.	1934–39, 1946–54, 1955–61	3.7	2.0	2.9	Jan. 14, 1938 Jan. 15, 1938	2.0	2.7	Nov. 13, 1950
Sweeper Cove, Adak I.	1943–60	3.7	2.2	2.6	Jan. 5, 1951 Jan. 6, 1951	2.2	2.9	Nov. 11, 1950
Massacre Bay, Attu I.	1943–61	3.3	1.5	1.9	Jan. 27, 1949	1.8	2.5	Nov. 12, 1950 Nov. 13, 1950

ᵃDiurnal range.

NOTE: Complete data from Disney, 1955, modified by U. S. Coast and Geodetic Survey, 1962; stations for which only mean range and extreme tides are given are from Harris and Lindsay, 1957. Pacific Coast data from Disney, 1955, modified by U. S. Coast and Geodetic Survey, 1962.

tides, extreme high tides, and extreme low tides for many United States stations.

Statistical studies of maximum water levels have been made for some regions, and are being made for other places (see Fig. 5.10; also Bruun, Gerritsen, and Morgan, 1958, p. 503).

It is difficult to assess in nature the relative effects of coupled waves, space and time resonant response of coastal configurations, and the wind stress on the water surface. Some, and/or all, and possibly other causes, such as mass transport due to wind waves, were responsible for the water levels reported in Table 12.2. Detailed studies of some of these effects are available in such publications as the *Proceedings of the Second Technical Conference on Hurricanes* (Alaka, 1962). In this section, only the effect of wind stress on the water surface will be described. Wind blowing over the water surface drags water along with it as a drift current. The motion of the water is complicated by the Coriolis force which deflects the current to the right of the wind in the Northern Hemisphere and to the left of the wind in the Southern Hemisphere (see Chapter 13 for a discussion of this, and for numerical examples). The amount of deflection depends upon the duration of the winds, the orientation of the coast with respect to the wind direction, and the depth of water (Figs. 13.10 and 13.12). Storms move, winds are not steady, water depths not constant, and coastlines are irregular. Hence, it is very difficult to arrive at solutions for natural conditions, and these are only approximate. In this section, only the simplest case will be described.

Consider a long, narrow, shallow rectangular tank, with a constant wind parallel to the side walls of the tank, blowing over the water surface. The shear stress on the water surface τ_0 is related to the wind speed through a dimensionless coefficient C as given by Eq. 13.8 and Fig. 13.11. As this relationship is still subject to considerable discussion, it should be used with caution. The water piles up at the end of the tank toward which the wind is blowing while the water level decreases at the other end of the tank. In order to satisfy the continuity condition after steady state has been reached, a sufficient head must be created to cause a hydraulic flow back which has the same total flow rate as transported by the surface current. Because of this, the more shallow the water, the greater must be the pile-up of water at the leeward shoreline, all other conditions being equal. The roughness of the bottom of the tank is taken into consideration through a bottom shear stress, τ_b(lb/ft²). This problem has been considered theoretically by many people (see, for example, Hellstrom 1941; Langhaar, 1951; Keulegan, 1951). The slope of the water surface is given by

$$\frac{dy_s}{dx} = \lambda \frac{\tau_0}{\rho_w g y_s}, \qquad (12.1)$$

where ρ_w is the mass density of the water (slugs/ft³), g is the gravitational constant (ft/sec²), y_s is the vertical distance from the bottom to the water surface, x is the horizontal coordinate in the direction toward which the wind is blowing, and λ is a dimensionless coefficient which depends upon the turbulence of the flow. Keulegan (1951) defined λ as

$$\lambda = 1 + \frac{\tau_b}{\tau_0} \qquad (12.2)$$

and used a value of 1.25. Hellstrom (1941) found that λ varied between 1.15 and 1.30 for tanks of moderate and large depth. Sibul and Johnson (1957) found that λ depended upon the roughness of the bottom and upon a Reynolds number based upon the wind speed, water depth, and air viscosity; their values ranged from about 0.7 to 2.0 for a smooth bottom; and from 1.0 to 1.8 for a rough bottom.

The equation of the water surface as given by Hellstrom (1951) is

$$y_s^2 = \frac{2\lambda\tau_0}{\rho_w g}(x + C_1), \qquad (12.3)$$

where C_1 must be obtained from considerations of continuity. The water surface is parabolic, and Eq. 12.2 can be expressed as

$$\zeta_s^2 = \frac{2\lambda\tau_0}{\rho_w g}\xi. \qquad (12.4)$$

Equations 12.3 and 12.4 are plotted in Fig. 12.7, in which the physical significance of C_1 can be seen. The three cases are: (a) the drawdown on the windward side is minor; (b) the drawdown reaches the bottom of the tank; and (c) a portion of the bottom is exposed. In this figure, F refers to the fetch, which is defined as the horizontal extent of the water surface prior to the wind blowing over it.

The wind setup, h, is given by

$$h^2 = \frac{2\lambda\tau_0}{\rho_w g}(x + C_1) - d, \qquad (12.5)$$

where h and d are defined as shown in Fig. 12.7.

Using Eq. 12.4, Eq. 13.10, the continuity conditions, and defining a new dimensionless coefficient k as

$$k = \frac{\lambda\tau_0}{\rho_w U_o^2}, \qquad (12.6)$$

where U_o is the wind speed (ft/sec), the solution can be put in tabular form with h/d as a function of x/F and $kU_o^2 F/gd^2$. This has been done by Bretschneider (1958) and the results are given in Tables 12.3 and 12.4. Table 12.3 is for cases *a* and *b* of Fig. 12.7, where no bottom is exposed; x_n refers to the horizontal distance from the windward still water line to the nodal line. Table 12.4 is for case *c* of Fig. 12.7, where a horizontal bottom of length x_0 is exposed. In both tables, positive values of h refer to increases in water level above the initial water

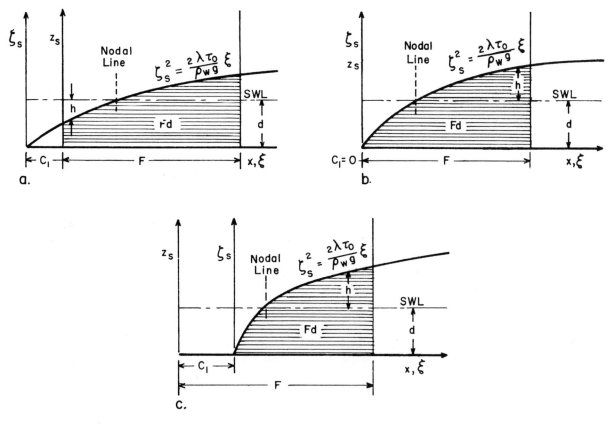

Fig. 12.7. (after Sibul and Johnson, 1957)

Table 12.3. PARAMETER RELATIONS FOR WIND SETUP IN RECTANGULAR CHANNEL OF CONSTANT DEPTH FOR NONEXPOSED BOTTOM
(from Bretschneider, 1958)

$\dfrac{kU_0^2 F}{gd^2}$	$\dfrac{x_n}{F}$	x/F 0	0.1	0.2	0.3	0.4	0.5	0.6	0.7	0.8	0.9	1.0
		Values of h/d corresponding to x/F and $kU_0^2 F/gd^2$										
0.201	0.492	−0.104	−0.082	−0.060	−0.039	−0.018	0.002	0.021	0.041	0.060	0.079	0.097
0.209	0.492	−0.109	−0.086	−0.063	−0.041	−0.019	0.002	0.022	0.043	0.063	0.082	0.101
0.218	0.491	−0.114	−0.089	−0.066	−0.043	−0.020	0.002	0.024	0.045	0.065	0.086	0.106
0.228	0.491	−0.119	−0.094	−0.069	−0.045	−0.021	0.002	0.025	0.047	0.068	0.089	0.110
0.239	0.490	−0.125	−0.098	−0.072	−0.047	−0.022	0.002	0.026	0.049	0.072	0.094	0.115
0.251	0.489	−0.132	−0.103	−0.076	−0.049	−0.023	0.003	0.027	0.052	0.075	0.098	0.121
0.265	0.488	−0.139	−0.109	−0.080	−0.051	−0.024	0.003	0.029	0.054	0.079	0.104	0.127
0.280	0.488	−0.147	−0.115	−0.084	−0.054	−0.025	0.003	0.031	0.058	0.084	0.109	0.134
0.296	0.488	−0.157	−0.122	−0.089	−0.057	−0.026	0.004	0.033	0.061	0.089	0.116	0.142
0.315	0.487	−0.167	−0.130	−0.095	−0.061	−0.028	0.004	0.035	0.065	0.094	0.123	0.150
0.337	0.486	−0.180	−0.140	−0.101	−0.065	−0.029	0.005	0.038	0.070	0.101	0.131	0.160
0.361	0.485	−0.194	−0.150	−0.109	−0.069	−0.031	0.006	0.041	0.075	0.108	0.140	0.171
0.390	0.484	−0.211	−0.163	−0.117	−0.074	−0.033	0.006	0.044	0.081	0.117	0.151	0.184
0.423	0.482	−0.230	−0.177	−0.127	−0.080	−0.035	0.008	0.049	0.088	0.126	0.163	0.199
0.463	0.480	−0.255	−0.195	−0.140	−0.087	−0.038	0.009	0.054	0.097	0.138	0.178	0.217
0.511	0.478	−0.286	−0.217	−0.154	−0.096	−0.041	0.011	0.060	0.108	0.153	0.196	0.238
0.571	0.476	−0.324	−0.244	−0.172	−0.106	−0.044	0.014	0.069	0.121	0.171	0.219	0.265
0.648	0.472	−0.377	−0.280	−0.195	−0.118	−0.048	0.018	0.080	0.138	0.194	0.247	0.298
0.750	0.467	−0.452	−0.329	−0.226	−0.134	−0.052	0.024	0.095	0.162	0.224	0.284	0.341
0.894	0.464	−0.587	−0.409	−0.274	−0.160	−0.059	0.032	0.115	0.192	0.265	0.334	0.399
0.930	0.458	−0.614	−0.421	−0.278	−0.159	−0.055	0.039	0.125	0.205	0.280	0.350	0.418
0.971	0.455	−0.659	−0.443	−0.290	−0.164	−0.055	0.043	0.132	0.215	0.292	0.365	0.434
1.015	0.452	−0.715	−0.467	−0.302	−0.169	−0.055	0.047	0.140	0.226	0.306	0.382	0.453
1.066	0.440	−0.794	−0.494	−0.315	−0.174	−0.054	0.053	0.150	0.239	0.322	0.401	0.475
1.125	0.444	−1.000	−0.526	−0.329	−0.178	−0.051	0.061	0.162	0.255	0.342	0.423	0.500

Table 12.4. PARAMETER RELATIONS FOR WIND SETUP IN RECTANGULAR CHANNEL OF CONSTANT DEPTH FOR EXPOSED BOTTOM
(from Bretschneider, 1958)

$\dfrac{kU_0^2F}{gd^2}$	$\dfrac{x_0}{F}$	$\dfrac{x_n}{F}$	x/F 0	0.1	0.2	0.3	0.4	0.5	0.6	0.7	0.8	0.9	1.0
			Values of h/d corresponding to x/F and kU_0^2F/gd^2										
1.125	0	0.444	−1.0	−0.526	−0.329	−0.178	−0.051	0.061	0.162	0.255	0.342	0.423	0.500
1.20	0.021	0.438	−1.0	−0.565	−0.345	−0.182	−0.047	0.072	0.179	0.276	0.367	0.452	0.533
1.40	0.070	0.428	−1.0	−0.712	−0.397	−0.198	−0.039	0.097	0.218	0.328	0.429	0.524	0.613
1.60	0.112	0.423	−1.0	−1.000	−0.466	−0.222	−0.038	0.116	0.251	0.373	0.485	0.589	0.687
1.80	0.145	0.423	−1.0	−1.0	−0.556	−0.253	−0.042	0.130	0.280	0.413	0.536	0.649	0.754
2.00	0.175	0.425	−1.0	−1.0	−0.681	−0.292	−0.050	0.141	0.305	0.450	0.582	0.704	0.817
2.20	0.200	0.428	−1.0	−1.0	−1.00	−0.338	−0.063	0.148	0.326	0.483	0.624	0.755	0.876
2.40	0.223	0.432	−1.0	−1.0	−1.0	−0.393	−0.079	0.153	0.345	0.513	0.664	0.802	0.931
2.60	0.244	0.436	−1.0	−1.0	−1.0	−0.459	−0.099	0.155	0.361	0.540	0.701	0.847	0.983
2.80	0.262	0.441	−1.0	−1.0	−1.0	−0.539	−0.121	0.155	0.376	0.566	0.736	0.890	1.033
3.00	0.279	0.446	−1.0	−1.0	−1.0	−0.643	−0.147	0.152	0.388	0.590	0.768	0.931	1.080
3.20	0.294	0.450	−1.0	−1.0	−1.0	−0.806	−0.177	0.148	0.399	0.612	0.799	0.969	1.126
3.40	0.308	0.455	−1.0	−1.0	−1.0	−1.0	−0.210	0.142	0.408	0.632	0.829	1.006	1.169
3.60	0.321	0.460	−1.0	−1.0	−1.0	−1.0	−0.248	0.134	0.416	0.651	0.856	1.041	1.210
3.80	0.333	0.465	−1.0	−1.0	−1.0	−1.0	−0.289	0.125	0.423	0.669	0.883	1.075	1.251
4.00	0.345	0.470	−1.0	−1.0	−1.0	−1.0	−0.336	0.114	0.429	0.686	0.908	1.108	1.290
4.20	0.355	0.475	−1.0	−1.0	−1.0	−1.0	−0.388	0.102	0.433	0.701	0.933	1.139	1.327
4.60	0.375	0.483	−1.0	−1.0	−1.0	−1.0	−0.517	0.074	0.440	0.730	0.978	1.299	1.399
5.00	0.392	0.492	−1.0	−1.0	−1.0	−1.0	−0.712	0.041	0.443	0.756	1.021	1.255	1.466
5.40	0.407	0.505	−1.0	−1.0	−1.0	−1.0	−1.0	0.0011	0.433	0.778	1.060	1.307	1.520
6.00	0.428	0.511	−1.0	−1.0	−1.0	−1.0	−1.0	−0.068	0.438	0.808	1.114	1.381	1.621
7.00	0.456	0.528	−1.0	−1.0	−1.0	−1.0	−1.0	−0.218	0.418	0.847	1.194	1.492	1.759
8.00	0.480	0.542	−1.0	−1.0	−1.0	−1.0	−1.0	0.433	0.386	0.877	1.263	1.593	1.885
9.00	0.500	0.555	−1.0	−1.0	−1.0	−1.0	−1.0	−1.0	0.342	0.897	1.323	1.683	2.000

depth d, and negative values of h refer to decreases in the water level. The value k has been obtained for one prototype case, Lake Okeechobee, Florida, and found to be about 3.3×10^{-6} (U.S. Army, Corps of Engineers, Beach Erosion Board, 1961, p. 52i).

The foregoing discussion has been in regard to the steady state condition. Little work has been done on the transient state; the results of one laboratory study, however, have shown that the maximum water levels reached during the initial transient condition are of the order of twice the steady state level (Tickner, 1961).

As mentioned previously, the prediction of the maximum water levels along the open coast and in bays, considering all mechanisms by which the water level is increased, is extremely complicated. Conner, Kraft, and Harris (1957) developed a semiempirical formula to be used in conjunction with hurricanes along the United States coast of the Gulf of Mexico. The maximum height of the combined tide above mean sea level for the open coast, $R_{\max}$, was expressed in terms of the pressure at the center of the hurricane, P_0 (in millibars). It was found that the best fit for 30 hurricanes was

$$R_{\max} = 0.154(1019 - P_0) \qquad (12.7)$$

The over-all correlation coefficient was only .68, with a spread of data of about plus or minus 40 per cent, so that this formula can be used only as a rule of thumb.

REFERENCES

Alaka, M. A. (ed.), *Proc. Second Technical Conf. Hurricanes,* June 27–30, 1961, Miami Beach, Fla., U.S. Dept. of Commerce, Weather Bureau, National Hurricane Research Project, Rept. No. 50, March, 1962.

Bretschneider, C. L., Engineering aspects of hurricane surge, *Proc. Technical Conf. Hurricanes,* sponsored by Amer. Meteor. Soc., Miami Beach, Florida, November, 1958.

Bruun, Per, F. Gerritsen and W. H. Morgan, Florida coastal problems, *Proc. Sixth Conf. Coastal Eng.,* Berkeley, Calif.: The Engineering Foundation Council on Wave Research, 1958, pp. 463–509.

Conner, W. C., R. H. Kraft and D. Lee Harris, Empirical methods for forecasting the maximum storm tide due to hurricanes and other tropical storms, *Monthly Weather Review,* 85, 4 (April, 1957), 113–16.

Defant, Albert, *Ebbe und Flut des Meeres, der Atmosphäre, und der Erdfeste.* Berlin-Göttingen-Heidelberg: Springer, 1953.

Disney, L. P., Tide heights along the coasts of the United States, *Proc. ASCE,* 81, separate no. 660, April, 1955.

Doodson, A. T., Oceanic tides, in *Advances in Geophysics,* vol. 5. New York: Academic Press, Inc., 1958, pp. 117–52.

Gilluly, James, and U. S. Grant, Subsidence in the Long Beach Harbor area, California, *Bull. Geological Soc. Amer.,* 60 (March, 1949), 461–529.

Harris, D. Lee and C. V. Lindsay, *An index of tide gages and tide gage records for the Atlantic and Gulf Coasts of the United States*, U.S. Dept. of Commerce, Weather Bureau, National Hurricane Research Project Rept. No. 7, 1957.

Hellstrom, B., *Wind effect on lakes and rivers*, Royal Swedish Inst. Eng. Res., Bull. No. 41, 1941.

International Union of Geodesy and Geophysics, Monthly and annual mean height of sea level, up to and including the year 1936, *Publ. Sci. Assn. Oceanog. Phys.*, No. 5, 1940.

——, Monthly and annual mean heights of sea level, 1937–1946, and unpublished data for earlier years, *Publ. Sci. Assn. Oceanog. Phys.*, No. 10, 1950.

Inui, Teturo, On deformation, wave patterns and resonance phenomenon of water surface due to a moving disturbance, *Proc. Physico-Math. Soc. Japan*, ser. 3, **18** (February, 1936), 60–113.

Keulegan, G. H., Wind tides in small closed channels, *J. Res. National Bureau of Standards*, **46**, 5 (May, 1951), 358–81.

Langhaar, H. L., Wind tides in inland waters, *First Midwestern Conf. Fluid Dynamics*, Ann Arbor, Mich.: J. W. Edwards Co., pp. 278–96, 1951.

Laplace, P. S., Recherches sur plusieurs points du système du monde, *Mem. Acad. Roy. Sci.* (Paris), **88** (1775), 75–185.

Marmer, H. A., *Tidal datum planes*, U.S. Dept. of Commerce, Coast and Geodetic Survey, Special Pub. No. 135, 1951.

——, Changes in sea level determined from tide observations, *Proc. Second Conf. Coastal Eng.*, Berkeley, Calif.: The Engineering Foundation Council on Wave Research, 1952, pp. 62–67.

Newton, Sir Isaac, *Principia*, Motte's translation revised by Florian Cajori. Berkeley, Calif.: Univ. Calif. Press, 1947.

Pattullo, June, Walter Munk, Roger Revelle, and Elizabeth Strong, The seasonal oscillation in sea level, *J. Mar. Res.*, **14**, 1 (1955), 88–155.

Proudman, J. and A. T. Doodson, Tides in oceans bounded by meridians, Parts I and II, *Phil. Trans. Roy. Soc.* (London), ser. A, **235**, 753 (May, 1936), 273–342, Part III, by A. T. Doodson, **237**, 779 (June, 1938), 311–73.

Schureman, Paul, *Manual of harmonic analysis and prediction of tides*, U.S. Dept. of Commerce, Coast and Geodetic Survey, Special Publ. No. 98, 1940. (Reprinted 1958 with corrections.)

Sibul, O. J. and J. W. Johnson, Laboratory study of wind tides in shallow water, *J. Waterways Harbors Div., Proc. ASCE*, **83**, WW1, Paper No. 1210, April, 1957.

Sterneck, Robert, Die Gezeiten des Ozean, *Sitzungsberichte, Akademie der Wissenschaften in Wien, Mathematisch-Naturwissenschaftliche Klasse*, **129**, 2a, 2 (1920), 131–50.

Tickner, E. G., *Transient wind tides in shallow water*, U.S. Army, Corps of Engineers, Beach Erosion Board, Tech. Memo. No. 123, January, 1961.

U.S. Army, Corps of Engineers, *Bibliography on tidal hydraulics*, Committee on Tidal Hydraulics, Rept. No. 2, February, 1954.

——, Beach Erosion Board, *Shore protection planning and design*, Tech. Rept. No. 4, 560 pp, 1961.

U.S. Dept. of Commerce, Coast and Geodetic Survey, *Extreme tides, letter of* October 26, 1962, No. 2221/274/706, from Captain Kenneth S. Ulm.

CHAPTER THIRTEEN

Currents

1. INTRODUCTION

There are several main classes of currents in the oceans. These currents may be classified in various ways, one being (1) wind-drift currents of relatively short duration; (2) currents related to surface waves; (3) tidal currents; and (4) major ocean currents generally considered to be a part of the oceanic circulation. In addition to these classes of currents, there are currents in nearshore areas where water from rivers and nearly enclosed bays is discharged into the ocean with considerable momentum. The effects of such currents generally are not important relative to the major currents, but rather large local effects may sometimes occur. As pointed out by Sverdrup (1943, p. 92), one of the main difficulties in studying ocean currents is that all types exist simultaneously, and a current-measuring program of long duration with measuring devices at various depths and many locations would be needed to obtain the necessary information to differentiate the effects of each type. (See Wiegel and

Johnson, 1960, for descriptions of current meters and the operational difficulties encountered in measuring currents.)

Measurements with refined current meters show that the movement of the water varies continually both in speed and direction. Rapid traverses with respect to depth show this fine structure as a function of position (Fig. 13.1). A study of several hundred continuous GEK observations in the Pacific Equatorial Countercurrent showed large-scale turbulent fluctuations which were proportional to the mean current and from one-half to two-thirds of the magnitude of the mean current (Knauss, 1961). Much larger-scale fluctuations have been observed in the major ocean currents; for example, in the Gulf Stream meanders occur and have been measured in detail south of Cape Hatteras (Webster, 1961). Figure 13.2 shows the shifting of the positions of the maximum surface current speed and the 20°C isotherm at a depth of 100 meters. These two measurements are both fairly reliable indicators of the center of the current. The

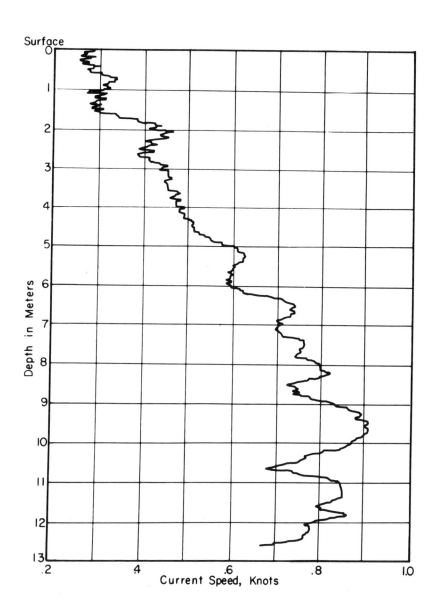

Fig. 13.1. Example of the variation of current speed with depth. Obtained with a Snodgrass meter in the Pacific Ocean, August 23, 1957 (courtesy of J. M. Snodgrass)

double amplitude of the meanders in the lateral direction was found to be of the order of 20 nautical miles with the instantaneous width of the Gulf Stream in this region being of the order of 30 to 50 nautical miles.

An example of changes that occur on a yearly scale is the California Current which moves in a southerly direction during the greater part of the year along the California coast. From the middle of November until the middle of February, however, an inshore flow to the north occurs, this being called the Davidson Current (Sverdrup *et al.*, 1942).

Long-term changes occur in the water masses which appear to be related to the variations in a major ocean current. Indirect measurements, such as temperature anomalies, are usually the only indications of these changes. One such set of data for the California Current is available (Reid, Roden, and Wyllie, 1958), and is shown in Fig. 13.3. It appears as if these long-term

changes may be related to the winds, an example of which is also shown in this figure.

There are apparently eddies of all sizes associated with the major currents, and these play an important part in lateral diffusion of momentum. Detailed measurements have been made of the detaching from the Gulf Stream of an eddy measuring 200 by 60 miles (Fuglister and Worthington, 1951).

In addition to turbulent fluctuations, eddies, etc., major currents vary in total transport. Figure 13.4 shows the variation in transport of the Florida Current as measured between Key West and Havana. An even more spectacular example has been reported by Knauss (1961) for the portion of the Pacific Equatorial Countercurrent beneath the thermocline. In the region in which the measurements were made, the transport was in excess of 60×10^6 cubic meters per second in 1958 and only 1×10^6 cubic meters per second in 1959.

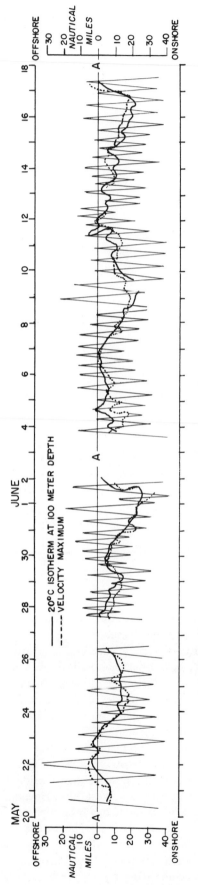

Fig. 13.2. Meanders of the Gulf Stream *Note:* Slanting lines across the face of the diagram represent the path of the ship in space and time (after Webster, 1961)

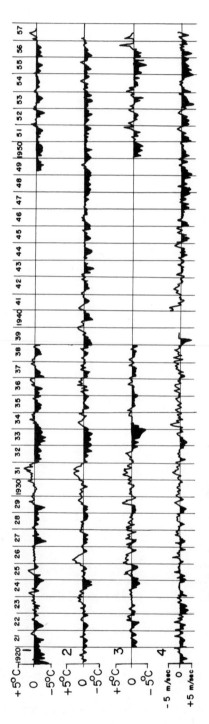

Fig. 13.3. Monthly differences from average sea surface temperature (degrees Centigrade at (1) 30°–35°N, 115°–120°W, (2) Scripps Pier, and (3) 25°–30°N, 110°–115°W and (4) monthly differences from average northerly wind component (in meters per second) at 30°N, 110°–130°W. (after Reid, Roden, and Wyllie, 1958)

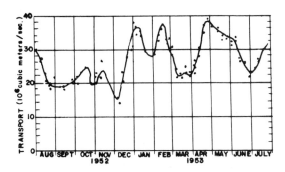

Fig. 13.4. Mass transport of the Florida current (after Wertheim, 1954)

The magnitude of the various currents not only varies between classes but the magnitude of any particular class will vary from location to location and also with time at a given location. A summary, however, to give the general order of magnitude of the various classes of currents is of interest. Typical examples of the magnitude and range of value of these currents are presented in Table 13.1, together with the reference for the source

Table 13.1. RELATIVE MAGNITUDE OF OCEANIC AND ESTUARINE CURRENTS

Type of current	Magnitude (knots)	Reference
Large-scale movements		
California Current, off Monterey, January, 1958	0.3–0.6	Jennings and Schwartzlose (1958)
Off Southern California, 1937	0.13–0.4	Tibby (1939)
Florida Current	1.7–3.5	Murray (1962)
Cromwell Current	2.4	Knauss (1961)
Pacific Equatorial Countercurrent	0.6–1.2	Knauss (1961)
Currents caused by wind stress		
Amrum Bank Light Ship	0.06–0.2[a]	Mandelbaum (1955)
San Francisco Lightship	0.06	Disney and Overshine (1925)
Inertial currents	very small	Stommel (1954)
Tidal currents		
Rotary San Francisco Lightship	0.05	Disney and Overshine (1925)
Reversing San Francisco Bay, Golden Gate		
Max. flood	3.3	U.S.C. & G.S.
Max. ebb	4.5	Tidal current charts (1947)
Hydraulic Cape Cod Canal (max.)	4.5	Wilcox (1958)
Wave-induced currents	small	

[a] Depends on wind speed.

of such information.

Current patterns may be represented in either of two manners, both of which are based upon hydrodynamical relationships. These representations are the "path" method of Lagrange and the "flow" method of Euler. The Lagrangian method follows the behavior of a given fluid particle (such as is represented to some extent by a current drogue) during its motion through space. The Euler method observes the flow characteristics in the vicinity of a given point as the particles pass (such as are recorded by an anchored current meter). Although the Eulerian or flow method lends itself readily to many practical hydraulic problems, it is not so descriptive of the fate of the individual particle. Because the flow patterns in one system can be predicted from observations in the other system, and vice versa, for only non-turbulent flow with relatively simple boundary conditions, the determination of the actual currents should involve both current meter and drogue observations. One type of observation might well be more extensive than the other, but the availability of both types of data would permit at least a qualitative transformation from one flow representation to the other.

Examples of the flow and path methods of representing current patterns are shown in Figs. 13.5 and 13.6 for a coastal area in the southwestern part of the Netherlands. Figure 13.5 shows the magnitude and direction of the current at various times before, at, and after highwater at the Hook of Holland (Arlman et al., 1958). These illustrations show instantaneous flow patterns which occurred at various times during a tidal cycle. Such Eulerian flow patterns or streak patterns would be obtained if an airplane passed over the area and took a vertical photograph with a time exposure of a large number of surface floats. Since the travel time for all such floats is a constant and equal to the time exposure, the distance that each float travels is a function of the velocity of each float. The instantaneous flow direction is obtained by placing an arrow head on the appropriate end of the particle path. This procedure of obtaining flow patterns from "aerial photographs" is a common technique in studying the rather complicated flow patterns in estuaries by the use of hydraulic models. It must be recognized that flow patterns obtained by this technique pertain only to the surface. Where large vertical density variations, and consequent velocity gradients occur, subsurface determinations of current speed and direction must be determined by suitable current-measuring devices.

For the same area to which the flow patterns shown in Fig. 13.5 pertain, polar diagrams of current speed and direction are shown in Fig. 13.6(a) for a relatively large number of selected stations. The rotary nature of the currents is quite evident; at some stations, however, the current is almost reversing in character. The polar diagrams shown in Fig. 13.6(a) are typical of the average

The figures at the arrows indicate the average rates, in knots, in the upper layer, under normal weather conditions and for days with mean range and fresh water discharge.

(a) Three hours before H.W. at Hook of Holland

(b) H.W. at Hook of Holland

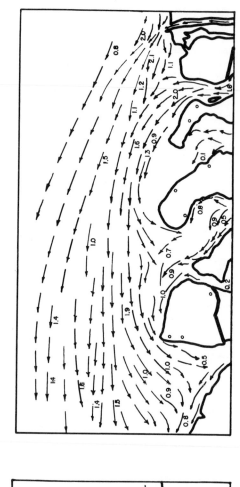

(c) Three hours after H.W. at Hook of Holland

(d) Six hours after H.W. at Hook of Holland

Fig. 13.5. Current patterns in the coastal area in the southwestern part of the Netherlands (from Arlman, et al., 1958)

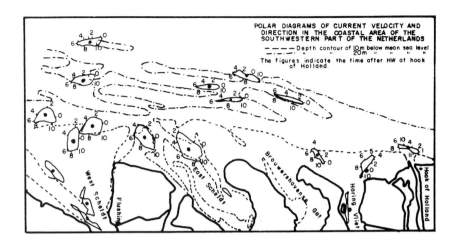

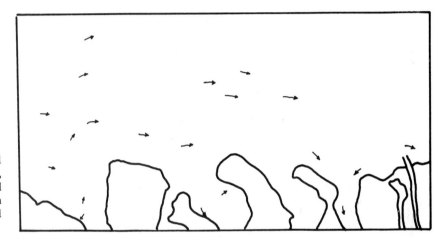

Fig. 13.6. (a) Polar diagrams of current and direction (after Arlman, *et al.*, 1958); (b) Current direction and velocity at H.W. as plotted from (a) above (after Johnson and Wiegel, 1959)

current conditions that might be obtained in this area from current meters anchored at the various locations and observations made over a period of time. The group of polar diagrams shown in this figure could readily be obtained from such flow patterns as shown in Fig. 13.5; however, to plot the flow patterns as shown in these latter figures from the polar diagrams shown in Fig. 13.6(a) would be relatively difficult, and subject to considerable error in their preparation. For example, an overlay on Fig. 13.6(a) with the current at each station shown for high water at the Hook of Holland will give the pattern shown in Fig. 13.6(b). The difficulty in arriving at the detailed pattern shown in Fig. 13.5 from that shown in Fig. 13.6(b) is obvious, even though the number of observation stations shown in Fig. 13.6(a) is relatively large. It must be recognized that the complications of tides, discharges of rivers, density differences in the water, and the wind and wave conditions that occur along the section of the coast for which the data shown in the figure cited apply, probably are much more severe than that which might occur along most sections of a shoreline, such as the coast of California. Never-

theless, the number of stations at which observations must be taken to define adequately the average current conditions for the area under study is apparent from a comparison of Figs. 13.5 and 13.6.

Although the data on currents as given in the plots shown in Figs. 13.5 and 13.6 are valuable in giving certain information on the currents in the area under study, they do not give the highly desirable data on the movement of a water particle that, for example, might be released from an outfall at an offshore location. For example, Fig. 13.7(a) shows the relatively complicated path of a surface float that was observed through several cycles for an area in the North Sea near Wilhelmshaven, Germany, by Kruger (1911; Thorade, 1933). Had similar data for the area shown in Figs. 13.5 and 13.6 been obtained, a rather complete measure of water movement would have resulted. Only by the "tracking" of floats released almost simultaneously at many points could the desired details on water movement be provided.

The current plots shown in Figs. 13.6(a) and 13.7 are representative of the basic data that might be obtained in an engineering survey of a particular area. The type

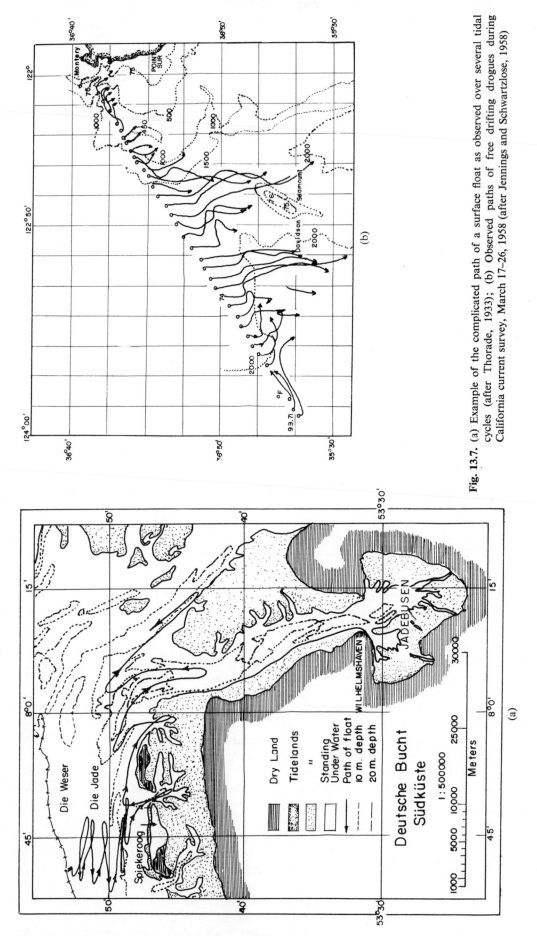

Fig. 13.7. (a) Example of the complicated path of a surface float as observed over several tidal cycles (after Thorade, 1933); (b) Observed paths of free drifting drogues during California current survey, March 17–26, 1958 (after Jennings and Schwartzlose, 1958)

316

of current pattern that is required—whether "path" or "flow"—dictates the type of instrument that must be used in the survey.

2. WIND DRIFT

When wind blows over the surface of the water, a shear stress exists at the interface and the air drags water along. Because of turbulent mixing, this current gradually deepens. When the wind fields are large and of long duration, such as the trade winds, the resultant current is one of the major components of the large oceanic circulation. We are considering here, however, the wind drift currents resulting from winds blowing over a rather limited area for a limited time, although the area may be of the order of many square miles and the duration may be of the order of a day or so.

An idea of how this type of current forms can be seen in a glass wall wind-water tank. As the wind blows over the surface, water is dragged along, forming the surface current. Waves are also formed. If a few small drops of colored nonmiscible fluid of the same specific gravity as the water are placed in the water, they will have nearly the same paths as the water, but their paths can be followed by eye. It is convenient to take motion pictures of their motions and then plot their paths. An example is shown in Fig. 7.36 (Johnson and Rice, 1952). If no wind drift were present, the particles would travel in a nearly closed orbit. It can be seen that they have an appreciable net movement in the direction of wind travel. The few particles well below the surface have rather strange trajectories. Some of these occur because their specific gravities are a little different from the specific gravity of the water, but others are due to some undetermined cause. In general, they move in a direction opposite to the surface current because this experiment was done in a closed system and there had to be a return flow to balance the surface current. In the laboratory, the distance over which the wind blows is so short that the depth of the surface current is very small. In the ocean, this depth can be relatively large.

If it were not for eddy viscosity and the mixing by wind waves, the depths of the wind drift currents would be small. Lamb (1945, p. 590) developed the equation for the velocity distribution $u(z)$ as a function of the wind velocity at the interface U_s and of time for the case of laminar flow in the water. The equation is

$$\frac{u(z)}{U_s} = 1 - \frac{2}{\sqrt{\pi}} \int_0^\theta e^{-\theta^2} \, d\theta, \qquad (13.1)$$

where θ is $z/\sqrt{4vt}$, z is the depth below the surface (ft), v is the kinematic viscosity of water (ft²/sec), and t is the time (sec). This equation is plotted in Fig. 13.8. In this equation, the effect of Coriolis force is not considered.

At the suggestion of Nansen, Ekman (1905) examined theoretically the problem of wind drift in homogeneous

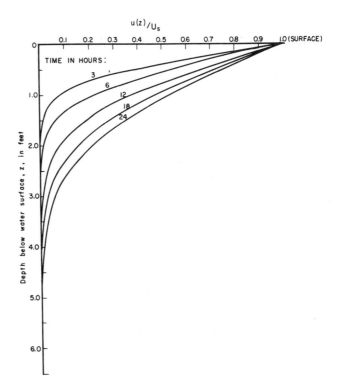

Fig. 13.8. Development of wind drift current due to viscosity, as a function of time

deep water, considering only frictional and Coriolis forces and assuming the eddy viscosity of the water to be constant (see also Proudman, 1953). In addition, the motions were considered in a plane of infinite extent (with only one value of latitude, however) with winds of constant velocity. For the case of the wind stress on the water surface (τ_0) in the direction of the y axis (associated with the velocity component v) he found that the steady state solution in the Northern Hemisphere was

$$u = U_s \, e^{-az} \cos (45° - az)$$
$$v = U_s \, e^{-az} \sin (45° - az) \qquad (13.2)$$

where U_s is the absolute velocity of the current at the water surface (ft/sec), z is the vertical coordinate taken positive downwards, and

$$a = +\sqrt{\frac{\rho_w \Omega \sin \phi}{\mu_e}} \qquad (13.3)$$

in which ρ_w is the mass density of the water (slugs/ft³), Ω is the angular velocity of the Earth (0.0000729 radians/sec), ϕ is the latitude, and μ_e is the eddy viscosity (1b-sec/ft²). At the surface the current is directed at an angle of 45° cum sole from the wind stress. For positions beneath the surface, the current speed decreases exponentially with depth and varies in direction. Ekman's schematic representation of a pure wind current is shown in Fig. 13.9. Projected on a horizontal plane, the end points of the vectors lie on a logarithmic spiral, commonly called the *Ekman spiral*. The wind also

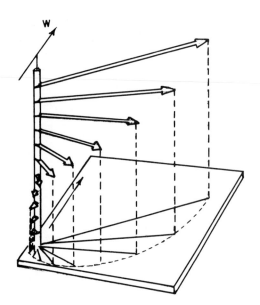

Fig. 13.9. Schematic representation of a wind current in deep water, showing the decrease in velocity and change of direction at regular intervals of depth (the Ekman spiral). *W* indicates direction of wind (from Sverdrup, *et al.*, 1942)

changes direction with elevation above the water surface; for all practical purposes, however, the direction of the wind stress at the interface is the same as the direction of the wind at normal anemometer height (30 ft, or so).

At some depth $z = D$ the direction of the current is opposed to the surface current, and is only one-twenty-third the surface value. This depth is defined by

$$D = \frac{\pi}{a} \qquad (13.4)$$

and is called the *frictional layer*, or sometimes the Ekman

layer. For the case of the water depth not large compared with D, the angle of deflection depends upon the water depth as well as the distance below the surface. He found that as long as D was less than the water depth, the effect was practically negligible. For an extreme case of the water depth being only $0.1D$ the direction at the surface, and at all depths, was essentially the same as the direction of the wind stress.

Although water is being transported in all directions (the direction depending upon the distance beneath the surface) in deep water, the net transport is at right angles to the wind stress. For the case of τ_0 being in the y direction,

$$S_x = \int_0^\infty u\,dz = \frac{U_s D}{\pi \sqrt{2}} \quad \text{and} \quad S_y = \int_0^\infty v\,dz = 0. \quad (13.5)$$

If the winds suddenly start from zero, and then continue to blow at a constant velocity, the variation of the surface current in deep water with time was found by Ekman to be as shown in Fig. 13.10 for the surface and for a depth of $0.5D$. The numbers shown on the curves refer to the number of pendulum hours after the start of the wind, where a pendulum hour is the sidereal hour divided by $\sin \phi$. In the higher latitudes, the length of time for the surface current to reach approximately its steady state velocity is about one day. For winds blowing only a pendulum hour or so, the current at the surface is nearly in the same direction as the wind stress; at about 8 pendulum hours after the start, the current makes an angle of about 60 degrees with the wind stress. Beneath the surface, the time-dependent solution varies much more from the steady state solution, both in magnitude and in direction, than is the case for the current at the surface. In addition, the velocity varies periodically in both magnitude and direction, with slowly

Fig. 13.10. Time dependent solution for Ekman drift (from Ekman, 1905)

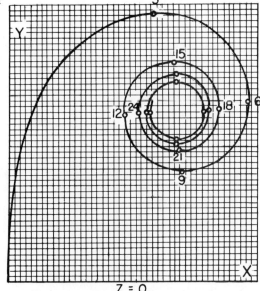

Z = 0

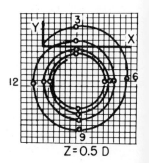

Z = 0.5 D

decreasing amplitudes, with a period of 12 pendulum hours. In the higher latitudes, this period is very near that of the tides and should be considered when reducing current meter records. In the lower latitudes, the pendulum hour is long, becoming infinitely long at the equator. Hence, part of the paradox of what happens to the current at the equator may be explainable by Fig. 13.10 which would indicate a current in the direction of wind stress on the surface, as it would take an infinite time to cause the current to flow at an angle to the wind stress.

Rossby and Montgomery (1953, p. 92) developed a boundary layer theory for the wind-induced currents, assuming that Prandtl's mixing length λ decreases linearly with distance beneath the surface as

$$\lambda = \frac{k(h-z)}{\sqrt{2}} \qquad (13.6)$$

where k is a nondimensional constant estimated to have a value of 0.65, and h is the depth of the drift current. The resulting form of the eddy viscosity is

$$\mu_e = \rho_a \lambda^2 \left| \frac{\mu}{z} \right| = \rho_a \lambda \sqrt{\frac{\tau_0}{\rho_a}} \cdot \sqrt{\frac{z}{h}} \qquad (13.7)$$

where

$$\tau_0 = C\rho U_0^2 \qquad (13.8)$$

in which C is a dimensionless drag coefficient, ρ_a is the mass density of air (slugs/ft³), and U_0 is the surface wind speed (ft/sec). The surface wind speed seems to have been taken as the speed at the shipboard anemometer, which depends upon the ship; hence, the coefficient C must be a value associated with the wind speed measured at a specified elevation. The numerical values for the angle between the current at the surface and the surface wind velocity depended upon the relationship between the surface wave heights and the wind speed. As is evident from Chapter 9, there is no simple relationship, the wave heights depending upon the fetch, duration, and stability of the air as well as upon the wind speed. Thus, the values given in Table 13.2 are useful only to show some of the variations that can be expected.

The depth of the current, h, was found to vary inversely with sin ϕ, and directly with the surface wind speed.

An electric analog computer has been developed for use in the solution of the Ekman wind drift problem for

the case of the eddy viscosity varying arbitrarily with depth (Wilson *et al.*, 1953), and its use should permit the development of a more realistic theory with respect to the distribution of eddy viscosity.

Both Ekman's and Rossby and Montgomery's theories show the current depth to become infinitely deep at the equator, which does not happen. Goldsbrough (1935) states that Ekman's solution (and presumably the solution of Rossby and Montgomery also) holds good only in the vicinity of the poles. He treated the problem of a steady space periodic wind blowing over the surface of a rotating sphere completely covered with deep homogeneous water. For the case of the periodicity of the winds chosen to resemble, roughly, the wind systems in the Atlantic Ocean, it was found that the angle between the surface wind and surface current directions depended upon the coordinates of the point in question, and could even be negative.

It should be emphasized that, although there has been much theoretical work on the Ekman drift, few systematic measurements have been made of drift currents. Hence, engineers should consider with caution theories that have been described, and those that will be described in future sections.

One set of measurements in areas where the effects of solid boundaries should be small was made by Stommel (1954). They were made over an interval of several months in the vicinity of Bermuda. It was found that the speed of the surface wind drift was of the order of one-twentieth to one-thirtieth of the surface wind speed, with the current and wind speeds being averages of either 1½- or 3-hr intervals. The direction of the drift varied considerably with very large-scale eddies possibly playing an important part in this variation. The current and wind speed directions were obtained by instantaneous measurements. The results are shown in Table 13.3. A more selective choice of data, based upon days when the average wind speeds were relatively steady for a 24-hr interval showed that the direction of the wind drift was to the right of the surface wind direction by an angle varying between 0 and 100 degrees for a large number of observations (Table 13.4).

The average deflection of the wind drift has been measured by others (Krummel, 1911; Bowden, 1953) in the higher latitudes and it has been considered that Ekman's simple theory predicted the direction correctly. On the other hand, the results of large-scale measurements, using drift cards designed to float at the water surface in such a manner that they wouldn't be acted upon directly by the wind, showed the drift current to be in the same direction as the gradient wind (Hughes, 1956). The speed of the current at the surface was found to be about one-fiftieth the gradient wind speed, which would be the equivalent of about one-thirty-third of the wind speed at a height of 30 ft above the water surface.

It is not possible to tell from Stommel's paper the

Table 13.2. ANGLE (DEGREES) BETWEEN THE DIRECTION OF THE CURRENT AT THE SURFACE AND THE SURFACE WIND DIRECTION
(after Rossby and Montgomery, 1935, p. 95)

Latitude ϕ (degrees N)	Surface wind speed, U_0 (meters per second)			
	5	10	15	20
15	35.0	38.7	41.1	43.0
30	38.6	42.8	45.7	48.0
45	40.6	45.4	48.4	50.9
60	42.0	46.8	50.2	52.7
75	42.6	47.7	51.1	53.8
90	42.8	48.0	51.4	54.1

Table 13.3. TOTAL NUMBER OF VANE READINGS (CUMULATIVE) FOR EACH 10° ANGLE OF DEVIATION OF CURRENT FROM WIND
(from Stommel, 1954)

Right		Left	
170	4	010	274
160	6	020	288
150	7	030	297
140	18	040	307
130	25	050	309
120	30	060	311
110	32	070	314
100	46	080	318
090	56	090	320
080	68	100	322
070	79	110	325
060	103	120	331
050	121	130	333
040	141	140	338
030	154	150	340
020	172[a]	160	341
010	192	170	345
000	223		

[a] Half the readings are on either side of this point.

velocity distribution with depth. The velocities were measured from a "submerged" drifting raft whose drift was controlled by a drogue suspended at depths from 120 to 530 ft, so that the surface velocities reported by him were relative to whatever current existed at the depth of the drogue. The current measurements were made a few feet below the surface.

Carruthers, Lawford, and Veley (1951) report wind drift measurements made from the North Goodwin Light Vessel, off the southeast coast of England. The measurements were made with a device that averages the velocity over a vertical distance of 5 ft which was apparently suspended with its top some 5 to 10 feet below the ocean surface. Although their data had considerable scatter, they concluded that the speed of the wind-drift current was from one-forty-fifth to one-fiftieth the wind speed.

The reason for the difference between the ratios of wind drifts to wind as obtained by the two sets of measurements may be that Stommel's were measured closer to the surface and that the wind drift decreases very rapidly with depth in the top few feet.

Table 13.4. 24-HR VECTORIAL MEANS OF CURRENTS AND WINDS FOR DAYS WITH STEADY WINDS
(from Stommel, 1954)

Day	Buoy	Wind blows towards	Wind speed (knots)	Current Direction	Current Speed (knots)	Angle of current relative to wind Method 1[f]	Angle of current relative to wind Method 2[g]
1953							
29 Oct.	3	018	20.0	060	1.09	042R	070R
29	4	018	20.0	070	0.70	052R	000
3 Nov.	3	225	6.1	230	0.51	005R	025R
6	3	305	11.2	345	0.64	040R	045R
8	3	230	4.4	235	0.72	005R	040R
9	3	270	4.5	270	0.57	000	055R
23[a]	2	285	11.2	315	0.45	030R	010R
24	2	335	6.0	335	0.56	000	005L
26	2	015	13.0	030	0.57	015R	005R
2–3 Dec.	2	150	13.0	170	0.50	020R	020R
5	8	270	3.0	(b)	0.08	(b)	(b)
6	8	260	4.0	(b)	0.18	(b)	(b)
18	7	140	13.0	220	0.61	080R	(e)
19	7	145	16.0	225	0.67	080R	(e)
22	7	030	17.0	075	0.62	045R	(e)
23	7	060	6.0	115	0.48	055R	(e)
1954							
1 Jan.	6	150	18.0	(c)	0.70	(c)	(c)
3–4	6	030	17.0	(c)	0.50	(c)	(c)
8	6	150	18.0	(c)	0.68	(c)	(c)
11[d]	6	025	18.0	(c)	0.72	(c)	(c)
26–27	9	270	4.0	010	0.48	100R	055R
9	14	270	23.0	(d)	1.45	(d)	045R

[a] Current data not available for entire 24 hours, but winds blew steady for previous 4 days.

[b] Mean direction meaningless because of large oscillatory motions.

[c] Angles have unknown systematic errors due to faulty bridling of Buoy 6.

[d] Current data not available for entire 24 hours.

[e] Wind vane broken.

[f] Method 1. Angle between mean wind direction at observatory on Bermuda (less than 20 miles away from buoy) and current direction given by magnetic compass on buoy.

[g] Method 2. Angle between wind vane on buoy which measures instantaneous angle between wind and buoy orientation.

Almost no measurements have been made of current distribution with depth in the open ocean, probably because of the difficulty in obtaining them. Some data obtained by Vine, Knauss, and Volkmann (1954) in the area between Cuba and Jamaica of what might be a well-established wind drift show a speed just under the surface of 1.4 knots which had decreased to about 0.2 knots at a depth of 600 feet below the surface. Knauss, however, doubts this (Personal communication, 1962).

Stommel's measurements are not surprising when one considers the irregularity of both wind speed and direction. Thus a gust of wind acts upon the water surface for a limited extent of time, and if the water was initially still, sets the water in motion in the direction of the wind velocity in accordance with Fig. 13.10. The wind direction in this gust may be the same as the mean wind direction, or in some other direction, the distribution of probabilities perhaps being quasi-Gaussian.

As the wind fluctuates in direction, it must act on a water current initially heading at some angle to it and because of the relatively great inertia of the water probably cannot change its direction too much. It will, however, change it to some extent. Thus, the variability of motions of the water within the wind drift must be similar to some extent to the variability of the motions of the air in the wind. As was stated earlier in this section, the wind blowing over the water surface generates both waves and currents. Currents should be treated in a manner similar to that given waves (Chapter 9), with mean values of speed and direction being presented together with the distributions of fluctuations from these means; in fact it would be desirable to use a power spectra type of analysis.

The relationship between the wind speed and the speed of the current at the surface has been the subject of considerable discussion and study. It is related to the

Table 13.5. U_s/U_0 As a Function of Latitude and Surface Wind Speed
(after Rossby and Montgomery, 1935, p. 95)

Latitude (degrees)	Surface wind speed, U_0 (meters per second)			
	5	10	15	20
15	0.0317	0.0291	0.0276	0.0266
30	0.0292	0.0268	0.0254	0.0245
45	0.0280	0.0256	0.0243	0.0234
60	0.0273	0.0249	0.0237	0.0228
75	0.0269	0.0246	0.0246	0.0226
90	0.0268	0.0245	0.0233	0.0225

shear stress of the air on the water surface (Rossby and Montgomery, 1953; Francis, 1954; Deacon, Sheppard, and Webb, 1956; Stommel, 1958; Francis, 1959). According to Rossby and Montgomery (1935, p. 65), the steady state speed of the drift current at the surface U_s is given by

$$\tau_0 = \frac{3}{2} \rho_w k^2 U_s^2 = C \rho_a U_0^2. \qquad (13.9)$$

Subject to the conditions stated previously, the ratio U_s/U_0 is given in Table 13.5 as a function of surface wind speed and latitude for a value of $C = 0.0025$. This value is consistent with the value reported by Francis (1959) for higher wind speeds, as can be seen in Fig. 13.11. It has been found that the stability of the air affects the value of C, but not enough data are available to evaluate the effect numerically (Deacon, Sheppard, and Webb, 1956). Wilson (1960) examined the results of 47 investigators and found that, for higher wind speeds, $C = 0.0024 \pm 0.0005$, and for light winds, $C = 0.0015 \pm 0.0008$.

3. WIND DRIFT AND THE EFFECT OF BOUNDARIES

When there is a boundary, such as a shore, winds blowing toward a shore cause an increase in water level, the

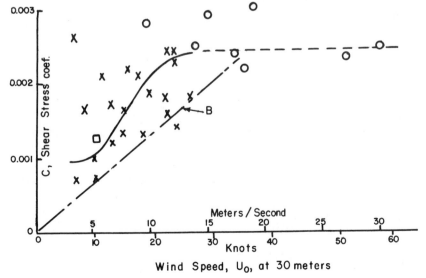

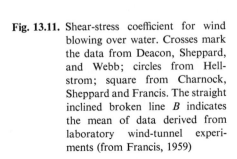

Fig. 13.11. Shear-stress coefficient for wind blowing over water. Crosses mark the data from Deacon, Sheppard, and Webb; circles from Hellstrom; square from Charnock, Sheppard and Francis. The straight inclined broken line B indicates the mean of data derived from laboratory wind-tunnel experiments (from Francis, 1959)

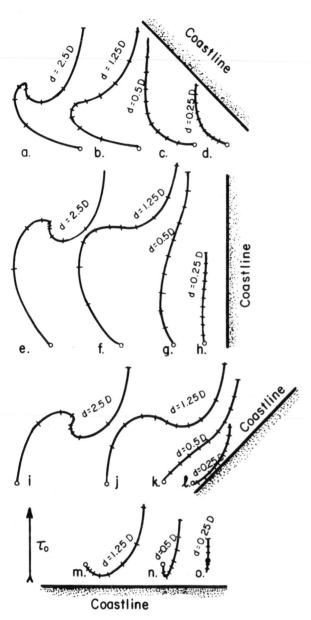

Fig. 13.12. Effect of coastline on Ekman drift (from Ekman, 1905)

on currents. Associated with the increase of decrease in water level along a coast are horizontal pressure gradients. Ekman studied these currents for the case of horizontal pressure gradient and Coriolis force only. As one example of this process, he considered a long rectangular channel with a hydraulic current flowing in it. He found that the total inclination would be directed approximately along the channel if d/D were small, and nearly perpendicularly to it if d were greater than D. The motion consists of a current in the direction of the channel and a circulation in the planes perpendicular to the axis of the channel.

Ekman (1905) also considered the case of a vertical straight boundary with a wind stress as well as the Coriolis force and horizontal pressure gradient. The results are shown graphically in Fig. 13.12. The direction of the shear stress τ_0 with respect to the orientation of the coasts is given by the arrow. The directions and magnitudes of the currents at different depths are given by an arrow running from the origin of each curve (an open circle on the curves) to the curve. Each dot on the curve is $\frac{1}{10} D$. When $d > D$ there are essentially three "currents": a bottom current of rapidly varying magnitude, an intermediate current of nearly constant magnitude but rapidly varying direction, and a surface current of nearly constant direction.

In the Northern Hemisphere, the net transport due to wind drift is directed at right angles to the wind according to Ekman's theory (see Sverdrup et al., 1942). This transport depends only upon the wind stress and the latitude and, as described, can develop only in the open ocean in regions where the wind blows with a constant speed and direction over wide areas. Near coasts, however, certain modifications occur and the secondary effect of the wind becomes important. For example, consider a wind in the Northern Hemisphere which is blowing approximately parallel to a coast with the coast on the right-hand side of an observer facing in the direction in which the wind blows (Fig. 13.13a). Assuming that the density of the sea water increases with depth, primarily because the temperature decreases with depth, the direct effect of the wind causes a transport of light and warm surface water toward the coast, Fig. 13.13(b). Because the coast represents an obstruction to this flow, the light and warm water piles up against the coast, Fig. 13.13(c), and at some distance from the coast a denser and colder subsurface water must rise to replace that which has been carried toward the coast. The distribution of density consequently is altered and, as a secondary effect, a current develops which flows in the direction of the wind or parallel to the coast. A steady state may be reached, Figs. 13.13(b), 13.13(c), depending upon the wind stress and upon the rapidity with which the water that rises toward the surface is heated.

If the coast lies to the left of the wind direction, Fig. 13.13(d), the light and warm surface water is transported

amount of the increase depending upon the wind velocity, the fetch, and duration of the winds, the water depth, and the roughness of the bottom (Sibul and Johnson, 1957). For all other conditions being equal, the shallower the water, the greater will be the increase in the water level. This increase in water level is often called a *wind-tide* or *setup*, and often causes serious damage in areas such as those along the Louisiana and Texas coasts in the United States as a result of hurricanes, and along the southern coast of the North Sea as a result of gales. Because this subject is largely connected with the rise of the water level, it was discussed in Chapter 12.

Ekman (1905) and others (Section 7 of this chapter; also Hidaka, 1947) have studied the effect of boundaries

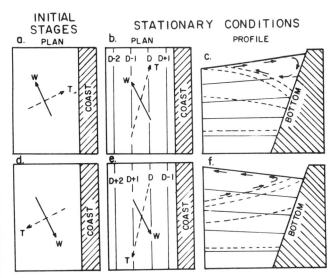

Fig. 13.13. Schematic representation of effect of wind towards producing currents parallel to a coast in the Northern Hemisphere and vertical circulation. W shows wind direction, and T, direction of transport. Contours of sea surface shown by lines marked D, $D + 1$, ... Top figures show sinking near the coast; bottom figures show upwelling (after Sverdrup, *et. al.*, 1942)

away from the coast and is replaced by denser and colder subsurface water, Fig. 13.13(f). This process is known as "upwelling" and leads to an altered distribution of density and gives rise to a current flowing in the direction of the wind, Fig. 13.13(e). The phenomenon of upwelling as illustrated in Figs. 13.13(d), 13.13(f) is typical of the California coast where the prevailing northwesterly winds, during the period from the middle of February to the end of July, cause an overturning of the upper layers in which water of relatively low temperatures is brought to the surface from depths that probably do not exceed 1000 feet. This period in which this phenomenon occurs is termed the "period of upwelling" (Reid *et al.*, 1958). Records of surface temperature during this period show that, on the coast, the lowest temperatures regularly occur in certain localities separated by regions with higher surface temperature. Sverdrup *et al.* (1942) plotted the surface temperatures as a function of latitude for the Pacific Coast from La Jolla, California, to the mouth of the Columbia River, and found that in the regions of intense upwelling (i.e., near Point Arguello and at Blunts Reef) the spring temperatures are lower than the winter temperatures, but in the regions of less intense upwelling, they are higher.

Toward the end of the summer, the upwelling gradually ceases and the more or less regular pattern of currents flowing alternately away from and toward the coast breaks down into a number of irregular eddies which may carry coastal water far out into the ocean. Other eddies carry oceanic waters in toward the coast in the regions between centers of upwelling. The period lasting from the end of July to the middle of November is

termed the *oceanic period*. During this period, a counter-current gradually develops in the surface layers which then becomes the Davidson Current during the winter months as previously discussed.

4. INERTIAL CURRENTS

Once the driving force (the wind) has stopped, the current generated by it will be under the influence of its inertia, fluid resistance at its boundaries, and the deflecting Coriolis force, and it will be a damped inertial current. Its path will be elliptical (varying from a horizontal line in the direction of original motion at the equator, to a circle at the North Pole) and its period, T in hours, given by

$$T = \frac{12}{\sin \phi}, \qquad (13.10)$$

which is half a pendulum day. The period varies from an infinite period at the equator to 12 hours at the North Pole.

5. WAVE-INDUCED CURRENTS

Wind blowing over the water surface transmits energy to the water. Part of this energy is in the form of surface currents, and part in the form of surface waves. The proportion of energy partition into these two forms of energy transmission mechanisms (direct transport and radiation) is unknown. In the region where currents and waves (called *wind waves* or a *sea*) are being generated, the motions of the water particles are complicated, as can be seen in Fig. 9.32. After the wind has died down, or the currents and waves have left the generating area, they are free (currents are now called *inertia currents* and the waves are called *swell*).

In this section, the motion of the water due to wave action alone is being considered. This would be the case of swell. The motions in the area of generation are too complicated to form a quantitative theory at present. Higher-order (second and higher) theory for Stokes waves predicts that the water particles subject only to uniform periodic waves will be transported in the direction of wave advance. Theoretical results are given in Chapter 2, together with the results for the case of null net transport. In a wave tank, and along a shore (neglecting three-dimensional effects), the amount of water transported shoreward by the waves must be compensated for by a hydraulic current flowing seaward. Comparison of theory and experiment shows fair agreement between them.

For the case in deep water, in the absence of return flow, some curves of velocity and transport are shown in Figs. 13.14 and 13.15. As this is a second-order effect, it might be expected that other second-order effects due to viscosity might modify the mass transport to a considerable extent in shallow water. This has been found

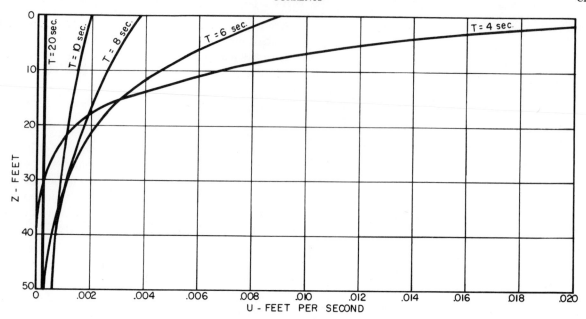

Fig. 13.14. Velocity of mass transport at various depths in deep water as a function of wave period for a wave one foot high, Eq. 2.117

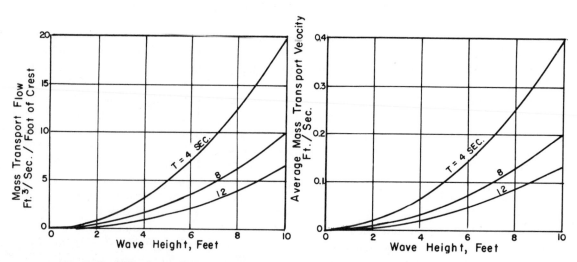

Fig. 13.15. Flow of mass transport and average velocity as a function of wave height and period for the region between the water surface and a depth of 50 ft (from Johnson and Wiegel, 1959)

to be the case. In deep water, the fact that water is viscous has little effect upon the mass transport as has been shown by laboratory measurements. The solution for the mass transport function, considering the viscosity of water, is given by Eq. 2.201, and the theoretical and experimental results shown in Figs. 2.41 and 2.42. As pointed out in Chapter 2, there are two outstanding features of this theory and its experimental verification: (1) The transport near the bottom is always in the direction of wave advance. (2) For relatively shallow water, the transport at the surface is in the opposite direction to the direction of wave advance. The first feature is of importance to the motion of sediments; both facts are of importance to the motion of sewage from sewer outfalls. The second fact might explain the drifting of

moored ships under certain circumstances and is certainly important in regard to the understanding of forces on marine pile structures.

Because of the mass transport of water, and because a wave upon breaking along a beach becomes a wave of translation, a certain amount of water piles up against the coast. A hydraulic head is thus established, and the water must return seaward. Sometimes it returns directly seaward, but in other places it flows along the beach as an alongshore current inside the breaker zone.

Along many beaches there are irregularities that cause the water to flow seaward as a narrow rip. Idealized sketches and an aerial photograph of this general process of nearshore circulation (Shepard and Inman, 1951) are shown in Fig. 13.16. As indicated in the sketch, a rip current consists of three main parts: the feeder current, which flows parallel to the shore inside the breakers; the rip current proper, which flows through the breakers in a narrow band; and the seaward portion, where the current widens and slackens. Rip currents are easy to detect from the air but more difficult to detect from the ground. A ground observer usually can distinguish a rip by a stretch of relatively unbroken water in the breaker line and patches of foam and discolored water offshore. Once rip currents have formed, they cut troughs in the sand and remain fairly stable until the wave conditions change.

Waves approaching a shoreline at an angle not only undergo the changes in velocity, height, and length for waves parallel to the shore, but they also are bent, or refracted because the inshore portion of the wave front travels at a lower velocity than does the portion in deeper water. Consequently, the waves "swing around" and tend to conform to the bottom contours. The characteristics of the bottom topography, the wave period, and the wave direction in deep water determine the pattern of the wave crests in shallow water. The result of refraction is a change in height and direction of the waves. With very irregular bottom conditions, the heights may differ greatly between closely adjacent points along a coast. The magnitude of the height and direction changes resulting from refraction can be estimated by use of a refraction diagram. The details of the construction of such diagrams, as well as some of their engineering applications, are described in Chapter 7.

Although waves tend to become parallel with the coast as a result of refraction, they usually break at a slight angle to the shore, with the result that a "littoral current" is induced and is effective in moving a mass of water slowly along the coast in the surf zone. The strength of the littoral current in terms of the wave and beach characteristics for a relatively straight shoreline has been studied by several investigators (Putnam *et al.*, 1949; Shepard, 1950a; Inman and Quinn, 1952). From an analysis of both laboratory and field data, Inman and Quinn proposed the following expression for the velocity of the littoral current:

$$V = \left[\left(\frac{1}{4x^2} + y \right)^{1/2} - \frac{1}{2x} \right]^2, \qquad (13.11)$$

where V = velocity of the littoral current (feet per second).
$\quad x = (108.3\, H_b i \cos \alpha)/T.$
$\quad y = C_b \sin x.$
$\quad C_b = \sqrt{2.28 g H_b}$ = wave velocity, (ft/sec).
$\quad \alpha$ = angle between breaker and shoreline (degrees).
$\quad H_b$ = height of breaker (feet).
$\quad T$ = wave period (seconds).
$\quad g$ = acceleration of gravity (ft/sec²).
$\quad i$ = slope of beach.

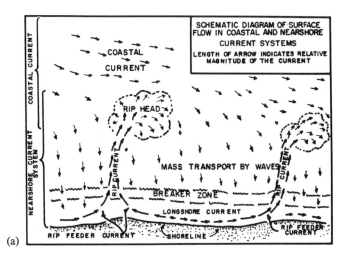

(a)

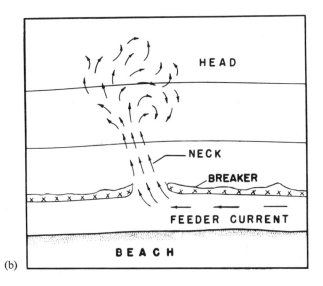

(b)

(c)

Fig. 13.16. (a) Schematic diagram of the two interrelated current systems nearshore as proposed by Shepard and Inman (1951); (b) Idealized rip current; (c) Aerial photo of rip current

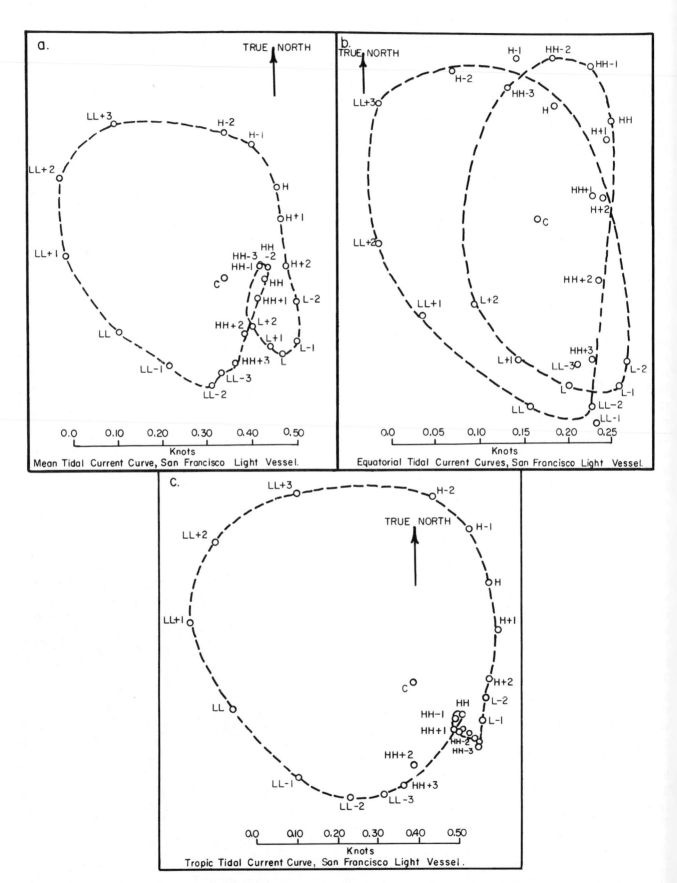

Fig. 13.17. Rotary currents at San Francisco Light Vessel (from Marmer, 1926)

One of the important effects of littoral currents is movement of sand along a coast as "littoral drift." The littoral current, combined with the agitating action of the breaking waves, is the primary factor in causing such sand movement. Studies indicate that the greatest percentage of sand transported along a coast occurs shoreward of the breaker zone (Johnson, 1956); there are indications, however, that sand is moved by waves along the bottom at depths up to as much as 170 ft (Trask, 1955; Inman, 1957).

6. TIDAL CURRENTS

The astronomical forces of the moon and sun cause tides in the ocean which have both vertical and horizontal motions. These tidal motions, combined with topographical features, give rise to three types of tidal currents: (a) the rotary type, illustrated by currents in the open ocean and along the sea coast; (b) the rectilinear or reversing type, illustrated by currents in most inland bodies of water, such as in parts of San Francisco Bay; (c) the so-called hydraulic type, illustrated by the currents in straits connecting two independently tidal bodies of water, such as the Cape Cod Canal in Massachusetts or Deception Pass in Washington. Since these three types of currents are of tidal origin, they are periodic. The tidal currents vary from locality to locality, depending upon the character of the tide, the water depth, and the configuration of the coast, but in any given locality they repeat themselves as regularly as the tides to which they are related. In the open ocean, the tidal currents usually are rotating due to the effect of the Coriolis force; i.e., from hour to hour the currents change in both direction and speed. In the Northern Hemisphere, the change in direction is clockwise, and the current completes one rotation in about 12 hours where the tide is semidiurnal and in about 24 hours where the tide is diurnal. The net transport of water in either case is zero. A typical example of a rotary current is the tidal current observed at the San Francisco Lightship (Fig. 13.17; Marmer, 1962). This current rotates and completes one rotation in about 24 hours. The diurnal inequality of the tide at this lightship is so great that the current is very largely diurnal; i.e., during the greater part of the month the current changes direction at the rate of about 15 degrees per hour, giving but one strength of flood and one strength of ebb in a day.

In recent years, the problem of circulation in estuaries has been studied extensively by such investigators as Tully (1949), Rhodes (1950), Redfield (1951), Simmons (1952), Stommel (1953a and 1953b), Ketchum (1953), Farmer and Morgan (1953), Pritchard (1955), Todd and Lau (1956), Stewart (1958), and others. Pritchard (1952 and 1955), for example, classifies estuaries into four types, each with a distinct density stratification and circulation pattern, and concludes that an estuary tends

to shift from type A (highly stratified) through type B (moderately stratified) to type C or D (vertically homogeneous) as a result of such factors as decreasing river flow, increasing tidal velocities, increasing width, or decreasing depth. The Mississippi River, Fig. 13.18(a), for example, is typical of the type A estuary; whereas San Francisco Bay, Fig. 13.18(b) is typical of type D. Redfield (1951), on the other hand, gives three types of estuarine circulation which are classified more as to the geometry of the estuary. These are (a) a narrow estuary with complete vertical mixing normal to the axis, (b) a deep estuary with vertical stratification, (c) a wide estuary with a symmetrical circulation due to the earth's rotation.

Where there is a definite salt water wedge, as in Fig. 13.18(a), the time of occurrence of slack water will vary between the top and bottom layers, as well as between flood and ebb flows. This condition will occur to some extent in San Francisco Bay at times when the fresh

Fig. 13.18. (a) This shows schematically the distribution of flow in an estuary having a well-defined salt-water wedge (similar to lower Mississippi River). Flow in the wedge is always upstream, and flow in the fresh-water strata is always downstream. The thickness of the interfacial layer varies with the fresh-water discharge. (b) This shows schematically the distribution of salinity in an estuary in which mixing caused by tidal turbulence does not permit formation of a well defined salt-water wedge (similar to San Francisco and Delaware Bays). The salt-water front advances and retreats with tidal action, but there is little difference in salinity from surface to bottom (after Simmons, 1954)

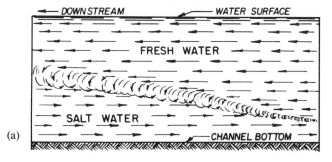

(a)

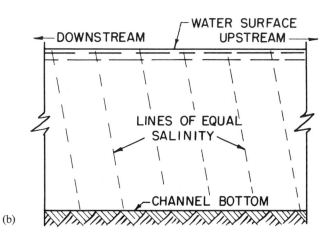

(b)

328 CURRENTS CHAP. 13

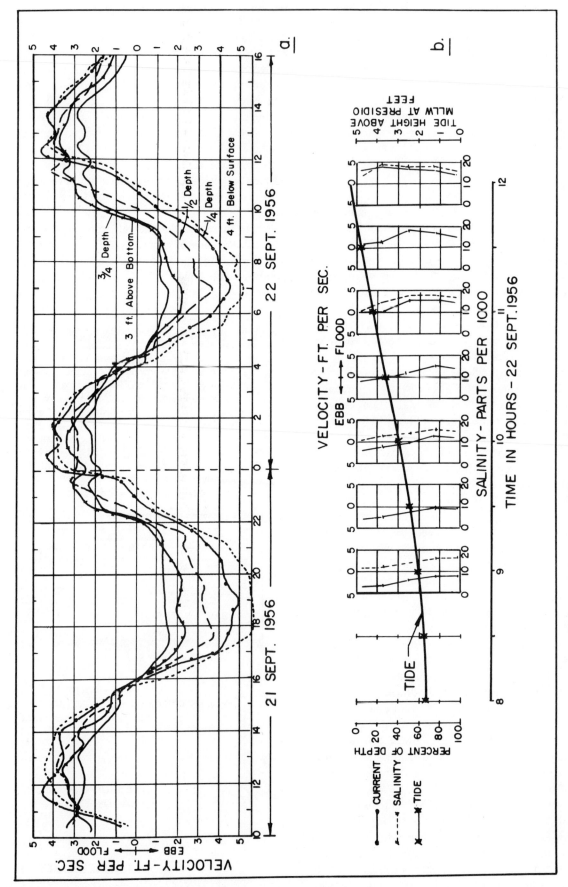

Fig. 13.19. Variation of velocity and salinity with depth and time in Carquinez Straits, California, September 21–22, 1956 (courtesy of the San Francisco District, Corps of Engineers)

water inflow is relatively small. Of particular interest on this point are the results of some of the observations taken in San Francisco Bay by the Corps of Engineers to provide data necessary to the operation of the new hydraulic model of the bay. For example, Fig. 13.19(a) shows the variation in velocity at five different depths from near surface to near bottom in Carquinez Straits over a two-day period when the fresh water flow from the Sacramento and San Joaquin Rivers was relatively small. Similar curves are available for periods during the winter months when the river flows were relatively large. It is obvious from Fig. 13.19(a) that there is considerable variation in velocity in the vertical at any particular time; however, to give a clearer indication of the velocity variation, both with depth and time, Fig. 13.19(b) has been prepared. This figure shows the vertical distribution of both velocity and salinity at various times during the rising tide on September 22, 1956. Although these distribution curves are roughly vertical, it is evident that the higher salinity generally occurs in the lower part of the channel. It is also of interest to note that at about midtide the high salinity flood currents are

moving upstream along the bottom while lower salinity ebb currents exist at the surface. As high tide is approached, the current is in flood throughout the depth, and near low tide the currents are in ebb throughout the depth.

Similar curves prepared for times when a relatively large fresh water flow was moving seaward showed that the salt wedge was almost eliminated, and the velocity distribution was almost vertical. The salinity distribution was relatively uniform throughout the depth, except at high tide, thus indicating that considerable mixing occurred with the large fresh water flows.

Another method of illustrating the variation of flow with depth are the plots presented in Figs. 13.20(a), 13.20(b), which show the predominance of ebb or flood with depth over one tidal cycle for two different periods —the first period was on September 21–22, 1956, when the fresh water flow was relatively small; the second was on March 3–4, 1958, when the fresh water flow was relatively large. The procedure in preparing such curves is illustrated in Fig. 13.20(c) where the ebb predominance at a particular depth is obtained by planimetering the

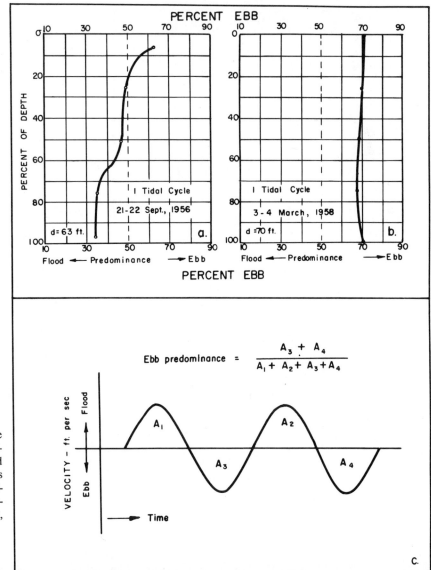

Fig. 13.20. Examples of ebb predominance curves for Carquinez Straits, California for (a) the dry season and (b) the wet season. (c) shows method of obtaining ebb predominance at a particular depth. (courtesy of the San Francisco District, Corps of Engineers.)

areas under the velocity curve obtained for the particular depth and then computing the predominance as indicated. This procedure is repeated for each of the five velocity curves shown in Fig. 13.19(a).

The significance of the predominance curves presented in Fig. 13.20 is that the quantity of flow at various depths is readily apparent. For example, Fig. 13.20(a) indicates that, in the upper 20 per cent of the depth, the ebb flow is in excess of 50 per cent of the total flow; whereas, in the lower 80 per cent of the depth, flood currents are predominant, i.e., a flat wedge moves upstream along the bottom.

Consider next the curve shown in Fig. 13.20(b). Here due to the large fresh water flow, the predominance curve is essentially vertical; i. e., that salt wedge has disappeared and the ebb flow predominates throughout the vertical with approximately uniform intensity. For a detailed discussion of flow predominance curves, especially as they apply to the problem of silt deposition in an estuary, the reader is referred to the work of Schultz and Simmons (1957).

The so-called hydraulic currents are those occurring in straits and canals which connect two bodies of water which have independent tides. They are reversing currents which are due primarily to a temporary difference in head between two bodies brought about by tidal action, rather than by action of a progressive or stationary type of wave passing through the channel as discussed by Brown (1932). The hydrodynamic problems involved in the problem of hydraulic currents are highly involved and even where solutions have been found they necessitate a great deal of complicated numerical work. For a discussion of the theory of hydraulic currents, and the application to specific cases, the reader is referred to the work of Rude (1928), Harwood (1936), Pillsbury (1956), and Wilcox (1958).

7. MAJOR OCEAN CURRENTS

a. *Introduction.* Far more is known of the surface currents of the ocean than of the deeper currents. Most of the observations of the surface currents were obtained from ship drift observations. An example of the type of data available on surface currents, together with the names of the major ocean currents, is given in Fig. 13.21 (see Schott, 1943, for a larger, more detailed chart). Charts of monthly mean surface currents for most areas are also available (U.S. Navy Hydrographic Office, see Fig. 15.1). With the better and more numerous instruments for measuring currents that have become available in recent years, the complexity of the ocean's currents is

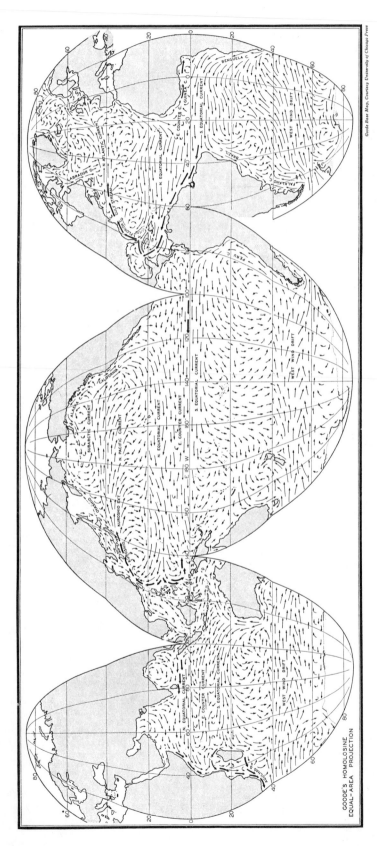

Fig. 13.21. Surface currents of the oceans in February–March (from Sverdrup, *et al.*, 1942) Goode Base Map, Courtesy of University of Chicago Press

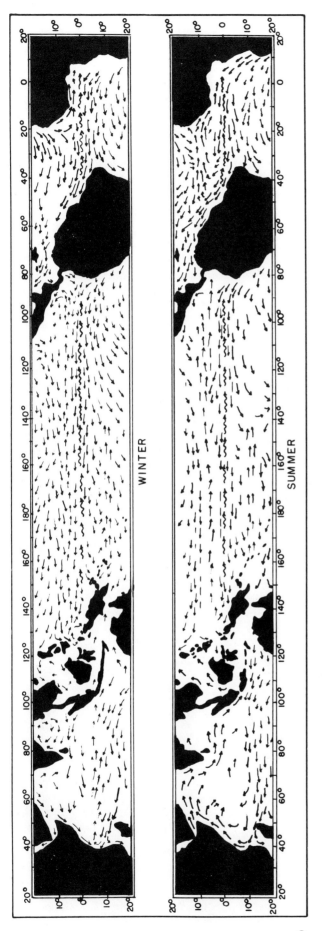

WINTER

SUMMER

(a)

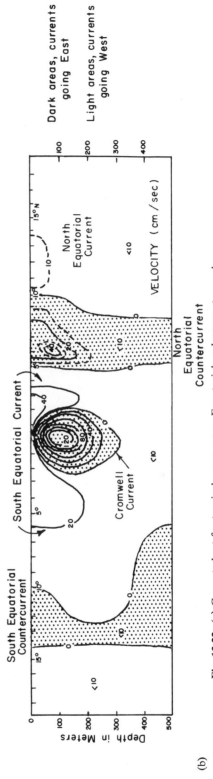

Dark areas, currents going East

Light areas, currents going West

(b)

Fig. 13.22. (a) Current chart for tropical ocean areas. Equatorial undercurrents are shown as wavy lines; (b) Equatorial zonal currents, Pacific Ocean (reprinted with permission from J.A. Knauss, *Equatorial Current Systems*, New York: John Wiley & Sons, Inc., 1963)

331

becoming more evident. Shallow, subsurface currents are being discovered, such as the Cromwell Current (Cromwell, Montgomery, and Stroup, 1954; Knauss, 1960), Fig. 13.22. Use of the Swallow float (Knauss, 1961) has resulted in the measurements of very deep (a thousand fathoms), slow-moving (1/20 ft per sec) currents.

Wind is the prime source of the energy which the currents need in order to maintain their motion. The situation is more complicated than the foregoing statement would indicate because the atmosphere-ocean is a coupled system. The mixing that occurs in the top layer of the water results in the storage of a considerable amount of heat in a thick layer, whereas on land heat must be transmitted down into the rock by the slow process of conduction (Sverdrup, 1943). Furthermore, the currents transport vast quantities of water (hence, heat); for example, Stommel (1958, p. 163) estimates the net transport of the Gulf Stream system to be 35×10^6 cu meters per second. This transport of warm water north results in the transfer of heat from an area of total annual surplus of energy received by the ocean to an area of annual loss of energy from the ocean to the atmosphere (Sverdrup, 1943, p. 230). The fact that the Atlantic Polar Front lies just to the east of the region of high energy transfer from the sea to the atmosphere in the Atlantic Ocean (associated with the Gulf Stream) and the Pacific Polar Front lies just to the east of a similar region (associated with the Kuroshio) in the Pacific Ocean strongly suggests that the ocean currents may be partly responsible for them. These fronts, in turn, influence the wind systems blowing over the oceans, and the winds, in turn, cause the movement of the water.

b. *Wind-induced surface currents.* There have been attempts to predict the main features of the circulations of the oceans, from a consideration of the wind field over the ocean, and the Coriolis deflecting force (Sverdrup, 1947; Reid, 1948a, b; Hidaka, 1949; Munk, 1950; Munk and Carrier, 1950); some of these attempts have included the effect of lateral stresses. For details of the theories, the reader is referred to the original papers, and to the books by Stommel (1958) and Defant (1961) which discuss the various theories.

Munk's (1950) result for a rectangular plane ocean can be seen in Fig. 13.23 for the case of zonal winds (winds blowing in a direction parallel to the latitudinal lines) in the Northern Hemisphere portion of the Pacific Ocean. Some of the main features of the observed oceanic features are predicted by this theory, such as the intensification of the currents along the western boundaries. It should be noted that the intensification of the currents along the western boundaries has been shown by Stommel (1958) to be due primarily to the fact that the Coriolis force varies with latitude. A countercurrent eastward of the western boundary currents is predicted as are the existence of a number of gyres. It was found that the boundaries between these gyres were determined by the latitudes at which the zonal wind shear stresses in the east-west direction went through their maximum or minimum values (that is, when $d\tau_x/dy = 0$, where x is positive to the east and y is positive to the north).

Munk also examined the effect of some meridional winds and decided that the currents resulting from the winds were responsible for the eastern boundary currents such as the California Current system, the Peru Current and the Benguela Current.

As a result of these studies, Munk suggested a nomenclature for the major circulation features associated with the surface currents. This is shown in Fig. 13.24, and the identification of these general features with specific currents is made in Table 13.6.

There are, however, major features of the circulation

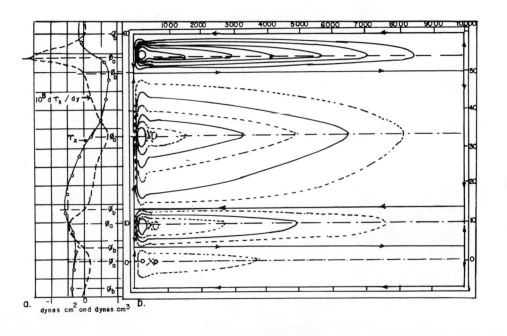

Fig. 13.23. (a) The mean annual zonal wind stress $\tau_x(y)$ over the Pacific Ocean and its curl $d\tau_x/dy$ (b) Mass transport streamlines $\psi(x, y)$, with transport between adjacent lines being 10 million metric tons per second

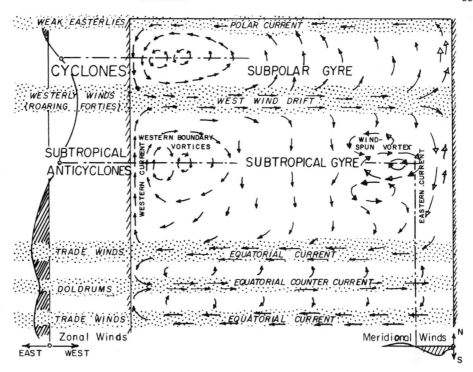

Fig. 13.24. Schematic presentation of circulation in a rectangular ocean resulting from zonal winds (filled arrowheads), meridional winds (open arrowheads), or both (half-filled arrowheads). The width of the arrows is an indication of the strength of the currents. The nomenclature applies to either hemisphere, but in the Southern Hemisphere the subpolar gyre is replaced largely by the Antarctic circumpolar current (west wind drift) flowing around the world. Geographic names of the currents in various oceans are summarized in Table 13.6 (from Munk, 1950)

of the ocean that are not predicted by Munk's work—for example, the Cromwell Current in the Equatorial Pacific which has a total transport of about 40×10^6 cu meters per second (Knauss, 1961), which is approximately the same as the transport by the Gulf Stream. Nor did extensions of Munk's work to higher-order approxima-

Table 13.6. The General Ocean Circulation
(from Munk, 1950)

Classification (see Fig. 13.24)	North Atlantic	South Atlantic	North Pacific	South Pacific	North Indian	South Indian
Polar current (z)	E. Greenland (3)	(p?)	(p)	(p (w))		(p)
Western c. (z)	Labrador (6)?	Falkland	Oyashio	(a)		(a)
W. bdry vortex (z)	p? (50°N, 50°W)	(a?)	(?)	(a)		(a)
Eastern c. (m)	Norwegian (3)	(a)	Alaska	(a)	A	(a)
Wind-spun vortex (z, m)	(a)	(a?)	Gulf of Alaska Aleutian (15)	(a)	S	(a)
					I	
West wind drift (z)	⌐ N. Atlantic (38)	West wind drift	⌐ N. Pacific (20)	West wind drift	A	West wind drift
	Gulf Stream (70)		Kuroshio Ext. (65)			
Western c. (z)	Florida (26)	Brazil (7)	Kuroshio (23)	E. Australia?		Agulhas (25)
W. bdry vortex (z)	⌐ Sargasso Sea	(p)	(p)	(p)		(p)
Eastern c. (m)	Canary?	Benguela (16)	California (15)	Peru (13)		(p (v))
Wind-spun vortex (x, m)	Azores vortex	(p, 30°S, 5°W)	E. Pac. vortex	(p, 30°S, 100°W)	————	(p, 30°S, 95°E)
Equatorial current (z)	N. Equat. c. (32)	S. Equat. c.	N. Equat. c. (45)	S. Equat. c.	N. Equat. c.	S. Equat. c.
Equat. counter current (z)	Equat. Cc (4)	(a)	Equat. Cc (25)	(a)	(a)	Equat. Cc

Subpolar Gyres (left bracket for Western c. through Wind-spun vortex); *Subtropical Gyres* (left bracket for West wind drift through Wind-spun vortex). Gulf Stream S. labels the bracket under North Atlantic; Kuroshio S. labels the bracket under North Pacific.

a: absent; p: present, but unnamed; m: related to meridional winds; S: system; v: variable; w: weak; z: related to zonal winds. Figures in parentheses give mass transport in 10^{12} g sec^{-1} estimated from oceanographic observations.

tion (Munk, Groves, and Carrier, 1950) or to triangular planes (Munk and Carrier, 1950) lead to the prediction of a countercurrent on the western side of the Gulf Stream, the countercurrents beneath the Gulf Stream, or the complexity of the North Atlantic currents in the Gulf Stream system (Fig. 15.13).

Although Munk's use of Sverdrup's integrated mass transport as the dependent variable permits the development of the theory for a baroclinic ocean, the results do not show the interesting details of velocity distribution with depth and the relationship of the velocity to such features as the thermocline. An example of simultaneous measurements of temperature and current speed variation with depth in the Pacific Equatorial Countercurrent show these to be intimately related (Knauss, 1961), as can be seen in Fig. 13.25.

c. *Jets.* One effect of the winds that blow to the west in the tropical latitudes north of the equator is to cause water to pile up in the Gulf of Mexico. Sverdrup (1943, p. 181) states that the average sea level at Cedar Keys, which is off the southwest tip of Florida, is 19 cm higher than the average sea level off St. Augustine, Florida. Neglecting all other forces this would result in a current

Fig. 13.25. Comparison of simultaneous velocity and temperature profiles through the counter-current made by attaching an electronic bathythermograph to the Roberts meter. The *BT* recorded continuously. The instruments were stopped at different depths (noted by dots on the temperature profile), and current observations were made. The uncertainty of any given current observation (15 cm/sec) is noted. Surface observations were made with drogues and have a smaller uncertainty (5 cm/sec) (from Knauss, 1961)

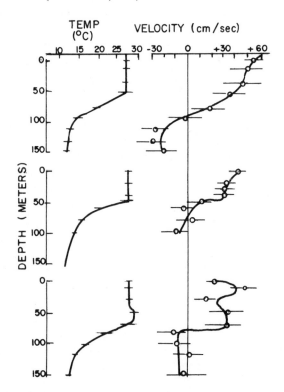

velocity $v = \sqrt{2gh} = 193$ cm/sec (3.75 knots), where h is the difference in sea level of 19 cm. This is about the speed of the Florida current as measured by Murray (1952). One is tempted to consider the Florida Current, and its extensions, the Gulf Stream and the North Atlantic Current as a jet. This has been done by Rossby (1936). Rather than develop a jet theory from the standpoint of the conservation of momentum, the conservation of vorticity was used. This was based upon a theoretical study by Taylor (1932) who showed that in a rotating or curved system the use of conservation of momentum or the conservation of vorticity led to conflicting conclusions, and the findings resulting from a conservation of vorticity were more appropriate. Details of Rossby's work will not be given here; for some information on jets, see Chapter 16. In general, use of the jet theory predicts a considerable mixing of the sea water, a gradual increase in volume flow downstream, a western countercurrent and an eastern compensating current.

One possibility that should be considered in studying the Gulf Stream system is that the general circulation of the North Atlantic Ocean as predicted by Munk (1950) is complicated by the geography. Thus, part of the transport of the South Atlantic Equatorial current is deflected northward by Brazil and joins a part of the transport of the North Pacific Equatorial current, with part of the combined flows moving into the Gulf of Mexico. A part of the North Atlantic Equatorial current, acting more in accordance with Munk's circulation theory, flows as the Antilles Current and joins the Florida Current soon after it leaves the Straits of Florida. The average transport through the Straits of Florida is about 27×10^6 cu meters per second although it varies considerably (Fig. 13.4). The Antilles Current transports about 12×10^6 cu meters per second (Sverdrup, 1943, p. 165). The total is 41×10^6 cu meters per second which is somewhat in excess of the total transport of the Gulf Stream as estimated by Stommel (1958); however, as Stommel pointed out, the estimates of the Gulf Stream transport are as great as 123×10^6 cu meters per second.

d. *Thermohaline circulation.* The fact that one portion of the ocean receives a greater net amount of heat than another part results in a slow circulation of the oceans' water. The process is complicated by the fact that the water density is dependent on the salinity as well as the temperature of the water; hence, the term *thermohaline.* This means that evaporation from the ocean surface, rainfall over the ocean, river runoff, and ground water flow into the ocean along its shores must be taken into consideration. The problem is further complicated by the shape of the oceans' boundaries and the distribution of net heat absorption areas of the oceans, Fig. 13.26.

V. Bjerknes has shown (see Sverdrup, 1943, p. 150) that for heat energy to be transformed into mechanical energy, the heating must take place under higher pressure

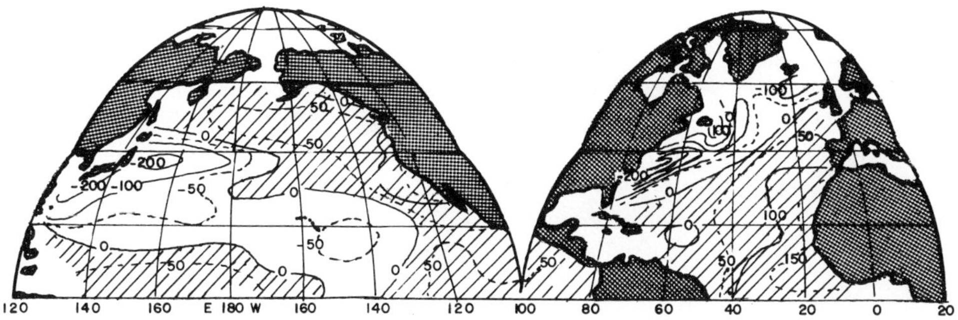

Fig. 13.26. Total annual surplus of energy received by the ocean water (gram-calories per square centimeter per day) when the exchange of energy with the atmosphere is taken into account. Areas with positive surplus are shaded (from H. U. Sverdrup, 1942)

than the cooling. Thus, the atmosphere is a relatively good heat engine as the air is heated close to the ground where the pressure is high and cooled at a great height where the pressure is low. The ocean, on the other hand, is a poor heat engine with both heating and cooling occurring at the surface. It does act as a heat engine to some extent as the warm surface water flows toward the higher latitudes, gradually cooling, and the cold water that sinks in the regions of high latitude flows toward the lower latitudes and is gradually heated by conduction; hence, some heating occurs at a greater pressure than some cooling. Although no theory seems to take it into consideration, this process is complicated by the fact that there is a net flow of heat from the earth to the bottom waters of about one-tenth of a gram calorie per square centimeter per day (Maxwell and Revelle, 1957).

Stommel (1958, p. 153) has considered thermohaline circulation in conjunction with the circulation due to winds in the North Atlantic and predicted a countercurrent beneath the Gulf Stream which appears to have been shown to exist (Swallow and Worthington, 1961). Wyrtki (1961) has developed the theory of the circulation due only to the heat distribution in the ocean, for a hemispheric ocean (no continents). He found that the meridional transport would be from 3 to 10×10^6 cu meters per second for the different oceans, these values being from 10 to 20 per cent of the total meridional transports. It was also concluded that the upward velocities through the thermocline would be from 2 to 5×10^{-5} cm per second.

e. *Geostrophic relationship.* There is a well-known set of equations relating the distribution of sea water density to strength and direction of currents. In their simplest form they are known as the hydrostatic and geostrophic equations; all acceleration terms are considered to be negligible, and all forces are neglected except the pressure term (related to the distribution of sea water density) and the deflecting Coriolis force. In a plane representing a limited extent of the ocean's surface, with the x coordinate positive toward the east, the y coordinate positive

toward the north, and the z coordinate positive downward, the equations become (Sverdrup, 1943, p. 92)

$$\rho_w g - \frac{\partial p}{\partial z} = 0 \qquad \text{(hydrostatic eq.)} \qquad (13.12)$$

$$u = -\frac{1}{\rho_w 2\Omega \sin \phi} \frac{\partial p}{\partial y} \qquad\qquad (13.13)$$
$$\text{(geostrophic eqs.)}$$
$$v = \frac{1}{\rho_w 2\Omega \sin \phi} \frac{\partial p}{\partial x} \qquad\qquad (13.14)$$

in the Northern Hemisphere, where ρ_w is the mass density of sea water, p is the pressure, and the other symbols are as defined previously. As pointed out by Sverdrup, there is no inference in the equations as to cause and effect.

The geostrophic equations have been found to be very useful in computing the relative velocity distributions of many of the currents in the ocean. The reason for the usefulness of these equations has been pointed out by Stommel (1958, p. 17), who calculated that in the Gulf Stream, for example, the forces associated with local accelerations, convective accelerations, lateral shearing stresses, and wind stress were each of the order of about one-tenth the Coriolis force and forces due to the horizontal pressure gradients.

The results, however, are only approximate. For example, Knauss (1961) in three sections across the Pacific Equatorial Countercurrent found that measured values of the current speed were from 20 to 50 per cent greater than predictions made using geostrophic equations. Calculations made for the current due to winds blowing on the region of the measurements, together with the appropriate values of the Coriolis force and horizontal pressure gradients due to the measured density field of the sea water, indicated that the resulting current would be about one-third greater than the "geostrophic current." This is in better agreement with the measured current.

As previously stated the geostrophic equations can be used to compute relative velocities. In order to obtain the actual velocities, the velocity must be known for one value of z. This is often taken as zero at some "level of no motion," with this being taken as the bottom by some, or at some intermediate depth by others. Stommel (1958, p. 19) pointed out that this level must be determined by direct measurement. This has not been done directly, but has been done indirectly for one line across the Gulf Stream (Small and Worthington, 1961). In a section from Cape Romain, South Carolina, to Bermuda the level of no motion was determined to be at about 1000 fathoms.

8. DATA RECORDING AND ANALYSIS

One of the main problems in recording data today is not merely trying to record some small signal, but rather to record it in a form that can be rapidly analyzed. This is relatively difficult because, until a considerable insight on the problem is available, it is not possible to group a series of components in such a manner as to get a maximum efficiency in the analyses. For example, for many problems where a large number of data must be reduced, it probably would be best to record the data on magnetic tape or on punched tape or cards for machine analysis. Neither of these systems, however, lends itself easily to scanning by eye. It is this scanning by eye that is necessary in the early stages of investigation, before the phenomena are well understood. It would be desirable to record the data simultaneously on both standard chart paper for visual scanning and on magnetic tape for machine scanning. This, of course, would mean a much higher expenditure for equipment.

If it is necessary to standardize upon data storing and data analysis techniques, it should be done in digital form because many measurement techniques give data only in this form, e.g., the motion of a float, the motion of dye, the readout of the Roberts radio current meter every so many revolutions of the runner, etc. In addition, data in digital form can be handled easily by machines in performing harmonic analyses.

The current as measured at any particular location will consist of all or a portion of various superimposed current components: tidal currents, oceanic circulation (sometimes called the *mean current*), wind drift currents, inertial currents, currents associated with waves, hydraulic currents, currents due to fresh water run-off from land, random currents of unknown cause, and for some installations, fluctuations which are not currents but are due to the motion of the moored station relative to the water.

A paper by Bowden (1954) on the direct measurement of subsurface currents in the ocean includes a discussion of the methods of analysis of ocean current records. The first step is to plot the observations of speed and direction as a function of time, if the records are not already in this form. It should then be possible to detect and correct for gross errors in the records. Then these curves should be

converted into the N-S, E-W velocity components, each being plotted as a function of time. It might be well to use a lunar hour time scale rather than a solar hour time scale. An example of a record obtained on a series of observations made by Swallow (1955) of a deep floating buoy are shown in Fig. 13.27(a). The dates and times of the fixes are given on the plot. The components are shown in Fig. 13.27(b). Then a harmonic analysis should be made to determine the tidal components, and these should be subtracted from the record. The method of plotting the tidal components depends upon whether they are reversing currents (as in an estuary) or rotary. An example of the plot of the rotary current obtained by Swallow (1955) is shown in Fig. 13.27(c). The mean drift can be obtained readily from the plot of the N-S and E-W components, which for the case shown in Fig. 13.27(b) is 2.4 cm/sec in the direction 300 degrees (true)—(this particular case is a deep current in the open ocean, hence the low speed). The residuals are shown in Fig. 13.27(c). Bowden (1954) emphasizes that in order for the harmonic analyses to be reliable in regard to both amplitude and phase, they must be made for a number of complete cycles. A thorough discussion of this problem has been given by Haurwitz (1952, 1953). If the records are short in length, coefficients may be obtained by Fourier analysis which have little more validity than if they were obtained from a random time series. The statistical tests that should be performed to check their relative validity have been given in detail by Haurwitz.

The possibilities in the analysis of the residuals and the mean current are so numerous that Bowden (1954) recommends that each set of records be considered on its own.

One further harmonic analysis that can be made of the record is for the period of inertial currents, which are half a pendulum day. The period is given by $T = 12/\sin \phi$. It is convenient to use 24 equally spaced intervals of the lunar day (24 hours, 50 minutes), 31 minutes.

With the current data and other data, such as wind speed and direction, wave height, period, and direction, put on punched cards, it should be possible first to obtain the rotary or reversing tidal currents and their correlations with the published data for standard U.S. Coast and Geodetic current stations; then to obtain correlations between variations in the tidal currents and the wind and wave conditions; and finally to obtain correlations between the residual drift currents and wind and wave conditions. The drift current direction could be plotted as a function of wind direction, and its speed could be plotted as a function of wind speed and duration. The same could be done for the mass transport velocity associated with waves. Once these correlations have been determined, maximum use could be made of the tidal current data and oceanic circulation data and the wind data available in the files of old weather

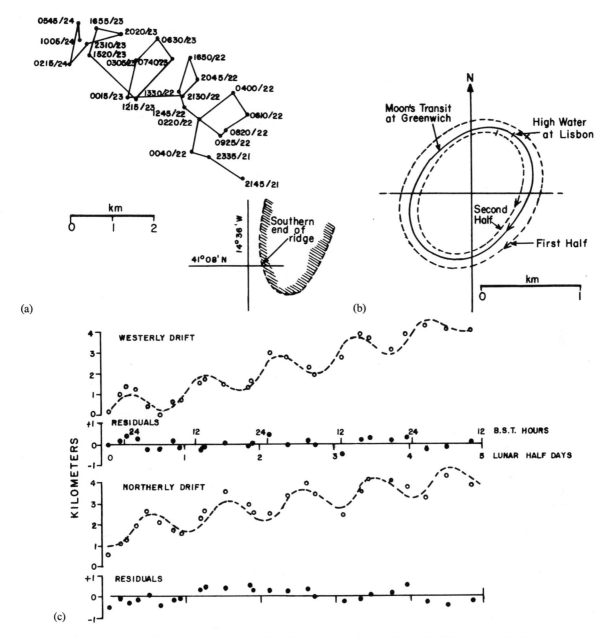

Fig. 13.27. Example of the partial analysis of an ocean current: (a) Track of float, June 21–24, 1955; (b) Tidal components of displacement combined to form an eclipse; (c) Northward and westward movements of float plotted against time (after Swallow, 1955)

maps. These data have been accumulated for many years and they can be used to develop a broad statistical picture of the currents to be expected. The "mean" currents can be obtained from the afore mentioned files and the wind patterns can be obtained from the weather maps. These data, together with the correlations obtained during the study plus the techniques available for predicting ocean waves from weather maps, permit the prediction of variations of the currents from the mean values. If the current data obtained during the study are used only to plot a current rose directly, then the data cannot be extended.

REFERENCES

Arlman, J. J., P. Santema, and J. N. Svasek, *Movement of bottom sediment in coastal waters by currents and waves; measurements with the help of radioactive tracers in the Netherlands*, U.S. Army, Corps of Engineers, Beach Erosion Board, Tech. Memo. No. 105, March, 1958.

Bowden, K. F., Measurement of wind currents in the sea by

the method of towed electrodes, *Nature*, **171**, 4356 (April, 1953), 735–37.

Brown, E. I., Flow of water in tidal canals, *Trans. ASCE*, **96**, (1932), 749–813.

Carruthers, J. N., A. L. Lawford, and V. F. C. Veley, Water movement at the North Goodwin Light Vessel, *Marine Observer* (January, 1951), pp. 36–46.

Charnock, H., J. R. D. Francis, and P. A. Sheppard, An investigation of wind structures in the trades, Anegada, 1953, *Phil. Trans. Roy. Soc.* (London), ser. A, **249**, 963 (1956), 179–234.

Cromwell, T., R. B. Montgomery, and E. D. Stroup, Equatorial undercurrent in Pacific Ocean revealed by new methods, *Science*, **119** (May 7, 1954), 648–49.

Deacon, E. L., P. A. Sheppard, and E. K. Webb, Wind profiles over the sea surface and the drag at the sea surface, *Australian J. Physics*, **9**, 4 (December, 1956), 511–41.

Defant, Albert, *Physical oceanography*, Vol. 1. London and New York: Pergamon Press, 1961.

Disney, L. P., and W. H. Overshine, *Tides and currents in San Francisco Bay*, U.S. Coast and Geodetic Survey Special Pub. No. 115, 1925.

Ekman, V. W., On the influence of the earth's rotation on ocean currents, *Arkiv. f. Matem., Astr. o. Fysik* (Stockholm), **2**, 11, 1905.

Farmer, H. G., and G. W. Morgan, The salt wedge, *Proc. Third Conf. on Coastal Eng*, Berkeley, Calif.: The Engineering Foundation, Council on Wave Research, 1953, pp. 54–64.

Francis, J. R. D., Wind stress on a water surface, *Quart. J. Roy. Met. Soc.* (London), **80**, 345 (July, 1954), 438–43.

Francis, J. R. D., Wind action on a water surface, *Proc. Inst. Civ. Engrs.*, **12** (February, 1959), 197–216.

Fuglister, F. C., and L. V. Worthington, Some results of a multiple ship survey of the Gulf Stream, *Tellus*, **3**, 1 (February, 1951), 1–14.

Goldsbrough, G. R., On ocean currents produced by winds, *Proc. Roy. Soc.* (London), ser. A, **148**, 863 (January, 1935), 47–58.

Harwood, E. C., Proposed improvements of the Cape Cod Canal, *Trans. ASCE*, **101** (1936), 1440–75.

Haurwitz, B., *On the reality of internal lunar tidal waves in the ocean*, Woods Hole Oceanog. Inst., Ref. No. 52–71, September, 1952. (Unpublished.)

———, Internal tidal waves in the ocean, Woods Hole Oceanog. Inst., Ref. No. 53–69, September, 1953. (Unpublished manuscript.)

Hidaka, Koji, Drift currents in an enclosed sea and the rotation of the Earth, *Trans. Amer. Geophys. Union*, **28**, 4, (August, 1947), 549–58.

———, Mass transport in ocean currents and lateral mixing, *Geophys. Notes*, Univ. Tokyo, **2**, 3 (1949), 1–4.

Hughes, P., A determination of the relation between wind and

sea-surface drift, *Quart. J. Roy. Met. Soc.* (London), **82**, 354 (October, 1956), 494–502.

Inman, D. L., *Wave generated ripples in nearshore sands*, U.S. Army, Corps of Engineers, Beach Erosion Board, Tech. Memo. No. 100 October, 1957.

———, and W. H. Quinn, Currents in the surf zone, *Proc. Second Conf. Coastal Eng.* Berkeley, Calif.: The Engineering Foundation Council on Wave Research, 1952, pp. 24–35.

Jennings, F. D., and R. A. Schwartzlose, *Deep sea current survey off Monterey*, Scripps Inst. Oceanog. March, 1958. (Unpublished.)

Johnson, J. W., Dynamics of nearshore sediment movement, *Bull. Amer. Soc. Petroleum Geologists*, 40, 9 (September, 1956), 2211–32.

———, and E. K. Rice, A laboratory investigation of wind-generated waves, *Trans. Amer. Geophys. Union*, **33**, 6 (December, 1952), 845–54.

———, and R. L. Wiegel, *Investigations of current measurement in estuarine and coastal waters*. State Water Pollution Control Board, Sacramento Calif.: No. 19, 1959.

Ketchum, B. H., Circulation in estuaries, *Proc. Third Conf. Coastal Eng.* Berkeley, Calif.: The Engineering Foundation Council on Wave Research, 1953, pp. 65–76.

Knauss, John A., Measurements of the Cromwell Current, *Deep Sea Res.* **6**, 4 (June, 1960), 265–86.

———, The structure of the Pacific Equatorial counter-current, *J. Geophys. Res.*, **66**, 1 (January, 1961), 143–55.

———, *Equatorial current systems, The sea: ideas and observations*, Vol. II. New York: John Wiley & Sons, Inc., 1963, 235–52.

Kruger, W., Meer und Küste bei Wangeroog und die Kräfte, die auf ihre Gestaltung einwirten, *Z. Bauwes.*, **61**, 1911.

Krummel, O., Handbuch der Ozeanographie, Vol. II. Stuttgart: J. Engelhovns Nachf., 1911.

Malkus, W., and K. Johnson, *Atlantis Cruise 198 and Caryn Cruise 78, a drift study of the Gulf System*, Woods Hole Oceanog. Inst. Ref. No. 54–67, August, 1954. (Unpublished.)

Mandelbaum, H., Wind-generated ocean currents at Arum Bank Lightship, *Trans. Amer. Geophys. Union*, **36**, 6, (December, 1957), 867–78.

Marmer, H. A., *Coastal currents along the Pacific Coast of the United States*, U.S. Coast and Geodetic Survey, Special Pub. No. 121, 1926.

Maxwell, A. E., and Roger Revelle, Heat flow through the Pacific Ocean Basin, *Travaux Scientifiques, Bur. Central Seismologique Intern.*, ser. A, Part 19 (Jan. 25, 1957), 395–405.

Munk, W. H., On the wind-driven ocean circulation, *J. Met.*, **7**, 2 (April, 1950), 79–93.

———, and G. F. Carrier, The wind-driven circulation in ocean basins of various shapes, *Tellus*, **2**, 3 (August 1950), 158–67.

————, G. W. Groves and G. F. Carrier, Note on the dynamics of the Gulf Stream, *J. Mar. Res.*, **9**, 3 (1950), 218–38.

Murray, Kenneth M., Short period fluctuations of the Florida current from Geomagnetic electrokinetograph observations, *Bull. Mar. Sci. Gulf and Caribbean*, **2**, 1 (November, 1952), 360–75.

Pillsbury, G. B., *Tidal hydraulics*, rev. ed. U.S. Army, Corps of Engineers, Waterways Experiment Station, 1956.

Pritchard, D. W., Estuarine circulation patterns, *Proc. ASCE*, **81**, separate no. 717 (June, 1955).

Proudman, J., *Dynamical oceanography*, London: Methuen & Co. Ltd., 1953.

Putnam, J. A., W. H. Munk, and M. A. Traylor, The prediction of longshore currents, *Trans. Amer. Geophys. Union*, **30**, 3 (June, 1949), 337–45.

Redfield, A. C., The flushing of harbors and other hydrodynamic problems in coastal water, in *Hydrodynamics in Modern technology*. Cambridge: Mass. Inst. Tech., 1951, pp. 127–35.

Reid, J. L., Jr., G. I. Roden, and J. G. Wyllie, *Studies of the California current system*, California Cooperative Oceanic Fisheries Investigations, Progress Report, July 1, 1956– January 1, 1958, 1958, pp. 27–56.

Reid, R. O., The equatorial currents of the eastern Pacific as maintained by the stress of the wind, *J. Mar. Res.*, **7**, (1948a). 74–99.

————, A model of the vertical structure of mass in equatorial wind-driven currents of a baroclinic ocean, *J. Mar. Res.*, **7**, 3 (1948b), 304–12.

Rhodes, R. F., *Effect of salinity on current velocities*, U.S. Army, Corps of Engineers, Committee on Tidal Hydraulics, Report No. 1, Evaluation of Present State of Knowledge of Factors Affecting Tidal Hydraulics and Related Phenomena, February, 1950, 41–100.

Robson, R., Radar method in ocean current detection, *Commonwealth Eng.* (Australia), **42** (January, 1955), 219–25.

Rossby, C. G., Dynamics of steady ocean currents in the light of experimental fluid mechanics, *Papers Physical Oceanog. Met.*, Mass. Inst. Tech. and Woods Hole Oceanog. Inst., **5**, 1 (August, 1936).

————, and R. B. Montgomery, The layer of frictional influence in wind and ocean currents, *Papers in Physical Oceanog. Met.*, Mass. Inst. Tech. and Woods Hole Oceanog. Inst., **3**, 3 (April, 1935).

Rude, G. T., Tides and their engineering aspects, *Trans. ASCE*, **92** (1928), 606–71.

Schott, G., Weltkarte zur Übersicht der Meeresströmungen, *Ann. Hydrogr. maritimen Meteorologie*, **71**, 7 (July, 1943), 281; Chart 22.

Schutz, B., Beiträge zur Kenntis der Gezeiten an der Flandrischen Küste und auf der unteren Schelde, B. Aerologische und Hydrographische Beobachtungen der Deutschen Marine-stationen während der Kriegszeit 1914–1918,

Hamburg: *Archiv Deut. Seewarte*, Vol. 47, No. 2, 1925, p. 27.

Shepard, F. P., *Longshore current observations in Southern California*, U.S. Army, Corps of Engineers, Beach Erosion Board, Tech. Memo. No. 13, 1950.

————, and D. L. Inman, Nearshore circulation, *Proc. First Conf. Coastal Eng.*, Berkeley, Calif.: The Engineering Foundation Council on Wave Research, 1951, pp. 50–59.

Sibul, O. J., and J. W. Johnson, Laboratory study of wind tide in shallow water, *J. Waterways Harbors Div., Proc. ASCE*, **83**, WW1, Paper No. 1210 (April, 1957).

Simmons, H. B., Salinity problems, *Proc. Second Conf. Coastal Eng.*, Berkeley, Calif.: The Engineering Foundation Council on Wave Research, 1952, pp. 68–85.

Stewart, H. B., Upstream bottom currents in New York Harbor, *Science*, **127**, 3306 (May, 1958), 1113–14.

Stommel, H., The roles of density currents in estuaries, *Proc. Minn. Intern. Hyd. Conv.* (1953a), 305–20.

————, Computation of pollution in a vertically mixed estuary, *Sewage Industrial Wastes*, **25**, 9 (September, 1953), 1065–71.

————, Serial observations of drift currents in the Central and North Atlantic Ocean, *Tellus*, **6**, 3 (August, 1954), 203–14.

————, *The Gulf Stream, a physical and dynamical description*. Berkeley, Calif.: Univ. Calif. Press, 1958.

Sverdrup, H. U., *Oceanography for meteorologists*. Englewood Cliffs, N.J.: Prentice-Hall, Inc., 1943.

————, Wind-driven currents in a baroclinic ocean; with application to the equatorial currents of the eastern Pacific, *Proc. Nat. Acad. Sci.*, **33**, 11 (November, 1947), 318–26.

————, Martin W. Johnson, and Richard H. Fleming, *The oceans, their physics, chemistry and general biology*. Englewood Cliffs, N.J.: Prentice-Hall, Inc., 1942.

Swallow, J. C., A neutral-buoyancy float for measuring deep currents, *Deep-Sea Res.*, **3**, 1 (October, 1955), 74–81.

————, and L. V. Worthington, An observation of a deep countercurrent in the Western North Atlantic, *Deep-Sea Res.* **8**, 1 (June, 1961), 1–19.

Taylor, G. I., The transport of vorticity and heat through fluids in turbulent motion, *Proc. Roy. Soc.* (London), ser. A, **135**, 828 (Apr. 1, 1932), 685–702.

Thorade, H., Method zum Studium der Meeresströmungen, in *Handbuch der biologischen Arbeitsmethoden*, Abt. II, *Physikalische Methoden*, **3**, 3 (1933), 2865–3095.

Tibby, Richard B., *Report on returns of drift bottles released off Southern California*, 1937, Calif. Div. Fish and Game, Fish Bull. No. 55, 1939.

Todd, D. K., and L. Lau, On estimating streamflow into a tidal estuary, *Trans. Amer. Geophys. Union*, **37**, 4 (August, 1956), 468–73.

Trask, Parker D., *Movement of sand around Southern California promontories*, U.S. Army, Corps of Engineers, Beach Erosion Board, Tech. Memo. No. 76, 1955.

Tully, J., Prediction of pulp mill pollution, *Western Pulp Paper* (Canada) (December, 1948).

Vine, A. C., J. A. Knauss, and G. H. Volkmann, *Current studies of the Eastern Cayman Sea*, Woods Hole Oceanog. Inst., Ref. No. 54–35, May, 1954. (Unpublished.)

von Arx, W. S., An electromagnetic method for measuring the velocities of ocean currents from a ship under way, *Papers Physical Oceanog. Met.*, Mass. Inst. Tech. and Woods Hole Oceanog. Inst., **11**, 3 (March, 1950).

Webster, Ferris, *Crawford Cruise 18, a study of Gulf Stream meanders*, Woods Hole Oceanog. Inst., Ref. No. 61–18, September, 1961. (Unpublished.)

Wertheim, Gunther K., Studies of the electric potential between Key West, Florida, and Havana, Cuba, *Trans. Amer. Geophys. Union*, **35**, 6 (1954), 872–82.

Wiegel, R. L., and J. W. Johnson, Ocean currents, measurement of currents and analysis of data, in *Waste Disposal in the Marine Environment*. London: Pergamon Press 1960, pp. 175–245.

Wilcox, B. W., Tidal movement in the Cape Cod Canal, Massachusetts, *J. Hyd. Div., Proc. ASCE*, Paper 1586, April, 1958.

Wilson, Basil W., Note on surface wind stress over water at low and high wind speeds, *J. Geophy. Res.*, **65**, 10, (October, 1960), 3377–82.

Wilson, K. G., A. B. Arons, and Henry Stommel, *A simple electrical analog for the solution of the Ekman wind drift problem with the coefficient of eddy viscosity varying arbitrarily with depth*, Woods Hole Oceanog. Inst., Ref. No. 55–52, September, 1955. (Unpublished.)

Wyrtki, Klaus, The thermohaline circulation in relation to the general circulation in the oceans, *Deep-Sea Res.*, **8**, 1 (June, 1961), 39–64.

CHAPTER FOURTEEN

Shores and Shore Processes

1. INTRODUCTION

Technically the *shore* is that strip of ground bordering any body of water which is alternately exposed, or covered by tides and/or waves (Fig. 14.1). A shore of unconsolidated material is usually called a *beach*. Shores are continually changing; at best they are in dynamic equilibrium with sediment moving offshore and then back onshore, along the shore in one direction, and perhaps in the other direction. Beaches will be the particular type of shore considered in most detail, because unconsolidated material is subject to greatest change within the normal life of a man-made structure.

2. CLASSIFICATION OF SHORES

In order to deal logically with a physical feature as large and as complex as shores, it is necessary to have a scheme of classification which clarifies the subject and accents the differences and similarities of individual shores.

Some of the more important classifications which have been devised are described herein.

a. *Genetic classification.* As an introduction, it is most illuminating to use the genetic and evolutionary classification of D. W. Johnson (1919), which describes how a particular coast as a whole came into being, and also its subsequent changes. Note that there are no implications as to whether the land or the sea moved, or whether the water level rose or fell.

(1) *Shorelines of submergence*, or those shorelines produced when the water surface comes to rest against a partially submerged land area.

(2) *Shorelines of emergence*, or those resulting when the water surface comes to rest against a partially emerged sea floor or lake.

(3) *Neutral shorelines*, or those whose essential features do not depend on either submergence of a former land surface or the emergence of a former subaqueous surface.

(4) *Compound shorelines*, or those whose essential

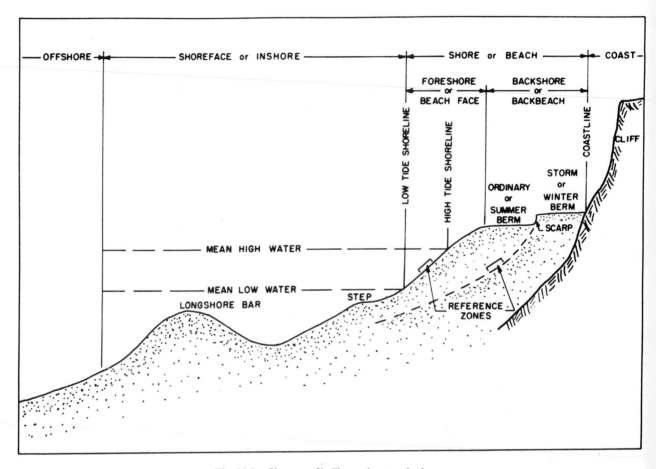

Fig. 14.1. Shore profile illustrating terminology

features combine elements of at least two of the preceding classes.

A shoreline of submergence is any shoreline formed by the partial submergence of a land mass dissected by river valleys, or any shoreline formed by the partial submergence of a region of glacial troughs (called *fjord* shorelines). A shoreline of emergence is one produced by the raising of a submarine /or sublacustrine) plain. In its early stages it will be simple in outline. Among neutral shorelines we may list deltas, alluvial plains, outwash plains, volcano shorelines, coral reefs, and fault shorelines.

When the ocean rises relative to land, an irregular shoreline is most likely to result, with rocky ridges becoming peninsulas, river valleys becoming elongated embayments, and the rough submerged section of land that formerly was close to shore becoming sea bottom. As these relative movements occur very slowly, a new shoreline is being modified by wave action while the movement is going on. As a result of wave refraction, the promontories usually are attacked most severely, and because the water is relatively deep immediately offshore, the full energy of the waves is spent directly against

these exposed sections of the coast. At first a notch is cut, and as the support of the rock above is removed, rock and earth slides occur, slowly forming a seacliff. The debris is moved by wave and current action from the toe of the cliffs, leaving a flat, seaward-sloping, wave-cut rock bench. The coarse debris of erosion is deposited immediately offshore; the finer sediments are moved farther offshore or are transported by littoral currents into the newly formed bays. Additional sediments are desposited in the submerged valley bays by streams which bring them from the hinterland. The net result is a tendency to straighten the shoreline and to shoal the areas offshore, with bay mouth beaches, spits, and other depositional features being formed. Eventually the entire shoreline will retreat, with both rocky ridges and lagoons disappearing, and a mature coast will evolve with straight low cliffs or gently sloping hinterland bordered with wide sand beaches and a flat shallow offshore area which gradually deepens seaward across the continental shelf.

When the land rises relative to the ocean, the newly formed coastal strip is likely to be smooth, with the old wave-cut cliffs and benches becoming terraces and bays forming river valleys. As is the case with shorelines of submergence, the relative movements occur very slowly with wave action continually modifying the shoreline; benches and cliffs are formed and the resulting material deposited as beaches. Owing to the differential in hardness of the coastal material, open bays are likely to form in certain locations. Eventually, shorelines of emergence are likely to appear similar to old shorelines of submergence. A fine example is shown in Fig. 14.2.

In the case of neutral shorelines, such as deltas, alluvial plains, outwash plains, volcano shorelines, coral reefs, fault shorelines, etc., it is even more difficult to generalize the processes than for shorelines of submergence and emergence, and it will not be attempted in this limited introduction.

b. *Descriptive classification of beaches.* Though a genetic classification of shores is desirable for engineers and geologists in evaluating many of the practical problems that confront them, a descriptive classification of readily accessible shores with respect to their constituent materials is also useful. A broad threefold classification is most convenient for this purpose. Such a classification has been presented by Trask (University of California, 1951, Sec. I-I):

(1) *Sand, gravel, shingle and cobble beaches,* or those beaches consisting of coarse materials.

(2) *Muddy, silty, or clayey beaches,* or those beaches consisting of fine-grained materials.

(3) *Bedrock and reefs,* or those shores which consist either of rock which has been worn away by the waves to form a more or less gently sloping underwater offshore bench, or those which consist of coral.

The coarse-grained beaches (sand, gravel, shingle, and cobble) have different sizes and shapes of the sand and gravel constituents. Beaches of this type, however, have the common property of exposure to the open ocean and the action of more or less large waves. As a result of this wave action, the individual constituents are of essentially the same size on any particular part of the beach. With sands, the principal characteristics are the average size and shape of the sand particles, the steepness of the beach, and the water content of the sand. With gravel, cobble, and shingle beaches, the chief characteristics are the smoothness, looseness, and steep slope of the material.

Beaches in areas protected from large waves, such as in bays, lagoons, or marshes, generally are composed of fine particles of silt and clay. They characteristically are soft to walk on when wet and the constituents generally are not so well sorted as in sand. They contain varying mixtures of sand and varying amounts of water. The water content depends principally upon the grain size, the depth of overburden, and the extent to which the beaches are exposed to air. As a general rule, the finer the grain size, the greater is the water content, and the poorer the bearing capacity. The greater the overburden of mud, the smaller the water content, and the greater is the strength. The longer the sediments are exposed to air, the greater is the evaporation and loss of water. A special characteristic of fine-grained beaches is the sensitivity of the clay: i.e., the degree to which the structure of the soil can be destroyed by reworking or repeated passage of vehicles. The fundamental differences between fine-grained, or muddy, beaches and coarse-grained, or sandy, beaches is the cohesive character of the fine-grained beaches. This tendency for clay and silt particles to adhere together causes fundamental differences in strength compared with sand.

Many shores consist of rock which has been worn away by the waves to form a more or less gently sloping underwater offshore bench. Some benches consist of firm rock and are smooth, but in others the bedrock consists of shale or clay which may be softened by the continual exposure to water. Such soft bedrock is not common.

Fig. 14.2. Two terraces near Little River, California, are fine examples of wavecut benches now raised above the sea. A new and very irregular seacliff is now being carved by the waves, and stacks, caves, and arches are all present. Note the cave open to the surface at the extreme left and the irregularities in the old bottom which show as hills on top of the point

Rocky shores frequently are covered with thin veneers of sand or other types of deposit. The thickness and location of this veneer of sediment changes with variations in waves and currents. The beach itself may be covered with sand or gravel during certain times of the year. The offshore bench may contain hummocks or ridges of varying height. Such features are encountered where the bedrock varies in hardness or has been badly fractured.

Coral reefs commonly found in tropical waters are special types of rocky features. They are characterized by a broad shallow bench of coral rock, varying in width from a few feet to a mile or more. The surface of this rock is irregular and cut here and there by channels of deeper water. In places, soft deposits of coral clay lie upon the benches which are difficult to traverse. Also, coral reefs vary greatly in character from one reef to another.

c. *Exposure and descriptive classification.* Any appraisal of beach characteristics involves the three problems of statics, dynamics, and stability. *Statics* relates to the physical state of the individual components of the beach, such as the configuration, grain size, and strength. *Dynamics* relates to the physical, chemical, and biological processes that control the beach and its constituent materials; i.e., the processes by which the components of the beach are formed or change once they have formed. *Stability* relates to the rate of change of the beach characteristics once they have formed. Stability is of particular concern in operating on shores because of the desirability of ascertaining whether the beach will or will not change significantly from one period of time to another, and if any change is likely, under what conditions, and to what extent it may occur. The most pertinent question would be the effect of storms upon the beach, especially storms originating from different quarters of the wind. A classification of this type was developed by W. N. Bascom (University of California, 1951, Sec. I-I). It is based on the ability of waves to create and maintain beaches from whatever materials are available at the particular shore. Bascom reasoned that, as waves are governed in size by the size of the body of water and are modified by the configuration of the shoreline and the inshore hydrography, there are three basic items which control the nature of any beach: (1) deep water wave dimensions, (2) wave refraction, and (3) beach material. Thus, this classification is based upon the exposure of the beach material to the prevailing storm waves. The classification was simplified by assuming that the deep water wave dimensions of the prevailing waves depend upon the size of the body of water, as oceans, small seas, etc., and that the effect of wave refraction is included in the general shape of beaches and with regard to shielding by offshore barriers, promontories, etc. The classification is composed of combinations of the following items:

(1) Wave size
    (a) Oceans
    (b) Large seas
    (c) Small seas
    (d) Lakes
(2) Types of beach orientation
    (a) Convex
    (b) Straight
    (c) Oblique
    (d) Headland-protected
    (e) Embayed (open, tight)
    (f) Reef-protected
    (g) Lee
(3) Size of beach materials
    (a) Mud
    (b) Find sand
    (c) Coarse sand
    (d) Gravel granules
    (e) Pebbles and shingle
    (f) Cobbles

Thus a beach would be classified as "fine sand beach oblique to the North Sea," "straight shingle beach on South Atlantic," "coarse sand open embayed beach with some reef protection on Mozambique Channel," etc.

3. THE SHORE AND ITS ENVIRONMENT

a. *Introduction.* Shores change in character rapidly compared with many other geographic features. Nevertheless, these changes may be relatively slow, such as the erosion by waves of resistant headlands (Fig. 14.3) or coral reefs; they may be rapid, as the erosion of a beach foreshore during the course of a storm which lasts perhaps but a few hours; they may occur at intermediate rates, like the building of dunes by winds along sections of beaches being furnished a surplus of sediments by littoral drift; shore changes may also alternate, receding in winter and advancing in summer, or between spring and neap tides. Whatever the physical causes and whatever the rate, all shores are changing constantly.

b. *Factors affecting beach characteristics.* Beaches are as variable as the weather and waves which act upon them, they are dependent primarily upon wave action for their origin, existence, texture, structure, and configuration.

The four primary factors that control the formation and configuration of a beach are the geomorphology of the land adjacent to the beach, the type and quantity of beach material available, and the waves impinging on the beach. Each factor gives the beach different characteristics, though they may be difficult to separate. The first two factors tend to change only gradually, whereas the second two tend to change periodically over intervals of years, seasonally throughout the year, and even at shorter intervals, perhaps daily or hourly.

Fig. 14.3. Surf at Cape Lookout, Oregon

The first factor, geomorphology, depends upon long-term geologic circumstances. The second factor (type of beach material available) is normally fairly constant for a given area. This factor is dependent upon the source of material. In the case of sand beaches, the most important single characteristic has been found to be the median diameter of the sand grains. The type of material largely determines the slope of the beach face; the finer the material, the flatter the beach. The third factor, quantity of material available, is largely dependent upon rivers and streams carrying sediments to the ocean, and the erosion of land in contact with the ocean with subsequent transport of material. The quantity of any given type of material available determines the width of the beach. Large amounts of materials cause wide beaches.

The fourth factor, the interaction of waves and beach, is responsible for the details of a beach. This phenomenon causes sand to move along the coast as well as onto or off a beach. It causes the rapid change in a beach and is the most difficult to predict or evaluate quantitatively. When waves enter shoaling waters along the coast and reach a depth of about one-half the wave length, the wave begin to "feel bottom"; they start to slow down, decrease in length, and after a bit, increase in height. When they reach a water depth of approximately 1 to $1\frac{1}{2}$ times their height, they break. On a steep beach, they break at the beach face; on a flat beach, they often break on an outer bar, then re-form as much lower waves and continue to shore, where they break again. In deep water, the water particles move in nearly closed circular orbits, with a small mass transport in

the direction of wave propagation. The orbits' diameters decrease exponentially with distance below the surface until, at a depth equal to about one-half the wave length, there is effectively no motion. When the total depth of water is less than about one-half the wave length, the waves begin to "feel bottom," i.e., there is motion at the bottom. In addition, the particle motion becomes elliptical rather than circular. The more shallow the water, the flatter the ellipse; and the greater, the bottom velocity. A mechanism is thus available for moving sediments, the distance from shore at which sediments can be put in motion depending upon the wave dimensions and sediment size; the motion, however, is relatively small. But the zone in which the waves break is one of great turbulence. The smaller materials are thrown into suspension, and once in suspension even currents of very low velocity are capable of transporting material in considerable amount.

Upon breaking, wave motion becomes translatory. The water moves forward as a foam line and then rushes up the beach face, carrying fine sand in suspension and moving coarser grains along the bottom. The uprush gradually slows down, due to gravity, friction, and percolation, depositing a thin layer of sand along the way. At its maximum landward limit on sandy beaches, a thin line of sand grains (called a *swash mark*) is deposited, the grains usually being of a larger size than the rest of those on the beach. When its motion up the beach face has ceased, the water which has not percolated into the sand returns as gravity flow down the beach face, moving sediment with it. This sediment consists mainly of grains greater in size than average (like those in the swash marks). When this backflow comes in contact with forward moving water of the next breaking wave, the coarse material is deposited, forming a low seaward-facing step which adjusts itself to changing wave conditions. During periods when storm waves are breaking along the beach, it changes form slightly, becoming a bar. Waves reaching the beach face break over this step and, becasue of the great agitation caused by the breakers, a great deal of sediment is thrown into suspension. As much of the material on the step consists of the coarser particles which are thus continually moved about, it is thought that this plays an important role in the way in which waves tend to sort out the larger particles and very slowly abrade them to uniform size.

The *berm* is the nearly horizontal depositional formation of beach material and is usually what people are referring to when they speak of "the beach." The width of the berm depends primarily upon the amount of sediment available, and to a lesser extent upon the amount being removed by the wind. The height of the berm depends directly upon the wave height and tidal range and indirectly upon beach material. It is readily apparent that whether a beach is aggrading or eroding depends primarily upon whether more material is being moved onshore or offshore. The primary factor that controls the relationship between waves and a given beach is the deep water steepness of the wave. When the wave is steep the beach face is cut back, the face tends to flatten somewhat, the tidal step disappears, and offshore bars form. These steep waves are usually *storm* waves, this term referring to the fact that they have been generated in a local storm. There is a range of wave steepnesses that is transitional in their effect upon shores, their effect depending upon the stability of the shore. Flat waves have the opposite effect from the steep waves. The beach face tends to steepen a little, sand is deposited, building up the berm, the bars tend to disappear, and a tidal step forms.

The relationship between the size of the bar and breaker height depends upon the deep water steepness of the wave and the offshore beach slope. For a given deep water wave steepness, the higher the breaker, the bigger the bar and the further it is seaward from the foreshore of the beach. It has also been found that bars tend to increase in size (all other conditions being equal) with decreasing beach steepness (Wiegel, 1950; Shepard, 1950).

Rip currents have a very definite effect on bars. The currents flowing seaward through the surf zone cut channels through the troughs. These rip currents are closely related to the distribution of wave heights along the coast (Shepard and Inman, 1950).

The height of a berm is dependent upon the height of the breakers. The higher the waves, the higher will be the berm. The height of the berm is also dependent upon

the tidal conditions; hence, if waves are building a berm during the high spring tides, the berm is built higher than during the lower neap tides, wave conditions being equal.

When steep waves erode a beach, scarps are often formed in the beach face. These scarps have been observed with heights of at least 6 ft (Fig. 14.4).

Often cusps form on a beach (Fig. 14.5). A *cusp* is a nearly semicircular cutout in the beach face which tends to be at right angles to the sea margin, tapering to a point seaward. The depth of the hollows seems to increase with increasing steepness of the beach face. The mechanics of cusp formation are not definitely known, but they are certainly associated with waves. Many of the mechanisms proposed in scientific works on the subject are invalid, as cusps have been formed in the laboratory under conditions of uniform waves and no tides.

It is evident that height and direction of waves at a particular section of shore are extremely important. Hence, the effect on waves of the hydrography of the region will be discussed here in some detail.

In some areas, winds are of considerable importance since they are responsible for the loss of sand from the beach and its movement inland as dunes. Studies made at the mouth of the Columbia River, with the beach sand being 0.19 mm median diameter, showed that sand began to be transported by winds of 9 miles per hour, the wind being measured 5 ft above the surface. The rate of transport increased exponentially with increasing wind speed being 0 at 13.4 ft/sec., 1000 lb per feet of beach per day at 30 ft/sec, and 2300 lb per feet per day at 40 ft/sec (O'Brien and Rindlaub, 1936).

c. *Effect of wave refraction.* The conditions that will exist at a particular section of shore depend upon the characteristics of the shore itself and also upon the

Fig. 14.4. Scarp at Oceanside, California

Fig. 14.5. Aerial view of cusps at El Segundo, California

location of the shore in relation to the geography and hydrography of the region as well as the meteorological conditions that exist.

The relationship between meteorological conditions and the waves that will be present at a beach have been explained in the chapters on wind waves and waves in shoaling water. In this section, the over-all effect of wave refraction for different conditions will be explained. It is refraction that causes waves to become nearly parallel to a beach before breaking, although far off-shore they may be moving at a considerable angle to the beach face.

There is an old seafaring expression "headlands draw the waves, " meaning that breakers are usually higher at the land points that jut into the sea than in the adjacent areas. This is probably owing to a combination of two factors: First, the underwater topography is very often a continuation of the landform, and so there is likely to be an underwater ridge at the end of a headland. Secondly, the refraction occurring over an underwater ridge causes convergence of wave energy and thus higher waves. In Fig. 14.6 the refraction diagram for such a condition is given together with an aerial photograph of the same location showing that this is actually what occurs.

How can refraction cause the height of waves to vary from one location to another? Consider a wave moving between two points along its crest any chosen distance apart. Now, suppose a line is drawn from each of these points in such a way that the lines are always at right angles to the crest of the wave as it moves forward, these lines being known as *orthogonals*. The wave power transmitted between two orthogonals remains constant for most practical purposes (neglecting friction) as the wave moves forward. Thus, if two adjacent orthogonals converge, the energy density increases; if they diverge, the energy density decreases. In Fig. 14.6, the waves converged on the point, causing high breakers, and diverged at Arena Cove, resulting in low breakers. As the intensity of wave action is one of the most important forces in regard to coastal processes, the importance of wave refraction is evident. A detailed example of this is given in Chapter 18.

d. *Effect of offshore hydrography.* It is commonly believed that the lee of an island affords good shelter from swell and breakers. This, although usual, is not always the case. The degree of shelter depends upon the local hydrography and the period and direction of the waves. Due to some combination of underwater ridges and canyons, there may be areas of higher swell and surf. A second difficulty may be understood by examining

Fig. 14.6. Arena Cove, California. Refraction diagrams for Arena Cove, California. Compare with the aerial photograph shown above for an illustration of how such diagrams can be utilized to determine points where low and high wave action can be expected

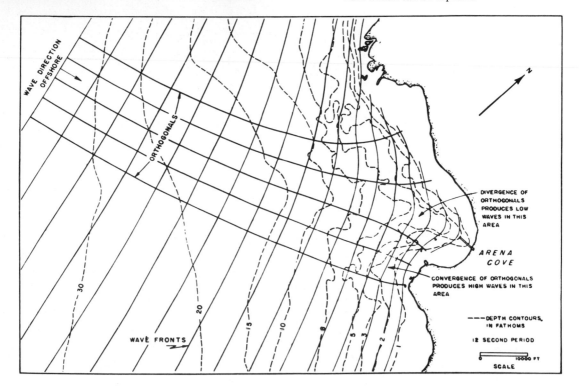

Fig. 14.7, which shows swell, after being refracted around both sides of the island, interacting on the far side to cause a varying surface. The swells may completely cancel in some section while reinforcing each other in other sections, causing a choppy surface.

Large islands or groups of islands generally offer greater areas of shelter than small islands. Consider the Channel islands off the coast of Southern California. Their sheltering effect is far-reaching and greatly influences the height of breakers along the coast. An example of this is discussed in detail in Chapter 18 (see Figs. 18.1 and 18.2).

It is not necessary for an obstacle to rise above the water surface to be capable of disturbing waves. Submerged reefs can be almost as beneficial as an island in offering shelter. A good example of this is shown in Fig. 14.8, where the underwater reef acts as a breakwater.

The energy or swell passing over a submarine canyon not in deep water tends to spread out (diverge), causing the waves to be lower in this area; conversely, the energy density of swell passing over submarine ridges tends to increase (convergence of orthogonals), causing higher waves. This has been stated previously and both the refraction drawing and an aerial photograph have been

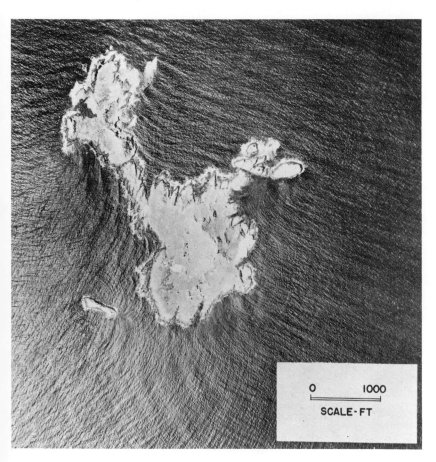

Fig. 14.7. Refraction and diffraction of both swell and wind waves at Farrallon Island, California. Note reflection from small island at lower left

Fig. 14.8. Effect of submerged reef on waves, at Halfmoon Bay, California

shown of a region where both a submarine canyon and ridge exist near each other (Fig. 14.6).

Often the net result of a submarine canyon, with its head extending into shallow water near shore, is a region of low waves with a region of high waves on either side, so that the angle the surf makes with beach upcoast is in the opposite direction, relatively, to the surf angle down coast. Because of these two factors, a current circulation system is set up, with water flowing along the beach from both sides, then offshore to the canyon head (Shepard and Inman, 1950; Inman, 1953). The resulting tendency of the littoral drift of sand along the coast in such an area is for quantities of the sand to deposit in the region of the canyon head. Shepard (1951) has described a series of detailed and precise measurements of the water depths in such an area, La Jolla Submarine Canyon, and has found rapid increases in water depth in sandy regions. The most logical explanation is that a large mass of sand has slid into the canyon. In one such case approximately 130,000 cubic yards of sand slid down the canyon wall, in another case about 140,000 cu yd, and in another case about 10,000 cu yd. Similar events have been documented at Redondo Beach, and Moss Landing (Monterey Canyon), California. Surveys of the canyon bottoms indicate that this material probably finds its way far offshore, although just how this occurs for sand is not known.

It appears that such submarine canyons act as an effective barrier to a substantial amount of the littoral drift of sand; furthermore, this results in a net loss of littoral material.

e. *Effect of promontories.* The effects of promontories are varied. It has been stated previously that because of the similarity of landforms, waves usually converge off headlands. These same promontories, however, generally offer sheltered areas near them because energy cannot be concentrated in one locality without the accompanying decrease of energy density in a nearby area.

There are many examples of this phenomenon. One is illustrated in Fig. 14.9 which shows refraction diagrams

and aerial photographs of Point Sur, California. With a northerly swell running there are high breakers at the head of the point and yet craft can easily land on the southern beach. In this case, the point is acting in a similar manner to a breakwater. Even when a westerly swell is running, however, the breakers are high at the head and yet quite small on the southern side.

A major effect of headlands is that they often are a substantial barrier to the littoral drift of sand. Sometimes a major headland acts only as a partial barrier, as has been shown by Trask (1955). It was found that a considerable amount of sand moved around the headland in water from 30 to 60 ft deep (the median diameter of the sand was about 0.15 mm).

f. *Bays.* A person thinking of a bay normally visualizes protection from waves. In general a bay does offer protection, but because of refraction and diffraction there may be certain areas where the breakers are quite high. The many variations of bay types can be grouped roughly into five classes: (1) closed, a small entrance compared to the size of the bay; (2) hooked, shaped like a hook, with a deep indentation at one end which gradually decreases towards the other end; (3) open, a big gentle indentation in the coastline; (4) finger, a long narrow bay, such as the Norwegian fjords; (5) cove, a small recess in the general coastline.

Closed bays: An excellent example of this type is San Francisco Bay. It affords the maximum protection from swell. The waves must enter through a relatively small opening and then refract greatly, causing the swell inside the bay to become much smaller than the swell outside. Protection is almost always afforded from swell from any direction.

Hooked bays: This type of bay normally offers a great gradation of protection, the most sheltered regions being within the hooked portion. The further a section of shore is from this portion, the higher will be the waves under most circumstances. There is hardly any chance of convergence of waves inside the hook, as this

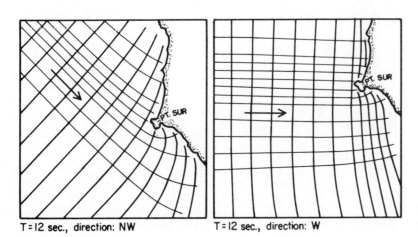

T = 12 sec., direction: NW   T = 12 sec., direction: W

Fig. 14.9. Wave refraction diagrams for Point Sur, California

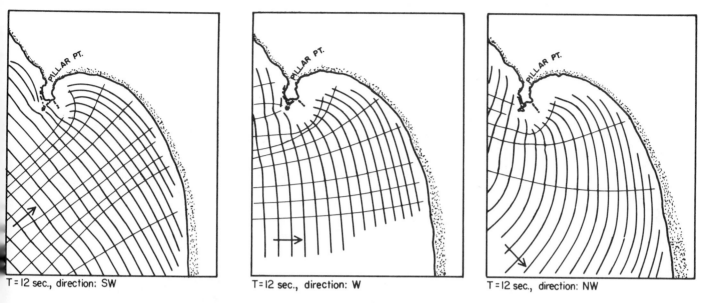

T = 12 sec., direction: SW   T=12 sec., direction: W   T=12 sec., direction: NW

Fig. 14.10. Wave refraction diagrams for Halfmoon Bay, California

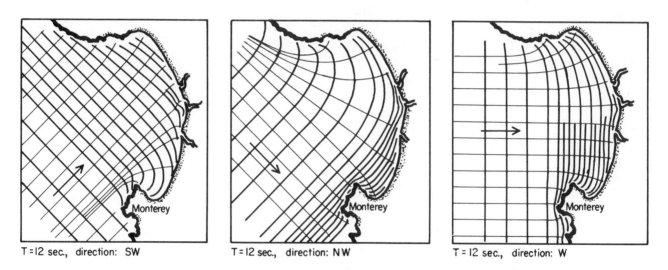

T =12 sec., direction: SW   T =12 sec., direction: N W   T = 12 sec., direction: W

Fig. 14.11. Wave refraction diagrams for Monterey Bay, California

has usually filled in with sand so that it is flat and even. An example of this type of bay is shown in Fig. 14.10.

Open bays: This type usually offers large sections of shore where there are low breakers. There may, however, be sections where, due to offshore configurations, convergence takes place. In most regions there are certain prevailing directions from which waves come. These shores generally offer protection from the prevailing swell along a good portion of their coast. An example of this type of bay is shown in Fig.14.11.

Finger bays: This type is much in evidence on subarctic coasts. They usually offer excellent protection from swell.

Coves: This is the most numerous and universal type. They are merely small indentations in the coastline.

Some lie to the side of headlands, some at the mouths of rivers, and some inside longer bays of other types. The type of breakers encountered in a cove is as varied as the coves themselves. Shelter may be had from waves of particular periods from certain directions, wherers for other conditions there may be convergence of the waves, thus creating a hazard rather than a haven.

g. *Ground water:* Some studies on ground water in beaches have been made in an attempt to find some sort of a satisfactory datum connected to the tide stage of ocean beaches (Isaacs and Bascom, 1949; Patrick, 1950). These studies were made in California, Oregon, and Washington. An example of the ground water level is shown in Fig. 14.12. In this figure are also shown the usual and emergency potable limits for drinking water.

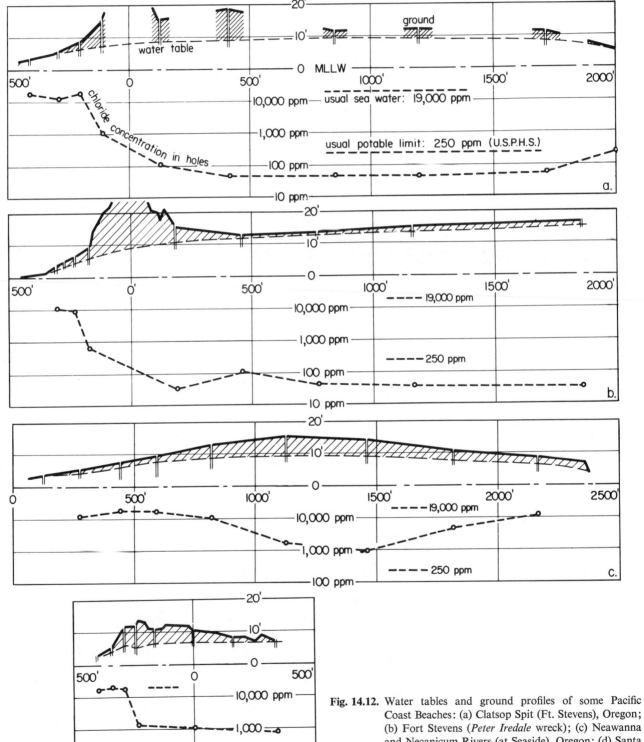

Fig. 14.12. Water tables and ground profiles of some Pacific Coast Beaches: (a) Clatsop Spit (Ft. Stevens), Oregon; (b) Fort Stevens (*Peter Iredale* wreck); (c) Neawanna and Necanicum Rivers (at Seaside), Oregon; (d) Santa Margarita River, California (after Patrick, 1950)

It was found that, as the tide rose, there was a tendency for salt water to flow into the beach and at the same time the effective fresh water head was reduced. At the maximum height of the tide, the trace of the water table at the seaward section became a curve that was concave upward, undoubtedly due to the relatively slow movement of water in the beach as compared with the change of tide level. As the tide receded, the curves gradually straightened and then turned downward. It was found on the beaches investigated that tide affected ground water in a narrow area of the beach face.

By introducing dye into the ground water under the beach it was found for these beaches that the rate of flow was about 1 ft per hour. The flow rate will of course vary greatly. Another study (Grant, 1948) made in Southern California showed that on one beach the flow seaward was at the rate of 4.5 ft/hr. the measurements were made 20 to 40 ft from the water line, and it was found that the water table lagged from 1 to 3 hr behind the tide and the amplitude of the water table was approximately that of the tide. Additional studies in the same region have been made by Emery and Foster (1948).

Some work by Grant (1948) indicates that a high water table accelerates beach erosion, and conversely, a low water table may result in a pronounced aggradation of the foreshore.

h. *Rivers and streams:* The major effect of rivers and streams upon a coast is that they often supply a major portion of the sediment that makes up the beaches. The area in which the stream flows into the ocean is one in which changes are often rapid.

4. DESCRIPTION OF BEACH MATERIALS

Beach sediments and soils may come from the erosion of the shoreline itself, such as the coast of New Jersey; they may come from nearby land areas, transported by local streams and rivers; they may be transported from areas hundreds of miles inland by major river systems, such as the Columbia River; they may come from offshore coral reefs; they may come from old deposits offshore, such as the glacial deposits off the west coast of Denmark. All sizes of debris are available as intertidal material. These sediments may be deposited locally or they may be transported many miles by littoral currents.

The large blocks and cobbles generally remain near their origin, whereas sand, silts, and clays are often moved large distances; in fact, owing to wave and current action, silts and clays tend to remain in suspension near the shore, eventually to be deposited in offshore areas. If, however, the currents are weak and wave action small, they may deposit on shore. This usually occurs only in well-protected bays. Beaches which are exposed to the ocean almost invariably are of sand, pebbles, and cobbles.

Many studies of the characteristics of beach material have been made, primarily for the purpose of determining the fundamental properties of beaches. Persons studying the characteristics of beaches also attempt to determine the mode of origin of the features they observe, but owing to the great number of variables and their complex interrelation, the fundamental laws governing the formation of beach materials are at present imperfectly known.

Table 14.1. SOIL CLASSIFICATION BY SIZE

1. Wentworth's size classification

Grade limits (diameters, mm)	Name
Above 256	Boulder
256–64	Cobble
64–4	Pebble
4–2	Granule
2–1	Very coarse sand
1–1/2	Coarse sand
1/2–1/4	Medium sand
1/4–1/8	Fine sand
1/8–1/16	Very fine sand
1/16–1/256	Silt
Below 1/256	Clay

2. U.S. Army, Corps of Engineers' classification

Grade limits (diameters, mm)	Name
Greater than 76.2	Cobbles
76.2–25.4	Coarse gravel
25.4–4.76	Medium gravel
4.76–2.00	Fine gravel
2.00–0.42	Coarse sand
0.42–0.074	Fine sand
If liquid limit is 27 or less and plasticity index (based upon —40 fraction) is less than 6	Silt
If liquid limit is 27 or less and plasticity index (based on —40 fraction) is 6 or greater	Clay

3. U.S. Bureau of Soils' classification

Grade limits (diameters, mm)	Name
2–1	Fine gravel
1–1/2	Coarse sand
1/2–1/4	Medium sand
1/4–1/10	Fine sand
1/10–1/20	Very fine sand
1/20–1/200	Silt
Below 1/200	Clay

4. Atterberg's size classification

Grade limits (diameters, mm)	Name
2000–200	Blocks
200–20	Cobbles
20–2	Pebbles
2–0.2	Coarse sand
0.2–0.002	Fine sand
0.02–0.002	Silt
Below 0.002	Clay

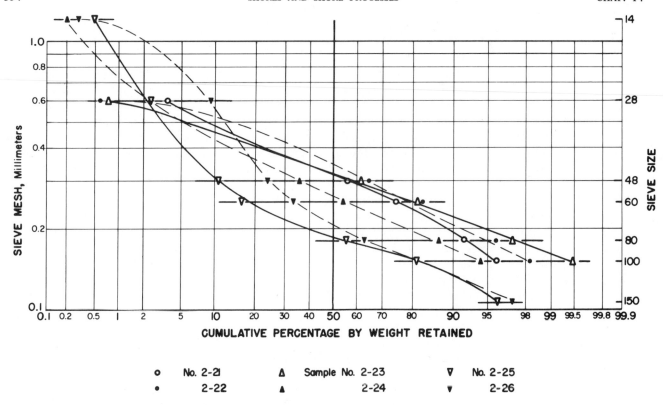

 (legend below)

o	No. 2-21	△	Sample No. 2-23	▽	No. 2-25
•	2-22	▲	2-24	▼	2-26

Fig. 14.13. Sieve analysis of sand samples from Santa Margarita River beach (Oceanside), California

a. *Classifications:* In describing beach materials quantitatively, it is convenient to be able to refer to a standard classification that relates descriptive terms to size. Some of the classifications most often used are given in Table 14.1.

In addition to size, beach materials are described by their shape, color, chemical composition, specific gravity, hardness, and cleavage. Together with a size classification, it is desirable to give an indication of the distribution of grain sizes in the sample. An example of such a distribution is given in Fig. 14.13. It is often possible to describe the distribution curve by means of a sorting coefficient which is defined as $S_0 = \sqrt{Q_1/Q_3}$, where Q_1 is the grain diameter having 75 per cent of the sample smaller and Q_3 is the grain diameter having 25 per cent of the sample smaller. Another coefficient of considerable use is the skewness coefficient, $S_k = Q_1\,Q_3/M^2$, where M is the median diameter. A sediment in adjustment with its marine environment will have values of S_0 and S_k close to unity.

Some photomicrographs of sand samples of various beaches are shown in Fig. 14.14. Other details of these samples are given in Table 14.2.

b. *Structure:* Emery and Stevenson (1950) found that the median grain size and the sorting coefficients determined by ordinary mechanical analyses of scoop samples of sand are composites because sand characteristically consists of alternating layers of coarser sand

and finer sand. On the beaches studied, it was found that the layers were from 1 to 20 mm in thickness. The laminae were most apparent where the coarse layers consisted of large light-colored quartz and feldspar grains and the fine layers consisted of dark heavy minerals, such as magnetite, ilmenite, hornblende, and olivine. In beaches of simple composition, such as the quartz-feldspar beach at Daytona, Florida, or the calcium carbonate beaches of coral reefs, the laminae have little contrast, but are nevertheless present.

A detailed study was made by Thompson (1937) of the structure of several beaches in California. It was found that the average thickness of the laminae was 0.3 in. (502 observations), with 80 per cent of the laminae being between 0.1 and 1.0 in. At the other extreme were layers a foot or more thick and layers only one grain of sand in thickness. In cross sections of the beach parallel to the shoreline, a typical lamina could be traced 100 ft or more before it disappeared; in sections normal to the shoreline, a single lamina seldom could be traced more than 25 ft before it was terminated by a surface on which rested other similar laminae with slightly differing dips. Results of sieve analyses showed that the grains of the individual lamina fell characteristically within a narrow size range, although they differed considerably from equally well-sorted grains of adjacent laminae.

Thompson (1937) discusses the variables dependent largely upon variations in the transporting capacity of

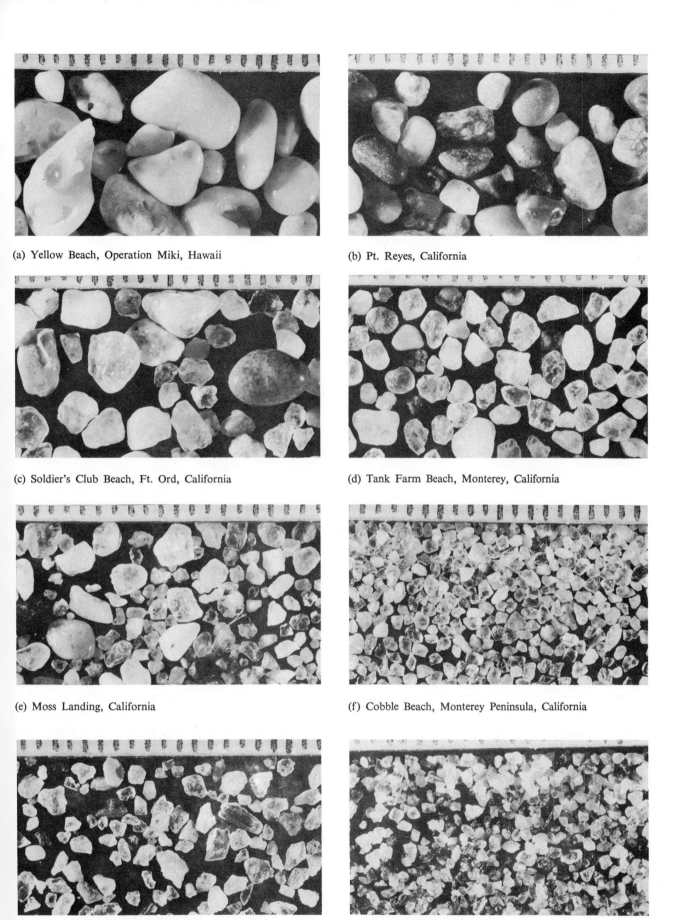

(a) Yellow Beach, Operation Miki, Hawaii

(b) Pt. Reyes, California

(c) Soldier's Club Beach, Ft. Ord, California

(d) Tank Farm Beach, Monterey, California

(e) Moss Landing, California

(f) Cobble Beach, Monterey Peninsula, California

(g) Santa Margarita River Beach, California

(h) Clatsop Spit, Oregon

Fig. 14.14. Comparison of photomicrographs of sand samples arranged from steepest to flattest beaches studied

Table 14.2. Characteristics of Sand Samples from Critical Locations on Various Beaches

Beach	Sample no.	Slope of beach face on range where sand samples were taken	Median sand diameter (on beach face, mm)	Sorting coefficient	Roundness	Sphericity	Density of sample
Operation MIKI							
Green	2–2	1/1.5	0.34	1.26			
	2–3		0.34	1.22			
Blue	2–4	1/6.0	0.56	1.19			
	2–5		0.70	1.28			
Yellow	2–7	1/6.5	0.60	1.25			
	2–8		0.76	1.29	0.8	0.85	2.64
Soldiers Club	2–33	1/8.7	0.66	1.53	0.5–	0.85	2.47
Beach, Fort Ord,	2–34		0.61	1.35	0.8		
Calif.	2–35		0.75	1.42			
Point Reyes	2–58		1.04	1.23	0.8	0.8	2.61
Beach, Calif.	2–60	1/8.3	0.78	1.17			
	2–61		0.84	1.17			
Tank Farm	2–39		0.44	1.24			
Beach, Monterey	2–40	1/11.1	0.50	1.28	0.7	0.75	2.58
Calif.	2–41		0.56	1.25			
Cobble Beach,	2–47		0.29	1.14			
Spanish Bay,		1/17					
Monterey, Calif.	2–48		0.27	1.10	0.55	0.75	2.47
(sand section)							
South Beach,	2–44	1/14.4	0.30	1.18	0.4	0.7	2.58
Moss Landing,	2–45		0.23	1.22			
Calif.							
Santa Margarita	2–24	1/20.0	0.28	1.27	0.4	0.6	2.62
River Beach,	2–25		0.21	1.20			
Oceanside, Calif.							
Test Site Beach,	2–64		0.17	1.15			
Clatsop Spit,	2–65	1/42	0.21	1.18			
Ore.	2–66		0.24				
	2–67		0.25	1.18	0.4–	0.6	2.70
				1.25	0.5		

waves carrying particles of different sizes and specific gravities. The general idea was illustrated by means of a study on one beach for four months. The detritus deposited on the beach ranged in size from cobbles 2 in. in diameter to very fine sand. It was found that, following an interval of unusually great wave activity, the surface of the beach was strewn with cobbles, pebbles, coarse sand, and fragments of shell; during periods of lesser waves, finer sediments were deposited. Successive changes in grain size of material observed on the surface of the beach corresponded to the similar changes in grain size of the constituents in the successive laminae that made up the beach.

Trask and Scott (1954) drilled a series of bore holes in the beach that formed upcoast of the Santa Barbara breakwater as a result of the breakwater interrupting the littoral drift of sand. Some of the holes were drilled in an area that was originally in 10 to 15 ft of water. The cores were examined and a series of mechanical analyses were made of samples taken from the cores. Laminae were evident, but the median diameter of samples of individual laminae not too far apart in depth (say 3 ft, or so) varied by a maximum of only 15 per cent. It is evident that the variation between laminae is strongly dependent upon the range of material available.

The studies just cited dealt with the structure associated with normal changes in wave and tidal conditions. There have been few studies of the effects of rare severe storms. McKee (1959) surveyed an atoll in the Pacific Ocean after a typhoon struck it. It was found that the deposits differed greatly in texture and structure from those normally found in the same area, with the sediments usually being much coarser. Gravel sheets from $\frac{1}{2}$ to 3 ft thick were deposited across two-thirds of many of the islets.

c. *Sorting.* Some information on sorting was described in the section on beach structure. A comprehensive study was made of the source, transportation, and deposition of beach sediments in the section of Southern California lying between Carpenteria and Point Fermin by Handin (1957). Samples taken from 88 stations were analyzed and it was found that the median diameter ranged from 0.16 to 1.68 mm, with the sorting coefficients varying from 1.15 to 4.95. Only five samples, however, had sorting coefficients greater than 2.5. Thus, the sand was well sorted.

Sorting was also studied by Bascom (University of California, 1951, Sec. I-I). Of the ocean beaches studied, most were composed of materials which fell within a rather limited size range, whereas on some beaches there was good sorting on two levels; these beaches had well-sorted fine sand below MLLW (mean lower low water) and cobbles which composed the berms. In addition, some beaches have a mixture of sand and cobbles, one such beach being in Kodiak Island, Alaska (Wiegel, 1949). It was found that wave action even in the most protected regions of the coast removed sand smaller than 0.17 mm median diameter from the beach face. Data from other sources show that the finer sands are deposited offshore. Bascom found that (1) the largest sands have the widest range of variation along the profile and the finest sands show the least variation; (2) the coarsest sand is always found at the plunge point, just seaward of the backrush, which observation indicates is the point of maximum agitation in the waves. Another group of large particles is found on top of the berm; apparently these are churned up at the plunge point and swept by maximum wave uprushes across the top of the berm, from whence they cannot return (those on the seaward face are carried back to the plunge point by the backrush; (3) the sand becomes increasingly finer to seaward; above water, the finest sand is in the dunes. These results, in graphical form, are shown in Fig. 14.15.

Periodic studies of two stations situated 5 miles apart on a long straight beach north of Point Reyes peninsula (Trask and Johnson, 1955) showed considerable variation in grain size during different times of the year. At the west and slightly more exposed end of the beach the median grain size ranged between 0.5 and 0.7 mm; compared with 0.6 and 0.67 mm at the east end. At the individual stations, for given times of the year and within a radius of 250 ft, the maximum grain size of individual samples was approximately 3 times the minimum grain size. The beach cut and filled as much as 5 ft between successive intervals of sampling.

Bore hole studies by Trask and Scott (1954) showed a definite sorting of material, with the finer sand being deposited in deeper water.

Inman (1953) studied the sediments on the beach and near shore in the vicinity of the La Jolla Submarine Canyon, California. Sand samples were taken at intervals throughout a year at a large number of stations. As an example, the median diameter of the sand along one range line was about 0.18 mm median diameter on the beach face, 0.12 mm in 50 ft of water, and 0.07 mm in 200 ft of water. Additional data on the sorting of beach sands has been presented by Miller and Zeigler (1958).

Two-dimensional laboratory studies by Scott (1954), using sand of median diameter of 0.31 mm, showed definite sorting with coarser grains being deposited on the berm and tops of the bars, and with finer sand being

Fig. 14. 15. Size distribution across a beach (after Bascom, 1951)

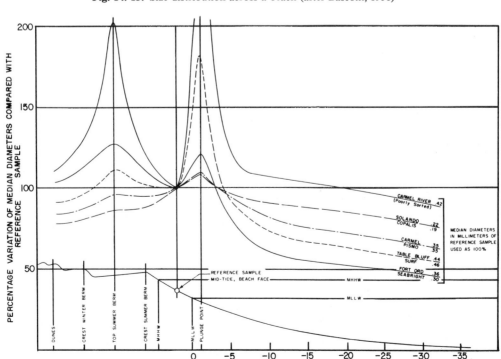

deposited in the troughs and seaward of the bar. Observations of the mechanics of this phenomenon in the ocean have been made by the Beach Erosion Board (U. S. Army, 1933).

Some samples were taken by the author from mid-beach face at approximately 1 mile intervals along the extensive beach extending from the mouth of the Columbia River south approximately 20 miles to a major headland, Tillamook Head. The same samples ranged from 0.22 to 0.30 mm median diameter, with no trend of size variation with distance from the source. No sand beaches existed on Tillamook Head, only cobbles. In this case, no lateral sorting occurred.

d. *Abrasion*. A study by Mason and Schulte (1942) showed that the abrasion of beach sands occurs at rates so low as to be of no practical importance in short-range protection problems, although from a geological standpoint they are important. It was found that the product of the abrasion of beach sands was smaller than 0.07 mm diameter, so that it would be removed from the beach and be deposited offshore.

e. *Sampling*. Any investigation of beach sediments is concerned with the extent to which measurements are indices of the conditions that will be encountered at other times. If samples are taken over a certain part of the beach, will the results be the same for other parts of the beach, and if not, what will be the difference? Some sort of a sampling program is needed to answer these questions. As in all types of measurements, two fundamental matters are involved (1) variations in the nature of the items measured, (2) variations inherent in the methods of measurement.

The fundamental problems of measurements, variations in constituent particles, and other properties of beaches have been described by Krumbein (1953) who gives particular consideration to the representativeness or reliability of individual samples and to the proper sampling interval required to give reliable results for given beaches. The work, done on some of the beaches along Lake Michigan, was concerned with the sampling of local sand populations to determine the inherent

variability of particle size, shape, roundness, mineral content, moisture content, penetrability, etc. Most of the work was done with sampling grids, which are also applicable to larger-scale studies. It was found on the beaches studied, that the berm line is an important line for changes in population attributes.

5. BEACH CONFIGURATION

a. *Beach profiles*. Profiles have been measured on many beaches. They vary greatly, depending upon the presence or absence of reefs, and other offshore hydrography. In general, however, they depend upon the size and amount of sediment available and the wave climate in the area. Some beach profiles are given in Fig. 14.16 (note distortion of vertical scale).

b. *Foreshore slope and sand size*. Sand samples were obtained and beach profiles measured for most of the beaches along the Pacific Coast of the United States as well as many other beaches (Wiegel, 1950, University of California, 1951; Bascom, 1951). In general, it was found that there was a good correlation between median sand diameters and the slope of the foreshore. The sorting of beach sediments along a profile, as shown in Fig. 14.15, explains a part of the scatter of the data. Bascom (1951) was able to reduce the scatter by using only those data obtained from the reference points shown in Fig. 14.15. The relationship obtained by Bascom is shown in Fig. 14.17. The other field studies confirmed the general trends given by Bascom, as can be seen in the figure.

The only factor besides median diameter that appears to affect the slope of the foreshore is the degree of exposure; the less exposed the beach, the steeper the beach face for a given median grain diameter. It was found that the beach face became flatter when eroding and steeper when building up. An example of this can be seen in Fig. 14.19 where the beach was subjected to storm waves a few days before to the photograph (*a*) and subjected to several months of long low waves before photographs (*b*) were taken. This observation

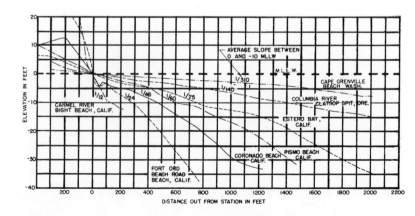

Fig. 14.16. Typical Pacific Coast beaches showing the wide range of beach slopes that are commonly found

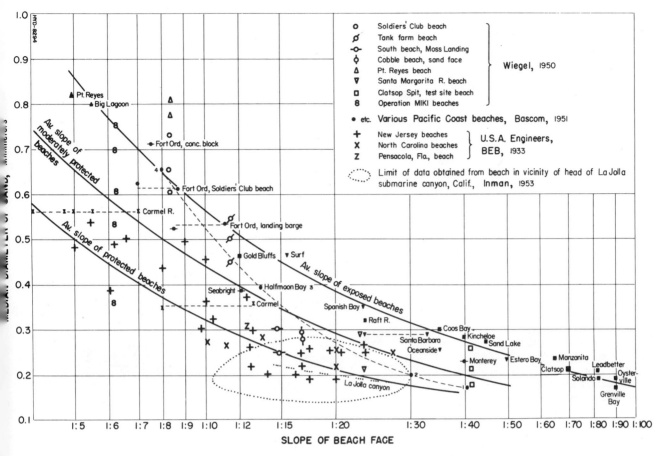

Fig. 14.17. Size-slope relationship of beaches at the reference point

was also made during the course of a model study by Scott (1954). Sand from a natural beach was used, 0.31 mm median diameter, and waves about 0.1 ft high with a 1 sec period were generated and allowed to act upon the beach. These tests should provide information on the limiting case of a well-protected beach. The slope of the equilibrium foreshore ranged from 1:3.5 to 1:4.2, which is well below the average slope of protected ocean beaches.

In some laboratory studies using coarse material (Bagnold, 1947), it was found that the slope of the foreshore depended only upon the size of the material, over the experimental range. Pebbles with a median diameter of 7 mm developed a foreshore of slope 1:2.5; pebbles with a median diameter of 3 mm developed a foreshore slope of 1:2.8; and sand with a median diameter of 0.5 mm developed a foreshore slope of 1:4.0. The last measurement is interesting, as it is just below the limiting curve for protected beaches of Fig. 14.17, and the waves used in the laboratory would certainly be less than even a well-protected beach along the ocean.

c. *Berm height.* A *berm* is the nearly horizontal deposit of beach material resulting from waves and tide action

(Fig. 14.18). Bascom measured berm heights for a portion of Monterey Bay from an exposed to a well-protected region, with the degree of protection gradually varying between the two extremes (University of California, 1951, Sec. I-I). The berm height was 16 ft in the most exposed area, 10 ft high about midway between the two extremes, and negligible in the most protected area.

A previous study (Krumbein, 1947) on a similar bay did not show such clear evidence. The berm height varied from 16 ft at the most exposed location to 10 ft at a more protected location, with a true berm not existing in a few locations in the more protected parts. It should be cautioned, however, that the sand beach was still very much in evidence, but that the beach sloped gradually up to the foot of sand dunes.

Two sets of laboratory experiments by Bagnold (1940) with pebbles of 7 mm and 3 mm median diameters showed that the height of the berm was directly related to the wave height, the berm height being about one-third higher than the wave height.

Bascom (1954b) has presented arguments to show that the stream outlets to the ocean are largely controlled by the berm height, and hence depend largely upon wave refraction.

An uprush crossing the crest of a berm. In this manner the berm grows vertically, and traces of previous berm crests are eliminated. Excess water returns to sea through the run-off channel at right

A new berm (under DUKW) has widened the beach at Seacliff, California. The old "winter" shoreline with its large cusps is now abandoned at the back of the beach. Run-off channel is just over the right hand tree

Fig. 14.18.

6. SOURCE OF MATERIAL

The source of sand on beaches is a matter of prime concern in regard to the character of the beach, especially with the slope, which in turn varies with the grain size. The grain size is a function of the sorting action of the waves and longshore currents, and it also is a function of the grain size of the sediment brought to the ocean by the rivers. If the rivers supply only fine sand to the ocean, the beach itself must consist of fine sand and hence be gently inclined. If the rivers supply coarse sand as well as fine sand, the adjoining beaches may consist of coarse or of fine sand, depending upon the distribution and sorting of the river sand by waves and currents. If the rivers bring down well-rounded sand, the adjoining beach sands are likely to consist of rounded grains. If they bring down angular sand grains, considerable wave action may be needed to round the grains. If they bring down large quantities of sand, the adjoining beaches will advance seaward because the waves and currents will supply more sand to the beach than the downrush can scour away. The amount of sand that a river can transport depends upon the river bed gradient, the flow of water in the river, and the material available. A detailed study of the phenomenon for a section of the California coast has been made by Johnson (1959).

For a given stream, when large amounts of water flow during storms, large quantities of sand will be moved onto the beach. An example of the change in a beach due to very heavy rainfall in the watershed for an interval of several days can be seen in Fig. 14.19. If the waves and currents bring little sand to the beaches, or the scouring action of the waves is high, more sand will be removed from the beaches than is brought to them. The front of the beach will then recede.

The artificial construction of breakwaters and jetties interferes with the movement of sand along shores, causing it to pile up on the side of the jetty facing the direction from which the sand comes, and to be removed from the beach for several miles on the other side of the jetty (see Chapter 18).

Handin (1951) made a detailed study of the physical and mineralogical properties of the beach sands at 88 stations along the coast between Ventura River and Santa Monica Bay, a distance of 75 miles, in order to determine the origin of the beach sand. The work showed that the sand migrated southeastward along the beach from the mouths of the main rivers, which are the principal sources of supply. Handin states that the sand moves around rocky headlands between the Ventura River and Santa Monica Bay, but does not move around point Fermin, which lies east of the bay. As the beach in Santa Monica Bay is not prograding, and as the supply of sand is continually brought to the bay by longshore currents, sand accordingly is either lost to the deeper water offshore, or is blown off the beach onto the adjoining land by wind. The beach front advances or retreats with (1) changing supply of sand from rivers, (2) varying waves, (3) storms. The grain size of the beaches is influenced by wave height and by grain size of the material brought to the ocean by the streams. The average grain size is 0.35 to 0.40 mm.

Trask (1952) investigated the distribution of mineral grains in the sand along the beaches west of Santa Barbara, California. It was found that there was a significant amount of augite, 2 to 3 per cent of the heavy mineral fraction, in the sand at Santa Barbara, and that this percentage increased progressively upcoast to a large fraction, about 55 per cent of the heavy minerals, at Morro Bay. It was further found that no augite was entering the littoral stream from rivers or by bluff erosion south of Pismo Beach, but that significant quantities were entering the littoral stream from rivers between Pismo Beach and Piedras Blancas. These studies indicated that a significant part of the sand moves around rocky promontories, such as Point Conception and Point Arguello. The data presented in this report indicate that 5 to 10 per cent of the sand on the beach at Santa Barbara was moved more than 125 miles along the coast. The results of this and other studies indicate that some 300,000 yards of sand a year move along the beach in the Santa Barbara area.

In some areas, the source of the beach sand is the shore itself. New Jersey is an example. Its coast is a plain of unconsolidated strata of gravel, sand, silt, and clay. The general recession of a large portion of the New Jersey shore through the years shows that it has provided a good portion of the littoral material (Wicker, 1951). Streams flow through relatively flat areas and are too sluggish to carry much sand, and the coast is effectively isolated from material from any major rivers, the Hudson and Delaware.

The opposite case exists along a portion of the west coast of Denmark (Bruun, 1955). The offshore slope is only 1:4500, and sand bars have been found to move shoreward from water at least as deep as 70 feet.

7. STABILITY OF BEACHES

Beaches are either prograding, remaining neutral, or retrograding. The specific condition may be due to long-term processes, annual processes, or short-term processes.

Even when beaches are retrograding in the long run, they build up under certain wave conditions; beaches prograding in the long run, or remaining neutral, erode under certain wave conditions. Each time water rushes up and down the foreshore of a beach, material is moved.

Owing to a combination of tide, waves, ground water, evaporation, and seepage, the porosity of the shore material changes. In addition, waves and wind are very powerful sorting tools. Sediment size varies considerably along any profile across a beach, along any contour along a beach, and with depth.

The Beach Erosion Board (U.S. Army, 1933), described one test made on an ocean beach that consisted of placing colored sand in a trench from the berm to mean sea level. Except for the sand at the plunge point during high tide, very little of the colored sand moved; instead it was gradually covered to depths up to 2 feet. It was not until the waves changed and erosion occurred that the sand moved. In regard to erosion of prograding beaches, it was found that the rearrangement of sand along the foreshore took place with great rapidity during some storms. During the period when a beach was undergoing rapid change, the slope of the foreshore varied considerably. Ninety-one observations taken at one point at Long Branch, New Jersey showed a variation of 3 to 12 degrees. It was cautioned that a cursory inspection of a beach immediately following a storm when rapid changes have been taking place may lead to conclusions that would not be substantiated a few days later.

Measurements of sand level were taken from the deck of a pier at 45 equally spaced stations about once a week for two years; twice a week for one year; and daily for one year (Shepard and La Fond, 1940). Wave, current, tide, and wind measurements were also made. It was found that at the stations which were nearly always above mean sea level, the sand underwent a single major cyclic change during the course of a year, with numerous smaller cycles superimposed upon it. At the outer stations (700–900 ft offshore), a less regular cycle took place; at the intermediate stations, many large short period cycles occurred during the course of a year. When the section above mean sea level was losing sand, the outer section was gaining sand. In general, the period of greatest cut inside and greatest fill outside was closely related to the season of the largest waves, and the period of shore fill and pier end cut was associated with the months of small waves. In addition, it was noted that the wave period was important as well as the wave height. However, no correlations among data were presented. Data were presented that showed the relationship between the changes in sand height of the section above mean sea level and the range of the tide. There seemed to be some correspondence between the two factors, particularly from October until the large fill in June. During this period, the spring tides were almost invariably accompanied by a cut, and the neap tides by a fill. The relationship was obscured by the effect of the waves, but the tide appeared to have had fully as important an effect as the waves at this section. The data on currents and mean sea level were discussed in regard to average sand levels rather than in relation to changes on the beach itself.

Measurements have also been made by Shepard (1946) of beach profiles, primarily in Southern California, but with a few measurements on Cape Cod, and Martha's Vineyard, Massachusetts. In addition, some grain size studies were made, as were studies of mean sea level and erosion of sea cliffs. Data were presented which showed that the fill occurred during the summer and fall seasons in Southern California and summer season in Massachusetts, these being the seasons of small waves. The data also showed erosion occurred during the winter months when large waves were prevalent. At the time the berm was being built, the slope of the foreshore became on the average somewhat steeper; during periods of erosion, scarps were often formed so that there were greatly oversteepened portions of the foreshore, even though the average slope was less. In addition, it was found that during the summer months excessive flat areas were developed close to low tide level, but they were removed rapidly by large waves. Coarse sand beaches developed berms much more rapidly than fine sand, and the berms also were cut back more rapidly.

Data taken during the period of 1945–49 were summarized by Shepard (1950) for many beaches in Southern California. The data substantiated the findings presented by Shepard (1946) and Shepard and La Fond (1940);

i.e., the widths of the berms decreased during the winter and spring months, these being the months when most of the large waves occurred, and the sand moved back onto the beaches during the summer and fall, these being the months of lower waves. An example of this can be seen in Fig. 14.19, by comparing the sand level at the cut-off piles in the foreground. These piles are exposed during the winter and covered during the summer. In comparing these two photographs, it is cautioned that other factors were at work because photograph (a) was taken just after a major flood had brought a great deal of sand down the river and deposited it at the mouth of the river.

Data, covering an interval of ten months, were obtained of waves, longshore currents, and beach conditions near the mouth of the Santa Margarita River, Oceanside, California (Wiegel, Patrick, and Kimberly, 1954). This beach is nearly straight with rather even parallel contours offshore, and the river mouth was closed during the entire interval. The wave and longshore current measurements are shown in Figs. 14.20 and 14.21.

Beach profiles were taken along ranges 500 ft apart (designated 7 + 50, 12 + 50, and 17 + 50 in the figures) at approximately two week intervals. Profiles through the surf were obtained by means of a DUKW (an amphibious vehicle) and leadline. In Fig. 14.22 are presented the profiles, measured to a depth of about 25 ft below *MLLW*. In Fig. 14.23 are shown (using a larger scale) the profiles of the beach face.

Fig. 14.19.

(a) Beach at Capitola, California, showing large amount of sand brought down Soquel Creek during floods in late December, 1955

(b) Beach at Capitola under normal summer conditions. Tide stage approximately the same as in above photograph; May, 1957

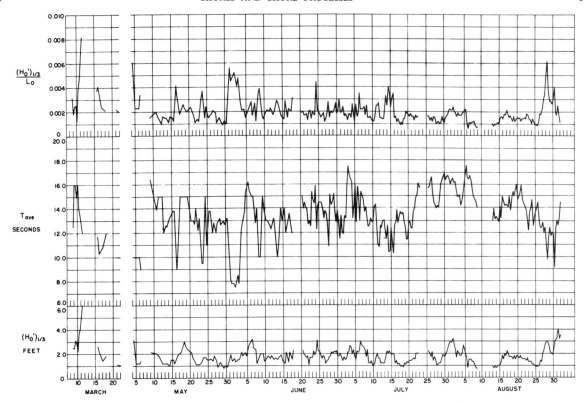

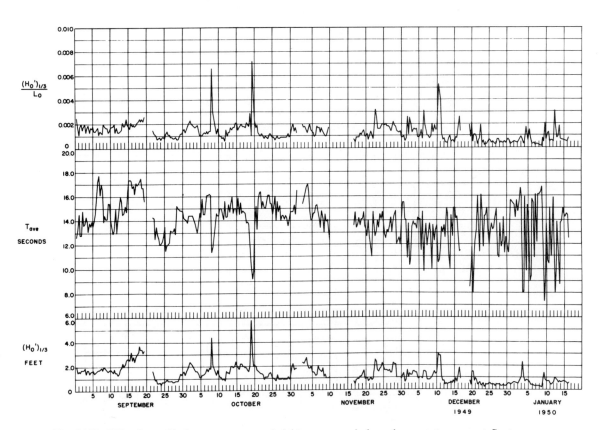

Fig. 14.20. "Unrefracted" deep water wave height, wave period, and wave steepness at Santa Margarita River Beach Oceanside, California (after Wiegel, Patrick, and Kimberley, 1954)

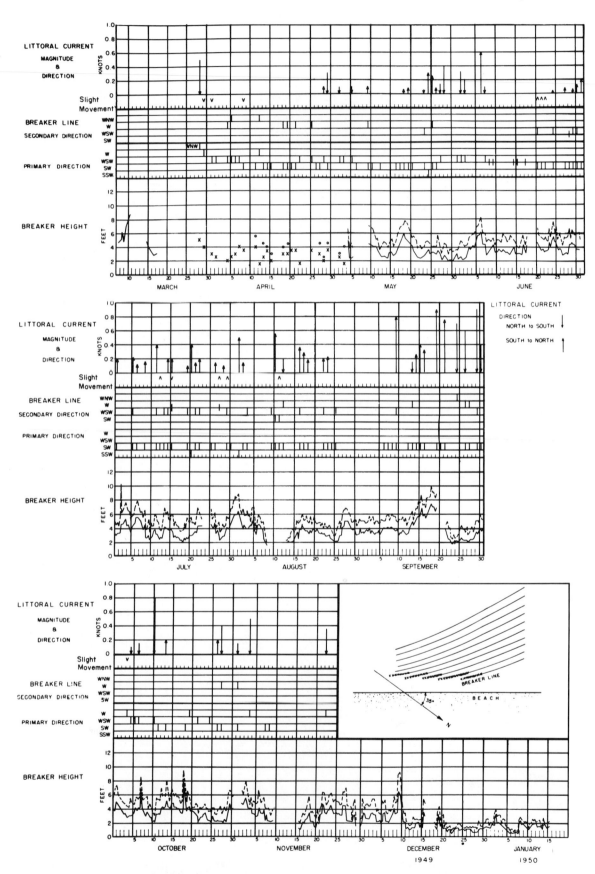

Fig. 14.21. Breaker height and direction, littoral current direction and magnitude at Santa Margarita River Beach Oceanside, California (after Wiegel, Patrick, and Kimberley, 1954)

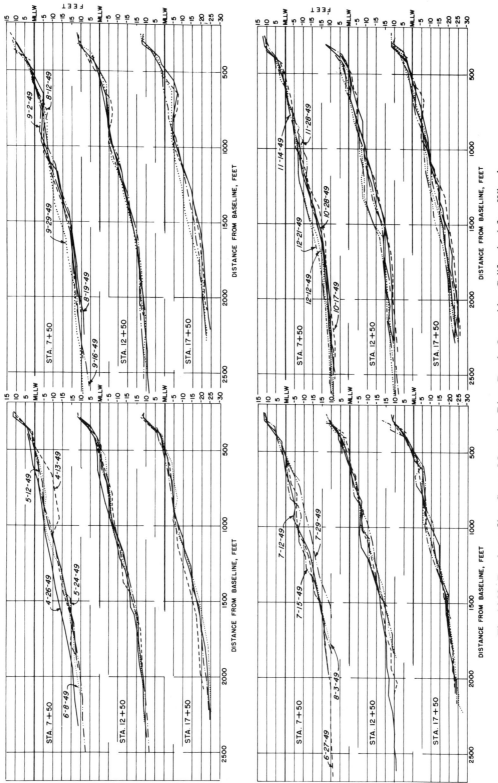

Fig. 14.22. Beach profiles, Santa Margarita River Beach Oceanside, California (after Wiegel, Patrick, and Kimberley, 1954)

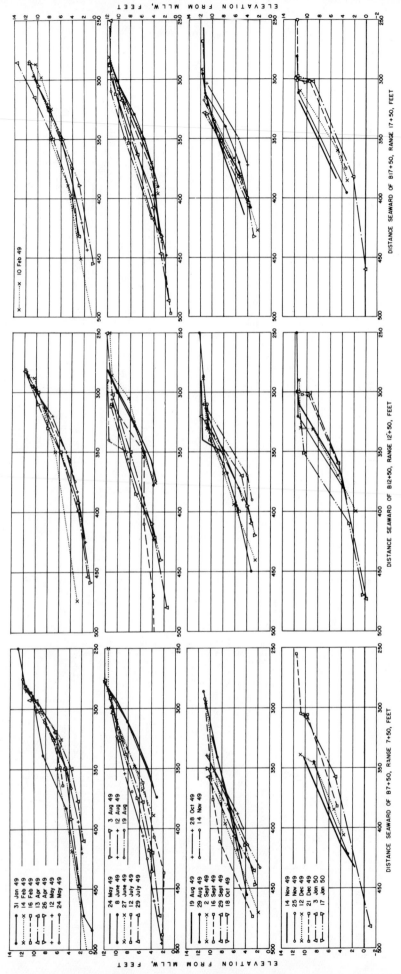

Fig. 14.23. Beach face profiles, Santa Margarita River Beach Oceanside, California (after Wiegel, Patrick, and Kimberley, 1954)

366

It can be seen that, in general, the width of the berm increased during the summer months and decreased during the winter months, but the three sets of profiles were not consistent. For example, considerable sand was deposited on the beach face on Range 7 + 50 between September 2 and September 16, 1949, while erosion occurred between September 16 and September 29; on Range 12 + 50 a slight amount of erosion occurred between September 2 and September 16, as well as between September 16 and September 29; and on Range 17 + 50 there was some slight deposition between September 2 and September 16 and between September 16 and September 29. Some of the apparent discrepancies can be explained by the presence of cusps on the beach during a portion of the time; thus, one profile may be across the ridge of a cusp, whereas another is across the hollow of a cusp. This condition makes it difficult to obtain an accurate picture of what is happening unless a large number of ranges are surveyed. Certain changes were of such magnitude that they were observed on all ranges; for example, between December 12 and December 21, a storm occurred near the coast; severe erosion took place along the entire beach, and a pronounced scarp was formed.

The three sets of offshore profiles were more consistent than were the series of onshore profiles. Although a distinct bar was often lacking, a step was usually present, but almost never was there a time when neither bar nor step was present. Also, although the profiles were constantly changing, they were always within remarkably narrow limits. The total vertical variation at any given distance offshore was less than plus or minus 4 feet from the mean. The beach never reaches a static equilibrium condition, but rather is in dynamic equilibrium.

Because of the relationship between wave steepness and beach profiles found in laboratory work (U.S. Army, Corps of Engineers, Beach Erosion Board, 1936; Johnson, 1949; Saville, 1950) a similar relationship was

sought from these data for ocean beaches. Although no definite relationship was found, it is believed that the 2-week interval between profile surveys was so large that completely adequate data were not obtained. Some laboratory studies showed that for waves steeper than $H_0/L_0 = 0.03$ the "storm wave" profile existed (that is, a bar formed), and for values less than 0.025 no bar was present, whereas for values of wave steepness in between, the profiles were unstable. Other studies (Iwagaki and Noda, 1963) have shown that the value of wave steepness which causes a bar to form is dependent upon the ratio of the deep water wave height to the mean diameter of the sand, H_0/d_{50}. For H_0/d_{50} less than about 300 the critical H_0/L_0 was about 0.025 or a little higher (up to 0.04), but as H_0/d_{50} increased, the critical value of H_0/L_0 decreased rapidly, being 0.02, 0.01, 0.008, and 0.004 for values of H_0/d_{50} of 430, 850, 1000, and 1300, respectively. In the case of Santa Margarita River data, the wave steepness, based upon the significant wave height and period, never approached the critical value of 0.03, and a bar or a step was usually present. The greatest steepness measured was about 0.008, and it was seldom above 0.003. The values of H_0/d_{50} were of the order of 5000, however, so that the field measurements are not inconsistent with the findings of Iwagaki and Noda.

Bascom (University of California, 1951, Sec I-I) made measurements of the growth and retreat of the berm and face of an exposed beach over an interval of several months (Fig. 14.24) and also over a short interval (Fig. 14.25). This example illustrates how large the variation in berm width can be.

Measurements of the beach above water can be made accurately. Measurements of the underwater portion cannot be made with the same accuracy if using the conventional methods of a fathometer or a lead line, which are about ± 0.5 ft and ± 1 ft maximum variations, respectively. The development of self-contained breathing apparatus (SCUBA), and its rapid spread

Fig. 14.24. Deposition and erosion of a beach (after Bascom, 1951)

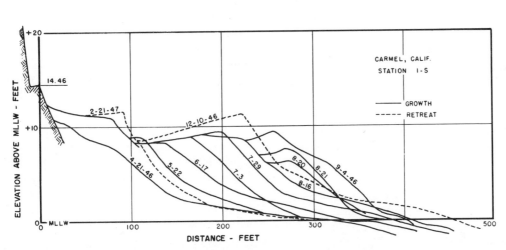

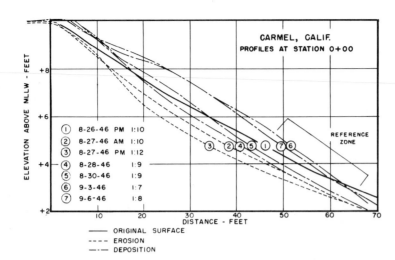

CARMEL, CALIF.
PROFILES AT STATION 0+00

① 8-26-46 PM 1:10
② 8-27-46 AM 1:10
③ 8-27-46 PM 1:12
④ 8-28-46   1:9
⑤ 8-30-46   1:9
⑥ 9-3-46   1:7
⑦ 9-6-46   1:8

REFERENCE ZONE

——— ORIGINAL SURFACE
- - - - EROSION
—·—·— DEPOSITION

Fig. 14.25. Slope changes with erosion and deposition (after Bascom, 1951)

as an engineering tool, has made it possible to obtain a large body of data on the ocean bottom near shore. Inman and Rusnak (1956) made use of SCUBA to install a series of rods in the sand bottom in water up to 75 ft deep off a beach at La Jolla, California. The median diameter of the sand ranged from about 0.18 mm on the beach face to 0.12 mm at the 30-ft depth, and 0.11 mm at the 75-ft depth. This area is rather special in that it is between two heads of a submarine canyon. The heights of the sand at the reference rods were measured by divers at intervals. The changes in elevation, obtained in this manner, are summarized in Fig. 14.26. Few data were obtained at the 18-ft water depth rod as the rod was lost during an interval of erosion, and few data were obtained at the 6-ft water depth rod as it was usually buried in the sand. In depths 30 ft and greater, the change throughout an interval of 3 years was only of the order of ±0.1 ft. This is an apparent conflict with other measurements (Wiegel, Patrick, and Kimberley, 1954), for example. There are two possible reasons for this, the most likely one being that the measurements of Inman and Rusnak were more precise than the other measurements and the other being the unusual hydrography.

Experiments made in the ocean by the Corps of Engineers (U.S. Army, 1938, 1948; Harris, 1954) are probably verifications of the findings of Inman and Rusnak for the deeper water portion of their work. In one case about 200,000 cu yards of sand was dredged from the Santa Barbara, Calif., harbor and dumped in about 22 ft of water 1000 ft from shore. It formed a mound 2200 feet long and 5 ft high. Surveys made nine years after it was dumped showed that the mound at that time was at no point more than a foot below its original level. This sand, however, was relatively coarse.

Measurements made by Inman and Rusnak of the portion of the beach exposed during low tide showed that changes of the order of several tenths of a foot occurred between high tide and low tide, even under normal wave conditions.

An indication has been obtained of the large areal movement of sand by using radioactive quartz particles to trace the movement (Inman and Chamberlain, 1958). The quartz particles were obtained from the site to be studied and subjected to slow neutron irradiation. The irradiated quartz particles (860 grams) were spread over an area of 3 sq ft on the bottom in 10 ft of water in the same area studied by Inman and Rusnak (1956) by divers using SCUBA. The techniques used in measuring the areal concentration of tracer grains provided a minimum sensitivity of 1 in 100,000. Background readings made before the test showed no radioactivity in the area. At the time of the tests, the waves were $1\frac{1}{4}$ ft high with a 14-sec period, and during the 24-hour duration of the tests the beach face prograded, while the area in which the tracer sand was placed eroded about 0.1 ft. At the end of $\frac{1}{4}$ hour, the tracer sand had spread to an elliptical shape with the long axis pointing toward shore. The count was 1000 per 100,000 at the center and 1 per 100,000 25 ft from the center. By the time $7\frac{1}{2}$ hours had elapsed, the 1 per 100,000 count extended 50 ft upcoast and downcoast from the center, 75 ft to sea, and 150 ft toward shore. At the end of 24 hours, irradiated grains were detected on the beach face about $\frac{1}{4}$ mile from the release point. The sampling showed that the particles dispersed over an area as great as 1 sq. mile within 24 hours.

The foregoing study and the studies of Shepard on sand deposited at the head of submarine canyons appear to conflict to some extent with the findings of Inman and Rusnak. A great deal of sand, however, can move along the bottom without an increase in the bottom level. The fact that large volumes of sand are lost into the submarine canyon shows only that an equilibrium exists in the region. There can be a large littoral drift without an appreciable change in bottom depth.

If a structure is placed offshore, it can upset this equilibrium to a considerable extent, often resulting in scouring in the immediate vicinity of the structure. This scouring can lead to a foundation failure of the struc-

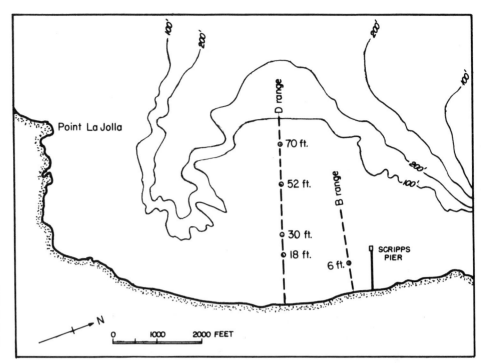

Index chart showing the location of ranges and stations. Six reference rods were placed at each station on *D* range. Single rods were placed at 20-foot intervals along a profile through the surf zone on *B* range. Depths are in feet below mean lower low water

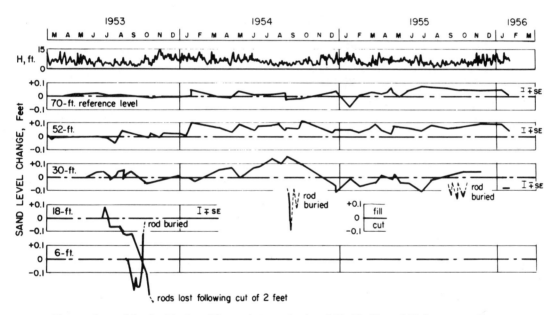

Changes in sand level with time. The stations at depths of 18, 30, 52, and 70 feet are on *D* range and the fluctuations in sand level are based on the mean of six reference rods at each station. The magnitude of two values of the standard error, $\mp$ SE, is shown as a measure of reliability of the observations. The station at a depth of 6 feet was on *B* range and consisted of a single reference rod which was completely covered with sand during the winter. The wave heights are for significant breakers at the point of wave convergence near *D* range (after Inman and Rusnak, 1956)

Fig. 14.26.

ture. The sediment that is deposited offshore is usually just in equilibrium with the forces exerted by waves and currents. A structure that is placed in water that is not too deep causes a partial standing wave to form and this causes an increase in the water motion at the bottom, and this can put the sand into suspension; any current then moves it away from the structure.

A series of laboratory experiments were made by the Beach Erosion Board to determine the relationship between varying wave conditions and beach configuration (U.S. Army, Corps of Engineers, 1936). In general, it was found that normal wave conditions tended to move the offshore sand toward a beach, storm conditions caused erosion of the beach and the formation of an offshore bar. This offshore bar was built of material taken from the beach and possibly some from the bottom. With an offshore bar formed by storm conditions, a return to normal conditions caused a movement of the bar shoreward. The time required to move a bar shoreward under normal conditions was 4 to 6 times as long as was required to form it under storm conditions.

Meyer (1936) performed experiments in a wave basin to investigate the effect of varying wave conditions. A dimensional analysis showed that the foreshore slope depended upon the size of the sand and the wave steepness. The beach profiles were measured as they changed with time and varying wave conditions. The data were reworked and presented in a dimensionless form by Johnson (1949). It was found that "storm" profiles developed when the wave steepness (ratio height to length) was greater than 0.06 and that "ordinary" profiles developed when the wave steepness was less than 0.02; a transition zone existed for wave steepness between 0.02 and 0.06. Laboratory experiments performed in a wave channel using sand of 0.30 mm median diameter by Saville (1950) also showed there were two primary types of beach profiles, "storm" profiles and "ordinary" profiles, and that the controlling factor was the deep-water wave steepness. For wave steepnesses greater than 0.03, sand was moved from the beach face, forming an offshore bar; for steepnesses less than 0.025, the offshore bar gradually moved onshore and sand was deposited on the beach; for steepness between 0.025 and 0.030, either type of beach could exist. The tests further showed that both the slope of the foreshore and the offshore slope depended upon the wave dimensions, but no definite relationship was established.

In some ways the foregoing results are in conflict with those of Rector (1954) and Watts and Dearduff (1954). In these studies, the beach face was found to build up for 0.46 and 3.44 mm median diameter sand and pebbles for the same waves that eroded a 0.22 mm sand beach. The size distributions of these materials, however, showed them to be poorly sorted compared with most ocean beaches, and as shown by Shay and Johnson (1950), this makes a difference in their response to waves. The results were in general agreement, however, with the findings of Iwagaki and Noda described previously.

Watts and Dearduff (1954) studied the effect of a tidal cycle on the interaction between waves and beach materials. The effect on 0.46 and 3.44 mm material was not great, but the effect on the 0.22 mm sand was considerable. For waves of relatively low steepness, the beach face built up for the 0.22 mm sand when no tidal cycle was used, but when the tidal cycle was introduced, this build-up did not occur.

In another set of tests, Watts (1954) varied the wave period about two mean wave periods in such a manner that one set of waves was fairly flat, while the other set of waves was fairly steep. Steep waves acting on all three materials and flat waves acting on the coarse materials (0.46 and 3.44 mm median diameter) produced

about the same results as were obtained using waves of constant period the same as the mean period; flat waves of constant period acting on 0.22 mm sand built up the beach face, whereas this did not occur for the case of the varying wave period.

One difficulty in using the results of many of these laboratory tests is that the beach was first formed by hand as an inclined plane, and then subjected to waves of a given character, then the beach was re-formed into an inclined plane by hand and subjected to another set of waves. Such a beach may be stable and unstable to different wave conditions than is true for a concave-shaped beach, which is closer to the form of a natural beach.

Two-dimensional studies were made of the relationship between sand beaches and waves (Johnson and Shay, 1950) to determine the behavior of various specially prepared, relatively fine sands. These tests indicated that a stable beach could be established for a particular sand only if the range of grain sizes in the mixture was relatively small, i.e., well sorted, as usually is the case for most of the beach sands in nature. Sand mixtures with a relatively large range of grain sizes resulted in a large range of profiles for the same wave conditions. Any attempt to predict a beach profile for a given wave condition appeared to be impossible as the beach was unstable for such mixtures. Further, the median diameter of the sand had to be as coarse as the sands found on natural beaches in order for a stable beach to form. Waves acting on finer sand caused continual erosion, with the sand gradually being moved seaward. These observations, together with the numerous measurements of natural beaches that show them to consist of well-sorted material, indicate that an artificial beach, or a beach fill, must be made with properly sorted material of appropriate size. In regard to the relationship among foreshore slope, median diameter of the sand, and wave steepness, it was found that the slope depended only upon the sand size for waves steeper than about 0.03, but that the wave steepness was also important for steepnesses less than 0.03, and the foreshore slope was steeper for the flatter waves. In addition to studying the effect of sand mixture on the development of equilibrium profiles, tests were also made to determine equilibrium profiles as developed under the influence of tides and waves. It was found that rising tides keep the original profile intact but move the beach position landward, eroding the beach, whereas a receding tide tends to flatten or smooth out the entire beach profile. Tests were also made of the effect of changing wave period. It was found that, if the materials used in these tests were transposed to prototype conditions, the profile changes which took place as a result of period change would require so long a time that they would probably be confused with those resulting from purely tidal action.

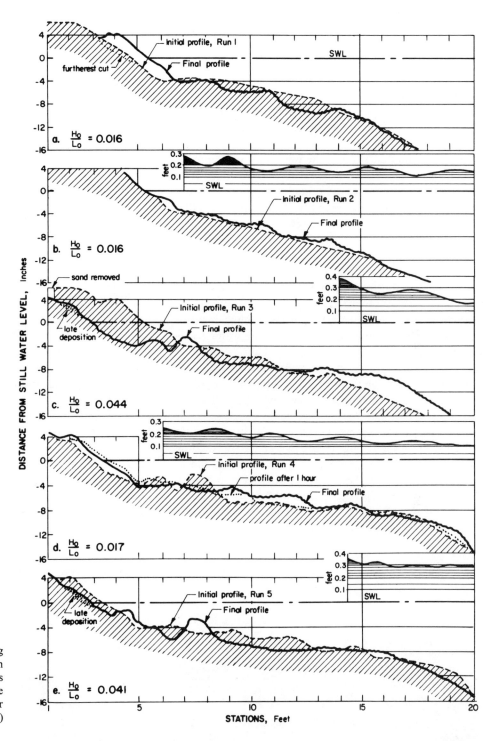

Fig. 14.27. Effect of changing wave steepness on beach profiles. Curves show trace of wave crests above still water level (after Scott, 1954)

Laboratory studies by Scott (1954) confirmed some of the observations of Saville (1950) and Johnson and Shay (1950). A 0.31 median diameter sand was used and about 12,000 waves of 0.016 steepness allowed to work on the beach, this being the number of waves necessary to create an equilibrium beach profile, with a well-developed berm (Fig. 14.27). Then the wave steepness was increased to 0.044 and 113,000 waves allowed to act on the beach to bring it into a new equilibrium. This resulted in an erosion of the foreshore and the movement of sand offshore. Decreasing the wave steepness and having 106,000 waves act upon the beach resulted in building up the width of the berm again, but not to the original width, as these waves were not able to bring ashore the sand lost to the "deeper water" by the higher steep waves. The wave steepness was again increased to 0.041

and 19,000 waves permitted to act on the beach. Again the berm was eroded, with material being carried farther seaward than during the first run with steep waves. The main results of this set of tests is to indicate that ocean beaches, which are subject to flat and steep waves, probably lose sand to deeper water, at least fine sand, This last statement is apparently not universally true, but depends upon the offshore area, as there is some evidence that sand moves onshore from relatively deep water along portions of the west coast of Denmark (Bruun, 1951).

Some advances have been made by Eagleson and Dean (1959), both theoretically and experimentally, on the relationship between wave characteristics, particle size, and bottom slope

Although there is much experimental data on the offshore and onshore movement of beach material, the mechanism (or mechanisms) controlling the direction of net movement of material is not fully understood. That an equilibrium condition can be reached is proof that the bottom modifies the motion of the water particles as well as vice versa. Hence it is not possible to determine the direction of net movement of sediment from a study or wave theory for an immovable bottom.

It is sometimes stated that the mass transport of waves would result in a shoreward movement of the sediment. Consider the null transport solution of a closed system for a slightly viscous fluid (see Chapter 2). In shallow water, the net movement of water at the bottom is shoreward, whereas above bottom it is seaward, so that a good deal of sand in suspension will move seaward. This could explain to a certain extent the observation of Scott (1954) that even on an eroding beach the sand

ripples shoreward of some null point moved shoreward, while the net movement (sand movement in suspension less the transport of sand due to the movement of the ripples) was offshore. Russell and Osorio (1958) found experimentally, however, that there are two regimes as far as the mass transport is concerned, one seaward of the bar and one landward of the bar.

One possible mechanism of the erosion of the beach face by storm waves is the effect of increased water level in the surf zone when high short waves (storm waves) are present (Munk, 1949; Fairchild, 1958). The profiles shown in Fig. 14.22 are roughly concave-shaped. These profiles were formed gradually over many tidal cycles and vary with the tidal cycle as shown by Inman and Rusnak (1956). Now, a local storm occurs, generating steep waves. The water level along the shore rises above the normal tide; the steep portion of the concave shape beach is unstable for this new water level and begins to erode from the bottom. That is, an undercutting occurs, then sand slumps down, a new undercutting occurs, etc. This would also explain the formation of scarps during intervals of erosion. If this concept is correct, erosion should occur mostly at high tide. Partial verification of this concept is the observation of Shepard and La Fond (1950) on an ocean beach that spring tides were almost invariably accompanied by a cut in the beach and neap tides by a fill.

In regard to the effect of high water it should be noted that one of the main reasons for the extensive damage due to the waves of the March 1962 storm on the Atlantic Coast of the U.S. was the fact that the storm lasted through five successive high tides with the water level being generally 4 to 5 ft above normal, (Cassidy, 1962).

Fig. 14.28. Waves breaking at an angle with beach generate longshore currents; north of Oceanside, California

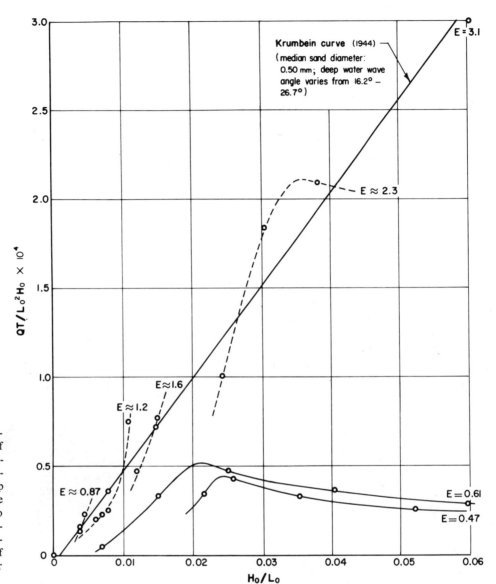

Fig. 14.29. Ratio of sand transport as a function of deep water wave steepness. Median sand diameter: 0.30 mm; deep water wave angle varied from 10.0° to 13.1°. E is energy content of waves in footpounds per foot of wave crest (after Saville, 1950)

In addition the storm occurred at the peak of the spring tides. In many areas the combination of high tides, high wind tides, and high waves eroded lines of dunes and attacked the land, roads, and buildings behind the dunes.

8. LONGSHORE DRIFT

Sand moves along beaches as well as on and off them. A glance at Fig. 14.28 will show why. Waves usually break at a small angle to a beach. Because of this, a wave-induced current is set up that flows along the beach. This current is called the *littoral* or *longshore* current. In addition, the water of the breakers rushes up the beach face at a slight angle, and then down the beach face in such a manner that there is a step type of motion of water along a beach. Studies made along the New

Jersey shore showed that sand moved parallel to the shore as well as on and off it, and that the longshore drift of sand consisted of two parts: the general drift due to longshore currents and the zigzag path on the foreshore due to the uprush and backwash of the waves (U.S. Army, 1933).

A laboratory study by Saville (1950) showed the same two regimes as the field studies. These studies were made using sand of 0.30 mm median diameter placed on a 1:10 slope. Waves were generated with an original angle of 10 degrees with the beach. The angle the breakers made with the beach depended upon the amount of refraction. Waves were allowed to break on the beach until equilibrium conditions were obtained. Measurements were made of the total drift along the beach and also of the drift past different positions along a beach profile, It was found that the sand transport along

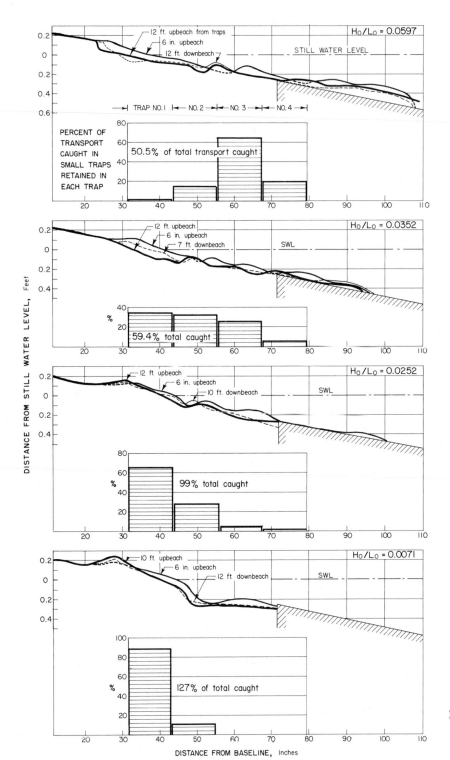

Fig. 14.30. Relationship between wave steepness and percentage of sand transport in different beach zones (after Saville, 1949)

incipient ordinary (or summer) beaches was much greater than along storm beaches, for waves of the same energy content, and for equilibrium beach conditions, but the transport dropped rapidly for waves less steep than 0.02. Peak transport occurred for a deep water wave steepness of from 0.02 to 0.025 (Fig. 14.29), with this transport almost entirely due to beach drifting. In this figure, Q is the sand transport rate and L_o, H_o, and T are the deep water wave length, height, and period, respectively,

and E is the wave energy. In Fig. 14.30 are shown the percentages of total sand trapped in different sections of the beach profile, with trap No. 4 being on the beach face. The location of the other traps with respect to the breaker varied, depending upon the wave steepness. Their locations can best be judged from the bar location in the beach profiles. It will be noted that more than 100 per cent of the total transport was trapped in one instance; this was for a very flat wave that refracted to such an extent that it broke parallel to the beach, so that the sand moved first in one direction and then in another direction. This is also the reason for the fact that very little net transport of sand occurs for flat waves. For "storm" beaches, the sediment transport occurred mainly in suspension in the littoral current. For unsteady conditions, it was found that greater longshore transport of sand occurred when the beaches were eroding than when they were prograding. Sand samples were taken along the profiles perpendicular to the beach and along lines parallel to the beach. Sieve analyses of the samples showed that no sorting took place along the lines parallel to the beach, but that sorting occurred along the profiles, with the finer sand moving seaward of the bar or step.

Other studies (Shay and Johnson, 1951; Johnson, 1953) were conducted for a series of offshore angles of wave approach (20, 30, 40, and 50 degrees). The results for sand of 0.30 mm median diameter are shown in Fig. 14.31. For given wave dimensions, the maximum movement of sand along the shore occurred when the angle of wave approach was approximately 30 degrees. It was found that a considerable length of time was required to obtain a quasi–steady-state condition of sand transport when the wave conditions changed (Fig. 14.32).

Fig. 14.31. (a) Relationship between rate of littoral transport and wave steepness for various angles of wave approach (mean sand size: 0.30 mm); (b) Dimensionless relationship between rate of littoral transport, wave power, wave steepness, and angle of wave approach (mean sand size: 0.30 mm)

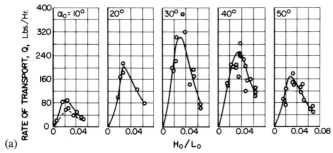

(a)

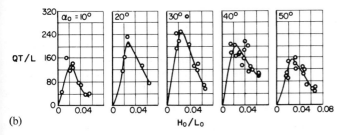

(b)

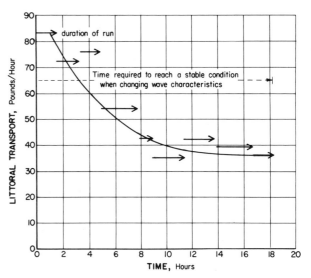

Fig. 14.32. Variation in rate of transport during period when beach is adjusting itself from an initial equilibrium condition ($H_0/L_0 = 0.0217$) to a final equilibrium condition. Wave steepness during adjustment: 0.039. First wave conditions: $T = 1.0$ sec, $H_0 = 0.111$ ft, $H_0/L_0 = 0.0217$; second wave conditions: $T = 0.86$ sec, $H_0 = 0.145$ ft, $H_0/L_0 = 0.039$ (after Johnson, 1953)

The angles given in Fig. 14.31 are the angles between the wave generator and the beach, so that the effects of refraction must be considered. One way of taking this into account would be to measure the angle the breaker makes with the beach. This is not easy to do in the laboratory even by photographic methods. Another method is to measure the littoral current by placing dye in the water and timing its travel. Results of such a procedure are shown in Fig. 14.33. In Fig. 14.33(a) the relationships among wave steepness, wave angle, and littoral current are given, and in Fig. 14.33(b) the relationship between sand transport and littoral current is shown. Q/T^5 is a dimensionless number when multiplied by a dimensional constant which depends upon the specific weight of the sand and gravity, and the relationship between wave length and wave period.

There have been some studies of the relationship between wave energy and sand transport along ocean beaches, but the difficulties involved in obtaining wave characteristics, especially the angle of wave approach, and the inherent difficulties in surveying water depths from a floating platform make the results difficult to assess (Watts, 1953; Caldwell, 1956; Savage, 1959). If no measurements of sand transport are available, the formula of Caldwell (1956) can be used to estimate the amount of transport. This formula is

$$Q = 210\,E^{0.8}$$

where Q is the number of cubic yards per day that moves along a sandy coast and E is the alongshore energy in millions of foot-pounds per foot of beach per day. E

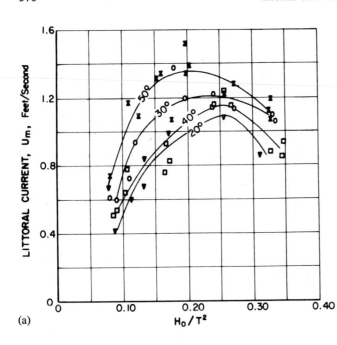

(a)

Fig. 14.33. (a) Littoral current as a function of wave steepness and wave angle; (b) Sand transport as a function of littoral current (after Shay and Johnson, 1951)

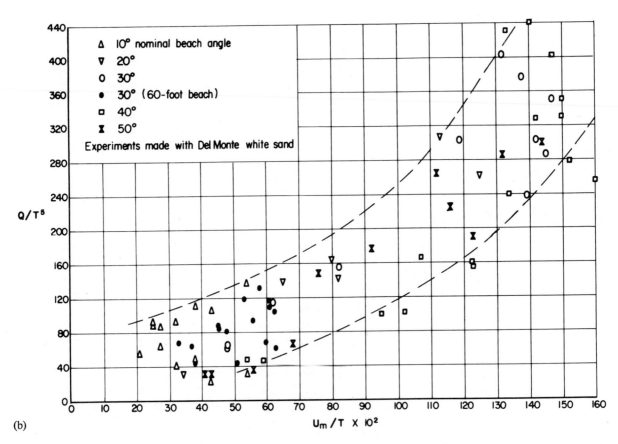

(b)

is given by

$$E = E_t \sin \phi \cos \phi$$

where E_t is the total wave energy, and ϕ is the angle between the wave crest and the beach and the point in question. The calculations by Caldwell were for waves passing the 12-ft contour.

In spite of these and many other studies, there is at present no general relationship between wave and sediment characteristics available for estimating the rate of transport along a given shore (Johnson, 1956).

The probable rates of transport along natural beaches can be best estimated from the amount trapped by shore structures (Fig. 14.5) or natural barriers, and from knowledge of the rate of depletion of sources of supply, such as flow through a river mouth, or eroding cliffs (Johnson, 1959). For an example, see Chapter 18.

Johnson (1956; 1957) has tabulated measured rates of drift for many locations, and his findings are given in Table 14.3.

Table 14.3. SUMMARY OF MEASURED RATES OF LITTORAL DRIFT ALONG COASTS
(from Johnson, 1956; 1957)

Location	Predominant direction of drift	Rate of drift (cubic yards per year)	Method of measure of rate of drift	Years of record	Reference
Atlantic Coast					
Suffolk Co., N.Y.	W	300,000	Accretion	1946–55	U.S. Army (1955a)
Sandy Hook, N.J.	N	493,000	Accretion	1885–1933	U.S. Army (1954b)
Sandy Hook, N.J.	N	436,000	Accretion	1933–51	U.S. Army (1954b)
Asbury Park, N.J.	N	200,000	Accretion	1922–25	U.S. Army (1954b)
Shark River, N.J.	N	300,000	Accretion	1947–53	U.S. Army (1954b)
Manasquan, N. J.	N	360,000	Accretion	1930–31	U.S. Army (1954b)
Barneget Inlet, N.J.	S	250,000	Accretion	1939–41	U.S. Army (1954b)
Absecon Inlet, N.J.	S	400,000	Erosion	1935–46	U.S. Army (1954b)
Ocean City, N.J.	S	400,000	Erosion	1935–46	U.S. Army (1953a)
Cold Spring Inlet, N.J.	S	200,000	Accretion	—	U.S. Congress (1953b)
Ocean City, Md.	S	150,000	Accretion	1934–36	U.S. Army (1948a)
Atlantic Beach, N.C.	E	29,500	Accretion	1850–1908	U.S. Congress (1948)
Hillsboro Inlet, Fla.	S	75,000	Accretion	—	U.S. Army (1955b)
Palm Beach, Fla.	S	150,000 to 225,000	Accretion	1925–30	U.S. Army (1947)
Gulf of Mexico					
Pinellas Co., Fla.	S	50,000	Accretion	1922–50	U.S. Congress (1954a)
Perdido Pass, Ala.	W	200,000	Accretion	1934–53	U.S. Army (1954c)
Galveston, Texas	E	437,500	Accretion	1919–34	U.S. Congress (1953c)
Pacific Coast					
Santa Barbara, Calif.	E	280,000	Accretion	1932–51	Johnson (1953b)
Oxnard Plain Shore, Calif.	S	1,000,000	Accretion	1938–48	U.S. Congress (1953d)
Port Hueneme, Calif.	S	500,000	Accretion	1938–48	U.S. Congress (1954b)
Santa Monica, Calif.	S	270,000	Accretion	1936–40	U.S. Army (1948b)
El Segundo, Calif.	S	162,000	Accretion	1936–40	U.S. Army (1948b)
Redondo Beach, Calif.	S	30,000	Accretion	—	U.S. Army (1948b)
Anaheim Bay, Calif.	E	150,000	Erosion	1937–48	U.S. Congress (1954c)
Camp Pendleton, Calif.	S	100,000	Accretion	1950–52	U.S. Army (1953a)
Great Lakes					
Milwaukee Co., Wis.	S	8,000	Accretion	1894–1912	U.S. Congress (1946)
Racine Co., Wis.	S	40,000	Accretion	1912–49	U.S. Congress (1953e)
Kenosha, Wis.	S	15,000	Accretion	1872–1909	U.S. Army (1953b)
Ill. state line to Waukegan	S	90,000	Accretion	—	U.S. Congress (1953f)
Waukegan to Evanston, Ill.	S	57,000	Accretion	—	U.S. Congress (1953f)
South of Evanston, Ill.	S	40,000	Accretion	—	U.S. Congress (1953f)
Hawaii					
Waikiki Beach, Hawaii	—	10,000	Suspended load samples	—	U.S. Congress (1953)
Outside of the United States					
Monrovia, Liberia	N	500,000	Accretion	1946–54	U.S. Army (1955)
Port Said, Egypt	E	910,000	Dredging	—	Vernon-Harcourt, 1900
Port Elizabeth, South Africa	N	600,000	Accretion	—	Matthews (1906)
Durban, South Africa	N	383,000	Dredging	1897–1904	Methven, C. W. (1905–1906)
Madras, India	N	740,000	Accretion	1886–1919	Vernon-Harcourt, L. F. 1881–82; Spring, F. J. E. 1919–20
Mucuripe, Brazil	N	427,000	Accretion	1946–50	—

Silvester (1962) combined the available information on shapes of bays, prevailing waves, and littoral drift to develop a criterion for the prediction of the direction of littoral drift. He then examined about 250 charts covering the major coastlines of the world, and prepared a series of maps which showed the most likely direction of littoral drift. These maps should be of considerable value to an engineer in the preliminary feasibility study of harbor development in a relatively unfamiliar part of the world.

In addition to the general drift of sand discussed so far, there are migrations of undulations of a sand coast. Bruun (1955) measured a series of these undulations in the vicinity of the Lime Inlet, Denmark. They were found to be between 900 and 6000 ft between crests and they migrated at rates varying from zero to 3000 ft per year. Bruun also presents some evidence of the migration of large humps of sand on the bottom near shore, and in a direction roughly parallel to the shore.

There are many interesting phenomena associated with littoral drift. One of these is the nodal point. A classic example is located in New Jersey (Wicker, 1951). The predominant waves moving toward the coast of New Jersey are from the northeast quadrant. The effect of the waves is modified by the sheltering characteristics of Long Island, Cape Cod, and the islands adjoining Nantucket Sound. As a result, waves from this quadrant dominate south of the area between Manasquin Inlet and Barnegat Inlet, whereas waves from the southeast quadrant dominate the area to the north. The nodal point lies between these two inlets, with sand moving upcoast north of the nodal point, and downcoast south of the nodal point (Fig. 14.34). Measurements of the littoral drift of sand presented in Table 14.3 show this clearly.

The *tombolo* is a sand feature that builds from the shore to a rock, island, or offshore breakwater, as a result of wave diffraction and refraction (Fig. 14.35).

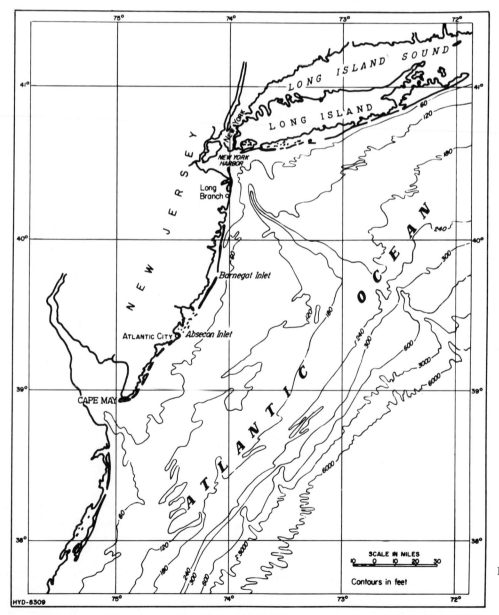

Fig. 14.34. New Jersey coastal area

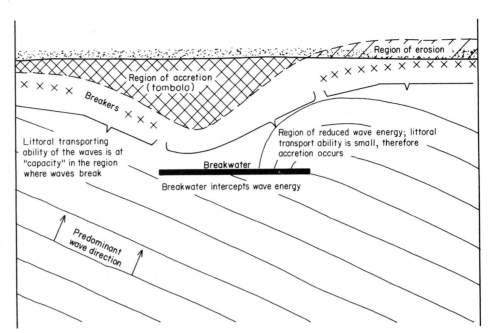

Fig. 14.35. Schematic representation of the transportation, deposition, and scour of littoral sediments at a detached breakwater (from Johnson, 1957)

The net effect is a local drift from both downcoast and upcoast in the immediate area of the offshore obstruction, with a deposit of sand in the shadow of the obstruction. Eventually the sand tombolo bridges the gap between shore and obstruction. Model studies have shown that there are several modifications of the pattern, depending upon the general littoral drift in the region (Sauvage de Saint Marc and Vincent, 1955).

9. ESTUARY ENTRANCES

The relationships among littoral drift, waves, ocean currents, tidal prisms, fresh water flow, the supply of sediment by the river, and the dimensions and configurations of the entrances to estuaries are complex. The configurations and dimensions change with time, an example of the variation in configuration with time for an ocean inlet being presented in Fig. 14.36. For details of the changes that occur over very long intervals of time see Abecasis (1955). This paper presents data on the entrance to the port of Aveiro, Portugal, from 1318 through 1954. It is the classic study of an estuarial inlet.

A model study has been made of the effect of waves, tides, and currents on an uncontrolled bay entrance through sand 0.23 mm median diameter, $S_o = 1.16$, and $S_k = 0.99$ (Saville, Caldwell, and Simmons, 1957). First the beach, without bay entrance, was allowed to attain equilibrium configuration, when subjected to waves from one direction, then another direction, etc. The wave and current program was such that there were roughly four units of sand transport in the downcoast direction to one unit in the upcoast direction. Then a channel was cut

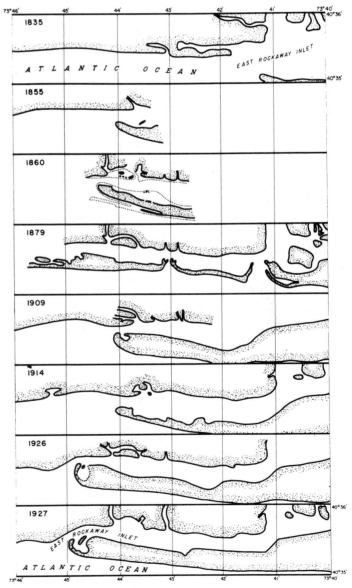

Fig. 14.36. Shoreline changes, East Rockaway Inlet, N.Y. (after U.S. Army, 1933)

379

through the beach to a bay, the channel length being 4 times the width, and the width being 10 times the depth. The bottom of the bay sloped from the entrance to a depth about 6 times the channel depth. For small prisms, it was found that the channel soon closed by a sand spit building in the downcoast direction. That this occurs in nature has been verified many times along the New Jersey coast (Wicker, 1951).

The bay area was increased in size until there was enough tidal flow through the inlet, with high enough velocities, to keep the channel open. It was found that the channel was unstable, changed in depth, migrated, with offshore bars forming with one or two channels through the bar, etc. The test was then repeated, but with the bay the same depth as the entrance. In this case, the channel was nearly closed by the end of 300 tidal cycles.

In spite of the complexities and the fact that the entrances are always changing, O'Brien (1931) observed that a simple relationship existed between the volume of the tidal prism between MLLW and MHHW and the area of

the inlet below mean sea level. These data are shown in Fig. 14.37. For the estuaries on the Atlantic Coast and Gulf Coast, the mean tidal range was used. It appears that Tillamook Bay does not follow the curve through the other points; this entrance, however, had been changing rapidly since construction of the jetties at the entrance, and the change in the entrance area had been in the direction of the curve. A tendency existed for entrance areas to be larger in fine sand than in coarse sand.

Data on other tidal prisms and entrance areas have been published (Bruun and Gerritsen, 1958; Bruun, Gerritsen, and Morgan, 1958). These have also been shown in Fig. 14.37. Note that some of the values are in conflict with the original values used by O'Brien. Some of these disagreements are for bays that have had dredging operations performed at their entrances, and as such are not representative of natural conditions. Other disagreements reflect the difficulty in computing tidal prisms, and the fact that tidal entrances vary from time to time from their mean conditions. The Wester-

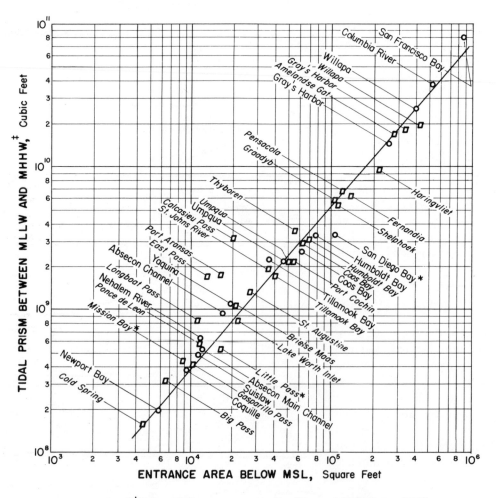

○ O'Brien (1931)   ▯ Bruun and Gerritsen (1958)

‡ Tidal prism for estuaries with diurnal tides based upon mean tide range

* Entrances dredged

Fig. 14.37. Relationship between tidal prism and entrance area of bays

schelde and Oosterschelde which were reported in the paper by Bruun and Gerritsen have been omitted, as they are not separate estuaries, but are so connected that they are not independent.

The harbors that are known to have dredged entrances are noted in Fig. 14.37. There are undoubtedly others in the figures that are dredged, and the reader is cautioned to consider this fact when using the mean curve

Another condition to be considered in using Fig.14.37 is the effect on the entrance area of the amount of littoral drift of sand. It is to be expected that the entrance in a region of large littoral drift would be different from the entrance in a region of small littoral drift. Another condition that would probably modify the relationship between tidal prism and entrance area is the presence or absence of jetties at the entrances.

REFERENCES

Abecasis, Carlos Krus, The history of a tidal lagoon inlet and its improvement (the case of Aveiro, Portugal), *Proc. Fifth Conf. Coastal Eng.*, Berkeley, Calif.: The Engineering Foundation Council on Wave Research, 1955, pp. 329–63.

Bagnold, R. A., Beach formation by waves: some model experiments in a wave tank, *J. Inst. Civil Engrs.* (London), Paper No. 5237 (November, 1940), 27–52.

Bascom, Willard J., The relationship between sand size and beach-face slope, *Trans. Amer. Geophys. Union*, 32, 6 (1951), 866–74.

——, Characteristics of natural beaches, *Proc. Fourth Conf. Coastal Eng.*, Berkeley, Calif.: The Engineering Foundation Council on Wave Research, 1954a, 163–80.

——, The control of stream outlets by wave refraction, *J. Geology*, 62, 6 (November, 1954), 600–605.

Bruun, Per, Migrating sand waves or sand humps, with special reference to investigations carried out on the Danish North Sea coast, *Proc., Fifth Conf. Coastal Eng.*, Berkeley, Calif.: The Engineering Foundation Council on Wave Research, 1955, pp. 269–95.

——, and F. Gerritsen, Stability of coastal inlets, *J. Waterways Harbors Division, ASCE*, 84, WW3, Paper No. 1644, (May, 1958).

——, ——, and W. H. Morgan, Florida coastal problems, *Proc. Sixth Conf. Coastal Eng.*, Berkeley, Calif.: The Engineering Foundation Council on Wave Research, 1958, pp. 463–509.

Caldwell, Joseph M., *Wave action and sand movement near Anaheim Bay*, California, U.S. Army, Corps of Engineers, Beach Erosion Board, Tech. Memo, No. 68, February, 1956.

Cassidy, W. F., Recovery operations after Atlantic Coast storm, *The Military Engineer*, 54, 360, July–Aug. 1962, 246–48.

Eagleson, P. S. and R. G. Dean, Wave-induced motion of bottom sediment particles, *J. Hyd. Div. ASCE*, 85, HY 10, Paper No. 2202 (October, 1959), 53–79.

Emery, K. O. and J. F. Foster, Water tables in marine beaches, *J. Mar. Res.*, 7, 3 (1948), 644–54.

——, and R. E. Stevenson, Laminated beach sand, *J. Sed. Pet.*, 20, 4 (1950), 220–23.

Fairchild, John C., Model study of wave set-up induced by hurricane waves at Narragansett pier, Rhode Island, U.S. Army, Corps of Engineers, Beach Erosion Board, *Ann. Bull.*, 12 (July, 1958), 9–20.

Grant, U. S., Influence of water table on beach aggradation and degradation, *J. Mar. Res.*, 7, 3 (November, 1948), 655–60.

Handin, J. W., *The source, transportation and deposition of beach sediment in Southern California*, U.S. Army, Corps of Engineers, Beach Erosion Board, Tech. Memo. No. 22, 1951.

Harris, Robert L., *Restudy of test-shore nourishment by offshore deposition of sand, Long Branch, New Jersey*, U.S. Army, Corps of Engineers, Beach Erosion Board, Tech. Memo. No. 62, November, 1954.

Inman, D. L., *Areal and seasonal variations in beach and nearshore sediment at La Jolla, California*, U.S. Army, Corps of Engineers, Beach Erosion Board, Tech. Memo. No. 39. March, 1953.

——, and T. K. Chamberlain, Tracing sand movement with irradiated quartz, *J. Geophys. Res.*, 64, 1 (January, 1959), 41–47.

——, and G. S. Rusnak, *Changes in sand level on the beach and shelf at La Jolla, California*, U.S. Army, Corps of Engineers, Beach Erosion Board, Tech. Memo. No. 82, July, 1956.

Isaacs, J. D., and W. N. Bascom, Water-table elevations in some Pacific Coast beaches, *Trans. Amer. Geophys. Union*, 30, 2 (April, 1949), 293–94.

Iwagaki, Y., and Hideaka Noda, Laboratory study of scale effect in two-dimensional beach processes, *Proc. Eighth Conf. Coastal Eng.*, Berkeley, Calif.: The Engineering Foundation Council on Wave Research, 1963, 194–210.

Johnson, Douglas W., *Shore processes and Shoreline Development*. New York: John Wiley & Sons, Inc., 1919.

Johnson, J. W., Scale effects in hydraulic models involving wave motion, *Trans. Amer. Geophys. Union*, 30, 4 (1949), 517–25.

——, Sand transport by littoral currents, *Proc. Fifth Hyd. Conf.*, Iowa Inst. Hyd. Res., 1953, pp. 89–109.

——, Dynamics of nearshore sediment movement, *Bull. Amer. Soc. Pet. Geologists*, 40, 9 (September, 1956), 2211–32.

——, The littoral drift problem at shoreline harbors, *J. Waterways Harbors Div.*, ASCE, 83, WW1, Paper 1211, April, 1957.

——, The supply and loss of sand to the coast, *J. Waterways Harbors Div.*, ASCE, 85, WW3, Paper 2177 (September, 1959), 227–51.

————, and E. A. Shay, *Sand studies in two-dimensional wave motion*, Univ. Calif. IER, Tech. Rept. 14–5, 1950. (Unpublished.)

Krumbein, W. C., *Shore currents and sand movement on a model beach*, U.S. Army Corps of Engineers, Beach Erosion Board, Tech. Memo. No. 7, 1944.

————, *Shore processes and beach characteristics*, U.S. Army, Corps of Engineers, Beach Erosion Board, Tech. Memo. No. 3, 1947.

————, Statistical design for sampling beach sand, *Trans. Amer. Geophys. Union*, **34**, 6 (December, 1953), 857–68.

McKee, Edwin D., Storm sediments on a Pacific atoll, *J. Sed. Pet.*, **29**, 3 (September, 1959), 354–64.

Mason, M. A., and C. A. Schulte, *Abrasion of beach sand*, U.S. Army, Corps of Engineers, Beach Erosion Board, Tech. Memo. No. 2, 1942.

Matthews, Wm., Discussion of harbours of South Africa, *Min. Proc. Inst. Civil Engrs.*, **160** (1906) 47–49.

Methven, C. W., The harbors of South Africa; with special reference to causes and treatment of sand bars, *Min. Proc. Inst. Civil Engrs.*, **166** (1905–1906), 40.

Meyer, Richard D., A model study of wave action on beaches, M.S. Thesis in Civil Engineering, Univ. Calif., 1936.

Miller, Robert L., and John M. Zeigler, A model relating dynamics and sediment pattern in equilibrium in the region of shoaling waves, breaker zone, and foreshore, *J. Geology*, **66**, 4 (July, 1958), 417–41.

Munk, W. H., Surf beats, *Trans. Amer. Geophys. Union*, **30**, 6 (December, 1949), 849–54.

O'Brien, Morrough P., Estuary tidal prisms related to entrance areas, *Civil Eng.*, **1**, 8 (May, 1931), 738–39.

————, and B. D. Rindlaub, The transport of sand by wind, *Civil Eng.*, **6**, 5 (May, 1936), 325–27.

Patrick, D. A., *Ground water adjacent to four Pacific Ocean beaches*, Univ. Calif. IER, Tech. Rept. 29–35, 1950. (Unpublished.)

Rector, Ralph L., *Laboratory study of equilibrium profiles beaches*, U.S. Army, Corps of Engineers, Beach Erosion Board, Tech. Memo. No. 41, August, 1954.

Russell, R. C. H., and J. D. C. Osorio, An experimental investigation of drift profile in a closed channel, *Proc. Sixth Conf. Coastal Eng.*, Berkeley, Calif.: The Engineering Foundation Council on Wave Research, 1958, pp. 171–93.

Sauvage de Saint Marc, M. G., and M. G. Vincent, Transport littoral formation de flèches et de tombolos, *Proc. Fifth Conf. Coastal Eng.*, Berkeley, Calif.: The Engineering Foundation Council on Wave Research 1955, pp. 296–328.

Savage, Rudolph P., *Laboratory study of the effect of groins on the rate of littoral transport: equipment development and initial tests*, U.S. Army, Corps of Engineers, Beach Erosion Board, Tech. Memo. No. 114, June, 1959.

Saville, Thorndike, Jr., *Preliminary report on model studies of sand transport along an infinitely long straight beach*, Univ. Calif. IER, Tech. Rept. 3–305, July, 1949. (Unpublished.)

————, Model study of sand transport along an infinitely long straight beach, *Trans. Amer. Geophys. Union*, **31**, 4 (August, 1950), 555–65.

————, Joseph M. Caldwell, and Henry B. Simmons, *Preliminary report: laboratory study of the effect of an uncontrolled inlet on the adjacent beaches*, U.S. Army, Corps of Engineers, Beach Erosion Board, Tech. Memo. No. 94, May, 1957.

Shay, E. A., and J. W. Johnson, *Model studies on the movement of sand transported by wave action along a straight beach*, Univ. Calif. IER, Tech. Rept. 14–7, February, 1951. (Unpublished.)

Shepard, F. P., *Beaches and wave action*, Univ. Calif., Scripps Inst. Oceanog., Wave Project Rept. No. 56, 1946. (Unpublished.)

————, Beach cycles in Southern California, Univ. Calif., Scripps Inst. Oceanog., Submarine Geology Rept. No. 11, 1950.

————, *Longshore-bars and longshore-troughs*, U.S. Army, Corps of Engineers, Beach Erosion Board, Tech. Memo. No. 15, January, 1950.

————, Mass movement in submarine canyon heads, *Trans. Amer. Geophys. Union*, **32**, 3, (June, 1951), 405–18.

————, and D. L. Inman, Nearshore water circulation related to bottom topography and wave refraction, *Trans. Amer. Geophys. Union*, **31**, 3 (April, 1950). 196–212,

————, and E. C. La Fond, Sand movement along the Scripps Institution Pier, California, *Amer. J. Sci.*, **238**, (1940), 272–85.

Silvester, Richard, Sediment movement around the coastlines of the world, *Conf. Civil Eng. Problems Overseas*, **1962**, Inst. Civil Eng., Paper No. 14, 1962.

Spring, F. J. E., Coastal sand travel near Madras Harbour, *Min. Proc. Inst. Civil Engrs.*, **210** (1919–20), 27–28.

Thompson, W. O., Original structure of beaches, bars and dunes, *Bull. Geograph. Soc. Amer.*, **48** (1937), 723–51.

Trask, Parker D., *Source of beach sand at Santa Barbara, California as indicated by mineral grain studies*, U.S. Army, Corps of Engineers, Beach Erosion Board, Tech. Memo. No. 28, 1952.

————, *Movement of sand around Southern California promontories*, U.S. Army, Corps of Engineers, Beach Erosion Board, Tech. Memo. No. 76, 1955.

————, and Charles A. Johnson, *Sand variation at Point Reyes Beach, California*, U.S. Army, Corps of Engineers, Beach Erosion Board, Tech. Memo. No. 65, October, 1955.

————, and Theodore Scott, *Bore hole studies of the naturally impounded fill at Santa Barbara, California*, U.S. Army, Corps of Engineers, Beach Erosion Board, Tech. Memo. No. 49, August, 1954.

U.S. Army, Corps of Engineers, Baltimore District, *Survey of Ocean City Harbor and Inlet and Sinepuxent Bay, Maryland*, 1948. (Unpublished.)

————, Beach Erosion Board, *Interim report*, 1933.

———, Beach Erosion Board, *Wave tank experiments on sand movement*, 2 vols., 1936. (Unpublished.)

———, Beach Erosion Board, *Beach Erosion Report on Cooperative Study at Palm Beach, Florida*, 1947. (Unpublished.)

———, Beach Erosion Board, *Shore protection planning and design*, Tech. Rept. No. 4, 1954.

———, Beach Erosion Board, *Study of Monrovia Harbor, Liberia, and adjoining shore line*, 1955 (a).

———, Beach Erosion Board, Sand by-passing at Hillsboro Inlet, Florida, *Beach Erosion Board Bulletin*, **9**, 2 (1955b), 1–6.

———, Los Angeles District, *Shoreline effects, harbor at Playa Del Rey, Calif.*, Enclosure 20, 1948, pp. 39–40. (Unpublished.)

———, Los Angeles District, *Interim report on harbor-entrance improvement, Camp Pendleton, California*, 1953, p. 11. (Unpublished)

———, Milwaukee District, *Preliminary analysis of cooperative beach erosion study, City of Kenosha, Wisconsin*, 1953, p. 17. (Unpublished).

———, Mobile District, *Beach erosion control report, cooperative study of Perdido Pass, Alabama*, 1954 p. 12. (Unpublished.)

———, New York District, *Atlantic Coast of New Jersey, Sandy Hook to Barnegat Inlet, Beach Erosion control report on cooperative study*, 1954, pp. 39–40. (Unpublished.)

———, New York District, *Atlantic Coast of Long Island, N.Y., Fire Island Inlet and shore westerly to Jones Inlet, Beach Erosion control report on cooperative study*, 1955, pp. A–16. (Unpublished.)

U.S. Congress, *Manasquan River and Inlet, New Jersey*, House Doc. 482, 70th Cong., 2d Sess., p. 14, 1928.

———, *Beach erosion at Santa Barbara, Calif.*, House Doc. No. 552, 75th Cong., 3d Sess., March 18, 1938.

———, *Beach erosion study, Lake Michigan Shore Line of Milwaukee County, Wisconsin*, House Doc. 526, 79th Cong., 2d Sess., p. 16, 1946.

———, *North Carolina Shore Line, Beach erosion study*, House Doc. 763, 80th Cong., 2d Sess. p. 20, 1948.

———, *Santa Barbara, Calif., Beach erosion control study*, House Doc., No. 761, 80th Cong., 2d Sess., Dec. 22, 1958.

———, *Ocean City, New Jersey, Beach erosion control study*, House Doc. 184, 83d Cong., 1st Sess., p. 23, 1953a.

———, *Cold Spring Inlet (Cape May Harbor), New Jersey*, House Doc. 206, 1st Sess., p. 30, 1953b.

———, *Gulf Shore of Galveston Island, Texas, Beach erosion control study*, House Doc. 218, 83d Cong., 1st Sess., p. 19, 1953c.

———, Appendix I, *Coast of California, Carpinteria to Point Mugu*, House Doc. 29, 83d Cong., 1st Sess., p. 25, 1953d.

———, *Racine County, Wisconsin*, House Doc. 88, 83d Cong., 1st Sess., p. 10, 1953e.

———, *Illinois shore of Lake Michigan*, House Doc. 28, 83d Cong., 1st Sess., p. 39, 1953f.

———, *Waikiki Beach, Island of Oahu, T. H., Beach erosion study*, House Doc. No. 227, 83d Cong., 1st Sess., p. 23, 1953g.

———, *Pinellas County, Florida*, House Doc. 380, 83d Cong., 2d Sess., p. 27, 1954a.

———, *Port Hueneme, California*, House Doc. 362, 83d Cong., 2d Sess., p. 25, 1954b.

———, *Anaheim Bay Harbor, California*, House Doc. 349, 83d Cong., 2d Sess., p. 37, 1954c.

University of California, *Manual on amphibious oceanography*, ed. R. L. Wiegel. IER, Contract N7onr–29535, Washington, D.C.: Pentagon Press, 1951.

Vernon-Harcourt, L. F., Harbours and estuaries on sandy coasts, *Min. Proc. Inst. Civil, Engrs.*, **70** (1881–82) 1–32.

———, Discussion on the Suez Canal, *Min. Proc. Inst. of Civil Engrs.*, **141** (1900), 197–202.

Watts, George M., *A study of sand movement at South Lake Worth Inlet, Florida*, U.S. Army, Corps of Engineers, Beach Erosion Board, Tech. Memo. No. 42, October, 1953.

———, *Laboratory study of effect of varying wave periods on beach profiles*, U.S. Army, Corps of Engineers, Beach Erosion Board, Tech. Memo. No. 53, September, 1954.

———, and Robert Dearduff, *Laboratory study of effects of tidal action on wave-formed beach profiles*, U.S. Army, Corps of Engineers, Beach Erosion Board, Tech. Memo. No. 52, December, 1954.

Wicker, C. F., History of New Jersey coastline, *Proc. First Conf. Coastal* Eng., Berkeley, Calif.: The Engineering Foundation Council on Wave Research, 1951, pp. 299–319.

Wiegel, R. L., *Characteristics of Kalsin Bay Beach, Kodiak Island, Alaska, on 12 Feb. 1949*, Univ. Calif. IER, Tech. Rept. 29–3, 1949. (Unpublished.)

———, *Trafficability studies of selected beaches on the Pacific Coast of the United States and Oahu, T. H.*, Univ. Calif. IER, Tech. Rept. 29–30, Sept. 15, 1950. (Unpublished.)

———, D. A., Patrick and H. K. Kimberley, Wave, longshore current, and beach profile records for Santa Margarita River Beach, Oceanside, California, 1949, *Trans. Amer. Geophys. Union*, **35**, 6 (December, 1954), 887–96.

CHAPTER FIFTEEN

Some Characteristics of the Oceans' Waters

Much of this chapter is descriptive. No attempt will be made to go into the details of the water masses of the oceans or the reasons for the characteristics of these masses. Instead, the reader is referred to the classic book on the oceans by Sverdrup, Johnson, and Fleming (1942). In addition, he is referred to papers on the detailed characterisitcs of specific areas; for example, the work of Iselin (1936) and Pyle (1962) on the western North Atlantic.

Some engineering problems require a knowledge only of average conditions, or the range of likely conditions. Hence, illustrations will be given of the general temperature, salinity, and density distributions in the ocean, together with references to reports containing large amounts of data of the type needed for a specific design study.

1. TEMPERATURE, SALINITY, AND DENSITY

The temperature of the surface water of the oceans is quite variable both seasonally and daily, but that of the bottom water changes hardly at all, except in coastal regions. Between the surface water and the bottom water is a zone in which temperature fluctuations gradually decrease. This zone is in the form of a permanent thermocline in the tropics and temperate zones. *Thermocline* is the term applied to the portion of a temperature-depth curve in which the temperature decreases rapidly with depth compared with the rest of the curve.

There are extensive data on surface temperatures, and the monthly means have been plotted by the U.S. Navy Hydrographic Office (1944) in the form shown in Fig. 15.1. Hutchins and Scharff (1947) have used these data to obtain maximum and minimum mean monthly surface temperatures (Fig. 15.2). In general, the difference between maximum and minimum temperatures is greatest in middle latitudes and along the western boundaries of the oceans. Monthly averages have been plotted by Böhneck (1936) for the Atlantic Ocean, by Fuglister (1947) for the western North Altantic Ocean, Gulf of Mexico, and the Caribbean, and monthly surface

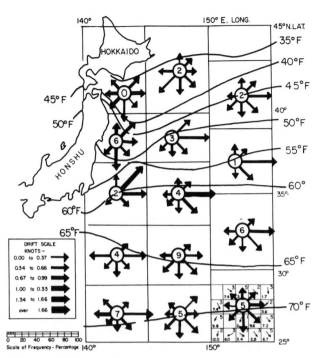

Large arrows show frequency of direction and average drift of current.
Numbers in large circles show per cent of observations in which no
appreciable current was observed.

In 1° quadrangles, the number in upper right hand corner shows the total
current observations. The lower left-hand number shows the resultant
drift in miles per day. The arrow indicates resultant set.

temperatures for each year have been published for the
Gulf of Alaska and the northeast Pacific Ocean by the
Bureau of Commercial Fisheries.

Between about 50° North and 50° South latitude
there are three general regions of water based upon
temperature characteristics: the mixed region (surface
layer), the depth of which is from 100 feet or less near
the equator to nearly 1000 feet deep in mid-latitudes
(30° to 40°); the main thermocline which extends from
the bottom of the surface layer to a depth of from 2000
feet near the equator to 3000 feet in the mid-latitudes;
and the deep water which extends from the bottom of
the main thermocline to the ocean bottom. In winter,
above about 50° latitude, the water temperature is
approximately isothermal, whereas in summer, a seasonal
thermocline develops. Examples of winter and summer
water temperature profiles are given in Fig. 15.3. Three
general types of thermoclines exist in the ocean: the

Fig. 15.1. Surface currents and temperatures near Japan (after
U.S. Navy Hydrographic Office, 1944)

Fig. 15.2. Maximum and minimum monthly mean sea surface
temperature charted from *World Atlas of Sea Surface
Temperatures* (after Hutchins and Sharff, 1947)

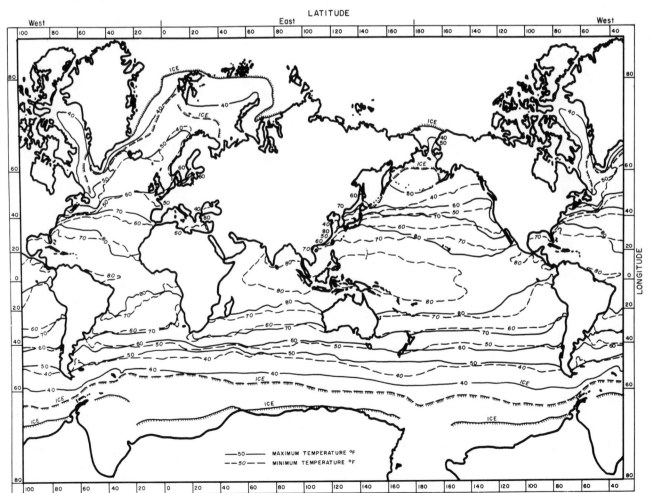

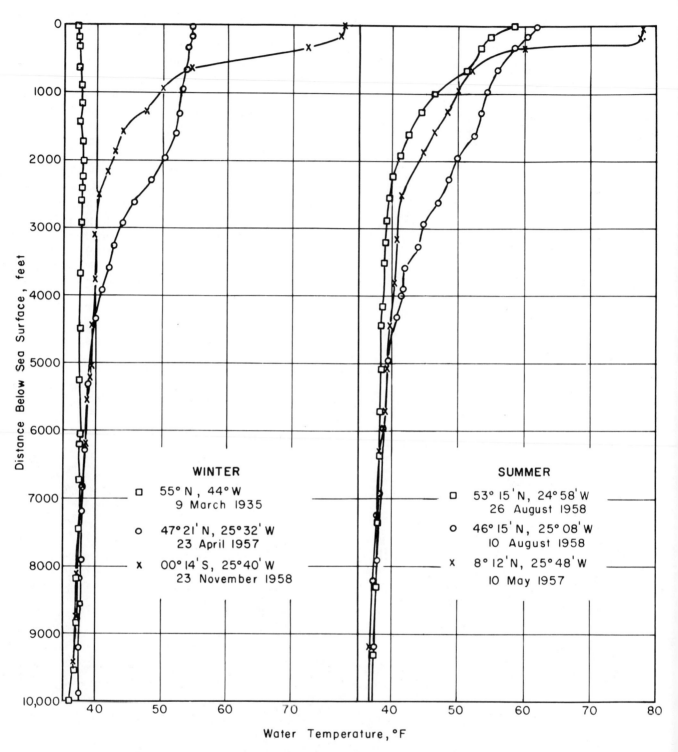

Fig. 15.3. Water temperature profiles (data from Fuglister, 1960, and NRC, 1946)

permanent (main) one, which is relatively deep and which does not exist in the high latitudes; the seasonal one, which develops in the spring and disappears in the late fall; and the daily one, which develops in the morning and disappears during the evening (Montgomery, 1954).

In general, the surface waters of the Atlantic and

Pacific Oceans between about 35° North and 35° South latitudes are relatively saline due to evaporation. In higher latitudes, the rainfall is in excess of evaporation and the surface waters are fresher. The surface waters in the lower latitudes are heated by solar radiation and they expand into the higher latitudes, where they gradually become cooler. Because of the relatively high salinity

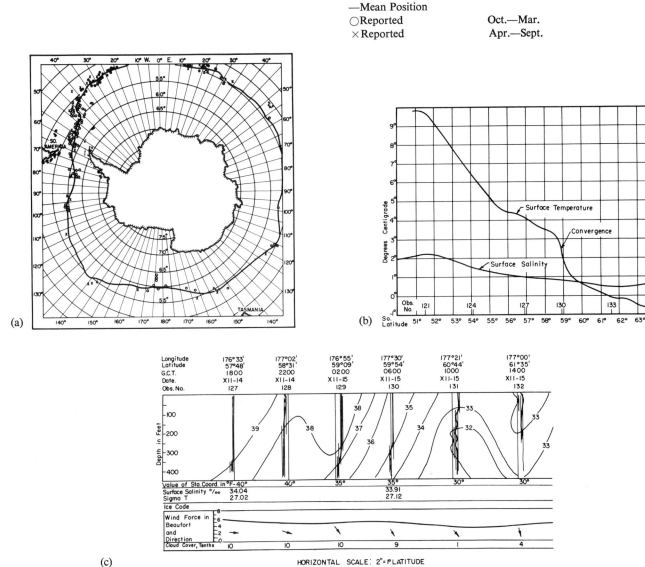

Fig. 15.4. The antarctic convergence: (a) Mean position of antarctic convergence (after Mackintosh, 1946); (b) Surface temperature and salinity across the antarctic convergence, 177°W long., Dec., 1947 (after Robinson, Cochrane and Burt, 1950); (c) Vertical section across antarctic convergence (after Robinson, Cochrane, and Burt, 1950)

of this water, when it cools sufficiently it sinks into the main thermocline; at the same time, arctic waters from slightly higher latitudes move in. This zone, which is one of mixing of the two waters, is called the *arctic convergence* in the north and the *antarctic convergence* in the south. Figure 15.4 shows some characteristics of the antarctic convergence. The arctic waters flow very slowly as bottom waters toward the equator to replace the surface waters that have moved toward the poles owing to the expansion just described and to currents such as the Gulf Stream system.

There are several major atlases which present data on temperature, salinity, density, oxygen content, etc., of the Atlantic, Pacific, Indian, and Southern Oceans (Schott, 1935; Wüst and Defant, 1936; U.S. Navy

Hydrographic Office, 1957; NORPAC Committee, 1960; Fuglister, 1960). One sample of a vertical temperature section in the antarctic has been shown in Fig. 15.4, and detailed sections are available (Deacon, 1937; U.S. Navy Hydrographic Office, 1957; Burling, 1961). Some of the data on vertical temperature sections are shown in abbreviated form in Figs. 15.5 and 15.6 for the Atlantic and Pacific Oceans. The NORPAC data for the North Pacific Ocean and the Bering Sea are novel in that they represent the closest approach yet made in obtaining synoptic data for the ocean; they were obtained by nineteen research vessels of fourteen institutions from three countries. Two vertical temperature sections of the Indian Ocean are shown in Fig. 15.7.

Charts of horizontal temperature distributions are

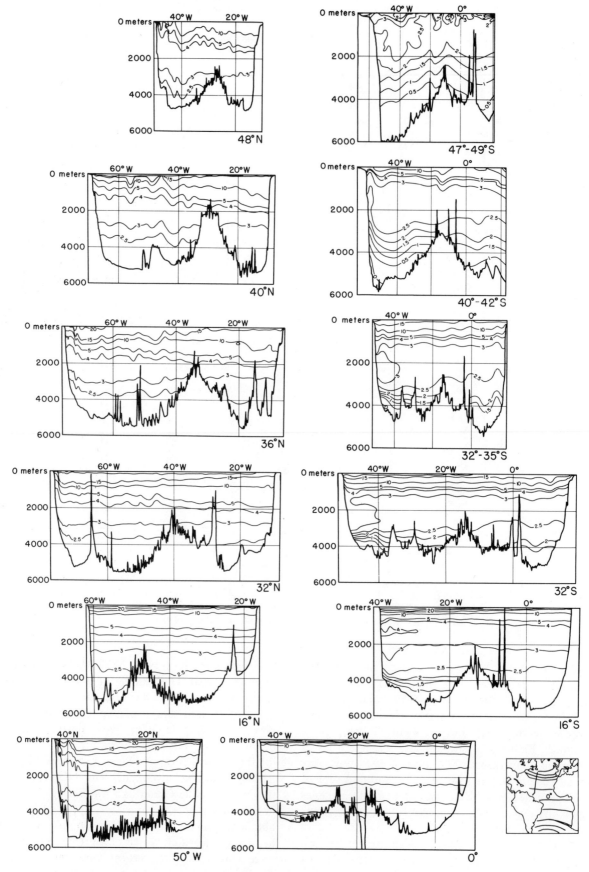

Fig. 15.5. Vertical east-west and north-south temperature sections, Atlantic Ocean; temperature °C (abbreviated from Fuglister, 1960 and Wüst and Defant, 1936)

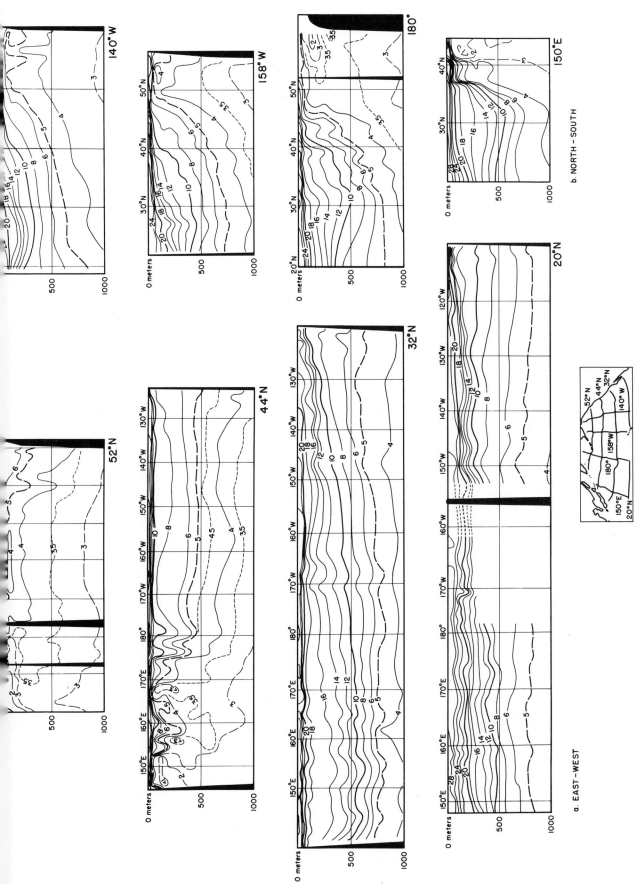

Fig. 15.6. Vertical east-west and north-south temperature sections, North Pacific Ocean; temperature °C (after NORPAC Committee, 1960)

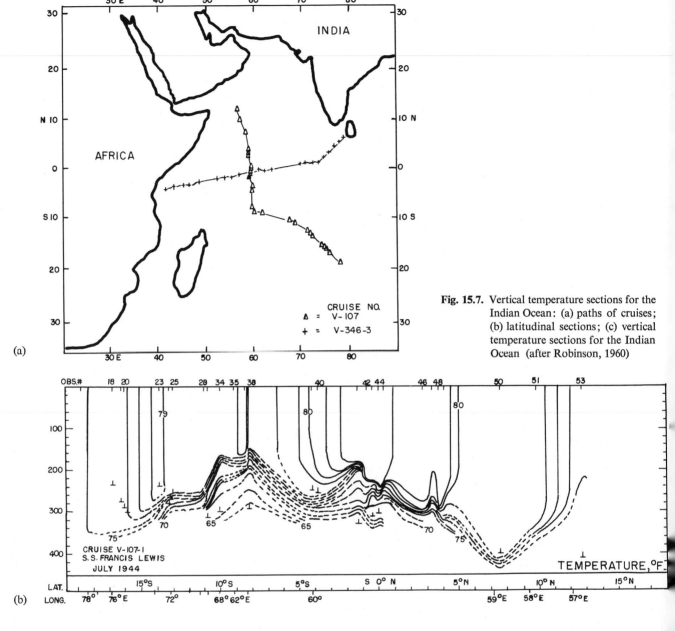

(a)

Fig. 15.7. Vertical temperature sections for the Indian Ocean: (a) paths of cruises; (b) latitudinal sections; (c) vertical temperature sections for the Indian Ocean (after Robinson, 1960)

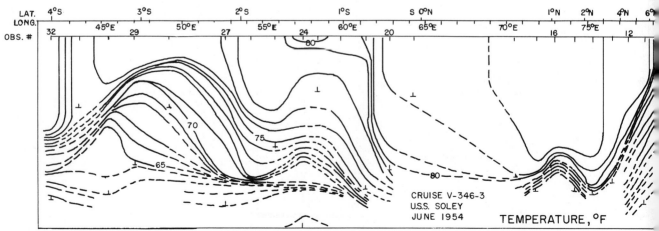

(b)

(c)

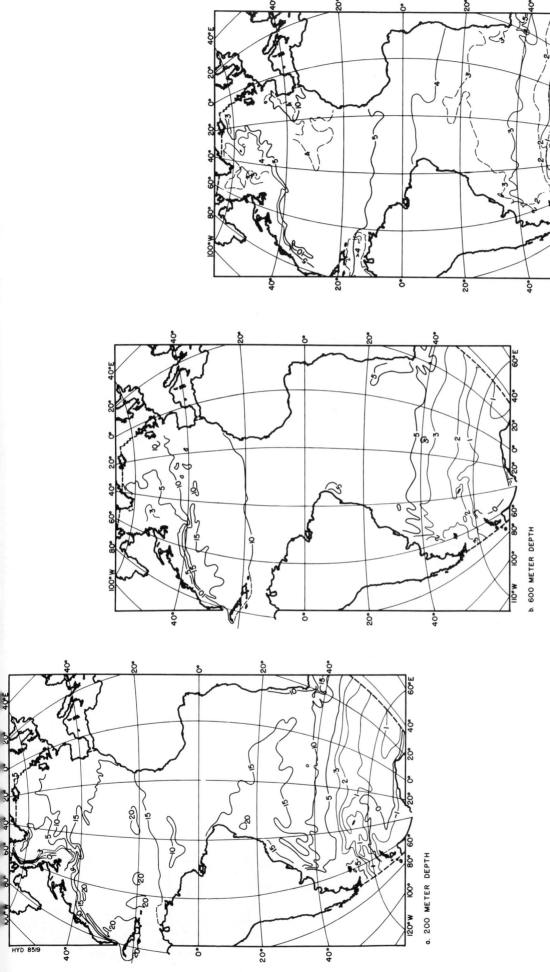

Fig. 15.8. Temperature in the Atlantic Ocean at depths of 200, 600, 1000, and 2000 meters; temperature °C (abbreviated from Wüst and Defant, 1936)

c. ——— 1000 METER DEPTH, ———— 2000 METER DEPTH

b. 600 METER DEPTH

a. 200 METER DEPTH

HYD 8519

also available, and examples for the Atlantic Ocean are shown in Figs. 15.8 and 15.9. Figure 15.8 shows some of the work from the atlas of Wüst and Defant (1936); Fig. 15.9 shows an abbreviated version of the chart of Fuglister (1953; 1957) of the average temperature at a depth of 200 meters. This chart was constructed using 40,000 observations, about 10 times the number used in the chart by Wüst and Defant. The new chart includes the Norwegian Sea, the Gulf of Mexico, the Mediterranean Sea, and Baffin Bay. Even with the extended coverage, the vast area from 10° North latitude south is based upon relatively few observations (Fig. 15.10). Fuglister made a close study of the two sets of charts and found that there was little difference between them except for greater detail in the later chart. Conditions at 200 meters depth appear to be very nearly constant over most of the North Atlantic Ocean. The complex areas on the charts exist in the regions of the major ocean currents, for example, the North Equatorial, Florida, Gulf Stream, Labrador, North Atlantic, and Norway currents.

Charts of horizontal temperature distributions for the

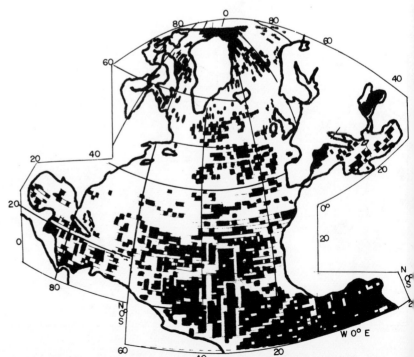

Fig. 15.9. Average sea temperature chart at 200 meters depth, North Atlantic Ocean, Norwegian Sea, Gulf of Mexico, Mediterranean Sea, and Baffin Bay; temperature °C (after Fuglister, 1953, 1957)

Fig. 15.10. Distribution of temperature observations in the North Atlantic Ocean at 200 meters depth (after Fuglister, 1953, 1957)

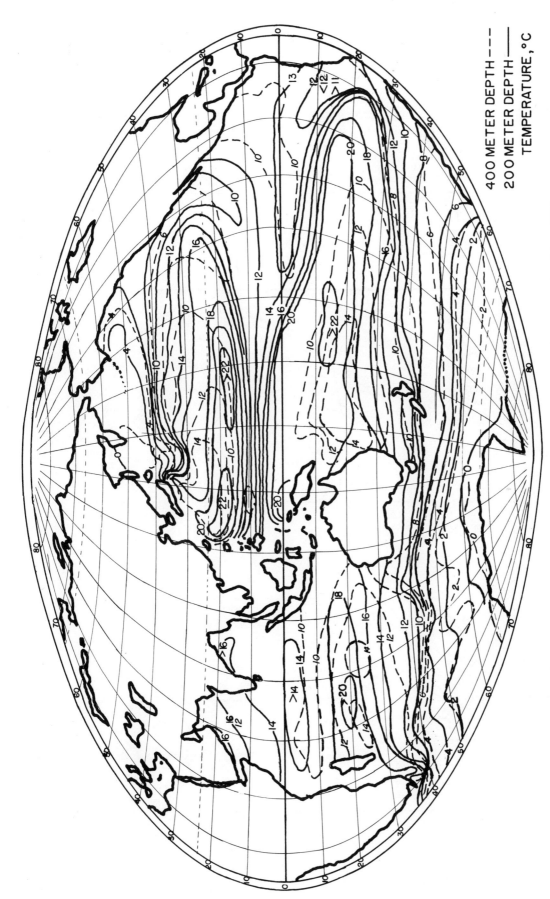

400 METER DEPTH ---
200 METER DEPTH ——
TEMPERATURE, °C

Fig. 15.11. Temperatures at 200 and 400 meters depth in the Pacific and Indian Oceans (after Schott, 1935)

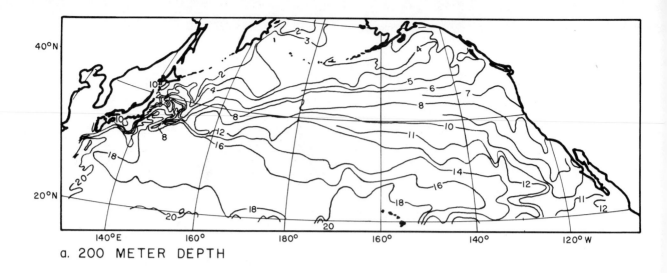

a. 200 METER DEPTH

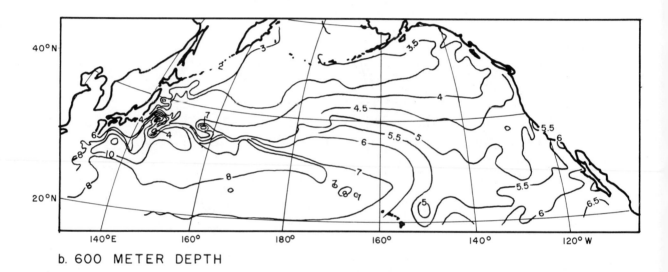

b. 600 METER DEPTH

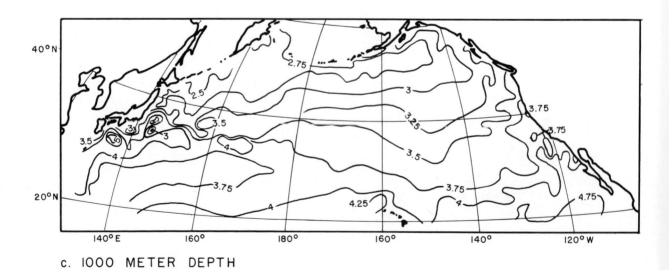

c. 1000 METER DEPTH

Fig. 15.12. Temperatures in the North Pacific at depths of 200, 600, and 1000 meters; temperature °C (abbreviated from NORPAC Committee, 1960)

Pacific and Indian Oceans for 200- and 400-meter depths are shown in Fig. 15.11. In Fig. 15.12, the NORPAC charts for the North Pacific Ocean are shown in abbreviated form. Horizontal temperature charts of monthly conditions for the Gulf of Alaska and the northeast Pacific Ocean for surface, 100-, 200-, 300-, and 400-ft depths are available (Robinson, 1957), together with annual means for these depths.

In addition to the atlases cited there are numerous reports and papers on the temperatures of the many seas and oceans of the world. Vaughan (1937) has listed the sources of most of the pre-World War II data. Since then, the number of observations has increased tremendously, with vast numbers of data cards being stored at Scripps Institution of Oceanography and Woods Hole Oceanographic Institute. The U.S. Navy Hydrographic Office had three-quarters of a million such cards in 1956 (Lyman, 1957). These types of data are now stored at the U.S. National Oceanographic Data Center. Many data are in the form of technical reports issued by oceanographic departments; for example, for the Gulf of Mexico see McLellan (1960). In addition, a considerable body of data obtained by the Russians in the Arctic is becoming available in translation (see for example, Treshnikov, 1960).

Regions of major ocean currents are of considerable interest from the standpoint of water temperatures at 200 meters depth. Fuglister (1951, 1952, 1953) has studied the Gulf Stream system in detail, apparently being able to define a portion of the system as three easterly currents separated and bounded by countercurrents, forming from two easterly currents in the vicinity of 50° West longitude (Fig. 15.13). Similar steep temperature gradients are found in the left side (facing downstream) of the Kuroshio, Fig. 15.12(a). An attempt to correlate the position of the Gulf Stream system with surface temperatures (Strack, 1953) showed close agreement between surface and relatively deep temperatures from November to May. A feeling for the relationship between the vertical temperature profile and the current can be obtained from Fig. 15.14. Not only do relatively steep temperature gradients exist in the vertical sections, but the depths of the mixed surface layer are deeper due to increased mixing because of the turbulence of the currents; Fig. 15.15 is a good example of this.

Fuglister (1951) studied the variations of temperatures from mean values at a depth of 200 meters in the North Atlantic Ocean and discovered that these were largely found in the regions of major ocean currents, and that they showed some characteristics not shown by the gradients of mean isotherms. It would appear that studies of such variations would be of considerable importance (Fuglister, 1957).

One of the remarkable discoveries in physical oceanography was made by Forchhammer in 1856. He found that regardless of the absolute concentration of the

Fig. 15.13. Schematic chart of temperature (°C) at a depth of 200 meters in the Gulf Stream area (after Fuglister, 1951)

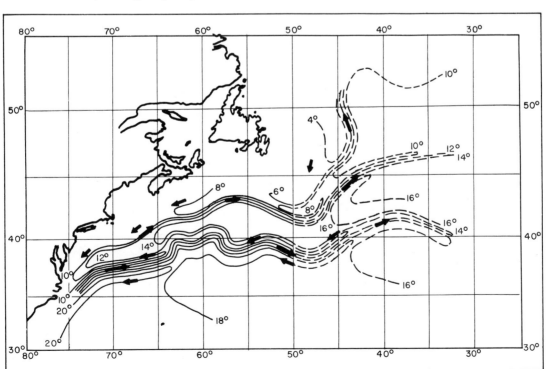

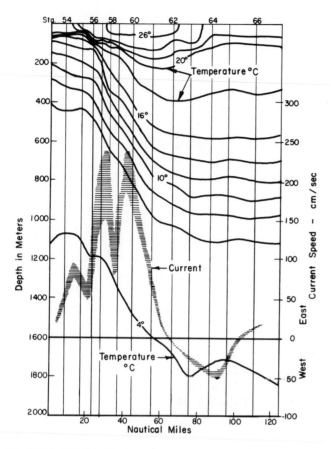

Fig. 15.14. Relationship between temperature gradients and currents, Gulf Stream system (after Worthington, 1954)

total solids in sea water (called the *salinity*, grams per kilogram, 0/00), the ratios between the more abundant constituents are nearly constant. Dittmar (1884) made a careful analysis of 77 water samples taken from all oceans and found results very close to those accepted at present. More recent values are shown in Table 15.1. (For a tabulation of the less abundant constituents, together with a thorough discussion of the chemistry of sea water, see National Academy of Science, 1932; 1959, and Sverdrup, Johnson, and Fleming, 1942.)

The uniformity of the relative composition is the result of mixing and circulation. It is difficult to measure the salinity directly, but the chlorinity can be measured by several techniques (Sverdrup, Johnson, and Fleming, 1942; Paquette, 1959). The empirical relationship between salinity and chlorinity, as established by the International Commission is

$$\text{salinity } (0/00) = 0.030 + 1.805 \times \text{chlorinity } (0/00).$$

$$(15.1a)$$

It has been recommended on the basis of much additional evidence that Eq. 15.16 be modified as follows (UNESCO, 1963)

$$\text{salinity } (0/00) = 1.80655 \times \text{chlorinity } (0/00) \quad (15.1b)$$

There are not so many data available on salinity and density distributions as there are on temperature distributions. Abbreviated examples of salinity distributions are given in Figs. 15.16–15.22.

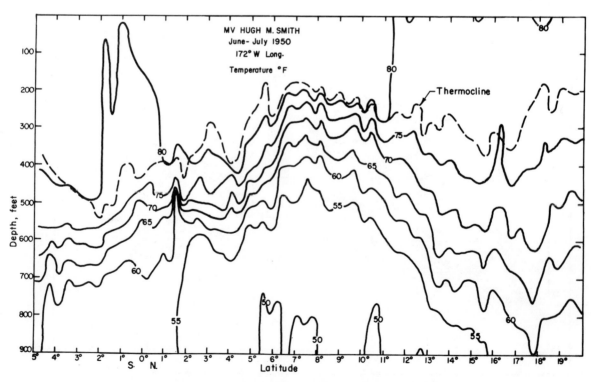

Fig. 15.15. Vertical temperature section (after Robinson, 1952)

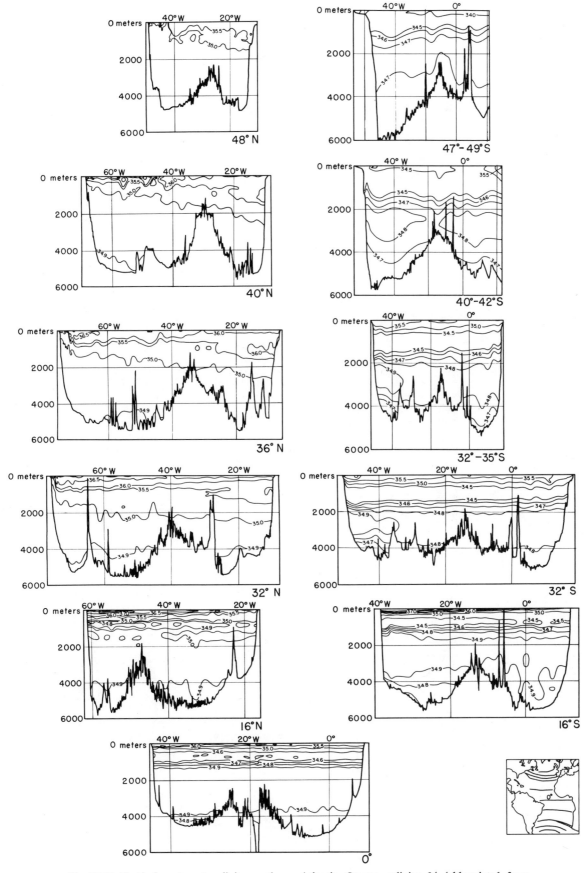

Fig. 15.16. Vertical east-west salinity sections, Atlantic Ocean; salinity ‰ (abbreviated from Fuglister, 1960, and Wüst and Defant, 1936)

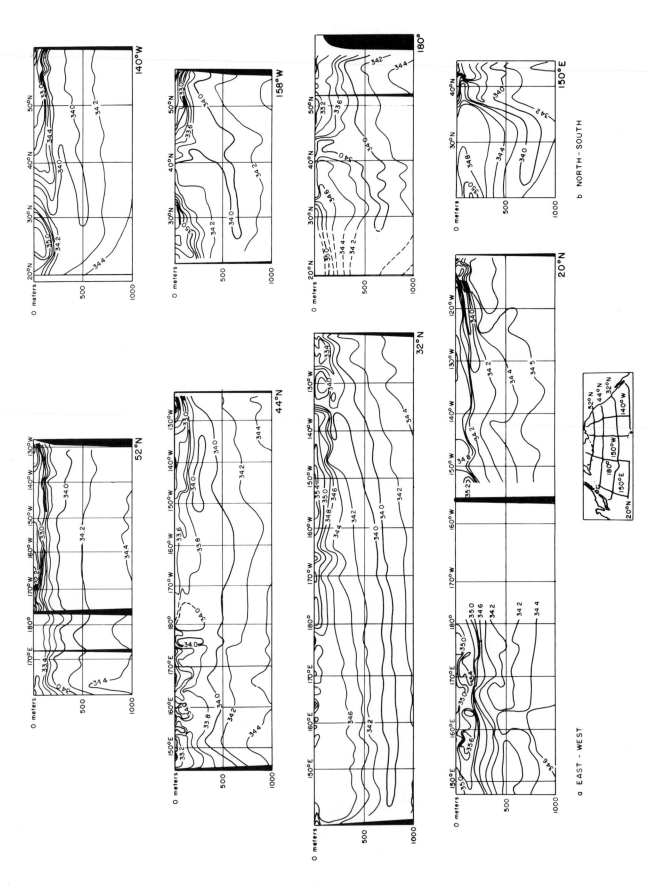

Fig. 15.17. Vertical east-west and north-south salinity sections, North Pacific Ocean; salinity ‰ (after NORPAC Committee, 1960)

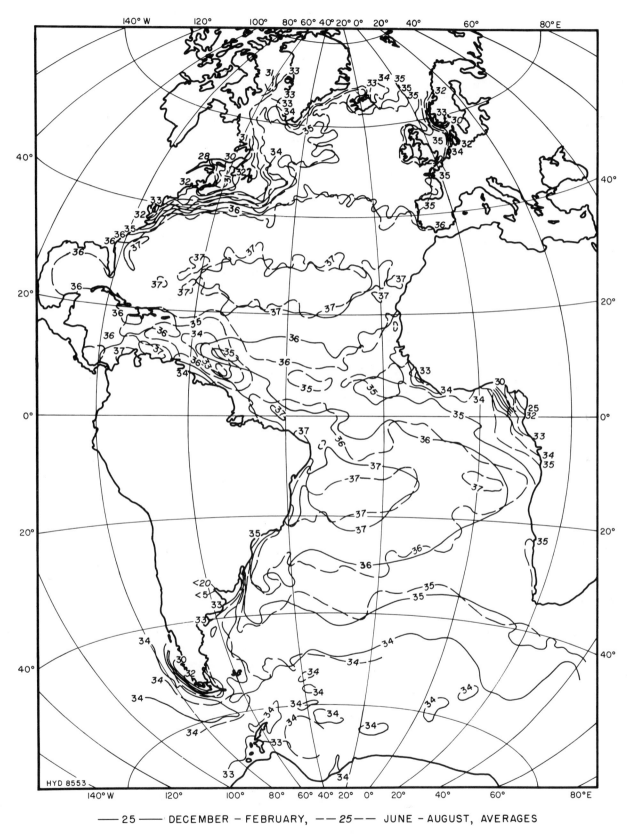

Fig. 15.18. Salinities, ‰, of surface waters, Atlantic Ocean (after Böhnecke, 1936)

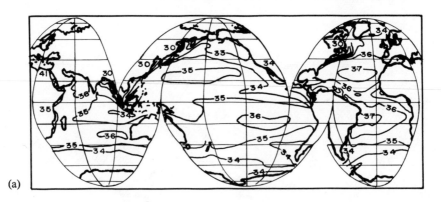

(a)

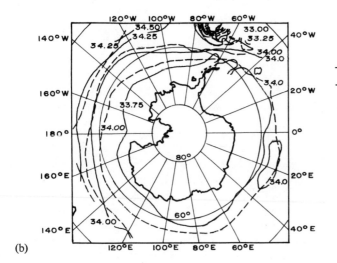

(b)

Fig. 15.19. (a) Surface salinities of the oceans in northern summer (after Sverdrup, Johnson, and Fleming, 1942); (b) Average sea surface salinity in the Antarctic Ocean, February and August (after U.S. Navy Hydrographic Office, 1957)

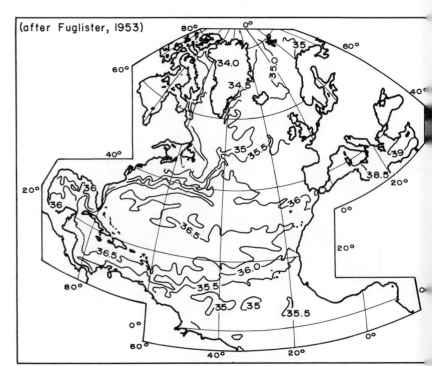

Fig. 15.20. Salinities at 200 meters, depth in the North Atlantic Ocean, Norwegian Sea, Gulf of Mexico, Mediterranean Sea, and Baffin Bay

Table 15.1. MAJOR CONSTITUENTS OF SEA WATER
(after Lyman and Fleming, 1940)

Ion	Dittmar's original values		Recalculated, 1940 atomic weights		1940 values	
	Cl = 19%	%	Cl = 19%	%	Cl = 19%	%
Cl⁻	18.971	55.29	18.971	55.26	18.980	55.04
Br⁻	0.065	0.19	0.065	0.19	0.065	0.19
SO₄⁼	2.639	7.69	2.635	7.68	2.649	7.68
CO₃⁼	0.071	0.21	0.071	0.21	—	—
HCO₃⁻	—	—	—	—	0.140	0.41
F⁻	—	—	—	—	0.001	0.00
H₃BO₃	—	—	—	—	0.026	0.07
Mg⁺⁺	1.278	3.72	1.292	3.76	1.272	3.69
Ca⁺⁺ }	0.411	1.20	0.411	1.20	0.400	1.16
Sr⁺⁺ }					0.013	0.04
K⁺	0.379	1.10	0.385	1.12	0.380	1.10
Na⁺	10.497	30.59	10.498	30.58	10.556	30.61
Total	34.311		34.328		34.482	

When temperature and salinity are both plotted against depth, there does not appear to be any simple relationship (Fig. 15.23a). As can be seen in Fig. 15.23(a), however, when the temperature-depth and salinity-depth curves are cross-plotted at constant depth for temperature-salinity (T-S) curves a remarkably consistent relationship is obtained. The T-S relationship has been used in correcting both temperature and salinity measurements, especially in deciding whether one determination was in error. In Fig. 15.23(b), the T-S curves corresponding to the data in Fig. 15.23(a) are plotted, together with two cross-hatched areas that represent the normal range within which T-S curves should fall for the east North Pacific central water mass and the west North Pacific water mass (Sverdrup, Johnson, and Fleming, 1942). The boundaries and T-S relationships of the principal water masses of the oceans are given in Fig. 15.24.

Density is the mass per unit volume (slugs/ft³ in the fps system and grams/cm³ in the cgs system), and is given the symbol $\rho_{s,t,p}$ as it depends upon the salinity, temperature, and pressure of the sample. In publications of oceanographic data specific gravities are usually reported although the term "density" is often used in referring to the data. This all right, as specific gravities in oceanography are referred to distilled water at 4°C, and as the density of distilled water at this temperature is unity, the values of specific gravity and density are numerically equal. Often the term $\sigma_{s,t,p}$ is used in reporting data. This is done to save space in tabulating data. The relationship between σ and ρ is

$$\sigma_{s,t,p} = (\rho_{s,t,p} - 1)\,1000. \qquad (15.2)$$

If $\rho_{s,t,p} = 1.02819$, $\sigma_{s,t,p} = 28.19$. The symbol $\sigma_{s,t,p}$ is used to refer to the density of a sea water sample meas-

Fig. 15.21. Salinities in the Atlantic Ocean at depths of 200, 600, 1000, and 2000 meters (abbreviated from Wüst and Defant, 1936)

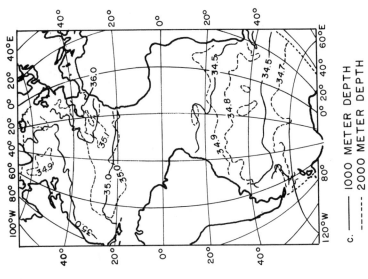

c. —— 1000 METER DEPTH
------ 2000 METER DEPTH

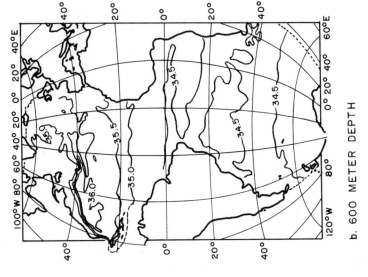

b. 600 METER DEPTH

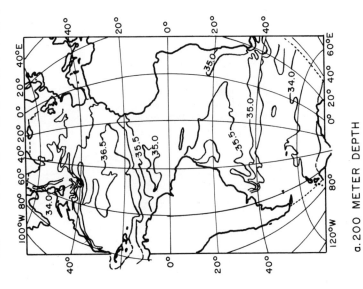

a. 200 METER DEPTH

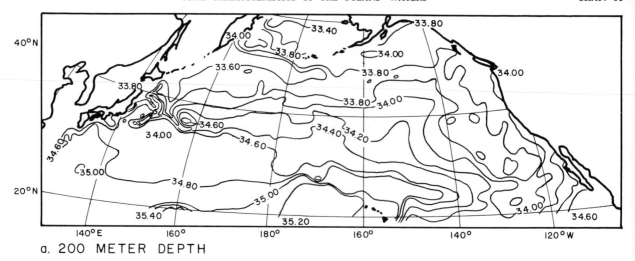

a. 200 METER DEPTH

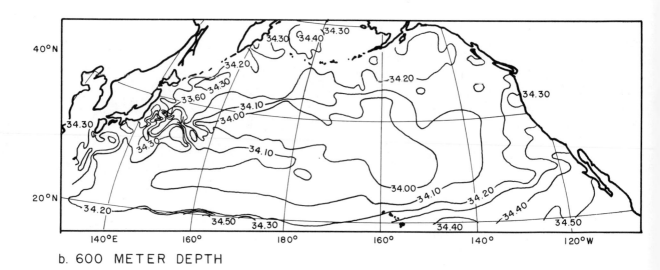

b. 600 METER DEPTH

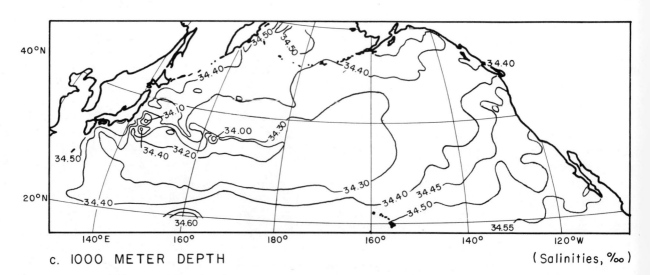

c. 1000 METER DEPTH   (Salinities, ‰)

Fig. 15.22. Salinities in the North Pacific at depths of 200, 600, and 1000 meters (abbreviated from NORPAC Committee, 1960)

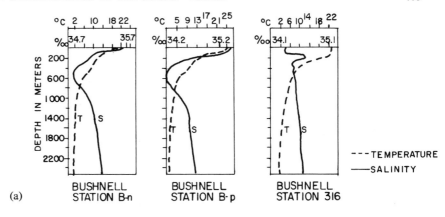

(a)

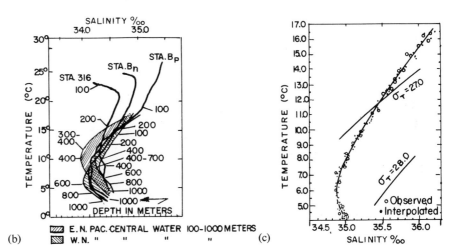

Fig. 15.23. Relationship between temperature and salinity: (a) Temperature and salinity versus depth, Hawaiian area (after Leipper and Anderson, 1948); (b) Temperature-salinity curves, Hawaiian area (after Leipper and Anderson, 1948); (c) Temperature-salinity relationship for all stations, Gulf of Mexico, 1958 and 1959 (After McLellan, 1960)

(b)   (c)

ured at the temperature and pressure at which it was collected (called *density in situ*), σ_t refers to the density at atmospheric pressure and temperature, and σ_0 refers to the density at atmospheric pressure and 0°C (Sverdrup, Johnson, and Fleming, 1942). The International Commission found the following relationship between σ_0 and the chlorinity, Cl.

$$\sigma_0 = -0.069 + 1.4708\,Cl - 0.001570\,Cl^2$$
$$+ 0.0000398\,Cl^3. \qquad (15.3)$$

Equation 15.3 must be used in conjunction with Eq. 15.1 if the salinity is given rather than the chlorinity. σ_t can be written as

$$\sigma_t = \sigma_0 - D, \qquad (15.4)$$

where D is a complicated function of both σ_0 and temperature (Knudsen, 1901). Tables are now available that can be used to find the density σ_t rather simply for any reasonable value of temperature and salinity (U.S. Navy Hydrographic Office, 1952). It has been found that the relationship between density and electrical conductivity is a more reliable relationship than the relationship between density and chlorinity which varies by approximately 0.03% rather than 0.004% which is the case for the density-conductivity relationship (UNESCO, 1963).

Examples of density (σ_t) distributions are shown in Figs. 15.25–15.29.

2. LONG-TERM, SEASONAL, AND DAILY THERMOCLINES AND MIXED LAYER DEPTH

In winter seasons, winds are stronger for longer intervals of time than in summer seasons, and in summer the water is subject to relatively greater solar heating. Hence the mixed layer depth normally increases in the winter season and decreases in the summer season, and surface thermoclines form in the summer season. An example of the summer thermocline in the regions above 50° latitude can be seen in Fig. 15.3.

Average water temperatures have been plotted for each month of the year for the surface and for depths of 100, 200, and 300 ft in the Hawaiian Island area (Leipper and Anderson, 1948). The mixed layer depth was from 150 to 350 ft in the winter season and from 100 to 250 ft in the summer season. A study of bathythermographs shows that it is not easy to determine the mixed layer depth; in fact, the term is too simple to describe actual conditions, but it is a rather useful simplification. An attempt to develop a subjective measure led to the conclusion that, if the depth is taken to be that at which a 1/50°F or greater change occurs in

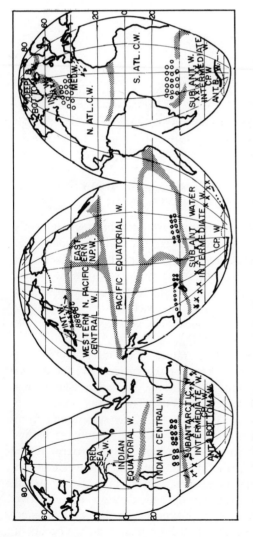

(a) Approximate boundaries of the upper water masses of the ocean. Circles indicate the regions in which the central water masses are formed, and crosses show the lines along which the Antarctic and Arctic intermediate waters sink.

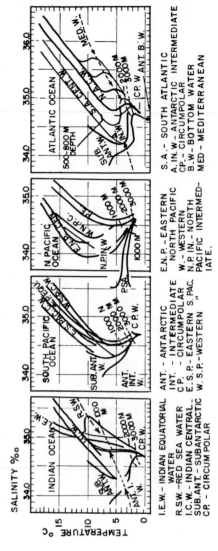

(b) Temperature-salinity relations of the principal water masses of the oceans

I.E.W. – INDIAN EQUATORIAL
        WATER
R.SW. – RED SEA WATER
I.C.W. – INDIAN CENTRAL
SUB.ANT. – SUBANTARCTIC
C.P. – CIRCUM POLAR

ANT. – ANTARCTIC
INT. – INTERMEDIATE
C.P. – CIRCUMPOLAR
E.S.P. – EASTERN S.PAC.
W.S.P. – WESTERN "

E.N.P. – EASTERN
        NORTH PACIFIC
W.– WESTERN
N.P.IN.– NORTH
        PACIFIC INTERMED-
        IATE.

S.A. – SOUTH ATLANTIC
A.IN.W – ANTARCTIC INTERMEDIATE
C.P.– CIRCUMPOLAR
B.W.– BOTTOM WATER
MED – MEDITERRANEAN

404

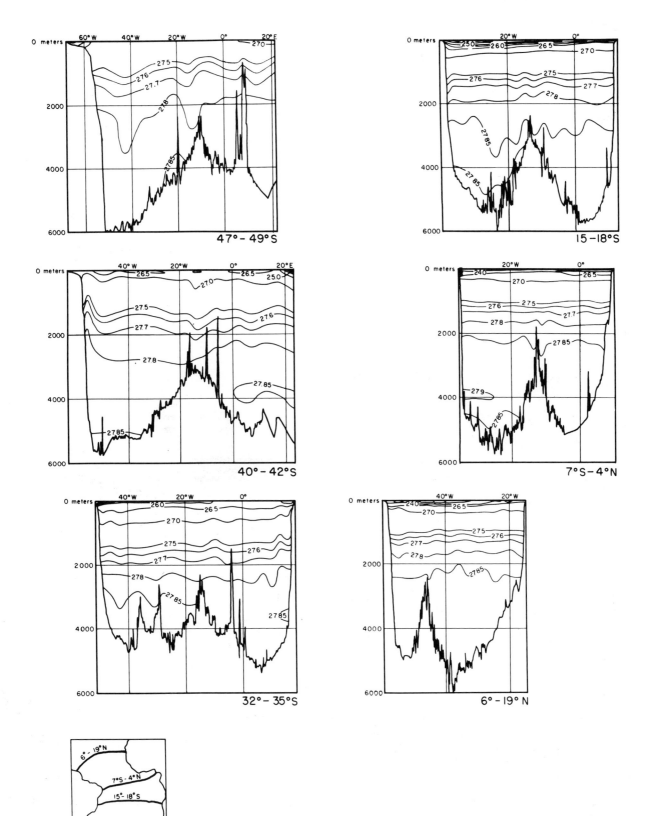

Fig. 15.25. Vertical east-west density (σ_t) sections, Atlantic Ocean (abbreviated from Wüst and Defant, 1936)

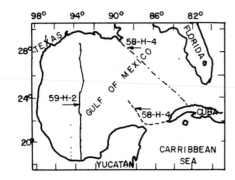

(a) Location of oceanographic station

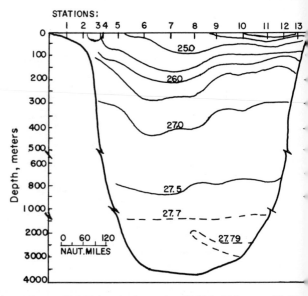

(b) σ_t along 93°30 W longitude, cruise 59-H-2, February 17–26, 1959

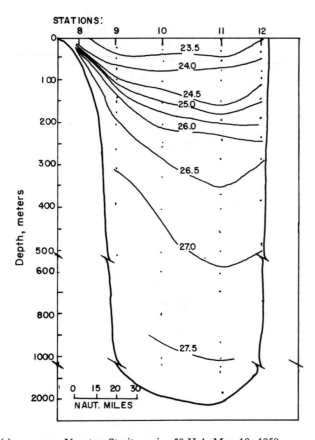

(c) σ_t across Yucatan Strait, cruise 58-H-4, May 18, 1958

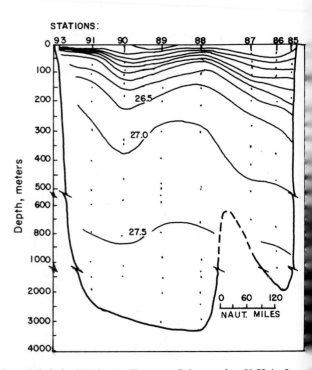

(d) σ_t Mississippi Delta to Havana, Cuba, cruise 58-H-4, June 25–28, 1958

Fig. 15.26. Vertical density (σ_t) sections, Gulf of Mexico, 1958 and 1959 (after McLellan, 1960)

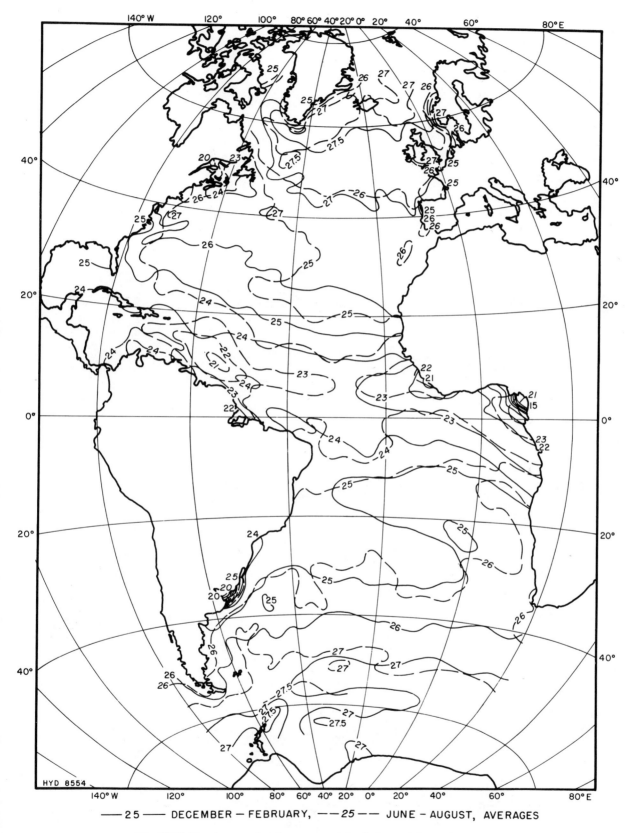

Fig. 15.27. Density, σ_t, of surface waters, Atlantic Ocean (after Böhnecke, 1936)

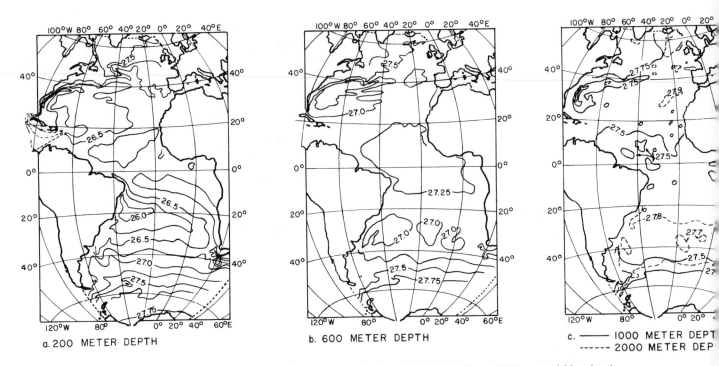

a. 200 METER DEPTH

b. 600 METER DEPTH

c. ——— 1000 METER DEPTH
   ----- 2000 METER DEPTH

Fig. 15.28. Densities in the Atlantic Ocean at depths of 200, 600, 1000, and 2000 meters (abbreviated from Wüst and Defant, 1936)

Fig. 15.29. Density (σ_t, 4°C) of surface water in the Pacific and Indian Oceans with values shown being for local summer conditions in both the northern and southern hemispheres (after Schott, 1935)

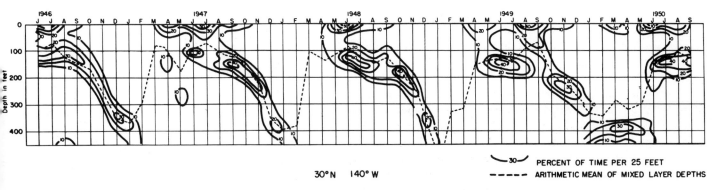

30°N   140°W

⌒ 30 ⌒   PERCENT OF TIME PER 25 FEET

- - - - -   ARITHMETIC MEAN OF MIXED LAYER DEPTHS

Fig. 15.30. Frequency distributions of the critical gradients (after Patullo, 1952)

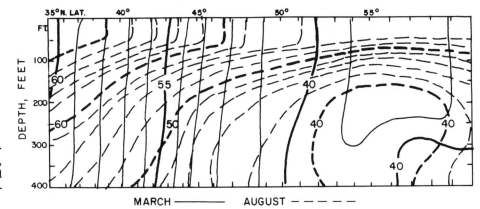

MARCH ———— AUGUST - - - - -

Fig. 15.31. Average vertical temperature (°F) section, at 145°W, March and August (after Robinson, 1957)

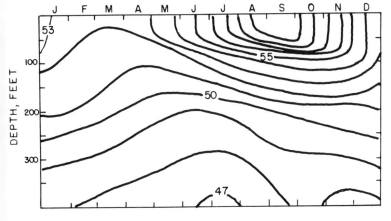

Fig. 15.32. Vertical time sections of average temperature, (°F), 37°N, 123°W (after Robinson, 1957)

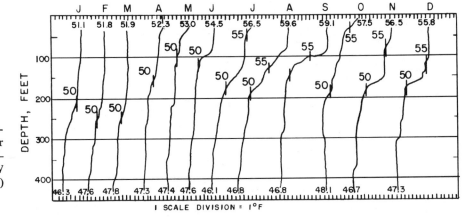

I SCALE DIVISION = 1°F

Fig. 15.33. Typical 450-foot bathythermograms for each month, 37°–38°N, 123°–124°W (after Robinson, 1957)

a 25-ft interval, a fairly consistent result is obtained (Pattullo, 1952), Fig. 15.30.

Average vertical temperature sections for winter and summer months are given in Fig. 15.31. The vertical isotherms show the mixed layer region which is deep in winter and shallow in summer. Time variations at a given location can be presented in two manners, Figs. 15.31 and 15.32. Figure 15.33 illustrates thermoclines nicely: for example, two distinct thermoclines are seen to exist in July with the upper thermocline moving deeper during August and September, but then developing into a step type of distribution in October. By December, the mixed layer has developed and deepened, and there is a very sharp thermocline. By March, the thermocline has practically disappeared and the mixed surface layer extends down to the permanent thermocline. Of interest in respect to the mixed layer region are long-term variations of water temperature. Measurements have been made at a weather station named "Extra" near Japan, Fig. 13.34(a). The variations in moving monthly averages and moving 12-monthly averages are shown in Fukuoka and Tsuiki (1953) Fig. 13.34(b). It is believed that these variations are associated with the movement of water masses.

Nearly continuous temperature measurements for approximately a year have been obtained off Bermuda at the bottom in water 50 meters deep and in water about 500 meters deep (near the top of the main thermocline) about ¾ mile and 1½ miles offshore, respectively (Stommel and Hodgson, 1956). For reference, the temperature and salinity measurements made in deep water are given in Fig. 15.35; the actual stations are represented by vertical lines; the rest of the isotherms are drawn by interpolation. Several things in the continuous record

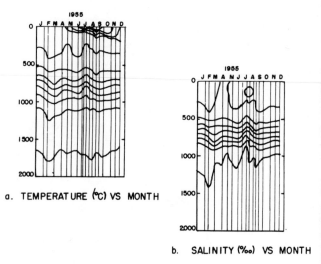

a. TEMPERATURE (°C) VS MONTH

b. SALINITY (‰) VS MONTH

Fig. 15.35. Temperature and salinity versus month, off Bermuda, 1955 (after Stommel and Hodgson, 1956)

are of interest, and examples are given in Fig. 15.36. A series of severe storms over the North Atlantic started on January 2, 1955, and by January 4, large temperature disturbances were recorded, with a long-period oscillation being evident during January 6–12. The period of these oscillations was a little less than 24 hr, and was perhaps an inertial period. An example of the record when nothing particular had occurred is shown in Fig. 15.36(b), July 26–July 30. There are practically continuous fluctuations of the order of ±0.2°C of relatively short periods (1/2 to 2 hr) at even the 500-meter depth and fluctuations of about ±0.6°C from periods in the range of 12 to 48 hr. It was found that the temperature records at 500-meters depth do not bear any resem-

Fig. 15.34. (a) Location of weather station "extra"; (b) Moving monthly and twelve month averages (after Fukuoka and Tsuiki, 1953)

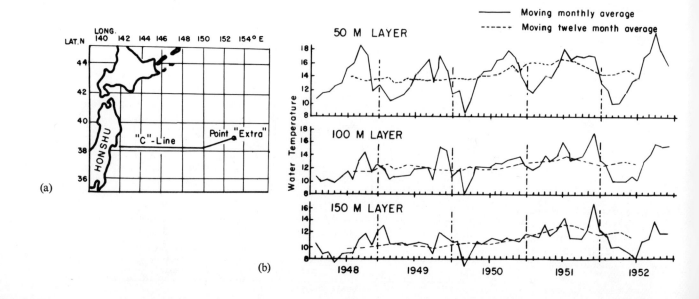

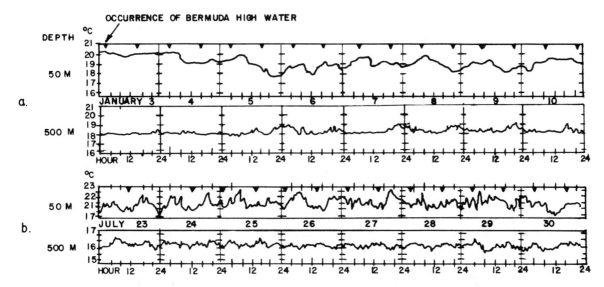

Fig. 15.36. Examples of continuous temperature records on the bottom in water 50 and 500 meters deep off Bermuda, 1955 (after Stommel and Hodgson, 1956)

blance to a tidal phenomenon, and that the phases were randomly distributed with respect to time of high water. One other point of importance can be made in comparing the type of temperature time history shown in Fig. 15.35 with the continuous records of Fig. 15.36; that is, most of the fluctuations in the deeper water of Fig. 15.35 are the result of sampling error.

The study just cited with respect to thermogram–tidal period noncorrelation was similar in results to a statistical study made of a series of bathythermograms from two deep sea U.S. Coast Guard stations in the northeastern Pacific Ocean (Rudnick and Cochrane, 1951) which showed minor fluctuations of tidal period. In addition, fluctuations of surface temperature were found to be only half due to diurnal solar heating, and about half random. An approximation to an energy spectrum was obtained, and it was found that, except for a peak for one series at an approximate diurnal period, the rest of the energy was distributed in a continuous spectrum down to a period of several hours which was the lower limit imposed by the sampling interval.

Some continuous measurements have been made of the water temperature 3 to 4 feet below the water surface in the Black Sea with simultaneous measurements of surface water waves (Komonkova and Kontoboitseva, 1957). It was found that the temperature varied in a manner similar to the waves with respect to time, with the wave periods and periods of temperature fluctuations being nearly identical. When a positive temperature gradient existed (temperature increasing with distance beneath the surface), the maximum temperature occurred in phase with the wave crest, and when a negative gradient existed, the maximum temperature occurred in phase with the wave trough; this is consistent with the motion of the water mass due to waves.

Many factors enter into the development and changes of a daily and seasonal thermocline, such as hours of daylight, cloud cover, winds, and currents (see, for example, Pollak, 1954). It is difficult to study the effect of the variables separately, or even to be consistent in the study of the effect of one variable, neglecting the others. Francis and Stommel (1953) examined the bathythermograms from 502 weather ship cruises in the North Atlantic. The data they chose had to meet the following conditions: (1) the records had to be between May and October to insure a surface thermocline; (2) there had to be winds below Beaufort force 4 for at least two days followed by winds of force 8 or greater for at least one day; (3) the ship had to be in the same place throughout the interval chosen. Only 15 series of bathythermograms were found to meet these conditions, and these are shown in Fig. 15.37. In general, a single gale can deepen the depth of the seasonal surface thermocline by about 20 to 30 ft on the average; but on some occasions the thermocline is less deep.

Part of the explanation for the variations in the foregoing studies can be made using the studies of nearshore temperatures by Arthur (1954); his findings are presented in Section 3.

Some advances have been made in the prediction of the development of daily and seasonal thermoclines and the mixing of the surface waters from theoretical and semiempirical standpoints (Rossby and Montgomery, 1935; Munk and Anderson, 1948; Schule, Simpson, and Shapiro, 1952; Martineau, 1952; Gilcrest, Jung, and Freeman, 1954). It appears, however, that the consensus of opinion at a conference on thermocline held in 1953 (Pollak, 1954) was that a satisfactory method of predicting the quantitative behavior of a thermocline did not exist at that time, although subjective techniques were

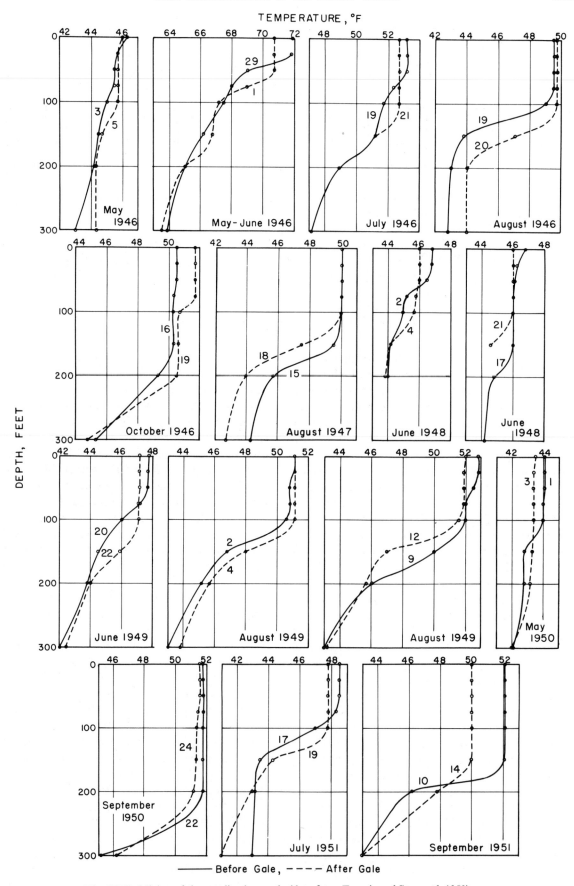

Fig. 15.37. Mixing of thermocline by a gale (data from Francis and Stommel, 1953)

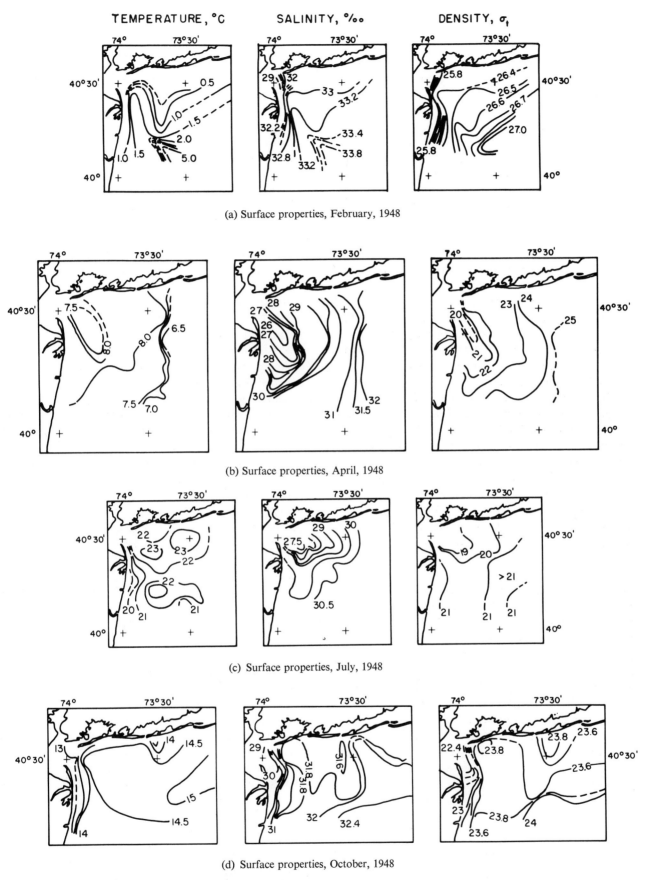

(a) Surface properties, February, 1948

(b) Surface properties, April, 1948

(c) Surface properties, July, 1948

(d) Surface properties, October, 1948

Fig. 15.38. Horizontal distribution of surface temperature, salinity, and density, New York Bight, 1948 (after Ketchum, Redfield, and Ayers, 1951)

quite useful. Some of the reasons for the difficulties are described in Chapter 16.

3. COASTAL WATER

Iselin (1936) has developed a classification of the waters off the east coast of the United States, based upon the salinity. These classes are coastal, slope (over the continental slope), Gulf Stream, and ocean waters, with the latter three being above the deep water. These various classes can be distinguished to a certain extent by their salinities, with the coastal waters (over the continental shelf) being relatively fresh owing to inflow of river water into the ocean along the coast. A study of the illustrations of salinities shows that this salinity characteristic is not always true, but it is common enough to be a useful concept. There are many other distinguishing characteristics of coastal water, such as upwelling, the absence of the major ocean currents, no permanent thermocline or deep water, large amounts of sediment in suspension and moving along the bottom, and surf. The main characteristic is that the water is at a boundary; the other characteristics are derivatives of this.

Along the coasts of the United States, the major ocean currents act as an effective barrier to the ocean waters. For example, numerous measurements of surface water temperatures have shown that at no season of the year do surface temperature gradients in the slope water shoreward of the Gulf Stream system reliably indicate those seaward of the system (Strack, 1953).

Coastal waters may be grouped roughly into two classes: coastal water into which considerable volumes of fresh water flows, and coastal water adjacent to arid regions. New York Bight is an example of the first type. Illustrations of the horizontal and vertical distributions of water temperature, salinity, and density are given in Figs. 15.38 and 15.39 (Ketchum, Redfield, and Ayers, 1951). One feature that is evident is the vertical orientation of the isotherms in February, changing to a horizontal orientation by April (thermocline structure), developing into a very steep thermocline structure by July, and then changing into a well-mixed vertical structure (vertical orientation of the isotherms) by October. At the same time, the water mass warmed from February to April, and April to July, and then cooled between July and October. Although the isotherms are oriented vertically in both February and October, the water is much warmer in October than in

February. Other characteristics are apparent; for example, the brackish water was in a band along the New Jersey coast during the February and October, 1948, surveys, but rather evenly distributed during the April and July, 1948, surveys. The location of the brackish water apparently depends more upon flow of fresh water into the bight than upon the season, as in August, 1949, when the river flow was much lower than during the comparable season of 1948, the brackish water was along the New Jersey coast. In studying the inshore regions of the oceans, it is necessary to include data on river flow as shown in Table 15.2 as well as ground water percolation into the inshore area.

Temperatures and salinities can change rapidly in inshore areas. For example, oceanographic data were collected along a series of ranges between Cape Hatteras and Cape Fear during the latter half of March and again during the latter half of April, 1948. One set of measurements is shown in Fig. 15.40 (Bumpus and Wehe, 1949).

An example of a semiarid region is the area off Southern California. Temperatures and salinities for the general region are shown in Fig. 15.41. Changes of surface temperature, salinity, and density between spring and fall are shown for the water immediately adjacent to the coast in Fig. 15.42.

As will be shown in Chapter 16, and in Chapter 17, Section 12 on the functional design of an ocean outfall sewer, the difference in temperature and (density) between the bottom and the surface, and especially the absence or presence of a strong thermocline, is of great importance in determining whether the sewage effluent will reach the surface or not. In areas such as the inshore water off Southern California, the temperature difference varies considerably from top to bottom and strong thermoclines may exist in the spring through fall seasons but not during the winter. Examples of this for the Point Loma area near San Diego, California, are given in Figs. 15.43 and 15.44. As will be made evident in Chapter 16, the effluent–sea water mixture will reach the surface during the winter months, but be likely to spread as a submerged field during the late spring and summer months.

There has been considerable discussion in the literature on the cause of subsurface temperature oscillations with periods of the order of hours. One of the possible causes has been stated to be internal waves. Measurements of temperatures along piers in coastal water seem

Table 15.2. MONTHLY AVERAGE FLOW OF THE HUDSON AND OTHER RIVERS ENTERING THE NEW YORK BIGHT BETWEEN SANDY HOOK AND FAR ROCKAWAY POINT (MILLIONS OF CUBIC FEET PER DAY) (after Ketchum, Redfield, and Ayers, 1951)

Month	J	F	M	A	M	J	J	A	S	O	N	D
1948	1,050	1,920	5,250	4,090	3,230	2,130	1,300	1,140	570	600	1,340	2,310
1949	5,510	3,730	3,070	2,810	1,660	720	570	570	790	940	1,330	2,400

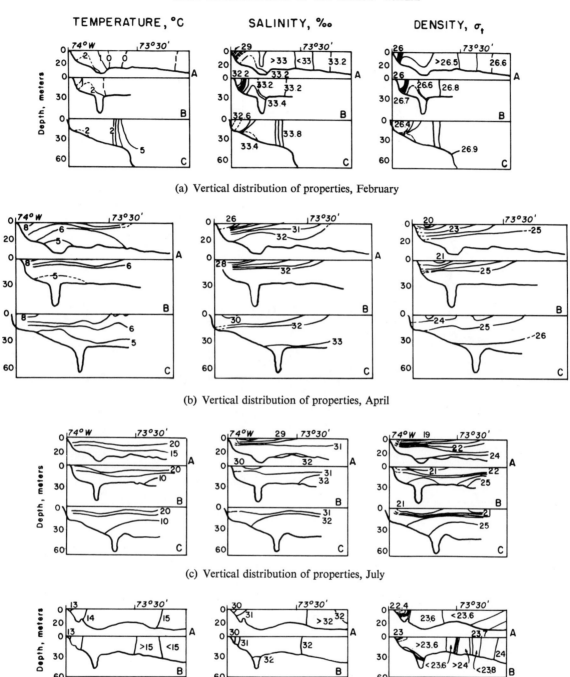

(a) Vertical distribution of properties, February

(b) Vertical distribution of properties, April

(c) Vertical distribution of properties, July

(d) Vertical distribution of properties, October

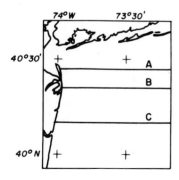

Fig. 15.39. Vertical distributions of temperature, salinity, and density, New York Bight, 1948 (after Ketchum, Redfield, and Ayers, 1951)

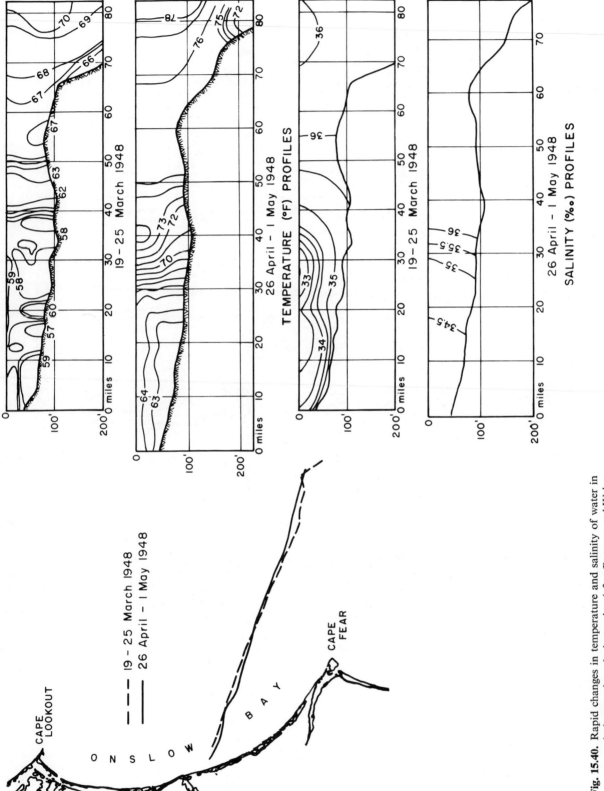

Fig. 15.40. Rapid changes in temperature and salinity of water in inshore regions during spring (after Bumpus and Wehe, 1949)

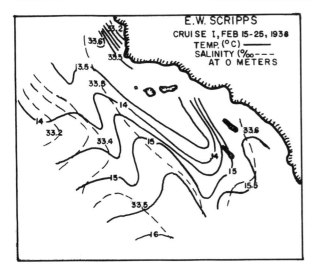

(a) Surface, February 15–25, 1938

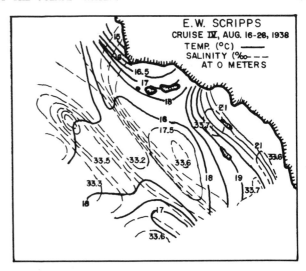

(b) Surface, August 16–26, 1938

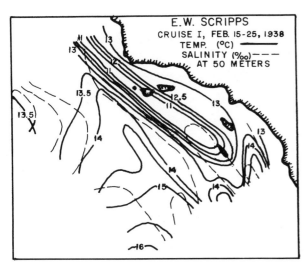

(c) Fifty meters' depth, February 15–25, 1938

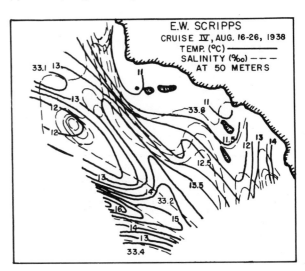

(d) Fifty meters' depth, August 16–26, 1938

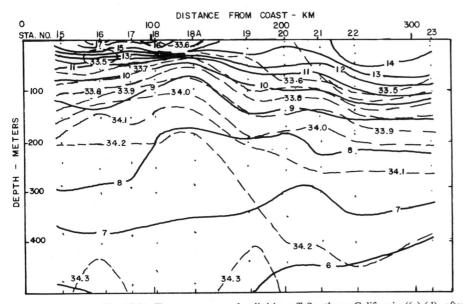

(e) Vertical section, May 5–13, 1937, showing distribution of temperature (——— °C) and salinity (——— ‰) from San Pedro towards S 50° W

Fig. 15.41. Temperatures and salinities off Southern California ((a)-(d), after Sverdrup, 1942; (e), after Sverdrup and Fleming, 1941)

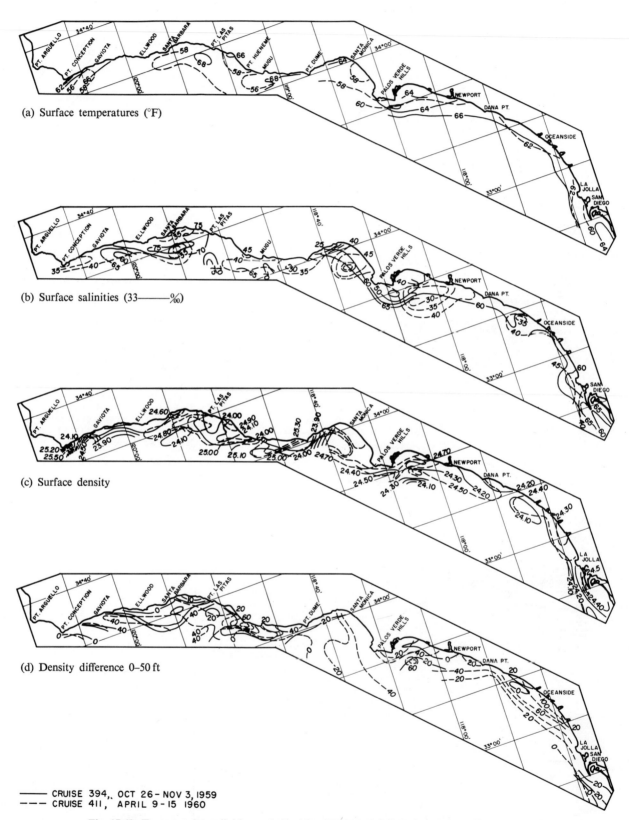

(a) Surface temperatures (°F)

(b) Surface salinities (33———‰)

(c) Surface density

(d) Density difference 0–50 ft

——— CRUISE 394, OCT 26 – NOV 3, 1959
- - - CRUISE 411, APRIL 9 – 15 1960

Fig. 15.42. Temperatures, salinities, and densities, spring and fall, inshore water off southern California (after Allan Hancock Foundation, 1960)

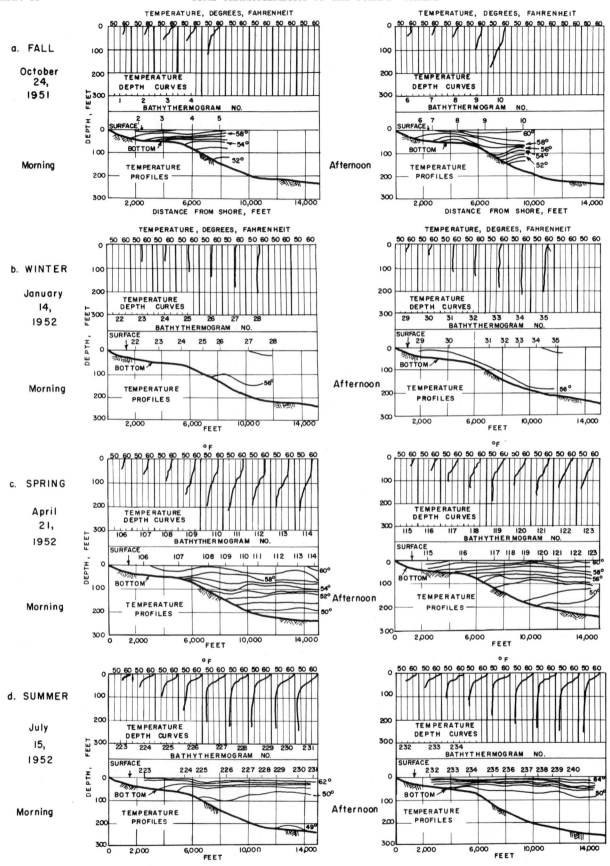

Fig. 15.43. Temperatures in the ocean off Point Loma, near San Diego, California (after Caldwell, Hyde, and Rawn, 1952)

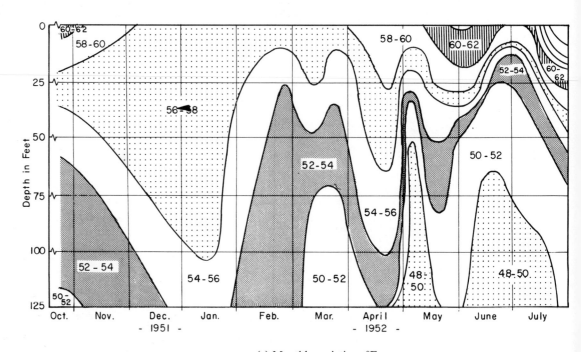

(a) Monthly variation, °F

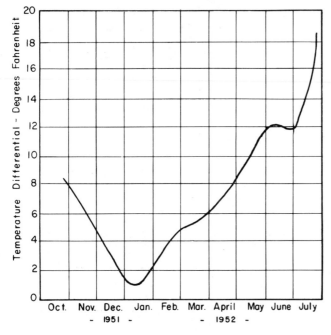

(b) Temperature difference between surface and bottom

Fig. 15.44. Monthly variation of temperature in the ocean off Point Loma, near San Diego, California (after Caldwell, Hyde, and Rawn, 1952)

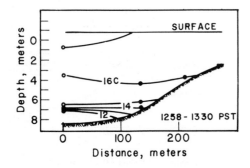
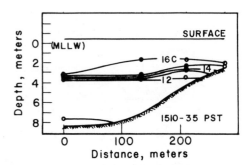

Fig. 15.45. Vertical temperature sections along pier at Oceanside, California, July 9, 1952 (after Arthur, 1954)

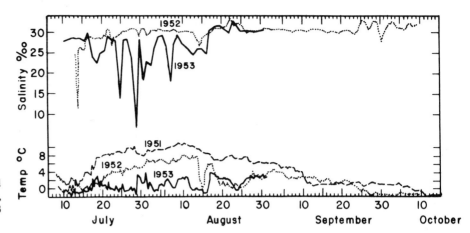

Fig. 15.46. Inshore surface temperatures and salinities during open water seasons, Point Barrow, Alaska, 1951–1953 (after Wilimovsky, 1953)

to have verified the belief that internal waves associated with a thermocline can cause these temperature oscillations (Arthur, 1954). Measurements of temperature time histories along two piers about 35 miles apart in Southern California have shown no obvious linear coherence; this would appear to rule out the tidal mechanism. At the same time there was linear coherence between two sets of measurements about $3\frac{1}{2}$ miles apart within a bight. It was also found that the water on the bottom, say, 35 feet deep, would change as much as 5°C and return to its original value several hours later, but that such pronounced changes occurred only during the summer months when a well-developed thermocline existed. This observation also tends to rule out the tide as a cause

and to confirm the assumption of waves as a cause. An example of an oscillation of the thermocline is shown in Fig. 15.45. Such a motion of the thermocline is important in the mixing of effluent from an ocean outfall sewer.

In working with data taken in coastal waters it is necessary to obtain measurements for several years so that erroneous conclusions will not be drawn because of insufficient data. An excellent example of this is the series of measurements made during open water periods off Point Barrow, Alaska, during 1951–1953 (Wilimovsky, 1953). Conditions during the month of July are entirely different for 1952 than for 1953, as can be seen in Fig. 15.46.

REFERENCES

Arthur, Robert S., Oscillations in sea temperature at Scripps and Oceanside piers, *Deep Sea Res.*, **2**, (1954), 107–121.

Böhnecke, Gunther, Atlas zu Temperatur, Salzgehalt, und Dichte an der Oberfläche des Atlantischen Ozeans, in *Deutschen Atlantischen Expedition auf dem Forschungs und Vermessungschiff "Meteor" 1925–1927*, Vol. 5. Berlin: Walter de Gruyter & Co., 1936.

Bumpus, D. F., and T. J. Wehe, *Hydrography of the Western Atlantic: coastal water circulation off the east coast of the United States between Cape Hatteras and Florida*, Woods Hole Oceanog. Inst., Ref. No. 49–6, January, 1949. (Unpublished.)

Bureau of Commercial Fisheries, *California fishery market news monthly summary*, U.S. Dept. of the Interior, Bureau of Commercial Fisheries, Biological Laboratory, San Diego, Calif.

Burling, R. W., *Hydrology of circumpolar waters south of New Zealand*, New Zealand Dept. of Sci. and Industr. Res. Bull. No. 143, 1961.

Caldwell, David H., Charles Gilman Hyde, and A. M. Rawn, *Report on the collection, treatment and disposal of the sewage of San Diego County, California*, Report to the Board of Supervisors, County of San Diego, California, September, 1952. (Unpublished.)

Deacon, G.E.R., The hydrology of the Southern Ocean, in *Discovery Reports*, Vol. 15. Cambridge: Cambridge Univ. Press, 1937.

Dittmar, W., Report on researches into the composition of ocean water, collected by H.M.S. Challenger, in *Challenger Reports, Phys. and Chem.*, Vol. 1, 1884, pp. 1–251.

Francis, J. R. D., and H. Stommel, How much does a gale mix the surface layers of the ocean?, *Quart. J. Roy. Met. Soc.*, **79**, 342, (October, 1953), 534–35.

Fuglister, F. C., Average monthly sea surface temperatures of the Western North Atlantic Ocean, *Papers in Physical Oceanography and Meteorology*, Mass. Inst. Tech. and Woods Hole Oceanog. Inst., **10**, 2 (May, 1947).

———, Multiple currents in the Gulf Stream System, *Tellus*, **3**, 4 (1951), 230–33.

———, *Recent temperature surveys of the Gulf Stream System*, Woods Hole Oceanog. Inst., Ref. No. 52–39, May, 1952. (Unpublished.)

———, *Average temperature and salinity at a depth of 200 meters in the North Atlantic*, Woods Hole Oceanog. Inst., Ref. No. 53–58, August, 1953. (Unpublished.)

———, The thermal structure in the deep sea, *Proc. Symposium: Aspects of Deep-Sea Research*, Nat. Acad. Sci.– Nat. Research Council, Pub. No. 473, 1957, 10–18.

————, *Atlantic Ocean Atlas: temperature and salinity profiles and data from the International Geophysical Year of 1957–1958*. Woods Hole, Mass.: Woods Hole Oceanog. Inst., 1960.

Fukuoka, J., and T. Tsuiki, On the variation of the oceanographic condition of the sea near the fixed point "Extra," *Rec. Oceanog. Works in Japan*, new ser., **1**, 1 (March, 1953), 23–27.

Gilcrest, Robert A., Glen H. Jung, and John C. Freeman, Jr., *Empirical relations between the weather and the ocean mixed layer*, The Agricultural and Mechanical College of Texas, Dept. of Oceanography, Contract N7, onr–487 T. O. 3, Tech. Rept. No. 8, March, 1954. (Unpublished.)

Hancock, Allan, Foundation, *Oceanographic survey of the continental shelf area of Southern California*, *Annual report 1959–60*, Allan Hancock Foundation, University of Southern California, submitted to the California State Water Pollution Control Board, June 30, 1960. (Unpublished.)

Hutchins, Louis W., and Margaret Scharff, Maximum and minimum monthly mean sea surface temperatures charted from the "World Atlas of Sea Surface Temperatures," *J. Mar. Res.*, **6**, 3 (December, 1947), 264–68.

Iselin, C. O'D., A study of the circulation of the western North Atlantic, *Papers in Physical Oceanog. Met.*, Mass. Inst. Tech. and Woods Hole Oceanog. Inst., **4**, 4, August, 1936.

Ketchum, Bostwick H., Alfred C. Redfield, and John C. Ayers, The oceanography of the New York Bight, *Papers in Physical Oceanog. Met.*, Mass. Inst. Tech., and Woods Hole Oceanog. Inst., **12**, 1, August, 1951.

Knudsen, Martin, *Hydrographical tables*. Copenhagen: G.E.C. Gad, 1901.

Komonkova, G. R., and N. V. Kontoboitseva, Temperature oscillations in the surface layer of the sea caused by wave movement, *Akademii Nauk SSSR*, Geophysics ser., Trans. published by Pergamon Press for *Amer. Geophys. Union*, No. 12, 1957, pp. 57–64.

Leipper, D. F., and E. R. Andersen, *Sea temperature in the Hawaiian Island area*, Scripps Inst. Oceanog., Oceanographic Report No. 12, June, 1948. (Unpublished.)

Lyman, John, and R. H. Fleming, Composition of sea water, *J. Mar. Res.*, **3** (1940), 134–46.

————, *Proc. Symposium: Aspects of Deep-Sea Research*, ed. William S. von Arx, National Academy of Sciences–National Research Council, Pub. No. 473, 1957, p. 10.

Mackintosh, N. A., The Antarctic convergence and the distribution of surface temperatures in antarctic waters, Vol. 23 in *Discovery Reports*, London: Cambridge Univ. Press, 1946, pp. 177–212.

Martineau, Donald P., *An objective method for forecasting the diurnal thermocline*, Woods Hole Oceanog. Inst., Ref. No. 53–21, April, 1953. (Unpublished.)

McLellan, Hugh J., The waters of the Gulf of Mexico as observed in 1958 and 1959, The Agricultural and Me-

chanical College of Texas, Contract N7onr–48702, Ref. 60–14T, September, 1960. (Unpublished.)

Montgomery, R. B., Summary of some of the existing knowledge of the origin and behavior of the thermocline, *Notes from the Conference on the Thermocline of 25–27 May, 1953*, The Johns Hopkins Univ., Chesapeake Bay Inst., Ref. 54–2, March, 1954, pp. 5–16.

Munk, W. H., and E. R. Anderson, Notes on a theory of the thermocline, *J. Mar. Res.*, **7** (1948), 276–95.

National Academy of Sciences–National Research Council, *Physics of the Earth—V, Oceanography*, Bull. National Research Council, No. 85, 1932.

————, *Physical and chemical properties of sea water*, Pub. No. 600, 1959.

NORPAC Committee, *Oceanic Observations of the Pacific: 1955, the NORPAC data*. Berkeley, Calif.: Univ. California Press, 1960.

————, *Oceanic Observations of the Pacific: 1955, the NORPAC Atlas*. Berkeley, Calif.: Univ. California Press, 1960.

Paquette, Robert G., Salinometers, *Conf. on Physical and Chemical Properties of Sea Water*, National Academy of Sciences–National Research Council, Pub. No. 600, 1959, pp. 128–45.

Pattullo, June, *Mixed layer depth determined from critical gradient frequency*, Scripps Inst. Oceanog., Ref. No. 52–25, May, 1952. (Unpublished.)

Pollak, M. J. (ed.), *Notes from the conference on the thermocline of 25–27 May, 1953*, The Johns Hopkins University, Chesapeake Bay Institute, Ref. No. 54–2, March, 1954.

Pyle, Robert L., *Serial atlas of the marine environment: Folio 1. Sea surface temperature regime in the Western North Atlantic 1953–1954*. New York: Amer. Geog. Soc., 1962.

Robinson, Margaret K., *Sea temperature in the Marshall Islands area*, Scripps Inst. Oceanog., Contract Nonr–233(05), Ref. No. 52–45, September, 1952. (Unpublished.)

————, Sea temperature in the Gulf of Alaska and in the Northeast Pacific Ocean, 1941–1952, *Bull. Scripps Inst. Oceanog.*, **7**, 1, October, 1957.

————, Indian Ocean vertical temperature sections, *Deep Sea Res.*, **6**, (1960), 249–58.

————, John D. Cochrane and Wayne V. Burt, *Analysis and interpretation of bathythermograms from the Antarctic development project 1947–48*, Scripps Inst. Oceanog. Contract N6ori–111, Task Order 6, June 16, 1950. (Unpublished.)

Rossby, C.G., and R.B. Montgomery, The layer of frictional influence in wind and ocean currents, *Papers in Physical Ocean. Met.*, Mass. Inst. Tech., and Woods Hole Oceanog. Inst., **3**, 3, April, 1935.

Rudnick, Philip, and John D. Cochrane, Diurnal fluctuations in bathythermograms, *J. Mar. Res.*, **10**, 3 (1951), 257–62.

Schott, Gerhard, *Geographie des Indischen und Stillen Ozeans*. Hamburg: C. Boysen, 1935.

Schule, J. J., Jr., L. S. Simpson and A. Shapiro, *Effects of weather upon the thermal structure of the ocean*, U.S. Navy Hydrographic Office, H. O. Misc. 15360, 1952.

Stommel, Henry, and Sloat F. Hodgson, *Consecutive temperature measurements at 500 meters off Bermuda*, Woods Hole Oceanog. Inst. Ref. No, 56–43, July, 1956. (Unpublished.)

Strack, S. L., *Surface temperature gradients as indicators of the position of the Gulf Stream*, Woods Hole Oceanog. Inst., Ref. No. 53–53, July, 1953. (Unpublished.)

Sverdrup, H. U., and R. H. Fleming, The waters off the coast of Southern California, March to July 1937, *Bull. Scripps Inst. Oceanog.*, **4**, 10 (1941), 261–378.

——, Martin W. Johnson, and Richard H. Fleming, *The oceans, their physics, chemistry and general biology*. Englewood Cliffs, N. J.: Prentice-Hall, Inc., 1942.

——, and Staff, Oceanographic observations on the "E.W. Scripps" in cruises of 1938, in *Records of Observations*, Scripps Inst. Oceanog., **1**, 1 (1942), 1–64.

Treshnikov, A. F., The Arctic discloses its secrets: new data on the bottom topography and the waters of the Arctic Basin, *Priroda* **2**, (1960). 25–32. E. R. Hope, trans. Defence Research Board, Canada, T357R, August, 1961.

United Nations Educational Scientific and Cultural Organization, *Second report of joint panel on the equation of state of sea water*, Berkeley, Calif, IUGG, Thirteenth Assembly Aug. 19, 1963.

U.S. Navy Hydrographic Office, *World atlas of sea surface temperatures*, H. O. Publication No. 225, 1944.

——, *Tables for sea water density*, H. O. Pub. No. 615, 1952.

——, *Oceanographic atlas of the Polar seas*, H. O. Pub. No. 705, 1957.

Vaughan, Thomas Wayland, *International aspects of oceanography. Oceanographic data and provisions for oceanographic research*. Washington, D.C.: National Academy of Sciences, 1937.

Wilimovsky, Norman J., *Inshore temperature and salinity data during open water periods, Point Barrow, Alaska, 1951–1953*, Natural History Museum, Stanford Univ., Tech. Rept. 4, Contract N6onr–25136, December, 1953. (Unpublished.)

Worthington, L. V., Three detailed cross-sections of the Gulf Stream, *Tellus*, **6**, 2 (1954) 116–23.

Wüst, Georg and Albert Defant, *Atlas zur Schlichtung and Zirkulation des Atlantischen Ozeans, Deutschen Atlantischen Expedition auf dem Forschungs und Vermessungsschiff "Meteor" 1925–1927*, Vol. 6. Berlin: Walter de Gruyter and Co., 1936.

CHAPTER SIXTEEN

Mixing Processes

1. INTRODUCTION

Mixing processes are of prime importance to the engineer and to the oceanographer. The transfer of heat down through the surface layers of the ocean is largely due to mixing by eddy diffusion and by large-scale stirring induced by the breaking of waves in a sea. The entrainment of water by currents is due to the mixing of the currents with the boundary waters. The disposal of sewage by coastal towns is usually done by the mixing of the effluent from ocean outfall sewers with the sea water. Some of these mixing processes are described in this chapter.

The principal difficulty in trying to apply the results presented in the separate sections of this chapter is that they have all been studied separately, whereas in nature many of the mixing processes occur simultaneously. For example, theories and experimental results are presented for turbulent jets in a receiving fluid that is not turbulent, whereas in reality it often is turbulent. The mixing due

to wind waves is studied in the absence of a thermocline; how does such a stable stratification affect mixing of sewage? How much greater is the mixing of a turbulent jet in a turbulent current than a turbulent jet in a nonturbulent body of receiving water?

2. MIXING OF TURBULENT JETS

a. *Introduction.* Ocean outfall sewers are one of many operations which require a knowledge of mixing processes. When sea water is distilled into fresh water, quantities of dense brine must be pumped back into the ocean and dispersed. The use of sea water as the cooling water in nuclear or steam-electric generating plant requires the mixing of large quantities of heated sea water in the sea. In Norway, the discharge of fresh water into the fjords from hydroelectric plants in winter requires the mixing of the fresh water and sea water so that the fresh water will not remain on the surface and freeze.

First, the effluent is discharged through some sort of

a port into the receiving water, and mixing occurs by the process associated with a turbulent jet. The jet mixing may be affected by the action of currents or stirring by seas, to an extent not well defined. Depending upon the amount of mixing that takes place close to the pipelines and upon the characteritics of the thermocline, the effluent–sea water mixture may or may not reach the surface. In one case, the sewage spreads laterally either as a submerged sewage mixture field or as a surface sewage mixture field. In both cases, the effects of turbulence in the sea water and mixing by the stirring processes associated with seas must be considered. If it were not for currents, the problem would then be one of a circulation system of ever-increasing size and pollution. The currents, however, transport the mixture away from the outfall, and at the same time, dilute it by further mixing, both laterally and vertically. Depending upon the relationships among the location of the outfall, the strength and direction of the current, the location of beaches, etc., the effect of the current may be largely beneficial, or detrimental.

Laboratory studies have been made of the individual processes of mixing, separately. In field studies it has not been possible to distinguish clearly between the effects of the various processes. Both methods have their advantages and disadvantages. In this chapter, the various mixing mechanisms will be treated separately.

b. *Mixing of a turbulent jet.* First, the mixing of a turbulent jet discharging into a fluid of the same density will be considered. A turbulent jet issuing from an outlet will entrain part of the surrounding fluid so that the volume flux will increase with increasing distance from the outlet, and the velocity will decrease. In Fig. 16.1, the jet is issuing from an outlet of thickness L_o as uniform flow (which is an approximation to the velocity distribution of highly turbulent flow). The boundary between the jet and the surrounding fluid is unstable

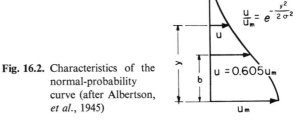

Fig. 16.2. Characteristics of the normal-probability curve (after Albertson, *et al.*, 1945)

(Landau and Lifshitz, 1959), high shear stresses exist, and mixing will occur with the transfer of momentum, temperature, and material from the jet to the receiving fluid. Two zones of flow are considered in this model. The zone of flow establishment consists of a core of uniform velocity, u_o, and a surrounding boundary layer of fluid with a velocity u. In the zone of established flow, mixing takes place throughout the jet.

In the development of a theory of the turbulent jet, Albertson, *et al.* (1948) assumed that the pressure distribution was hydrostatic throughout the zone of motion, that the momentum flux was constant, that the flow was dynamically similar in every section within the mixing region. It was also assumed that the velocity profiles in the mixing region could be approximated by the Gaussian normal probability function.

$$\frac{u}{u_m} = \exp\left(\frac{-y^2}{2\sigma^2}\right) \qquad (16.1)$$

in which u is the velocity component in the direction of the jet axis, u_m is the axial velocity along the centerline of the jet at any distance x from the outlet, y is the coordinate normal to x, and where σ mathematically is the standard deviation, and physically is the distance from the centerline of the jet to the point of maximum velocity gradient (Fig. 16.2). Equation 16.1 has been used by almost every person investigating this problem although all experimental evidence shows it to be a poor assumption except in the central region of a jet. Furthermore, such an assumption forces one to an erroneous conclusion regarding the possibility of circulation in the receiving fluid. This will be discussed later.

In the zone of flow establishment, it was assumed that u_m would correspond to u_o until the point where the eddies generated by the mixing of the jet with the receiving fluid penetrated to the center of the jet. This is shown in Fig. 16.3 for a slot type of opening.

σ must be determined experimentally. The value of σ is not used directly, rather, another coefficient, C, is obtained which is related to σ through the expression $\sigma/x = C$. For the case of a circular orifice $\sigma/x = C_2$ is used, together with the momentum flux assumption that

$$\frac{M}{M_o} = \frac{\int_0^\infty u^2 \, dA}{u_o^2 \, A_o} = 1 \qquad (16.2)$$

Fig. 16.1. Schematic representation of jet mixing (after Albertson, *et al.*, 1945)

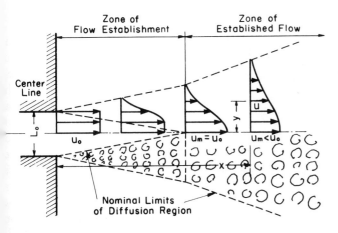

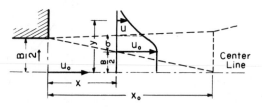

Fig. 16.3. Schematic representation of zone of flow establishment (after Albertson, *et al.*, 1945)

to give

$$\frac{x_o}{D_o} = \frac{1}{2C_2},$$

where D_o is the diameter of the orifice. For the case of a slot, the relationship $\sigma/x = C_1$ is used, together with Eq. 16.2, to give

$$\frac{x_o}{B_o} = \frac{1}{C_1\sqrt{\pi}},$$

where B_o is the thickness of the slot.

The values of C_1 and C_2 are obtained from the data plotted in the manner of Fig. 16.4, as x_o/D_o is the value of x/D_o at the intersection of the horizontal and sloping lines through the data. It was found that $C_1 = 0.109$ and $C_2 = 0.081$. These data were for air jets in air with the densities of the jet and receiving air being identical. C_2 calculated from experiments of fresh water jets in saline water ($\rho_s/\rho_o = 1.01$ where ρ_s is the density of the receiving fluid and ρ_o is the density of the jet at the orifice) was found to be 0.078 (Forstall and Gaylord, 1955). Corrsin and Uberoi (1949) found in their experiments with air jets that $C_2 = 0.072$ and that

$$\frac{1}{2}C_2^2 = 96\left[1 + 0.19\left(\frac{\rho_s}{\rho_o} - 1\right)\right]^{-2}$$

so that C_2 is essentially constant for the range of interest in sewage issuing into sea water, $1 < \rho_s/\rho_o < 1.025$.

In the zone of flow establishment, the equations for the velocity distributions are

$$\frac{u}{u_o} = \exp\left[-\frac{(y + \frac{1}{2}\sqrt{\pi}\,C_1 x - \frac{1}{2}B_o)^2}{2C_1^2 x^2}\right],$$

for

$$\qquad\qquad (16.3a)$$

$$y > \frac{B_o}{2} - C_1 x,$$

and $\dfrac{u}{u_o} = 1,$ for $y < \dfrac{B_o}{2} - C_1 x,$ $\quad$ (16.3b)

for the case of a slot,

and $\qquad \dfrac{u}{u_o} = \exp\left[-\dfrac{(r + C_2 x - \frac{1}{2}D_o)^2}{2C_2^2 x^2}\right],$

for

$$\qquad\qquad (16.4a)$$

$$y > \frac{D_o}{2} - C_2 x$$

and $\dfrac{u}{u_o} = 1,$ for $y < \dfrac{D_o}{2} - C_2 x$ $\quad$ (16.4b)

for the case of a circular orifice, with r being the radial coordinate. After putting in the experimentally determined values of C_1 and C_2, these equations can be expressed as

$$\log_{10}\frac{u}{u_o} = -18.4\left(0.096 + \frac{y - \frac{1}{2}B_o}{x}\right)^2 \quad (16.5)$$

for the slot, and

$$\log_{10}\frac{u}{u_o} = -33\left(0.081 + \frac{r - \frac{1}{2}D_o}{x}\right)^2 \quad (16.6)$$

for the circular orifice.

For the zone of established flow, the equations are

$$\frac{u}{u_o} = \sqrt{\frac{1}{C_1\sqrt{\pi}}\frac{B_o}{x}}\exp\left[-\frac{1}{2C_1^2}\frac{y^2}{x^2}\right] \quad (16.7)$$

for the slot, and

$$\frac{u}{u_o} = \frac{1}{2C_2}\frac{D_o}{x}\exp\left[-\frac{1}{2C_2^2}\frac{r^2}{x^2}\right] \quad (16.8)$$

for the circular orifice.

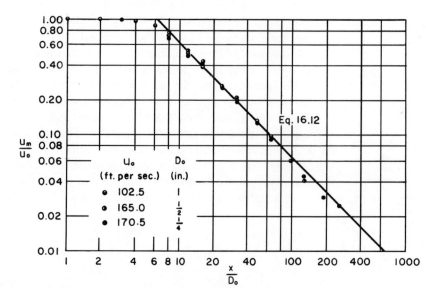

Fig. 16.4. Centerline velocity for flow from an orifice (after Albertson, *et al.*, 1945)

After substituting the numerical values of C_1 and C_2 into Eqs. 16.7 and 16.8, they can be expressed as

$$\log_{10} \frac{u}{u_o} \sqrt{\frac{x}{B_o}} = 0.36 - 18.4 \frac{y^2}{x^2} \qquad (16.9)$$

fro the slot, and

$$\log_{10} \frac{u}{u_o} \frac{x}{D_o} = 0.79 - 33 \frac{r^2}{x^2} \qquad (16.10)$$

for the circular orifice. The equations for the velocity along the centerline in the zone of established flow, u_m, can be expressed as

$$\frac{u_m}{u_o} \sqrt{\frac{x}{B_o}} = 2.28 \qquad (16.11)$$

for the slot, and

$$\frac{u_m}{u_o} \frac{x}{D_o} = 6.2 \qquad (16.12)$$

for the circular orifice. It can be seen in Fig. 16.4 that Eq. 16.12 fits the data. A similar fit of data to Eq. 16.11 was found for the slot.

Velocity distributions of values of x and r are shown in Fig. 16.5 for the zone of flow establishment, and Fig. 16.6 for the zone of established flow for the circular orifice. It can be seen that Eqs. 16.6 and 16.10 describe the velocity distributions adequately only in the central

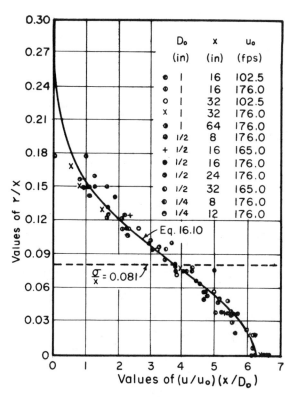

Fig. 16.6. Distribution of longitudinal velocity in zone of established flow from orifice (after Albertson, *et al.*, 1948)

region of the jet. Similar results were found for the jets issuing from a slot.

General flow patterns were developed by Albertson *et al.* (1948), and these are shown in Fig. 16.7. No specific expressions for the lateral velocity components (v) were obtained; however, these values were calculated by graphical means and the results compared with the measurements. Although the shape of the curves and the data distribution were similar, the measured lateral components were about 100 per cent greater than the computed values. Even so, the lateral velocity components were only of the order of one-tenth the longitudinal velocity components.

Measurements of the direction of particle velocities (Fig. 16.8) showed that the angle was zero at the centerline of the jet, gradually sloping up to about 7 degrees (water particles moving away from the centerline) at a value of y/x or r/x of about 0.15, then dropping rapidly to minus 90 degrees (heading directly toward the centerline for y/x and r/x of about 0.27.

Flow rates, Q, were given by Albertson *et al.* (1948) in terms of the discharge rate Q_o as

$$\frac{Q}{Q_o} = 1 + C_1 \sqrt{\pi} \left(\sqrt{2} - 1 \right) \frac{x}{B_o} = 1 + 0.080 \frac{x}{B_o}$$
$$(16.13)$$

for the slot, and

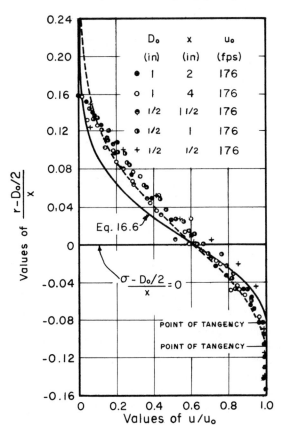

Fig. 16.5. Distribution of longitudinal velocity in zone of flow establishment (after Albertson, *et al.*, 1948)

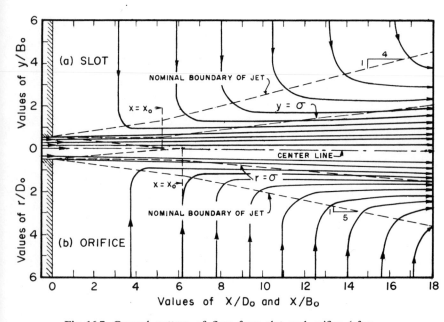

Fig. 16.7. General pattern of flow from slot and orifice (after Albertson, *et al.*, 1948)

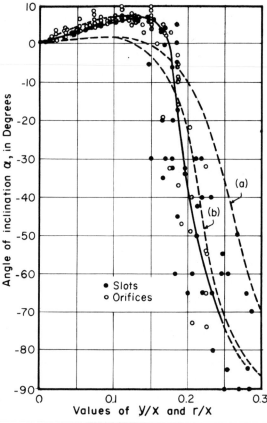

Fig. 16.8. Direction of flow within jets from both slots and orifices (Albertson, *et al.*, 1948)

$$\frac{Q}{Q_0} = 1 + 2\left(\sqrt{2\pi} - 2\right) C_2 \frac{x}{D_0} + 4\left(3 - \sqrt{2\pi}\right) C_2^2 \frac{x^2}{D_0^2}$$

$$= 1 + 0.083 \frac{x}{D_0} + 0.0128 \frac{x^2}{D_0^2} \qquad (16.14)$$

for the circular orifice in the zone of flow establishment. In the region of established flow, the equations were found to be

$$\frac{Q}{Q_0} = \sqrt{2C_1 \sqrt{\pi}\, \frac{x}{B_0}} = 0.62 \sqrt{\frac{x}{B_0}} \qquad (16.15)$$

for the slot, and

$$\frac{Q}{Q_0} = 4C_2 \frac{x}{D_0} = 0.32 \frac{x}{D_0} \qquad (16.16)$$

for the circular orifice.

Similar results have been found by other investigators, and the coefficients have been compared by Abraham (1960).

Other investigators (Hinze and van der Hegge Zijnen, 1948; Forstall and Gaylord, 1955; Abraham, 1960) used equations for material transfer in terms of the concentration c of some tracer material of concentration c_0 in the issuing jet. The equations are based upon the assumption that the distribution of c/c_0 with respect to x and r (for a circular orifice) would be analogous to the distribution of axial momentum. The necessary constant was obtained in the same manner as for the equation of the velocity distribution in the axial direc-

tion. Thus, for the zone of flow establishment for a jet issuing from an orifice

$$\frac{c}{c_0} = \exp\left\{-\frac{[r + C_3 x - (D_0/2)]^2}{2C_3^2 x^2}\right\}$$

for $\qquad\qquad\qquad\qquad\qquad\qquad\qquad\qquad (16.17)$

$$r > \frac{D_0}{2} - C_3 x$$

and $\dfrac{c}{c_0} = 1,$ for $r < \dfrac{D_0}{2} - C_3 x.$ $\qquad (16.18)$

Several investigators have found C_3 experimentally to be as follows: 0.092 (Corrsin and Uberoi, 1949); 0.085 (Keagy *et al.*, 1949); 0.096 (Forstall and Gaylord, 1955).

In the zone of established flow ($x/D_0 > 6$ or 7),

$$\frac{c}{c_m} = \exp\left[-\frac{1}{2C_3^2} \frac{r^2}{x^2}\right] \qquad (16.19a)$$

$$\frac{c_m}{c_0} = \frac{1}{2C_3} \frac{D_0}{x}. \qquad (16.19b)$$

The various experiments cited have shown that the mixing of momentum occurs in a different manner than either the mixing of material or temperature, at least in the main part of the jet. In this region, the distribution of temperature and matter were found to be the same by Hinze and van der Hegge Zijnen (1948), but greater than the distribution of momentum in the axial direction. This does not appear to be the case in the region normally considered to be the "edge" of the jet. A physical reason for this can be obtained by looking at the flow direction data of Fig. 16.8. Material and temperature criteria are functions of location and are independent of the direction of flow, whereas the measurements of momentum distribution have been made only for the axial component. Thus in the central region of the jet θ/θ_m (temperature) and c/c_m should be greater than u/u_m for a given x and r.

At some value of y or r from the jet centerline there must be a transition from turbulent to laminar flow. This would affect the mixing rates as calculated from the equations obtained for turbulent mixing. Folsom and Ferguson (1949), using the work of Corrsin (1943) and other measurements, concluded that the flow in the annular viscous ring would be less than 5 per cent of the total fluid flowing so that it should be relatively unimportant. The problem of decay of turbulence with increasing x was also considered, but there were not enough data available to draw conclusions. There were some data for an air jet up to values of x of 40 orifice diameters in which the flow was still turbulent.

c. *Jet Discharging into denser fluid.* The density of sewage effluent is very nearly that of fresh water. It is usually discharged from a pipe on the ocean bottom. Because of this, the effect of buoyancy of the effluent in sea water on the mixing processes must be considered. For turbulent jets, dimensional analysis shows that (Rawn, Bowerman, and Brooks, 1961)

$$\frac{u}{u_o} = f_1\left(\frac{x}{D_o}, \frac{y}{D_o}, \frac{\Delta\rho}{\rho_o}, N_F\right) \quad (16.20)$$

$$\frac{c}{c_o} = f_2\left(\frac{x}{D_o}, \frac{r}{D_o}, \frac{\Delta\rho}{\rho_o}, N_F\right) \quad (16.21)$$

as the effect of Reynolds number can be neglected. Here $\Delta\rho$ is the difference in density between the issuing jet and the surrounding water. The Froude number is given by

$$N_F = \frac{u_o}{\sqrt{(\Delta\rho/\rho_o)g D_o}}. \quad (16.22)$$

In the previous section, the nonbuoyancy case was considered. Here the buoyancy-only case will be considered first. This is the equivalent of a gravitational convection problem, i.e., there is no initial jet momentum. Because of the buoyant force, the resulting jet that forms will gain momentum as it rises.

Rouse, Yih, and Humphreys (1952) assumed that the initial laminar zone was small compared with the subsequent zone of turbulent convection and that the change in density of the jet as it rises is small enough to be neglected compared with the buoyant force due to the difference in density of the jet, ρ, and the density of the receiving fluid, ρ_s. Similarity is assumed, together with a Gaussian distribution of vertical velocity and density.

In applying the solution to the case of a jet of density ρ_o (which is less than ρ_s), it is assumed that the bouyancy flux can be given by

$$\frac{\pi}{4} D_o^2 g(\rho_s - \rho_o) u_o \quad (16.23)$$

for small values of N_F and that the term $\Delta\rho/\rho_o$ in Eq. 16.22 can be used in place of $\Delta\rho/\rho_s$ as $\Delta\rho/\rho_o \approx \Delta\rho/\rho_s$. Using the numerical constants obtained from the experimental work of Rouse et al., Abraham (1960) found that u and c could be expressed as

$$\frac{u}{u_o} = 4.35 N_F^{-2/3}\left(\frac{x}{D_o}\right)^{-1/3} \exp\left[-96\left(\frac{r}{x}\right)^2\right], \quad (16.24)$$

$$\frac{c}{c_o} = 9.35 N_F^{2/3}\left(\frac{x}{D_o}\right)^{-5/3} \exp\left[-71\left(\frac{r}{x}\right)^2\right]. \quad (16.25)$$

The experimental coefficients were obtained by Rouse et al. for plumes above a single gas burner. They also developed the theory and obtained coefficients experimentally for the case of a line source.

Morton, Taylor, and Turner (1956) also studied the problem. As an approximation they assumed that the velocity and buoyant forces were constant across the jet discharging vertically, that the jet had no initial momentum, and that the source was a point (they also considered a line source). Abraham (1960) assumed that the buoyancy flux could again be given by Eq. 16.23 and found that the equations due to Morton et al., using their experimentally determined experimental coefficients could be expressed as

$$\frac{u_m}{u_o} = 3.65 N_F^{-2/3}\left(\frac{x}{D_o} + 2\right)^{-1/3}, \quad (16.26)$$

$$\frac{u}{u_m} = \exp\left[-80\left(\frac{r}{x}\right)^2\right], \quad (16.27)$$

$$\frac{c_m}{c_o} = 9.7 N_F^{2/3}\left(\frac{x}{D_o} + 2\right)^{-5/3}, \quad (16.28)$$

$$\frac{c}{c_m} = \exp\left[-80\left(\frac{r}{x}\right)^2\right]. \quad (16.29)$$

The term $2 + x/D_o$ rather than x/D_o was found experimentally by Morton et al., to allow for the displacement along the x axis of the virtual source of diameter D_o, as the theoretical source at $x = 0$ is a point rather than an orifice of diameter D_o. The experimental data of Morton et al. were for fresh water being discharged vertically into salt water; hence experimental coefficients include the effect of small initial momentum. The experimental work of Abraham (1960) tended to confirm the coefficients due to Morton et al., and these results were also obtained by discharging a jet of fresh water vertically into a tank full of salt water.

In sewer outfalls, both the initial momentum and buoyancy are important. A practical solution to this problem was obtained by Abraham (1960) by an approach that was suggested by the equations of F. H. Schmidt

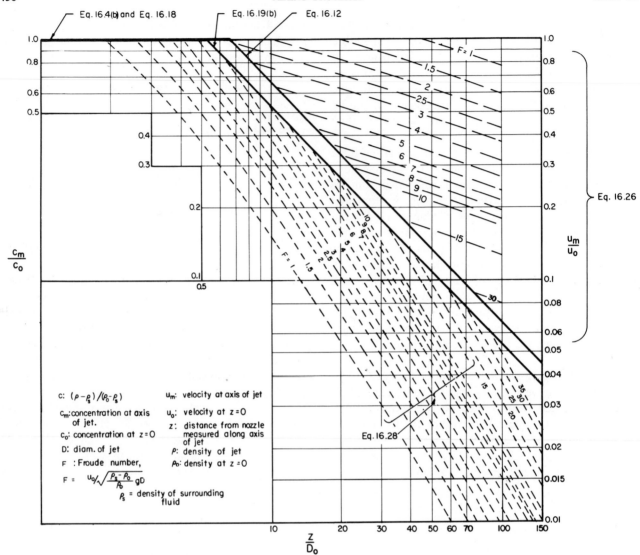

Fig. 16.9. Concentration and velocity along jet axis (adapted from Abraham, 1960)

(1957) for the general case. This was to match the solutions of the nonbuoyant and buoyancy-only cases as shown in Fig. 16.9. As the experimental data of Abraham were obtained for jets of fresh water discharging vertically into salt water,

$$c = \frac{(\rho - \rho_s)}{\rho_o - \rho_s}. \qquad (16.30)$$

In the theoretical study of F. H. Schmidt (1957), no assumption was made regarding the profile of the axial component of velocity. He found a solution of the form

$$\frac{u}{u_m} = (1 - \gamma\eta^\gamma)\, e^{-\eta^\gamma}, \qquad (16.31)$$

for a line source, and

$$\frac{u}{u_m} = \left(1 - \frac{\gamma}{2}\eta^\gamma\right) e^{-\eta^\gamma} \qquad (16.32)$$

for an axially symmetric jet, where γ is a constant that must be determined empirically, and η is y/R for the line

source and r/R for the axially symmetric jet. R was defined in general terms by F. H. Schmidt as being a measure of the "width" of the jet, but was not specifically defined in terms of some known parameter. The most likely values of γ in Eqs. 16.31 and 16.32 have been found by Schmidt to be 1.5 and 1.8, respectively. These values were obtained by plotting the data of Rouse *et al.* on graphs of u/u_m versus η for several values of γ and obtaining the best fit, as shown in Fig. 16.10 and 16.11.

One of the most important findings in Schmidt's work is that when a Gaussian distribution is not imposed on the velocity distribution one finds that a circulation is set up. This circulation is certainly the case for the studies made by the author, and must be the case for any jet discharging into a finite body of receiving fluid, even if the volume of the receiving fluid is large, such as the ocean or the atmosphere. This means, of course, that there will be a continual contamination of the receiving fluid. In practice, this result is modified by the presence

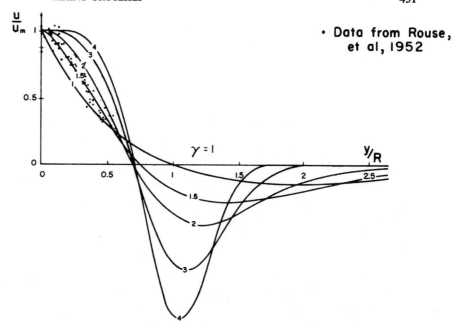

Fig. 16.10. Distribution of u/u_m as a function of y/R for a two-dimensional jet (after F. H. Schmidt, 1957)

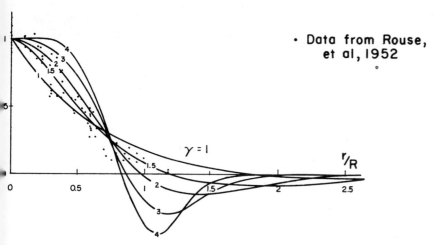

Fig. 16.11. Distribution of u/u_m as a function of r/R for an axially symmetric jet (after F. H. Schmidt, 1957)

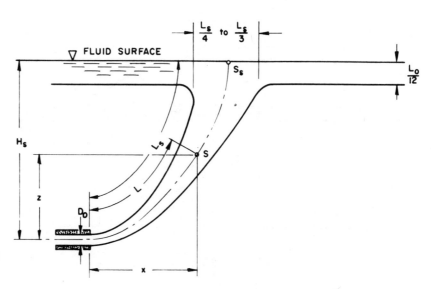

Fig. 16.12. Sketch of single rising column (after Rawn, *et al.*, 1960)

of currents which move the receiving water so that the local area does not necessarily become highly contaminated.

d. *Jet Discharging horizontally into denser fluid.* The case in which the initial momentum and the buoyancy force are normal to each other has been studied empirically by Rawn, Bowerman, and Brooks (1960) for the case of fresh water discharging into sea water. The characteristics of this system are shown in Fig. 16.12. Model studies were made with a turbulent jet so that the effect of Reynolds number was minimized. The two variables were H_s/D_o and a Froude number,

$$N_F = \frac{Q}{\pi/4} D_o^2 \sqrt{g D_o (\rho_s - \rho_o)/\rho_o} = \frac{u_o}{\sqrt{g D_o (\rho_s - \rho_o)/\rho_o}},$$

where ρ_s is the density of the receiving water and ρ_o is the density of the effluent. Curves plotted by Rawn *et al.* from their data are given in Fig. 16.13. The dilution at the intersection of the axis of the jet and the water surface, S_s, is the fraction of effluent in the sample. More recent work by Frankel and Cumming (1963) is given in Fig. 17.25. These results also were for other angles of discharge.

e. *Jet discharging into fluid with a stable density gradient.* Often the ocean has a density gradient, with the density increasing with increasing distance beneath the water surface (see Chapter 15). Sewage effluent discharging from a pipe on the bottom has a density that is about that of fresh water. As it rises, it mixes with the receiving water, and the mixture gradually becomes more dense as it rises. If the density of the mixture becomes equal to the surrounding fluid (which decreases with increasing distance above the bottom) the mixture begins to move in a horizontal direction.

Morton, Taylor, and Turner (1956) studied the case of a vertical convection plume from a point source in a uniformly and stably stratified fluid, when there is no initial momentum. The solution is

$$z = 0.410 \alpha^{-1/2} F_o^{1/4} G^{-3/8} z_1 \qquad (16.33)$$

$$b = 0.819 \alpha^{1/2} F_o^{1/4} G^{-3/8} R \qquad (16.34)$$

$$u = 1.158 \alpha^{-1/2} F_o^{1/4} G^{1/8} U \qquad (16.35)$$

$$\frac{\rho_z - \rho_m}{\rho_s} = 0.819 g^{-1} \alpha^{-1/2} F_o^{1/4} G^{5/8} \Delta, \qquad (16.36)$$

where the horizontal distributions within the main part of the jet (before it starts to spread) are Gaussian; b is an effective jet radius (the horizontal distance from the plume centerline at any height where the axial velocity and buoyancy amplitude are $1/e$ of those on the axis); α has been found experimentally to be 0.093; ρ_z is the receiving fluid density at any level z above the origin of the plume; ρ_m is the fluid density along the axis of the plume at any level z; ρ_s is the density of the receiving fluid at the origin of the plume;

$$G = -\frac{g}{\rho_s} \frac{d\rho_z}{dz} \qquad (16.37)$$

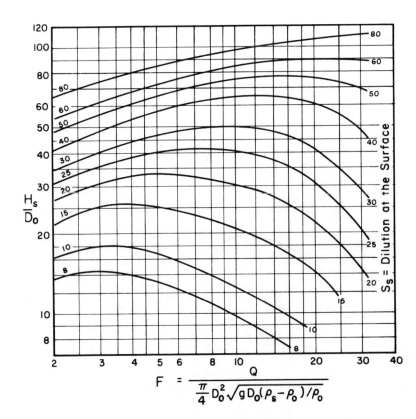

Fig. 16.13. Dilution at the surface, S_s, as a function of H_s/D_o and N_F for horizontal discharge (after Rawn, *et al.*, 1960)

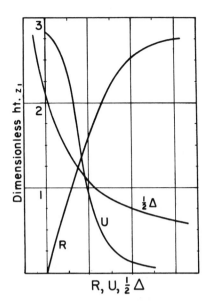

Fig. 16.14. Variation with height of the horizontal extent (R), the vertical velocity (U), and the buoyancy (Δ), in non-dimensional units for a turbulent plume in a uniformly and stably stratified fluid (after Morton, et al., 1956)

and F_o can be given approximately as

$$F_o = \frac{\pi}{2}\left(\frac{\pi}{4}\,D_o^2\right)u_o g\,\frac{\rho_s - \rho_o}{\rho_s} \approx \frac{\pi^2}{8}\,D_o^2 u_o g\,\frac{\rho_s - \rho_o}{\rho_o}$$

(16.38)

in a manner similar to Eq. 16.23. Specific results can be obtained from Eqs. 16.33–16.36 using Eqs. 16.37 and 16.38, together with the dimensionless plots of R, U, and $\frac{1}{2}\Delta$ a function of z_1 as given in Fig. 16.14. It can be seen that the buoyancy force goes to zero at a height, z_1, of 2.125 and the momentum goes to zero at a height of 2.8. Morton *et al.* concluded that a plume would spread out horizontally between these two values of z_1, and the experiments they performed appeared to substantiate this conclusion.

The case of a jet with initial momentum has been studied by Rawn *et al.* (1960) and Hart (1961). The jet of fresh water (ρ_o) discharged vertically into receiving water of constant density, ρ_s, can be described by Eqs. 16.26–16.29 until it reaches the bottom of the thermocline. Equations 16.33–16.36 might be used to approximate the jet characteristics in the thermocline. Because of either the air-water boundary or the thermocline the rising jet ultimately must spread. The mixture may reach the surface and spread out. It may never reach the surface, but will spread out in the thermocline because the mixture is less dense than the surface layer. The mixture may be more dense than the surface water but still reach the surface due to its momentum, then plunge beneath the surface to about the top of the thermocline where it spreads out as a submerged field. The three cases are illustrated in Fig. 16.15, where the Froude number is

the velocity of the jet leaving the orifice divided by $\sqrt{gD_o(\rho_s - \rho_o)/\rho_o}$. It can be calculated, using the relationship

$$\rho_h = \rho_s + 0.51\,c_m(\rho_o - \rho_s),$$

(16.39)

where c_m can be estimated from Fig. 16.9 for a value of z equal to the vertical distance from point of issue of the jet to the center of the thermocline. ρ_h is the average mass density through the cross section of the jet and ρ_w is the mass density of the receiving water either above the thermocline or at the surface. It was found that $(\rho_h - \rho_w)/\rho_o$ was the most significant factor in determining the type of sewage field that would develop. Within the limits of the experimental conditions, it was found that, when this parameter was negative, the sewage field spread at the surface (Fig. 16.15A), when the parameter was between 0 and about 10 to 15×10^{-4} the intermediate type of field occurred (Fig. 16.15B), and when the parameter was greater than about 15×10^{-4}, a submerged sewage field occurred (Fig. 16.15C).

Abraham (1962) has developed a step type of procedure for calculating the type field that will occur by considering the thermocline to be a series of layers of fluid of slightly different densities. The results of his calculations show close agreement with Hart's experimental data except in a couple of cases, and these were for experimental runs which were difficult to analyze.

f. Horizontal jet at a free surface. A vertical jet encountering a free surface spreads laterally, with horizontal momentum. How does this deflected jet (or perhaps it should be called a surface current) mix with the surrounding fluid? A study of the mixing of a jet discharged horizontally at the free surface is of help in understanding this phenomenon. Horikawa (1958) made laboratory studies of a single jet discharging horizontally from a nozzle at the free surface between water and air, and a series of jets discharging from nozzles spaced along a supply pipe. The jet and the receiving water were the same density. Figure 16.16 shows the definitions for the horizontal distribution of velocity, where the component of velocity parallel to the jet axis x is u. The results for both a small-scale and large-scale experiment are shown in Fig. 6.31. It is interesting that the data and curve for a submerged jet are nearly the same as the lateral velocity distribution resulting from a jet discharging horizontally at a free surface. This holds true even for the decrease of centerline velocity with distance from a single nozzle, Fig. 16.17, and from a line of nozzles.

In these and other tests (Williams, 1960), a circulation was set up with the current decreasing rather rapidly with distance beneath the water surface, and reversing direction at some distance beneath the surface. It was found that although the lateral surface distribution of the axial component of a single surface jet has the same distribution as a submerged jet, this is not the case for

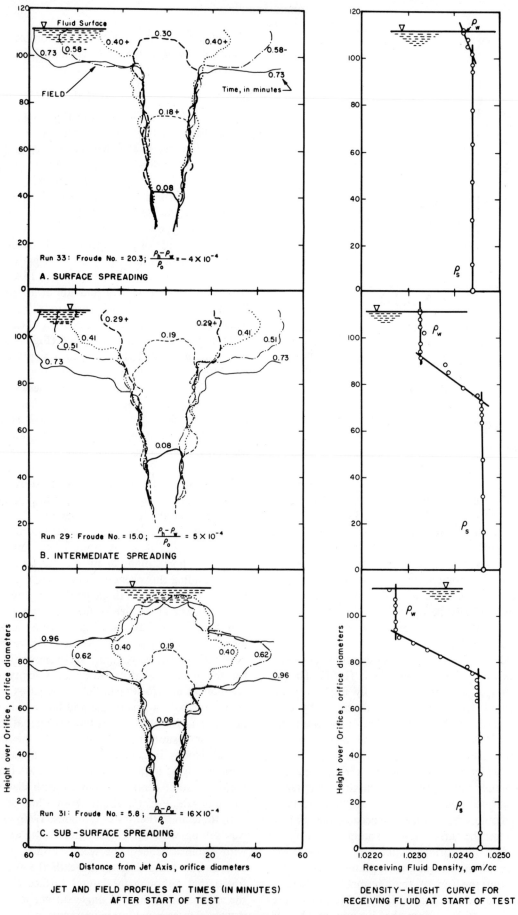

Fig. 16.15. Jet and field profiles and density-height curve (adapted from Hart, 1961)

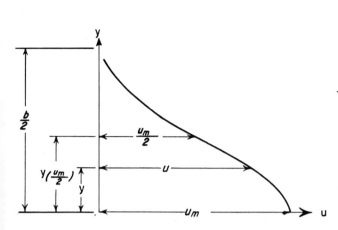

Fig. 16.16. Definition sketch (after Horikawa, 1958)

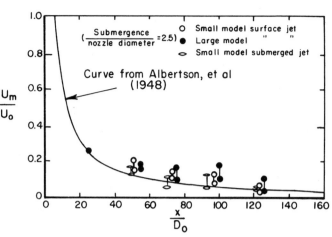

Fig. 16.17. Effect of distance from origin on centerline velocity (after Horikawa, 1958)

the distribution with distance beneath the surface for a series of jets along a line. The distribution appears to be similar to the solution of F. D. Schmidt for $\gamma = 1$ (Fig. 16.10).

3. EDDY DIFFUSION

The classic equation for eddy diffusion is Fick's equation

$$\frac{\partial c}{\partial t} = D \frac{\partial^2 c}{\partial s^2} = D \nabla^2 c, \qquad (16.40)$$

where $\partial c / \partial t$ is the time rate of change of the diffusate of concentration c, s is a direction, and D is the coefficient of eddy diffusivity. This equation may be expressed as

$$\frac{\partial c}{\partial t} = D_x \frac{\partial^2 c}{\partial x^2} + D_y \frac{\partial^2 c}{\partial y^2} + D_z \frac{\partial^2 c}{\partial z^2}, \qquad (16.41)$$

Table 16.1. VALUES AND CONDITIONS OF OBSERVATION OF HORIZONTAL EDDY DIFFUSIVITY
(after Pearson, 1956)

Place of Observation	Reported by	Condition at sea	Sea layer observed	Sea depth, (feet)	Tracer employed	Horizontal eddy diffusivity	
						(cm²/sec)	(ft²/sec)
N. W. North Atlantic Ocean	Neuman-Sverdrup	Strong currents	Surface	—	—	4×10^8	0.0062×10^8
Atlantic Equat. Current	Montgomery-Sverdrup	Moderate currents	0–200m	—	—	4×10^7	0.0062×10^7
South Atlantic Ocean	Sverdrup	Weak currents	2500–4000m	—	—	1×10^8	0.0016×10^8
California Current	Sverdrup	Weak currents	200–400m	—	—	2×10^6	0.0031×10^6
Pacific Ocean	Sverdrup	—	Surface	—	Dyes	5×10^2	0.0078×10^2 (a)
Pacific Ocean	Sverdrup	—	Surface	—	Dyes	5×10^7	0.0078×10^7 (b)
Bikini Lagoon	Munk et al.	Calm-drift 0.1 knots	Surface	52	Radiosiotopes	1.5×10^5	0.0023×10^5
Bikini Lagoon	Munk et al.	Calm-drift 0.03 knots	50m	150	Radiosiotopes	0.5×10^5	0.0076×10^5
Bikini Lagoon	von Arx	—	Surface	52	Dyes	0.7×10^4	0.0011×10^4 (c)
Bikini Lagoon	von Arx	—	Surface	52	Dyes	1.8×10^4	0.0027×10^4 (c)
New York Bight	Ketchum	Moderate wind, Wind force 1–3	Surface	60	Iron waste	2.5×10^3	0.0039×10^3
New York Bight	Ketchum	Moderate wind 1–3	Surface	60	Iron waste	1.9×10^3	0.0029×10^3
New York Bight	Ketchum	Moderate wind 1–2	Surface	26	Iron waste	6.8×10^3	0.0011×10^3
Atlantic Ocean	Bourret and Broida	Wind speed 15 mph	Surface	15	Mimeograph paper	2.7×10^2	0.0042×10^2

a Reported value for small-scale phenomena, dye spot $r \approx 10^3$.
b Reported value for large-scale phenomena, $r \approx 10^8$ cm.
c Reported value based on characteristic radius of dye spot ≈ 100 m.

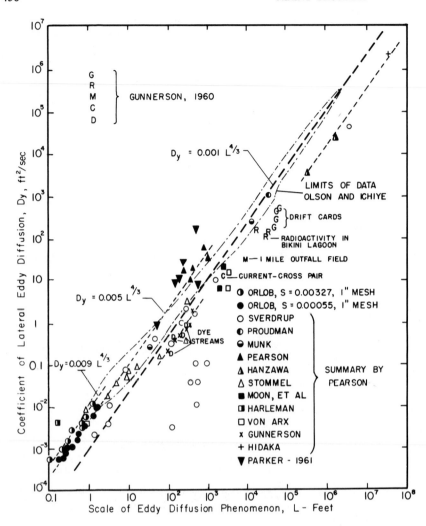

Fig. 16.18. Eddy diffusion as a function of scale; summary of reported investigations (after Orlob, 1959)

where x, y, and z are the three space coordinates. There have been numerous papers on this subject (see, for example, Taylor, 1915; Richardson, 1926; Batchelor and Townsend, 1956). There are several limitations to the use of Eq. 16.40 in the study of turbulent diffusion in the ocean, one of these limitations being the tremendous range of values of D reported by various authors.

Most investigators of eddy diffusion in the ocean have limited their studies to the horizontal component of eddy diffusivity, D_y. Values obtained by several investigators have been assembled by Pearson (1956) and are shown in Table 16.1. It can be seen that the values of D_y range from 5×10^2 to 4×10^8, a spread of almost 10^6. It would seem that the equation just is not satisfactory. In Fig. 16.18, many of the values of D_y are presented in relationship to the size of the phenomenon being considered, with the slopes of the lines being drawn in conformance with the so-called four-thirds law. This "law" states that the coefficient of eddy diffusivity D_y is proportional to the $\frac{4}{3}$ power of the scale of the disturbance (Richardson, 1926).

The reasoning often given for the increase in the coefficient of eddy diffusivity with scale is that, if one

studies the separation of two particles, initially close together, the rate at which they tend to separate increases with increasing distance between them. It has been reasoned that when they are close together they tend to move apart by the action of small-scale turbulence, at rates small compared with the mean flow velocity. As the distance between them increases, each particle may be entrained in separate larger eddies, and move away from the other faster, etc. Batchelor and Townsend (1956) have obtained a four-thirds relationship from theoretical studies, using the two-particle separation approach, but with the restriction that the distance between the particles had to be small compared with the length scale of the turbulence. Because of this, they state that the fact that their theory predicts a four-thirds law and that the measurements of Richardson apparently show a four-thirds relationship is partly fortuitous. This seems to be in agreement with the experimental findings of Orlob (1959). His measurements were made in a laboratory channel which effectively limited the maximum size of eddies. It was found that close to the source the width of the dispersion pattern increased about in proportion to the three-halves power of the distance from the source, and that relatively far from the source the width of the dispersion pattern increased about in proportion to the one-half power of the distance from the source (a parabolic shape), with a transition region between. This indicates that the coefficient of eddy diffusivity is proportional to the 4/3 power of the scale to the source, but becoming a constant at some distance from the source. Laboratory studies made by Masch (1961) of the mixing at the surface by wind-generated waves and currents have shown results similar to those reported by Orlob.

The form used by Richardson for the diffusivity of neighboring particles is (Richardson and Stommel, 1948)

$$\frac{\partial c}{\partial t} = \frac{\partial}{\partial l}\left[F(l)\frac{\partial c}{\partial l}\right], \qquad (16.42)$$

where

$$F(l_o) = \frac{\text{mean of } (l_1 - l_o)^2 \text{ for all pairs}}{2(t_1 - t_o)} \quad (16.43)$$

in which l_o is the initial distance between two particles at time t_o and l_1 is the distance between the same two particles at time t_1. They point out that this is in conflict with a common method of obtaining an eddy coefficient

$$D_y = \frac{\sigma_1^2 - \sigma_o^2}{2(t_1 - t_o)} \qquad (16.44)$$

(see Munk, Ewing, and Revelle, 1949; Gunnerson, 1960) as the standard deviation of two particles $\sigma = l/2$, so that

$$D_y = \frac{l_1^2 - l_o^2}{8(t_1 - t_o)} \qquad (16.45)$$

rather than the $F(l_o)$ given by Eq. 16.43.

Tests made by Richardson and Stommel using floats 2 cm in diameter which were nearly completely immersed in the ocean near shore in water 2 meters deep, using a

series of small initial displacements (0–52 cm) and a series of large initial displacements (117–320 cm). They measured the separations at the end of 30 secs. It was found that $F(l) = 0.07\, l^{1.4}$, which is very close to the four-thirds power relationship. No indication was given of the scale of the turbulence, so that it is not possible to decide whether or not these measurements were made for values of l less than, or greater than, the scale of the turbulence. This is true of all of the measurements made in the ocean. In the ocean, however, the scale of the turbulence can be quite large, as was indicated in Chapter 13, so that for many engineering purposes the mixing probably follows the four-thirds law.

Another difficulty that exists in trying to interpret published values of coefficients of eddy diffusivity is that there are several ways of extracting the coefficient from the measured data, with some of the methods being subject to considerable error. Stommel and Woodcock (1951) used two different methods to compute the coefficient of vertical eddy diffusivity D_z in the ocean and tried to correlate the results with the wind speed as shown in Fig. 16.19. The first method depends upon a knowledge of the heat flux at the sea surface and at some distance beneath the surface. The second method (symbol x in the figure) consisted of using Eq. 16.41 for the vertical direction, with sea water temperature being the property that was measured. They discussed the problem, but were not able to draw any firm conclusions because each method was subject to a different kind of weakness.

4. MIXING BY WIND WAVES

When a fan is turned on in a wind-wave tunnel, and the air is moving at a relatively low speed (15 ft/sec, or so), waves are not generated over the entire area at once.

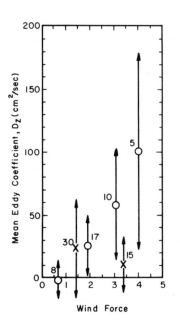

Fig. 16.19. Mean eddy coefficients as functions of wind force computed by two different methods (after Stommel and Woodcock, 1951)

They start at the line of contact between the wind and the water due to separation of the air as it leaves the Venturi section. The wind does drag a very thin (a thousandth of a foot or so) layer of surface water with it. If dye is placed in the water it is dragged along with the surface layer, but without much mixing. Occasionally it separates into a series of parallel streaks. When the waves, first generated at the upwind end of the tank, reach the dye great mixing occurs extremely rapidly. It is evident that the main mixing that occurs in the surface region of water is due to the waves rather than directly to the winds.

In many papers, the eddy coefficient in the surface region of the ocean has been considered to be a function of the wind strength, and computed values have shown considerable scatter (Stommel and Woodcock, 1951), as can be seen in Fig. 16.19. It has been proposed that one reason for the scatter is that eddy diffusivity is related more closely to the wave spectrum generated by the winds than to the wind itself (Wiegel, 1963). Thus not only are the wind strength and variability important, but the fetch and duration are also important. Some data relating the mixing of sea water to the state of the sea have been obtained by Ichiye (1953), but only averages have been published so that the relative amount of scatter of the data is unknown to the author.

Wave theory has been developed almost entirely for irrotational fluid motion. That wind waves are not irrotational in the generating area is readily apparent to an observer. Figure 9.32 is an example of the type of motion that does occur, which shows that a surface current (shear flow) and wind waves are generated simultaneously. Furthermore, it is a highly turbulent shear flow, as can be made evident by introducing dye into the water as previously described. There is some indication that swell is not important to the mixing process (Munk, 1947) as apparently it is nearly irrotational. Seas are important, as Munk's experiment showed. Waves were generated in a tank by an oscillating plunger. Two layers of water of slightly different densities were distinguished by mixing dye with the lower layer. The "swell" did not cause mixing of the two layers. When light winds were blown over the surface (13 ft/sec), no mixing occurred, but when the winds were increased to about 23 ft/sec, breaking (whitecaps) occurred and mixing took place. Similarly, destruction of a thermocline in the laboratory has been accomplished by generating breaking waves by wind, and the mixing was rapid (Wiegel, 1954).

Phillips (1958a) has developed a theory for a steady state energy spectrum for wind-generated waves, this condition coming into existence when the energy transfer from the wind to the water in the form of waves is balanced by the dissipation of mechanical energy by the turbulence generated by breaking waves. Some measurements (Burling, 1955; Phillips, 1958b) have shown that a portion of the energy spectrum of waves is predicted by Phillips' theory (Fig. 9.14). If this can be accepted, then the portion of the energy spectrum of waves that matches the curve due to Phillips should be useful in predicting the amount of mixing that will take place, and the depth of mixing. Then the mixing can be related to the winds, fetches, and duration through an eddy coefficient.

One mechanism by which wind waves can cause mixing was studied by Reynolds (1900) in connection with the possible partial calming of the sea by rain. He found that drops of water allowed to fall on the water surface not only caused a splash, but under certain circumstances formed into a vortex ring which descended into the water with a gradually decreasing speed and increasing size. Thus water is carried downward, and because of continuity, water must move in to replace the water being accumulated by the vortex ring. The numerous drops of water associated with breaking waves might act in this manner.

Several studies have been made of the mixing by wind waves (Johnson, 1960; Johnson and Hwang, 1961; Masch, 1961; 1962). Masch made one study in a wave tank which was wide enough (4 ft) that the wave spectrum was essentially two-dimensional, as is the case in nature. Polyethelene spheres (average diameter 2.8 mm) with a specific gravity of 0.97 were used so that they floated on the surface in order to preclude vertical mixing. Masch states that the wind did not act directly upon the spheres. As has been described in previous section, wind blowing over the water surface creates a surface current as well as waves. Both the current and the waves were measured. Masch found no evidence of the four-thirds law, except near the source. Over most of the range of the horizontal plume D_y was found to be essentially a constant which was found to correlate with the sum of the surface current u_s (approximately 0.027 U, where U was the wind speed) and the water particle orbit speed $\bar{q}_w$ (where $\bar{q}_w = \pi H_{1/3}/T_s$, $H_{1/3}$ being the significant wave height and T_s being the average period of the significant waves). The results, shown in Fig. 16.20, can be expressed as

$$D_y = 0.0038[u_s + \bar{q}_w]^{3.2}. \qquad (16.46)$$

Due to the close association of the significant wave height and the peak of the energy spectrum (Chapter 9), this finding confirms the supposition presented previously. Because of the close tie between the surface current and waves generated by the wind it would be enlightening to perform similar experiments in channels of different slopes in order to be able to vary the current to a greater extent.

In real problems, vertical as well as lateral mixing occurs. This is more difficult to study in the laboratory

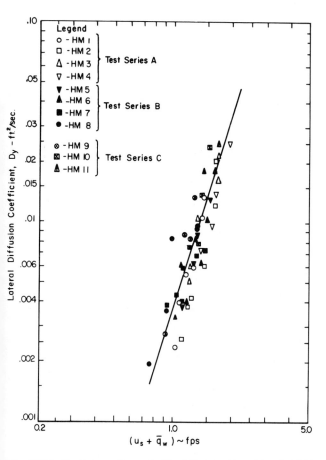

Fig. 16.20. Coefficient of lateral eddy diffusivity (from Masch, 1961)

than is lateral mixing at the surface. Wind drags the surface water to the downwind end of the tank and piles it up, producing a hydraulic head which causes a flow to occur in the opposite direction. Thus, the mixing and transport of fluid away from the source by advection, is not due to the wind waves and surface current alone (Johnson, 1960; Johnson and Hwang, 1961; Masch, 1962).

5. MIXING BY CURRENTS

Mixing by currents has been covered essentially in Section 3 of this chapter. The mixing of a concentrate from a line source will be considered here. Sewage discharged from a series of ports near the end of an ocean outfall sewer is mixed first as the jet rises towards the surface. As has been pointed out previously, the mixture may or may not reach the surface. The case where the mixture reaches the surface will be considered here. Usually there will be a current, which may consist of the combination of general ocean currents, tidal currents, wind drift, and wave mass transport.

The problem of the further mixing of the sewage–sea water mixture by currents has been treated by Brooks

(1960). The analysis is restricted to a current flowing continuously in one direction, and it is not clear how the effect of rotary tidal currents can be handled in addition, although these are always present in the ocean. Brooks considers a line source of length b and a concentration C_o of coloforms in the sewage mixture entering a current of mean advective velocity U moving in the direction x. The line source is in the horizontal direction y normal to x. The width of the sewage at any distance x from the source is L, where L is defined

$$L = 2\sqrt{3}\,\sigma. \qquad (16.47)$$

This definition of L with respect to the standard deviation of the concentrate distribution in the y direction, σ, is used so that L will equal b at the origin. The coloforms will die with time in the hostile environment, with k being the die-off constant. The concentration of live coloforms at any point, C, is considered to be given by

$$C = C_n\,e^{-kx/U}, \qquad (16.48)$$

where C_n is the concentration of total of live and dead coloforms at that point. The general equation is

$$\frac{\partial}{\partial y}\left(-D_y\frac{\partial C}{\partial y}\right) + U\frac{\partial C}{\partial x} + kC = 0. \qquad (16.49)$$

Solutions to three cases were obtained. These were for $D_y = $ constant, D_y increasing linearly with L/b, and D_y increasing with the four-thirds power of L/b. The field width L is given by

$$\frac{L}{b} = \left(1 + 2\beta\frac{x}{b}\right)^{1/2}, \qquad \text{for } D_y = \text{const} \qquad (16.50)$$

$$\frac{L}{b} = 1 + \beta\frac{x}{b}, \qquad \text{for } D_y = D_{y_o}\frac{L}{b} \qquad (16.51)$$

$$\frac{L}{b} = \left(1 + \frac{2}{3}\beta\frac{x}{b}\right)^{3/2}, \quad \text{for } D_y = D_{y_o}\left(\frac{L}{b}\right)^{4/3} \qquad (16.52)$$

where

$$\beta = \frac{12 D_{y_o}}{Ub} \qquad (16.53)$$

and D_{y_o} is the value of D_y at $x = 0$.

The concentrations of live coloforms along the center-line of the sewage field $y = 0$ are

$$C_{y=0} = C_o\,e^{-kx/U}\operatorname{erf}\left\{\sqrt{\frac{3}{4\beta x/b}}\right\}, \qquad \text{for } D_y = \text{const} \qquad (16.54)$$

$$C_{y=0} = C_o\,e^{-kx/U}\operatorname{erf}\left\{\sqrt{\frac{3/2}{[1 + \beta(x/b)]^2 - 1}}\right\},$$
$$\text{for } D_y = D_{y_o}\frac{L}{b} \qquad (16.55)$$

$$C_{y=0} = C_o\,e^{-kx/U}\operatorname{erf}\left\{\sqrt{\frac{3/2}{[1 + \frac{2}{3}\beta(x/b)]^3 - 1}}\right\},$$
$$\text{for } D_y = D_{y_o}\left(\frac{L}{b}\right)^{4/3} \qquad (16.56)$$

where erf is the standard error function. The travel time t is x/U.

REFERENCES

Abraham, G., Discussion of jet discharge into a fluid with a density gradient, *J. Hyd. Div., Proc. ASCE*, **88**, HYD2 (March, 1962), 195–98.

————, Jet diffusion in liquid of greater density, *J. Hyd. Div., Proc. ASCE*, **86**, HY6 (June, 1960), 1–13.

Albertson, M. L., Y. B. Dai, R. A. Jensen, and Hunter Rouse, Diffusion of submerged jets, *Proc. ASCE*, **74**, 10, (December, 1948), 1157–96.

Batchelor, G. K., and A. A. Townsend, *Turbulent diffusion, Surveys in Mechanics*. London: Cambridge Univ. Press, 1956, pp. 352–99.

Brooks, Norman H., Diffusion of sewage effluent in an ocean current, *First Inter. Conf. on Waste Disposal in the Marine Environment*. London and New York: Pergamon Press, 1960, pp. 246–67.

Burling, R. W., Wind generation of waves on water, Ph.D. dissertation, Imperial College, Univ. London, 1955.

Corrsin, S., *Investigation of flow in an axially symmetrical heated jet of air*, NACA Wartime Report ACR No. 3L23, W–94, December, 1943.

————, and M. S. Uberoi, *Further experiments on the flow and heat transfer in a heated turbulent air jet*, NACA Tech. Note N 1865, April, 1949.

Folsom, R. G., and C. K. Ferguson, Jet mixing of two liquids, *Trans. ASME*, **71** (January, 1949), 73–77.

Forstall, W., and E. W. Gaylord, Momentum and mass transfer in a submerged water jet, *J. Applied Mech.*, **22**, 2 (June, 1955), 161–64.

Frankel, Richard J., and James D. Cumming, *Turbulent mixing phenomena of ocean outfalls*, Univ. Calif., IER Tech. Rept. No. HEL–3–1, February, 1963. (Unpublished.)

Gunnerson, Charles G., Discussion of eddy diffusion in homogeneous turbulence, *J. Hyd. Div., Proc. ASCE*, **86**, HY4 (April, 1960), 101–109.

Hanzawa, M., On the eddy diffusion of pumices ejected from Myojin Reef in the Southern Sea of Japan, *Rec. Oceanog. Works in Japan*, new series, **1**, 1 (March, 1953), 18–22.

Hart, W. E., Jet discharge into a fluid with a density gradient, *J. Hyd. Div., Proc. ASCE*, **87**, HY6 (November, 1961), 171–200.

Hinze, J. O., and B. G. van der Hegge Zijnen, Heat and mass transfer in the turbulent mixing zone of an axially symmetrical jet, *Proc. Seventh Intern. Congr. Applied Mechanics*, **2**, Part I (1948), 286–99.

Horikawa, K., *Three-dimensional model studies of hydraulic breakwaters*, Univ. Calif. IER., Tech. Rept. 104–8, October, 1958. (Unpublished.)

Ichiye, T., On the effect of waves on the vertical distribution of water temperatures, *Rec. Oceanog. Works in Japan*, new ser., **1**, 1 (March, 1953), 63–70.

Johnson, J. W., The effect of wind and wave action on the

mixing and dispersion of wastes, *Proc. First Intern. Conf. Waste Disposal in the Marine Environment*. London and New York: Pergamon Press, 1960, pp. 328–43.

————, and H. C. Hwang, *Mixing and dispersion by wind waves*, Univ. Calif. IER Tech. Rept. 138–5, January, 1961.

Keagy, W. R., A. E. Weller, F. A. Reed, and W. T. Reid, Batelle Memorial Inst., The RAND Corp., Santa Monica, Calif., February, 1949.

Landau, L. D., and E. M. Lifshitz, *Fluid mechanics*. J. B. Sykes and W. H. Reid, trans. London and New York: Pergamon Press, 1959.

Masch, Frank D., *Mixing and dispersive action of wind waves*, Univ. Calif. IER, Tech Rept. 138–6, November, 1961.

————, *Observations on vertical mixing in a closed wind wave system*, Univ. Calif. IER Tech. Rept. 138–7, January, 1962.

Morton, B. R., Sir Geoffrey Taylor, and J. S. Turner, Turbulent gravitational convection from maintained and instantaneous sources, *Proc. Roy. Soc.* (London), ser. A, **234**, 1196, (January 24, 1956), 1–23.

Munk, Walter H., A critical wind speed for air-sea boundary processes, *J. Mar. Res.*, **6**, 3 (1947), 203–18.

————, G. C. Ewing and R. R. Revelle, Diffusion in Bikini Lagoon, *Trans. Amer. Geophys. Union*, **30**, 1 (February, 1949), 59–66.

Orlob, G. T., Eddy diffusion in homogeneous turbulence, *J. Hyd. Div., Proc. ASCE*, **85**, HY9 (September, 1959), 75–101.

Parker, F. L., Eddy diffusion in reservoirs, *J. Hyd. Div., Proc. ASCE*, **87**, HY3 (May, 1961), 151–71.

Pearson, E. A., *An investigation of the efficacy of submarine outfall disposal of sewage and sludge*, California, State Water Pollution Control Board, Pub. No. 14, 1956.

Phillips, O. M., On the equilibrium range in the spectrum of wind-generated waves, *J. Fluid Mech.*, **4**, Part 4 (August, 1958a), 426–34.

————, On some properties of the spectrum of wind-generated ocean waves, *J. Mar. Res.*, **16**, 3 (October, 1958b), 231–40.

Rawn, A. M., F. R. Bowerman, and Norman H. Brooks, Diffusers for disposal of sewage in sea water, *J. Sanitary Eng. Div., Proc. ASCE*, **86**, SA 2, Paper No. 2424 (March, 1960), 65–105.

Reynolds, Osborne, On the action of rain to calm the sea, in *Papers on Mechanical and Physical Subjects*. London: Cambridge Univ. Press, 1900, pp. 86–88.

Richardson, Lewis F., Atmospheric diffusion shown on a distance-neighbour graph, *Proc. Roy. Soc.* (London), ser. A, **110** (1926), 709–37.

————, and Henry Stommel, Note on eddy diffusion in the sea, *J. Meteorology*, **5**, 5 (October, 1948), 238–40.

Rouse, Hunter, C. S. Yih, and H. W. Humphreys, Gravitational convection from a boundary source, *Tellus*, **4**, 3 (August, 1952), 201–10.

Schmidt, F. H., On the diffusion of heated jets, *Tellus*, **9**, 3 (August, 1957), 378–83.

Schmidt, W., Turbulente Ausbreitung eines Stromes erhitzter Luft, *Z.A.A.M.*, **21** (1941), 265; 351.

Stommel, Henry, and Alfred H. Woodcock, Diurnal heating of the surface of the Gulf of Mexico in the spring of 1942, *Trans. Amer. Geophys. Union*, **32**, 4 (August, 1951), 565–71.

Taylor, G. I., Eddy motion in the atmosphere, *Phil. Trans., Roy. Soc.* (London), ser. A, **215** (1915), 1–26.

Tollmien, W., Momentum transfer theory for a jet, translation, NACA TM 1085, 1945. (Originally *Z.A.M.M.*, **6**, 1926, 468–78.)

Wiegel, R. L., *Final report: wave instrumentation*, Univ. Calif. IER, Tech. Rept. 3–372, June, 1954. (Unpublished.)

————, Some engineering aspects of wave spectra, *Ocean Conf. Wave Spectra: Proceedings of a Conference*, Englewood Cliffs, N.J.: Prentice-Hall, Inc., 1963, 309–21.

Williams, John A., *Verification of the Froude modeling law for hydraulic breakwater*, Univ. Calif. IER, Tech. Rept. 104–11, August, 1960. (Unpublished.)

CHAPTER SEVENTEEN

Functional Design

1. INTRODUCTION

Functional design is concerned with those ideas and details that are necessary to insure that a structure, equipment, or process does the job for which it was conceived. In regard to marine structures, this is often more difficult than the structural design. The complex oceanographic processes and their interrelationships with man-made structures are only partially understood. Furthermore, these processes are in a state of dynamic equilibrium and sometimes the building of a structure creates a problem more serious than the problem it was intended to solve. A documented example of this is presented in Chapter 18.

This chapter gives some details on operations and limitations of existing structures as an aid to understanding a few of the considerations involved in functional design. In addition, some items concerned with the design of structures that did not seem to fit into other chapters are presented here. No detailed functional design will be given, however, as each job presents its own problems.

The primary factors one has to deal with in the design of marine structures are those associated with winds, waves, tides, and currents. In addition, biological activities, such as fouling and boring organisms, must be considered for certain structures as must certain chemical (corrosion), electrolytic, and thermal (ice, expansion) phenomena.

Wind is often of direct importance from a structural rather than a functional design standpoint. It is often important indirectly through the waves, surface currents, and storm tides which it generates. Wind blowing over the sea surface drags surface water along with it. At a boundary such as a coast this water piles up, forming a hydraulic head, which supports a flow of water seaward. The more shallow the water, the higher the hydraulic head (the "wind tide") that must be built up to support the necessary return flow. In addition to the rise in the water surface due to wind stress there is often a rise due to a lower atmospheric pressure at the storm center which is accentuated by dynamical effects.

The effect of the offshore water depth can be seen by comparing the effect of a storm of hurricane force (say, 100 miles per hour) acting over a distance of approximately 50 miles in a depth of 100 ft. If the depth were constant, the wind tide would be about 6 ft; if the bottom slope were only 5 ft per mile, however, the wind tide would be only about 3 ft. This would be rather typical in the Gulf of Mexico region. Offshore of Southern California the maximum wind tide would be more of the order of $\frac{1}{2}$ ft.

Examples of extreme storm tides can be given: (1) Galveston, September 8, 1900, 10 ft above astronomical mean high water, (2) Hook of Holland, February 1, 1953, 10 ft above astronomical high water.

The effect of waves may be direct or indirect. As an example of a direct effect in the determination of the height of a breakwater, it is necessary to predict the most probable height of waves to which the structure might be subject during the design life. At the same time, the effect of the breakwater upon the wave-induced littoral drift of sand along the coast must be considered. The engineer is in an unfortunate position insofar as wave data are concerned. Although astronomical tide data are available, with acceptable accuracy, for nearly every coastal location there seldom are reliable wave or wind tide data.

The effect of tides is much more important in some areas than in others; for example, the tides are seldom over a foot or two in a harbor such as Apra, Guam, M.I., and range from 6 feet at San Francisco, to 15 feet at Ketchikan, Alaska, and may be as great as 18 feet at Dover, England, and more than 40 feet in some parts of the Bay of Fundy. It is evident that the design of quayside mooring systems would be more difficult for a port which had a great tidal range than for a port which had a small range; on the other hand, owing to the strong currents usually associated with a large tidal range, a sewer outfall might be more effective in such an area.

There are several main types of currents in the ocean: the general oceanic currents, the tidal currents, local wind-induced surface currents, wave-induced mass transport, and littoral currents. These currents are important for many reasons. They refract the waves; they are responsible for some of the mixing of fresh water discharged into the ocean by rivers, and of sewage and industrial wastes deposited into the ocean; they transport the mixed waters, sediments, and flotsam; they cause scour and deposits; and because of the hydraulic forces due to currents, they are important in the mooring of ships and barges.

Anyone who has been at sea, or along the seacoast, is aware of the problems presented by the growth of marine organisms, and the corrosion of metals. Some of the effects have been described in Chapter 1, together with a few possible solutions.

A seafaring man once said that people get tired and materials get tired but the sea never gets tired. This statement should always be kept in mind when designing any structure for use in the ocean.

2. SOME MARINE STRUCTURES AND THEIR FUNCTIONS

Some marine structures together with some of their functions are as follows:

a. *Floating structures.* Floating structures are vessels, buoys, nets, floating dredge pipelines, and floating platforms; submerged tanks are included in this category. The function depends upon the structure; the primary function of the system, however, is keeping a structure at a particular location, within certain predetermined limits. This fixing of location must be accompanied by its ability to perform its task; for example, a buoy must be seen, or its bell heard.

b. *Pile-supported structures*
(1) *Offshore platforms:* Offshore platforms are used as a base for coring and well-drilling operations by oil-producing companies, the installation of ocean outfall sewers, as a base for radar equipment, and as light stations. Their decks must be high enough above the water so that even under the worst oceanographic and meteorological conditions the deck is safe from a standpoint of personnel and equipment. Drilling platforms must remain fairly static so that the drill pipe does not bend excessively during drilling. They must have facilities for transferring equipment to and from serivce vessels.
(2) *Piers:* A pier is a structure which extends into the water from shore to serve as a landing place or a recreational facility rather than to afford coastal protection. They must often extend through the surf zone. It should normally be high enough to be above the biggest wave that might occur during times of maximum water heights.

c. *Offshore islands.* The functions of an offshore island are the same as for offshore platforms; in addition, they may be for recreational purposes. They usually should have a lee area which can be used for small supply ships.

d. *Breakwaters.* The purpose of a breakwater is to cause reduction of wave heights in its lee. Breakwaters do not have to be of either the rubble mound or masonry type. They may be submerged or moored "vessels," either rigid or flexible, compressed air or water jets.

e. *Seawalls.* The primary functions of seawalls are to maintain a fixed boundary between land and sea, and to prevent serious overtopping by water.

f. *Jetties*. The functions of jetties are to confine and direct river or tidal flow at the entrance to an estuary, with the main purpose being the scouring of a channel to maintain a project depth. In addition, jetties often act functionally as breakwaters. Depths must be great enough to allow vessels to pass and also deep enough so that swell does not peak up and become even more dangerous.

g. *Groins*. The function of a groin is to prevent or retard the erosion of an existing beach, to widen an existing beach, or provide a beach where none exists.

h. *Underwater pipe lines and cables*. The function of underwater pipelines is to transmit fluids, pulverized solids, or solids in suspension in fluids, and the function of underwater cables is to transmit messages or electric power.

i. *Sewer outfalls*. Sewer outfalls serve as the point of entry to the ocean of effluent from pipe lines. They should, insofar as possible, cause the effluent to enter the ocean in a manner that will lead to the most effective dispersion of the wastes. It must be dispersed to such an extent that it presents neither a health nor an aesthetic problem.

j. *Miscellaneous*. In addition to the structures mentioned previously there are many others: marine slipways, dry docks, ferry slips, quays, moles, etc.

In regard to quays, moles, and other structures which may have vessels moored alongside, one of the primary functions is that they be so located that seiching or surging is not a serious problem.

3. FLOATING STRUCTURES

Floating structures are vessels, buoys, nets, floating dredge pipelines, floating platforms, etc. Submerged tanks will also be included in this category, although the problems connected with these tanks are somewhat different from the problems associated with structures floating at the free surface. The functional design of ships is not included here, as there is a vast literature on this category. Regardless of the end product, the type of structure considered in this section must be able to maintain a position, within certain predetermined limits, for a specified length of time. To do this, the structure may be moored or it may be maneuvered under power (National Academy of Sciences, 1961). The functions may differ widely from one another. For example, a marker buoy is moored and normally must maintain its position under all conditions of waves, currents, and winds, and it must either be seen, or its bell heard (U.S. Coast Guard, 1957; 1959). On the other hand, a dump barge may have to maneuver over a particular location in order to dump rock in the construction of a jetty, but it is not expected to work under any but the mildest of sea conditions. (Details of moored structures

will not be covered here as these are presented in Chapter 19.)

In operating equipment along seacoasts, many factors must be considered. The following list is a modification of Santema's (1955):

a. Type of work

b. Equipment available or possible, including dynamic characteristics, such as period of pitch, etc., of a vessel

c. Current direction, speed, and variability

d. Wind direction, speed, variability

e. Wave length, height, and direction

f. Orientation of vessel with respect to wind, waves and currents; types of moorings and characteristics of moored or freely floating vessel

g. Fog, mist, temperature, and ice

h. Relative position of operation site with respect to hazards such as shoals, reefs, beaches, piers, etc.

i. Distance to port or other safe haven

j. Ship traffic intensity

k. Duration of intervals of adverse conditions

l. Human factors

There is very little information available relating the physical factors with the ability of floating structures to perform their functions. One reason for this is that often human factors predominate. With the same physical conditions and equipment, one crew will work while another crew will claim that they cannot work. Santema (1955) cites a case where the number of operating days for a dredging operation rose suddenly from the day the dredger of a competing firm started to work in the vicinity, the prestige of the crew coming into play. Likely limiting wind wave heights (probably the significant wave height) for floating equipment used in the Netherlands in dredging and dumping operations are given in Table 17.1.

Table 17.1 WAVE HEIGHTS LIMITING OPERATIONS
OF DREDGES AND BARGES
(after Santema, 1955)

Equipment and kind of work	Limiting wave height (feet)
1. Dredging with	
a. Seaworthy suction hopper dredge, with rigid suction tube and cutters	2–3
b. Seaworthy suction hopper dredge, with flexible suction tube	$4\frac{1}{2}$–6
c. Suction dredges of the nonpropelled, low pontoon type, rigid suction tube	$1\frac{1}{2}$–3
d. Bucket dredge nonpropelled, low pontoon type, hard bottom	$1\frac{1}{2}$
e. Seaworthy tin dredges	$4\frac{1}{2}$–6
2. Mooring barges alongside a dredge with barge discharge, or alongside a barge-unloading dredge	$1\frac{1}{2}$–$2\frac{1}{4}$
3. Dumping stones, sand, or clay with dump barge with bottom doors (up to 400 tons)	$1\frac{1}{2}$–3
4. Dumping stones and clay with self-tipping barges (up to 600 tons)	$1\frac{1}{2}$–$2\frac{1}{4}$
5. Transport and sinking fascine mattresses	1–$1\frac{1}{2}$
6. Pumping stones in layers on fascine mattresses from barges	$1\frac{1}{4}$–$2\frac{1}{2}$

Experience obtained in laying a $12\frac{3}{4}$ inch pipeline in the Gulf of Mexico has shown that barges can work only in waves up to 4 ft in height (Aldridge, 1956). Glenn (1950) has given a list of various types of floating structures and equipment, together with the wave height that limits their operation. No periods are given, but as the statistics are for off the Louisiana coast of the Gulf of Mexico where offshore oil operations are undertaken, they are probably for wind waves with periods in the range of 4–8 secs. (Glenn's values are given in Table 17.2.)

Dredges are one of the most important types of floating structures. Dredging channels through offshore bars, deepening tidal inlets, and by-passing sand along open coasts are more difficult than operating in relatively well-protected inland waters. There is greater exposure to wave action currents (with the tidal currents often being swift and reversing), fog, ice, etc. These factors increase the hazard of loss or damage to the working plant, increase the insurance cost, and decrease the number of working days per year.

There are three general classes of dredges—hydraulic pipeline, bucket or dipper, and sea-going hopper.

The hydraulic pipeline dredge usually has a scow type of hull (square end, flat bottom) which contains a large centrifugal pump with the necessary power plant. This type of dredge seldom has its own propelling equipment and must be towed into position. A hinged "ladder" is mounted at one end of the hull which carries the cutter shaft and intake pipe, and the discharge pipeline goes off the stern end of the hull, with the pipe being kept afloat by a series of pontoons. A spud is mounted at each side of the stern, and one is pushed to the bottom to act as a pivot. The bow is swung around this pivot, using a cable leading from a drum and power winch on the dredge to an anchor. As it pivots, the cutting head stirs up the bottom, mixing the sediment with water, with the pump removing it. After a complete swing, the other spud is pushed into the bottom and the first spud lifted, this resulting in a steplike forward progress of the dredge. The spuds are subject to damage when lifted out of the mud when high waves exist, and the integrity of the floating pipeline is sensitive to waves and currents.

The bucket, or dipper, dredge also has a scow type of hull. It has a hinged boom with a controlled bucket, or

Table 17.2. GENERALIZED PERFORMANCE DATA FOR MARINE OPERATIONS
(after Glenn, 1950)

Type of operation	Wave heights[a] (feet) for		
	Safe, efficient operation	Marginal operation	Dangerous and/or inefficient operation
Deep sea tug			
Handling oil and water barge	0–2	2–4	>4
Towing oil and water barge	0–4	4–6	>6
Handling derrick barge	0–2	2–3	>3
Handling and towing LST-type vessel	0–3	3–5	>5
Crew boats, 60–90 ft in length			
Underway	0–8	8–15	>15
Loading or unloading crews at platform	0–3	3–5	>5
Supervisor's boats, fast craft, 30–50 ft in length			
Underway at cruising speed	0–2	2–4	>4
Loading or unloading personnel at platform or floating equipment	0–2	2–4	>4
LCT-type vessel and cargo luggers			
Underway	0–4	4–5	>5
Loading or unloading at platform	0–3	3–4	>4
Loading or unloading at floating equipment	0–4	4–5	>5
Buoy laying (using small derrick barge)	0–2	2–3	>3
Platform building			
Using ship-mounted derrick	0–4	4–6	>6
Using large derrick barge	0–3	3–5	>5
Pipeline construction	0–3	3–4	>4
Gravity-meter exploration using surface vessel (limiting conditions caused by instrument becoming noisy)	0–4	4–6	>6
Seismograph exploration using craft under 100 ft in length	0–6	6–8	>8
Large amphibious aircraft (PBY)			
Sea landings and take-offs	0–1.5	1.5–3	>3
Boat-to-plane transfer operations in water	0–1	1–2	>2
Small amphibious aircraft	0–1	1–2	>2

[a] Wave heights used are those of the average maximum waves. Height limits given are not rigid and will vary to some extent with locality, local wind conditions, experience of personnel, etc.

dipper, at its end. The material removed from the bottom is usually deposited by the bucket in a barge which stands alongside the dredge. One of the main criteria from the operational standpoint is the ability of the barge to maintain its position relative to the dredge under adverse conditions.

The seagoing hopper dredge has a ship's hull, and is self-propelled. Some have one suction pipe in a well amidship; others have one suction pipe on each side of the ship. The pumps discharge into hoppers within the ship, the sediment settles out, and the water passes overboard through scuppers. When a full load has been obtained, the ship proceeds to the dumping area, and the load is dumped by opening gates in the bottom of the hoppers. Because of its greater draft, it must operate in deeper water than the other two types.

Blackman (1951) lists the following criteria for the operations of hydraulic pipeline and bucket dredges:

a. If possible, the work should be scheduled for the season of favorable weather, waves, and current conditions.

b. Dredges should have as high a freeboard as possible to increase the working range of wave conditions. For the same reason they should be large.

c. A high dredging capacity is necessary to minimize the working time that the dredge will be exposed to possible hazardous conditions.

d. It should operate close to a safe haven to which it may be moved when conditions make it inoperable.

e. Adequate towboat service must be available at all times.

f. It should dredge from shoreward seaward, to have a deep channel between it and shore, if dredging a harbor entrance. In addition, if a jetty is being constructed as a part of a harbor entrance improvement job, the jetties should be built first.

g. It should be used to open the minimum cut necessary to permit a hopper dredge to finish the job.

In choosing the type of dredge to use, consideration must be given to secondary problems, such as the possible use of the dredged material. For example, it might be cheaper to maintain a channel using a sea-going hopper dredge, but the draft of such a dredge might be so great that the material is dumped at sea rather than in a location that could use the fill because this was not a part of the original project.

4. OFFSHORE PLATFORMS

Most offshore platforms that have been built were for oil drilling, and numerous papers have been written on the subject (see, for example, Howe and Collipp, 1957; Rechtin, Steele, and Scales, 1957). A glance at a world map of potential offshore oil production areas emphasizes the tremendous future of such structures. A plat-

form has been built in 50 ft of water 7 miles at sea off Louisiana to mine sulphur (McNamara, 1960). A platform was built about 1800 ft offshore from Pakning, Indonesia, in 60 ft of water. It was designed to accommodate simultaneously two 50,000 ton deadweight tankers (Silveston, 1959). Other structures either have been built or are being planned for handling ship cargoes. Radar platforms have been constructed (U.S. Congress, 1961a); 1961b). Similar platforms have been considered to replace some lightships and some have already been constructed (U.S. Coast Guard, 1957; *Life* 1962). They are also being used for oceanographic purposes (LaFond, 1959). The types of mobile platforms alone are numerous, and some of these are shown in Fig. 17.1. The various types present different functional problems.

These structures must be high enough to be above the combination of the highest waves on top of the highest astronomical and meteorological tides. The astronomical tides are well known for most regions, but the meteorological tides (also called *storm surges*) are

Fig. 17.1. Classification of mobile units (after Howe and Collipp, 1957)

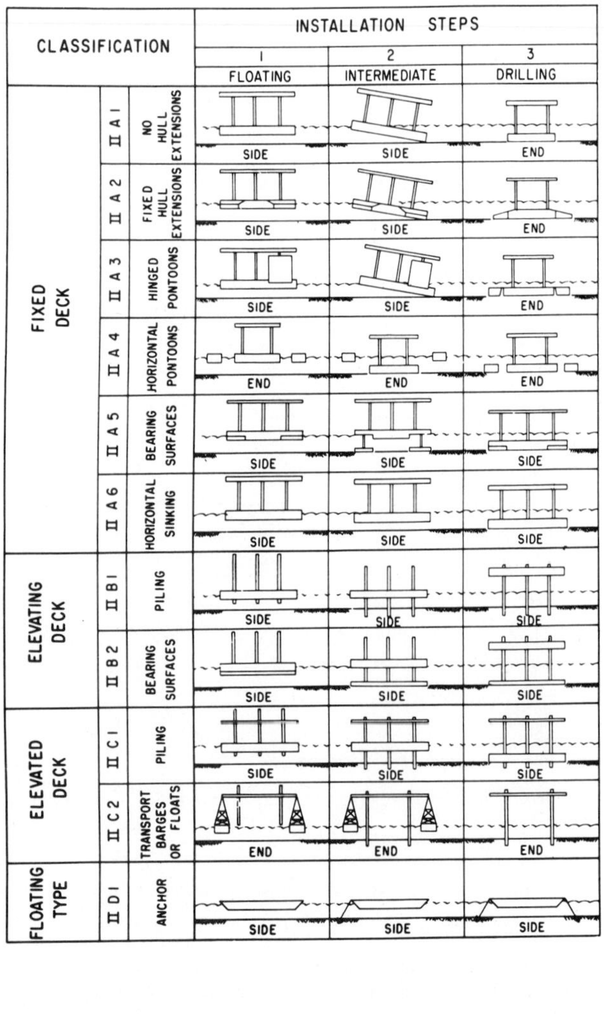

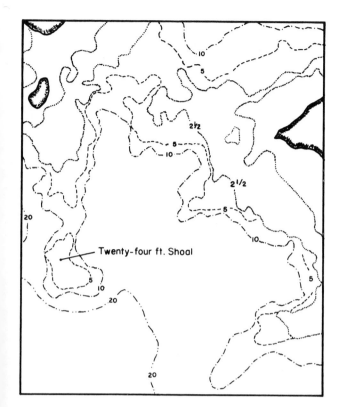

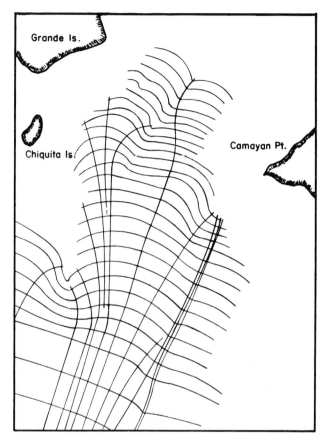

Fig. 17.2. Wave refraction over shoal

not as well known. They consist largely of two parts, the pile-up of water along a coast due to the tangential drag of the wind on the water surface and the increase in water depth due to the normal pressure forces of a hurricane, or other disturbance, moving over the water surface. These phenomena have been described in detail in Chapters 5 and 12. The highest waves can be estimated by statistical means as described in Chapter 9.

The refraction of waves can cause a considerable increase in wave heights in some locations. The bottom contours may be such that the wave train splits as shown in Fig. 17.2. On some occasions, two sections of the original wave train may be intersecting in a manner that results in an approximation to a standing wave with an amplitude equal to the sum of the amplitudes of the two wave sections. The resulting wave may even be so steep that the instability of the type described by Taylor occurs, and a mass of water is projected upwards. An engineer must be careful in selecting a site for a platform on a submarine ridge, as such a location is often a region in which such a phenomenon occurs. An example of such a condition is shown in Fig. 17.2.

Platforms placed in areas which apparently have stable bottoms have in some instances caused scouring to occur. In the Gulf of Mexico, scouring started sometime after five platforms had been placed. The maximum observed depth of the scour amounted to 10 feet (Posey

and Sybert, 1961). Layers of heavy pervious material were dumped at the sites to prevent further scouring, these layers being designed in accordance with the criteria proposed by Terzaghi (1941; 1948) for other types of hydraulic structures. This prevented further scouring. In another case, the platform had a large pontoon bearing on the bottom and a channel was formed under the pontoon. Graded rock dumped on the bottom can prevent the scouring. Mats of new flexible plastic filters, weighted with rocks, are also useful. Techniques also have been developed for pouring asphalt on the bottom which prevents scouring.

5. PIERS

Sometimes piles supporting a pier cause a change in the beach configuration. A pile structure can cause the beach to build seaward in the shape of a tombolo as is the case at Huntington Beach, California.

Suppose a pier has a cut-off wall at its beach end. What will be the effect of a wave breaking against the cut-off wall? Under certain conditions, a jet of water will shoot vertically up against the underside of the pier deck and past its sides. Measurements at Dieppe (DeRouville, *et al.*, 1938) showed these jets to have velocities as great as 250 ft/sec. This mixture of air and water, in addition to doing structural damage, may cause water damage and corrosion to supplies and equipment on the pier deck, which would be a functional failure.

6. ISLANDS

There have been few islands constructed. One of these has been fully described in an article by Blume and Keith (1959). In determining the elevation of the island, the same items must be considered as for an offshore platform. The problem is complicated by the run-up of waves on the sides of the island, unless the sides consist of vertical caissons or sheet piling. Methods for calculating the run-up are given in the section on seawalls.

Islands must be supplied. Unless they are connected to shore by a causeway, as is the case of Rincon Island (Blume and Keith, 1959), they need an adequate lee area if possible. The slope of a man-made island will usually be relatively steep so that diffraction will control the distribution of wave heights around the island rather than refraction effects. If the island is circular in shape, an approximation of heights can be obtained from Fig. 11.28 and Table 11.3. The actual heights will be lower owing to the wave energy dissipation characteristics of rubble mound revetments.

If the island is angular, consideration must be given to the Mach-stem effect for certain ranges of angles of approach of the incident waves (see Chapter 3).

Model studies would be of considerable value in the design of an island as was found to be the case for Rincon Island.

7. BREAKWATERS

There are many types of breakwaters. (For details on the various types see the papers and reports listed in the bibliography by Haferkorn, 1932.) They may be connected to shore, or detached. They all have one thing in common: they have a lee area in which the waves are lower than the incident waves. Breakwaters may cause waves to break, they may just reflect waves, or they may do either, depending upon the tide stage and the wave dimensions. The relative amount of wave energy dissipated and reflected by breakwaters depends upon their slope and porosity, the depth of water at the toe of the breakwater, the wave height, and the wave steepness. Some of the data for an impervious plane slope are given in Fig. 2.33. It is usually preferable for a breakwater to cause the dissipation of as much wave energy as possible to prevent possible dangerous wave heights off the breakwater, but this is not always economically feasible. Each breakwater must be considered on its own from this standpoint (California Institute of Technology, 1952).

Some breakwaters can be designed to take overtopping by waves, although it is probably never desirable.

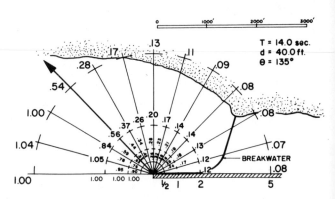

Fig. 17.3. Diffraction diagram overlay on a breakwater plan

In other cases, where the lee side of the breakwater is a quay, it is imperative to prevent overtopping.

As breakwaters must provide an area of relative calm, diffraction of waves plays a major part in the functional design of a breakwater. An example of the use of a diffraction plot to determine the effectiveness of a breakwater at a given location is shown in Fig. 17.3.

Breakwaters are usually constructed in conjunction with a harbor, and, as such, the problem of harbor surging must be considered at the same time. The entrance width between two arms of a breakwater, or the distance between a single breakwater and the coast is of great importance with respect to harbor surging. Details of harbor oscillations and the effect of entrance width are given in Chapter 5.

Littoral drift and breakwaters are inseparable. It is often necessary to prevent erosion on the downdrift side of a breakwater; it is sometimes necessary to prevent too great an accumulation of sand on the updrift side; and it is almost always necessary to prevent the sanding up of the harbor entrance. A detailed example of the construction of breakwater and its effect on the sand in the region is given in Chapter 18. Johnson (1957) has made an extensive study of numerous breakwaters and their effect on littoral drift. Offshore breakwaters have been built in the hope that they might not interfere with littoral drift. Because of diffraction and refraction of the waves, however, a tombolo forms in the manner shown in Fig. 14.35.

During the construction of breakwaters, the movement of sand can present a serious problem. When the breakwater at Zeebrugge, Belgium, was being constructed, the severe erosion which took place on the bottom in the immediate vicinity of the construction work caused the plans to be modified (Verschoore, 1935). About 450,000 cubic meters of rock were dumped on the bottom to stop the erosion, and the breakwater was founded on this mattress. In other circumstances, deposition might occur instead of erosion; an example of this sort is described in the section on jetties.

8. SEAWALLS

Waves acting against breakwaters and seawalls can break seaward of the structure, against the structure, or be reflected unbroken by the structure. The forces associated with these three conditions have been treated in Chapter 11. One of the main functions of such structures is to be high enough so that the waves will not overtop them, or if they do overtop them, to be constructed in such a manner that they will not fail structurally, Fig. 17.4. One of the most disastrous failures of coastal defences took place in the Netherlands on February 1, 1953, as a result of a severe gale in the North Sea. The main cause of failure of the dikes was erosion by the rapid flow of water down their back side (Wemelsfelder, 1954). In this case, a functional failure led to a structural failure. To prevent erosion, the back can be protected by grass, asphalt, or concrete (Edelman, 1961), the choice being an economic one which depends upon the region in which they are constructed. A considerable amount of work has been done on the proper use of asphalt, which is a very good solution, economics permitting (Van Asbeck, 1961). Theoretical work backed by practical experience in the Netherlands has shown that the ground water pressure is one of the controlling factors in determining the thickness of the asphalt, and that considerable care must be exercised to calculate the correct thickness. For protection against overtopping which only lasts for minutes, such as is the case of a tsunami, grass has been found to be adequate to prevent scouring.

How high should a seawall be? There is no definitive answer. It depends in part on how long the structure is supposed to serve its purpose and upon the risk one is willing to take on the chance of a storm exceeding a given intensity. It also depends upon how much overtopping can be accepted without serious trouble. The matter of overtopping depends upon several things: the mean water level; the slope, toe depth, roughness, porosity, and crest width of the seawall; the bottom slope in front of the seawall; and the characteristics of the waves. The effect of the winds in causing an increase in the water level has been covered in Chapters 5 and 12, including a discussion of the probabilities of storm tides exceeding a given value as a function of the proposed life of the structure and the percentage possibility of this storm occurring during the life of the structure. Superimposed upon this increased water level will be waves, and these will run up the structure. Most of the tests have been made in the laboratory using uniform periodic waves, and the problem in nature is complicated by the necessity of predicting most probable maximum waves.

Suppose economic considerations dictate that a certain amount of overtopping can be permitted. How can this be calculated? Laboratory tests have been made with uniform periodic waves (Saville and Caldwell, 1953; Sibul, 1955) and with wind waves (Sibul and Tickner, 1956; Paape, 1961). The results for the uniform periodic waves differed from the results for wind waves. The direct effect of the wind, rather than the effect of the irregularity of the waves was not determined, and as the direct wind effect is difficult to scale, it is uncertain how to apply the results to prototype conditions.

Laboratory tests with uniform periodic waves running up a plane impervious slope with a zero crown width showed that the greatest amount of overtopping, q (cubic

Fig. 17.4. Waves breaking on the seawall at Flushing, Netherlands, and overtopping the wall

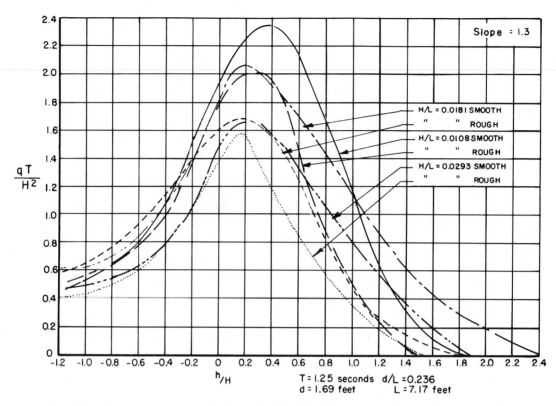

Fig. 17.5. Wave overtopping as a function of height of structure above still water for constant d/L ratio and variable wave steepness (after Sibul, 1955)

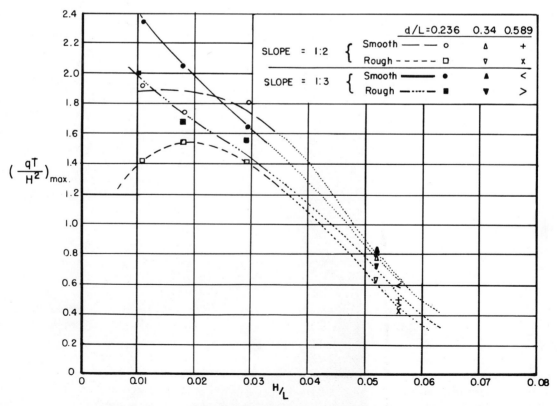

Fig. 17.6. Maximum value of $(qT/H^2)_{max}$ as a function of wave steepness (after Sibul, 1955)

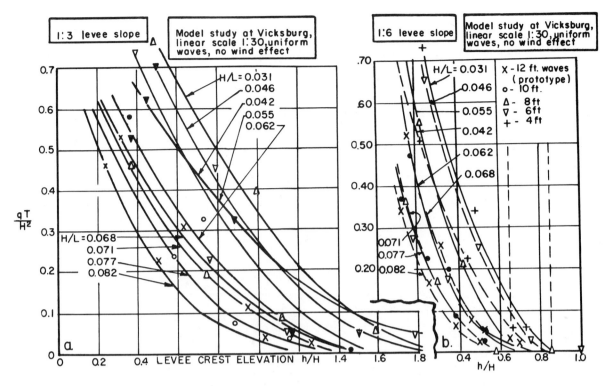

Fig. 17.7. Model study at Waterways Experiment Station, Vicksburg, Mississippi (after Sibul and Tickner, 1956)

feet per second per foot length of structure), occurred for values of h/H between 0 and 0.4, depending upon the wave steepness H/L, and the roughness or smoothness of the (Fig. 17.5). In this figure, h refers to the elevation of the crest of the structure, above or below the still-water level, T is the wave period in seconds, H is the wave height in feet, and L is the wave length in feet. The effect of wave steepness can be seen in Fig. 17.6 in which the maximum value of qT/H^2 of Fig. 17.5 and a similar plot for a seawall slope of 1:2 have been plotted as a function of wave steepness. Other tests for a model of an actual seawall with a slope of 1:3, but with a crown of finite width, showed similar results for the one wave steepness that could be compared (Sibul and Tickner, 1956). The trend, with respect to the effect of increasing wave steepness in the amount of overtopping, continued for values of H/L up to 0.082, which was the maximum value tested, Fig. 17.7.

A seawall subject to wind waves has greater overtopping than when it is subject to mechanically generated uniform waves. Some results of model tests (presented in terms of the prototype) are shown in Fig. 17.8. $H_{1/3}$ refers to the average of the highest one-third of the waves during the test, T_s and L_s refer to the average period and length of these highest one-third waves, U is the wind speed in the model, and h is the height of the structure above the mean water level at the time, that is, above the level of the storm surge. The curves shown in Fig. 17.8(b) show the difference in overtopping as meas-

ured in the uniform wave tests, Fig. 17.7(a), and in the wind wave tests, Fig. 17.8(a). Results for the case of a structure with a slope of 1:6 were similar with respect to the effect of $h/H_{1/3}$, but did not rise nearly so steeply as a function of wind speed, U.

Laboratory tests have been made with both wind waves and uniform waves acting on a plane impermeable structure, using a series of slopes of from 1:2 to 1:8 (Paape, 1961). It is not possible to compare the numerical results of these tests with the tests cited previously, as they were for a different range of conditions (h/H_{50} of greater than 1.5 for seawall slopes steeper than 1:3). The trends, however, were similar. A comparison of the overtoppings by uniform waves having the same height as the median wind wave height (H_{50}) showed the wind waves to cause a much greater overtopping.

How high must a structure be to prevent overtopping? There have been several laboratory studies made on wave run-up on structures of different slopes. Most of these tests were made with uniform periodic waves, on both smooth impervious and rough porous slopes. Tests made with solitary waves have been described in Chapter 3. Some tests have also been made for composite slopes, such as a 1:2 seawall founded on a 1:30 slope beach.

An example of the relationships among wave steepness, slope of the bottom, and run-up is given in Fig. 17.9. In this figure, R is the run-up (the vertical distance between the maximum height the water runs up the

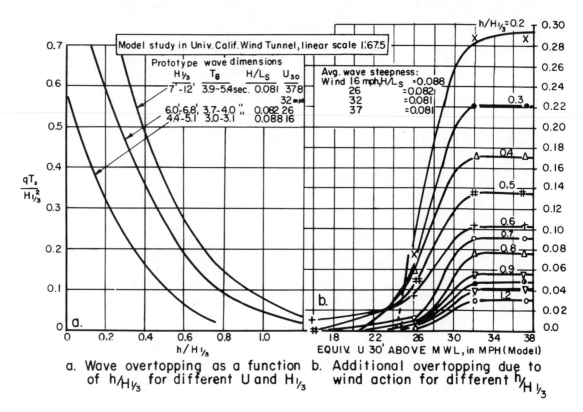

a. Wave overtopping as a function of $h/H_{1/3}$ for different U and $H_{1/3}$

b. Additional overtopping due to wind action for different $h/H_{1/3}$

Fig. 17.8. Overtopping by wind waves (after Sibul and Tickner, 1956)

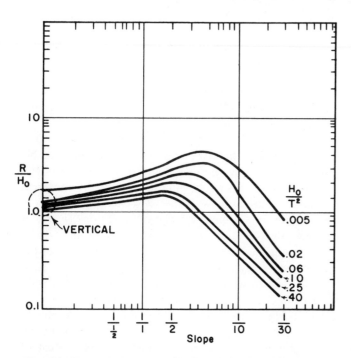

Fig. 17.9. Wave run-up on smooth slopes (after Savage, 1958)

structure and the still-water level), H_o is the wave height in deep water, and T is the wave period. There are three main features shown by the curves in this figure. First, there is a slope for each wave steepness that causes a maximum run-up (which is in accordance with the results of Granthem, 1953). Second, the flatter the wave, the smaller the run-up, which is in conflict with some of

the data of Granthem (1953) and Sibul (1955). Sibul found a maximum value of R/H for a particular value of wave steepness with R/H being smaller for both steeper and flatter waves, with the location and magnitude of the maximum being a function of the slope of the structure and d/L_o. Third, the run-up for the case of a vertical wall shows values of R/H_o greater than unity.

The reason for this difference is twofold. First, the data are referred to deep water wave heights, and except for a limited range of values of d/L_o, the wave height in some transitional or shallow water depth is greater than the deep water wave height so that even if R/H were equal to unity, R/H_o would be greater than unity. Second, these tests were made in water of uniform depth (1.25 ft, and waves with periods up to 4.70 sec were used), so that there were values of d/L_o as small as 1/76. Because of this, some of the tests were with highly nonlinear waves, and the wave crests were considerably more than half the wave height above the still water level. The resulting standing wave characteristics could probably be described using equations such as given in Chapter 2. Because of the nonlinearity of the waves in very shallow water, use of H_o obtained from linear theory can lead to doubtful results. In order to compute the run-ups, it would be necessary to have similar values of d/L_o in the prototype as in the model.

There was a considerable amount of scatter in the laboratory data. An example is given in Fig. 17.10 (Saville, 1956). These data are for a 1:1½ slope seawall founded on a 1:10 slope beach with the water depth at

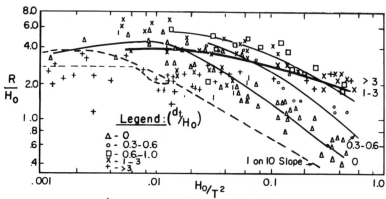

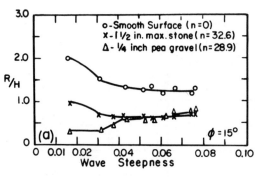

Fig. 17.10. Run-up as a function of structure depth and wave steepness (after Saville, 1956)

I on 1½ slope, with I on 10 beach slope from toe of structure.

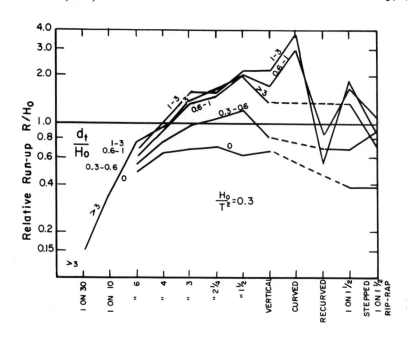

Fig. 17.11. Variation of relative run-up with structure type (after Saville, 1956)

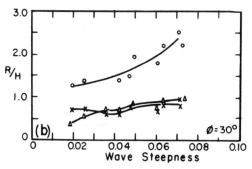

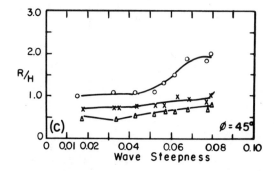

Fig. 17.12. Wave run-up as a function of wave steepness and porosity of structure for slope angles of 15°, 30°, and 45° with a constant relative depth, $d/L = 0.218$ (after Granthem, 1953)

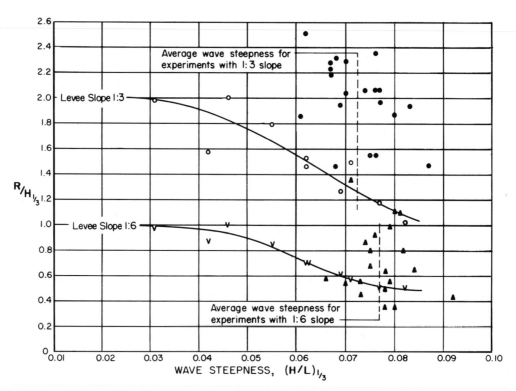

Fig. 17.13. Wave run-up as a function of wave steepness on levees with uniform side slopes of 1 : 3 and 1 : 6 (after Sibul and Tickner, 1955)

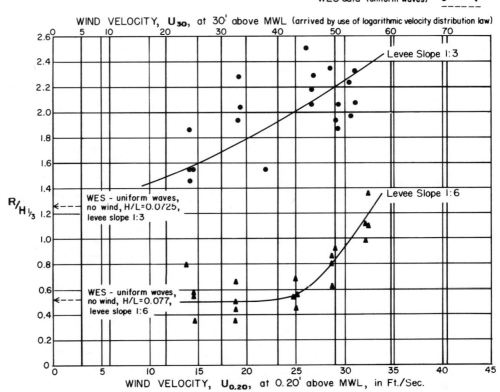

Fig. 17.14. Wave run-up as a function of wind velocity (after Sibul and Tickner, 1955)

the toe of the seawall designated by d_t. The 1:10 slope curve refers to a beach with no seawall.

A comparison of the effectiveness of beaches of different slopes, seawalls of different slopes, and seawalls of different shapes is given in Fig. 17.11. Except for the points representing the 1:30 and 1:10 slopes, the structures were all founded on a 1:10 slope. The data shown are for a deep water wave steepness of $H_0/T^2 = 0.3$ ($H_0/L_0 \approx 0.06$), which is approximately the steepness of the mean wave height in a generating area. Apparently a concave curved wall is the worst type that can be built, unless it recurves, in which case it is very good.

Run-up of waves on rough porous slopes is considerably less than on smooth impervious slopes, as is shown in Fig. 17.12.

Run-up of wind waves on structures shows trends similar to those found in the overtopping of the structures by wind waves. Laboratory studies of the run-up of such waves on plane impervious slopes of 1:3 and 1:6 have been made by Sibul and Tickner (1955). These data are shown in Fig. 17.13, where they are compared with the values of R/H obtained using uniform periodic waves. The value of $H_{1/3}$ for the wind waves is the significant wave height for a series of 100 consecutive waves. The runup is appreciably higher than would be estimated from uniform wave tests of the same steepness as the steepness of the significant wave. There did not appear to be any trend of $R/H_{1/3}$ as a function of $(H/L)_{1/3}$, but there was a relationship between $R/H_{1/3}$ and the speed of the wind used in generating the waves, as is shown in Fig. 17.14.

9. JETTIES

There are numerous types of jetties, based upon the type of construction (Hickson and Rodolf, 1951a): (1) random stone (Fig. 17.15); (2) stone and concrete (Fig. 17.16); (3) caisson type—usually built of reinforced concrete, floated into position, settled on a prepared foundation, filled with sand or stone for stability, and then capped (Fig. 17.17); (4) sheet pile; (5) crib types—built of timber, floated to the site, sunk onto a prepared foundation by loading with stone; (6) asphalt-filled rubble mound.

One of the main purposes of a jetty is to confine the flow of sea water as it moves in and out of an estuary, in order to scour the bottom to maintain a navigable depth. There is a rather well-established relationship of the cross-sectional area between a pair of jetties and the tidal prism within the estuary. This relationship has been discussed in Chapter 14 and given in Fig. 14.37. In order to maintain a navigable depth naturally when the tidal prism is small, a narrow entrance must be built, and this may present a serious hazard. The practical solution is to build the entrance wide enough to take the ships, to utilize as much natural scouring as possible, and to use dredges to do the rest of the job.

Jetties usually disrupt littoral drift just as do breakwaters. The problem of sediment movement, scour, and deposition is often more serious as jetties are usually built at the entrance to estuaries, and the rivers feeding into the estuaries carry sediment loads. During the construction of the south jetty at the mouth of the Columbia River, the regime was altered to such an extent that Clatsop Spit formed, with a length of about 3 miles and a width of about $\frac{1}{2}$ mile (Hickson and Rodolf, 1951b).

The formation and movement of bars at the entrances, and within jetties, have been discussed briefly in Chapter 14. The interaction between jetties, tides, waves, currents, and sand form the subject of many hydraulic model studies. Although the results of such model studies are probably unreliable with respect to a quantitative solution of the problem, they are very useful in helping

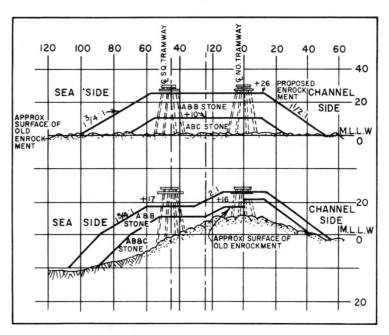

Fig. 17.15. Random stone (rubble mound) jetty, mouth of Columbia River (after Hickson and Rodolf, 1951)

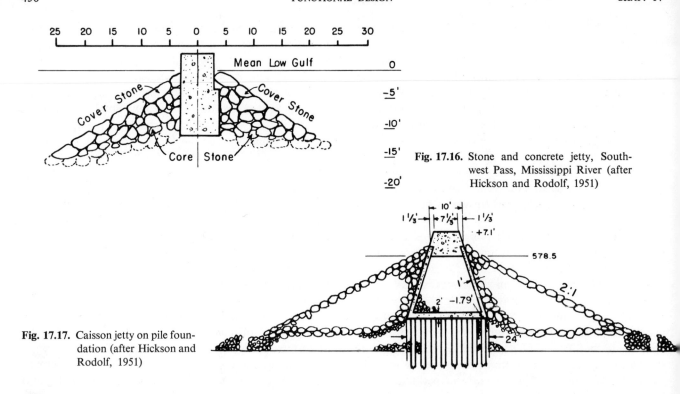

Fig. 17.16. Stone and concrete jetty, Southwest Pass, Mississippi River (after
Hickson and Rodolf, 1951)

Fig. 17.17. Caisson jetty on pile foundation (after Hickson and
Rodolf, 1951)

the engineer to formulate a qualitative solution to a specific problem.

10. GROINS

Groins may be classified in three categories: (a) permeable or impermeable, (b) high or low, (c) fixed or adjustable (U.S. Army, Corps of Engineers, Beach Erosion Board, 1961). Stone, steel, timber, asphalt (Williams, 1955), concrete, or combinations of these, as well as other materials, have been used in their construction. There are also many types of groins, based upon their planform, such as corner, inclined, T, Z, and angular as well as the standard groin (Bruun, 1953). Several types of low-cost groins have been described, many of which have been successful and many of which have failed (Brater, 1954). One of the most important points in constructing groins is being sure that they extend far enough inland so that they are not flanked and that they are placed deep enough in the sand so that they do not fail during the winter when storms normally move large quantities of sand offshore. Some asphalt groins have failed owing to erosion of sand from underneath them, causing the groins to settle, crack, and then have pieces moved away by the waves (Asphalt Institute 1962).

There is an extensive literature on the construction and operation of groin systems, much of which is controversial. Most prototype observations have been made of only the net result, with little or no observations of wave and current conditions associated with the movement of the beach sand. In fact there is little

published information on prototype groin usefulness (Savage, 1959).

One of the primary problems connected with the design of a groin system is the determination of the source of the littoral sediments and the quantity moving. Then it is necessary to decide upon the groin lengths and spacings. Structural design and construction procedures are probably of secondary importance. In deciding upon the lengths of the groins, the seasonal onshore-offshore movement of sand must be kept in mind.

When there is not an adequate amount of sand in a region for which a groin system is contemplated, an artificial beach may be considered in conjunction with the groins if it is either economically justified, or of enough importance for other reasons. When sand is imported to the area, the relationships among beach material, beach slope, and exposure of the beach to waves and currents must be used if satisfactory results are to be obtained (see Fig. 14.17, together with the discussion on sorting of sand by waves in Chapter 14). The coarser the material (for a given specific gravity and roughness, and given exposure to waves) the less the quantity that will be needed to produce a beach of a given width. The coarser the material, however, the steeper the beach face; hence the faster the backwash of the wave, and the more dangerous will be the beach from the standpoint of people swimming or wading in the surf.

Nagai (1956) made laboratory studies on groin length, orientation, and spacing. It was found that the groin should extend about 40 per cent of the distance from

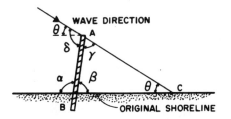

Fig. 17.18. Angles of groins to the shoreline (after Nagai, 1956)

the water line to the predominant line of plunging type breakers. When the groins extended further seaward, scour occurred in the vicinity of the groins. Another investigator (Horikawa, 1958) found that the groins should extend from 40 to 60 per cent of the distance, rather than 40 per cent. The reason for this can be seen in the results of laboratory studies on current around groins (Horikawa and Sonu, 1958; Nagai and Kubo, 1958).

Nagai's studies of the orientation of the groins with respect to the direction of wave advance and beach alignment (Fig. 17.18) showed that best results were obtained when the angle δ (the angle between the direction of wave advance and the groins) was from 100 to 110 degrees, and should not be less than 95 degrees nor greater than 120 degrees. When δ was less than 90 degrees, the sand on the updrift side of the groins was transported seaward, causing scouring inside the groin system near the groins; when δ was 90 degrees, severe scouring took place at the offshore ends and along the updrift sides of the groins. When δ was greater than 120 degrees, scouring took place along the updrift sides of the groins and eroded the shore at the base of the updrift side of the groins. Until tests are made with variable waves, in amplitude, frequency, and direction,

as they exist in prototype it is difficult to see whether or or not a variation of δ from 90 degrees would be an improvement.

The range of angles, α, which describes the alignment of the groins with respect to the shore, was determined analytically by assuming that the area of the triangle ABC (Fig. 17.18) should be a maximum. This resulted in the expression

$$(180° - \alpha) = \beta = \frac{\pi - \theta}{2}, \qquad (17.1)$$

where θ is the angle between the direction of wave advance and the shoreline. It was found that, for all practical purposes, θ could be taken as the angle occurring in the region of the breakers. This results in $\alpha = 105, 110, 115,$ and 120 degrees for $\theta = 30, 40, 50,$ and 60 degrees, respectively. Experiments verified this conclusion within the range of θ given above. When $\theta \geq 60$ degrees, it is not possible to satisfy both Eq. 17.1 and the condition that δ range from 100 to 110 degrees. It was found experimentally that the least unfavorable effects occurred for α between 90 and 110 degrees. The authors state that observations of prototype installations also show the same general results, which is that sand accumulation by jetties is more effective when $\theta < 60$ degrees than when $\theta > 60$ degrees.

The laboratory studies of Horikawa and Sonu (1958) showed that the best angle depended upon the wave steepness with $\alpha \approx 105$ degrees being better for steep storm waves, and $\alpha \approx 90$ degrees being better for relatively flat waves.

Some illustrations of the erosion and deposition for several values of θ and α are shown in Fig. 17.19.

Tests on sand accretion between groins was found to be as shown in Fig. 17.20 for θ between 35 and 50 degrees.

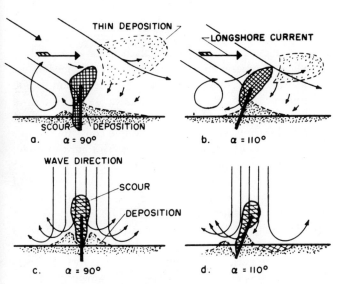

Fig. 17.19. Erosion and deposition of sand at a single groin (after Nagai, 1956)

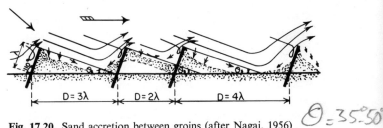

Fig. 17.20. Sand accretion between groins (after Nagai, 1956) $\theta = 35$-$50°$

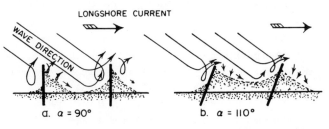

Fig. 17.21. Sand accretion as a function of groin spacing (after Nagai, 1956)

The effect of one groin upon another for the case of $\alpha = 110$ can be seen by comparing Fig. 17.20 with Fig. 17.19d. The deposition of sand against the protected side of the groin as shown in Figs. 17.20 and 17.21 appears to be in conflict with observations of prototype installations in the United States (Saville, 1962).

The effect of groin spacing could be predicted using the experimentally determined relationship

$$\theta_s \approx 0.4\theta, \qquad (17.2)$$

where θ_s is the angle between the sand deposited and the original shoreline. For the case of $\alpha = 110$ degrees, θ from 35 to 55 degrees, and Eq. 17.2 it was found that

$$D \approx 3\lambda, \qquad (17.3)$$

where λ is the length of the groin as measured seaward from the original shoreline. Experimental results are shown in Fig. 17.21 for $D = 2\lambda$, 3λ, and 4λ. For waves coming nearly straight on to shore (θ from 60 to 90 degrees), it was found that a spacing of $D \approx 4\lambda$ was better. Laboratory tests by Kressner (1928) indicated that a spacing of 2λ to 3λ was best, which is lower than a spacing of 4λ that he found most beneficial with tests made of a current flowing along the shore in the absence of waves. Other laboratory studies (Horikawa and Sonu, 1958) indicated that the groin spacing depended upon the length of the groins with respect to the distance from the shoreline to the breaker line. The greater the relative groin length, the smaller should be the distance between groins. The most desirable distance between groins was found to be from 2 to 3 times the groin length.

Surveys have been made for a series of equilibrium beaches along the southwest corner of England which, if used judiciously, might be of value in the determination of groin spacing (Hoyle and King, 1955). They were beaches with the shoreline being in the form of an arc of a circle sustained between two headlands. Two of the beaches were sheltered and three were exposed, with the projected lengths of the beaches (chords) being from 525 ft to 96,500 ft. The mid-ordinates (perpendicular distance from the middle of the chord to the arc) varied from 35 to 5280 ft, with the ratios of mid-ordinates to chords being about 0.25 for four of the beaches and 0.50 for the other beach. This would indicate that, for waves of variable direction as occur in nature, a given spacing of about four to one would be desirable for groins that are normal to the general coastline. This is in agreement with the findings of Nagai previously described.

The height of groins depends to some extent on their purpose. If it is not necessary to allow sand to pass a groin, it should be high enough to prevent overtopping by storm waves at high tide (Horton, 1951; U.S. Army, Corps of Engineers, Beach Erosion Board, 1961). This means that the shore portion of the groin should be at the elevation of the winter berm. If it is desirable for sand to pass the shore section, the top should not exceed the level of mean high tide (Wicker, 1958). The shore portion should be approximately horizontal, and the intermediate portion should have a slope parallel to the normal slope of the foreshore, and should extend to about the mean lower low water line. The outer section should be nearly horizontal, and the top should be at an elevation of about MLLW.

In the tests described in this section, the effect of beach slope (essentially sand size and wave conditions, see Chapter 14) was neglected. Considering the critical effect of slope on currents, cusp, bars, and other near-shore features, one would expect that it would have a considerable influence on the operation of a groin system.

One way of determining the effectiveness of a groin system is to measure the long-time rate of drift before and after the installation of the groins. Prototype measurements along the northern coast of New Jersey (U.S. Army Corps of Engineers, New York District, 1956) indicate that groin systems might have reduced the drift of sand by 12 per cent; too many variables, however, are associated with the prototype to enable us to accept or reject this figure. Laboratory measurements (Savage, 1959) on a 1:20 slope beach showed that one low short groin reduced the average transport rate by 12 per cent, one high short groin reduced the transport rate by about 25 per cent, and a high long groin reduced the transport rate by about 60 per cent, but this last figure was in doubt, as it was believed that the test was not run long enough for equilibrium conditions to be obtained.

Several laboratory experiments (Johnson, 1948; Nagai, 1956; Shimano, Hom-ma, Horikawa, and Sakou, 1957) have shown scouring at the tips of groins. This should be taken into consideration in the design and construction of groins. The Japanese have used brushwood mattresses in the past to minimize the scouring (Horikawa, 1958).

Should groins be permeable or impermeable? Part of the conflicting viewpoints on this may be explained by the lack of precise definition of the word "permeable" as used when describing groins. The Japanese seem to find permeable groins beneficial (Horikawa, 1958; Horikawa and Sonu, 1958). Most of their groins are not very permeable, however, a typical example consisting of two parallel rows of 1-ft-wide sheet piles with gaps of about 3/4 in. between piles, and the space between the two rows of piles being filled with rock. This amount of permeability was arrived at after two years of tests.

Laboratory studies by Johnson (1948) on a 1:10 slope beach using impermeable groins and permeable groins of much greater spacing than used by the Japanese showed the permeable groins to be ineffective. The permeable groins consisted of 3/4 in. × 1/4 in. vertical slats with the spacing between the slats varying so that permeabilities between 25 and 75 per cent could be tested. In the Great Lakes, permeable groins (vertical wood) have not been successful except where the wave action was mild or the beach material was coarse (Brater, 1954).

11. UNDERWATER PIPELINES AND CABLES

One major use of underwater pipelines is the transportation of oil either from tankers to shore or from offshore platforms to shore; another is the movement of sea water for use as cooling water in nuclear and steam electric power stations. In both these types of uses, as well as in other uses, the continuous operation of the pipelines is usually of paramount importance. Pipelines on the bottom of the sea or bay are difficult to inspect except during summer months, and are nearly always difficult and expensive to repair. In many locations they cannot be repaired during long intervals of time because of severe sea conditions.

A knowledge of the types of failures that have occurred and of the reasons for these failures will be of use to the design engineer. Information of this sort is difficult to obtain. Scientific works are filled with announcements and details of construction projects of this and other types of marine structures and installations, but little, if any, information is reported on failures of structures. The city of Vancouver, B.C., has used submerged pipelines across Burrard Inlet, a deep tidal arm of the Pacific Ocean, to supply water to the city. A list of the causes of breaks in the various pipes follows (Powell, 1935):

1. Sixteen breaks (between 1894 and 1935) caused by pipes being struck by ships or ships' anchors.
2. Many pipes damaged by sagging when pipes went over boulders or ledges even though divers were used to inspect for such conditions during construction and to correct the defect by supporting the suspended portion of the pipes with bags of sand, gravel, and cement.
3. Sagging due to erosion of material from under the pipes.
4. Abrasion.
5. Corrosion of bolts and nuts (later use of nickel steel bolts and nuts has stopped this cause of failure).

The large number of failures due to ships and ships' anchors emphasizes the desirability of burying submerged pipelines. In addition, a great amount of damage was done in the Gulf of Mexico to unburied pipelines during Hurricane Carla in September, 1961. In some areas of the world, local government regulations require the burial of pipelines in water less than a specified depth. Burying pipes also protects them from corrosion, marine life, movement by waves and currents, and spanning due to scouring action of waves and currents.

The most common types of pipes used in modern practice are steel and concrete. Steel pipes are normally coated with concrete reinforced with steel mesh over a proper type of enamel, or coated with an asphalt mastic (Aldridge, 1956; Timothy, 1956). Heavy aggregate, such as barite, is commonly used to increase the weight of the coating. The coat serves as a protection against corrosion and marine growths and provides additional weight for the submerged pipe to provide the necessary negative buoyancy.

One method of laying pipes is to perform most of the construction on a barge. The coated pipes are welded together on the deck, and the pipeline is laid off the stern of the barge. Protection of the welded joints is obtained by enamel, plastic tape, asphalt mastic, or quick-drying concrete. Care must be used in laying the pipe to make sure that neither the pipe nor the coating cracks. This is of particular importance in deeper water. For example, a $12\frac{3}{4}$-in. diameter steel pipe can be bent on a radius of about 300 ft before the elastic limit is reached (Timothy, 1956). When the water depth is such that this radius is about reached, the pipe must be handled in a way to keep the radius above this limit. This can be done by ramps, additional barges, and booms or floats. Concrete coatings make the job of laying the pipe more difficult, as it cracks more easily than the steel pipe. The concrete cover on a $12\frac{3}{4}$-in. diameter pipe will crack when the radius of curvature is about 1600 ft (water depth from 12 to 14 ft). Asphalt mastic has been developed for coating steel pipes so that the smaller radius of curvature controls (Timothy, 1956). Such a coating also permits a more rapid job to be done at sea. An example of a pipeline laying setup is given in Fig. 17.22.

Another method of installing pipelines is to do the construction work on shore and either push, or more commonly, pull the pipeline to sea. A 20-in. diameter pipe covered with enamel, fiberglass, asbestos felt, wire mesh and gunited with concrete, the total unit weighing 215 lb per foot was pulled about 3200 ft to sea, ending in water 50 ft deep (Sneddon, 1953). A much longer pipe (7 miles) was installed in the Gulf of Mexico to transport liquid sulphur (McNamara, 1960). The pipeline was made up on shore in sixteen 2000-ft sections, and each section was welded to the previous one after a 2000-ft pull. The pipeline was a complex one, the inner pipe (6 in. O.D.) carrying the molten sulphur. This pipe was placed inside a $7\frac{5}{8}$-in. O.D. pipe, with the annulus being used to run hot water under pressure to maintain the sulphur between 280°F and 320°F. This pipe was covered with a $2\frac{1}{4}$-in. layer of calcium silicate insulation wrapped with an aluminum foil jacket and placed inside a 14-in. bell-and-spigot 54.6-lb pipe which was covered with high-temperature coal tar and fiberglass. A 4-in. return water line was attached to one side of 14-in. pipe and a $6\frac{5}{8}$-in. cement-lined pipe was attached to the side to carry fresh water to the offshore mine. A barge with two

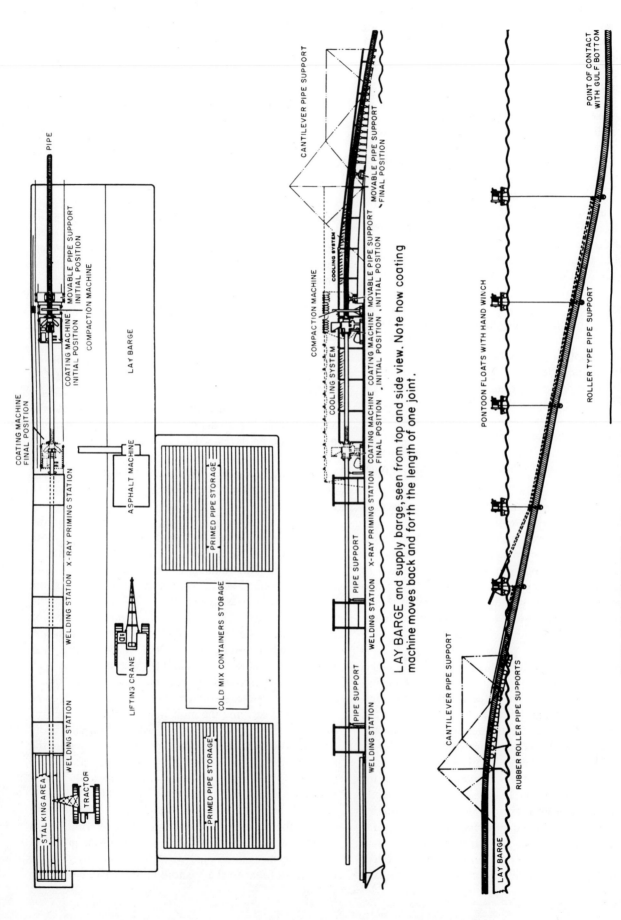

LAY BARGE and supply barge, seen from top and side view. Note how coating machine moves back and forth the length of one joint.

PIPELAYING METHOD for deep water uses pontoons to partially support pipe between barge and Gulf bottom.

Fig. 17.22. One type of marine pipeline-laying procedure (from Timothy, 1956)

250-ton winches was used to pull the sections. Considerable difficulties were encountered in the pulling phase of the job.

A 30-in. pipeline 19 miles long has been installed in the Persian Gulf off Iran (*Oil Gas J.*, 1959). The pipe was glass-wrapped and coated with 2½ in. of concrete. All work was done on shore and the pulling was done with 400-ft sections (8 hr per pull) using a barge.

A very large concrete pipeline, 12 ft in diameter, extending 5 miles into the sea was installed as a part of the Hyperion sewage disposal system (Los Angeles, California) using a very large mobile platform which was constructed specifically for the $20,297,500 job (Miller and Albriton, 1959). Eight 24-ft pipe sections were joined onshore and the 192-ft unit attached to the underside of a pontoon strongback mounted on a special barge. The barge was then towed to the mobile platform and maneuvered under the platform where the pipe section–pontoon unit was transferred to the bridge crane on the underside of the platform deck. After the pipe-pontoon

Fig. 17.23. Scattergood Steam Plant, Los Angeles, California: (a) Scattergood Steam Plant layout; (b) Vertical longitudinal section and hydraulic gradient circulating water system (from Mariner and Hunsucker, 1959)

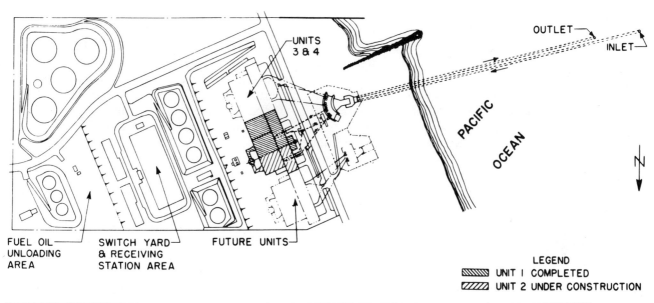

(a)

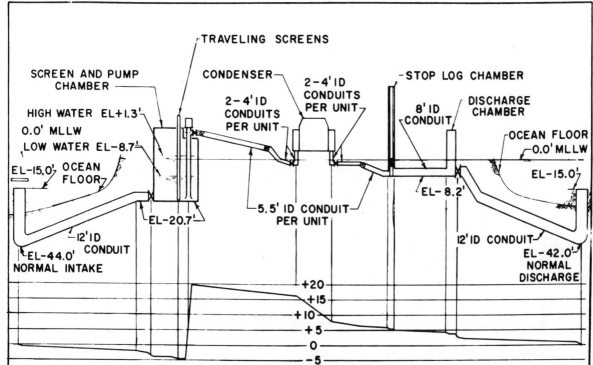

(b)

unit was lifted from the barge, the barge was towed away and the unit lowered into the water. The pontoon was flooded, lowered to the bottom, and attached to the end of the pipeline. The pontoon was then dewatered, brought to the surface, and placed back on the barge. After rock had been placed along the sides of the pipes, hydraulic jacks on the mobile platform lifted the support legs to free the platform from the bottom, the platform winched forward 192 ft, and the legs jacked to the bottom again, ready for another unit.

Numerous power stations have been built or are being built which utilize sea water for cooling (Spencer and Bruce, 1960). The volumes of water for the large stations are very large. For example, a plant designed to produce 800 megawatts(mw) uses nearly 800 cubic feet per second of cooling water; if it were fresh water, this quantity could supply a city of more than 3,000,000 people (Weight, 1958). The electric output of a power station of this size is between the outputs of the hydro-electric plants at Bonneville Dam, 518 mw, and Hoover Dam, 1320 mw (Mosonyi, 1957).

These power stations encounter several major problems, such as the possibility of re-circulation of heated water into the intake pipe (water is discharged from 15° to 20°F above intake temperature), marine growths decreasing the capacity of the pipes, and fish getting into the system. An example of a power station is shown in Fig. 17.23. In this installation, the intake is approximately 350 ft seaward of the discharge. Mixing of the discharge water is accomplished by jet mixing, waves, and currents. Few field studies have been made on the distribution of temperatures in the vicinity of a discharge port. One study (Pacific Gas and Electric Co., 1961) was made for a limited time which showed that the water temperatures of the sea water more than one or two thousand feet from the discharge were essentially the same as the ambient sea water temperatures. The studies were made during the winter, however, when there was a likelihood of greater mixing by wind waves than would be the case during the summer.

One method that has been used successfully to kill marine growths and prevent their building up inside a pipe is the use of hot water shock. Water at about 105°F pumped through the pipes for approximately 4 hours once a month has been found to be satisfactory in Southern California (Weight, 1958).

Apparently, water moving vertically into an intake is not utilized by fish as a danger signal and large numbers of them have been drawn into cooling water circulation systems. A modification of an inlet, Fig. 17.24, using a cap to cause the high-speed position of the water flow to be horizontal was found to result in far fewer fish being sucked into the pipe (Weight, 1958).

Underwater cables will not be discussed here, rather the reader is referred to a paper by Zajac (1957).

12. SEWER OUTFALLS

Nearly every coastal community in the world discharges sewage, raw or treated, into the sea. Over 140 such installations in the United States alone have been listed by Pearson (1956). The effects of such discharges may be studied from the standpoint of both public health and public nuisance (grease, oil slicks, visible floating solids, odors, etc.). The engineer designing an ocean outfall sewer must consider many factors, most of which are given in Table 17.3. Some details on 75 ocean outfall sewers, including their costs, are given in Table 17.4. For more details on some of the problems connected with disposal of wastes in the sea, the reader is referred to the proceedings of a conference on this subject (Pearson, 1960) and to the report by the Committee on Sewerage and Sewage Treatment of the ASCE (1961).

The principles of dilution are paramount in the engineering solution of the problem. The mixing of the receiving waters may be considered in two parts: first, mixing in the immediate vicinity of the discharge point or line, second, the movement and further dilution of the sewage sea water mixture. A general discussion of these phenomena has been given by Pearson (1956); specific

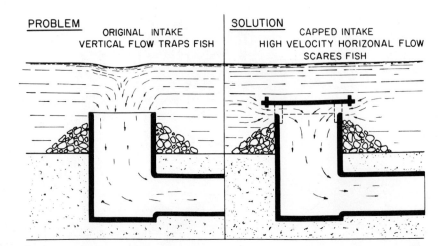

PROBLEM
ORIGINAL INTAKE
VERTICAL FLOW TRAPS FISH

SOLUTION
CAPPED INTAKE
HIGH VELOCITY HORIZONTAL FLOW
SCARES FISH

Fig. 17.24. Solution to fish control problem (from Weight, 1958)

Table 17.3. Factors to be Considered in the Design of Marine Waste Disposal Systems
(from Pearson, 1961)

I. *Beneficial uses*

1. Water contact sports
2. Marine recreation, boating
3. Marine working environment
4. Fishery, propagation, migration, etc.

5. Economic fishery, propagation, harvesting
6. Industrial commercial use, coding, etc.
7. Waste disposal
8. Other

II. *Water quality criteria to protect beneficial uses*

1. Public health
   (a) Coliform
   (b) Other
2. Fishery
   (a) Toxic substances
   (b) Antagonistic substances
   (c) Oxygen depressants
   (d) Stimulants, fertilizers
   (e) Transparency, turbidity
   (f) Suspended and settleable debris

3. Nuisance
   (a) Grease and oil films
   (b) Floating debris
   (c) Settleable debris
   (d) Odors
4. Aesthetic
   (a) Sleek areas
   (b) Colors
   (c) Turbidity—transparency
   (d) Floating debris
   (e) Plankton bloom
   (f) Other
5. Economic and other

III. *Oceanographic characteristics of outfall sites*

1. General water circulation system
2. Current
   (a) Surface and subsurface
   (b) Strength and direction as a function of time
   (c) Effect of wind, wave, tide, littoral drift
3. Eddy diffusivity or dispersion characteristics

4. Density structure, salinity-temperature-depth relationship
5. Wave and swell effects
6. Submarine topography
7. Submarine geology

IV. *Waste dispersion considerations*

1. Initial mixing—diffuser
   (a) Jet mixing
   (b) Buoyancy—gravitational mixing
   (c) Density gradients—thermoclines
   (d) Diffuser orientation
   (e) Waste dilution—flow continuity
   (f) Port selection, area—spacing

2. Waste transport—dispersion
   (a) Current regiment
   (b) Eddy diffusion
   (c) Mixing depth, effective
   (d) Rational dispersion equations
       (1) Concentration dilution only—conservative waste
       (2) Concentration including decay—nonconservative waste, i.e., bacteria, radioisotopes, BOD, etc.

V. *Economic analyses*

1. Various types of treatment and effluent characteristics.
2. Length, depth, and cost of outfall systems for each type of effluent to meet water quality criteria requirements.
3. Selection of optimum and least-cost combination of treatment and outfall system to protect beneficial uses.

studies of the mixing in the immediate vicinity of the discharge point or line have been made by Rawn and Palmer (1930); Rawn, Bowerman, and Brooks (1960); Abraham (1960; 1962; 1963), and Hart (1961); a specific study of the dilution of sewage by currents has been made by Brooks (1961), and specific studies of the mixing by wind waves have been made by Johnson (1960) and Masch (1961; 1962). Details of the mixing processes are given in Chapter 16.

The curves of Figs. 16.9 and 16.13 should be used with caution to determine the concentration of the effluent at the center of the plume as it reaches the surface for the case of effluent discharging into a nonturbulent stationary body of receiving water of uniform density. The reason for the caution is that very little dilution occurs in that portion of the rising plume between the level at which the plume deflects to a horizontal flow and the surface. After it has completely deflected to the horizontal the rate of dilution increases again. The results of the experiments by Frankel and Cumming (1963), which include data for a greater range of discharge angles (Fig. 17.25), have been reexamined, and the following procedure developed in order to take this surface effect into consideration: the ratio of the vertical distance between the nozzle and the water surface to inside diameter of the nozzle, H_s/D_0, should be multiplied by 0.75, and the value of c_m/c_0 read for this value. For example, suppose $H_s/D_0 = 80$; then the appropriate z/D_0 is $0.75 \times 80 = 60$, and the value of c_m/c_0 at the surface is 0.012 for a horizontal jet with an initial Froude number of 8.0.

As is evident from the data given in Chapter 15, the receiving sea water is rarely of even approximately uniform density, with thermoclines often existing.

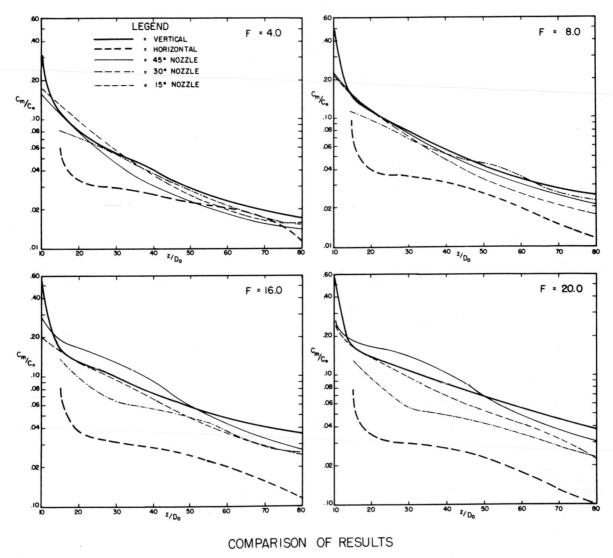

COMPARISON OF RESULTS

Fig. 17.25. Effect of angle of discharge on mixing (from Frankel and Cumming, 1963)

Sewage effluent has a density very nearly equal to that of fresh water. It is discharged at or near the bottom and rises to the surface largely because it is less dense than the surrounding sea water (it may also have a vertical component of momentum upon leaving some types of diffusers). As it rises, it mixes with the surrounding sea water, and the mixture becomes denser than the issuing effluent (Rawn and Palmer, 1930; Rawn, Bowerman, and Brooks, 1960; Abraham, 1960; 1962). Often the mixture reaches the surface and spreads out, resulting in a large patch of discolored water. If a thermocline and a mixed surface layer exist in the sea above the sewer outfall, the mixture may never reach the surface because the mixture may be more dense than the mixed surface layer. These conditions have been discussed in Chapter 16, together with an intermediate case, and examples of the three types of plumes given in Fig. 16.15. The advantage sometimes claimed for having the mixture remain beneath the surface may be of the "out-of-sight, out-of-mind" type. It would appear that more mixing would occur if the effluent mixture reached the surface because of the effect of wind waves and surface currents on the mixing processes.

In some regions where ocean outfall sewers are in operation or are proposed, thermoclines exist during certain seasons of the year, and do not exist during other seasons; furthermore, thermoclines and the mixed surface layers change with varying conditions of cloud cover, currents, winds, and waves. One of the main criteria in studying the changes in a thermocline and the surface mixed layer, and the effects of these changes upon the sewage mixture, as well as the direct effect on the mixing of the sewage, is the "eddy coefficient." This has been described in Chapter 16.

Effluent can be mixed more effectively if it is discharged from a series of ports along a length of pipe. In some cases, a multi-pipe system is used in the disposal area to discharge the effluent over a wide area rather than along a line (Fig. 17.26). This involves the design of manifold systems, which are beyond the scope of this book (see, for example, Keller, 1949; McNown, 1954; Acrivos, Babcock, and Pigford, 1959; and Rawn, Bowerman, and Brooks, 1960).

The viability of the pathogens must be considered in the functional design of an ocean outfall sewer. Much of the evidence on the die-off rates of various bacteria, etc., is conflicting and the engineer is cautioned against drawing any quick conclusions on the problem. Some data compiled by Pearson (1956) are given in Fig. 17.27. (The names shown with the various curves in this figure refer to the investigators.)

The method, or methods, by which water in the vicinity of sewer outfalls can be rated is a subject of considerable argument (Pearson, 1956). One common method is to determine the number of coliforms per 100 ml. Using this number as a base, many governments have passed legal standards with respect to the number that are permissible for swimming. Some of the standards refer to the arithmetic average of the number of coliforms for a given number of samples, some to the mean, some to the geometric average, some to monthly average, etc. The limits for "safe swimming" vary by a factor

of at least 10, from about 240 to 2400/100 ml (Garber, 1955).

There is one very disturbing item connected with the limits set by the various agencies. Aside from the problem of aesthetics, is swimming in "polluted" water harmful? A rather thorough study of this has been made by the Public Health Laboratory in England (Moore, 1960). Many claims of infection resulting from bathing in polluted waters were investigated very carefully over a three-year period, in Great Britain, and there was little evidence to support the claims that the infections were contracted while swimming, even though one area was grossly polluted, with median presumptive coliform counts of from 10,000 to 20,000 per 100 ml for samples taken over a long interval of time. It must be cautioned, however, that the incidence is low in Great Britain for paratyphoid or typhoid fever, etc. The danger of swimming in polluted areas in regions where such diseases are common might be entirely different. For example, waters off the beach at Rio de Janeiro, Brazil, have coliform counts of about 10,000/100 ml, and the incidence of enteric illness is 1000 to 10,000 times higher than it is in California. A beach in California would be considered to be polluted if the coliform count was about one-tenth that of Rio de Janeiro. If the California standard is acceptable, then the incidence of disease carriers in Rio

Fig. 17.26. Diffuser for 90-in. outfall (after Rawn, Bowerman, and Brooks, 1960)

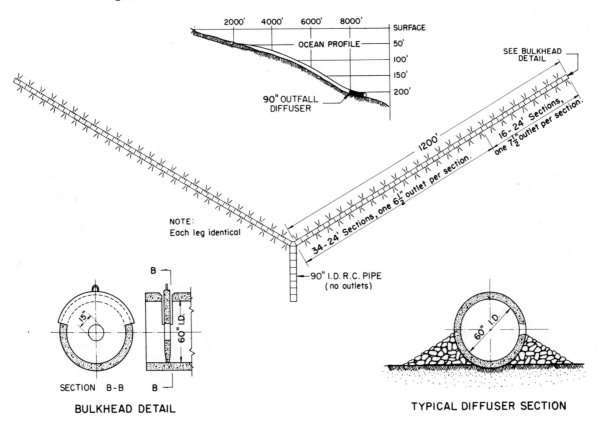

Table 17.4. SUBMARINE OUTFALL CONSTRUCTION COST SUMMARY
(from Pearson, 1956)

Location	Conduit[1] material	Length[2] of outfall (ft)	Diameter I. D. (in)	Depth below MLLW at outlet (ft)	Anchorage principally surf section
Capitola S. D., Calif.	Conc. stl. cyl.	1,000	12	14	Trench
Carlsbad, Calif.	C.I.	980	6	14	Pile bents
Carmel S. D., Calif.	Stl.	2,170	14	~0	Trench
East Bay M.U.D., Calif.	R.C.	5,960	108	40	Trench
Los Angeles City, Calif.	R.C.	5,280	144	40	Trench piles
Los Angeles Co. S.D. Calif., No 1	R.C.	5,000	60	110	Trench
Los Angeles Co. S.D. Calif., No 2	R.C.	4,300	72	110	Trench
Monterey, Calif., No 1	C.I.	1,000	24	50	Trench (Part)
Monterey, Calif., No 2	Stl.	1,045	24	29	Trench, conc. blocks
No. San Mateo Co. S.D., Calif.	C.I.	272	24	~0	Piles
Oceanside, Calif.	W.I.	2,000	12	—	Piles, conc. blocks
Orange County S.D., Calif.	R.C.	7,000	72	48	Trench
Pacific Grove, Calif.	C.I.	624	20	0	Trench
Pismo Beach, Calif.	W.I.	1,375	12	~15	Trench
San Diego, Calif., No 1	W.I.	1,435	12	15	Trench
San Francisco No. Pt., Calif.	C.I.	4@~900'	48	~9	Under wharf
Santa Ana (J.O.S. No 1), Calif.	C.I.	3,000	42	35	—
Santa Barbara, Calif.	R.C.	3,100	42	35	Trench, piles
Santa Cruz No 1 & 2	W.I.	2@2,000	15	34	Trench
Santa Cruz Ext. No 3	Stl-Conc cyl	1,200	36	36	Trench
Seaside, S.D., Calif.	Stl.	850	24	25	Trench, chain
Sharp Park, S.D., Calif.	C.I.	216	12	~+3	Pile bents
South Laguna S.D., Calif.	C.I., Stl.	1,900	16,18	55	Piles
Sunset Beach S.D., Calif.	C.I.	1,300	6	15	Chain
Ventura, Calif. No 1	Stl.	1,900	18	20	—
Ventura; Calif. No 2	Stl.	2,560	18 7/8	22	Trench
Watsonville, Calif. No 1	W.I.	1,500	16	18	Chains
Watsonville, Calif. No 2	W.I.	1,500	16	20	Chains
Anacortes, Wn. No 1	R.C.	701	24	38	Pile bents
Anacortes, Wn. No 2	C.I.	472	12	30	Sack conc. crib
Seattle, Wn. W. Dawson St.	C.I.	724	24	—	Rock cover
Seattle, W. Dawson Ext.	C.I.	127*	24	30	—
Seattle, Murray Ave. Ext.	C.I.	216*	24	30	—
Seattle, W. Barton St. Ext.	C.I.	395*	24	—	—
Seattle, Washington St. Ext.	C.I.	888*	16	30	—
Seattle, 32nd Ave. W.	C.I.	760	16	—	—
Seattle, 32nd Ave. W. Ext.	C.I.	792*	16	30	—
Tacoma, Wn., Edwards St.	R.C.	627	36	40	Pile bents
Tacoma, Division St.	R.C.	224	36	36	Pile bents
Key West, Fla.	C.I.	4,570	24	33	Trench
Miami Beach, Fla.	C.I.	7,000	36	40	Trench
Miami, Fla.	R.C.	4,500	90	20	Trench
Baltimore, Md., Patapsco R.	R.C.	525	60	13	Pile bents
Boston, Mass., MDC. Nut. Is.	C.I.	22,000	12	35	Trench
Gloucester, Mass.	C.I.	6,032	24	32	Trench, piles
Swampscott, Mass.	C.I.	3,754	20	44	Trench
Allenhurst, N. J.	W.I.	1,100	20	—	—
Asbury Park, N. J., No 1	W.I.	1,055	23	21	Trestle
Manasquon, N. J.	W.I.	1,250	12	—	—
Passaic Valley S.D., N. J.	R.C.	25,600	144	—	Tunnel
Seabright, N. J.	W.I.	1,000	15	18	Trench
Sea Girt, N. J.	W.I.	1,100	12	—	None
Nassau Co., N. Y.	R.C.	12,600	84	12	Pile bents
New Rochelle, N. Y.	C.I.	6,830	36	25	Trench, piles
New York City, Bowery Bay	R.C.*	1,403	72	27	Piles
New York City, Coney Is.	R.C.*	8,220	72	12	Trench, piles
New York City, Jamaica	R.C.	7,400	84	~20	Trench, piles
Westchester Co., N. Y., Blind Brook	C.I.	6,200	30	42	Trench
Westchester Co., Mamaroneck	C.I.	13,240	54	51	Pile bents
Westchester Co., New Rochelle S.D.	R.C.	9,900	54	46	Pile bents
Westchester Co., No. Yonkers	C.I.	1,025	60	45	Piles
Westchester Co., So. Yonkers	C.I.	300	48	18	Piles
Bristol, R. I.	R.C.*, C.I.	1,504	30	20	Trench
Norfolk, Va., Lamberts Pt.	R.C.	3,000	54	10	Pile bents
Waukegan, Ill.	C.I.	365	36	10	—
Ashtabula, Ohio	R.C.	3,550	42	20	—
Cleveland, Ohio, West	St.	2,600	60 & 72	30	Trench
Cleveland, Ohio, East No 1	R.C.	3,200	84	30	—
Cleveland, Ohio, East No 2	R.C.	700	108	9	Piles
Honolulu, T. H., No 1	C.I.	2,773	48	60	Trench
Toronto, Ont.	R.C.,* Stl.	5,330	60	19	Trench
Vancouver, B. C.	R.C.	3,040	66	80	Trench, piles

[1] C. I.—cast iron, R. C.—reinforced concrete, Stl.—steel, W. I.—wrought iron.

[2] Outfall contract length—usually from shore structure seaward.

[3] 1948 to date ENR Index for closest of 21 cities in index. Prior to 1948 yearly average index for 20 cities in ENR index.

Special diffuser	Special notes	Date of construction	Total cost ($)	Cost/foot when built ($)	E. N. R.[3] Index when built ($)	Adjusted cost/foot ENR = 700 index ($/ft)
4 pairs 3″ φ ports @ 9′ ctrs		1955	64,958	65.00	680	67.00
None		1929	6,500	6.60	206	22.40
None		1950	16,500	7.60	500	10.60
36–16″ φ ports @ 3′ ctrs		1951	1,229,158	206.00	540	267.00
Tapered–96″–78″–60″–42″ φ 3 pairs slots @ 100′ ctrs		1948	3,517,410	666.00	500	933.00
2 ports in end Not included in bid		1937	528,059	105.50	235	314.00
Not included in bid		1947	700,000	162.80	400	284.80
None		1934	15,000	15.00	198	53.00
Bonnet diffuser (?)		1952	86,000	82.30	560	102.90
None		1955	22,000	80.90	680	83.30
None		1923	32,000	16.00	214	52.30
None		1954	1,873,052	268.00	650	288.50
None		1952	20,111	32.20	560	40.20
None		1925	48,000	34.90	207	108.00
None		1928	22,000	15.30	206	52.00
None		1951	557,375	155.00	540	201.00
None		1927	300,000	100.00	206	340.00
None		1926	228,424	73.70	208	248.00
None		1927	90,000	22.50	208	75.70
6 pairs 6″ φ ports @ 8′ ctrs		1954	188,000	156.70	620	175.80
4 port bonnet type		1952	50,800	59.70	560	74.60
None		1953	22,000	101.90	570	125.00
Reducer—venturi		1955	130,000	68.40	680	70.50
None		1936	3,950	3.00	206	10.20
None		1927	37,000	19.50	206	66.20
None		1938	33,260	13.00	236	38.60
Double outlets 12″ φ turned down		1926	26,726	17.80	208	59.90
Double outlets 12″ φ turned down		1949	60,000	40.00	480	58.30
None		1955	28,865	41.20	680	42.40
None		1955	7,557	16.00	680	16.50
None		1929	5,892	8.10	207	27.40
None	*Total 851′	1948	2,948	23.20	510	31.90
None	*Total 738′	1948	4,979	23.00	510	31.60
None	*Total 853′	1946	14,789	37.40	346	75.70
None	*Total 1190′	1948	20,768	23.40	510	32.10
None		1928	6,065	8.00	206	27.20
None	*Total 1552′	1948	17,350	21.90	510	30.10
None		1947	46,000	73.40	400	128.50
None		1945	18,000	80.40	308	182.90
4–10″ φ ports @ 50′ ctrs		1954	243,420	53.20	460	81.00
5–16″ φ ports @ 65′ ctrs		1938	585,260	83.60	236	248.20
12–24″ & 36″ φ ports @ 15′ ctrs		1954	1,677,461	372.50	460	567.00
7–nozzles @ 20′ ctrs		1938	33,592	64.00	236	190.00
Special S-shaped port		1947	576,000	26.20	400	45.90
None		1927	241,532	40.10	206	136.20
None		1949	189,386	50.50	510	69.40
—		1939	25,000	22.70	236	67.40
—		1934	16,880	16.00	198	56.60
—		1944	50,000	40.00	299	93.60
Terminal chamber		1913	8,443,840	330.00	100	2,310.00
—		1939	18,000	18.00	236	53.40
—		1946	60,000	54.60	346	110.50
Terminal chamber—4 ports		1952	1,816,402	144.00	640	157.50
Crowsfoot, 3–200′, 24″ φ branches		1925	165,000	24.20	207	81.90
57 pairs 5″×16″ ports @ 3′ ctrs	*C.I. diffuser	1939	314,141	224.00	236	665.00
18 pairs 4″×14″ ports @ 18′	*C.I. diffuser	1936	380,125	46.20	206	157.00
36 pairs 5″×16″ ports @ 18′		1939	507,255	68.60	236	203.50
None		1928	275,366	44.40	206	151.00
None		1930	864,000	65.20	203	225.00
None		1953	1,176,000	118.70	680	122.10
None		1933	147,266	143.70	170	591.00
None		1913	65,977	220.00	100	1,540.00
None	*Surf half R.C.	1952	82,300	54.60	610	62.60
None		1948	175,000	58.40	450	90.90
—		1936	22,000	60.30	206	205.00
—		1925	125,000	35.20	207	119.00
—		1916	70,000	26.90	130	145.00
—		1918	205,000	64.00	189	237.00
—		1932	70,000	100.00	157	446.00
—		1927	286,281	103.20	206	350.70
120–4″ φ ports @ 8′ ctrs 60″–24″ φ diffuser	*1780′ R.C. 3550′ Stl.	1911	134,000	25.10	100	175.70
3–36″ φ branches		1931	243,200	79.90	189	296.00

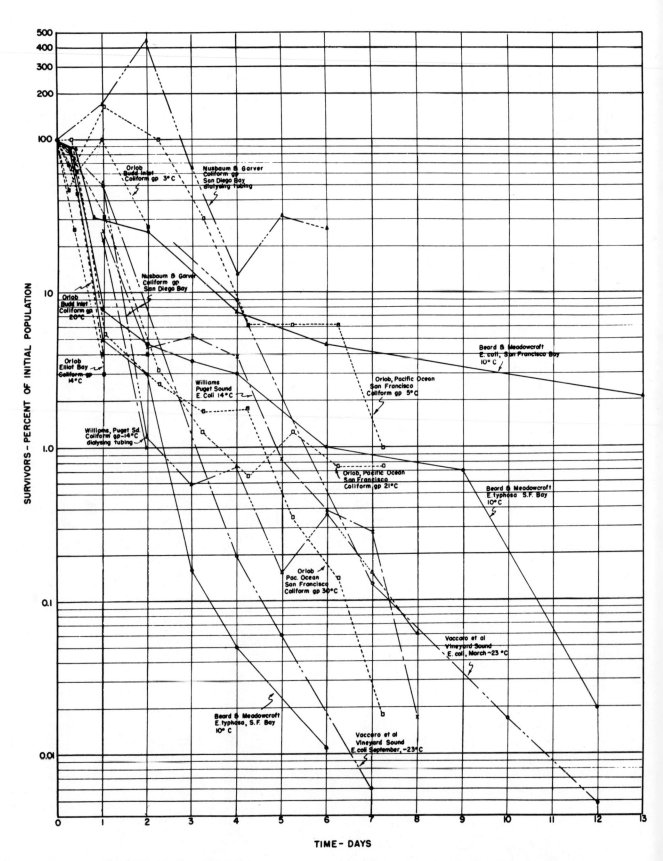

Fig. 17.27. Survival of bacteria in raw sea water; summary of reported investigations (after Pearson, 1956)

is of the order of 10,000 to 100,000 times that of California (Peixoto, daMotta, Selleck and Pearson, 1962). The results of a survey on the problem of coliform standards for recreational waters (American Society of Civil Engineers, Public Health Activities Committee, 1963) indicate that there is practically no evidence to support the existing numerical values used in classifying the quality of bathing waters. It appears that a considerable amount of work must be done by public health authorities before the engineer can design an ocean outfall sewer with complete confidence.

REFERENCES

Abraham, G., Jet diffusion in liquid of greater density, *J. Hyd. Div., Proc. ASCE*, **86**, HY6, Paper 2500 (June, 1960), 1–13.

————, Discussion of jet discharge into a fluid with a density gradient, *J. Hyd. Div., Proc. ASCE*, **88**, HY2 (March, 1962), 195–98.

————, Jet diffusion in stagnant ambient fluid, doctoral thesis, Delft University (also Delft Hydraulics Lab. Pub. No. 29), 1963.

Acrivos, A., B. D. Babcock, and R. L. Pigford, Flow distributions in manifolds, *Chem. Eng. Sci.*, **10** (1959), 112–24.

Aldridge, Clyde, What's involved in planning and constructing an offshore pipeline, *Oil Gas J.* **54**, 59 (June 18, 1956), 174–77.

American Society of Civil Engineers, Public Health Activities Committee, Coliform standards for recreational waters, *Jour. Sanitary Engineering Division*, **89**, SA4 (August, 1963), 57–94.

Asphalt Institute, *Asphalt in hydraulic structures*, Manual Series No. 12, April, 1962.

Blackman, Berkeley, Dredging at inlets on sandy coasts, *Proc. First Conf. Coastal Eng.*, Berkeley, Calif.: The Engineering Foundation Council on Wave Research, 1951, pp. 169–74.

Blume, John A., and James M. Keith, Rincon offshore island and open causeway, *J. Waterways Harbors Div., Proc. ASCE*, **85**, WW3, Paper No. 2170 (September, 1959), 61–92.

Brater, E. F., Low-cost shore protection used on the Great Lakes, *Proc. Fourth Conf. Coastal Eng.*, Berkeley, Calif.: The Engineering Foundation Council on Wave Research, 1954, pp. 214–26.

Brooks, Norman H., Diffusion of sewage effluent in an ocean current, *Proc. First Intern. Conf. Waste Disposal Marine Environment*. London: Pergamon Press, 1960, pp. 246–67.

Bruun, Per, Coastal protection: review of methods for defence, *Dock and Harbour Authority*, **34**, 397 (November, 1953), 217–22; 398 (December, 1953), 233–37.

California Institute of Technology, *Wave protection aspects of harbor design*, Hydro. Lab., Report No. E–11, August, 1952.

Committee on Sewerage and Sewage Treatment, Marine disposal of wastes, *J. Sanitary Eng. Div., Proc. ASCE*, **87**, SA1 (January, 1961), 23–56.

DeRouville, A., P. Bessen, and P. Petry, État actuel des études internationales sur les efforts dus aux lames, *Annales Ponts Chaussées*, **108**, Part 2, 7 (July, 1938), 5–113.

Edelman, T., Safety of seawalls, *Proc. Seventh Conf. Coastal Eng.*, Berkeley, Calif.: The Engineering Foundation Council on Wave Research, 1961, pp. 817–18.

Evenson, Don, Univ. Calif., personal communication, 1961.

Frankel, Richard J., and James D. Cumming, Turbulent mixing phenomena of ocean outfalls, Univ. Calif. IER, Tech. Rept. No. HEL–3–1, February, 1963. (Unpublished.)

Garber, W. F., A critical review of the bacteriological standards for bathing waters, *27th Annual Conv. Calif. Sewage and Industrial Wastes Assn.*, April, 1955.

Glenn, A. H., Progress report on solution of wave, tide, current, and hurricane problems in coastal operations, *Oil Gas J.*, **49**, 7 (June 22, 1950), 174–77.

Granthem, Kenneth N., Wave run-up on sloping structures, *Trans. Amer. Geophys. Union*, **34**, 5 (October, 1953), 720–24.

Haferkorn, H. E., *Breakwaters, a bibliography*. U.S. Army, Corps of Engineers, The Engineer School Library, 1932.

Hart, William E., Jet discharge into a fluid with a density gradient, *J. Hyd. Div., Proc. ASCE*, **87**, HY6 (November, 1962), 171–200.

Hickson, R. E., and F. W. Rodolf, Design and construction of jetties, *Proc. First Conf. Coastal Eng.*, Berkeley, Calif.: The Engineering Foundation Council on Wave Research, 1951a, pp. 227–45.

————, and F. W. Rodolf, History of Columbia River jetties, *Proc. First Conf. Coastal Eng.*, Berkeley, Calif.: The Engineering Foundation Council on Wave Research, 1951b, pp. 283–93.

Horikawa, Kiyoshi, *Japanese construction practice on groins*. Seminar on Groins at Princeton, N.J., sponsored by the ASCE, October, 1958. (Unpublished.)

————, and C. Sonu, An experimental study on the effect of coastal groins, *Coastal Eng. Japan*, Vol. 1, Committee of Coastal Eng., Japan Society of Civil Engineers, 1958, pp. 59–74.

Horton, Donald F., Design and construction of groins, *Proc. First Conf. Coastal Eng.*, Berkeley, Calif.: The Engineering Foundation Council on Wave Research, 1951, pp. 246–53.

Howe, R. J., and B. G. Collipp, Offshore mobile units—present and future, *Mech. Eng.*, **79**, 4 (April, 1957), 335–38.

Hoyle, J. W., and G. T. King, The longitudinal stability of beaches, *J. Inst. Municipal Engrs.*, **82**, 5 (November, 1955), 181–91.

Johnson, J. W., *The action of groins on beach stabilization*, Univ. Calif. IER, Tech. Rept. 3–283, April, 1948. (Unpublished.)

———, The littoral drift problem at shoreline harbors, *J. Waterways Harbors Div., Proc. ASCE*, **83**, WW1, Paper No. 1211, April, 1957.

———, Mixing and dispersion by wind waves, *Proc. First Intern. Conf. Waste Disposal Marine Environment*. London: Pergamon Press, 1960, pp. 328–43.

Keller, J. D., The manifold problem, *J. Applied Mechanics*, **16**, 1 (March, 1949), 77–85.

Kressner, B., Tests with scale models to determine the effect of currents and breakers upon a sandy beach, and the advantages of installation of groins, *Bautechnik*, **6** (June 12, 1928), 374–86. G. P. Specht, trans. Univ. Calif., Berkeley.

LaFond, E. C., How it works—the NEL oceanographic tower, *Proc. U.S. Naval Inst.*, **85**, 11 (November, 1959), 146–48.

Life, Lighthouse gone to sea, **52**, 10 (March 9, 1962), 43–48.

McNamara, E. J., Seven-mile hot pipeline in Gulf of Mexico, *Civil Eng.*, **30**, 4 (April, 1960), 47–49.

McNown, John S., Mechanics of manifold flow, *Trans. ASCE*, **119** (1954), 1103–18; discussions, 1119–42.

Mariner, L. T., and W. A. Hunsucker, Ocean cooling water systems for two thermal plants, *J. Power Div., Proc. ASCE*, **85**, PO4 (August, 1959), 65–85.

Masch, F. D., Mixing and dispersive action of wind waves, Ph.D. thesis in Civil Eng., Univ. Calif., IER Tech. Rept. 138–6, November, 1961.

———, *Observations on vertical mixing in a closed wind wave system*, Univ. Calif. IER, Tech. Rept. 138–7, January, 1962. (Unpublished.)

Miller, David R., and William R. Albriton, Spectacular underwater pipeline, *Western Construction*, **34**, 5 (May, 1959), 63–76.

Moore, B., The risk of infection through bathing in sewage-polluted waters, *Proc. First Intern. Conf. Waste Disposal Marine Environment*. London and New York: Pergamon Press, 1960, 29–38.

Mosonyi, Emil, *Water power development, Vol. 1, low-head power plants*. Hungarian Acad. of Sci., 1957.

Nagai, Shoshichiro, Arrangements of groins on a sandy beach, *J. Waterways Harbors Div., Proc. ASCE*, **82**, WW2, Paper No. 1063, September, 1956.

———, and Hirokazu Kubo, Motion of sand particles between groins, *J. Waterways Harbors Div., Proc. ASCE*, **84**, WW5, Paper No. 1876, December, 1958.

National Academy of Sciences, *Experimental drilling in deep water at La Jolla and Guadalupe sites*, National Research Council, Div. Earth Sci., AMSOC Committee, Pub. No. 914, 1961.

Oil Gas J., World's longest pipeline pull made in Persian Gulf off Iran, **57**, 50 (Dec. 7, 1959), 98–99.

Paape, A., Experimental data on the overtopping of seawalls by waves, *Proc. Seventh Conf. Coastal Eng.*, Berkeley, Calif.: The Engineering Foundation Council on Wave Research, 1961, pp. 674–81.

Pacific Gas and Electric Co., Application No. 43808 to the Public Utilities Commission, State of California, Oceanographic Data, 1961.

Pearson, Erman A., *An investigation of the efficacy of submarine outfall disposal of sewage and sludge*, State of California, Water Pollution Control Board, Publication No. 14, 1956.

———, (ed.), *Proc. First intern. Conf. Waste Disposal Marine Environment*. London and New York: Pergamon Press, 1960.

———, Marine waste disposal, *Engineering J.*, **44**, 11 (November, 1961), 3–8.

Peixoto, E. C., A. S. daMotta, R. E. Selleck, and E. A. Pearson, Marine waste disposal considerations at Rio de Janeiro, Brazil, *Seventh Congr., Interamerican Assoc. Sanitary Engrs.*, June 10–15, 1962, Washington, D.C. (in press).

Posey, C. J., and J. H. Sybert, Erosion protection of production structures, *Proc. Ninth Congr., Intern. Assoc. Hyd. Res.*, Belgrade, **4**, Paper No. 18, 1961.

Powell, W. H., Maintaining Vancouver's submerged pipe lines, *Eng. News-Record*, **114**, 4 (Jan. 24, 1935), 124–25.

Rawn, A. M., F. R. Bowerman, and Norman H. Brooks, Diffusers for disposal of sewage in sea water, *J. Sanitary Eng. Div., Proc. ASCE*, **86**, SA2, Part 1 (March, 1960), 65–105.

———, and H. K. Palmer, Predetermining the extent of a sewage field in sea water, *Trans. ASCE*, **94** (1930), 1036–86.

Rechtin, E. C., J. E. Steele, and R. E. Scales, Engineering problems related to the design of offshore mobile platforms, *Trans., Soc. Naval Architects Mar. Engrs.*, **65**, (1957), 633–81.

Santema, P., About the estimation of the number of days with favorable meteorological and oceanographical conditions for engineering operations on the sea coast and in estuaries, *Proc. Fifth Conf. Coastal Eng.*, Berkeley, Calif.: The Engineering Foundation, Council on Wave Research, 1955, pp. 405–410.

Savage, Rudolph P., Wave run-up on roughened and permeable slopes, *J. Waterways Harbors Div., Proc. ASCE*, **84**, WW3, May, 1958.

———, *Laboratory study of the effect of groins on the rate of littoral transport*, U.S. Army, Corps of Engineers, Beach Erosion Board, Tech. Memo. No. 114, June, 1959.

Saville, Thorndike, Jr., Wave run-up on shore structures, *J. Waterways Harbors Div., Proc. ASCE*, **82**, WW2, April, 1956.

———, Personal communication, July 20, 1962, (Unpublished letter.)

———, and J. M. Caldwell, Experimental study of wave overtopping on shore structures, *Proc. Minn. Intern. Hyd. Conv.*, Minneapolis, 1953, pp. 261–69.

Sibul, O. J., Flow over structures by wave action, *Trans. Amer. Geophys. Union*, **36**, 1 (February, 1955), 61–69.

———, and E. G. Tickner, *A model study of the run-up of wind-generated waves on levees with slopes of 1:3 and 1:6*, U.S. Army, Corps of Engineers, Beach Erosion Board, Tech. Memo. No. 67, 1955.

———, and E. G. Tickner, *Model study of overtopping of wind-generated waves on levees with slopes of 1:3 and 1:6*, U.S. Army, Corps of Engineers, Beach Erosion Board, Tech. Memo. No. 80, April, 1956.

Shimano, T., M. Hom-ma, K. Horikawa, and T. Sakou, Functions of groins: fundamental study on beach sediment affected by groins (1), *Proc. Fourth Conf. Coastal Eng. in Japan*, Japan Society of Civil Engineers, 1957.

Silveston, Barnett, Oil terminal for Pakning, Sumatra, *J. Waterways Harbors Div., Proc. ASCE*, Vol. 85, No. WW1, pp 89–95, March, 1959.

Sneddon, Richard, West Coast's largest submarine line laid in record time, *Petroleum Engr.*, **25**, 10 (September, 1953), D7–D8.

Spencer, R. W., and John Bruce, Cooling water for steam electric stations on tidewater, *J. Power Div., Proc. ASCE*, **86**, PO3 (June, 1960), 1–26.

Terzaghi, Karl, *Investigations of filter requirements for under-drains*, U.S. Army, Corps of Engineers, Waterways Experiment Station, Tech. Memo. 183-1, 1941.

———, *Laboratory investigations of filters for Enid and Grenada dams*, U.S. Army, Corps of Engineers, Waterways Experiment Station, Tech. Memo. 3-245, 1948.

Timothy, P. H., New method for coating offshore pipelines, *Oil Gas J.*, **54**, 35 (Jan. 2, 1956), 70–75.

U.S. Army, Corps of Engineers, New York District, *Beach Erosion Control Report on Coop. Study of the Atlantic Coast of New Jersey (Sandy Hook to Barnegat Inlet)*, House Document No. 361, 84th Cong., 2d Sess., 1956.

———, Beach Erosion Board, *Shore Protection Planning and Design*, Tech. Rept. No. 4, 1961.

U.S. Coast Guard, *Changes in Coast Guard buoy design*, Civil Eng. Rept. No. 25, January, 1957a.

———, *A study of the technical and economic feasibility of erecting lighthouses in deep waters*, Civil Eng. Rept. CG–250–28, December, 1957b.

———, *Performance characteristics of buoys*, Civil Eng. Rept. No. 33, February, 1959.

U.S. Congress, *Inquiry into the collapse of Texas Tower No. 4, Hearings before the Preparedness Investigating Subcommittee of the Committee on Armed Services*, Senate, 87th Cong., 1st Sess., 1961a.

———, *Investigation of the preparedness program. Report by Preparedness Investigating Subcommittee of the Committee on Armed Services, United States Senate on the collapse of Texas Tower No. 4*, S. Res. 43, 87th Cong., 1st Sess., 1961b.

Van Asbeck, Baron W. F., Modern design and construction of dams and dikes built with the use of asphalt, *Proc. Seventh Conf. Coastal Eng.*, Berkeley, Calif.: The Engineering Foundation Council on Wave Research, 1961, pp. 819–35.

Verschoore, E., 1st Question. Layout of outer protective works, maintenance of depths in harbours on sandy shores and before mouths of estuaries. Results obtained, *14th Intern. Congr. Navigation*, Brussels, Report No. 66, 1935.

Weight, Robert H., Ocean cooling water system for 800 MW power station, *J. Power Div., Proc. ASCE*, **84**, PO6, Paper 1888, December, 1958.

Wemelsfelder, P. J., The disaster in the Netherlands caused by the storm flood of February 1, 1953, *Proc. Fourth Conf. Coastal Eng.*, Berkeley, Calif.: The Engineering Foundation Council on Wave Research, 1954, pp. 258–71.

Wicker, C. F., *Summary statement concerning importance of a groin design criteria*. Seminar on Groins at Princeton, N.J., sponsored by the ASCE, October, 1958. (Unpublished.)

Williams, R. K., Jr., The asphalt groins at Ocean City, Maryland, *Asphalt Inst. Quart.* (April, 1955), 6–8.

Zajac, E. E., Dynamics and kinematics of the laying and recovery of submarine cable, in *Submarine cable: oceanography; marine biology and cable mechanics*, Bell Telephone System, Tech. Pub., **36**, no. 5, Monograph 2847 (September, 1957), 1129–1207.

CHAPTER EIGHTEEN

A Case History

1. INTRODUCTION

A case history of a structure and a region is a useful technique to emphasize some of the effects of man on his environment. The history given here does this, and at the same time shows how some of the information and techniques given in this book have been used. The case history given here is essentially a paper written by the author for the Task Committee on Sand Bypassing of the American Society of Civil Engineers (Wiegel, 1959). It is concerned with the construction of a breakwater at Santa Barbara, California, and the subsequent problems encountered. A breakwater constructed at Santa Barbara, California interrupted the littoral drift of sand. The man-made harbor filled with sand and the downcoast beaches were nearly completely stripped of sand. At first, the harbor was dredged periodically, and later nearly continuously, with the sand being dumped downcoast to maintain these beaches. Although costly, and not always satisfactory in all details, this method has in general adequately performed its function.

Santa Barbara is located on a sandy lowland on the southern coast of California in an area afforded considerable protection from waves of the Pacific Ocean by offshore islands (Fig. 18.1). The coastal region in this area is generally rugged with mountains extending almost to the shore. There are many rock headlands separated by coves with narrow cobble, gravel, or sand beaches. There are no large rivers, but there are many steep streams which drain narrow basins averaging 5 miles in width which, when in flood, carry large quantities of sediments and debris (U.S. Congress, 1938).

The major source of sand for the beaches in the Santa Barbara area is probably the streams feeding into the ocean (U.S. Congress, 1938; 1948; Handin, 1951; Trask, 1952; 1955). Some sand comes from the erosion of the coastal cliffs, but it is a common observation that this erosion has been small in this area in recent historical times. It has been determined by Trask (1952a; 1955) that some sand does move into this area from the region upcoast from Point Concepcion and Point Arguello. Other possible sources are sand blown by the wind and

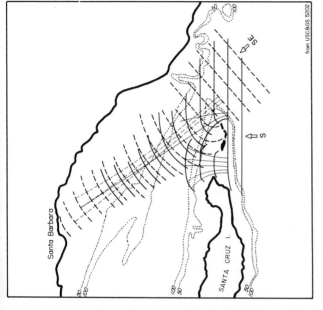

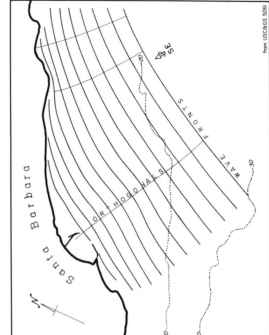

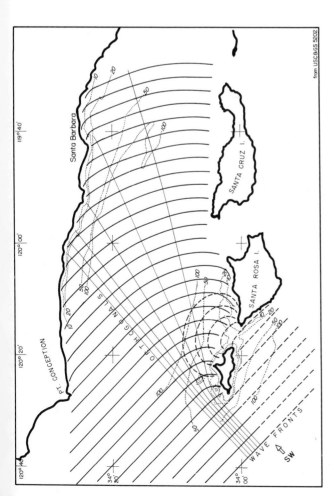

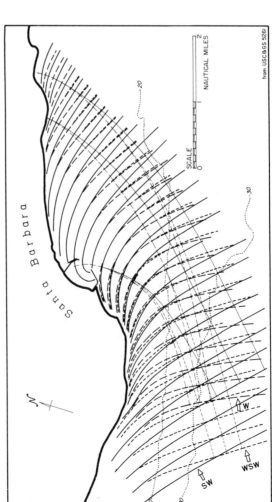

Fig. 18.1. Refraction diagrams for Santa Barbara and vicinity for waves of 10-second period: (a) waves from the southwest; (b) waves from the south and southeast; (c) expanded refraction drawing, waves from the west, west-southwest, and southwest; (d) expanded refraction drawing, waves from the southeast (after O'Brien, 1950)

Fig. 18.2.

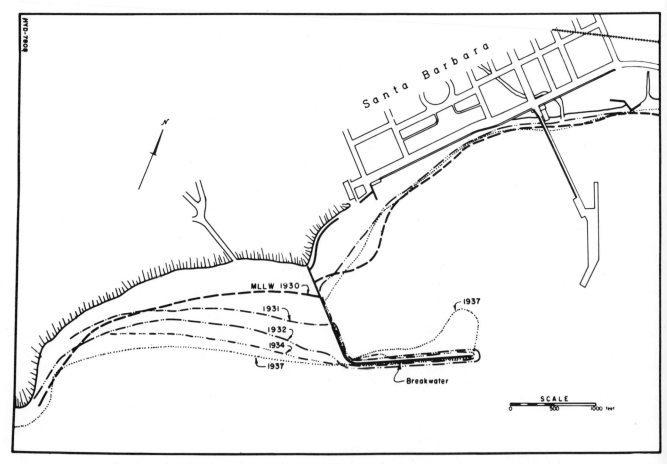

Fig. 18.3. Shoreline changes upcoast of the Santa Barbara breakwater (after Johnson, 1957)

474

sand from offshore areas. Observations by Bascom (1951) during times of high winds indicate that sand blown by the wind is a relatively unimportant source of sand for these beaches. Because of the relative importance of the streams as a source of sand for this region, the amount of sand should generally be strongly dependent upon the amount of rainfall in the area, and in particular, it should be dependent upon the number of storms that produce high rates of runoff, as it is this condition that results in a large quantity of sand being carried to the littoral region. A good correlation between littoral drift of sand and rainfall in this area has been shown (U.S. Congress, 1938; 1948), but it is cautioned that other effects of the same storms, such as the intensity and direction of the waves resulting from them, must be considered. From the long-range viewpoint, the program of flood control and soil erosion prevention within the United States will probably seriously affect the amount of sediment moving to the ocean, and make more serious the problem of coastal erosion.

The median diameter of the sand on the beach face and the slope of the beach face is best presented in relationship to other beaches on the Pacific Coast of the United States as shown in Fig. 14.17.

The waves in this region are either generated between the Channel Islands and the coast (local wind waves) or are generated in the ocean seaward of the islands, perhaps many thousands of miles away. Local observations lead to the conclusion that the predominant waves are from a westerly direction entering Santa Barbara Channel between Point Concepcion and San Miguel Island. This results in a wave-generated longshore drift from west to east. Current measurements along the shoreline (U.S. Congress, 1938) confirm this. Waves seldom exceed a height of 3 ft except during times of storms from the southeast with the average wave period being about 12 sec, but ranging from 8 sec to 16 sec (U.S. Congress, 1938; Johnson, 1953).

The net result of this is a gradual drift of sand from west to east, except during short intervals of time.

The refraction of waves arriving at Santa Barbara has been studied by O'Brien (1950) using graphical methods. He concluded from the refraction drawings that waves from the general westerly direction would almost always break at about the same angle with the beach (an observed fact) even though they were originally from the W, WSW, or SW (see Fig. 18.1 for an example). In addition, the waves would be less high at Santa Barbara than along the coast west of the city. Refraction diagrams drawn for waves from the S and SE directions show that only the SE waves would be relatively high at Santa Barbara.

As a result of the origin of the waves, the offshore islands, and the refraction of the waves, the wave pattern at Santa Barbara is nearly the same for all times of the year, with the exception of the waves from the occasional SE storms.

During 1927–28 a detached rubble mound breakwater with a concrete cap was constructed off Point Castillo at Santa Barbara, Fig. 18.2(a). The breakwater was about 1425 ft in length roughly parallel to shore, with an arm heading towards shore which was about 400 ft long. Because waves refract in shoaling water as well as diffracting in the lee of a breakwater, the waves from the west moved in an easterly direction through the gap between the shore and the breakwater whereas the waves in the vicinity of the easterly tip of the breakwater swung around in such a manner that they moved in a westerly direction in the gap area. This interaction of the waves caused sand to deposit in this area, causing the harbor to shoal. Because of this shoaling, in 1930 the breakwater was extended to shore by an additional 600 ft of structure, Fig. 18.2(b).

This breakwater proved to be an effective trap of the sand that was moving in the littoral stream. This sand filled the area west of the breakwater, moving along the breakwater, swinging around its tip, and eventually moving into the harbor, Fig. 18.3. Thus, one end result was the filling of the harbor with sand (Fig. 18.4). From measurements made of the rate of fill west of the breakwater, an annual drift of about 270,000 cu yd (about 740 cu yd per day) was computed (U.S. Congress, 1938). The waves acting on the beach immediately east of the breakwater continued to move sand eastward; this sand was not replaced, however, because the sand that would have been normally deposited on it was trapped by the breakwater. This problem gradually worked its way eastward. Thus, a second end result was the erosion of the beaches to the east of the breakwater.

Engineering studies were made of the problems and a decision was reached to dredge the harbor and to dump the sand east of the harbor in such a manner that this sand would be moved by the waves along the coast to build up the beaches and to maintain them in a desirable condition.

2. PERIODIC DREDGING AND SAND BYPASSING

The harbor was first dredged in the fall of 1935 by hopper dredges, with a little over 200,000 cu yd of material being moved. Material was dumped in about 22 ft of water approximately 1 mile east of the breakwater and about 1000 ft from shore. It formed a mound about 2200 ft long and 5 ft high. It was expected that the waves would move the sand onshore and eastward. This did not occur, as is evident in the chart of the May, 1937, survey (Fig. 18.5). Surveys made in 1946 showed that the mound at that time was at no point more than a foot below its 1937 depth.

As a result of this loss of material, insofar as beach

Fig. 18.4. Aerial photograph of Santa Barbara harbor looking upcoast, January 31, 1947

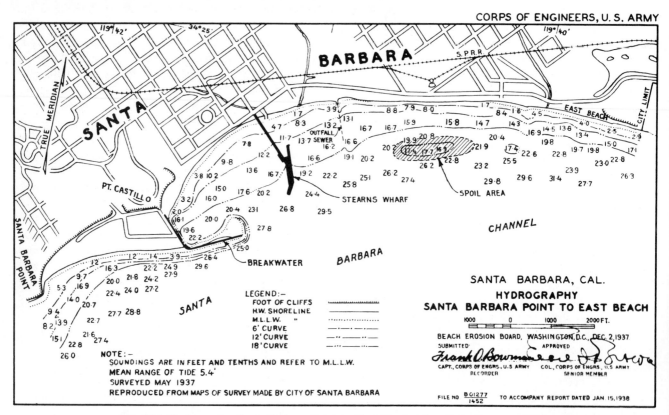

Fig. 18.5. Hydrography, Santa Barbara Point to East Beach (after U.S. Congress, 1938)

maintenance was concerned, it was decided to pump the material by pipeline in future dredgings and deposit it on the "feeder beach" just east of the outfall sewer line. The disposal areas for the 1938, 1940, 1942, and 1945 dredgings are shown in Fig. 18.6. Surveys of the effect of the sand moving from the feeder beach eastward showed that the shorter wider feeder beaches of the 1940 and 1942 dredgings were about as effective as the

longer narrower feeder beach of 1938, but that the shortest widest feeder beach (1945) was not so effective (U.S. Congress, 1948).

For complete details of the effectiveness of this method of sand bypassing, the reports of the U.S. Army Corps of Engineers to Congress should be consulted. A representative example is shown in Fig. 18.6. The net loss of material from the western side of East Beach occurs

476

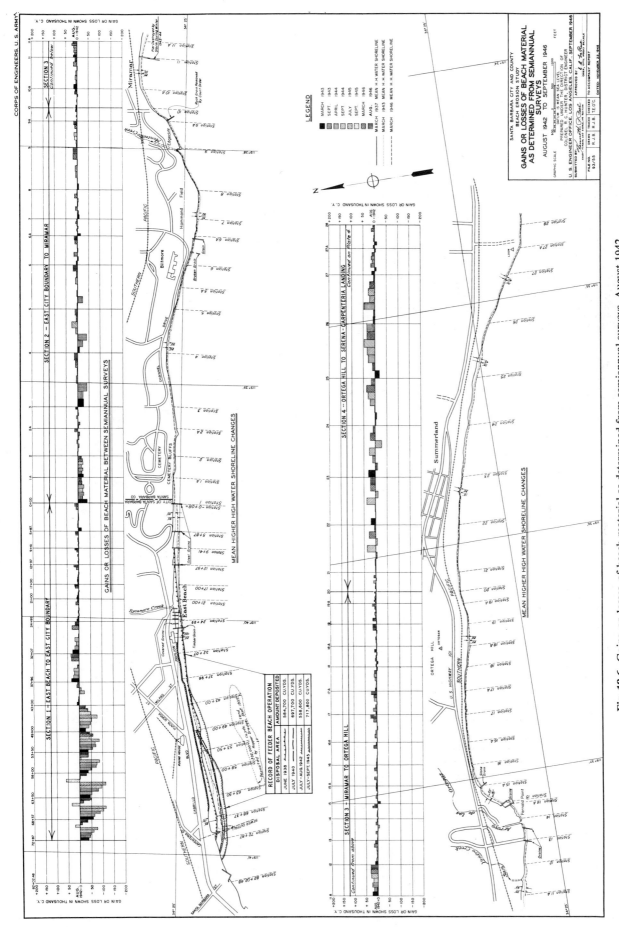

Fig. 18.6. Gains or losses of beach material as determined from semiannual surveys, August 1942 to September 1946 (after U. S. Congress, 1948)

Table 18.1. SANTA BARBARA HARBOR MAINTENANCE DREDGING

Year (a)	Contract unit price (b)	Pay quantity (c)	Contract payment (d) (b×c)	Government cost (e)	Nonpay yardage (f)	Gross yardage (g) (c+f)	Total cost (h) (d+e)	Cost per yard (i) (h/g)
Oct–Nov., 1952	0.2647	1,070,000	$283,670	$17,660	104,000	1,174,000	$301,330	0.2567
May–June, 1949	0.1838	704,106	$129,414.68	$18,624.91	82,419	838,152	$160,703.59	0.1917
Navy (Part of Area "C")	0.22	51,627	$ 11,357.94	$ 1,306.06				
May–June, 1947	0.1739	675,044	$100,000.15	$ 9,999.85	68,933	642,977	$110,000.00	0.1711
June–Sept., 1945	0.2204[j]	717,773	$158,171.16	$11,823.34	0	717,773	$170,000.00	0.2368
July–Aug., 1942	0.218	558,610[k]	$121,776.98	$ 9,531.97	41,500	600,110[k]	$131,308.95	0.2188
Inside breakwater west of Sec.B	—	38,690	2,400.00	Paid for by local interests $100/hr for 24 hr.				0.6203
June–July, 1940	0.178	646,067	$114,999.93	$ 9,532.05	51,652	697,719	$124,531.98	0.1785
May–June, 1933	0.21	532,000	$122,220.00	$ 5,884.11	24,427	606,427	$128,104.11	0.2112
Sept., 1935						202,000	$ 30,000.00	0.1485

  [j] Bid price 24.5¢ based on 420,500; additional 300,000 at 16.5¢ by supplemental agreement.

  [k] Includes about 103,000 cu yd dredged for the Navy in Area "C" (vicinity of Coast Guard pier and approach).

    The preceding data were obtained from U.S. Army Engineer District, Los Angeles, Ltr. SPLGP–R, Oct. 28, 1958.

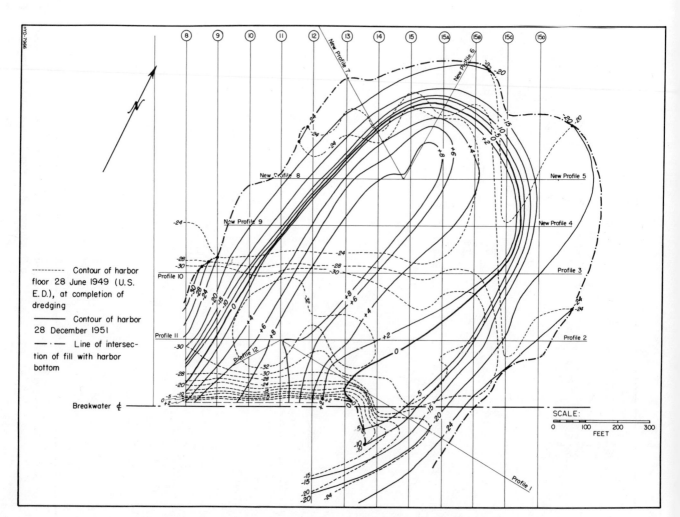

Fig. 18.7. Comparison of sand deposit 2½ years after dredging and dredged condition at end of breakwater inside harbor, Santa Barbara, California (from Wiegel, 1959)

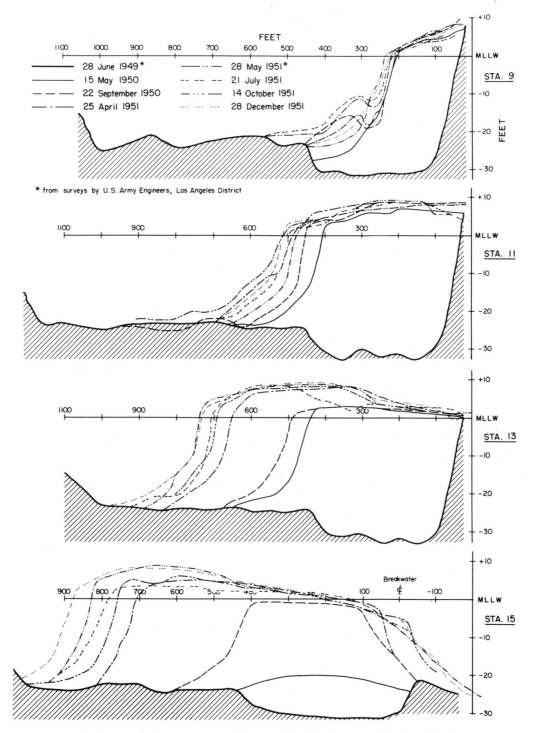

Fig. 18.8. Profiles of fill at end of breakwater, inside harbor, Santa Barbara, California (after Wiegel, 1959)

because this is the feeder beach. It is evident that waves moved the sand eastward causing an increase in the amount of sand all along the beach, except in a few locations. The determination of the reasons for the regions of opposite trend requires considerable attention in the future.

It should be emphasized that waves deposited the sand not only on the beach but for an appreciable distance offshore. Hence, a large portion of the sand necessary to replenish the beach is not noticed by the layman or the casual observer.

A summary of the dredging operations is shown in Table 18.1. With the exception of the 1935 dredging data, the data presented in Table 18.1 were obtained from the U.S. Army Engineer District, Los Angeles.

One of the most interesting and certainly one of the most important phenomena associated with the sand bypassing problem is the formation of the sand deposit at the end of the breakwater. The harbor in this area was dredged every two years or so, a portion of it to about 20 ft below MLLW. An example of the bottom contours just after a dredging are shown in Fig. 18.7, together with the bottom contours approximately $2\frac{1}{2}$ years after the dredging. Some of the profiles are shown in Fig. 18.8. One remarkable characteristic of the spit is the abrupt change in slope called the *breakoff line*. The relationship between the area defined by the breakoff and the volume of this fill was found to be remarkably consistent (Bascom, 1951; Trask, 1952b).

The buildup of this sand spit is remarkably consistent. In Fig. 18.9 are shown the MLLW profiles for seven cases. There are two modes: one perpendicular to the breakwater, and one making an angle of about 60 degrees with the breakwater. Why there should be two modes is not known, but it may be associated with the effects of the occasional SE storms, or with the strength of the littoral current, the latter possibility being indicated by a model study described later in this chapter. The change of the size and shape of the MLLW contour with time between two sets of dredgings is shown in Fig. 18.10.

It is not known why the change in orientation toward the end of the buildup should have occurred. Some detailed cross sections of the development of this sand spit are shown in Fig. 18.8.

Figure 18.11 shows the rate of fill for a two-year interval, with the surveys being taken often. The amount of material accumulated within the harbor (primarily the sand spit) for the years 1932 through 1951 are tabulated in Table 18.2 and shown in Fig. 18.12.

Lapsley (1937) made a model study of the effect of the breakwater on the sand movement at Santa Barbara. This study was made with a model of 1:750 horizontal scale and 1:100 vertical scale. Clean quartz sand with a median diameter of 0.35 mm and specific gravity of 2.62 was used. The wave steepness used during the tests corresponded to a prototype condition of continuous local storm waves. In spite of these discrepancies between prototype and model conditions, the

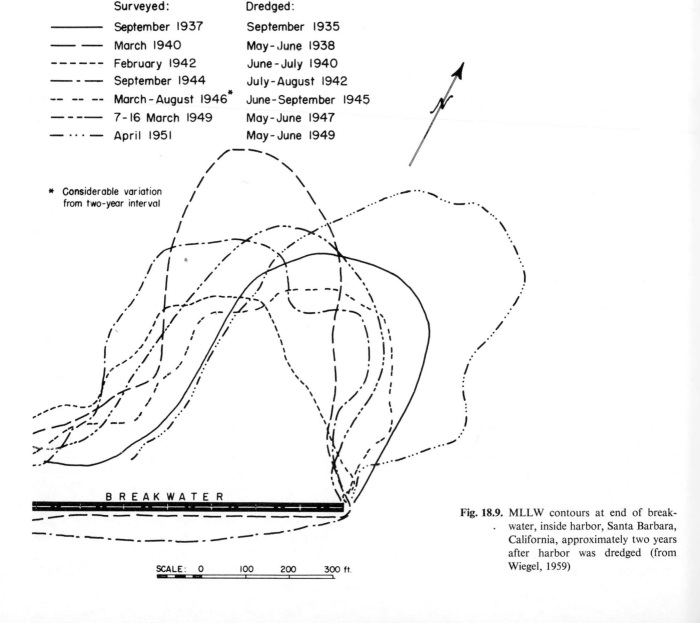

Surveyed:	Dredged:
———— September 1937	September 1935
—— —— March 1940	May-June 1938
- - - - - - February 1942	June-July 1940
—— · —— September 1944	July-August 1942
—— —— —— March-August 1946*	June-September 1945
—— · —— 7-16 March 1949	May-June 1947
—— ··· —— April 1951	May-June 1949

* Considerable variation from two-year interval

BREAKWATER

SCALE: 0   100   200   300 ft.

Fig. 18.9. MLLW contours at end of breakwater, inside harbor, Santa Barbara, California, approximately two years after harbor was dredged (from Wiegel, 1959)

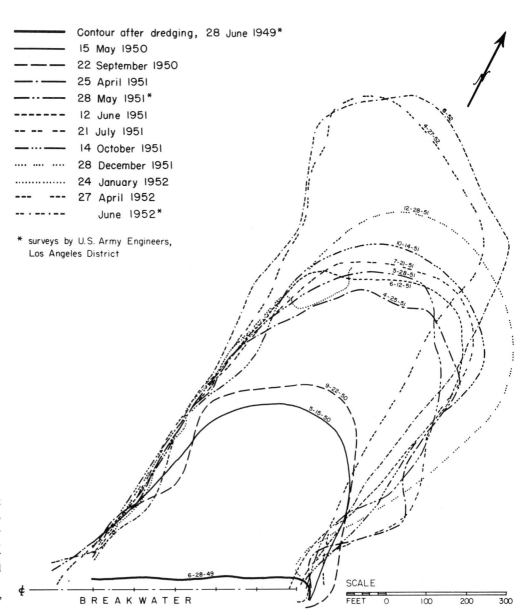

Contour after dredging, 28 June 1949*
15 May 1950
22 September 1950
25 April 1951
28 May 1951*
12 June 1951
21 July 1951
14 October 1951
28 December 1951
24 January 1952
27 April 1952
June 1952*

* surveys by U.S. Army Engineers,
Los Angeles District

Fig. 18.10. MLLW contours at end of breakwater, inside harbor, Santa Barbara, California, at various times after harbor was dredged (May-June, 1949) (from Johnson, 1957, and Wiegel, 1959)

results of the tests were informative. It was found that sand could be transported at the rate of 2.6 cu in. per minute in the model. Taking the average rate of sand transport in the prototype as 740 cu yd per day results

in a time scale such that 1 min in the model corresponds to 4.2 days in the prototype.

It was found that, after the sand had moved to the end of the breakwater, the current parallel to the breakwater was so strong that it swept the sand toward Stearns wharf. The model breakwater was constructed of a strip of heavy sheet metal, impervious and vertical. In subsequent tests, small rocks were cemented to the breakwater to decrease the current strength. It was then found that the sand swung around the end of the breakwater and deposited landward of the tip, as observed in the prototype.

Lapsley stated that he was never able to reproduce in the model the sand condition inside the breakwater that extended from the spit to the base of the breakwater arm. It has recently been found that this condition resulted from the movement of sand through the "porous" breakwater, and this movement has now been prevented by grouting the breakwater.

Although the model did reproduce certain charac-

Table 18.2. Estimated Annual Volumes of Accretion in Santa Barbara Harbor, 1932–51, Inclusive (from J. W. Johnson, 1953)

Year	Accretion (cubic yards)	Year	Accretion (cubic yards)
1932	225,000	1942	245,000
1933	265,000	1943	210,000
1934	390,000	1944	235,000
1935	200,000	1945	295,000
1936	225,000	1946	400,000
1937	205,000	1947	330,000
1938	235,000	1948	370,000
1939	280,000	1949	330,000
1940	310,000	1950	300,000
1941	260,000	1951	283,000

Average, 1932–51, inclusive = 279,650 cu yds per year

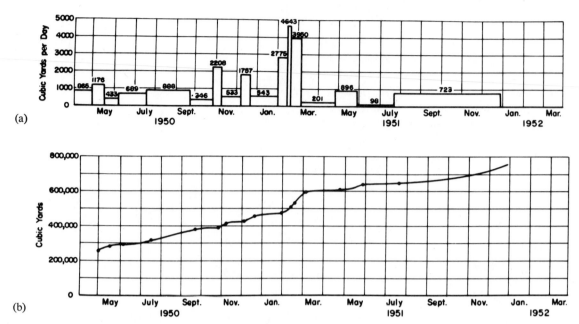

(a)

(b)

Fig. 18.11. (a) Rates of harbor accretion from June 28, 1949, to December 28, 1951; (b) Cumulative volume of harbor accretion from April 28, 1950 to December 28, 1951 (after Wiegel, 1959)

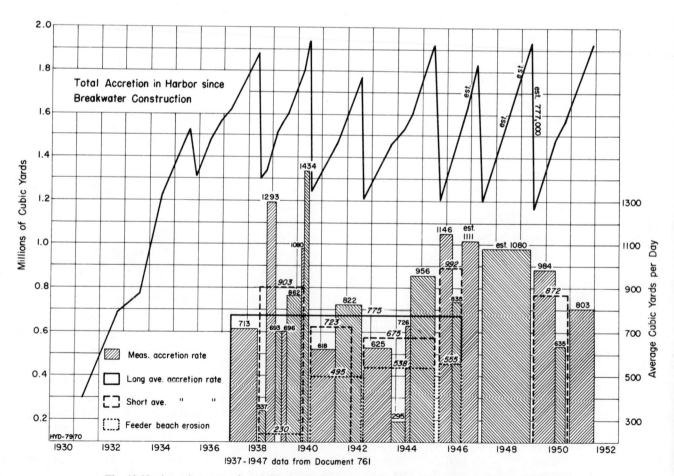

Fig. 18.12. Accretion rates of sand fill at end of breakwater inside harbor, Santa Barbara, California (after Johnson, 1957, and Wiegel, 1959)

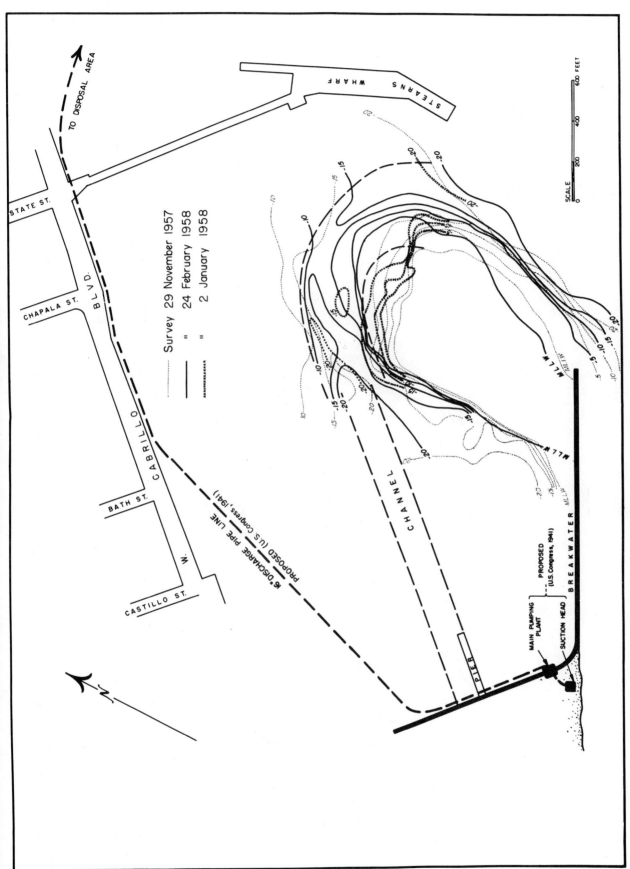

Fig. 18.13. Effect of southeast storm on sand deposit at end of breakwater, inside harbor, Santa Barbara, California (from Wiegel, 1959)

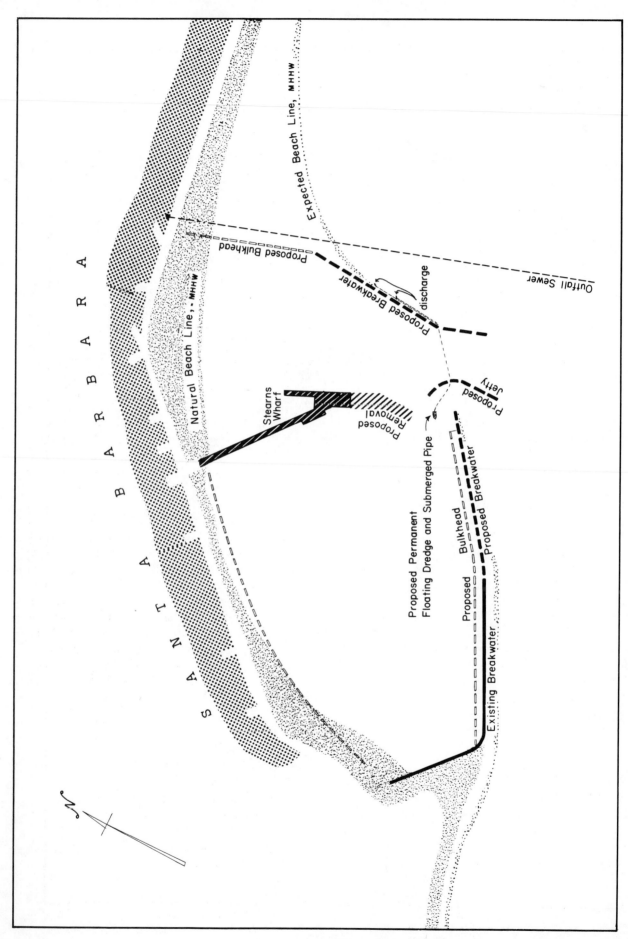

Fig. 18.14. Proposed harbor improvement, Santa Barbara, California (after Penfield, 1948)

teristics of the sand movement, it did not predict the amount of sand that would be moved, the width of the beach, etc.

3. CONTINUOUS DREDGING AND SAND BYPASSING

The last large dredging of the harbor was done during November and December, 1952. The city of Santa Barbara decided to use the sand spit inside the end of the breakwater as a protection to the inner harbor from waves from a southerly direction and to maintain an entrance into the inner harbor by means of nearly continuous dredging with a small dredge. The channel was to be maintained to a minimum depth of 15 ft below MLLW in the location shown in Fig. 18.13. Dredging operations were started in August, 1956.

The dredge had a capacity of about 1600 cu yd of material per 8-hr shift, and experience has shown that because of wave and weather conditions it pumps only about 72 per cent of the time. Although this capacity is adequate on a yearly basis, it is not adequate on a short-term basis. For example, in Fig. 18.11, it can be seen that between the surveys of February 15, and February 20, 1951, an average of 4643 cu yd per day were added to the sand spit, with most of this material probably being added during a SE storm lasting about one day.

This short-term difficulty is even more pronounced if one considers that large amounts of material may be shifted a few hundred feet without even considering a net increase of material. For example, one SE storm occurred during February 4–5, 1958 (City of Santa Barbara, 1958). The Public Works Department of Santa Barbara made spot checks of the water depth in the area both shortly before and shortly after the storm. It was stated that the storm had caused an unbelievable change to occur in the entrance channel. Many thousands of cubic yards of sand were moved into the channel. The westerly bank of the entrance channel shifted easterly from 100 to 400 ft. This nearly closed the entrance channel. The bottom contours obtained from a survey made shortly after the storm are shown in Fig. 18.14,

together with two sets of bottom contours surveyed a few months before the storm. It should be emphasized that the spit is developed primarily by waves from a westerly direction, and as such is probably unstable with respect to waves from a southerly direction.

4. PROPOSED FUTURE METHODS OF SAND BYPASSING

Other methods of sand bypassing have been proposed for Santa Barbara. One of these methods utilizes a fixed dredging plant; another utilizes a floating dredging plant.

The proposed fixed dredging plant has been described in a report to the U.S. Congress (1941). The general layout is shown in Fig. 18.13. The initial cost of the installation was estimated to be $106,000 (in 1941) and the annual operating cost was estimated to be $67,700 (in 1941), based upon transporting 300,000 cu yd of sand per year (22.6 cents per cubic yard). This system was conceived to operate with the existing breakwater, and it was expected that it would prevent the sand from moving around the breakwater by trapping the sand in the location shown in the plan drawing. One drawback to the system was the long distance through which the sand had to be pumped (from 5000 to 6000 ft, 16-in. diameter pipe, carrying about 10 per cent solids by volume at 13 ft/sec) with the high pumping costs associated with this distance (about a 1000 horsepower pump).

The floating dredging plant system was conceived to work in conjunction with a modified harbor as shown in Fig. 18.14. In this plan, a second breakwater would be constructed to the east of existing breakwater and the existing breakwater lengthened. The sand would be pumped through a submerged pipe only across the entrance to the proposed harbor, cutting down the distance through which the sand would have to be moved. Under this system, the waves and currents would be relied upon to continue to move the sand to the east.

At the present time neither of these plans has been implemented.

REFERENCES

Bascom, Willard N., *Investigation of coastal sand movements, Santa Barbara, California, Part I*, Univ. Calif. IER, Tech. Rept. series 14, issue 8, January, 1951. (Unpublished.)

City of Santa Barbara, Public Works Department, *Progress Report, Harbor maintenance dredging program*, May, 1958. (Unpublished.)

Handin, John W., *The source, transportation, and deposition of beach sediment in Southern California*, U.S. Army, Corps of Engineers, Beach Erosion Board, Tech. Memo. No. 22, March, 1951.

Johnson, J. W., Sand transport by littoral currents, *Proc.*

Fifth Hyd. Conf., Bulletin 34, State Univ. Iowa Studies in Engineering, 1953, pp. 89–109.

———, The littoral drift problem at shoreline harbors, *J. Waterways Harbors Div.*, ASCE, **83**, Paper 1211, April, 1957.

Lapsley, William W., Sand movement and beach erosion, unpublished M.S. thesis in Civil Engineering, Univ. Calif. 1937.

O'Brien, M. P., Wave refraction at Long Beach and Santa Barbara, California, *Bull.* Beach Erosion Board, Corps of Engineers, U.S. Army, **4**, 1 (January, 1950), 1–12.

Penfield, Wallace C., and Tom D. Cooke, *Report on a proposed improvement plan for Santa Barbara Harbor*. Santa Barbara, Calif.: The Santa Barbara Harbor Development Association, June, 1948. (Unpublished.)

Trask, Parker D., *Source of beach sand at Santa Barbara, California, as indicated by mineral grain studies*, Beach Erosion Board, Corps of Engineers, U.S. Army, Tech. Memo. No. 28, October, 1952a.

———, *Santa Barbara studies: Compilation of basic data*, Univ. California IER, Tech. Rept. series 14, issue 2, July, 1952b. (Unpublished.)

———, *Movement of sand around Southern California promontories*, Beach Erosion Board, U.S. Army Corps of Engineers, Tech. Memo. No. 76, June, 1955.

U.S. Army, Corps of Engineers, Beach Erosion Board, *Supplementary Report, Beach erosion study at Santa Barbara, California*, Feb. 18, 1942. (Unpublished.)

U.S. Congress, *Beach erosion at Santa Barbara, Calif.*, House Document No. 552, 75th Cong. 3d Sess. Mar. 18, 1938.

———, *Santa Barbara, Calif.: Letter from the Secretary of War*, House Document No. 348, 77th Cong., 1st Sess. Aug. 6, 1941.

———, *Santa Barbara, Calif., Beach erosion control study*, House Document No. 761, 80th Cong., 2d Sess., Dec. 22, 1948.

Wiegel, R. L., Sand bypassing at Santa Barbara, California, *J. Waterways Harbors Div., Proc. ASCE*, **85**, WW2 (June, 1959), 1–30.

CHAPTER NINETEEN

Moorings

1. INTRODUCTION

The motion of a freely floating vessel in a seaway is extremely complex. A floating body can have three rectilinear and three rotational motions. The rectilinear motions are heave—the vertical motion; surge—the horizontal motion along the longitudinal axes (through the bow and stern); and sway—the horizontal motion along the transverse axis. The rotational motions are yaw—about the vertical axis; roll—about the longitudinal axis; and pitch—about the transverse axis. A freely floating ship, or one underway has three natural periods, with the restoring force being gravity (or buoyancy): heave, roll, and pitch. When a ship is moored, it has three additional natural periods, with the elasticity and cable weight (catenary effect) being the restoring force: surge, sway, and yaw. The addition of mooring lines has little effect on the three natural periods of heave, roll, and pitch.

Studies have been made (Froude, 1861; Kriloff,

1898) of the motion of rolling and pitching of ships. More recently, Haskind (1946) and Havelock (1956) have studied heaving. Other investigators have studied these problems, and we have recent summaries of the present state of knowledge (Weinblum and St. Denis, 1950; Weinblum 1955; 1957). Weinblum and St. Denis present equations and graphs of various ship motion functions, among which are the magnification (response) factors for various degrees of freedom, which are shown to be functions of the ratio of the wave period to the natural period of the ship, and the exciting factors, which are shown to be functions of the ratio of the wave length to the ship length (see Fig. 19.1 for an example). Both these sets of factors were, of course, dependent upon several other wave and ship characteristics. The authors state that little is known of the surge, sway, and yaw motions of floating vessels and that the knowledge of the total motion of a vessel, including phase relationships between the various motions, is almost nil, although it is known that certain couplings between the

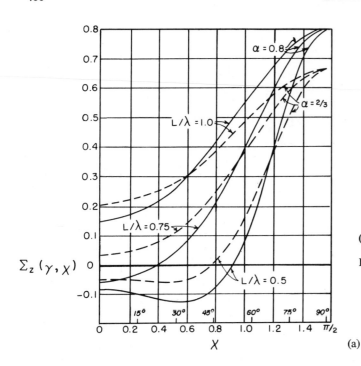

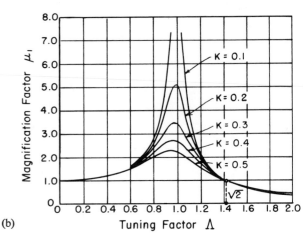

$\Sigma_z(\gamma, \chi)$

(b)

(a)

Fig. 19.1. (a) Heaving force function $\Sigma_z(\gamma, \chi)$ for two wall-sided vessels plotted against heading angle χ with L/λ as parameter, Waterline coefficient $\alpha = 2/3$ and 0.8; (b) Magnification factor

$$\mu_1 = \frac{\sqrt{1 + k^2\Lambda^2}}{\sqrt{(1 - \Lambda^2)^2 + k^2\Lambda^2}}$$

(from Weinblum & St. Denis, 1950)

motions in the different degrees of freedom existed. Of particular importance to studies of mooring problems are the possibilities of induced roll, sway, and yaw even when the vessel is encountering only head seas (Grim, 1952). Additional work on coupling effects has been done by Havelock (1955).

The aforementioned studies were concerned with periodic waves of uniform amplitude. A few studies of ship motion in nonuniform waves have been published (St. Denis and Pierson, 1953; Fuchs, 1955; Sibul, 1955b). Some comprehensive measurements of the motions of a ship at sea have been made, and the results compared with theory (Williams, 1953; Jasper, 1956; Cartwright and Rydill, 1957; Warnsinck and St. Denis, 1957). The tests last mentioned were concerned in particular with slamming of ships (see also Szebehely and Todd, 1955), which results from the ship motion (pitch and heave) getting out of phase with the wave motion. Slamming is of particular importance because of the very high impact loadings on the ship under these conditions.

Basic to all theories and to the prediction of prototype characteristics from model studies is a knowledge of wave-making resistance and skin friction resistance of ships, and the phenomenon of added masses (Saunders, 1957). There are hundreds of papers on the resistance characteristics (see, for example, U.S. Navy, David Taylor Model Basin, 1957) and many papers on added mass (Wendel, 1950; Landweber and Yih, 1956; Gerritsma, 1957; Porter, 1960; Cummins, 1962).

The details of ship motion are beyond the scope of this chapter and the reader is referred to the papers cited for a working knowledge of the subject. Before

dealing with mooring problems, it is necessary for an engineer to become familiar with the simpler problem of a ship without moorings.

Mechanical details of moorings, such as the relative value of chain, wire rope, and rope, fendering systems, and the data on strength of mooring materials are not considered here, as the details necessary to use this information are numerous. (See U.S. Navy, *Mooring Guide*, U.S. Navy, Bureau of Yards and Docks, 1954 for this information.)

2. LIMITED THEORY OF MOORED SHIPS

The problem of a moored ship is more complicated than the problem of unrestrained ship motion. There are three additional natural periods, and the restoring force (the mooring system) is nonlinear. A theoretical solution of the problem where the ship may be moored at any heading to the waves is not available at the present time. There are approximate solutions for several of the six components of motion, and computer solutions are obtainable. This is not true for the case of all six motions, including correct phase and couplings among the motions.

Wilson has studied the case of a ship moving only in surge, with the forcing functions being a long standing wave, and his results are presented in a series of papers (1950; 1959; Abramson and Wilson, 1955). The generalized geometry of the dockside mooring rope suspension, as presented by Wilson, is shown in Figs. 19.2 and 19.3. Referring to Fig. 19.2 for an explanation of the symbols (where in the generalized case the lowest point of cable sag is at C_1 which is outside of A_1B), the equations

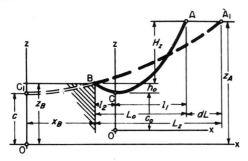

Fig. 19.2. Geometry of mooring rope suspension (from Wilson, 1950)

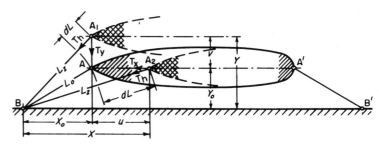

Fig. 19.3. Geometry of longitudinal and transverse ship motion (from Wilson, 1950)

of the catenary were given as

$$z = c \cosh \frac{x}{c} \qquad (19.1a)$$

$$s = c \sinh \frac{x}{c} \qquad (19.1b)$$

where s is the length of the hypothetical cable from point C_1 to any point (x, z) on the catenary. Further,

$$z_B + H_z = c \cosh \frac{x_B + L_z}{c} \qquad (19.2a)$$

$$z_B = \cosh \frac{x_B}{c} \qquad (19.2b)$$

and thus,

$$H_z = c \left(\cosh \frac{x_B + L_z}{c} - \cosh \frac{x_B}{c} \right) \qquad (19.3a)$$

$$s = c \left(\sinh \frac{x_B + L_z}{c} - \sinh \frac{x_B}{c} \right). \qquad (19.3b)$$

After certain simplifications were made, Wilson gave the horizontal component of rope tension T_h, at points A_1 and B as

$$T_h = \frac{W_c L_z^2}{12 \left[s^2 - (H_z^2 + L_z^2) \right]}, \qquad (19.4)$$

where W_c is the weight per unit length of the cable.

Wilson's next step was to determine the relationships between cable tensions and elongation for several types and sizes of standard mooring cables (coir rope and steel wire rope). In order to utilize these data, a further simplification was made wherein it was shown that, for common dockside moorings, the following equation was of sufficient accuracy:

$$L_z^2 = s^2 - H_z^2. \qquad (19.5)$$

Utilizing this approximation, and referring to Fig. 19.3 for an explanation of the symbols, the relationships between the components of cable tension and the components of ship movements were developed. These relationships have been shown in Fig. 19.4, together with the best-fit exponential curves, which were of the form:

$$T_{x,y} = k(u \text{ or } v)^n, \qquad (19.6)$$

where k is a constant and n is a numerical exponent.

For the case of a ship with many mooring lines, this was expressed for the longitudinal directions as

$$\sum T_x = C u^n, \qquad (19.7)$$

where C is a constant which depends upon the number, size, and condition of the cables and n depends upon the tension in the cables.

A harbor oscillation has a relatively long period (usually in the range of a minute or more—sometimes much longer) and usually has a low amplitude. It is possible to describe the water particle velocities and accelerations rather simply compared with the case of seas and swell. These approximations were used by Wilson (1950), together with an approximation which expresses the wave force on a vessel. It is not necessary here to go through the various steps, but merely to present the final results:

$$\frac{d^2 u}{dt^2} + \frac{N_x}{M_x'} \frac{du}{dt} + \frac{C}{M_x'} u^n$$

$$= \frac{Ag}{\sigma D} \frac{\sinh kd - \sinh kS}{\cosh kd} \frac{\sin k\lambda/2}{k\lambda/2} \qquad (19.8)$$

$$\sin kb \left(\sigma \sin \phi - \frac{N_x}{M_x'} \cos \phi \right)$$

where N_x is a coefficient of linear damping, M_x' is the virtual mass of the ship for surge motion, A is the amplitude of the standing wave, g is the gravitational constant, k is the wave number $(2\pi/L)$, σ is $2\pi/T$, T is the wave period, d is the water depth, D is the draft of the

Fig. 19.4. Relationship between rope tension and ship movement (from Wilson, 1950)

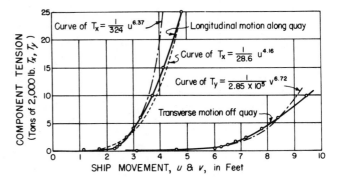

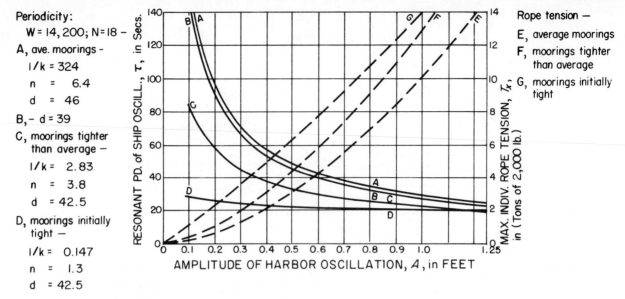

Periodicity:
W = 14,200; N = 18 –
A, ave. moorings –
  1/k = 324
  n = 6.4
  d = 46
B, – d = 39
C, moorings tighter
  than average –
  1/k = 2.83
  n = 3.8
  d = 42.5
D, moorings initially
  tight –
  1/k = 0.147
  n = 1.3
  d = 42.5

Rope tension —
E, average moorings
F, moorings tighter
  than average
G, moorings initially
  tight

Fig. 19.5. Influence of rope tightness on the resonance in longitudinal ship motion (from Wilson, 1950)

ship, S is the clearance beneath the ship, λ is the block length of the ship, b is the distance of the mass center from the side of the dock.

After certain other simplifications were made, Wilson (1950), showed the relationship between resonant period of ship oscillation and maximum individual cable tension, T_x, versus harbor oscillation amplitudes, A, for several conditions of cable tension (Fig. 19.5). The effect of cable tension on the periodicity of the system is apparent; the higher the tension, the lower the resonant period. The effect of amplitude is relatively unimportant for cables with high tension, but very important for cables with low tension: the greater the harbor oscillation amplitude, the lower the resonant period of the systems. Equation 19.8 has been treated by Abramson and Wilson (1955) using the Ritz method (which is useful for certain nonlinear differential equations), and solutions were obtained.

Solutions for a particular case are lengthy and the reader is referred to the later papers of Wilson (1959a; 1959b), Joosting (1957), and O'Brien and Kuchenreuther (1958b) for details. It is enlightening to look at the solution in two forms. In Fig. 19.6, the solution is given for one particular mooring arrangement, with the independent variables being the harbor oscillation amplitude and period. In Fig. 19.7, the solution for one harbor oscillation period is given, with the positions on the curves A, B, C, D, E referring to the same letters in Fig. 19.6. Figure 19.7 states that the ship with a given arrangement of mooring lines set at given initial tensions is subject to a 45-sec harbor oscillation, and the amplitude of the harbor oscillation gradually increases to a maximum of 1.0 ft, and then gradually decreases. Looking at Fig. 19.6, it can be seen that, as the harbor oscillation amplitude increases, an amplitude curve is reached that is tangent to the vertical harbor oscilla-

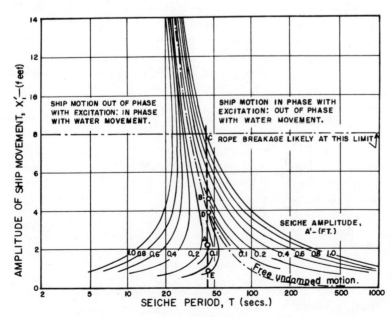

Fig. 19.6. Response characteristics to surging of the moored U.S.S. Norton Sound (after Wilson, 1959)

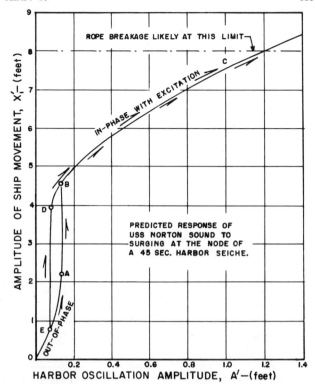

Fig. 19.7. Surging response of U.S.S. Norton Sound to nodal stimulus of 45-second harbor oscillation (after Wilson, 1959)

tion period line (Point *A*), and all values of harbor oscillation amplitudes above this point of tangency are less than this particular value in the region below the curve representing free undamped motion. In this region, the surging motion is out of phase with the excitation but in phase with the water motion. At this point, the motion suddenly increases to the value associated with the same value of the harbor oscillation amplitude, but in the region above the curve representing free undamped motion (Point *B*). Further increase

in harbor oscillation amplitude results in a slow increase in the surging motion of the ship, in phase with the excitation. As the harbor oscillation amplitude decreases, the ship motion decreases along the same line in Fig. 19.7 that it followed when increasing. Now, however, it does not jump down at Point *B*, but at some other place, Point *D*. The curve from Point *D* to Point *E* is conjecture; however, it must be something like this.

If the harbor oscillation period were higher, say about 100 sec, a similar jump would occur, but it would occur for a very low harbor oscillation, and the jump would be so small as to be practically unnoticeable.

The results of the analysis of the same problem by O'Brien and Kuchenreuther (1958b) are shown in Fig. 19.8. It is interesting that, for the standing waves of small steepness, it was found that the restoring force was proportional to the wave slope (Fig. 19.9). This was also true for the case reported by Deacon, Russell, and Palmer (1957).

3. MODEL STUDIES

Although the simpler cases of moored ships are amenable to analytical analyses of the sort outlined in the preceding section, model studies will probably be used to study the more difficult cases. These include ships moored in cross seas, mooring lines attached to buoys which in turn are moored to anchors, docked ships subject to lateral waves with the docks, including camel systems to absorb impact loadings.

a. *Model laws.* Gravity, inertia, elastic, surface tension, and viscous forces exist in a moored floating body system, neglecting compressibility effects. From a practical standpoint, it is not possible to model the system in compliance with all modeling laws simultaneously. The vessel and the waves were modeled in accordance

Fig. 19.8. Restoring force versus wave period for looser lines (from O'Brien and Kuchenreuther, 1958b)

Fig. 19.9. Graph of restoring force with slope of water surface for initial tensions from 15-20 KIPS (from O'Brien and Kuchenreuther, 1958b)

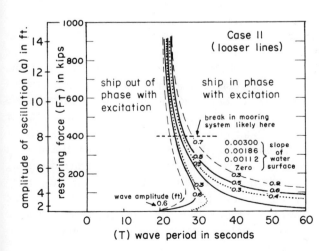

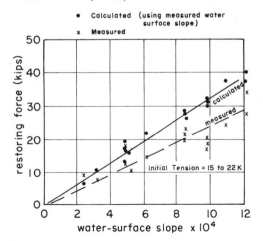

with the Froude modulus. The effect of the Reynolds modulus was neglected in the development of the model (however, turbulence existed in the flow) in conformity with standard naval architectural practice. The effect of surface tension was assumed to be of little importance, as the model was to be made large enough that the waves would be well into the "gravity regime" and the cables were to be large enough that surface tension forces would not interfere with their motions.

The most serious question raised in using a model of this sort is in regard to the effect of neglecting the Reynolds modulus (viscous forces). A brief analysis by Dr. Kitter (California Research Corporation, La Habra), however, indicated that the order of magnitude of viscous forces on the mooring cable would be small compared to gravity and inertia forces, and for the range of variables to be considered, the drag coefficient would be relatively insensitive to Reynolds number.

The following model laws were used in the design of the ship model, model mooring cable, and the cable force meter:

Let A = cable area (cross section).

C = a constant of the system; for the case considered herein it is related to the angle the mooring cable makes with the force meter.

E = modulus of elasticity of cable.

g = acceleration of gravity (32.2 ft/sec²).

I = moment of inertia of cable (cross section).

$\bar{L}$ = cable length.

P = applied force.

S = geometric scale ratio (1:80).

$\bar{T}$ = time

W = weight of a quantity.

$-m$ = refers to a model quantity.

$-p$ = refers to a prototype quantity.

$-r$ = refers to a ratio between model and prototype quantities.

$\Delta\bar{L}$ = axial deflection of cable.

γ = unit weight of a quantity.

Φ = force ratio.

For Froude similitude, set the ratio of model to prototype gravity forces equal to the ratio of model to prototype inertia force

(1) Gravity forces

$$\frac{\gamma_m \bar{L}_m^3}{\gamma_p \bar{L}_p^3} = \Phi_G = \gamma_r \bar{L}_r^3. \qquad (19.9)$$

(2) Inertia forces

$$\Phi_I = \frac{\gamma_m \bar{L}_m^3 g \bar{L}_m \bar{T}_p^2}{\gamma_p \bar{L}_p^3 g \bar{L}_p \bar{T}_m^2} = \gamma_r \bar{T}_r^{-2} \bar{L}_r^4 = \Phi_G. \qquad (19.10)$$

From Eq. 19.9 and 19.10,

$$\gamma_r \bar{T}_r^{-2} \bar{L}_r^4 = \gamma_r \bar{L}_r^3$$

$$\sqrt{\bar{L}_r} = \bar{T}_r. \qquad (19.11)$$

Next, the elastic force similitude can be obtained if the ratio of model to prototype gravity forces equals the ratios of model to prototype elastic forces:

(3) Axial elongation forces in the cable

$$\Delta\bar{L} = \frac{P\bar{L}}{AE}; \qquad AE = \frac{P}{\Delta\bar{L}/\bar{L}}; \quad \text{let } K = AE$$

$$\frac{K_m}{K_p} = \Phi_{AE} = \Phi_G = \gamma_r \bar{L}_r^3;$$

then $$K_r = \gamma_r \bar{L}_r^3. \qquad (19.12)$$

(4) Forces causing bending in the cable

$$\Delta = \frac{CP\bar{L}^3}{EI}, \qquad \frac{P_m}{P_p} = \Phi_B = \frac{\Delta_m (EI)_m \bar{L}_p^3}{\Delta_p (EI)_p \bar{L}_m^3} = \Phi_G = \gamma_r \bar{L}_r^3$$

$$\frac{\Delta_m}{\Delta_p} = \bar{L}_r; \quad \text{and} \quad (EI)_r = \gamma_r \bar{L}_r^5. \qquad (19.13)$$

These relationships will be referred to in the following paragraphs on the design and construction of the ship model, the mooring cable model, and the mooring cable force meter.

b. *Ship model.* Before designing a ship model, it is necessary to choose a scale ratio that will satisfy three criteria:

1. That the effects of surface tension will not be important.

2. That the model will be small enough for use in a model basin, yet large enough that minor variations, when magnified to prototype, will not prove to be excessive.

3. That the model will be large enough that a proper scale mooring system can be constructed and the expected mooring cable forces will be of sufficient magnitude to be measured accurately.

An example will be given. A scale model (1:80 scale) of a modified LSM was to be built and tested (Wiegel, Beebe, and Dilley, 1958). The model was scaled from a drawing. Transverse cross section drawings for many stations along the ship were prepared to assure accurate ship dimensions. The superstructure, except for the forward superstructure deck, was left off the model, as were the several decks. The weight of the completely balanced model was to be 3.78 lb, whereas the model shell was designed to weigh 1.21 lb, leaving 2.58 lb to be added for static and dynamic balancing.

After the basic model shell was completed, weights were added to bring the model up to its correct scale weight. These weights were placed in such a manner that the center of gravity of the model was properly located, and so that the natural periods of roll and pitch would be correct. This static and dynamic balancing was performed as follows: The model was suspended fore and aft by suspension bars. The weight of the forward suspension was counterbalanced on the scale before placing the model on the scale. Since the distances from the aft suspension bar to the center of gravity, from the aft suspension to the forward suspension bar, and the total weight of the model were known, it was possible to calculate the weight needed to balance the model when the longitudinal center of gravity was correctly positioned; this was done by taking moments about the aft suspension bar. Small pieces of lead were used as weights for adjusting the balance.

The model was placed on a special device and moments were taken about a given point (varied for each trial to minimize the effects of human error) and the necessary weight was calculated and set on the scale. The movable weights then were adjusted until the vertical center of gravity was properly located; this was completed after about twenty trials. After locating the vertical center of gravity the period of roll was adjusted.

The model was placed in the model basin, inclined, and released. The natural period of roll was observed and recorded. The transverse position of the center of gravity was obtained by adjusting weights until zero list was observed. Then lead weights were moved in a direction parallel to the axis of pitch, until the correct period of roll was obtained. By moving the weights parallel to the axis of pitch, the weights remained in the same position with respect to the vertical, so that the vertical center of gravity was unchanged. A few trials with the device used to locate the vertical center of gravity confirmed the fact that the vertical center of gravity had not shifted.

The last step in the dynamic balancing of the model was the adjustment of the period of pitch. The experimental method used is illustrated in Fig. 19.10 and the calculations required for this method are presented below. The period of pitch of a vessel is proportional to the radius of gyration of the vessel in the pitching direction, and is given by the equation (Rossell and Chapman, 1949)

$$T_P = \frac{1.108 K_1}{\sqrt{GM}}, \qquad (19.14)$$

where T_P = natural period of pitch
K_1 = radius of gyration of the model
GM = longitudinal metacentric height of the model
$1.108 = 2\pi/\sqrt{g}$ (where $g = 32.2$ ft/sec²).

Severe damping of the pitching motion of the structure precluded accurate direct measurement of the period of pitch so that it was necessary to determine the radius of gyration of the model about the pitching axis by using the bifilar pendulum method as given by Timoshenko (1937). For the model suspended as shown in Fig. 19.10, it can be shown that the period of oscillation of the bifilar pendulum setup is a function of the radius of gyration of the vessel about the yaw axis. In addition, the radius of gyration about the yaw axis, and the radius of gyration about the pitch axis are essentially equal. It can be shown that (Timoshenko, 1937)

$$T_N = \frac{2\pi K_1}{a} \sqrt{\frac{l}{g}}, \qquad (19.15)$$

where T_N = natural period of oscillation of suspended model in air
l = length of suspending lines
a = equal distance from CG to suspending lines.

In the experimental adjustment of the period of pitch $l = 7.125$ ft, $a = 1.0625$ ft, $g = 32.2$ ft/sec², and $K_1 = 63.4$ ft/80 = 0.792 ft. (The radius of gyration of the prototype was 63.4 ft.) Substituting these values in Eq. 10.15, it was found that the period of the pendulum would have to be 2.2 sec for the radius of gyration in the pitching direction to be correct. The weights in the model were then adjusted until this period of the pendulum was obtained. The longitudinal metacentric height was adjusted by other means.

Following this final adjustment of the weights, the complete balancing procedure was repeated, and the model was found to be balanced, and thus both geomet-

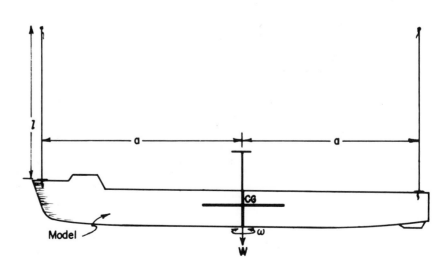

Fig. 19.10. Bifilar pendulum set-up (from Wiegel, Beebe, and Dilley, 1958)

rically and dynamically similar to the prototype, within the limits discussed in the introductory remarks.

c. *Mooring cable model.* Inspection of the model relationships previously derived indicates that, for geometric similarity, the diameter of the model cable must be scaled according to the relationship

$$d_m = \frac{d_p}{80}. \qquad (19.16)$$

If the same unit weight properties are to be maintained, it is necessary to scale the cable according to the relationship

$$d_m = \frac{d_p}{80b}$$

where $b > 1$, since the cable is stranded and the resulting unit weight for a given diameter is less than that of a solid wire. If the axial deflection forces of the mooring system are to be maintained in scale, the diameter of the mooring cable must scale according to the rule

$$(AE)_r = \gamma_r \bar{L}_r^3, \quad \text{but } E_r = \frac{\text{elastic modulus of model}}{\text{elastic modulus of prototype}}$$

$$E_r = \frac{3 \times 10^7}{10^7} = 3 \qquad (19.17)$$

$$(AE)_r = 3d_r^2 = \gamma_r \bar{L}_r^3; \qquad d_r = \frac{\gamma_r}{3} \bar{L}_r^{3/2}.$$

If the bending properties of the cable are to be in scale, it is necessary to scale the bending resistance of the cable cross section according to the relationship, as the moment of inertia is proportional to the fourth power of the diameter

$$(EI)_r = \gamma_r \bar{L}_r^5 \quad \text{or} \quad d_r^4 = \gamma_r \bar{L}_r^5$$

or

$$d_r = \sqrt[4]{\frac{\gamma_r}{3}} \bar{L}_r^{1.25}. \qquad (19.18)$$

It was not possible to satisfy the four model criteria by choosing a proper cable diameter. Therefore, the cable diameter was chosen so that the moment of inertia of the section would correspond to, and be in scale with, that of the prototype stranded cable. This meant it was necessary to choose a diameter somewhat smaller than the outside diameter of the prototype stranded cable, since a geometric scale would mean that the model cable extrapolated to prototype conditions would correspond to a steel rod in bending stiffness. Split lead shot were added to the mooring cable at 1-in. intervals to bring the cable up to the proper weight requirements. An arbitrary spacing of 1 in. between shot having been assumed, the required shot diameter was 0.050 in., which was readily available. The axial deflection characteristics of the cable were calculated and found to be less than those required; hence, some axial flexibility of the cable system was built into the force meters. This kept the total mooring system deflection in scale. The only

criterion not satisfied was that of geometric scale of the diameter, which would result in the viscous drag and inertia forces being even more out of scale. As was assumed in the case of the ship model, however, these forces would be small compared to other forces acting on the cable. Small-diameter stranded cables have been found subsequently which simplify the modeling of the mooring lines.

d. *Force meter.* The force meter served two purposes: it was used as a sensing element in a force recording system, and it was used in the modeled mooring system to satisfy scale factor requirements for cable elongation due to axial loads. In designing the meter, it was necessary to satisfy the similitude requirements for cable deflection (as determined from model laws) and to insure sufficient sensitivity so that the meter could measure the smallest expected forces.

In order to provide the proper deflection characteristics, it was necessary to use other than a simple structural shape. The deflection characteristics of the beam used were calculated through application of the principles of virtual work. The force measuring system was designed to provide maximum possible recorder sensitivity to applied cable force. Strain gauges were used for the force measuring elements; force was measured indirectly by measuring the strains induced in the moments caused by the force in the mooring cable. The strain measurements were recorded with a commercially available oscillograph.

Details of the force meter design, construction, and calibration are not included as they have been published elsewhere (Beebe, 1956).

e. *Tests and results.* The model was placed in the model basin at the desired orientation with respect to the wave generator. The waves and the forces acting on the mooring lines were measured and recorded on oscillographs, and at the same time, vertical and horizontal motion pictures were taken, with the motion pictures being synchronized with the oscillograph by means of timing devices and event markers. The motion pictures were analyzed frame by frame and the results plotted in the manner of Fig. 19.11. In these experiments, two bow and two stern mooring cables were used, with each cable making a horizontal angle of 12 degrees with the centerline of the ship. The cable scope (ratio of cable length to water depth) was 6. The anchor positions were adjusted so that the initial tension in the cables would be 3 kips (3000 lb), prototype, for one series of tests, 6 kips for a second series of tests, and 10 kips for a third series of tests.

Because of the complexity of the motions and force, a large number of results will be given in graphical form in order to give the engineer a feeling for the subject.

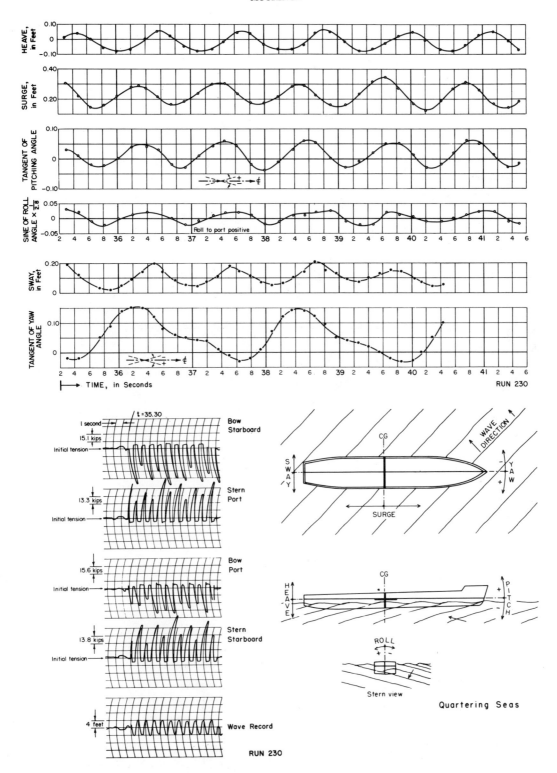

Fig. 19.11. Sample data with model LSM in quartering seas (from Wiegel, Beebe, and Dilley, 1958)

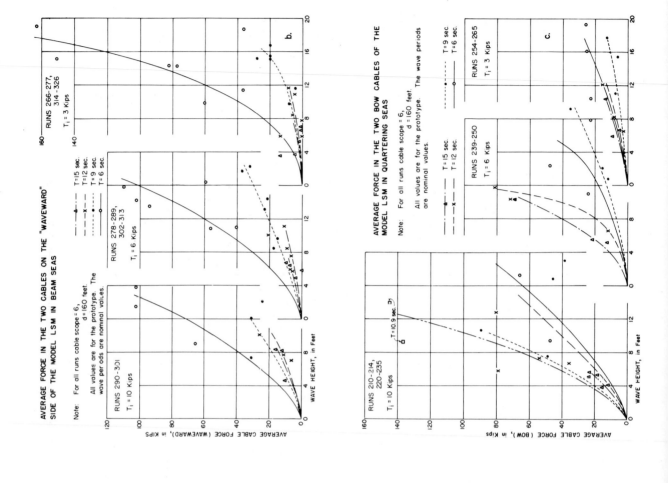

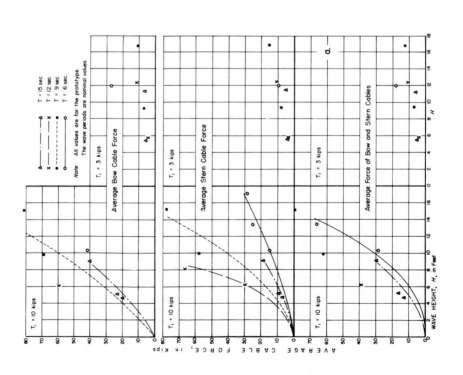

Fig. 19.12. Mooring cable forces of LSM in head, beam and quartering seas (from Wiegel, Beebe, and Dilley, 1958)

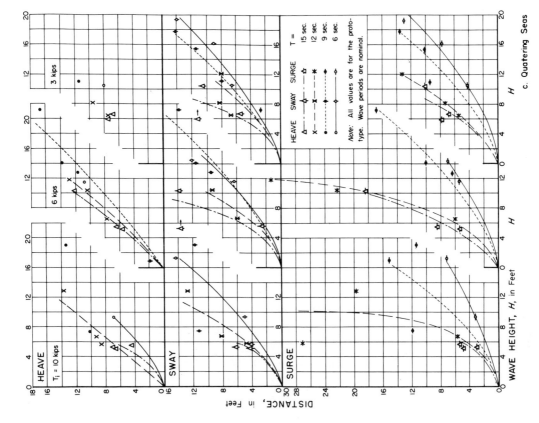

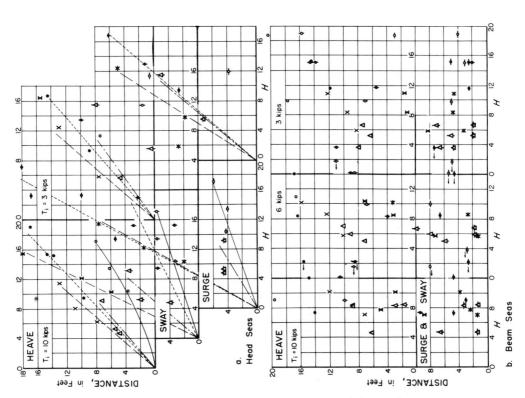

Fig. 19.13. Model LSM heave, sway, and surge in head, quartering, and beam seas (from Wiegel, Beebe, and Dilley, 1958)

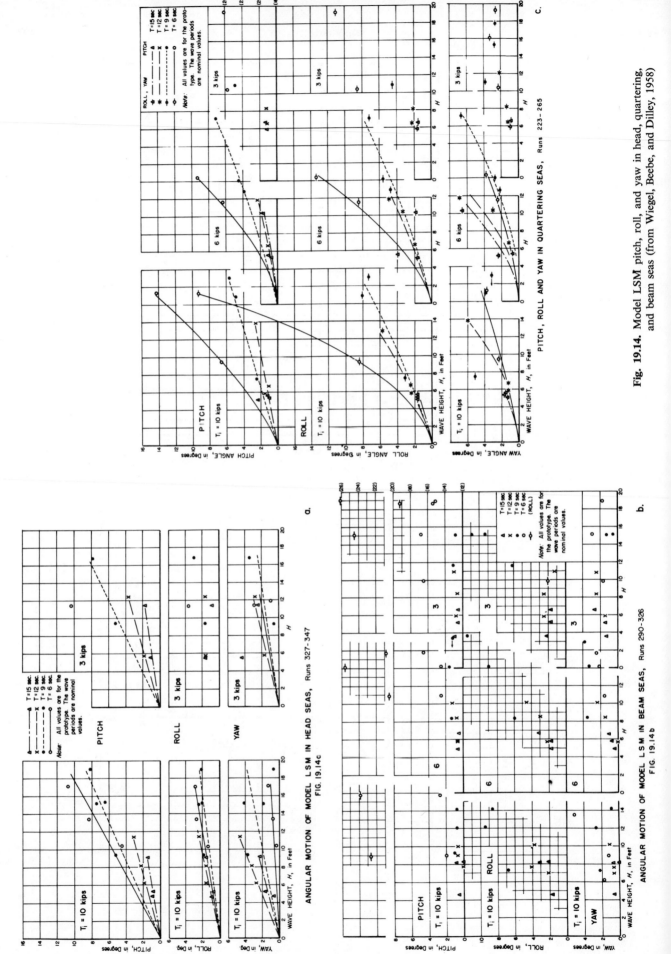

Fig. 19.14. Model LSM pitch, roll, and yaw in head, quartering, and beam seas (from Wiegel, Beebe, and Dilley, 1958)

Some of the results of the tests for the model are given in Figs. 19.12, 19.13, and 19.14 with all values being in terms of the prototype. The values of T_i appearing in the figure are the initial tensions in the mooring lines. The "least count" accuracy of the data was about ± 5 per cent. The scatter of the data was considerable, and much of this was due to the difficulty imposed by the least count when working with a model of this scale. The trends were apparent however, and the scatter of data of heave, surge, and pitch were probably within the necessary limits from an operational standpoint.

The most important fact evident in these data was in regard to the relationship between surging motion (and hence cable force) and the wave characteristics. The data showed a resonance condition occurring for certain combinations of wave and mooring characteristics. A check of the surging characteristics of the test was undertaken. It was found that the natural period of surge (for the 2 mil model wire with lead shot) was about 16 sec for the case of a cable scope of 6, a water depth of 200 ft, and an initial tension of 10 kips (Fig. 19.15). As this determination of the natural period was done with a new set of model cables, and owing to the difficulty in obtaining exact initial tensions, the natural period of surge during runs 1–96 might have been a few seconds less. It was evident from the data in Figs. 19.11 and 19.12 that the relationship between both surging and cable force and the wave height was non-linear, at least when near resonance. Both the surging motion and the cable force increased quite rapidly for the greater wave heights. It is believed that this is the same type of response as is shown in Fig. 19.7.

It is apparent that this difficulty would not be encountered when operations were conducted in an area where the prototype wave period was in the vicinity of 5 to 8 sec, nor would it become apparent for low waves (say, in the neighborhood of 3 feet) owing to the inability of a person to judge the extent of such small motion.

It was found that, for the case of low initial cable tension, the highest mooring cable forces were associated with the shortest wave period; however, as the initial tension increases, the highest forces were associated with the longer periods—those near, or at, resonance.

It was found that the pitching angle was an almost linear function of wave height with the shortest period (within the limits of the experiments) being responsible for the greatest pitch. The pitching angle appeared to be independent of the initial cable tension. Heave was greatest for the longest wave periods and appeared to be independent of the initial cable tension. Surging motion was dependent upon initial cable tension to a great degree. In addition, it was not linearly related to wave height.

As a part of one series of tests, data were obtained

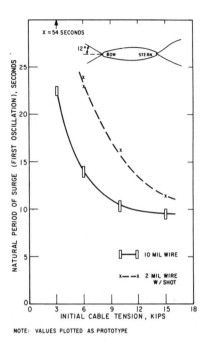

Fig. 19.15. Natural period of surge—1:80 scale model LSM (from Wiegel, Beebe, and Dilley, 1958)

which showed the effect of one mooring cable breaking. The tension in the diagonally opposing cable dropped off considerably and the vessel yawed slightly, slacking off the remaining two cables (from an "initial tension" standpoint). This caused an increase in the natural period of surge. Hence, the resonance occurred at a greater wave period. For example, a resonance or near-resonance condition occurred for a wave period of 9.2 sec for the 10-kip initial cable tension with four mooring lines, but shifted to 13.5 sec when the starboard bow cable was dropped. It was not possible to tell conclusively from these whether dangerously high forces will be experienced when one cable breaks, because of the spread of wave periods tested, with the resulting probability that the peak resonance condition might have been encountered for the 13.5-sec wave but not for the 9.2-sec wave, vice versa, or neither.

It was found that the vessel rolled, yawed, and swayed even though it was in "head seas." It is not possible to fix the vessel heading exactly, and any induced motion would be emphasized by the elastic mooring cables. In addition, there may be an inherent instability as discussed by Grim (1952).

Mooring line forces in quartering and beam seas are shown in Fig. 19.12.

In regard to motions in quartering seas, it can be seen that the surge and pitch amplitudes are of the same order of magnitude as for the head seas tests in the model basin (Figs. 19.13 and 19.14). The difference in water depth (160 ft rather than 200 ft) should not be of any significance as both depths are relatively deep as far as the water particle velocities and accelerations within the surface layer of the same thickness as the vessel's draft are concerned. The heaving amplitude is greater than for head seas; in fact, the heaving exceeded the wave height on many occasions. A portion of this can be attributed to the effect of angular distortion; however, it is mainly due to the fact (Weinblum and St. Denis, 1950) that the heaving force function increases with increasing angle between the ship's bow and the wave direction—see Fig. 19.1(a). This, combined with the magnification factor—Fig. 19.1(b)—for a damping coefficient of 0.4 (approximately the value for an average ship), leads to a heaving amplitude which can be in excess of the wave amplitude. In addition, there is considerable sway, roll, and yaw. An example of the type of data obtained has been given in Fig. 19.11. It was apparent from the tests that the sway and yaw records may be either fairly uniform or quite nonuniform. The record in Fig. 19.11 cannot be said to be typical; rather, there was an entire spectrum of variations. These two motions appeared to get in and out of phase for certain wave periods; maximum values of sway and yaw are often as much as 25 per cent greater than the average values. The mooring cable forces were of the same order of magnitude as in head seas and the trends appeared to be the same as in head seas.

In beam seas, the surging motion decreased considerably, although it did not go to zero. This was possibly due to a small error in the positioning of the vessel, unequal initial cable tensions, or both. The same was true for pitching. The yawing was appreciable, although the magnitude was not as great as for the case of quartering seas. The rolling motion appeared to be of about the same order of magnitude as for quartering seas. The sway increased. It was evident that large motions could be expected for certain combinations of wave heights and periods. The reason for the apparent "resonance" condition occuring for sway in beam seas was not apparent considering the long natural period. It was evident from Fig. 19.13(a) that it was nearly independent of initial cable tension. It well may be explained, however, if curves similar to Fig. 19.1 were available. Certainly the "swaying force function" must be very large. The ratio of the maximum values of sway to wave height never approached the maximum value of the ratios of surge to wave height.

The maximum cable forces occurred for lower wave periods in beam seas than in either head or quartering seas. Thus it might be necessary to head the vessel differently, depending upon the wave period (say, local seas or swell).

The bow cable forces for the model in head seas were considerably higher than the stern cable forces for the shortest wave periods used. For longer wave periods, this relationship was reversed; the longest waves tested

showed the relationship often reversed again. The curve was found to drop abruptly at about $6\frac{1}{2}$ sec to a minimum value and then to increase with increasing wave period.

f. *Other model studies.* When there is no strongly nonlinear effect, the mooring line forces can be presented in a relatively simple manner. A floating drydock(AFDL-1) was moored as shown in Fig. 19.16, and the results are as shown in Fig. 19.17. It is significant that in these tests, as has also been the case in other tests, the forces in the cables reached their maximum values with the second or third wave, demonstrating that the system is highly damped and that a long train of uniform waves is not required to develop maximum resonant amplification.

When the model was in head seas, it was noticed that a definite swaying motion occurred when the wave period was at the natural period of sway. It is believed that this was a coupling phenomenon similar to the type studied by Grim (1952).

As can be seen in Fig. 19.16, the mooring lines had concrete clumps attached to them in such a manner that when the motions became large the clumps were lifted off the bottom. It was found that the natural periods of surge and yaw (but apparently not sway) of the moored ship depended upon whether these clumps were lifted or not. The natural period of surge might be

Fig. 19.16. Anchoring system of AFDL-1 (from Wiegel, Clough, Dilley, and Williams, 1959)

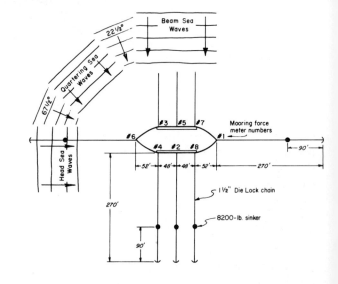

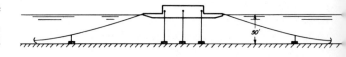

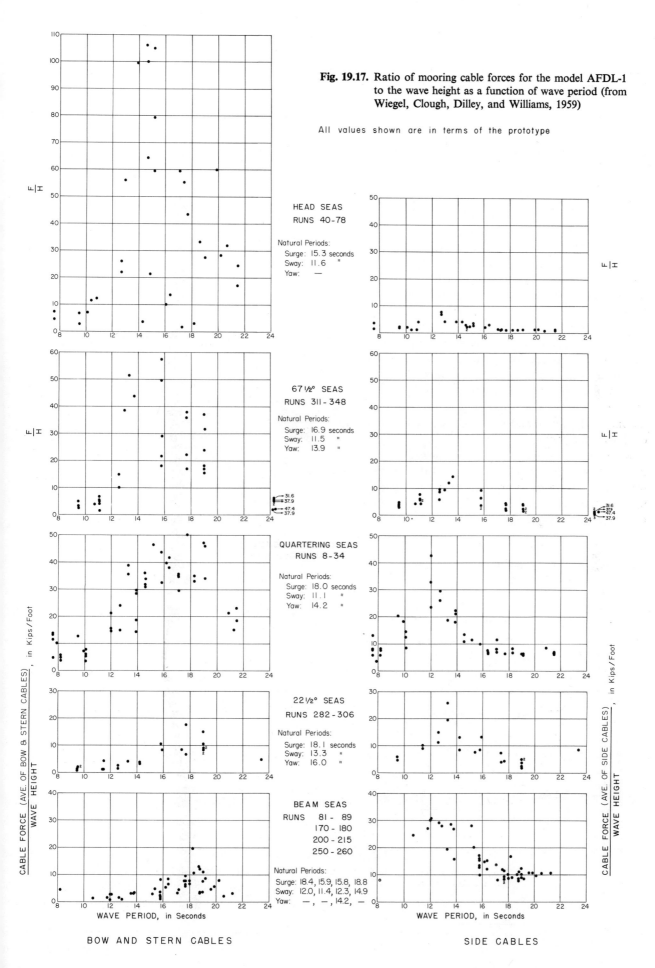

Fig. 19.17. Ratio of mooring cable forces for the model AFDL-1 to the wave height as a function of wave period (from Wiegel, Clough, Dilley, and Williams, 1959)

All values shown are in terms of the prototype

HEAD SEAS
RUNS 40-78

Natural Periods:
Surge: 15.3 seconds
Sway: 11.6 "
Yaw: —

67 1/2° SEAS
RUNS 311-348

Natural Periods:
Surge: 16.9 seconds
Sway: 11.5 "
Yaw: 13.9 "

QUARTERING SEAS
RUNS 8-34

Natural Periods:
Surge: 18.0 seconds
Sway: 11.1 "
Yaw: 14.2 "

22 1/2° SEAS
RUNS 282-306

Natural Periods:
Surge: 18.1 seconds
Sway: 13.3 "
Yaw: 16.0 "

BEAM SEAS
RUNS 81- 89
       170-180
       200-215
       250-260

Natural Periods:
Surge: 18.4, 15.9, 15.8, 18.8
Sway: 12.0, 11.4, 12.3, 14.9
Yaw: —, —, 14.2, —

BOW AND STERN CABLES

SIDE CABLES

501

about 18 to 19 sec (prototype) for small motions and 15 to 16 sec for large motions. As a result of this, when waves with periods of about 18 sec were run with the model in head seas, the motion would be built up for several consecutive waves (due to resonance) until its motion was sufficient to lift the clumps. The natural period of surge would then be only 15 or 16 sec and the motion was destroyed as it was no longer in resonance. Then the vessel would be in resonance again and the motion would again build up and be destroyed.

When the ratio of the forces in the bow cables to forces in the stern cables is examined for a model in head seas, say, it is found that for short wave periods the bow forces are much higher than the stern forces, whereas at times, for long periods the reverse is true. Observations show that the model shifts in the direction of wave advance for short-period waves, and often shifts in the opposite direction for long-period waves. In regard to short-period waves, it is believed that this is due to wave diffraction. When waves become long with respect to the water depth, the mass transport at the surface is in the opposite direction to the wave advance (Fig. 2.41), and this might cause the model to drift into the waves.

When a ship oscillates about a displaced mean position, the effective initial tension in the mooring line shifts, so that some of the lines have a higher initial tension than others. Natural periods can be obtained for approximately this condition by mounting the mooring

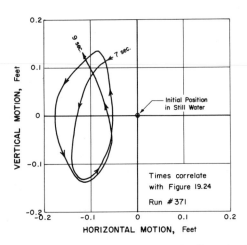

Fig. 19.19. Motion of the ship's bow at the leading edge (intersection of bow and shipboard) as a function of time; Shell model "donut" (from Tickner, Wiegel, and Swanstrom, 1958)

lines on long beams extending from a towing carriage, and measuring the natural period at different forward speeds. The results of such a test are shown in Fig. 19.18. At relatively low initial tensions, the effect of towing speed on the natural period of surge is pronounced.

The motion of the point of cable attachment is of prime importance. This motion depends upon the forcing function and response functions of the system. An example of the motion of the bow of one model vessel is shown in Fig. 19.19, with the motions being given in terms of the 1:60 scale model (Tickner, Wiegel, and Swanstrom, 1958). Sometimes when moored vessels are subject to large waves, green water will come over the bow and wash down the deck. An example of this is shown in Fig. 19.20 where the elevation of the bow above the datum is plotted, together with the water surface, as a function of time. When the two time histories coincide, water washes over the bow.

When a ship is moored to buoys which are anchored, consideration must be given to the natural periods of the moored buoy system. Model studies have shown that something close to impact forces has occurred in the lines between the buoys and the vessel when the buoys were undergoing certain resonant motions (Wiegel, Swanstrom, and Tickner, 1959).

The motions that occur in a moored ship may be relatively simple, or extremely complicated, with forces on some lines occurring only occasionally. An example of this can be combined with some information on the forces exerted on the camels of a ship moored alongside a dock (Wiegel, Dilley, and Williams, 1959). Examples of the mooring line and camel forces are shown in Fig. 19.21.

For long-period waves above 50 sec (prototype), the ship seemed to move with a regular cyclic motion

Fig. 19.18. Effect of nonsymmetrical mooring line tension on natural period of surge

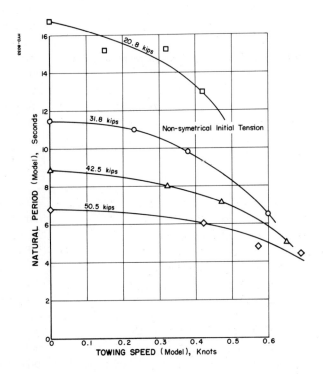

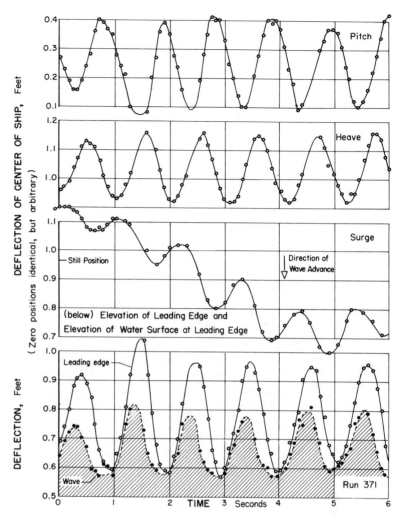

Fig. 19.20. Motions of model and water surface at swamping threshold; Shell Model "Donut;" initial mooring line tension—5 kips (prototype); head seas values are in terms of the model (from Tickner, Wiegel, and Swanstrom, 1959)

which caused force peaks on each line corresponding to each wave crest. These long waves caused multiple peaked force records, as can be seen in Fig. 19.21(a). Observations of the model motion showed that the line was pulled taut and then the vessel experienced a small amplitude high-frequency motion. This high-frequency motion was primarily roll for ship headings other than head and stern seas. In general, the multiple peaks occurred for long, high (in these tests the highest wave was only 5 ft, prototype) waves.

Waves of intermediate periods (20 to 50 sec) generally caused forces for each wave. These forces, however, exhibited large variations in amplitude from one wave to the next. It appeared that the ship would get into resonant motion and then out of it and then back again, etc. For the 4 and 5-ft waves at these intermediate periods, the force records became more like the records for the long waves, indicating that height plays an important part in the regularity of the motion and the resulting forces.

The short-period waves induced irregular motions in the ship and forces. With increased wave height, the motions and forces became more regular.

It should be emphasized that, in the nonlinear systems described in this section, the past history of the motion affects the motions and forces even during "steady state" conditions, this term meaning that the waves were of uniform height and periodic. This results in considerable variation in motions and forces from run to run even when wave conditions appear to be nearly identical. Forces may vary by a factor of 2 or so in a pair of symmetrical mooring lines.

4. COMPARISON OF MODEL AND PROTOTYPE

In any model work, it is necessary to have many comparisons between model and prototype measurements before the reliability of the model studies for the prediction of prototype action can be determined. These comparisons are not available for studies of wave-induced mooring forces.

One study has been made which should be of value in this respect. Personnel of the U.S. Navy Civil Engineering Laboratory set up an AFDL with one bow and one stern mooring line running to steel pile anchors. Measurements were made of the natural periods of

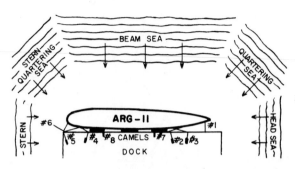

Force meter positions and wave directions, ARG-11 Model tests.

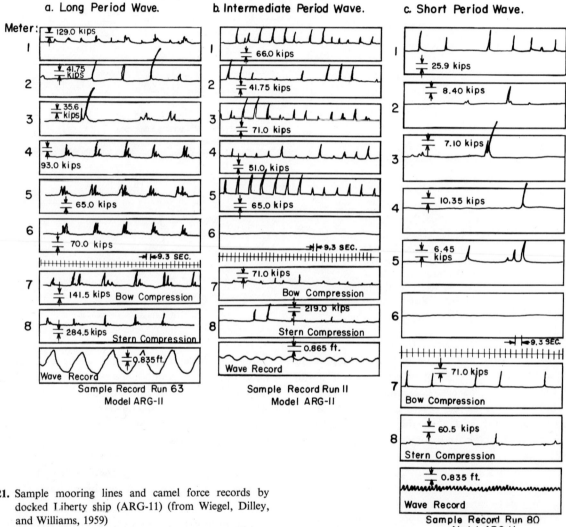

Fig. 19.21. Sample mooring lines and camel force records by docked Liberty ship (ARG-11) (from Wiegel, Dilley, and Williams, 1959)

surge and sway for several initial tensions in the mooring lines (O'Brien and Kuchenreuther, 1958).

The AFDL-1 described briefly in the previous section was modified slightly so that it was similar to the moored prototype, and then the natural periods of surge and sway were measured. The results of the natural periods of surge are shown in Fig. 19.22. The comparison appears to be good, and both sets of measurements agree favorably with calculations made by O'Brien and Kuchenreuther. The natural periods of sway were found to be between 170 and 215 secs, which confirmed approximately

those obtained from the prototype measurements and analytical analysis.

It appears from these limited comparisons that model studies should be fairly reliable in predicting the motions and mooring forces of a prototype system.

5. WIND AND CURRENT EFFECTS

The forces due to wind acting on individual ships, and ships tied up side by side have been studied in a wind tunnel (Ayers and Stokes, 1953). Longitudinal

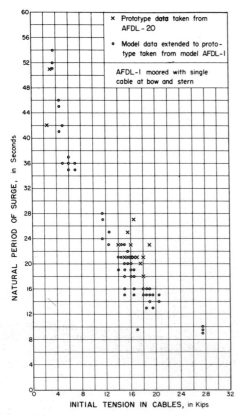

Fig. 19.22. Average cable tension versus period of 1st cycle (from Wiegel, Clough, Dilley, and Williams, 1958)

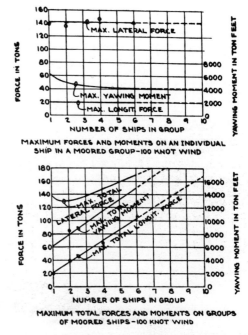

Fig. 19.24. Maximum forces and yawing moment on destroyers (after Ayers and Stokes, 1953)

and lateral forces and yawing moment were measured for wind speeds of 75, 100, and 125 knots, prototype. It was found that the wind-induced forces varied with the square of the wind speed.

An example of the forces measured on one ship model is shown in Fig. 19.23, with values being in terms of the prototype. The effect of the number of ships side by side on the maximum values is shown in Fig. 19.24.

Fig. 19.23. Forces and yawing moment on one destroyer—airspeed 100 knots. Forces and yawing moment on two destroyers—airspeed 100 knots (after Ayers and Stokes, 1953)

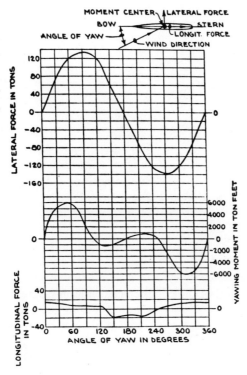

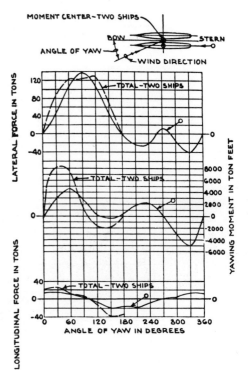

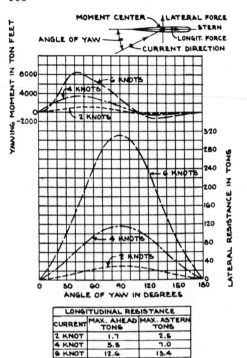

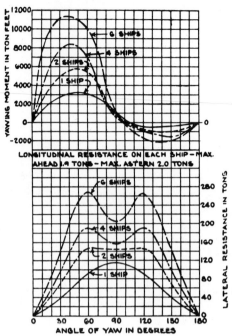

Fig. 19.25. Forces and yawing moment on one destroyer—water currents of 2, 4, and 6 knots. Lateral force and yawing moment on 1, 2, 4, and 6 destroyers — current, 4 knots; water depth, 25 feet (after Ayers and Stokes, 1953)

When several ships were moored side by side, the maximum values of the longitudinal and lateral forces and the yawing moment occurred at different wind angles than for the single ship setup.

The forces induced by currents were measured in a circulating water channel on models of several classes of ships. An example of the results for a single ship is shown in Fig. 19.25, and the results for groups of ships side by side are shown in Fig. 19.26. It was found that the forces were roughly proportional to the square of the wind speed. The depth of water affected the lateral resistance due to the currents (Fig.19.26).

Details for other classes of ships are available in U.S. Navy *Mooring Guide* (U.S. Navy, Bureau of Yards and Docks, 1954).

6. SUBMERGED BUOYANT STRUCTURES

A moored submerged structure is simpler to treat than a floating structure since there is no variation in the buoyant force as the waves pass over the structure. We are not considering here structures that are moored so far beneath the surface that they are effectively beneath the region of wave action.

A right circular cylinder, with its longitudinal axis horizontal, has been studied under a variety of wave conditions for several mooring setups. A 1:49 scale model was tested in head, quartering, and beam seas (Wiegel, Dilley, Whisenand, and Williams, 1956; Dilley, Whisenand, and Wiegel, 1956). When the tank (190 ft long by 22 1/2 ft diameter, prototype) was moored by vertical neutrally buoyant tubing (30 ft long by 1 1/2 ft O.D., prototype) it was possible to compute the maximum vertical force on the mooring lines due to the waves by use of Eq. 11.8b, using $C_M = 2.0$. The maximum vertical force for this size cylinder and the practical range of wave conditions were found to occur when $\partial v/\partial t$ = maximum and $v = 0$. The fact that Eq. 11.8b predicted the measured values is interesting in that the tank oscillated to and fro considerably when subjected to the waves. In fact, when the tank was in head

Fig. 19.26. Variations in lateral resistance with depth of water— one destroyer in a 4-knot current (after Ayers and Stokes, 1953)

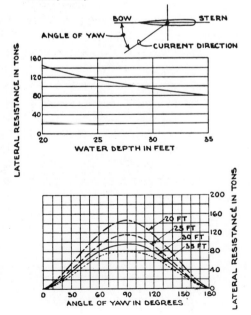

seas and subject to waves of 21 sec, prototype (the natural period of surge), the surging motion of the tank was as much as 6 times the wave height and 4 times the total horizontal motion of the water particles at the surface.

When neutrally buoyant no-sag bow, stern, and breast mooring lines were added, and some of the initial buoyant load taken by these lines, it was found that essentially no tank motion occurred regardless of wave height (up to 40 ft, prototype). At some point in the transference of the initial buoyant force from the vertical to the lateral mooring lines (the percentage of load transference depending upon the wave height and period), the vertical mooring lines would just slacken at one point in the cycle. This resulted in an impact type of loading upon takeup with high forces and considerable vibration.

When every other section of tubing of which the mooring "lines" were constructed was replaced by a length of chain, the forces were about the same as before. When the lateral lines were allowed to sag, the average maximum forces increased with increasing sag in a nonlinear manner, and some very high forces were measured. In the sagging line condition, the vertical mooring lines exhibited jerking for almost all the wave conditions and the lateral mooring lines exhibited high-frequency loading and unloading.

7. ANCHORING

A floating structure may be moored to a dock or similar structure, to a buoy which is in turn anchored, or it may be connected directly to an anchor by a mooring line or chain. For permanent moorings, the "anchor" may be a pile driven into the bottom which will stand a large horizontal load (Proceedings U.S. Naval Institute, 1946, describe one that will handle a 200,000-lb horizontal load).

The other extreme from a pile is the sea anchor. A sea anchor can be made of canvas or other material, in the shape of a truncated cone, or a plane, or a series of these, and is attached to the end of a line running from a ship into the water. When the ship moves, a force is developed by the resistance of the sea anchor to movement. One device with a 7-ft base diameter was found to produce a tension of 3600 lb at a speed of 6 knots in smooth water, with the same resistance achieved at 5 knots in rough water (Miller, 1900; 1906). Tests showed that a conical sea anchor with a length equal to its opening diameter moved erratically when towed through water. It would dive, broach, and yaw, although the strain in the cable would not vary much. If the length were 4 times the diameter, the motion would be stable, although the strain would be only half the strain of the sea anchor with a length equal to its diameter. Two conical sea anchors coupled in tandem, with lengths

equalling diameters, were more stable; one 7 ft and one 5 ft in tandem developed a tension of 15,000 lb while being pulled at 7 knots. In testing a multiplane sea anchor consisting of a series of sheets of 4-ft-square canvas spaced 8 ft apart, it was found that a series of five would pull in a stable manner, and were easier to launch and recover. When they were placed closer together, they would produce less resistance; when placed farther apart, they would produce more resistance.

There are many types of anchors, such as the concrete clump, the mushroom, the navy stockless anchor, the Danforth anchor, and the Admiralty stock anchor. The characteristics of a good anchor for general use are (Dove, 1950; Farrell, 1950): (1) speed of burial to develop good resistance; (2) capability of withstanding all the forces likely to act upon it without permanent set; (3) dead weight enough to be useful in poor holding grounds and heavy enough to drag the chain or cable from the locker or drum when let go; (4) stability, or freedom from rolling; (5) burial shallow enough not to make breaking out difficult; (6) nonfouling of cable or chain; (7) capability of digging into hard beds; (8) ease of manufacture.

Most anchors are designed to withstand a considerably greater horizontal force without pulling out, than a vertical force. Tests on one type of anchor (Dove, 1950) showed that the maximum vertical pull was only about two-fifths that of the maximum horizontal pull; this is necessary to permit the recovery of an anchor. Nearly all operators agree that the mooring line should lie flat at the anchor. One of the criteria in determining the length of cable necessary to assure this condition is the scope, that is, the ratio of length of the mooring line to the water depth. For single anchors a cable scope of from 5 to 8 is often recommended (Dove, 1950; Ogg and Danforth, 1954); the scope to be used, however, depends upon the water depth and many other factors (Dove, 1950). The results of Dove on the effect of cable scope are shown in Fig. 19.27 for a 5-ton Admiralty standard stockless anchor (Model No.1) attached to a $3\frac{1}{2}$-in. steel cable subject to the drag of a 5-knot current. The graph can be used in several ways. For example, suppose the water depth is 700 ft and only 1500 ft of cable are available. The intersection of these two inputs shows that the cable will make an angle θ_1 of just under 5 degrees. The intersection of the 1500-ft cable length and the 5-degree horizontal holding pull at ship curve shows that an 82,000 lb horizontal pull can be maintained at the water surface on the cable.

Anchor chain is effective as an anchor. Tests made pulling several sizes of chain alone through the holding ground showed that the ratio of holding pull to chain weight ranged from 1.36 to 1.24 (Dove, 1950). Wire rope has been found to have a holding pull of only about 60 per cent of its weight (Farrell, 1950).

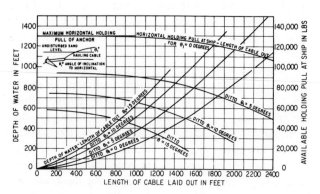

Fig. 19.27. Relation between horizontal holding pull of anchor and horizontal holding pull at ship (from Dove, 1950)

There have been several model studies of anchor characteristics and the effect of different types of bottom materials. One series of tests was made with model anchors $\frac{1}{8}$, $\frac{1}{5}$, $\frac{1}{3}$, and $\frac{1}{2}$ scale (Lucking, 1936). These tests were made on both sand and shingle beaches, and in prepared beds in a towing channel in sand and mud. For these types of bottoms, it was found that the holding pull was proportional to the third power of the geometric scale ratio. It was found that the absolute weight was important in getting the anchor to dig into the bottom. It was also found that the length of cable or chain per se was important as well as the scope, because of its own drag, and because it affected the rate of change of cable angle as the anchor burrowed into the bottom.

The effect of fluke area has been studied by means of model tests in a tank filled with fine sand and with clay (Leahy and Farrin, 1935). Tests with models of a U.S. Navy standard anchor showed that the holding pull was given by $P = KA^{1.53}$ where P is the holding pull in pounds, A is the fluke area in square inch, and K is an empirically determined coefficient which was 1.20 for the fine sand and 2.04 for the blue clay used in the experiments. Tests with models of another type of anchor showed similar results with the exponent being almost the same, but with different values for K.

Improvements have been made in one type of anchor

Table 19.1. ANCHOR CHAIN TESTS—U.S.S. TRENTON
(after Land, 1934)

	Stress at moment of break-out electrical method (lb) (1)	Force required to lift scope of chain off bottom (lb) (2)	Angular lift of shank at break-out (theoretical) max. degree (3)
Series 1: 11,000-lb standard anchor on $1\frac{5}{8}$ in. die-lock chain			
Test 1 90-fath. chain, 8-fath. water	68,000	52,480	0.7
2 75-fath. chain, 8-fath. water	69,000	36,608	3.5
3 45-fath. chain, 15-fath. water	37,000	8,704	17.2
4 120-fath. chain, 15-fath. water	62,000	55,296	0.7
5 75-fath. chain, 15-fath. water	44,000	22,208	6.5
Series 2: 11,000-lb standard anchor on $2\frac{1}{2}$ in. wrought-iron chain			
Test 1 90-fath. chain, 8-fath. water	55,000	129,000	0.0
2 54-fath. chain, 8-fath. water	74,000	33,284	6.5
3 75-fath. chain, 15-fath. water	52,000	54,479	0.0
4 45-fath. chain, 15-fath. water	55,000	21,352	13.0
5 120-fath. chain, 15-fath. water	96,000	135,000	0.0
6 30-fath. chain, 15-fath. water	39,000	10,990	24.4
Series 3: 14,000-lb standard anchor on $1\frac{5}{8}$ in. die-lock chain			
Test 1 90-fath. chain, 8-fath. water	90,000	52,480	2.5
2 45-fath. chain, 8-fath. water	58,000	13,568	9.7
3 75-fath. chain, 15-fath. water	58,000	22,208	8.0
4 120-fath. chain, 15-fath. water	72,000	55,296	1.7
5 45-fath. chain, 15-fath. water	79,000	8,704	19.4
Series 4: 11,000-lb experimental anchor on $1\frac{5}{8}$ in. die-lock chain			
Test 1 90-fath. chain, 8-fath. water	79,000	52,480	2.0
2 45-fath. chain, 8-fath. water	62,000	13,568	10.0
3 30-fath. chain, 8-fath. water	72,000	6,400	17.5
4 120-fath. chain, 15-fath. water	65,000	55,296	1.1
5 75-fath. chain, 15-fath. water	62,000	22,208	8.3
6 45-fath. chain, 15-fath. water	66,000	8,704	18.9
7 30-fath. chain, 15-fath. water	62,000	4,480	29.5

NOTE: Values in Column 1 were taken from test recorder sheets, after check and corrections.

as a result of a series of model tests, with the best model being able to hold 10 times its weight compared with only 2 times the weight of the original (Dove, 1950). As a result of these tests, it was concluded that (1) it is important to arrange for the optimum separation between fluke and shank; (2) the shank should be sharp on the side nearest the fluke; (3) the fluke should have a minimum cross-sectional area compatible with strength; (4) there is an optimum fluke angle; (5) holding pull can be increased by increasing the fluke area; (6) prevention of clogging of fluke angle limit stop is important; and (7) stability is an important factor governing holding pull.

In general, it was found that the comparison between model and prototype results were fairly good, although on occasions there were large differences. It is likely that many of the difficulties result from the problems of trying to model a particular bottom material. In fact, much work remains to be done in classifying bottoms with respect to their anchoring characteristics. The techniques of soil mechanics have barely been tapped. For example, for one type of anchor, the statement was made that fine sand and shingle bottoms proved best and soft mud worst, with no data on shear strength, median diameter of material, plasticity, etc.

Several sets of tests have been made with prototype anchors (Coombes, 1931; Lucking, 1932; Land, 1934; Thorpe and Farrell, 1948; Nunez, 1951). Land's results are given in Table 19.1.

Almost all tests described here have been performed with steady loadings on the anchor lines, whereas the reaction of a soil to a load depends upon its dynamic characteristics as well. When a ship is anchored in an open roadstead, the waves cause an oscillatory loading on the lines, and this must affect the holding pull of an anchor. An extensive search has revealed almost no information on this (Wiegel, 1955), although Thorpe and Farrell (1955) have considered the problem of wind gusts on a moored ship (but not on the anchor which is at the end of a cable). Theoretical studies have been made relating the motions of a cable and the forces in the cable to a periodic force at the surface end of the cable which are of value to the design engineer (Whicker, 1958), and studies have been made of the hydrodynamic forces on a mooring cable (Landweber, 1947).

There is also little information on the relative value of chain, wire rope, or rope on anchoring, although Farrell (1950) found that the use of chain made a marked improvement in the behavior of stockless anchors by limiting surges and improving stability, although it did not improve the behavior of the average holding pull to a great extent.

For details of anchors, cable strengths, mooring arrangements, and seamanship connected with anchoring see these three publications (Sharpey-Schafer, 1954; U.S. Navy, Bureau of Yards and Docks, 1954; Thorpe and Farrell, 1955).

REFERENCES

Abramson, H. Norman, and Basil W. Wilson, A further analysis of the longitudinal response of moored vessels to sea oscillations, *Proc. Midwestern Conf. Fluid Solid Mechanics*. Purdue Univ., Indiana: (September, 1955), 236–51.

Ayers, James R., and Ralph C. Stokes, Berthing of U.S. Navy Reserve Fleet, *Perm. Intern. Assoc. Nav. Congr.*, XVIIIth Congr. (Rome), Sect. II, Question 2, 1953, p. 69–85.

Beebe, K. E., Mooring cable force meter for models, *Proc. First Conf. Coastal Eng. Instr.*, Berkeley, Calif.: The Engineering Council on Wave Research, 1956, pp. 134–43.

Cartwright, D. E., and L. S. Rydill, The rolling and pitching of a ship at sea. A direct comparison between calculated and recorded motions of a ship in sea waves, *Trans., Inst. Naval Architects*, **99** (1957), 100–35.

Coombes, L. P., Tests of anchors for use on flying boats, Air Ministry (Great Britain), *Aeronautical Res. Committee, R and M* No. 1449 (May, 1931), 1086–98.

Cummins, W. E., The impulse response function and ship motion, *Schiffstechnik*, **9**, 47 (June, 1962), 101–9.

Deacon, G. E. R., R. C. H. Russell, and J. E. G. Palmer, Origin and effect of long period waves in ports, Sect. II,

Comm. 1, *XIXth Congr., Perm. Intern. Assoc. of Nav. Congr.*, London (1957), pp. 75–93.

Dilley, R. A., S. F. Whisenand, and R. L. Wiegel, *Model study of a submerged buoyant tank (Mk II) in waves*, Univ. Calif. IER, Tech. Rept. 91–5, September, 1956. (Unpublished.)

Dove, H. L., Investigations on model anchors, *Quart. Trans. Inst. Naval Architects*, **92**, 4 (October, 1950), 351–62.

Farrell, K. P., Improvements in mooring anchors, *Quart. Trans. Inst. Naval Architects*, **92**, 4 (October, 1950), 335–50.

Froude, W., On the rolling of ships, *Trans., Inst. Naval Architects*, **2** (1861), 180–227.

Fuchs, R. A., A linear theory of ship motion in irregular waves, *Proc. First Conf. Ships and Waves*, Berkeley, Calif.: The Engineering Foundation Council on Wave Research and SNAME, 1955, pp. 186–93.

Gerritsma, J., *Experimental determination of damping, added mass and added mass moment of inertia of a shipmodel*, Laboratorium voor Scheepsbouwkunde Technische Hogeschool, Pub. No. 8, 1957.

Grim, Otto, Rolling, stability and safety in a seaway, *Forschunyshefte fur Schijfbautechnik*, Vol. 1, 1952, a translation and summary of the original paper by E. V. Lewis,

Experimental Towing Tank, Stevens Inst. Tech., Note No. 234, 1953.

Haskind, M. D., The hydrodynamical theory of the oscillations of a ship in waves, *Prikladnaya Matematika i Mekhanika;* a translation by the SNAME, 1946.

Havelock, T. H., The coupling of heave and pitch due to speed of advance, *Trans. Inst. Naval Architects*, **97** (1955), 246.

———, The damping of heave and pitch: a comparison of two-dimensional and three-dimensional calculations, *Trans. Inst. Naval Architects*, **98** (1956), 464–67.

Jasper, N. H., Statistical distribution patterns of ocean waves and of wave-induced ship stresses and motions, with engineering applications, *Trans. Soc. Naval Architects Mar. Engrs.*, **64** (1956), 375–415.

Joosting, W. C. Q., Investigation into long period waves in ports, Sect. II, Comm. 1, *XIXth Congr., Perm. Intern. Assoc. of Nav. Congr.*, London (1957), pp. 205–27.

Kriloff, A., A general theory of the oscillations of a ship on waves, *Trans. Inst. Naval Architects*, **40** (1898), 135–90.

Land, E. S., Development in ground tackle for naval ships, *Trans. Soc. Naval Architects Mar. Engrs.*, **42** (1934), 164–67.

Landweber, L., Hydrodynamic forces on an anchor cable, U.S. Navy, David Taylor Model Basin, Report No. R–317, November, 1947.

Landweber, L., and C. S. Yih, Forces, moments, and added masses of Rankine bodies, *J. Fluid Mechanics*, 1, Part 3 (September, 1956), 319–36.

Leahy, William H., and James M. Farrin, Jr., Determining anchor holding power from model tests, *Trans. Soc. Naval Architects Mar. Engrs.*, **43** (1935), 105–14.

Lucking, D. F., Tests of full scale anchors in various sea beds, Air Ministry (Great Britain), *Aeronautical Res. Committee*, R and M No. 1546, November, 1932.

———, The experimental development of anchors for seaplanes, *Trans. Inst. Naval Architects*, **78** (1936), 201–12.

Miller, Spencer, The coaling of the U.S.S. Massachusetts at sea, *Trans. Soc. Naval Architects Mar. Engrs.*, **8** (1900), 155–65.

———, A new sea anchor for coaling at sea, *Trans. Soc. Naval Architects Mar. Engrs.*, **14** (1906), 269–75.

Nunez, Antonio Villanueva, Experiments with modern types of anchors, *Ingenieria Naval*, **19**, 188 (February, 1951), 64–79. Translation by Philip Chaplin, Defence Scientific Information Service (Canada), April, 1953.

O'Brien, J. T., and D. I. Kuchenreuther, Free oscillations in surge and sway of a moored floating dry dock, *Proc. Sixth Conf. Coastal Eng.*, Berkeley, Calif.: The Engineering Foundation Council on Wave Research, 1958a, 878–94.

———, and D. I. Kuchenreuther, *Forces induced by waves on the moored U.S.S. Norton Sound* (AVM–1); U.S. Naval Civil Eng. Lab., Tech. Memo. No. M–129, April, 1958b.

Ogg, R. D., and R. S. Danforth, *Anchors and anchoring.* Berkeley, Calif.: Danforth Anchors, 1954.

Porter, William R., *Pressure distributions, added-mass, and damping coefficients for cylinders oscillating in a free surface,* Univ. Calif. IER, Tech. Rept. 82–16, July, 1960.

Proceedings U.S. Naval Institute, New ship anchorage developed, **72**, 7 (July, 1946), 993.

Rossell, Henry E. and Lawrence B. Chapman (ed.), *Principles of Naval Architecture*, Vol. 2. New York: Society of Naval Architects and Marine Engineers, 1949.

Saunders, Harold E., *Hydrodynamics in ship design*, Vol. 2. New York: Society of Naval Architects and Marine Engineers, 1957.

Sharply-Schafer, J. M., Anchor dragging, *J. Inst. Navigation*, **7**, 3 (July, 1954), 290–300.

Sibul, Osvald, Laboratory studies of the motion of freely floating bodies in non-uniform and uniform long crested waves, *Proc. First Conference Ships and Waves*, Berkeley, Calif.: The Engineering Foundation Council on Wave Research, and SNAME, 1955, pp. 366–97.

St. Denis, Manley, and Willard J. Pierson, Jr., On some recent developments in the theory of ship motions, *Proc. First Conf. Ships and Waves*, Berkeley Calif.: The Engineering Foundation Council on Wave Research, and SNAME, 1955, pp. 160–84.

Szebehely, V. G., and M. A. Todd, *Ship slamming in head seas*, U.S. Navy, David Taylor Model Basin, Rept. No. 913, February, 1955.

Thorpe, T., and K. P. Farrell, Permanent moorings, *Trans. Inst. Naval Architects*, **90** (1948), 111–53.

Tickner, E. G., R. L. Wiegel and R. D. Swanstrom, *Model study of the dynamics of the Shell Model "Donut" moored in water gravity waves*, Univ. Calif. IER, Tech. Rept. No. 122–1, August, 1958. (Unpublished.)

Timoshenko, S., *Vibration Problems in Engineering*, 2d ed. Princeton, N. J.: D. Van Nostrand Co., Inc., 1937.

U.S. Navy, Bureau of Yards and Docks, *Mooring Guide*, Vols. 1, 2, Tech. Publ. NAVLOCKS TP–PW–2, March, 1954.

———, David Taylor Model Basin, *Trans. Eleventh General Meeting Amer. Towing Tank Conf.*, Report No. 1099, September, 1957.

Warnsinck, W. H. and M. St. Denis, Destroyer seakeeping trials, *Proc. symposium on the behavior of ships in a seaway* Wageningen, Netherlands: H. Veenman & Zonen, 1957, pp. 439–67.

Weinblum, George P., Progress of theoretical investigations of ship motions in a seaway, *Proc. First Conf. Ships Waves*, Berkeley, Calif.: The Engineering Foundation Council on Wave Research and SNAME, 1955, pp. 129–59.

———, Contribution of ship theory to the seaworthiness problem, *Symposium on Naval Hydrodynamics*, National Academy of Science–National Research Council, Pub. No. 515, 1957, pp. 61–98.

————, and M. St. Denis, On the motions of ships at sea, *Trans. SNAME*, **58** (1950), 184–248.

Wendel, Kurt, Hydrodynamic masses and hydrodynamic moments of inertia, *Jahrb, d. STG*, **44**, 1950, E. N. Labouvic and A. Borden, Trans. U.S. Navy, David Taylor Model Basin, Trans. No. 260, July, 1956.

Whicker, L. Folger, *Theoretical analysis of the effect of ship motion in mooring cables in deep water*, U.S. Navy, David Taylor Model Basin, R. and D. Report No. 1221, March, 1958.

Wiegel, R. L., *Ship mooring literature survey*, Univ. Calif. IER Tech. Rept. 91–2, August, 1955. (Unpublished.)

————, K. E. Beebe, and R. A. Dilley, Model studies of the dynamics of an LSM moored in waves, *Proc. Sixth Conf. on Coastal Eng.*, Berkeley, Calif.: The Engineering Foundation Council on Wave Research, 1957, pp. 844–77.

————, R. W. Clough, R. A. Dilley, and J. B. Williams, Model study of floating drydock mooring forces, *Intern. Shipbuilding Progress*, **6**, 56 (April, 1959), 147–59.

————, R. A. Dilley, and J. B. Williams, Model study of mooring forces of a docked ship, *J. Waterways Harbors Div., Proc. ASCE*, **85**, WW2, Paper No. 2071 (June, 1959), 115–34.

————, ————, S. F. Whisenand, and J. B. Williams, *Model study of a submerged buoyant tank in waves*, Univ. Calif., IER, Tech. Rept. 91–4, March, 1956. (Unpublished.)

————, R. D. Swanstrom, and E. G. Tickner, *Model study of the dynamics of the Shell Model "Delta" moored in water gravity waves*, Univ. Calif. IER, Tech. Rept. No. 122–2, January, 1959. (Unpublished.)

Williams, A. J., An investigation into the motions of ships at sea, *Trans. Inst. Naval Architects*, **95** (1953), 70–85.

Wilson, Basil W., Ship response to range action in harbor basins, *Proc. ASCE*, **76**, separate No. 41, November, 1950.

————, The energy problem in the mooring of ships exposed to waves, *Bull. Perm. Intern. Assoc. Navigation Congr.*, No. 50, 1959a. pp. 7–71.

————, A case of critical surging of a moored ship, *J. Waterways Harbors Div., Proc. ASCE*, **85**, WW3, Paper No. 2318 (December, 1959b), 157–76.

Appendixes

APPENDIX 1. TABLE OF FUNCTIONS OF d/L_o

d/L_o	d/L	$2\pi d/L$	tanh $2\pi d/L$	sinh $2\pi d/L$	cosh $2\pi d/L$	K	$4\pi d/L$	sinh $4\pi d/L$	cosh $4\pi d/L$	n	C_G/C_o	H/H'_o	M
0	0	0	0	0	1	1	0	0	1	1	0		
.0001000	.003990	.02507	.02506	.02507	1.0003	.9997	.05014	.05016	1.001	.9998	.02506	4.467	7,855
.0002000	.005643	.03546	.03544	.03547	1.0006	.9994	.07091	.07097	1.003	.9996	.03543	3.757	3,928
.0003000	.006912	.04343	.04340	.04344	1.0009	.9991	.08686	.08697	1.004	.9994	.04338	3.395	2,620
.0004000	.007982	.05015	.05011	.05018	1.0013	.9987	.1003	.1005	1.005	.9992	.05007	3.160	1,965
.0005000	.008925	.05608	.05602	.05611	1.0016	.9984	.1122	.1124	1.006	.9990	.05596	2.989	1,572
.0006000	.009778	.06144	.06136	.06148	1.0019	.9981	.1229	.1232	1.008	.9988	.06128	2.856	1,311
.0007000	.01056	.06637	.06627	.06642	1.0022	.9978	.1327	.1331	1.009	.9985	.06617	2.749	1,124
.0008000	.01129	.07096	.07084	.07102	1.0025	.9975	.1419	.1424	1.010	.9983	.07072	2.659	983.5
.0009000	.01198	.07527	.07513	.07534	1.0028	.9972	.1505	.1511	1.011	.9981	.07499	2.582	874.3
.001000	.01263	.07935	.07918	.07943	1.0032	.9969	.1587	.1594	1.013	.9979	.07902	2.515	787.0
.001100	.01325	.08323	.08304	.08333	1.0035	.9966	.1665	.1672	1.014	.9977	.08285	2.456	715.6
.001200	.01384	.08694	.08672	.08705	1.0038	.9962	.1739	.1748	1.015	.9975	.08651	2.404	656.1
.001300	.01440	.09050	.09026	.09063	1.0041	.9959	.1810	.1820	1.016	.9973	.09001	2.357	605.8
.001400	.01495	.09393	.09365	.09407	1.0044	.9956	.1879	.1890	1.018	.9971	.09338	2.314	562.6
.001500	.01548	.09723	.09693	.09739	1.0047	.9953	.1945	.1957	1.019	.9969	.09663	2.275	525
.001600	.01598	.1004	.1001	.1006	1.0051	.9949	.2009	.2022	1.020	.9967	.09977	2.239	493
.001700	.01648	.1035	.1032	.1037	1.0054	.9946	.2071	.2086	1.022	.9965	.1028	2.205	463
.001800	.01696	.1066	.1062	.1068	1.0057	.9943	.2131	.2147	1.023	.9962	.1058	2.174	438
.001900	.01743	.1095	.1091	.1097	1.0060	.9940	.2190	.2207	1.024	.9960	.1087	2.145	415
.002000	.01788	.1123	.1119	.1125	1.0063	.9937	.2247	.2266	1.025	.9958	.1114	2.119	394
.002100	.01832	.1151	.1146	.1154	1.0066	.9934	.2303	.2323	1.027	.9956	.1141	2.094	376
.002200	.01876	.1178	.1173	.1181	1.0069	.9931	.2357	.2379	1.028	.9954	.1161	2.070	359
.002300	.01918	.1205	.1199	.1208	1.0073	.9928	.2410	.2433	1.029	.9952	.1193	2.047	343
.002400	.01959	.1231	.1225	.1234	1.0076	.9925	.2462	.2487	1.031	.9950	.1219	2.025	329
.002500	.02000	.1257	.1250	.1260	1.0079	.9922	.2513	.2540	1.032	.9948	.1243	2.005	316
.002600	.02040	.1282	.1275	.1285	1.0082	.9919	.2563	.2592	1.033	.9946	.1268	1.986	304
.002700	.02079	.1306	.1299	.1310	1.0085	.9916	.2612	.2642	1.034	.9944	.1292	1.967	292
.002800	.02117	.1330	.1323	.1334	1.0089	.9912	.2661	.2692	1.036	.9942	.1315	1.950	282
.002900	.02155	.1354	.1346	.1358	1.0092	.9909	.2708	.2741	1.037	.9939	.1338	1.933	272
.003000	.02192	.1377	.1369	.1382	1.0095	.9906	.2755	.2790	1.038	.9937	.1360	1.917	263
.003100	.02228	.1400	.1391	.1405	1.0098	.9903	.2800	.2837	1.040	.9935	.1382	1.902	255
.003200	.02264	.1423	.1413	.1427	1.0101	.9900	.2845	.2884	1.041	.9933	.1404	1.887	247
.003300	.02300	.1445	.1435	.1449	1.0104	.9897	.2890	.2930	1.042	.9931	.1425	1.873	240
.003400	.02335	.1467	.1456	.1472	1.0108	.9893	.2934	.2976	1.043	.9929	.1446	1.860	233
.003500	.02369	.1488	.1477	.1494	1.0111	.9890	.2977	.3021	1.045	.9927	.1466	1.847	226
.003600	.02403	.1510	.1498	.1515	1.0114	.9887	.3020	.3065	1.046	.9925	.1487	1.834	220
.003700	.02436	.1531	.1519	.1537	1.0117	.9884	.3061	.3109	1.047	.9923	.1507	1.822	214
.003800	.02469	.1551	.1539	.1558	1.0121	.9881	.3103	.3153	1.049	.9921	.1527	1.810	208
.003900	.02502	.1572	.1559	.1579	1.0124	.9878	.3144	.3196	1.050	.9919	.1546	1.799	203
.004000	.02534	.1592	.1579	.1599	1.0127	.9875	.3184	.3238	1.051	.9917	.1565	1.788	198
.004100	.02566	.1612	.1598	.1619	1.0130	.9872	.3224	.3280	1.052	.9915	.1584	1.777	193
.004200	.02597	.1632	.1617	.1639	1.0133	.9869	.3263	.3322	1.054	.9912	.1602	1.767	189
.004300	.02628	.1651	.1636	.1659	1.0137	.9865	.3302	.3362	1.055	.9910	.1621	1.756	184
.004400	.02659	.1671	.1655	.1678	1.0140	.9862	.3341	.3403	1.056	.9908	.1640	1.746	180
.004500	.02689	.1690	.1674	.1698	1.0143	.9859	.3380	.3444	1.058	.9906	.1658	1.737	176
.004600	.02719	.1708	.1692	.1717	1.0146	.9856	.3417	.3483	1.059	.9904	.1676	1.727	172
.004700	.02749	.1727	.1710	.1736	1.0149	.9853	.3454	.3523	1.060	.9902	.1693	1.718	169
.004800	.02778	.1745	.1728	.1754	1.0153	.9849	.3491	.3562	1.062	.9900	.1711	1.709	165
.004900	.02807	.1764	.1746	.1773	1.0156	.9846	.3527	.3601	1.063	.9898	.1728	1.701	162
.005000	.02836	.1782	.1764	.1791	1.0159	.9843	.3564	.3640	1.064	.9896	.1746	1.692	159
.005100	.02864	.1800	.1781	.1809	1.0162	.9840	.3599	.3678	1.066	.9894	.1762	1.684	156
.005200	.02893	.1818	.1798	.1827	1.0166	.9837	.3635	.3715	1.067	.9892	.1779	1.676	153
.005300	.02921	.1835	.1815	.1845	1.0169	.9834	.3670	.3753	1.068	.9889	.1795	1.669	150
.005400	.02948	.1852	.1832	.1863	1.0172	.9831	.3705	.3790	1.069	.9887	.1811	1.662	147
.005500	.02976	.1870	.1848	.1880	1.0175	.9828	.3739	.3827	1.071	.9885	.1827	1.654	145
.005600	.03003	.1887	.1865	.1898	1.0178	.9825	.3774	.3864	1.072	.9883	.1843	1.647	142
.005700	.03030	.1904	.1881	.1915	1.0182	.9822	.3808	.3900	1.073	.9881	.1859	1.640	140
.005800	.03057	.1921	.1897	.1932	1.0185	.9818	.3841	.3937	1.075	.9879	.1874	1.633	137
.005900	.03083	.1937	.1913	.1949	1.0188	.9815	.3875	.3972	1.076	.9877	.1890	1.626	135
.006000	.03110	.1954	.1929	.1967	1.0192	.9812	.3908	.4008	1.077	.9875	.1905	1.620	133
.006100	.03136	.1970	.1945	.1983	1.0195	.9809	.3941	.4044	1.079	.9873	.1920	1.614	130
.006200	.03162	.1987	.1961	.2000	1.0198	.9806	.3973	.4079	1.080	.9871	.1935	1.607	128
.006300	.03188	.2003	.1976	.2016	1.0201	.9803	.4006	.4114	1.081	.9869	.1950	1.601	126
.006400	.03213	.2019	.1992	.2033	1.0205	.9799	.4038	.4148	1.083	.9867	.1965	1.595	124
.006500	.03238	.2035	.2007	.2049	1.0208	.9796	.4070	.4183	1.084	.9865	.1980	1.589	123
.006600	.03264	.2051	.2022	.2065	1.0211	.9793	.4101	.4217	1.085	.9863	.1994	1.583	121
.006700	.03289	.2066	.2037	.2081	1.0214	.9790	.4133	.4251	1.087	.9860	.2009	1.578	119
.006800	.03313	.2082	.2052	.2097	1.0217	.9787	.4164	.4285	1.088	.9858	.2023	1.572	117
.006900	.03338	.2097	.2067	.2113	1.0221	.9784	.4195	.4319	1.089	.9856	.2037	1.567	116

d/L_o	d/L	$2\pi d/L$	tanh $2\pi d/L$	sinh $2\pi d/L$	cosh $2\pi d/L$	K	$4\pi d/L$	sinh $4\pi d/L$	cosh $4\pi d/L$	n	C_G/C_o	H/H_o'	M
.007000	.03362	.2113	.2082	.2128	1.0224	.9781	.4225	.4352	1.091	.9854	.2051	1.561	114
.007100	.03387	.2128	.2096	.2144	1.0227	.9778	.4256	.4386	1.092	.9852	.2065	1.556	112
.007200	.03411	.2143	.2111	.2160	1.0231	.9774	.4286	.4419	1.093	.9850	.2079	1.551	111
.007300	.03435	.2158	.2125	.2175	1.0234	.9771	.4316	.4452	1.095	.9848	.2093	1.546	109
.007400	.03459	.2173	.2139	.2190	1.0237	.9768	.4346	.4484	1.096	.9846	.2106	1.541	108
.007500	.03482	.2188	.2154	.2205	1.0240	.9765	.4376	.4517	1.097	.9844	.2120	1.536	106
.007600	.03506	.2203	.2168	.2221	1.0244	.9762	.4406	.4549	1.099	.9842	.2134	1.531	105
.007700	.03529	.2218	.2182	.2236	1.0247	.9759	.4435	.4582	1.100	.9840	.2147	1.526	104
.007800	.03552	.2232	.2196	.2251	1.0250	.9756	.4464	.4614	1.101	.9838	.2160	1.521	102
.007900	.03576	.2247	.2209	.2265	1.0253	.9753	.4493	.4646	1.103	.9836	.2173	1.517	101
.008000	.03598	.2261	.2223	.2280	1.0257	.9750	.4522	.4678	1.104	.9834	.2186	1.512	100
.008100	.03621	.2275	.2237	.2295	1.0260	.9747	.4551	.4709	1.105	.9832	.2199	1.508	98.6
.008200	.03644	.2290	.2250	.2310	1.0263	.9744	.4579	.4741	1.107	.9830	.2212	1.503	97.5
.008300	.03666	.2304	.2264	.2324	1.0266	.9741	.4607	.4772	1.108	.9827	.2225	1.499	96.3
.008400	.03689	.2318	.2277	.2338	1.0270	.9737	.4636	.4803	1.109	.9825	.2237	1.495	95.2
.008500	.03711	.2332	.2290	.2353	1.0273	.9734	.4664	.4834	1.111	.9823	.2250	1.491	94.1
.008600	.03733	.2346	.2303	.2367	1.0276	.9731	.4691	.4865	1.112	.9821	.2262	1.487	93.0
.008700	.03755	.2360	.2317	.2381	1.0280	.9728	.4719	.4896	1.113	.9819	.2275	1.482	91.9
.008800	.03777	.2373	.2330	.2396	1.0283	.9725	.4747	.4927	1.115	.9817	.2287	1.478	90.9
.008900	.03799	.2387	.2343	.2410	1.0286	.9722	.4774	.4957	1.116	.9815	.2300	1.474	89.9
.009000	.03821	.2401	.2356	.2424	1.0290	.9718	.4801	.4988	1.118	.9813	.2312	1.471	88.9
.009100	.03842	.2414	.2368	.2438	1.0293	.9715	.4828	.5018	1.119	.9811	.2324	1.467	88.0
.009200	.03864	.2428	.2381	.2452	1.0296	.9712	.4855	.5049	1.120	.9809	.2336	1.463	87.1
.009300	.03885	.2441	.2394	.2465	1.0299	.9709	.4882	.5079	1.122	.9807	.2348	1.459	86.1
.009400	.03906	.2455	.2407	.2479	1.0303	.9706	.4909	.5109	1.123	.9805	.2360	1.456	85.2
.009500	.03928	.2468	.2419	.2493	1.0306	.9703	.4936	.5138	1.124	.9803	.2371	1.452	84.3
.009600	.03949	.2481	.2431	.2507	1.0309	.9700	.4962	.5168	1.126	.9801	.2383	1.448	83.5
.009700	.03970	.2494	.2443	.2520	1.0313	.9697	.4988	.5198	1.127	.9799	.2394	1.445	82.7
.009800	.03990	.2507	.2456	.2534	1.0316	.9694	.5014	.5227	1.128	.9797	.2406	1.442	81.8
.009900	.04011	.2520	.2468	.2547	1.0319	.9691	.5040	.5257	1.130	.9794	.2417	1.438	81.0
.01000	.04032	.2533	.2480	.2560	1.0322	.9688	.5066	.5286	1.131	.9792	.2429	1.435	80.2
.01100	.04233	.2660	.2598	.2691	1.0356	.9656	.5319	.5574	1.145	.9772	.2539	1.403	73.1
.01200	.04426	.2781	.2711	.2817	1.0389	.9625	.5562	.5853	1.159	.9751	.2643	1.375	67.1
.01300	.04612	.2898	.2820	.2938	1.0423	.9594	.5795	.6125	1.173	.9731	.2743	1.350	62.1
.01400	.04791	.3010	.2924	.3056	1.0456	.9564	.6020	.6391	1.187	.9710	.2838	1.327	57.8
.01500	.04964	.3119	.3022	.3170	1.0490	.9533	.6238	.6651	1.201	.9690	.2928	1.307	54.0
.01600	.05132	.3225	.3117	.3281	1.0524	.9502	.6450	.6906	1.215	.9670	.3014	1.288	50.8
.01700	.05296	.3328	.3209	.3389	1.0559	.9471	.6655	.7158	1.230	.9649	.3096	1.271	47.9
.01800	.05455	.3428	.3298	.3495	1.0593	.9440	.6856	.7405	1.244	.9629	.3176	1.255	45.3
.01900	.05611	.3525	.3386	.3599	1.0628	.9409	.7051	.7650	1.259	.9609	.3253	1.240	43.0
.02000	.05763	.3621	.3470	.3701	1.0663	.9378	.7242	.7891	1.274	.9588	.3327	1.226	41.0
.02100	.05912	.3714	.3552	.3800	1.0698	.9348	.7429	.8131	1.289	.9568	.3399	1.213	39.1
.02200	.06057	.3806	.3632	.3898	1.0733	.9317	.7612	.8368	1.304	.9548	.3468	1.201	37.4
.02300	.06200	.3896	.3710	.3995	1.0768	.9287	.7791	.8603	1.319	.9528	.3535	1.189	35.9
.02400	.06340	.3984	.3786	.4090	1.0804	.9256	.7967	.8837	1.335	.9508	.3600	1.178	34.4
.02500	.06478	.4070	.3860	.4184	1.0840	.9225	.8140	.9069	1.350	.9488	.3662	1.168	33.1
.02600	.06613	.4155	.3932	.4276	1.0876	.9195	.8310	.9310	1.366	.9468	.3722	1.159	31.9
.02700	.06747	.4239	.4002	.4367	1.0912	.9164	.8478	.9530	1.381	.9448	.3781	1.150	30.8
.02800	.06878	.4322	.4071	.4457	1.0949	.9133	.8643	.9760	1.397	.9428	.3838	1.141	29.8
.02900	.07007	.4403	.4138	.4546	1.0985	.9103	.8805	.9988	1.413	.9408	.3893	1.133	28.8
.03000	.07135	.4483	.4205	.4634	1.1021	.9073	.8966	1.022	1.430	.9388	.3947	1.125	27.9
.03100	.07260	.4562	.4269	.4721	1.1059	.9042	.9124	1.044	1.446	.9369	.4000	1.118	27.1
.03200	.07385	.4640	.4333	.4808	1.1096	.9012	.9280	1.067	1.462	.9349	.4051	1.111	26.3
.03300	.07507	.4717	.4395	.4894	1.1133	.8982	.9434	1.090	1.479	.9329	.4100	1.104	25.6
.03400	.07630	.4794	.4457	.4980	1.1171	.8952	.9588	1.113	1.496	.9309	.4149	1.098	24.8
.03500	.07748	.4868	.4517	.5064	1.1209	.8921	.9737	1.135	1.513	.9289	.4196	1.092	24.19
.03600	.07867	.4943	.4577	.5147	1.1247	.8891	.9886	1.158	1.530	.9270	.4242	1.086	23.56
.03700	.07984	.5017	.4635	.5230	1.1285	.8861	1.0033	1.180	1.547	.9250	.4287	1.080	22.97
.03800	.08100	.5090	.4691	.5312	1.1324	.8831	1.018	1.203	1.564	.9230	.4330	1.075	22.42
.03900	.08215	.5162	.4747	.5394	1.1362	.8801	1.032	1.226	1.582	.9211	.4372	1.069	21.90
.04000	.08329	.5233	.4802	.5475	1.1401	.8771	1.047	1.248	1.600	.9192	.4414	1.064	21.40
.04100	.08442	.5304	.4857	.5556	1.1440	.8741	1.061	1.271	1.617	.9172	.4455	1.059	20.92
.04200	.08553	.5374	.4911	.5637	1.1479	.8711	1.075	1.294	1.636	.9153	.4495	1.055	20.46
.04300	.08664	.5444	.4964	.5717	1.1518	.8681	1.089	1.317	1.654	.9133	.4534	1.050	20.03
.04400	.08774	.5513	.5015	.5796	1.1558	.8652	1.103	1.340	1.672	.9114	.4571	1.046	19.62
.04500	.08883	.5581	.5066	.5876	1.1599	.8621	1.116	1.363	1.691	.9095	.4607	1.042	19.23
.04600	.08991	.5649	.5116	.5954	1.1639	.8592	1.130	1.386	1.709	.9076	.4643	1.038	18.85
.04700	.09098	.5717	.5166	.6033	1.1679	.8562	1.143	1.409	1.728	.9057	.4679	1.034	18.49
.04800	.09205	.5784	.5215	.6111	1.1720	.8532	1.157	1.433	1.747	.9037	.4713	1.030	18.15
.04900	.09311	.5850	.5263	.6189	1.1760	.8503	1.170	1.456	1.766	.9018	.4746	1.026	17.82

APPENDIX 1. (Continued)

d/L_o	d/L	$2\pi d/L$	tanh $2\pi d/L$	sinh $2\pi d/L$	cosh $2\pi d/L$	K	$4\pi d/L$	sinh $4\pi d/L$	cosh $4\pi d/L$	n	C_G/C_o	H/H_o'	M
.05000	.09416	.5916	.5310	.6267	1.1802	.8473	1.183	1.479	1.786	.8999	.4779	1.023	17.50
.05100	.09520	.5981	.5357	.6344	1.1843	.8444	1.196	1.503	1.805	.8980	.4811	1.019	17.19
.05200	.09623	.6046	.5403	.6421	1.1884	.8415	1.209	1.526	1.825	.8961	.4842	1.016	16.90
.05300	.09726	.6111	.5449	.6499	1.1926	.8385	1.222	1.550	1.845	.8943	.4873	1.013	16.62
.05400	.09829	.6176	.5494	.6575	1.1968	.8356	1.235	1.574	1.865	.8924	.4903	1.010	16.35
.05500	.09930	.6239	.5538	.6652	1.2011	.8326	1.248	1.598	1.885	.8905	.4932	1.007	16.09
.05600	.1003	.6303	.5582	.6729	1.2053	.8297	1.261	1.622	1.906	.8886	.4960	1.004	15.84
.05700	.1013	.6366	.5626	.6805	1.2096	.8267	1.273	1.646	1.926	.8867	.4988	1.001	15.60
.05800	.1023	.6428	.5668	.6880	1.2138	.8239	1.286	1.670	1.947	.8849	.5015	.9985	15.36
.05900	.1033	.6491	.5711	.6956	1.2181	.8209	1.298	1.695	1.968	.8830	.5042	.9958	15.13
.06000	.1043	.6553	.5753	.7033	1.2225	.8180	1.311	1.719	1.989	.8811	.5068	.9932	14.91
.06100	.1053	.6616	.5794	.7110	1.2270	.8150	1.3231	1.744	2.011	.8792	.5094	.9907	14.70
.06200	.1063	.6678	.5834	.7187	1.2315	.8121	1.336	1.770	2.033	.8773	.5119	.9883	14.50
.06300	.1073	.6739	.5874	.7256	1.2355	.8093	1.348	1.795	2.055	.8755	.5143	.9860	14.30
.06400	.1082	.6799	.5914	.7335	1.2402	.8063	1.360	1.819	2.076	.8737	.5167	.9837	14.11
.06500	.1092	.6860	.5954	.7411	1.2447	.8035	1.372	1.845	2.098	.8719	.5191	.9815	13.92
.06600	.1101	.6920	.5993	.7486	1.2492	.8005	1.384	1.870	2.121	.8700	.5214	.9793	13.74
.06700	.1111	.6981	.6031	.7561	1.2537	.7977	1.396	1.896	2.144	.8682	.5236	.9772	13.57
.06800	.1120	.7037	.6069	.7633	1.2580	.7948	1.408	1.921	2.166	.8664	.5258	.9752	13.40
.06900	.1130	.7099	.6106	.7711	1.2628	.7919	1.420	1.948	2.189	.8646	.5279	.9732	13.24
.07000	.1139	.7157	.6144	.7783	1.2672	.7890	1.432	1.974	2.213	.8627	.5300	.9713	13.08
.07100	.1149	.7219	.6181	.7863	1.2721	.7861	1.444	2.000	2.236	.8609	.5321	.9694	12.92
.07200	.1158	.7277	.6217	.7937	1.2767	.7833	1.455	2.026	2.260	.8591	.5341	.9676	12.77
.07300	.1168	.7336	.6252	.8011	1.2813	.7804	1.467	2.053	2.284	.8572	.5360	.9658	12.62
.07400	.1177	.7395	.6289	.8088	1.2861	.7775	1.479	2.080	2.308	.8554	.5380	.9641	12.48
.07500	.1186	.7453	.6324	.8162	1.2908	.7747	1.490	2.107	2.332	.8537	.5399	.9624	12.34
.07600	.1195	.7511	.6359	.8237	1.2956	.7719	1.502	2.135	2.357	.8519	.5417	.9607	12.21
.07700	.1205	.7569	.6392	.8312	1.3004	.7690	1.514	2.162	2.382	.8501	.5435	.9591	12.08
.07800	.1214	.7625	.6427	.8386	1.3051	.7662	1.525	2.189	2.407	.8483	.5452	.9576	11.95
.07900	.1223	.7683	.6460	.8462	1.3100	.7634	1.537	2.217	2.432	.8465	.5469	.9562	11.83
.08000	.1232	.7741	.6493	.8538	1.3149	.7605	1.548	2.245	2.458	.8448	.5485	.9548	11.71
.08100	.1241	.7799	.6526	.8614	1.3198	.7577	1.560	2.274	2.484	.8430	.5501	.9534	11.59
.08200	.1251	.7854	.6558	.8687	1.3246	.7549	1.571	2.303	2.511	.8413	.5517	.9520	11.47
.08300	.1259	.7911	.6590	.8762	1.3295	.7522	1.583	2.331	2.537	.8395	.5533	.9506	11.36
.08400	.1268	.7967	.6622	.8837	1.3345	.7494	1.594	2.360	2.563	.8378	.5548	.9493	11.25
.08500	.1277	.8026	.6655	.8915	1.3397	.7464	1.605	2.389	2.590	.8360	.5563	.9481	11.14
.08600	.1286	.8080	.6685	.8989	1.3446	.7437	1.616	2.418	2.617	.8342	.5577	.9469	11.04
.08700	.1295	.8137	.6716	.9064	1.3497	.7409	1.628	2.448	2.644	.8325	.5591	.9457	10.94
.08800	.1304	.8193	.6747	.9141	1.3548	.7381	1.639	2.478	2.672	.8308	.5605	.9445	10.84
.08900	.1313	.8250	.6778	.9218	1.3600	.7353	1.650	2.508	2.700	.8290	.5619	.9433	10.74
.09000	.1322	.8306	.6808	.9295	1.3653	.7324	1.661	2.538	2.728	.8273	.5632	.9422	10.65
.09100	.1331	.8363	.6838	.9372	1.3706	.7296	1.672	2.568	2.756	.8255	.5645	.9411	10.55
.09200	.1340	.8420	.6868	.9450	1.3759	.7268	1.684	2.599	2.785	.8238	.5658	.9401	10.46
.09300	.1349	.8474	.6897	.9525	1.3810	.7241	1.695	2.630	2.814	.8221	.5670	.9391	10.37
.09400	.1357	.8528	.6925	.9600	1.3862	.7214	1.706	2.662	2.843	.8204	.5682	.9381	10.29
.09500	.1366	.8583	.6953	.9677	1.3917	.7186	1.717	2.693	2.873	.8187	.5693	.9371	10.21
.09600	.1375	.8639	.6982	.9755	1.3970	.7158	1.728	2.726	2.903	.8170	.5704	.9362	10.12
.09700	.1384	.8694	.7011	.9832	1.4023	.7131	1.739	2.757	2.933	.8153	.5716	.9353	10.04
.09800	.1392	.8749	.7039	.9908	1.4077	.7104	1.750	2.790	2.963	.8136	.5727	.9344	9.962
.09900	.1401	.8803	.7066	.9985	1.4131	.7076	1.761	2.822	2.994	.8120	.5737	.9335	9.884
.1000	.1410	.8858	.7093	1.006	1.4187	.7049	1.772	2.855	3.025	.8103	.5747	.9327	9.808
.1010	.1419	.8913	.7120	1.014	1.4242	.7022	1.783	2.888	3.057	.8086	.5757	.9319	9.734
.1020	.1427	.8967	.7147	1.022	1.4297	.6994	1.793	2.922	3.088	.8069	.5766	.9311	9.661
.1030	.1436	.9023	.7173	1.030	1.4354	.6967	1.805	2.956	3.121	.8052	.5776	.9304	9.590
.1040	.1445	.9076	.7200	1.037	1.4410	.6940	1.815	2.990	3.153	.8036	.5785	.9297	9.519
.1050	.1453	.9130	.7226	1.045	1.4465	.6913	1.826	3.024	3.185	.8019	.5794	.9290	9.451
.1060	.1462	.9184	.7252	1.053	1.4523	.6886	1.837	3.059	3.218	.8003	.5803	.9282	9.384
.1070	.1470	.9239	.7277	1.061	1.4580	.6859	1.848	3.094	3.251	.7986	.5812	.9276	9.318
.1080	.1479	.9293	.7303	1.069	1.4638	.6833	1.858	3.128	3.284	.7970	.5820	.9269	9.254
.1090	.1488	.9343	.7327	1.076	1.4692	.6806	1.869	3.164	3.319	.7954	.5828	.9263	9.191
.1100	.1496	.9400	.7352	1.085	1.4752	.6779	1.880	3.201	3.353	.7937	.5836	.9257	9.129
.1110	.1505	.9456	.7377	1.093	1.4814	.6752	1.891	3.237	3.388	.7920	.5843	.9251	9.068
.1120	.1513	.9508	.7402	1.101	1.4871	.6725	1.902	3.274	3.423	.7904	.5850	.9245	9.009
.1130	.1522	.9563	.7426	1.109	1.4932	.6697	1.913	3.312	3.459	.7888	.5857	.9239	8.950
.1140	.1530	.9616	.7450	1.117	1.4990	.6671	1.923	3.348	3.494	.7872	.5864	.9234	8.891
.1150	.1539	.9670	.7474	1.125	1.5051	.6645	1.934	3.385	3.530	.7856	.5871	.9228	8.835
.1160	.1547	.9720	.7497	1.133	1.5108	.6619	1.944	3.423	3.566	.7840	.5878	.9223	8.780
.1170	.1556	.9775	.7520	1.141	1.5171	.6592	1.955	3.462	3.603	.7824	.5884	.9218	8.726
.1180	.1564	.9827	.7543	1.149	1.5230	.6566	1.966	3.501	3.641	.7808	.5890	.9214	8.673
.1190	.1573	.9882	.7566	1.157	1.5293	.6539	1.977	3.540	.3.678	.7792	.5896	.9209	8.621

APPENDIX 1. (Continued)

d/L_o	d/L	$2\pi d/L$	tanh $2\pi d/L$	sinh $2\pi d/L$	cosh $2\pi d/L$	K	$4\pi d/L$	sinh $4\pi d/L$	cosh $4\pi d/L$	n	C_G/C_o	H/H_o'	M
.1200	.1581	.9936	.7589	1.165	1.5356	.6512	1.987	3.579	3.716	.7776	.5902	.9204	8.569
.1210	.1590	.9989	.7612	1.174	1.5418	.6486	1.998	3.620	3.755	.7760	.5907	.9200	8.518
.1220	.1598	1.004	.7634	1.182	1.5479	.6460	2.008	3.659	3.793	.7745	.5913	.9196	8.468
.1230	.1607	1.010	.7656	1.190	1.5546	.6433	2.019	3.699	3.832	.7729	.5918	.9192	8.419
.1240	.1615	1.015	.7678	1.198	1.5605	.6407	2.030	3.740	3.871	.7713	.5922	.9189	8.371
.1250	.1624	1.020	.7700	1.207	1.5674	.6381	2.041	3.782	3.912	.7698	.5926	.9186	8.324
.1260	.1632	1.025	.7721	1.215	1.5734	.6356	2.051	3.824	3.952	.7682	.5931	.9182	8.278
.1270	.1640	1.030	.7742	1.223	1.5795	.6331	2.061	3.865	3.992	.7667	.5936	.9178	8.233
.1280	.1649	1.036	.7763	1.231	1.5862	.6305	2.072	3.907	4.033	.7652	.5940	.9175	8.189
.1290	.1657	1.041	.7783	1.240	1.5927	.6279	2.082	3.950	4.074	.7637	.5944	.9172	8.146
.1300	.1665	1.046	.7804	1.248	1.5990	.6254	2.093	3.992	4.115	.7621	.5948	.9169	8.103
.1310	.1674	1.052	.7824	1.257	1.6060	.6228	2.104	4.036	4.158	.7606	.5951	.9166	8.061
.1320	.1682	1.057	.7844	1.265	1.6124	.6202	2.114	4.080	4.201	.7591	.5954	.9164	8.020
.1330	.1691	1.062	.7865	1.273	1.6191	.6176	2.125	4.125	4.245	.7575	.5958	.9161	7.978
.1340	.1699	1.068	.7885	1.282	1.6260	.6150	2.135	4.169	4.288	.7560	.5961	.9158	7.937
.1350	.1708	1.073	.7905	1.291	1.633	.6123	2.146	4.217	4.334	.7545	.5964	.9156	7.897
.1360	.1716	1.078	.7925	1.300	1.640	.6098	2.156	4.262	4.378	.7530	.5967	.9154	7.857
.1370	.1724	1.084	.7945	1.308	1.647	.6073	2.167	4.309	4.423	.7515	.5969	.9152	7.819
.1380	.1733	1.089	.7964	1.317	1.654	.6047	2.177	4.355	4.468	.7500	.5972	.9150	7.781
.1390	.1741	1.094	.7983	1.326	1.660	.6022	2.188	4.402	4.514	.7485	.5975	.9148	7.744
.1400	.1749	1.099	.8002	1.334	1.667	.5998	2.198	4.450	4.561	.7471	.5978	.9146	7.707
.1410	.1758	1.105	.8021	1.343	1.675	.5972	2.209	4.498	4.607	.7456	.5980	.9144	7.671
.1420	.1766	1.110	.8039	1.352	1.681	.5947	2.219	4.546	4.654	.7441	.5982	.9142	7.636
.1430	.1774	1.115	.8057	1.360	1.688	.5923	2.230	4.595	4.663	.7426	.5984	.9141	7.602
.1440	.1783	1.120	.8076	1.369	1.696	.5898	2.240	4.644	4.751	.7412	.5986	.9140	7.567
.1450	.1791	1.125	.8094	1.378	1.703	.5873	2.251	4.695	4.800	.7397	.5987	.9139	7.533
.1460	.1800	1.131	.8112	1.388	1.710	.5847	2.261	4.746	4.850	.7382	.5989	.9137	7.499
.1470	.1808	1.136	.8131	1.397	1.718	.5822	2.272	4.798	4.901	.7368	.5990	.9136	7.465
.1480	.1816	1.141	.8149	1.405	1.725	.5798	2.282	4.847	4.951	.7354	.5992	.9135	7.432
.1490	.1825	1.146	.8166	1.415	1.732	.5773	2.293	4.901	5.001	.7339	.5993	.9134	7.400
.1500	.1833	1.152	.8183	1.424	1.740	.5748	2.303	4.954	5.054	.7325	.5994	.9133	7.369
.1510	.1841	1.157	.8200	1.433	1.747	.5723	2.314	5.007	5.106	.7311	.5994	.9133	7.339
.1520	.1850	1.162	.8217	1.442	1.755	.5699	2.324	5.061	5.159	.7296	.5995	.9132	7.309
.1530	.1858	1.167	.8234	1.451	1.762	.5675	2.335	5.115	5.212	.7282	.5996	.9132	7.279
.1540	.1866	1.173	.8250	1.460	1.770	.5651	2.345	5.169	5.265	.7268	.5996	.9132	7.250
.1550	.1875	1.178	.8267	1.469	1.777	.5627	2.356	5.225	5.320	.7254	.5997	.9131	7.221
.1560	.1883	1.183	.8284	1.479	1.785	.5602	2.366	5.283	5.376	.7240	.5998	.9130	7.191
.1570	.1891	1.188	.8301	1.488	1.793	.5577	2.377	5.339	5.432	.7226	.5999	.9129	7.162
.1580	.1900	1.194	.8317	1.498	1.801	.5552	2.387	5.398	5.490	.7212	.5998	.9130	7.134
.1590	.1908	1.199	.8333	1.507	1.809	.5528	2.398	5.454	5.544	.7198	.5998	.9130	7.107
.1600	.1917	1.204	.8349	1.517	1.817	.5504	2.408	5.513	5.603	.7184	.5998	.9130	7.079
.1610	.1925	1.209	.8365	1.527	1.825	.5480	2.419	5.571	5.660	.7171	.5998	.9130	7.052
.1620	.1933	1.215	.8381	1.536	1.833	.5456	2.429	5.630	5.718	.7157	.5998	.9130	7.026
.1630	.1941	1.220	.8396	1.546	1.841	.5432	2.440	5.690	5.777	.7144	.5998	.9130	7.000
.1640	.1950	1.225	.8411	1.555	1.849	.5409	2.450	5.751	5.837	.7130	.5998	.9130	6.975
.1650	.1958	1.230	.8427	1.565	1.857	.5385	2.461	5.813	5.898	.7117	.5997	.9131	6.949
.1660	.1966	1.235	.8442	1.574	1.865	.5362	2.471	5.874	5.959	.7103	.5996	.9132	6.924
.1670	.1975	1.240	.8457	1.584	1.873	.5339	2.482	5.938	6.021	.7090	.5996	.9132	6.900
.1680	.1983	1.246	.8472	1.594	1.882	.5315	2.492	6.003	6.085	.7076	.5995	.9133	6.876
.1690	.1992	1.251	.8486	1.604	1.890	.5291	2.503	6.066	6.148	.7063	.5994	.9133	6.853
.1700	.2000	1.257	.8501	1.614	1.899	.5267	2.513	6.130	6.212	.7050	.5993	.9134	6.830
.1710	.2008	1.262	.8515	1.624	1.907	.5243	2.523	6.197	6.275	.7036	.5992	.9135	6.807
.1720	.2017	1.267	.8529	1.634	1.915	.5220	2.534	6.262	6.342	.7023	.5991	.9136	6.784
.1730	.2025	1.272	.8544	1.644	1.924	.5197	2.544	6.329	6.407	.7010	.5989	.9137	6.761
.1740	.2033	1.277	.8558	1.654	1.933	.5174	2.555	6.395	6.473	.6997	.5988	.9138	6.738
.1750	.2042	1.282	.8572	1.664	1.941	.5151	2.565	6.465	6.541	.6984	.5987	.9139	6.716
.1760	.2050	1.288	.8586	1.675	1.951	.5127	2.576	6.534	6.610	.6971	.5985	.9140	6.694
.1770	.2058	1.293	.8600	1.685	1.959	.5104	3.586	6.603	6.679	.6958	.5984	.9141	6.672
.1780	.2066	1.298	.8614	1.695	1.968	.5081	2.597	6.672	6.747	.6946	.5982	.9142	6.651
.1790	.2075	1.304	.8627	1.706	1.977	.5058	2.607	6.744	6.818	.6933	.5980	.9144	6.631
.1800	.2083	1.309	.8640	1.716	1.986	.5036	2.618	6.818	6.891	.6920	.5979	.9145	6.611
.1810	.2092	1.314	.8653	1.727	1.995	.5013	2.629	6.890	6.963	.6907	.5977	.9146	6.591
.1820	.2100	1.320	.8666	1.737	2.004	.4990	2.639	6.963	7.035	.6895	.5975	.9148	6.571
.1830	.2108	1.325	.8680	1.748	2.013	.4967	2.650	7.038	7.109	.6882	.5974	.9149	6.550
.1840	.2117	1.330	.8693	1.758	2.022	.4945	2.660	7.113	7.183	.6870	.5972	.9150	6.530
.1850	.2125	1.335	.8706	1.769	2.032	.4922	2.671	7.191	7.260	.6857	.5969	.9152	6.511
.1860	.2134	1.341	.8718	1.780	2.041	.4899	2.681	7.267	7.336	.6845	.5967	.9154	6.492
.1870	.2142	1.346	.8731	1.791	2.051	.4876	2.692	7.345	7.412	.6832	.5965	.9155	6.474
.1880	.2150	1.351	.8743	1.801	2.060	.4854	2.702	7.421	7.488	.6820	.5963	.9157	6.456
.1890	.2159	1.356	.8755	1.812	2.070	.4832	2.712	7.500	7.566	.6808	.5961	.9159	6.438

APPENDIX 1. (Continued)

d/L_o	d/L	$2\pi d/L$	tanh $2\pi d/L$	sinh $2\pi d/L$	cosh $2\pi d/L$	K	$4\pi d/L$	sinh $4\pi d/L$	cosh $4\pi d/L$	n	C_G/C_o	H/H_o	M
.1900	.2167	1.362	.8767	1.823	2.079	.4809	2.723	7.581	7.647	.6796	.5958	.9161	6.421
.1910	.2176	1.367	.8779	1.834	2.089	.4787	2.734	7.663	7.728	.6784	.5955	.9163	6.403
.1920	.2184	1.372	.8791	1.845	2.099	.4765	2.744	7.746	7.810	.6772	.5952	.9165	6.385
.1930	.2192	1.377	.8803	1.856	2.108	.4743	2.755	7.827	7.891	.6760	.5950	.9167	6.368
.1940	.2201	1.383	.8815	1.867	2.118	.4721	2.765	7.911	7.974	.6748	.5948	.9169	6.351
.1950	.2209	1.388	.8827	1.879	2.128	.4699	2.776	7.996	8.059	.6736	.5946	.9170	6.334
.1960	.2218	1.393	.8839	1.890	2.138	.4677	2.787	8.083	8.145	.6724	.5944	.9172	6.317
.1970	.2226	1.399	.8850	1.901	2.148	.4655	2.797	8.167	8.228	.6712	.5941	.9174	6.300
.1980	.2234	1.404	.8862	1.913	2.158	.4633	2.808	8.256	8.316	.6700	.5938	.9176	6.284
.1990	.2243	1.409	.8873	1.924	2.169	.4611	2.819	8.346	8.406	.6689	.5935	.9179	6.268
.2000	.2251	1.414	.8884	1.935	2.178	.4590	2.829	8.436	8.495	.6677	.5932	.9181	6.253
.2010	.2260	1.420	.8895	1.947	2.189	.4569	2.840	8.524	8.583	.6666	.5929	.9183	6.237
.2020	.2268	1.425	.8906	1.959	2.199	.4547	2.850	8.616	8.674	.6654	.5926	.9186	6.222
.2030	.2277	1.430	.8917	1.970	2.210	.4526	2.861	8.708	8.766	.6642	.5923	.9188	6.206
.2040	.2285	1.436	.8928	1.982	2.220	.4504	2.872	8.803	8.860	.6631	.5920	.9190	6.191
.2050	.2293	1.441	.8939	1.994	2.231	.4483	2.882	8.897	8.953	.6620	.5917	.9193	6.176
.2060	.2302	1.446	.8950	2.006	2.242	.4462	2.893	8.994	9.050	.6608	.5914	.9195	6.161
.2070	.2310	1.451	.8960	2.017	2.252	.4441	2.903	9.090	9.144	.6597	.5911	.9197	6.147
.2080	.2319	1.457	.8971	2.030	2.263	.4419	2.914	9.187	9.240	.6586	.5908	.9200	6.133
.2090	.2328	1.462	.8981	2.042	2.274	.4398	2.925	9.288	9.342	.6574	.5905	.9202	6.119
.2100	.2336	1.468	.8991	2.055	2.285	.4377	2.936	9.389	9.442	.6563	.5901	.9205	6.105
.2110	.2344	1.473	.9001	2.066	2.295	.4357	2.946	9.490	9.542	.6552	.5898	.9207	6.091
.2120	.2353	1.479	.9011	2.079	2.307	.4336	2.957	9.590	9.642	.6541	.5894	.9210	6.077
.2130	.2361	1.484	.9021	2.091	2.318	.4315	2.967	9.693	9.744	.6531	.5891	.9213	6.064
.2140	.2370	1.489	.9031	2.103	2.329	.4294	2.978	9.796	9.847	.6520	.5888	.9215	6.051
.2150	.2378	1.494	.9041	2.115	2.340	.4274	2.989	9.902	9.952	.6509	.5884	.9218	6.037
.2160	.2387	1.500	.9051	2.128	2.351	.4253	2.999	10.01	10.06	.6498	.5881	.9221	6.024
.2170	.2395	1.506	.9061	2.142	2.364	.4232	3.010	10.12	10.17	.6488	.5878	.9223	6.011
.2180	.2404	1.511	.9070	2.154	2.375	.4211	3.021	10.23	10.28	.6477	.5874	.9226	5.999
.2190	.2412	1.516	.9079	2.166	2.386	.4191	3.031	10.34	10.38	.6467	.5871	.9228	5.987
.2200	.2421	1.521	.9088	2.178	2.397	.4171	3.042	10.45	10.50	.6456	.5868	.9231	5.975
.2210	.2429	1.526	.9097	2.192	2.409	.4151	3.052	10.56	10.61	.6446	.5864	.9234	5.963
.2220	.2438	1.532	.9107	2.204	2.421	.4131	3.063	10.68	10.72	.6436	.5861	.9236	5.951
.2230	.2446	1.537	.9116	2.218	2.433	.4111	3.074	10.79	10.84	.6425	.5857	.9239	5.939
.2240	.2455	1.542	.9125	2.230	2.444	.4091	3.085	10.91	10.95	.6414	.5854	.9242	5.927
.2250	.2463	1.548	.9134	1.244	2.457	.4071	3.095	11.02	11.07	.6404	.5850	.9245	5.915
.2260	.2472	1.553	.9143	2.257	2.469	.4051	3.106	11.15	11.19	.6394	.5846	.9248	5.903
.2270	.2481	1.559	.9152	2.271	2.481	.4031	3.117	11.27	11.31	.6383	.5842	.9251	5.891
.2280	.2489	1.564	.9161	2.284	2.493	.4011	3.128	11.39	11.44	.6373	.5838	.9254	5.880
.2290	.2498	1.569	.9170	2.297	2.506	.3991	3.138	11.51	11.56	.6363	.5834	.9258	5.869
.2300	.2506	1.575	.9178	2.311	2.518	.3971	3.149	11.64	11.68	.6353	.5830	.9261	5.858
.2310	.2515	1.580	.9186	2.325	2.531	.3952	3.160	11.77	11.81	.6343	.5826	.9264	5.848
.2320	.2523	1.585	.9194	2.338	2.543	.3932	3.171	11.90	11.93	.6333	.5823	.9267	5.838
.2330	.2532	1.591	.9203	2.352	2.556	.3912	3.182	12.03	12.07	.6323	.5819	.9270	5.827
.2340	.2540	1.596	.9211	2.366	2.569	.3893	3.192	12.15	12.19	.6313	.5815	.9273	5.816
.2350	.2549	1.602	.9219	2.380	2.581	.3874	3.203	12.29	12.33	.6304	.5811	.9276	5.806
.2360	.2558	1.607	.9227	2.393	2.594	.3855	3.214	12.43	12.47	.6294	.5807	.9279	5.796
.2370	.2566	1.612	.9235	2.408	2.607	.3836	3.225	12.55	12.59	.6284	.5804	.9282	5.786
.2380	.2575	1.618	.9243	2.422	2.620	.3816	3.236	12.69	12.73	.6275	.5800	.9285	5.776
.2390	.2584	1.623	.9251	2.436	2.634	.3797	3.247	12.83	12.87	.6265	.5796	.9288	5.766
.2400	.2592	1.629	.9259	2.450	2.647	.3779	3.257	12.97	13.01	.6256	.5792	.9291	5.756
.2410	.2601	1.634	.9267	2.464	2.660	.3760	3.268	13.11	13.15	.6246	.5788	.9294	5.746
.2420	.2610	1.640	.9275	2.480	2.674	.3741	3.279	13.26	13.30	.6237	.5784	.9298	5.736
.2430	.2618	1.645	.9282	2.494	2.687	.3722	3.290	13.40	13.44	.6228	.5780	.9301	5.727
.2440	.2627	1.650	.9289	2.508	2.700	.3704	3.301	13.55	13.59	.6218	.5776	.9304	5.718
.2450	.2635	1.656	.9296	2.523	2.714	.3685	3.312	13.70	13.73	.6209	.5772	.9307	5.710
.2460	.2644	1.661	.9304	2.538	2.728	.3666	3.323	13.85	13.88	.6200	.5768	.9310	5.701
.2470	.2653	1.667	.9311	2.553	2.742	.3648	3.334	14.00	14.04	.6191	.5764	.9314	5.692
.2480	.2661	1.672	.9318	2.568	2.755	.3629	3.344	14.15	14.19	.6182	.5760	.9317	5.684
.2490	.2670	1.678	.9325	2.583	2.770	.3610	3.355	14.31	14.35	.6173	.5756	.9320	5.675
.2500	.2679	1.683	.9332	2.599	2.784	.3592	3.367	14.47	14.51	.6164	.5752	.9323	5.667
.2510	.2687	1.689	.9339	2.614	2.798	.3574	3.377	14.62	14.66	.6155	.5748	.9327	5.658
.2520	.2696	1.694	.9346	2.629	2.813	.3556	3.388	14.79	14.82	.6146	.5744	.9330	5.650
.2530	.2705	1.700	.9353	2.645	2.828	.3537	3.399	14.95	14.99	.6137	.5740	.9333	5.641
.2540	.2714	1.705	.9360	2.660	2.842	.3519	3.410	15.12	15.15	.6128	.5736	.9336	5.633
.2550	.2722	1.711	.9367	2.676	2.856	.3501	3.421	15.29	15.32	.6120	.5732	.9340	5.624
.2560	.2731	1.716	.9374	2.691	2.871	.3483	3.432	15.45	15.49	.6111	.5728	.9343	5.616
.2570	.2740	1.722	.9381	2.707	2.886	.3465	3.443	15.63	15.66	.6102	.5724	.9346	5.608
.2580	.2749	1.727	.9388	2.723	2.901	.3447	3.454	15.80	15.83	.6093	.5720	.9349	5.600
.2590	.2757	1.732	.9394	2.739	2.916	.3430	3.465	15.97	16.00	.6085	.5716	.9353	5.592

APPENDIX 1. (Continued)

d/L_o	d/L	$2\pi d/L$	tanh $2\pi d/L$	sinh $2\pi d/L$	cosh $2\pi d/L$	K	$4\pi d/L$	sinh $4\pi d/L$	cosh $4\pi d/L$	n	C_G/C_o	H/H_o'	M
.2600	.2766	1.738	.9400	2.755	2.931	.3412	3.476	16.15	16.18	.6076	.5712	.9356	5.585
.2610	.2775	1.744	.9406	2.772	2.946	.3394	3.487	16.33	16.36	.6068	.5707	.9360	5.578
.2620	.2784	1.749	.9412	2.788	2.962	.3376	3.498	16.51	16.54	.6060	.5703	.9363	5.571
.2630	.2792	1.755	.9418	2.804	2.977	.3359	3.509	16.69	16.73	.6052	.5699	.9367	5.563
.2640	.2801	1.760	.9425	2.820	2.992	.3342	3.520	16.88	16.91	.6043	.5695	.9370	5.556
.2650	.2810	1.766	.9431	2.837	3.008	.3325	3.531	17.07	17.10	.6035	.5691	.9373	5.548
.2660	.2819	1.771	.9437	2.853	3.023	.3308	3.542	17.26	17.28	.6027	.5687	.9377	5.541
.2670	.2827	1.776	.9443	2.870	3.039	.3291	3.553	17.45	17.45	.6018	.5683	.9380	5.534
.2680	.2836	1.782	.9449	2.886	3.055	.3274	3.564	17.64	17.67	.6010	.5679	.9383	5.527
.2690	.2845	1.788	.9455	2.904	3.071	.3256	3.575	17.84	17.87	.6002	.5675	.9386	5.520
.2700	.2854	1.793	.9461	2.921	3.088	.3239	3.587	18.04	18.07	.5994	.5671	.9390	5.513
.2710	.2863	1.799	.9467	2.938	3.104	.3222	3.598	18.24	18.27	.5986	.5667	.9393	5.506
.2720	.2872	1.804	.9473	2.956	3.120	.3205	3.610	18.46	18.49	.5978	.5663	.9396	5.499
.2730	.2880	1.810	.9478	2.973	3.136	.3189	3.620	18.65	18.67	.5971	.5659	.9400	5.493
.2740	.2889	1.815	.9484	2.990	3.153	.3172	3.631	18.86	18.89	.5963	.5655	.9403	5.486
.2750	.2898	1.821	.9490	3.008	3.170	.3155	3.642	19.07	19.10	.5955	.5651	.9406	5.480
.2760	.2907	1.826	.9495	3.025	3.186	.3139	3.653	19.28	19.30	.5947	.5647	.9410	5.474
.2770	.2916	1.832	.9500	3.043	3.203	.3122	3.664	19.49	19.51	.5940	.5643	.9413	5.468
.2780	.2924	1.837	.9505	3.061	3.220	.3106	3.675	19.71	19.74	.5932	.5639	.9416	5.462
.2790	.2933	1.843	.9511	3.079	3.237	.3089	3.686	19.93	19.96	.5925	.5635	.9420	5.456
.2800	.2942	1.849	.9516	3.097	3.254	.3073	3.697	20.16	20.18	.5917	.5631	.9423	5.450
.2810	.2951	1.854	.9521	3.115	3.272	.3057	3.709	20.39	20.41	.5910	.5627	.9426	5.444
.2820	.2960	1.860	.9526	3.133	3.289	.3040	3.720	20.62	20.64	.5902	.5623	.9430	5.438
.2830	.2969	1.866	.9532	3.152	3.307	.3024	3.731	20.85	20.87	.5895	.5619	.9433	5.432
.2840	.2978	1.871	.9537	3.171	3.325	.3008	3.742	21.09	21.11	.5887	.5615	.9436	5.426
.2850	.2987	1.877	.9542	3.190	3.343	.2992	3.754	21.33	21.35	.5880	.5611	.9440	5.420
.2860	.2996	1.882	.9547	3.209	3.361	.2976	3.765	21.57	21.59	.5873	.5607	.9443	5.414
.2870	.3005	1.888	.9552	3.228	3.379	.2959	3.776	21.82	21.84	.5866	.5603	.9446	5.409
.2880	.3014	1.893	.9557	3.246	3.396	.2944	3.787	22.05	22.07	.5859	.5600	.9449	5.403
.2890	.3022	1.899	.9562	3.264	3.414	.2929	3.798	22.30	22.32	.5852	.5596	.9452	5.397
.2900	.3031	1.905	.9567	3.284	3.433	.2913	3.809	22.54	22.57	.5845	.5592	.9456	5.392
.2910	.3040	1.910	.9572	3.303	3.451	.2898	3.821	22.81	22.83	.5838	.5588	.9459	5.386
.2920	.3049	1.916	.9577	3.323	3.471	.2882	3.832	23.07	23.09	.5831	.5584	.9463	5.380
.2930	.3058	1.922	.9581	3.343	3.490	.2866	3.843	23.33	23.35	.5824	.5580	.9466	5.375
.2940	.3067	1.927	.9585	3.362	3.508	.2851	3.855	23.60	23.62	.5817	.5576	.9469	5.371
.2950	.3076	1.933	.9590	3.382	3.527	.2835	3.866	23.86	23.88	.5810	.5572	.9473	5.366
.2960	.3085	1.938	.9594	3.402	3.546	.2820	3.877	24.12	24.15	.5804	.5568	.9476	5.361
.2970	.3094	1.944	.9599	3.422	3.565	.2805	3.888	24.40	24.42	.5797	.5564	.9480	5.356
.2980	.3103	1.950	.9603	3.442	3.585	.2790	3.900	24.68	24.70	.5790	.5560	.9483	5.351
.2990	.3112	1.955	.9607	3.462	3.604	.2275	3.911	24.96	24.98	.5784	.5556	.9486	5.547
.3000	.3121	1.961	.9611	3.483	3.624	.2760	3.922	25.24	25.26	.5777	.5552	.9490	5.342
.3010	.3130	1.967	.9616	3.503	3.643	.2745	3.933	25.53	25.55	.5771	.5549	.9493	5.337
.3020	.3139	1.972	.9620	3.524	3.663	.2730	3.945	25.82	25.83	.5764	.5545	.9496	5.332
.3030	.3148	1.978	.9624	3.545	3.683	.2715	3.956	26.12	26.14	.5758	.5541	.9499	5.328
.3040	.3157	1.984	.9629	3.566	3.703	.2700	3.968	26.42	26.44	.5751	.5538	.9502	5.323
.3050	.3166	1.989	.9633	3.587	3.724	.2685	3.979	26.72	26.74	.5745	.5534	.9505	5.318
.3060	.3175	1.995	.9637	3.609	3.745	.2670	3.990	27.02	27.04	.5739	.5530	.9509	5.314
.3070	.3184	2.001	.9641	3.630	3.765	.2656	4.002	27.33	27.35	.5732	.5527	.9512	5.309
.3080	.3193	2.007	.9645	3.651	3.786	.2641	4.013	27.65	27.66	.5726	.5523	.9515	5.305
.3090	.3202	2.012	.9649	3.673	3.806	.2627	4.024	27.96	27.98	.5720	.5519	.9518	5.300
.3100	.3211	2.018	.9653	3.694	3.827	.2613	4.036	28.28	28.30	.5714	.5515	.9522	5.296
.3110	.3220	2.023	.9656	3.716	3.848	.2599	4.047	28.60	28.62	.5708	.5511	.9525	5.292
.3120	.3230	2.029	.9660	3.738	3.870	.2584	4.058	28.93	28.95	.5701	.5507	.9528	5.288
.3130	.3239	2.035	.9664	3.760	3.891	.2570	4.070	29.27	29.28	.5695	.5504	.9531	5.284
.3140	.3248	2.041	.9668	3.782	3.912	.2556	4.081	29.60	29.62	.5689	.5500	.9535	5.280
.3150	.3257	2.046	.9672	3.805	3.934	.2542	4.093	29.94	29.96	.5683	.5497	.9538	5.276
.3160	.3266	2.052	.9676	3.828	3.956	.2528	4.104	30.29	30.31	.5678	.5494	.9541	5.272
.3170	.3275	2.058	.9679	3.851	3.978	.2514	4.116	30.64	30.65	.5672	.5490	.9544	5.268
.3180	.3284	2.063	.9682	3.873	4.000	.2500	4.127	30.99	31.00	.5666	.5486	.9547	5.264
.3190	.3294	2.069	.9686	3.896	4.022	.2486	4.139	31.35	31.37	.5660	.5483	.9550	5.260
.3200	.3302	2.075	.9690	3.919	4.045	.2472	4.150	31.71	31.72	.5655	.5479	.9553	5.256
.3210	.3311	2.081	.9693	3.943	4.068	.2459	4.161	32.07	32.08	.5649	.5476	.9556	5.252
.3220	.3321	2.086	.9696	3.966	4.090	.2445	4.173	32.44	32.46	.5643	.5472	.9559	5.249
.3230	.3330	2.092	.9700	3.990	4.114	.2431	4.185	32.83	32.84	.5637	.5468	.9562	5.245
.3240	.3339	2.098	.9703	4.014	4.136	.2418	4.196	33.20	33.22	.5632	.5465	.9565	5.241
.3250	.3349	2.104	.9707	4.038	4.160	.2404	4.208	33.60	33.61	.5627	.5462	.9568	5.237
.3260	.3357	2.110	.9710	4.061	4.183	.2391	4.219	33.97	33.99	.5621	.5458	.9571	5.234
.3270	.3367	2.115	.9713	4.085	4.206	.2378	4.231	34.37	34.38	.5616	.5455	.9574	5.231
.3280	.3376	2.121	.9717	4.110	4.230	.2364	4.242	34.77	34.79	.5610	.5451	.9577	5.227
.3290	.3385	2.127	.9720	4.135	4.254	.2351	4.254	35.18	35.19	.5605	.5448	.9580	5.223

APPENDIX 1. (Continued)

d/L_o	d/L	$2\pi d/L$	tanh $2\pi d/L$	sinh $2\pi d/L$	cosh $2\pi d/L$	K	$4\pi d/L$	sinh $4\pi d/L$	cosh $4\pi d/L$	n	C_G/C_o	H/H'_o	M
.3300	.3394	2.133	.9723	4.159	4.277	.2338	4.265	35.58	35.59	.5599	.5444	.9583	5.220
.3310	.3403	2.138	.9726	4.184	4.301	.2325	4.277	35.99	36.00	.5594	.5441	.9586	5.217
.3320	.3413	2.144	.9729	4.209	4.326	.2312	4.288	36.42	36.43	.5589	.5438	.9589	5.214
.3330	.3422	2.150	.9732	4.234	4.350	.2299	4.300	36.84	36.85	.5584	.5434	.9592	5.210
.3340	.3431	2.156	.9735	4.259	4.375	.2286	4.311	37.25	37.27	.5578	.5431	.9595	5.207
.3350	.3440	2.161	.9738	4.284	4.399	.2273	4.323	37.70	37.72	.5573	.5427	.9598	5.204
.3360	.3449	2.167	.9741	4.310	4.424	.2260	4.335	38.14	38.15	.5568	.5424	.9601	5.201
.3370	.3459	2.173	.9744	4.336	4.450	.2247	4.346	38.59	38.60	.5563	.5421	.9604	5.198
.3380	.3468	2.179	.9747	4.361	4.474	.2235	4.358	39.02	39.04	.5558	.5417	.9607	5.194
.3390	.3477	2.185	.9750	4.388	4.500	.2222	4.369	39.48	39.49	.5553	.5414	.9610	5.191
.3400	.3468	2.190	.9753	4.413	4.525	.2210	4.381	39.95	39.96	.5548	.5411	.9613	5.188
.3410	.3495	2.196	.9756	4.439	4.550	.2198	4.392	40.40	40.41	.5544	.5408	.9615	5.185
.3420	.3504	2.202	.9758	4.466	4.576	.2185	4.404	40.87	40.89	.5539	.5405	.9618	5.182
.3430	.3514	2.208	.9761	4.492	4.602	.2173	4.416	41.36	41.37	.5534	.5402	.9621	5.179
.3440	.3523	2.214	.9764	4.521	4.630	.2160	4.427	41.85	41.84	.5529	.5399	.9623	5.176
.3450	.3532	2.220	.9767	4.547	4.656	.2148	4.439	42.33	42.34	.5524	.5396	.9626	5.173
.3460	.3542	2.225	.9769	4.575	4.682	.2136	4.451	42.83	42.84	.5519	.5392	.9629	5.171
.3470	.3551	2.231	.9772	4.602	4.709	.2124	4.462	43.34	43.35	.5515	.5389	.9632	5.168
.3480	.3560	2.237	.9775	4.629	4.736	.2111	4.474	43.85	43.86	.5510	.5386	.9635	5.165
.3490	.3570	2.243	.9777	4.657	4.763	.2099	4.486	44.37	44.40	.5505	.5383	.9638	5.162
.3500	.3579	2.249	.9780	4.685	4.791	.2087	4.498	44.89	44.80	.5501	.5380	.9640	5.159
.3510	.3588	2.255	.9782	4.713	4.818	.2076	4.509	45.42	45.43	.5496	.5377	.9643	5.157
.3520	.3598	2.260	.9785	4.741	4.845	.2064	4.521	45.95	45.96	.5492	.5374	.9646	5.154
.3530	.3607	2.266	.9787	4.770	4.873	.2052	4.533	46.50	46.51	.5487	.5371	.9648	5.152
.3540	.3616	2.272	.9790	4.798	4.901	.2040	4.544	47.03	47.04	.5483	.5368	.9651	5.149
.3550	.3625	2.278	.9792	4.827	4.929	.2029	4.556	47.59	47.60	.5479	.5365	.9654	5.147
.3560	.3635	2.284	.9795	4.856	4.957	.2017	4.568	48.15	48.16	.5474	.5362	.9657	5.144
.3570	.3644	2.290	.9797	4.885	4.987	.2005	4.579	48.72	48.73	.5470	.5359	.9659	5.141
.3580	.3653	2.296	.9799	4.914	5.015	.1994	4.591	49.29	49.30	.5466	.5356	.9662	5.139
.3590	.3663	2.301	.9801	4.944	5.044	.1983	4.603	49.88	49.89	.5461	.5353	.9665	5.137
.3600	.3672	2.307	.9804	4.974	5.072	.1972	4.615	50.47	50.48	.5457	.5350	.9667	5.134
.3610	.3682	2.313	.9806	5.004	5.103	.1960	4.627	51.08	51.09	.5453	.5347	.9670	5.132
.3620	.3691	2.319	.9808	5.034	5.132	.1949	4.638	51.67	51.67	.5449	.5344	.9673	5.130
.3630	.3700	2.325	.9811	5.063	5.161	.1938	4.650	52.27	52.28	.5445	.5342	.9675	5.127
.3640	.3709	2.331	.9813	5.094	5.191	.1926	4.661	52.89	52.90	.5441	.5339	.9677	5.125
.3650	.3719	2.337	.9815	5.124	5.221	.1915	4.673	53.52	53.53	.5437	.5336	.9680	5.123
.3660	.3728	2.342	.9817	5.155	5.251	.1904	4.685	54.15	54.16	.5433	.5333	.9683	5.121
.3670	.3737	2.348	.9819	5.186	5.281	.1894	4.697	54.78	54.79	.5429	.5330	.9686	5.118
.3680	.3747	2.354	.9821	5.217	5.312	.1883	4.708	55.42	55.43	.5425	.5327	.9688	5.116
.3690	.3756	2.360	.9823	5.248	5.343	.1872	4.720	56.09	56.10	.5421	.5325	.9690	5.114
.3700	.3766	2.366	.9825	5.280	5.374	.1861	4.732	56.76	56.77	.5417	.5322	.9693	5.112
.3710	.3775	2.372	.9827	5.312	5.406	.1850	4.744	57.43	57.44	.5413	.5319	.9696	5.110
.3720	.3785	2.378	.9830	5.345	5.438	.1839	4.756	58.13	58.14	.5409	.5317	.9698	5.107
.3730	.3794	2.384	.9832	5.377	5.469	.1828	4.768	58.82	58.83	.5405	.5314	.9700	5.105
.3740	.3804	2.390	.9834	5.410	5.502	.1818	4.780	59.52	59.53	.5402	.5312	.9702	5.103
.3750	.3813	2.396	.9835	5.443	5.534	.1807	4.792	60.24	60.25	.5398	.5309	.9705	5.101
.3760	.3822	2.402	.9837	5.475	5.566	.1797	4.803	60.95	60.95	.5394	.5306	.9707	5.099
.3770	.3832	2.408	.9839	5.508	5.598	.1786	4.815	61.68	61.68	.5390	.5304	.9709	5.097
.3780	.3841	2.413	.9841	5.541	5.631	.1776	4.827	62.41	62.42	.5387	.5301	.9712	5.095
.3790	.3850	2.419	.9843	5.572	5.661	.1766	4.838	63.13	63.14	.5383	.5299	.9714	5.093
.3800	.3860	2.425	.9845	5.609	5.697	.1756	4.851	63.90	63.91	.5380	.5296	.9717	5.091
.3810	.3869	2.431	.9847	5.643	5.731	.1745	4.862	64.66	64.67	.5376	.5294	.9719	5.090
.3820	.3879	2.437	.9848	5.677	5.765	.1735	4.875	65.45	65.46	.5372	.5291	.9721	5.088
.3830	.3888	2.443	.9850	5.712	5.798	.1725	4.885	66.20	66.21	.5369	.5288	.9724	5.086
.3840	.3898	2.449	.9852	5.746	5.833	.1715	4.898	67.00	67.01	.5365	.5286	.9726	5.084
.3850	.3907	2.455	.9854	5.780	5,866	.1705	4.910	67.80	67.81	.5362	.5284	.9728	5.082
.3860	.3917	2.461	.9855	5.814	5.900	.1695	4.922	68.61	68.62	.5359	.5281	.9730	5.081
.3870	.3926	2.467	.9857	5.850	5.935	.1685	4.934	69.45	69.46	.5355	.5279	.9732	5.079
.3880	.3936	2.473	.9859	5.886	5.970	.1675	4.946	70.28	70.29	.5352	.5276	.9735	5.077
.3890	.3945	2.479	.9860	5.921	6.005	.1665	4.958	71.12	71.13	.5349	.5274	.9737	5.076
.3900	.3955	2.485	.9862	5.957	6.040	.1656	4.970	71.97	71.98	.5345	.5271	.9739	5.074
.3910	.3964	2.491	.9864	5.993	6.076	.1646	4.982	72.85	72.86	.5342	.5269	.9741	5.072
.3920	.3974	2.497	.9865	6.029	6.112	.1636	4.993	73.72	73.72	.5339	.5267	.9743	5.071
.3930	.3983	2.503	.9867	6.066	6.148	.1627	5.005	74.59	74.59	.5336	.5265	.9745	5.069
.3940	.3993	2.509	.9869	6.103	6.185	.1617	5.017	75.48	75.48	.5332	.5262	.9748	5.067
.3950	.4002	2.515	.9870	6.140	6.221	.1608	5.029	76.40	76.40	.5329	.5260	.9750	5.066
.3960	.4012	2.521	.9872	6.177	6.258	.1598	5.041	77.32	77.32	.5326	.5258	.9752	5.064
.3970	.4021	2.527	.9873	6.215	6.295	.1589	5.053	78.24	78.24	.5323	.5255	.9754	5.063
.3980	.4031	2.532	.9874	6.252	6.332	.1579	5.065	79.19	79.19	.5320	.5253	.9756	5.062
.3990	.4040	2.538	.9876	6.290	6.369	.1570	5.077	80.13	80.13	.5317	.5251	.9758	5.060

APPENDIX 1. (Continued)

d/L_o	d/L	$2\pi d/L$	tanh $2\pi d/L$	sinh $2\pi d/L$	cosh $2\pi d/L$	K	$4\pi d/L$	sinh $4\pi d/L$	cosh $4\pi d/L$	n	C_G/C_o	H/H_o'	M
.4000	.4050	2.544	.9877	6.329	6.407	.1561	5.089	81.12	81.12	.5314	.5248	.9761	5.058
.4010	.4059	2.550	.9879	6.367	6.445	.1552	5.101	82.08	82.08	.5311	.5246	.9763	5.056
.4020	.4069	2.556	.9880	6.406	6.483	.1542	5.113	83.06	83.06	.5308	.5244	.9765	5.055
.4030	.4078	2.562	.9882	6.444	6.521	.1533	5.125	84.07	84.07	.5305	.5242	.9766	5.053
.4040	.4088	2.568	.9883	6.484	6.561	.1524	5.137	85.11	85.11	.5302	.5240	.9768	5.052
.4050	.4098	2.575	.9885	6.525	6.601	.1515	5.149	86.14	86.14	.5299	.5238	.9777	5.050
.4060	.4107	2.581	.9886	6.564	6.640	.1506	5.161	87.17	87.17	.5296	.5236	.9772	5.049
.4070	.4116	2.586	.9887	6.603	6.679	.1497	5.173	88.20	88.20	.5293	.5234	.9774	5.048
.4080	.4126	2.592	.9889	6.644	6.718	.1488	5.185	89.28	89.28	.5290	.5232	.9776	5.046
.4090	.4136	2.598	.9890	6.684	6.758	.1480	5.197	90.39	90.39	.5287	.5229	.9778	5.045
.4100	.4145	2.604	.9891	6.725	6.799	.1471	5.209	91.44	91.44	.5285	.5227	.9780	5.044
.4110	.4155	2.610	.9892	6.766	6.839	.1462	5.221	92.55	92.55	.5282	.5225	.9782	5.043
.4120	.4164	2.616	.9894	6.806	6.879	.1454	5.233	93.67	93.67	.5279	.5223	.9784	5.041
.4130	.4174	2.623	.9895	6.849	6.921	.1445	5.245	94.83	94.83	.5277	.5221	.9786	5.040
.4140	.4183	2.629	.9896	6.890	6.963	.1436	5.257	95.96	95.96	.5274	.5219	.9788	5.039
.4150	.4193	2.635	.9898	6.932	7.004	.1428	5.269	97.13	97.13	.5271	.5217	.9790	5.037
.4160	.4203	2.641	.9899	6.974	7.046	.1419	5.281	98.30	98.30	.5269	.5215	.9792	5.036
.4170	.4212	2.647	.9900	7.018	7.088	.1411	5.294	99.52	99.52	.5266	.5213	.9794	5.035
.4180	.4222	2.653	.9901	7.060	7.130	.1403	5.305	100.7	100.7	.5263	.5211	.9795	5.034
.4190	.4231	2.659	.9902	7.102	7.173	.1394	5.317	101.9	101.9	.5261	.5209	.9797	5.033
.4200	.4241	2.665	.9904	7.146	7.215	.1386	5.329	103.1	103.1	.5258	.5208	.9798	5.031
.4210	.4251	2.671	.9905	7.190	7.259	.1378	5.341	104.4	104.4	.5256	.5206	.9800	5.030
.4220	.4260	2.677	.9906	7.234	7.303	.1369	5.353	105.7	105.7	.5253	.5204	.9802	5.029
.4230	.4270	2.683	.9907	7.279	7.349	.1361	5.366	107.0	107.0	.5251	.5202	.9804	5.028
.4240	.4280	2.689	.9908	7.325	7.392	.1353	5.378	108.3	108.3	.5248	.5200	.9806	5.027
.4250	.4289	2.695	.9909	7.371	7.438	.1345	5.390	109.7	109.7	.5246	.5198	.9808	5.026
.4260	.4298	2.701	.9910	7.412	7.479	.1337	5.402	110.9	110.9	.5244	.5196	.9810	5.025
.4270	.4308	2.707	.9911	7.457	7.524	.1329	5.414	112.2	112.2	.5241	.5195	.9811	5.024
.4280	.4318	2.713	.9912	7.503	7.570	.1321	5.426	113.6	113.6	.5239	.5193	.9812	5.023
.4290	.4328	2.719	.9913	7.550	7.616	.1313	5.438	115.0	115.0	.5237	.5191	.9814	5.022
.4300	.4337	2.725	.9914	7.595	7.661	.1305	5.450	116.4	116.4	.5234	.5189	.9816	5.021
.4310	.4347	2.731	.9915	7.642	7.707	.1298	5.462	117.8	117.8	.5232	.5187	.9818	5.020
.4320	.4356	2.737	.9916	7.688	7.753	.1290	5.474	119.2	119.2	.5230	.5186	.9819	5.019
.4330	.4366	2.743	.9917	7.735	7.800	.1282	5.486	120.7	120.7	.5227	.5184	.9821	5.018
.4340	.4376	2.749	.9918	7.783	7.847	.1274	5.499	122.2	122.2	.5225	.5182	.9823	5.017
.4350	.4385	2.755	.9919	7.831	7.895	.1267	5.511	123.7	123.7	.5223	.5181	.9824	5.016
.4360	.4395	2.762	.9920	7.880	7.943	.1259	5.523	125.2	125.2	.5221	.5179	.9826	5.015
.4370	.4405	2.768	.9921	7.922	7.991	.1251	5.535	126.7	126.7	.5218	.5177	.9828	5.014
.4380	.4414	2.774	.9922	7.975	8.035	.1244	5.547	128.3	128.3	.5216	.5176	.9829	5.013
.4390	.4424	2.780	.9923	8.026	8.088	.1236	5.560	129.9	129.9	.5214	.5174	.9830	5.012
.4400	.4434	2.786	.9924	8.075	8.136	.1229	5.572	131.4	131.4	.5212	.5172	.9832	5.011
.4410	.4443	2.792	.9925	8.124	8.185	.1222	5.584	133.0	133.0	.5210	.5171	.9833	5.010
.4420	.4453	2.798	.9926	8.175	8.236	.1214	5.596	134.7	134.7	.5208	.5169	.9835	5.009
.4430	.4463	2.804	.9927	8.228	8.285	.1207	5.608	136.3	136.3	.5206	.5168	.9836	5.008
.4440	.4472	2.810	.9928	8.274	8.334	.1200	5.620	137.9	137.9	.5204	.5166	.9838	5.007
.4450	.4482	2.816	.9929	8.326	8.387	.1192	5.632	139.6	139.6	.5202	.5165	.9839	5.006
.4460	.4492	2.822	.9930	8.379	8.438	.1185	5.644	141.4	141.4	.5200	.5163	.9841	5.005
.4470	.4501	2.828	.9930	8.427	8.486	.1178	5.657	143.1	143.1	.5198	.5161	.9843	5.005
.4480	.4511	2.834	.9931	8.481	8.540	.1171	5.669	144.8	144.8	.5196	.5160	.9844	5.004
.4490	.4521	2.840	.9932	8.532	8.590	.1164	5.681	146.6	146.6	.5194	.5158	.9846	5.003
.4500	.4531	2.847	.9933	8.585	8.643	.1157	5.693	148.4	148.4	.5192	.5157	.9847	5.002
.4510	.4540	2.853	.9934	8.638	8.695	.1150	5.705	150.2	150.2	.5190	.5156	.9848	5.001
.4520	.4550	2.859	.9935	8.693	8.750	.1143	5.717	152.1	152.1	.5188	.5154	.9849	5.000
.4530	.4560	2.865	.9935	8.747	8.804	.1136	5.730	154.0	154.0	.5186	.5152	.9851	5.000
.4540	.4569	2.871	.9936	8.797	8.854	.1129	5.742	155.9	155.9	.5184	.5151	.9852	4.999
.4550	.4579	2.877	.9937	8.853	8.910	.1122	5.754	157.7	157.7	.5182	.5150	.9853	4.998
.4560	.4589	2.883	.9938	8.910	8.965	.1115	5.766	159.7	159.7	.5181	.5148	.9855	4.997
.4570	.4599	2.890	.9938	8.965	9.021	.1109	5.779	161.7	161.7	.5179	.5146	.9857	4.997
.4580	.4608	2.896	.9939	9.016	9.072	.1102	5.791	163.6	163.6	.5177	.5145	.9858	4.996
.4590	.4618	2.902	.9940	9.074	9.129	.1095	5.803	165.6	165.6	.5175	.5144	.9859	4.995
.4600	.4628	2.908	.9941	9.132	9.186	.1089	5.815	167.7	167.7	.5173	.5143	.9860	4.994
.4610	.4637	2.914	.9941	9.183	9.238	.1083	5.827	169.7	169.7	.5172	.5141	.9862	4.994
.4620	.4647	2.920	.9942	9.242	9.296	.1076	5.840	171.8	171.8	.5170	.5140	.9863	4.993
.4630	.4657	2.926	.9943	9.301	9.354	.1069	5.852	173.9	173.9	.5168	.5139	.9864	4.992
.4640	.4666	2.932	.9944	9.353	9.406	.1063	5.864	176.0	176.0	.5167	.5138	.9865	4.991
.4650	.4676	2.938	.9944	9.413	9.466	.1056	5.876	178.2	178.2	.5165	.5136	.9867	4.991
.4660	.4686	2.944	.9945	9.472	9.525	.1050	5.888	180.4	180.4	.5163	.5135	.9868	4.990
.4670	.4695	2.951	.9946	9.533	9.585	.1043	5.900	182.6	182.6	.5162	.5134	.9869	4.989
.4680	.4705	2.957	.9946	9.586	9.638	.1037	5.912	184.8	184.8	.5160	.5132	.9871	4.989
.4690	.4715	2.963	.9947	9.647	9.699	.1031	5.925	187.2	187.2	.5158	.5131	.9872	4.988

APPENDIX 1. (Continued)

d/L_o	d/L	$2\pi d/L$	tanh $2\pi d/L$	sinh $2\pi d/L$	cosh $2\pi d/L$	K	$4\pi d/L$	sinh $4\pi d/L$	cosh $4\pi d/L$	n	C_G/C_o	H/H_o'	M
.4700	.4725	2.969	.9947	9.709	9.760	.1025	5.937	189.5	189.5	.5157	.5129	.9873	4.988
.4710	.4735	2.975	.9948	9.770	9.821	.1018	5.949	191.8	191.8	.5155	.5128	.9874	4.987
.4720	.4744	2.981	.9949	9.826	9.877	.1012	5.962	194.2	194.2	.5154	.5127	.9875	4.986
.4730	.4754	2.987	.9949	9.888	9.938	.1006	5.974	196.5	196.5	.5152	.5126	.9876	4.986
.4740	.4764	2.993	.9950	9.951	10.00	.1000	5.986	199.0	199.0	.5150	.5125	.9877	4.985
.4750	.4774	2.999	.9951	10.01	10.07	.09942	5.999	201.4	201.4	.5149	.5124	.9878	4.984
.4760	.4783	3.005	.9951	10.07	10.12	.09882	6.011	203.9	203.9	.5147	.5122	.9880	4.984
.4770	.4793	3.012	.9952	10.13	10.18	.09820	6.023	206.5	206.5	.5146	.5121	.9881	4.983
.4780	.4803	3.018	.9952	10.20	10.25	.09759	6.036	209.0	209.0	.5144	.5120	.9882	4.983
.4790	.4813	3.024	.9953	10.26	10.31	.09698	6.048	211.7	211.7	.5143	.5119	.9883	4.982
.4800	.4822	3.030	.9953	10.32	10.37	.09641	6.060	214.2	214.2	.5142	.5117	.9885	4.982
.4810	.4832	3.036	.9954	10.39	10.43	.09583	6.072	216.8	216.8	.5140	.5116	.9886	4.981
.4820	.4842	3.042	.9955	10.45	10.50	.09523	6.085	219.5	219.5	.5139	.5115	.9887	4.980
.4830	.4852	3.049	.9955	10.52	10.57	.09464	6.097	222.2	222.2	.5137	.5114	.9888	4.980
.4840	.4862	3.055	.9956	10.59	10.63	.09405	6.109	225.0	225.0	.5136	.5113	.9889	4.979
.4850	.4871	3.061	.9956	10.65	10.69	.09352	6.121	228.3	228.3	.5134	.5112	.9890	4.979
.4860	.4881	3.067	.9957	10.71	10.76	.09294	6.134	230.6	230.6	.5133	.5111	.9891	4.978
.4870	.4891	3.073	.9957	10.78	10.83	.09236	6.146	233.5	233.5	.5132	.5110	.9892	4.978
.4880	.4901	3.079	.9958	10.85	10.90	.09178	6.159	236.4	236.4	.5130	.5109	.9893	4.977
.4890	.4911	3.086	.9958	10.92	10.96	.09121	6.171	239.6	239.6	.5129	.5107	.9895	4.977
.4900	.4920	3.092	.9959	10.99	11.03	.09064	6.183	242.3	242.3	.5128	.5106	.9896	4.976
.4910	.4930	3.098	.9959	11.05	11.09	.09010	6.195	245.2	245.2	.5126	.5105	.9897	4.976
.4920	.4940	3.104	.9960	11.12	11.16	.08956	6.208	248.3	248.3	.5125	.5104	.9898	4.975
.4930	.4950	3.110	.9960	11.19	11.24	.08901	6.220	251.3	251.3	.5124	.5103	.9899	4.975
.4940	.4960	3.117	.9961	11.26	11.31	.08845	6.232	254.5	254.5	.5122	.5102	.9899	4.974
.4950	.4969	3.122	.9961	11.32	11.37	.08793	6.245	257.6	257.6	.5121	.5101	.9900	4.974
.4960	.4979	3.128	.9962	11.40	11.44	.08741	6.257	260.8	260.8	.5120	.5100	.9901	4.973
.4970	.4989	3.135	.9962	11.47	11.51	.08691	6.269	264.0	264.0	.5119	.5099	.9902	4.973
.4980	.4999	3.141	.9963	11.54	11.59	.08637	5.282	267.3	267.3	.5118	.5098	.9903	4.972
.4990	.5009	3.147	.9963	11.61	11.65	.08584	6.294	270.6	270.6	.5116	.5097	.9904	4.972
.5000	.5018	3.153	.9964	11.68	11.72	.08530	6.306	274.0	274.0	.5115	.5096	.9905	4.971
.5010	.5028	3.159	.9964	11.75	11.80	.08477	6.319	277.5	277.5	.5114	.5095	.9906	4.971
.5020	.5038	3.166	.9964	11.83	11.87	.08424	6.331	280.8	280.8	.5113	.5094	.9907	4.971
.5030	.5048	3.172	.9965	11.91	11.95	.08371	6.343	284.3	284.3	.5112	.5093	.9908	4.970
.5040	.5058	3.178	.9965	11.98	12.02	.08320	5.356	287.9	287.9	.5110	.5092	.9909	4.970
.5050	.5067	3.184	.9966	12.05	12.09	.08270	6.368	291.4	291.4	.5109	.5092	.9909	4.969
.5060	.5077	3.190	.9966	12.12	12.16	.08220	6.380	295.0	295.0	.5108	.5091	.9910	4.969
.5070	.5087	3.196	.9967	12.20	12.24	.08169	6.393	298.7	298.7	.5107	.5090	.9911	4.968
.5080	.5097	3.203	.9967	12.28	12.32	.08119	6.405	302.4	302.4	.5106	.5089	.9912	4.968
.5090	.5107	3.209	.9968	12.35	12.39	.08068	6.417	306.2	306.2	.5105	.5088	.9913	4.967
.5100	.5117	3.215	.9968	12.43	12.47	.08022	6.430	310.0	310.0	.5104	.5087	.9914	4.967
.5110	.5126	3.221	.9968	12.50	12.54	.07972	6.442	313.8	313.8	.5103	.5086	.9915	4.967
.5120	.5136	3.227	.9969	12.58	12.62	.07922	6.454	317.7	317.7	.5102	.5086	.9915	4.966
.5130	.5146	3.233	.9969	12.66	12.70	.07873	6.467	321.7	321.7	.5101	.5085	.9916	4.966
.5140	.5156	3.240	.9970	12.74	12.78	.07824	6.479	325.7	325.7	.5100	.5084	.9917	4.965
.5150	.5166	3.246	.9970	12.82	12.86	.07776	6.491	329.7	329.7	.5098	.5083	.9918	4.965
.5160	.5176	3.252	.9970	12.90	12.94	.07729	6.504	333.8	333.8	.5097	.5082	.9919	4.965
.5170	.5185	3.258	.9971	12.98	13.02	.07682	6.516	337.9	337.9	.5096	.5082	.9919	4.964
.5180	.5195	3.264	.9971	13.06	13.10	.07634	6.529	342.2	342.2	.5095	.5081	.9920	4.964
.5190	.5205	3.270	.9971	13.14	13.18	.07587	6.541	346.4	346.4	.5094	.5080	.9921	4.964
.5200	.5215	3.277	.9972	13.22	13.26	.07540	6.553	350.7	350.7	.5093	.5079	.9922	4.963
.5210	.5225	3.283	.9972	13.31	13.35	.07494	6.566	355.1	355.1	.5092	.5078	.9923	4.963
.5220	.5235	3.289	.9972	13.39	13.43	.07449	6.578	359.6	359.6	.5092	.5077	.9924	4.963
.5230	.5244	3.295	.9973	13.47	13.51	.07404	6.590	364.0	364.0	.5091	.5077	.9924	4.962
.5240	.5254	3.301	.9973	13.55	13.59	.07358	6.603	368.5	368.5	.5090	.5076	.9925	4.962
.5250	.5264	3.308	.9973	13.64	13.68	.07312	6.615	373.1	373.1	.5089	.5075	.9926	4.962
.5260	.5274	3.314	.9974	13.73	13.76	.07266	6.628	377.8	377.8	.5088	.5074	.9927	4.961
.5270	.5284	3.320	.9974	13.81	13.85	.07221	6.640	382.5	382.5	.5087	.5074	.9927	4.961
.5280	.5294	3.326	.9974	13.90	13.94	.07177	6.652	387.3	387.3	.5086	.5073	.9928	4.961
.5290	.5304	3.333	.9975	13.99	14.02	.07134	6.665	392.2	392.2	.5085	.5072	.9929	4.960
.5300	.5314	3.339	.9975	14.07	14.10	.07091	6.677	397.0	397.0	.5084	.5071	.9930	4.960
.5310	.5323	3.345	.9975	14.16	14.19	.07047	6.690	402.0	402.0	.5083	.5070	.9931	4.960
.5320	.5333	3.351	.9976	14.25	14.28	.07003	6.702	406.9	406.9	.5082	.5070	.9931	4.959
.5330	.5343	3.357	.9976	14.34	14.37	.06959	6.714	412.0	412.0	.5082	.5069	.9932	4.959
.5340	.5353	3.363	.9976	14.43	14.46	.06915	6.727	417.2	417.2	.5081	.5068	.9933	4.959
.5350	.5363	3.370	.9976	14.52	14.55	.06872	7.639	422.4	422.4	.5080	.5068	.9933	4.959
.5360	.5373	3.376	.9977	14.61	14.64	.06829	6.752	427.7	427.7	.5079	.5067	.9934	4.958
.5370	.5383	3.382	.9977	14.70	14.73	.06787	6.764	433.1	433.1	.5078	.5066	.9935	4.958
.5380	.5393	3.388	.9977	14.79	14.82	.06746	6.776	438.5	438.5	.5077	.5066	.9935	4.958
.5390	.5402	3.394	.9977	14.88	14.91	.06705	6.789	444.0	444.0	.5077	.5065	.9936	4.958

APPENDIX 1. (Continued)

d/L_o	d/L	$2\pi d/L$	tanh $2\pi d/L$	sinh $2\pi d/L$	cosh $2\pi d/L$	K	$4\pi d/L$	sinh $4\pi d/L$	cosh $4\pi d/L$	n	C_G/C_o	H/H_o'	M
.5400	.5412	3.401	.9978	14.97	15.01	.06664	6.801	449.5	449.5	.5076	.5065	.9936	4.957
.5410	.5422	3.407	.9978	15.07	15.10	.06623	6.814	455.1	455.1	.5075	.5064	.9937	4.957
.5420	.5432	3.413	.9978	15.16	15.19	.06582	6.826	460.7	460.7	.5074	.5063	.9938	4.957
.5430	.5442	3.419	.9979	15.25	15.29	.06542	6.838	466.4	466.4	.5073	.5063	.9938	4.956
.5440	.5452	3.426	.9979	15.35	15.38	.06501	6.851	472.2	472.2	.5073	.5062	.9939	4.956
.5450	.5461	3.432	.9979	15.45	15.48	.06461	6.863	478.1	478.1	.5072	.5061	.9940	4.956
.5460	.5471	3.438	.9979	15.54	15.58	.06420	6.876	484.3	484.3	.5071	.5060	.9941	4.956
.5470	.5481	3.444	.9980	15.64	15.67	.06380	6.888	490.3	490.3	.5070	.5060	.9941	4.955
.5480	.5491	3.450	.9980	15.74	15.77	.06341	6.901	496.4	496.4	.5070	.5059	.9942	4.955
.5490	.5501	3.456	.9980	15.84	15.87	.06302	6.913	502.5	502.5	.5069	.5059	.9942	4.955
.5500	.5511	3.463	.9980	15.94	15.97	.06263	6.925	508.7	508.7	.5068	.5058	.9942	4.955
.5510	.5521	3.469	.9981	16.04	16.07	.06224	6.937	515.0	515.0	.5067	.5058	.9942	4.954
.5520	.5531	3.475	.9981	16.14	16.17	.06186	6.950	521.6	521.6	.5067	.5057	.9943	4.954
.5530	.5541	3.481	.9981	16.24	16.27	.06148	6.962	528.1	528.1	.5066	.5056	.9944	4.954
.5540	.5551	3.488	.9981	16.34	16.37	.06110	6.975	534.8	534.8	.6065	.5056	.9944	4.954
.5550	.5560	3.494	.9982	16.44	16.47	.06073	6.987	541.4	541.4	.5065	.5056	.9945	4.953
.5560	.5570	3.500	.9982	16.54	16.57	.06035	7.000	548.1	548.1	.5064	.5055	.9945	4.653
.5570	.5580	3.506	.9982	16.65	16.68	.06997	7.012	554.9	554.9	.5063	.5054	.9946	4.953
.5580	.5590	3.512	.9982	16.75	16.78	.05960	7.025	562.0	562.0	.5063	.5053	.9947	4.953
.5590	.5600	3.519	.9982	16.85	16.88	.05923	7.037	569.1	569.1	.5062	.5053	.9947	4.953
.5600	.5610	3.525	.9983	16.96	16.99	.05887	7.050	576.1	576.1	.5061	.5053	.9947	4.952
.5610	.5620	3.531	.9983	17.06	17.09	.05850	7.062	583.3	583.3	.5061	.5052	.9948	4.952
.5620	.5630	3.537	.9983	17.17	17.20	.05814	7.074	590.7	590.7	.5060	.5051	.9949	4.952
.5630	.5640	3.543	.9983	17.28	17.31	.05778	7.087	598.0	598.0	.5059	.5051	.9949	4.952
.5640	.5649	3.550	.9984	17.38	17.41	.05743	7.099	605.0	605.0	.5059	.5050	.9950	4.951
.5650	.5659	3.556	.9984	17.49	17.52	.05707	7.112	613.2	613.2	.5058	.5050	.9950	4.951
.5660	.5669	3.562	.9984	17.60	17.63	.05672	7.124	620.8	620.8	.5057	.5049	.9951	4.951
.5670	.5679	3.568	.9984	17.71	17.74	.05637	7.136	628.5	628.5	.5057	.5049	.9951	4.951
.5680	.5689	3.575	.9984	17.82	17.85	.05602	7.149	636.4	636.4	.5056	.5048	.9952	4.951
.5690	.5699	3.581	.9985	17.94	17.97	.05567	7.161	644.3	644.3	.5056	.5048	.9952	4.950
.5700	.5709	3.587	.9985	18.05	18.08	.05532	7.174	652.4	652.4	.5055	.5047	.9953	4.950
.5710	.5719	3.593	.9985	18.16	18.19	.05497	7.186	660.5	660.5	.5054	.5047	.9953	4.950
.5720	.5729	3.600	.9985	18.28	18.31	.05463	7.199	668.8	668.8	.5054	.5046	.9954	4.950
.5730	.5738	3.606	.9985	18.39	18.42	.05430	7.211	677.2	677.2	.5053	.5046	.9954	4.950
.5740	.5748	3.612	.9985	18.50	18.53	.05396	7.224	685.6	685.6	.5053	.5045	.9955	4.950
.5750	.5758	3.618	.9986	18.62	18.64	.05363	7.236	694.3	694.3	.5052	.5045	.9955	4.949
.5760	.5768	3.624	.9986	18.73	18.76	.05330	7.249	703.2	703.2	.5052	.5044	.9956	4.949
.5770	.5778	3.630	.9986	18.85	18.88	.05297	7.261	711.9	711.9	.5051	.5044	.9956	4.949
.5780	.5788	3.637	.9986	18.97	19.00	.05264	7.274	720.8	720.8	.5051	.5043	.9957	4.949
.5790	.5798	3.643	.9986	19.09	19.12	.05231	7.286	729.9	729.9	.5050	.5043	.9957	4.949
.5800	.5808	3.649	.9987	19.21	19.24	.05198	7.298	739.0	739.0	.5049	.5043	.9957	4.948
.5810	.5818	3.656	.9987	19.33	19.36	.05166	7.311	748.1	748.1	.5049	.5042	.9955	4.948
.5820	.5828	3.662	.9987	19.45	19.48	.05134	7.323	757.5	757.5	.5048	.5042	.9958	4.948
.5830	.5838	3.668	.9987	19.58	19.60	.05102	7.336	767.0	767.0	.5048	.5041	.9959	4.948
.5840	.5848	3.674	.9987	19.70	19.73	.05070	7.348	776.7	776.7	.5047	.5041	.9959	4.948
.5850	.5858	3.680	.9987	19.81	19.84	.05040	7.361	786.5	786.5	.5047	.5040	.9960	4.948
.5860	.5867	3.686	.9987	19.94	19.96	.05009	7.373	796.4	796.4	.5046	.5040	.9960	4.948
.5870	.5877	3.693	.9988	20.06	20.09	.04978	7.386	806.5	806.5	.5045	.5040	.9960	4.947
.5880	.5887	3.699	.9988	20.19	20.21	.04947	7.398	816.5	816.5	.5045	.5039	.9961	4.947
.5890	.5897	3.705	.9988	20.32	20.34	.04916	7.411	826.7	826.7	.5045	.5039	.9961	4.947
.5900	.5907	3.712	.9988	20.45	20.47	.04885	7.423	837.1	837.1	.5044	.5038	.9962	4.947
.5910	.5917	3.718	.9988	20.57	20.60	.04855	7.436	847.6	847.6	.5044	.5038	.9962	4.947
.5920	.5927	3.724	.9988	20.70	20.73	.04824	7.448	858.2	858.2	.5043	.5037	.9963	4.947
.5930	.5937	3.730	.9989	20.83	20.86	.04794	7.460	868.9	868.9	.5043	.5037	.9963	4.946
.5940	.5947	3.737	.9989	20.97	20.99	.04764	7.473	879.8	879.8	.5043	.5037	.9963	4.946
.5950	.5957	3.743	.9989	21.10	21.12	.04735	7.485	890.8	890.8	.5042	.5036	.9964	4.946
.5960	.5967	3.749	.9989	21.23	21.25	.04706	7.498	901.9	901.9	.5042	.5036	.9964	4.946
.5970	.5977	3.755	.9989	21.35	21.37	.04677	7.510	913.4	913.4	.5041	.5036	.9964	4.946
.5980	.5987	3.761	.9989	21.49	21.51	.04648	7.523	925.0	925.0	.5041	.5035	.9965	4.946
.5990	.5996	3.767	.9989	21.62	21.64	.04619	7.535	936.5	936.5	.5040	.5035	.9965	4.946
.6000	.6006	3.774	.9990	21.76	21.78	.04591	7.548	948.1	948.1	.5040	.5035	.9965	4.945
.6100	.6106	3.836	.9991	23.17	23.19	.04313	7.673	1,074	1,074	.5036	.5031	.9969	4.944
.6200	.6205	3.899	.9992	24.66	24.68	.04052	7.798	1,217	1,217	.5032	.5028	.9972	4.943
.6300	.6305	3.961	.9993	26.25	26.27	.03806	7.923	1,379	1,379	.5029	.5025	.9975	4.942
.6400	.6404	4.024	.9994	27.95	27.97	.03576	8.048	1,527	1,527	.5026	.5023	.9977	4.941
.6500	.6504	4.086	.9994	29.75	29.77	.03359	8.173	1,771	1,771	.5023	.5020	.9980	4.940
.6600	.6603	4.149	.9995	31.68	31.69	.03155	8.298	2,008	2,008	.5021	.5018	.9982	4.940
.6700	.6703	4.212	.9996	33.73	33.74	.02964	8.423	2,275	2,275	.5019	.5017	.9983	4.939
.6800	.6803	4.274	.9996	35.90	35.92	.02784	8.548	2,579	2,579	.5017	.5015	.9985	4.939
.6900	.6902	4.337	.9997	38.23	38.24	.02615	8.674	2,923	2,923	.5015	.5013	.9987	4.938

APPENDIX 1. (Continued)

d/L_o	d/L	$2\pi d/L$	tanh $2\pi d/L$	sinh $2\pi d/L$	cosh $2\pi d/L$	K	$4\pi d/L$	sinh $4\pi d/L$	cosh $4\pi d/L$	n	C_G/C_o	H/H_o'	M
.7000	.7002	4.400	.9997	40.71	40.72	.02456	8.799	3,314	3,314	.5013	.5012	.9988	4.938
.7100	.7102	4.462	.9997	43.34	43.35	.02307	8.925	3,757	3,757	.5012	.5011	.9989	4.937
.7200	.7202	4.525	.9998	46.14	46.15	.02167	9.050	4,258	4,258	.5011	.5010	.9990	4.937
.7300	.7302	4.588	.9998	49.13	49.14	.02035	9.175	4,828	4,828	.5010	.5009	.9991	4.937
.7400	.7401	4.650	.9998	52.31	52.32	.01911	9.301	5,473	5,473	.5009	.5008	.9992	4.937
.7500	.7501	4.713	.9998	55.70	55.71	.01795	9.426	6,204	6,204	.5008	.5007	.9993	4.936
.7600	.7601	4.776	.9999	59.30	59.31	.01686	9.552	7,034	7,034	.5007	.5006	.9994	4.936
.7700	.7701	4.839	.9999	63.15	63.16	.01583	9.677	7,976	7,976	.5006	.5005	.9995	4.936
.7800	.7801	4.902	.9999	67.24	67.25	.01487	9.803	9,042	9,042	.5005	.5004	.9996	4.936
.7900	.7901	4.964	.9999	71.60	71.60	.01397	9.929	10,250	10,250	.5005	.5004	.9996	4.936
.8000	.8001	5.027	.9999	76.24	76.24	.01312	10.05	11,620	11,620	.5004	.5004	.9996	4.936
.8100	.8101	5.090	.9999	81.19	81.19	.01232	10.18	13,180	13,180	.5004	.5004	.9996	4.936
.8200	.8201	5.153	.9999	86.44	86.44	.01157	10.31	14,940	14,940	.5003	.5003	.9997	4.935
.8300	.8301	5.215	.9999	92.05	92.05	.01086	10.43	17,340	17,340	.5003	.5003	.9997	4.935
.8400	.8400	5.278	1.000	98.01	98.01	.01020	10.56	19,210	19,210	.5003	.5003	.9997	4.935
.8500	.8500	5.341	1.000	104.4	104.4	.009582	10.68	21,780	21,780	.5002	.5002	.9998	4.935
.8600	.8600	5.404	1.000	111.1	111.1	.009000	10.81	24,690	24,690	.5002	.5002	.9998	4.935
.8700	.8700	5.467	1.000	118.3	118.3	.008451	10.93	28,000	28,000	.5002	.5002	.9998	4.935
.8800	.8800	5.529	1.000	126.0	126.0	.007934	11.06	31,750	31,750	.5002	.5002	.9998	4.935
.8900	.8900	5.592	1.000	134.2	134.2	.007454	11.18	36,000	36,000	.5002	.5002	.9998	4.935
.9000	.9000	5.655	1.000	142.9	142.9	.007000	11.31	40,810	40,810	.5001	.5001	.9999	4.935
.9100	.9100	5.718	1.000	152.1	152.1	.006574	11.44	46,280	46,280	.5001	.5001	.9999	4.935
.9200	.9200	5.781	1.000	162.0	162.0	.006173	11.56	52,470	52,470	.5001	.5001	.9999	4.935
.9300	.9300	5.844	1.000	172.5	172.5	.005797	11.69	59,500	59,500	.5001	.5001	.9999	4.935
.9400	.9400	5.906	1.000	183.7	183.7	.005445	11.81	67,470	67,470	.5001	.5001	.9999	4.935
.9500	.9500	5.969	1.000	195.6	195.6	.005113	11.94	76,490	76,490	.5001	.5001	.9999	4.935
.9600	.9600	6.032	1.000	203.5	203.5	.004914	12.06	86,740	86,740	.5001	.5001	.9999	4.935
.9700	.9700	6.095	1.000	222.8	222.8	.004489	12.19	98,350	98,350	.5001	.5001	.9999	4.935
.9800	.9800	6.158	1.000	236.1	236.1	.004235	12.32	111,500	111,500	.5001	.5001	.9999	4.935
.9900	.9900	6.220	1.000	251.4	251.4	.003977	12.44	126,500	126,500	.5000	.5000	1.000	4.935
1.000	1.000	6.283	1.000	267.7	267.7	.003735	12.57	143,400	143,400	.5000	.5000	1.000	4.935

APPENDIX 2. CIRCULAR FUNCTIONS of x/L and t/T

x/L t/T	$2\pi x/L$ $2\pi t/T$	$4\pi x/L$ $4\pi t/T$	$6\pi x/L$ $6\pi t/T$	sin $2\pi x/L$ $2\pi t/T$	sin $4\pi x/L$ $4\pi t/T$	sin $6\pi x/L$ $6\pi t/T$	cos $2\pi x/L$ $2\pi t/T$	cos $4\pi x/L$ $4\pi t/T$	cos $6\pi x/L$ $6\pi t/T$
.00	.0000	.0000	.0000	.0000	.0000	.0000	1.0000	1.0000	1.0000
.01	.0628	.1257	.1885	.06276	.1254	.1874	.9980	.9921	.9298
.02	.1257	.2513	.3770	.1254	.2487	.3681	.9921	.9686	.9298
.03	.1885	.3770	.5655	.1874	.3681	.5358	.9823	.9298	.8443
.04	.2513	.5027	.7540	.2487	.4818	.6846	.9686	.8763	.7290
.05	.3142	.6283	.9425	.3091	.5878	.8090	.9510	.8090	.5878
.06	.3770	.7540	1.1310	.3681	.6848	.9048	.9298	.7290	.4258
.07	.4398	.8796	1.3195	.4258	.7705	.9686	.9048	.6374	.2487
.08	.5027	1.0053	1.5080	.4818	.8443	.9980	.8763	.5358	.06276
.083333	.5236	1.0472	1.5708	.5000	.8660	1.0000	.8660	.5000	.0000
.09	.5655	1.1310	1.6965	.5358	.9048	.9921	.8443	.4258	−.1254
.10	.6283	1.2566	1.8850	.5878	.9510	.9510	.8090	.3091	−.3091
.11	.6912	1.3823	2.0735	.6374	.9823	.8763	.7705	.1874	−.4818
.12	.7540	1.5080	2.2619	.6846	.9980	.7705	.7290	.06276	−.6374
.125000	.7854	1.5708	2.3562	.7071	1.0000	.7071	.7071	.0000	−.7071
.13	.8168	1.6336	2.4504	.7290	.9980	.6374	.6846	−.06276	−.7705
.14	.8796	1.7593	2.6389	.7705	.9823	.4818	.6374	−.1874	−.8763
.15	.9425	1.8850	2.8274	.8090	.9510	.3091	.5878	−.3091	−.9510
.16	1.0053	2.0106	3.0159	.8443	.9048	.1254	.5358	−.4258	−.9921
.166667	1.0472	2.0944	3.1416	.8660	.8660	.0000	.5000	−.5000	−1.0000
.17	1.0681	2.1363	3.2044	.8763	.8443	−.06276	.4818	−.5358	−.9980
.18	1.1310	2.2619	3.3929	.9048	.7705	−.2487	.4258	−.6374	−.9686
.19	1.1938	2.3876	3.5814	.9298	.6846	−.4258	.3681	−.7290	−.9048
.20	1.2566	2.5133	3.7699	.9510	.5878	−.5878	.3091	−.8090	−.8090
.21	1.3195	2.6389	3.9584	.9686	.4818	−.7290	.2487	−.8763	−.6846
.22	1.3823	2.7646	4.1469	.9823	.3681	−.8443	.1874	−.9298	−.5358
.23	1.4451	2.8903	4.3354	.9921	.2487	−.9298	.1254	−.9686	−.3681
.24	1.5080	3.0159	4.5239	.9980	.1254	−.9823	.06276	−.9921	−.1874
.25	1.5708	3.1416	4.7124	1.0000	.0000	−1.0000	.0000	−1.0000	.0000
.26	1.6336	3.2673	4.9009	.9980	−.1254	−.9823	−.06276	−.9921	.1874
.27	1.6965	3.3929	5.0894	.9921	−.2487	−.9298	−.1254	−.9686	.3681
.28	1.7593	3.5186	5.2779	.9823	−.3681	−.8443	−.1874	−.9298	.5358
.29	1.8221	3.6442	5.4664	.9686	−.4818	−.7290	−.2487	−.8763	.6846
.30	1.8850	3.7699	5.6549	.9610	−.5878	−.5878	−.3091	−.8090	.8090
.31	1.9478	3.8956	5.8434	.9298	−.6846	−.4258	−.3681	−.7290	.9048
.32	2.0106	4.0212	6.0319	.9048	−.7705	−.2487	−.4258	−.6374	.9686
.33	2.0735	4.1469	6.2204	.8763	−.8443	−.06276	−.4818	−.5358	.9980
.333333	2.0944	4.1888	6.2832	.8660	−.8660	.0000	−.5000	−.5000	1.0000
.34	2.1363	4.2726	6.4088	.8443	−.9048	.1254	−.5358	−.4258	−.9921
.35	2.1991	4.3982	6.5973	.8090	−.9510	.3091	−.5878	−.3091	.9510
.36	2.2619	4.5239	6.7858	.7705	−.9823	.4818	−.6374	−.1874	.8763
.37	2.3248	4.6496	6.9743	.7290	−.9980	.6374	−.6846	−.06276	.7705
.375000	2.3562	4.7124	7.0686	.7071	−1.0000	.7071	−.7071	.0000	.7071
.38	2.3876	4.7752	7.1628	.6846	−.9980	.7705	−.7290	.06276	.6374
.39	2.4504	4.9009	7.3513	.6374	−.9823	.8763	−.7705	.1874	.4818
.40	2.5133	5.0265	7.5398	.5878	−.9510	.9510	−.8090	.3091	.3091
.41	2.5761	5.1522	7.7283	.5358	−.9048	.9921	−.8443	.4258	.1254
.416667	2.6180	5.2360	7.8540	.5000	−.8660	1.0000	−.8660	.5000	.0000
.42	2.6389	5.2779	7.9168	.4818	−.8443	.9880	−.8763	.5358	−.06276
.43	2.7018	5.4035	8.1053	.4258	−.7705	.9686	−.9048	.6374	−.2487
.44	2.7646	5.5292	8.2938	.3681	−.6846	.9048	−.9298	.7290	−.4258
.45	2.8274	5.6549	8.4823	.3091	−.5878	.8090	−.9510	.8090	−.5878
.46	2.8903	5.7805	8.6708	.2487	−.4818	.6846	−.9686	.8763	−.7290
.47	2.9531	5.9062	8.8593	.1874	−.3681	.5358	−.9823	.9298	−.8443
.48	3.0159	6.0318	9.0478	.1254	−.2487	.3681	−.9921	.9686	−.9298
.49	3.0788	6.1575	9.2363	.06276	−.1254	.1874	−.9980	.9921	−.9823
.50	3.1416	6.2832	9.4249	.0000	.0000	.0000	−1.0000	1.0000	−1.0000

APPENDIX 2. (Continued)

x/L t/T	$2\pi x/L$ $2\pi t/T$	$4\pi x/L$ $4\pi t/T$	$6\pi x/L$ $6\pi t/T$	$\sin \dfrac{2\pi x/L}{2\pi t/T}$	$\sin \dfrac{4\pi x/L}{4\pi t/T}$	$\sin \dfrac{6\pi x/L}{6\pi t/T}$	$\cos \dfrac{2\pi x/L}{2\pi t/T}$	$\cos \dfrac{4\pi x/L}{4\pi t/T}$	$\cos \dfrac{6\pi x/L}{6\pi t/T}$
.51	3.2044	6.4088	9.6133	−.06276	.1254	−.1874	−.9980	.9921	−.9823
.52	3.2673	6.5345	9.8018	−.1254	.2487	−.3681	−.9921	.9686	−.9298
.53	3.3301	6.6602	9.9903	−.1874	.3681	−.5358	−.9823	.9298	−.8443
.54	3.3929	6.7858	10.1788	−.2487	.4818	−.6846	−.9686	.8763	−.7290
.55	3.4558	6.9115	10.3673	−.3091	.5878	−.8090	−.9510	.8090	−.5878
.56	3.5186	7.0372	10.5558	−.3681	.6846	−.9048	−.9298	.7290	−.4258
.57	3.5814	7.1628	10.7442	−.4258	.7705	−.9686	−.9048	.6374	−.2487
.58	3.6442	7.2885	10.9327	−.4818	.8443	−.9880	−.8763	.5358	−.06276
.583333	3.6652	7.3304	10.9956	−.5000	.8660	−1.0000	−.8660	.5000	.0000
.59	3.7071	7.4142	11.1212	−.5358	.9048	−.9921	−.8443	.4258	.1254
.60	3.7699	7.5398	11.3097	−.5878	.9510	−.9510	−.8090	.3091	.3091
.61	3.8327	7.6655	11.4982	−.6374	.9823	−.8763	−.7705	.1874	.4818
.62	3.8956	7.7912	11.6867	−.6846	.9980	−.7705	−.7290	.06276	.6374
.625000	3.9270	7.8540	11.7810	−.7071	1.0000	−.7071	−.7071	.0000	.7071
.63	3.9584	7.9168	11.8752	−.7290	.9880	−.6374	−.6846	−.06276	.7705
.64	4.0212	8.0425	12.0637	−.7705	.9823	−.4818	−.6374	−.1874	.8763
.65	4.0841	8.1681	12.2522	−.8090	.9510	−.3091	−.5878	−.3901	.9510
.66	4.1469	8.2938	12.4407	−.8443	.9048	−.1254	−.5358	−.4258	.9921
.666667	4.1888	8.3776	12.5664	−.8660	.8660	.0000	−.5000	−.5000	1.0000
.67	4.2097	8.4195	12.6292	−.8763	.8443	.06276	−.4818	−.5358	.9980
.68	4.2726	8.5451	12.8177	−.9048	.7705	.2487	−.4258	−.6374	.9686
.69	4.3354	8.6708	13.0062	−.9298	.6846	.4258	−.3681	−.7290	.9048
.70	4.3982	8.7965	13.1947	−.9510	.5878	.5878	−.3091	−.8090	.8090
.71	4.4611	8.9221	13.3832	−.9686	.4818	.7290	−.2487	−.8763	.6846
.72	4.5239	9.0478	13.5717	−.9823	.3681	.8443	−.1874	−.9298	.5358
.73	4.5867	9.1735	13.7602	−.9921	.2487	.9298	−.1254	−.9686	.3681
.74	4.6496	9.2991	13.9487	−.9980	.1254	.9823	−.06276	−.9921	.1874
.75	4.7124	9.4248	14.1372	−1.0000	.0000	1.0000	.0000	−1.0000	.0000
.76	4.7752	9.5504	14.3257	−.9980	−.1254	.9823	.06276	−.9921	−.1874
.77	4.8381	9.6761	14.5142	−.9921	−.2487	.9298	.1254	−.9686	−.3681
.78	4.9009	9.8018	14.7027	−.9823	−.3681	.8443	.1874	−.9298	−.5358
.79	4.9637	9.9274	14.8912	−.9686	−.4818	.7290	.2487	−.8763	−.6846
.80	5.0265	10.0531	15.0796	−.9510	−.5878	.5878	.3091	−.8090	−.8090
.81	5.0894	10.1788	15.2681	−.9298	−.6846	.4258	.3681	−.7290	−.9048
.82	5.1522	10.3044	15.4566	−.9048	−.7705	.2487	.4258	−.6374	−.9686
.83	5.2150	10.4301	15.6451	−.8763	−.8443	.06276	.4818	−.5358	−.9980
.833333	5.2360	10.4720	15.7080	−.8660	−.8660	.0000	.5000	−.5000	−1.0000
.84	5.2779	10.5558	15.8336	−.8443	−.9048	−.1254	.5358	−.4258	−.9921
.85	5.3407	10.6814	16.0221	−.8090	−.9510	−.3091	.5878	−.3091	−.9510
.86	5.4035	10.8071	16.2106	−.7705	−.9823	−.4818	.6374	−.1874	−.8763
.87	5.4664	10.9327	16.3991	−.7290	−.9980	−.6374	.6846	−.06276	−.7705
.875000	5.4978	10.9956	16.4934	−.7071	−1.0000	−.7071	.7071	.0000	−.7071
.88	5.5292	11.0584	16.5876	−.6846	−.9980	−.7705	.7290	.06276	−.6374
.89	5.5920	11.1841	16.7761	−.6374	−.9823	−.8763	.7705	.1874	−.4818
.90	5.6549	11.3097	16.9646	−.5878	−.9510	−.9510	.8090	.3091	−.3091
.91	5.7177	11.4354	17.1531	−.5358	−.9048	−.9921	.8443	.4258	−.1254
.916667	5.7596	11.5192	17.2780	−.5000	−.8660	−1.0000	.8660	.5000	.0000
.92	5.7805	11.5611	17.3416	−.4818	−.8443	−.9980	.8763	.5358	.06276
.93	5.8434	11.6867	17.5301	−.4258	−.7705	−.9686	.9048	.6374	.2487
.94	5.9062	11.8124	17.7186	−.3681	−.6846	−.9048	.9298	.7290	.4258
.95	5.9690	11.9381	17.9071	−.3091	−.5878	−.8090	.9510	.8090	.5878
.96	6.0319	12.0637	18.0956	−.2487	−.4818	−.6846	.9686	.8763	.7290
.97	6.0947	12.1894	18.2841	−.1874	−.3681	−.5358	.9823	.9298	.8443
.98	6.1575	12.3150	18.4726	−.1254	−.2487	−.3681	.9921	.9686	.9298
.99	6.2204	12.4407	18.6611	−.06276	−.1254	−.1874	.9980	.9921	.9823
1.00	6.2832	12.5664	18.8496	.0000	.0000	.0000	1.0000	1.0000	1.0000

APPENDIX 3. WAVE SPEED AND LENGTH AS A FUNCTION OF WAVE PERIOD

T (sec)	C_o (ft/sec)	C_o (knots)	L_o (feet)	T (sec)	C_o (ft/sec)	C_o (knots)	L_o (feet)	T (sec)	C_o (ft/sec)	C_o (knots)	L_o (feet)	T (sec)	C_o (ft/sec)	C_o (knots)	L_o (feet)
0.5	2.56	1.52	1.28	7.0	35.8	21.2	251	13.5	69.1	40.9	933	20.0	102.4	60.6	2047
0.6	3.07	1.82	1.84	7.1	36.3	21.5	258	13.6	69.6	41.2	947	20.1	102.9	60.9	2068
0.7	3.58	2.12	2.51	7.2	36.8	21.8	265	13.7	70.1	41.5	961	20.2	103.4	61.2	2088
0.8	4.09	2.42	3.28	7.3	37.4	22.1	273	13.8	70.6	41.8	975	20.3	103.9	61.5	2109
0.9	4.61	2.73	4.15	7.4	37.9	22.4	280	13.9	71.1	42.1	989	20.4	104.4	61.8	2130
1.0	5.12	3.03	5.12	7.5	38.4	22.7	288	14.0	71.6	42.4	1004	20.5	104.9	62.1	2151
1.1	5.63	3.33	6.19	7.6	38.9	23.0	296	14.1	72.2	42.7	1018	20.6	105.4	62.4	2172
1.2	6.14	3.64	7.37	7.7	39.4	23.3	304	14.2	72.7	43.0	1032	20.7	105.9	62.7	2193
1.3	6.65	3.94	8.65	7.8	39.9	23.6	312	14.3	73.2	43.3	1047	20.8	106.4	63.0	2214
1.4	7.17	4.24	10.0	7.9	40.4	23.9	320	14.4	73.7	43.6	1062	20.9	107.0	63.3	2235
1.5	7.68	4.55	11.5	8.0	40.9	24.2	328	14.5	74.2	43.9	1076	21.0	107.5	63.6	2257
1.6	8.19	4.85	13.1	8.1	41.4	24.5	336	14.6	74.7	44.2	1091	21.1	108.0	63.9	2278
1.7	8.70	5.15	14.8	8.2	42.0	24.8	344	14.7	75.2	44.5	1106	21.2	108.5	64.2	2300
1.8	9.21	5.45	16.6	8.3	42.5	25.1	353	14.8	75.7	44.8	1121	21.3	109.0	64.5	2322
1.9	9.72	5.76	18.5	8.4	43.0	25.4	361	14.9	76.2	45.1	1137	21.4	109.5	64.8	2344
2.0	10.2	6.06	20.5	8.5	43.5	25.7	370	15.0	76.8	45.4	1152	21.5	110.0	65.1	2366
2.1	10.7	6.36	22.6	8.6	44.0	26.1	379	15.1	77.3	45.8	1167	21.6	110.5	65.4	2388
2.2	11.2	6.67	24.8	8.7	44.5	26.4	388	15.2	77.8	46.1	1183	21.7	111.1	65.8	2410
2.3	11.8	6.97	27.1	8.8	45.0	26.7	397	15.3	78.3	46.4	1199	21.8	111.6	66.0	2432
2.4	12.3	7.27	29.5	8.9	45.6	27.0	406	15.4	78.8	46.7	1214	21.9	112.1	66.4	2455
2.5	12.8	7.58	32.0	9.0	46.1	27.3	415	15.5	79.3	47.0	1230	22.0	112.6	66.7	2477
2.6	13.3	7.88	34.6	9.1	46.6	27.6	424	15.6	79.8	47.3	1246	22.1	113.1	67.0	2500
2.7	13.8	8.18	37.3	9.2	47.1	27.9	433	15.7	80.4	47.6	1262	22.2	113.6	67.3	2522
2.8	14.3	8.48	40.1	9.3	47.6	28.2	442	15.8	80.9	47.9	1277	22.3	114.1	67.6	2545
2.9	14.8	8.79	43.0	9.4	48.1	28.5	452	15.9	81.4	48.2	1293	22.4	114.6	67.9	2568
3.0	15.4	9.1	46.1	9.5	48.6	28.8	461	16.0	81.9	48.5	1310	22.5	115.2	68.2	2591
3.1	15.9	9.4	49.2	9.6	49.1	29.1	471	16.1	82.4	48.8	1326	22.6	115.7	68.5	2614
3.2	16.4	9.7	52.4	9.7	49.6	29.4	481	16.2	82.9	49.1	1343	22.7	116.2	68.8	2637
3.3	16.9	10.0	55.8	9.8	50.2	29.7	491	16.3	83.4	49.4	1359	22.8	116.7	69.1	2661
3.4	17.4	10.3	59.2	9.9	50.7	30.0	502	16.4	83.9	49.7	1376	22.9	117.2	69.4	2684
3.5	17.9	10.6	62.7	10.0	51.2	30.3	512	16.5	84.4	50.0	1393	23.0	117.7	69.7	2707
3.6	18.4	10.9	66.4	10.1	51.7	30.6	522	16.6	85.0	50.3	1410	23.1	118.2	70.0	2731
3.7	18.9	11.2	70.1	10.2	52.2	30.9	533	16.7	85.5	50.6	1427	23.2	118.7	70.3	2755
3.8	19.4	11.5	73.9	10.3	52.7	31.2	543	16.8	86.0	50.9	1444	23.3	119.2	70.6	2779
3.9	20.0	11.8	77.9	10.4	53.2	31.5	554	16.9	86.5	51.2	1461	23.4	119.8	70.9	2803
4.0	20.5	12.1	81.9	10.5	53.7	31.8	564	17.0	87.0	51.5	1479	23.5	120.3	71.2	2827
4.1	21.0	12.4	86.1	10.6	54.2	32.1	575	17.1	87.5	51.8	1496	23.6	120.8	71.5	2851
4.2	21.5	12.7	90.3	10.7	54.8	32.4	586	17.2	88.0	52.1	1514	23.7	121.3	71.8	2875
4.3	22.0	13.0	94.7	10.8	55.3	32.7	597	17.3	88.5	52.4	1531	23.8	121.8	72.1	2899
4.4	22.5	13.3	99.1	10.9	55.8	33.0	608	17.4	89.0	52.7	1549	23.9	122.3	72.4	2924
4.5	23.0	13.6	104	11.0	56.3	33.3	620	17.5	89.6	53.0	1567	24.0	122.8	72.7	2948
4.6	23.5	13.9	108	11.1	56.8	33.6	631	17.6	90.1	53.3	1585	24.1	123.3	73.0	2973
4.7	24.0	14.2	113	11.2	57.3	33.9	642	17.7	90.6	53.6	1603	24.2	123.8	73.3	2997
4.8	24.6	14.5	118	11.3	57.8	34.2	654	17.8	91.1	53.9	1621	24.3	124.4	73.6	3022
4.9	25.1	14.8	123	11.4	58.3	34.5	665	17.9	91.6	54.2	1639	24.4	124.9	73.9	3047
5.0	25.6	15.2	128	11.5	58.9	34.8	677	18.0	92.1	54.5	1658	24.5	125.4	74.2	3072
5.1	26.1	15.5	133	11.6	59.4	35.1	689	18.1	92.6	54.8	1677	24.6	125.9	74.5	3097
5.2	26.6	15.8	138	11.7	59.9	35.4	701	18.2	93.1	55.1	1695	24.7	126.4	74.8	3123
5.3	27.1	16.1	144	11.8	60.4	35.7	713	18.3	93.6	55.4	1714	24.8	126.9	75.1	3148
5.4	27.6	16.4	149	11.9	60.9	36.1	725	18.4	94.2	55.8	1732	24.9	127.4	75.4	3173
5.5	28.1	16.7	155	12.0	61.4	36.4	737	18.5	94.7	56.1	1751	25.0	128.0	75.7	3199
5.6	28.7	17.0	161	12.1	61.9	36.7	750	18.6	95.2	56.4	1770	25.1	128.5	76.0	3225
5.7	29.2	17.3	166	12.2	62.4	37.0	762	18.7	95.7	56.7	1789	25.2	129.0	76.4	3250
5.8	29.7	17.6	172	12.3	63.0	37.3	775	18.8	96.2	57.0	1809	25.3	129.5	76.7	3276
5.9	30.2	17.9	178	12.4	63.5	37.6	787	18.9	96.7	57.3	1828	25.4	130.0	77.0	3302
6.0	30.7	18.2	184	12.5	64.0	37.9	800	19.0	97.2	57.6	1847	25.5	130.5	77.3	3328
6.1	31.2	18.5	191	12.6	64.5	38.2	813	19.1	97.8	57.9	1867	25.6	131.0	77.6	3354
6.2	31.7	18.8	197	12.7	65.0	38.5	826	19.2	98.3	58.2	1886	25.7	131.5	77.9	3380
6.3	32.2	19.1	203	12.8	65.5	38.8	839	19.3	98.8	58.5	1906	25.8	132.0	78.2	3407
6.4	32.8	19.4	210	12.9	66.0	39.1	852	19.4	99.3	58.8	1926	25.9	132.6	78.5	3433
6.5	33.3	19.7	216	13.0	66.5	39.4	865	19.5	99.8	59.1	1946	26.0	133.1	78.8	3460
6.6	33.8	20.0	223	13.1	67.0	39.7	879	19.6	100.3	59.4	1966	26.1	133.6	79.1	3486
6.7	34.3	20.3	230	13.2	67.6	40.0	892	19.7	100.8	59.7	1986	26.2	134.1	79.4	3513
6.8	34.8	20.6	237	13.3	68.1	40.3	906	19.8	101.3	60.0	2006	26.3	134.6	79.7	3540
6.9	35.3	20.9	244	13.4	68.6	40.6	919	19.9	101.8	60.3	2027	26.4	135.1	80.0	3567

Index